Birkhäuser

Systems & Control: Foundations & Applications

More information about this series at https://link.springer.com/bookseries/4895

Ali Saberi • Anton A. Stoorvogel
Meirong Zhang • Peddapullaiah Sannuti

Synchronization of Multi-Agent Systems in the Presence of Disturbances and Delays

Birkhäuser

Ali Saberi
Electrical Engineering & Computer Science
Washington State University
Pullman, WA, USA

Meirong Zhang
School of Engineering & Applied Science
Gonzaga University
Spokane, WA, USA

Anton A. Stoorvogel
Electrical Engineering, Mathematics &
Computer Science
University of Twente
Enschede, Overijssel, The Netherlands

Peddapullaiah Sannuti
Department of Electrical and Computer
Engineering
Rutgers The State University of New Jersey
Piscataway, NJ, USA

ISSN 2324-9749 ISSN 2324-9757 (electronic)
Systems & Control: Foundations & Applications
ISBN 978-3-030-88147-4 ISBN 978-3-030-88148-1 (eBook)
https://doi.org/10.1007/978-3-030-88148-1

Mathematics Subject Classification: 9302

This book is published under the imprint Birkhäuser, www.birkhauser-science.com, by the registered company Springer Nature Switzerland AG
The registered company address is: Gewerbestrasse 11, 6330 Cham, Switzerland

This book is dedicated to:

My family: Ann, Ingmar, Ula, Dmitri, and Mirabella

(Ali Saberi)

My family: Zuan, Andrew, and Caleb

(Meirong Zhang)

My grandchildren: Jayanth, Keshav, and Prajna

(Pedda Sannuti)

and

to the glory of man

Contents

This Is from Pedda

As I begin to retire,
I recognize the spirit of true friendship and cooperation of Ali and Anton for about four decades.

I recognize the way they spring concepts,
the concepts that flow from their hearts,
the concepts that sustain our work,
the concepts that bind coherently all we do.

I recognize their warmth, not physical warmth,
the warmth that resides deep in them,
the warmth that flows in their arteries and veins,
the warmth that fuels our work,
the warmth that binds us three together.

Space: the final frontier,
to explore ··· to boldly go where no one has gone before.
Space of Systems and Signals,
our deep and varied exploration together for about four decades,
a saga of many tales, many avenues, and many characters alike.

Ali a name,
no, Ali the soul of a hard-winged angel,
the soul with full of creative concepts.
Name and Soul,
they are the same.

Anton a name,
no, Anton a falcon in spread-winged flight,
a flight of mathematics.
Name and Falcon,
the same.

Concepts form foundation and framework,
mathematics the superstructure.

Where do I fit in?
To be fair, all I have been doing recently is simply solidifying, cementing, and smoothing the union of foundation, framework, and superstructure.
Thanks a lot guys.

Who Are We?

Does all the work we did for about four decades define who we are or what we are?

We are only what our experience shaped us to be or
we are simply that which others know of us, nothing more and nothing less.
We are what we are.
Our stay on Earth is just a tiny tiny fragment in the inexorable eternity called time.
We are just tiny tiny beings on a tiny dust—a spec of dust in a giant ocean of sand.
Billions and billions of humans were swallowed in the grand parade of history.
Among all the humans that ever lived, our legacy exists only through what we had accomplished.
Whatever legacy we leave behind is however a diminishing asset fading away with time.

Chapter 1
Notation and Preliminaries

1.1 Linear Algebra

We denote by $\mathbb{R}$ and $\mathbb{C}$ the real and complex numbers, respectively. Moreover, $\mathbb{C}^-$, $\mathbb{C}^0$, and $\mathbb{C}^+$ denote the open left half complex plane, the imaginary axis, and the open right half plane, respectively. Similarly, $\mathbb{C}^\ominus$, $\mathbb{C}^\bigcirc$, and $\mathbb{C}^\oplus$ denote the open unit disc, the unit circle, and the area outside the unit circle, respectively.

For a vector x, we denote by x' its transpose and by x^* its conjugate transpose. Moreover, $\mathrm{col}\{x_1, \ldots, x_N\}$ is a vector constructed by stacking scalars or vectors $x_1, x_2, \ldots, x_N$. Finally, $\|x\|$ denotes the 2-norm of the vector x.

A' and A^* denote the transpose and the complex conjugate of the matrix A. I_N and 0_N depict the N-dimensional identity and zero matrix, respectively, which will be denoted by I and 0 if the dimensions are clear from the context. We will denote by $\mathrm{blkdiag}\{A_1, \ldots, A_N\}$, a block diagonal matrix with $A_1, \ldots, A_N$ as the diagonal elements. Finally, $\|A\|$ denotes the induced 2-norm of the matrix A.

We denote the eigenvalues of a matrix $A \in \mathbb{R}^{n\times n}$ by $\lambda_i(A)$ for $i = 1, \ldots, n$. The spectral radius of a matrix A is denoted by $\lambda_{\max}(A)$. The singular values of a matrix $A \in \mathbb{R}^{n\times n}$ are denoted by $\sigma_i(A)$ for $i = 1, \ldots, n$. We denote by $\sigma_{\min}(A)$ and $\sigma_{\max}(A)$ the smallest and largest singular value of the matrix A, respectively.

A matrix P is positive definite if it is symmetric and all its eigenvalues are positive. Similarly the matrix is positive semidefinite if it is symmetric and all its eigenvalues are nonnegative. A matrix P is negative definite or negative semidefinite if the matrix $-P$ is positive definite or positive semidefinite, respectively.

We denote the Kronecker product between A and B by $A \otimes B$. The Kronecker product is bilinear and associative:

$$A \otimes (B + C) = A \otimes B + A \otimes C,$$
$$(A + B) \otimes C = A \otimes C + B \otimes C,$$

A. Saberi et al., *Synchronization of Multi-Agent Systems in the Presence of Disturbances and Delays*, Systems & Control: Foundations & Applications,
https://doi.org/10.1007/978-3-030-88148-1_1

$$(kA) \otimes B = A \otimes (kB) = k(A \otimes B),$$
$$(A \otimes B) \otimes C = A \otimes (B \otimes C),$$

where A, B, and C are matrices and k is a scalar. The following properties of the Kronecker product will be useful:

$$(A \otimes B)(C \otimes D) = (AC \otimes BD),$$
$$(A \otimes B)^{-1} = A^{-1} \otimes B^{-1},$$
$$(A \otimes B)' = A' \otimes B' \text{ and } (A \otimes B)^* = A^* \otimes B^*.$$

Finally, the H_2 and H_∞ norm of a transfer function T are indicated by $\|T\|_2$ and $\|T\|_\infty$, respectively.

1.2 Multi-Agent Systems and Graphs

Our notations are basically the same as in [69]. For continuous-time agents, each agent $i \in \{1, \ldots, N\}$ has access to the quantity:

$$\zeta_i = \sum_{j=1}^{N} a_{ij}(y_i - y_j), \tag{1.1}$$

where $a_{ij} \geq 0$, $a_{ii} = 0$ for $i, j \in \{1, \ldots, N\}$. The topology of the network can be described by a directed graph (digraph) $\mathcal{G}$ with nodes corresponding to the agents in the network and edges given by the coefficients a_{ij}. In particular, $a_{ij} > 0$ implies that an edge exists from agent j to i. Agent j is then called a *parent* of agent i, and agent i is called a *child* of agent j. The weight of the edge equals $a_{ij} \geq 0$. We do not have self-loops in the graph, i.e., $a_{ii} = 0$. In this context, the matrix $A = [a_{ij}]$ is referred to as the adjacency matrix.

The *weighted in degree* of a vertex i is given by:

$$d_{\text{in}}(i) = \sum_{j=1}^{N} a_{ij}.$$

Similarly, the *weighted out degree* of a vertex i is given by:

$$d_{\text{out}}(i) = \sum_{j=1}^{N} a_{ji}.$$

A graph is called *balanced* if for every node we have:

$$d_{\text{in}}(i) = d_{\text{out}}(j).$$

Based on the adjacency matrix and the weighted in degree, we can associate a Laplacian matrix L with a graph:

$$L = \text{diag}\{d_{\text{in}}(1), d_{\text{in}}(2), \ldots, d_{\text{in}}(N)\} - A.$$

Based on the above definition, it is easily verified that a Laplacian matrix has the property that all the row sums are zero. In terms of the coefficients of L, (1.1) can be rewritten as:

$$\zeta_i = \sum_{j=1}^{N} \ell_{ij} y_j. \tag{1.2}$$

We denote by $\mathbf{1}_N$ the column vector in $\mathbb{R}^n$ with all elements equal to 1. We will use $\mathbf{1}$ if the dimension is obvious from the context. Note that [164, Corollary 2.37] yields the following lemma.

Lemma 1.1 *The graph $\mathcal{G}$ describing the communication topology of the network is balanced if and only if the associated Laplacian matrix L has the property that $L + L' \geq 0$. Moreover, in that case, $\mathbf{1}$ is both a left and a right eigenvector of the Laplacian matrix L.*

A *directed tree* is a directed subgraph of $\mathcal{G}$, consisting of a subset of the nodes and edges, such that every node has exactly one parent, except a single-root node with no parents. In that case, there exists a directed path from the root to every other agent. A *directed spanning tree* is a directed tree that contains all the nodes of $\mathcal{G}$. In that case, the root node with no parents is called a *root agent.*

A directed graph may contain many directed spanning trees, and thus there may be several choices for the root agent. The set of all possible root agents for a graph $\mathcal{G}$ is denoted by $\Pi_{\mathcal{G}}$.

Example 1.2 Figure 1.1 illustrates a directed graph containing multiple directed spanning trees.

We recall from [1] and more specifically [91, Lemma 3.3] the following crucial connection between the graph and its associated Laplacian matrix.

Lemma 1.3 *The graph $\mathcal{G}$ describing the communication topology of the network contains a directed spanning tree if and only if the associated Laplacian matrix L has a simple eigenvalue at the origin. Moreover, in that case, the associated right eigenvector is given by* $\mathbf{1}$.

Definition 1.4 Let $\mathbb{G}^N$ denote the set of directed graphs with N nodes that contain a directed spanning tree.

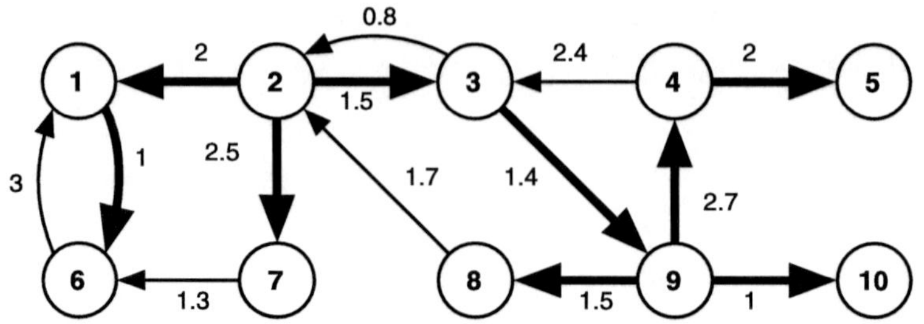

Fig. 1.1 The depicted directed graph contains multiple directed spanning trees, rooted at nodes 2, 3, 4, 8, and 9. One of these, with root node 2, is illustrated by bold arrows

Definition 1.5 Let $\mathcal{L}_N \subset \mathbb{R}^{N\times N}$ be the family of all possible Laplacian matrices associated with a graph with N agents.

Definition 1.6 For any given $\beta > 0$, let $\mathbb{G}^N_\beta$ denote the set of directed graphs with N nodes that contain a directed spanning tree and for which the corresponding Laplacian matrix has the property that its nonzero eigenvalues have a real part larger than or equal to β.

Similarly for any given $\alpha > 0$, let $\mathbb{G}^N_\alpha$ denote the set of directed graphs with N nodes that contain a directed spanning tree and for which the corresponding Laplacian matrix has the property that its norm is strictly less than or equal to α.

Definition 1.7 For any given $\alpha \geq \beta > 0$, let $\mathbb{G}^N_{\alpha,\beta}$ denote the set of directed graphs with N nodes that contain a directed spanning tree and for which the corresponding Laplacian matrix L has the property that its nonzero eigenvalues have a real part larger than or equal to β and $\|L\| \leq \alpha$.

Definition 1.8 For any given $\alpha \geq \beta > 0$ and $\theta \in [0, \frac{\pi}{2}]$, let $\mathbb{G}^N_{\alpha,\beta,\theta}$ denote the set of directed graphs with N nodes that contain a directed spanning tree and for which the corresponding Laplacian matrix has the property that its nonzero eigenvalues have a real part larger than or equal to β and an argument between $-\theta$ and θ while $\|L\| \leq \alpha$.

Sometimes we need to consider balanced graphs.

Definition 1.9 For any given $\alpha \geq \beta > 0$, let $\mathbb{G}^{b,N}_{\alpha,\beta}$ denote the set of undirected graphs with N nodes which are strongly connected and for which the corresponding Laplacian matrix L is balanced, i.e.:

$$L + L' \geq 0$$

and $L + L'$ has the property that its nonzero eigenvalues are larger than or equal to β while $\|L\| \leq \alpha$.

Finally, we consider the case of undirected graphs.

Definition 1.10 Let $\mathbb{G}^{u,N}$ denote the set of directed graphs with N nodes that contain a directed spanning tree.

Remark 1.11 Many problem definitions in this book assume that the network graph is in some arbitrary set $\mathbb{G}$. We will always restrict attention to graphs which contain a directed spanning tree so we will always assume $\mathbb{G} \subset \mathbb{G}^N$ for some N.

However, we will run into difficulties if the set $\mathbb{G}$ contains a sequence of network graphs whose corresponding Laplacian matrices contain a nonzero eigenvalue which converges to zero because then in the limit we will have a network without a directed spanning tree. Similarly, a problem arises if in this sequence an eigenvalue goes off to infinity.

To avoid this issue, we will say $\mathbb{G}$ is contained in the interior of $\mathbb{G}^N$ whenever there exists a $\alpha > \beta > 0$ such that $\mathbb{G} \subset \mathbb{G}^N_{\alpha,\beta}$.

Definition 1.12 For any given $\alpha \geq \beta > 0$, let $\mathbb{G}^{u,N}_{\alpha,\beta}$ denote the set of undirected graphs with N nodes which are strongly connected and for which the corresponding (symmetric) Laplacian matrix has the property that its nonzero eigenvalues are larger than or equal to β while $\|L\| \leq \alpha$. $\mathbb{G}^{u,N}_{\alpha,0}$ and $\mathbb{G}^{u,N}_{\infty,\beta}$ will be denoted by $\mathbb{G}^{u,N}_{\alpha}$ and $\mathbb{G}^{u,N}_{\beta}$, respectively

Remark 1.13 It is worth noting that these sets of directed graphs are subject to the relationship:

$$\mathbb{G}^N_{\alpha,\beta,\theta} \subset \mathbb{G}^N_{\alpha,\beta} \subset \mathbb{G}^N_{\beta} \subset \mathbb{G}^N.$$

We should also note that for any $\alpha > \beta > 0$, there exists a θ such that:

$$\mathbb{G}^N_{\alpha,\beta} \subset \mathbb{G}^N_{\alpha,0,\theta}.$$

For undirected and balanced graphs, we have:

$$\mathbb{G}^{u,N}_{\alpha,\beta} \subset \mathbb{G}^{u,N}$$

and:

$$\mathbb{G}^{u,N}_{\alpha,\beta} \subset \mathbb{G}^{b,N}_{\alpha,\beta}.$$

For discrete-time agents, we use:

$$\zeta_i = \sum_{j=1}^{N} d_{ij}(y_i - y_j), \tag{1.3}$$

where the $d_{ij} \geq 0$ forms a matrix D which is a row-stochastic matrix, i.e.:

$$\sum_{j=1}^{N} d_{ij} = 1.$$

The topology of the network can again be described by a directed graph (digraph) $\mathcal{G}$ with nodes corresponding to the agents in the network and edges given by the coefficients d_{ij}. In particular, $d_{ij} > 0$ implies that an edge exists from agent j to i. Agent j is then called a *parent* of agent i, and agent i is called a *child* of agent j. The weight of the edge equals the magnitude of d_{ij}. Clearly, in this case, all weights are less than or equal to 1, and, additionally, the weighted in degree has to be less than or equal to 1. Note that the diagonal elements d_{ii} do not affect (1.3) and are chosen to be such that the matrix becomes a row stochastic matrix.

We have the following result (see [91] for the sufficiency part).

Lemma 1.14 *Assume that the graph $\mathcal{G}$ describing the communication topology of the network contains a directed spanning tree and $d_{ii} > 0$ for $i = 1, \ldots, N$ implies that the row stochastic matrix D has a simple eigenvalue at* 1 *and all other eigenvalues have amplitude strictly less than* 1. *Moreover, in that case, the associated right eigenvector is given by* $\mathbf{1}$.

Remark 1.15 Note that the condition that $d_{ii} > 0$ for $i = 1, \ldots, N$ is not a necessary condition for the property that the row stochastic matrix D has a simple eigenvalue at 1 and all other eigenvalues have amplitude strictly less than 1. This can be seen from the example:

$$D = \tfrac{1}{4}\begin{pmatrix} 2 & 1 & 1 \\ 2 & 0 & 2 \\ 1 & 1 & 2 \end{pmatrix}.$$

A necessary condition is that the graph contains a spanning tree and the row stochastic matrix is aperiodic.

Proof Based on the row stochastic matrix D, we can associate a Laplacian matrix $L = I - D$ where both the Laplacian matrix and the row stochastic matrix D are connected to the same graph but with different weighting of the edges. Based on Lemma 1.3, we know that L has a simple eigenvalue at 0 with associated right eigenvector given by $\mathbf{1}$. Clearly this implies that D has a simple eigenvalue at 1 with associated right eigenvector given by $\mathbf{1}$.

From Geršgorin's circle criterion, any eigenvalue of the matrix D is contained in a disc with center d_{ii} and radius $1 - d_{ii}$ for some i. Since $d_{ii} \neq 0$, we find that the only eigenvalue on the unit disc can be 1. Therefore, all eigenvalues unequal to 1 have amplitude strictly less than 1. ∎

We can then define a set of network graphs as follows.

Definition 1.16 For $\beta \in (0, 1]$, let $\bar{\mathbb{G}}_{\beta}^{N}$ denote the set of digraphs with N nodes which contain a directed spanning tree and for which the corresponding matrix D has the property that its eigenvalues unequal to 1 have absolute value smaller than β.

Remark 1.17 For $\beta = 1$, $\bar{\mathbb{G}}_{1}^{N}$ is the set of all directed graphs with N nodes which contain a direct spanning tree. In this case, we shall drop the subscription 1 and simply denote it as $\bar{\mathbb{G}}^{N}$ but it implies that $\beta = 1$.

Remark 1.18 It is worth noting that these two sets are subject to the relationship:

$$\bar{\mathbb{G}}_{\beta}^{N} \subset \bar{\mathbb{G}}^{N}.$$

Remark 1.19 Many problem definitions in this book assume that the network graph is in some arbitrary set $\mathbb{G}$. We will always restrict attention to graphs which contain a directed spanning tree so we will always assume $\mathbb{G} \subset \bar{\mathbb{G}}^{N}$ for some N.

However, as noted for the continuous time in Remark 1.11, we will run into difficulties if the set $\mathbb{G}$ contains a sequence of network graphs whose corresponding row stochastic matrices contain an eigenvalue unequal to 1 which converges to 1 because then in the limit we will have a network without a directed spanning tree.

To avoid this issue, we will say $\mathbb{G}$ is contained in the interior of $\bar{\mathbb{G}}^{N}$ whenever there exists a $\beta \in (0, 1)$ such that $\mathbb{G} \subset \bar{\mathbb{G}}_{\beta}^{N}$.

Finally, we consider the case of undirected graphs.

Definition 1.20 For any given $\beta > 0$, let $\bar{\mathbb{G}}_{\beta}^{u,N}$ denote the set of undirected graphs with N nodes which are strongly connected and for which the corresponding matrix D has the property that its nonzero eigenvalues have absolute value smaller than β.

Remark 1.21 As before, for $\beta = 1$, we simply use $\bar{\mathbb{G}}^{u,N}$ to denote the set $\bar{\mathbb{G}}_{1}^{u,N}$.

Regarding regulated output synchronization of heterogeneous multi-agent systems, we give the following definition and remark.

Definition 1.22 We say that a matrix pair (A, C) contains the matrix pair (S, R) if there exists a matrix Π such that $\Pi S = A\Pi$ and $C\Pi = R$.

Remark 1.23 Definition 1.22 implies that for any initial condition $\omega(0)$ of the system $\dot{\omega} = S\omega$, $y_r = R\omega$, there exists an initial condition $x(0)$ of the system $\dot{x} = Ax$, $y = Cx$, such that $y(t) = y_r(t)$ for all $t \geq 0$.

1.3 Passivity

We define the concept of passivity and several generalizations for continuous- and discrete-time systems.

1.3.1 Continuous Time

Consider a general, strictly proper system Σ:

$$\Sigma : \begin{cases} \dot{x} = Ax + Bu, \\ y = Cx + Du, \end{cases} \tag{1.4}$$

where $x \in \mathbb{R}^n$, $u \in \mathbb{R}^m$, and $y \in \mathbb{R}^p$. We first define passive and passifiable systems.

Definition 1.24 The system (1.4) is called *passive* if the system is square (i.e., $m = p$) and for initial condition $x(0) = 0$, for any input u, and for any $T \geq 0$ we have:

$$\int_0^T y'(t)u(t)\,\mathrm{d}t \geq 0.$$

The system is called *passifiable via static output feedback* if the system is square and there exists a matrix H such that for initial condition $x(0) = 0$, for any input v, and for any $T \geq 0$ we have:

$$\int_0^T y'(t)u(t)\,\mathrm{d}t \geq \int_0^T y'(t)Hy(t)\,\mathrm{d}t.$$

The system is called *passifiable via static input feedforward* if the system is square and there exists a matrix R such that for initial condition $x(0) = 0$, for any input v and for any $T \geq 0$ we have:

$$\int_0^T y'(t)u(t)\,\mathrm{d}t \geq \int_0^T u'(t)Ru(t)\,\mathrm{d}t.$$

The positive real lemma (see, for example, [2, 161]) gives an easy characterization when systems are passive.

Lemma 1.25 *Assume that (A, B) is controllable and (A, C) is observable with B and C full-column and full-row rank, respectively. The system (1.4) is passive if and only if there exists a matrix $P > 0$ such that:*

$$G(P) = \begin{pmatrix} PA + A'P & PB + C' \\ B'P - C & -D - D' \end{pmatrix} \leq 0, \tag{1.5}$$

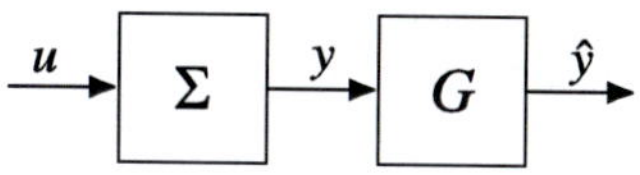

Fig. 1.2 A G-passive system

Remark 1.26 For strictly proper systems, the condition (1.5) reduces to:

$$\begin{aligned} PA + A'P &\leq 0, \\ PB &= C'. \end{aligned} \tag{1.6}$$

Classical passivity requires the system to be square. For non-square systems, G-passivity and G-passifiability have been introduced in [26]. Given a prespecified $m \times p$-matrix G, a system (1.4) is called *G-passive* if the cascade of the system (1.4) with post-compensator $G \in \mathbb{R}^{m\times p}$ as shown in Fig. 1.2 is passive. Similarly, given a prespecified $m \times p$-matrix G, a system (1.4) is called *G-passive* if the cascade of the system (1.4) with post-compensator $G \in \mathbb{R}^{m\times p}$ as shown in Fig. 1.2 is passifiable via static output feedback.

From the positive real lemma, we almost immediately find a characterization of G-passivity.

Lemma 1.27 *Assume that (A, B) is controllable and (A, C) is observable with B and GC full-column and full-row rank, respectively. The system (1.4) is G-passive if and only if there exists a matrix $P > 0$ such that*

$$G(P) = \begin{pmatrix} PA + A'P & PB + C'G' \\ B'P - GC & -GD - D'G' \end{pmatrix} \leq 0. \tag{1.7}$$

Remark 1.28 For strictly proper systems, the condition (1.7) reduces to

$$\begin{aligned} PA + A'P &\leq 0, \\ PB &= C'G'. \end{aligned} \tag{1.8}$$

From the literature on squaring down, it follows that such G should be designed, because in general G may introduce invariant zeros in $\mathbb{C}^+$ and in this case the system can never be passive or passifiable.

In this book, we consider the more general concepts of *squared-down passive*, *squared-down passifiable via static output feedback* and *squared-down passifiable via static input feedforward* for a non-square system (1.4) based on the idea of *squaring down* in [95]. Since this book is considering MAS with strictly proper systems, we define these concepts for strictly proper systems.

A system (1.4) is called *squared-down passive* if there exists a pre-compensator G_1 and a post-compensator G_2 such that the interconnection in Fig. 1.3 with input $\hat{u}$ and output $\hat{y}$ is passive. Assuming G_1 and G_2 are such that (A, BG_1) is controllable and (A, G_2C) is observable, then this is equivalent to the existence of a positive

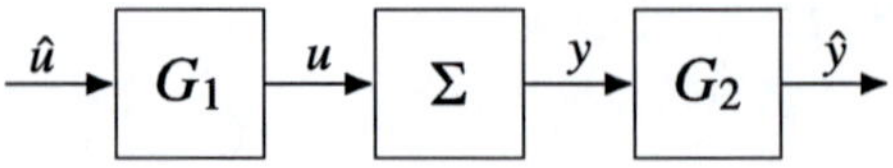

Fig. 1.3 A squared-down passive system

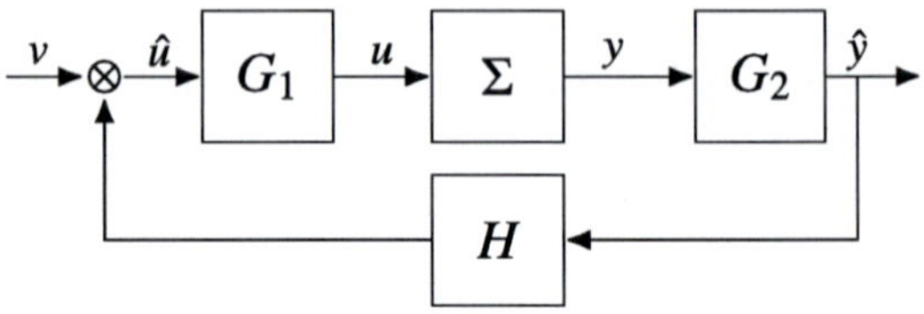

Fig. 1.4 A squared-down passive system via static output feedback

definite matrix P, such that:

$$\begin{aligned} PA + A'P &\leq 0, \\ PBG_1 &= C'G_2'. \end{aligned} \tag{1.9}$$

Remark 1.29 Note that when $G_1 = I$, squared-down passivity is similar to G-passivity as used in [26]. However, in [26], a more strict version of passivity is used which requires the system to be asymptotically stable. Our version can also be used for neutrally stable systems.

For a square system, i.e., $G_1 = G_2 = I$, squared-down passivity becomes conventional passivity.

Similar to the definition of G-passifiability, a system (1.4) is called *squared-down passifiable via static output feedback* if there exists a pre-compensator $G_1 \in \mathbb{R}^{m\times q}$, a post-compensator $G_2 \in \mathbb{R}^{q\times p}$, and a static output feedback:

$$\hat{u} = -H\hat{y} + v = -HG_2y + v, \qquad u = G_1\hat{u} \tag{1.10}$$

such that interconnection of the system (1.4) and the pre-feedback (1.10) is passive with respect to the new input v and output $\hat{y}$ as shown in Fig. 1.4.

Assume G_1 and G_2 are such that (A, BG_1) is controllable and (A, G_2C) is observable with BG_1 and G_2C full-column and full-row rank, respectively. In that case, squared-down passifiable via static output feedback is equivalent to the existence of a matrix H and a positive definite matrix P such that:

$$\begin{aligned} P(A - BG_1HG_2C) + (A - BG_1HG_2C)'P &\leq 0, \\ PBG_1 &= C'G_2'. \end{aligned} \tag{1.11}$$

A system (1.4) is called *squared-down passifiable via static input feedforward* if there exists a pre-compensator $G_1 \in \mathbb{R}^{m\times q}$, a post-compensator $G_2 \in \mathbb{R}^{q\times p}$, and an input feedforward matrix R, which make the interconnection of (1.4) and:

$$z = R\hat{u} + \hat{y} = R\hat{u} + G_2y, \qquad u = G_1\hat{u}$$

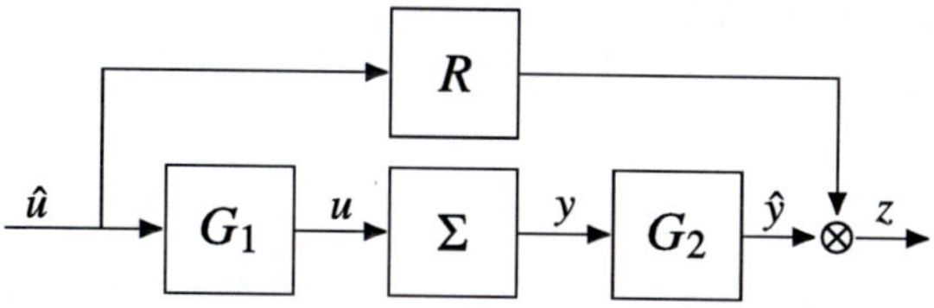

Fig. 1.5 Squared-down passive system via static input feedforward

passive with respect to the new input $\hat{u}$ and the new output z, as shown in Fig. 1.5. If (A, BG_1) is controllable and (A, G_2C) is observable with BG_1 and G_2C full-column and full-row rank, respectively, then this is equivalent to the existence of a positive definite matrix P such that:

$$G(P) = \begin{pmatrix} PA + A'P & PBG_1 - C'G_2' \\ G_1'B'P - G_2C & -R - R' \end{pmatrix} \leq 0. \tag{1.12}$$

Note that systems, which are squared-down passifiable via static input feedforward, are always neutrally stable which follows directly from (1.12).

Remark 1.30 If $G_1 = G_2 = I$ with the choice of $R = aI$, our squared-down passifiability via static input feedforward is reduced to input feedforward passivity as given in [86, Equation (4)].

1.3.2 Discrete Time

Consider a general, strictly proper system Σ:

$$\Sigma : \begin{cases} x(k+1) = Ax(k) + Bu(k), \\ y(k) \quad = Cx(k) + Du(k), \end{cases} \tag{1.13}$$

where $x \in \mathbb{R}^n$, $u \in \mathbb{R}^m$, and $y \in \mathbb{R}^p$. We first define passive and passifiable systems.

Definition 1.31 The system (1.13) is called *passive* if the system is square (i.e., $m = p$) and for initial condition $x(0) = 0$, for any input u, and for any $T \geq 0$ we have:

$$\sum_{k=0}^{T} y'(k)u(k) \geq 0.$$

The system is called *passifiable via static output feedback* if the system is square and there exists a matrix H such that for initial condition $x(0) = 0$, for any input v, and for any $T \geq 0$ we have:

$$\sum_{k=0}^{T} y'(k)u(k) \geq \sum_{k=0}^{T} y'(k)Hy(k).$$

The system is called *passifiable via static input feedforward* if the system is square and there exists a matrix R such that for initial condition $x(0) = 0$, for any input v, and for any $T \geq 0$ we have:

$$\sum_{k=0}^{T} y'(k)u(k) \geq \sum_{k=0}^{T} u'(k)Ru(k).$$

The positive real lemma (see, for example, [37]) gives an easy characterization when systems are passive.

Lemma 1.32 *Assume that* (A, B) *is controllable and* (A, C) *is observable with* B *and* C *full-column and full-row rank, respectively. The system* (1.4) *is passive if and only if there exists a matrix* $P > 0$ *such that:*

$$G(P) = \begin{pmatrix} A'PA - P & A'PB + C' \\ B'PA - C & B'PB - D - D' \end{pmatrix} \leq 0. \tag{1.14}$$

In this book, we consider the more general concepts of *squared-down passive*, *squared-down passifiable via static output feedback*, and *squared-down passifiable via static input feedforward* for a non-square system (1.4) based on the idea of *squaring down* in [95]. Since this book is considering MAS with strictly proper systems, we define these concepts for strictly proper systems. However, for discrete-time systems, this implies that the system can never be *squared-down passive* and *squared-down passifiable via static output feedback*. However, strictly proper, discrete-time agents can be squared-down passifiable via static input feedforward.

Consider a linear system Σ given by:

$$\begin{aligned} x(k+1) &= Ax(k) + Bu(k), \\ y(k) &= Cx(k), \end{aligned} \tag{1.15}$$

where $x(k) \in \mathbb{R}^n$, $u(k) \in \mathbb{R}^m$, and $y(k) \in \mathbb{R}^p$.

The linear system (1.15) is called *squared-down passifiable via static input feedforward* if there exists a pre-compensator G_1, a post-compensator G_2, and an input feedforward matrix R such that the system (1.13) is squared-down passive with respect to the new input $\hat{u}$ and the new output z, shown in Fig. 1.6.

Assume G_1 and G_2 are such that (A, BG_1) is controllable and (A, G_2C) is observable with BG_1 and G_2C full-column and full-row rank, respectively. In that case, squared-down passifiable via static output feedback is equivalent to the

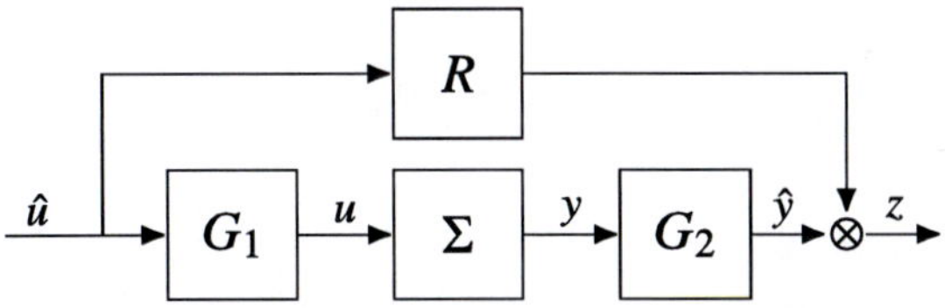

Fig. 1.6 A squared-down passifiable system via static input feedforward

existence of matrix R and a positive definite matrix P such that:

$$G(P) = \begin{pmatrix} A'PA - P & A'PBG_1 - C'G_2' \\ G_1'B'PA - G_2C & -R - R' + G_1'B'PBG_1 \end{pmatrix} \leq 0. \tag{1.16}$$

Note that systems which are squared-down passifiable via static input feedforward are always neutrally stable which follows directly from (1.16).

Part I
Synchronization of Homogeneous Systems

The first part of this book focuses on state synchronization of non-introspective homogeneous multi-agent systems. A multi-agent system is homogeneous when the dynamics of all agents are identical. For homogeneous systems, we will consider state synchronization.

We determine control strategies for both continuous- and discrete-time systems that achieve state synchronization under full state coupling as well as partial state coupling. It is shown that both synchronization problems can be solved through a connection with certain robust stabilization problems.

For full state coupled networks, control strategies can be generated via algebraic Riccati equation (ARE) based methods and/or asymptotic time-scale eigenstructure assignment (ATEA) based methods. We note that our design methodologies, at a first glance, might look very similar. However, often there are crucial differences that make all the difference in their behavior.

For partial state coupled networks, as expected, the use of dynamic controllers is generally required which will be often constructed via high-gain observers coupled with the state feedback gains designed via ARE or ATEA based methods. We will see that a structure related to passivity in some cases enables the use of static controllers for partial state coupled networks. Most of our work here is concentrated on linear time-invariant agents with only the standard communication between agents. It will be shown that the presence of an additional communication channel enables us to remove certain constraints on the poles and zeros of an agent.

In the following, we first consider linear time-invariant systems. Chapters 2 and 3 consider continuous- and discrete-time systems, respectively.

Frequently, multi-agent systems are subject to delays. Delays can be classified as input or communication delays. Chapters 6 and 7 consider input-delays for continuous- and discrete-time systems, respectively. Chapter 8 considers the case of communication delays for continuous-time systems.

There are several types of extensions of the above. Chapter 4 considers state synchronization for continuous-time systems subject to actuator saturation. Chapter 5 considers a certain class of nonlinear systems and Chap. 11 considers time-varying graphs. Combining these two chapters, we consider nonlinear systems with time-varying graphs in Chap. 12.

Finally, Chap. 13 considers the effect of disturbances. In that case, we can no longer guarantee exact synchronization and we have to allow for almost synchronization. The chapter considers both stochastic disturbances whose effect is measured by the H_2 norm and disturbances with bounded power whose effect is measured by the H_∞ norm.

Chapter 2
Synchronization of Continuous-Time Linear MAS

2.1 Introduction

This chapter considers synchronization problems for homogeneous linear continuous-time multi-agent systems (MAS). A multi-agent system is homogeneous when the dynamics of all agents are identical. For homogeneous systems, we will primarily consider state synchronization where the differences between the states of different agents converge to zero. We also address the case of formation where the differences between states of different agents converge to, a priori given, vectors.

In the literature, introspective and non-introspective synchronization problems are considered. In the introspective case, each agent has access to his own state or part of his own state in addition to the information through the network. The main focus of this chapter is the non-introspective case where the agents have access only to the information through the network. However, we will briefly address the introspective case as well.

The non-introspective problem was first studied under full-state coupling. Full-state coupling is said to exist if an agent has the access to the difference between its own *state* and the *state* of each of any other agent. Early work can be found in, for instance, [79, 90, 143, 149, 158, 166, 171].

On the other hand, non-introspective partial-state coupling is said to exist if an agent has only access to the difference between its own *output* and the *output* of other agents. Several papers have focused on conditions to achieve consensus by using static output-feedback protocols, for example, [64, 145, 167, 168]. It is shown that only a restricted class of agents satisfy these solvability conditions. These conditions are related to passivity. See, for instance, [3, 4, 17, 175, 192] and references therein.

When using dynamic output-feedback protocols, synchronization can be achieved for a larger class of agents. The seminal work [165, 166] showed that the synchronization problem can be solved via solving a simultaneous stabilization

A. Saberi et al., *Synchronization of Multi-Agent Systems in the Presence of Disturbances and Delays*, Systems & Control: Foundations & Applications,
https://doi.org/10.1007/978-3-030-88148-1_2

problem. The solvability of these simultaneous stabilization problems requires a specific and stringent requirement on the right half plane (RHP) poles and zeros of the agent model, that is, the simultaneous presence of RHP poles and zeros is not allowed. Therefore, in the literature on MAS with partial-state coupling and purely decentralized protocols, three kinds of restrictions can be observed:

Agents are at most weakly unstable.
In [117], general multi-input multi-output (MIMO) agents with all eigenvalues in the closed left half plane are considered, and a low-gain-based protocol is proposed.

Agents are at most weakly non-minimum-phase.
More specifically, [33, 179] require the agents to be minimum-phase, [16] requires agents to be weakly minimum-phase, and [128, 129] require agents to be weakly non-minimum-phase with the restriction that the Jordan blocks associated with zeros on the imaginary axis are at most size 2.

Note that the above conditions for dynamic protocols are closely related to our conditions for static protocols. After all, passivity implies, among other conditions, that the agents are neutrally stable and weakly minimum-phase.

Historically, the problem of partial-state coupling was first addressed by using an additional communication channel between the protocols of each agent relying on the same network topology. See, for instance, [49, 114]. This extra communication is mathematically very convenient making the solvability condition weaker and the analysis simpler. However, from a practical point of view, it is not very realistic.

This chapter considers the full-state coupling case in Sect. 2.4. In Sect. 2.5, we consider the partial-state coupling case. This latter section primarily focuses on dynamic protocols but also considers static protocols connected to passivity in Sect. 2.5.7. The extra communication channel between protocols is discussed in Sect. 2.5.8. Section 2.5.9 shows how conditions can be relaxed if agents are introspective, i.e., besides communication through the network, agents have access to part of their own state. Finally, we study the formation problem in Sect. 2.6 and show its connection to the synchronization problem as studied in the rest of this chapter.

The write-up of this chapter is partially based on [63] and [130].

2.2 Multi-agent Systems

Consider a multi-agent system (MAS) composed of N identical linear time-invariant continuous-time agents of the form:

$$\begin{aligned} \dot{x}_i &= Ax_i + Bu_i, \\ y_i &= Cx_i, \end{aligned} \qquad (i = 1, \ldots, N) \tag{2.1}$$

where $x_i \in \mathbb{R}^n$, $u_i \in \mathbb{R}^m$, and $y_i \in \mathbb{R}^p$ are, respectively, the state, input, and output vectors of agent i. We make the following standard and necessary assumption for the agent dynamics.

Assumption 2.1 *(A, B) is stabilizable, and (A, C) is detectable.*

The communication network provides each agent with a linear combination of its own output relative to that of other neighboring agents. In particular, each agent $i \in \{1, \ldots, N\}$ has access to the quantity:

$$\zeta_i = \sum_{j=1}^{N} a_{ij}(y_i - y_j), \tag{2.2}$$

where $a_{ij} \geq 0$, $a_{ii} = 0$ for $i, j \in \{1, \ldots, N\}$. The topology of the network can be described by a graph $\mathcal{G}$ with nodes corresponding to the agents in the network and edges given by the nonzero coefficients a_{ij}. In particular, $a_{ij} > 0$ implies that an edge exists from agent j to i. The weight of the edge equals the magnitude of a_{ij}.

As clarified in Sect. 1.2, we can express the communication in the network in terms of the Laplacian matrix L associated with this weighted graph $\mathcal{G}$. In particular, ζ_i can be rewritten as:

$$\zeta_i = \sum_{j=1}^{N} \ell_{ij} y_j. \tag{2.3}$$

We need the following assumption on the network graph.

Assumption 2.2 *The graph $\mathcal{G}$ describing the communication topology of the network contains a directed spanning tree.*

As noted in Lemma 1.3, Assumption 2.2 guarantees that the Laplacian matrix L has a simple eigenvalue at the origin, with corresponding right eigenvector $\mathbf{1}$. As will be clarified in Corollary 2.6, this condition is basically necessary to be able to achieve state synchronization.

2.3 Problem Formulation

Since all the agents in the network are identical, we can pursue state synchronization among agents. State synchronization is defined as follows:

Definition 2.1 Consider a homogeneous network described by (2.1) and (2.3). The agents in the network achieve state synchronization if:

$$\lim_{t\to\infty} \left[x_i(t) - x_j(t)\right] = 0, \qquad \forall i, j \in \{1, \ldots, N\}. \tag{2.4}$$

Note that if C has full-column rank then, without loss of generality, we can assume that $C = I$ and the quantity ζ_i becomes

$$\zeta_i = \sum_{j=1}^{N} a_{ij}(x_i - x_j) = \sum_{j=1}^{N} \ell_{ij} x_j, \tag{2.5}$$

which means agents in the network have access to the relative information of the full state of their neighboring agents compared to their own state.

Definition 2.2 Consider a MAS described by (2.1) and (2.3). The case where $C = I$ will be referred to as *full-state coupling*; otherwise, it is referred to as *partial-state coupling*. In the latter case, C does not have full-column rank.

We formulate two state synchronization problems, one for a network with full-state coupling and the other for partial-state coupling.

Problem 2.3 (Full-State Coupling) Consider a MAS described by (2.1) and (2.5). Let $\mathbf{G}$ be a given set of graphs such that $\mathbf{G} \subseteq \mathbb{G}^N$. The *state synchronization* problem with a set of network graphs $\mathbf{G}$ is to find, if possible, a linear static protocol of the form:

$$u_i = F\zeta_i, \tag{2.6}$$

for $i = 1, \ldots, N$ such that for any graph $\mathcal{G} \in \mathbf{G}$ and for all the initial conditions of agents, the state synchronization among agents is achieved.

Note that in MAS, it is often the case that the network graph and its associated weights are not known to the agents. Therefore, in the above, we design a protocol which will work for any graph in the set $\mathbf{G}$. Note that the graph $\mathcal{G}$ determines the coefficients of the Laplacian matrix L. In the extreme case where $\mathbf{G}$ contains only a single graph, the agents know the graph and the protocol (2.6) can be designed based on a known Laplacian matrix. However, the more interesting case is the situation where the set $\mathbf{G}$ is large, and we need to design a protocol based on minimal knowledge of the graph and its Laplacian matrix. For instance, in this book, we will often rely only on a lower bound for the real part of the nonzero eigenvalues of the Laplacian matrix. Sometimes, we will also require an upper bound for the Laplacian, but this still covers almost all possible graphs since these lower and upper bounds can be arbitrary small or large, respectively.

Problem 2.4 (Partial-State Coupling) Consider a MAS described by (2.1) and (2.3). Let $\mathbf{G}$ be a given set of graphs such that $\mathbf{G} \subseteq \mathbb{G}^N$. The *state synchronization* problem with a set of network graphs $\mathbf{G}$ is to find, if possible, a linear time-invariant dynamic protocol of the form:

$$\begin{cases} \dot{\chi}_i = A_c \chi_i + B_c \zeta_i, \\ u_i = C_c \chi_i + D_c \zeta_i, \end{cases} \tag{2.7}$$

for $i = 1, \ldots, N$ where $\chi_i \in \mathbb{R}^{n_c}$, such that for any graph $\mathcal{G} \in \mathbf{G}$ and for all the initial conditions for the agents and their protocol, state synchronization among agents is achieved.

Additionally, whenever the protocol (2.7) is required to be stable (i.e., the matrix A_c must be asymptotically stable), it is referred to as a partial-state synchronization via a stable protocol.

2.4 Protocol Design for MAS with Full-State Coupling

In this section, we establish a connection between the state synchronization among agents in the network and a robust stabilization problem. Then we will exploit this connection to design appropriate protocols based on full-state coupling.

2.4.1 Connection to a Robust Stabilization Problem

The MAS system described by (2.1) and (2.5) after implementing the linear static protocol (2.6) is described by:

$$\dot{x}_i = Ax_i + BF\zeta_i,$$

for $i = 1, \ldots, N$. Let:

$$x = \begin{pmatrix} x_1 \\ \vdots \\ x_N \end{pmatrix}.$$

Then, the overall dynamics of the N agents can be written as:

$$\dot{x} = (I_N \otimes A + L \otimes BF)x, \tag{2.8}$$

where $\otimes$ denotes the Kronecker product and I_N denotes the $N \times N$ identity matrix.

A key observation, which is rooted in the seminal work [165, 166], is that synchronization for the system (2.8) is equivalent to the asymptotic stability of the following $N-1$ subsystems:

$$\dot{\tilde{\eta}}_i = (A + \lambda_i BF)\tilde{\eta}_i, \quad i = 2, \ldots, N, \tag{2.9}$$

where λ_i, $i = 2, \ldots, N$ are the $N-1$ nonzero eigenvalues of the Laplacian matrix L.

Theorem 2.5 *The MAS* (2.8) *achieves state synchronization if and only if the system* (2.9) *is globally asymptotically stable for* $i = 2, \ldots, N$.

Combining this theorem with Lemma 1.3 immediately yields the following necessary condition for the state synchronization.

Corollary 2.6 *Consider a MAS described by* (2.1) *and* (2.5). *If there exists a protocol of the form* (2.6) *that achieves state synchronization, then the associated graph must have a directed spanning tree or A must be asymptotically stable.*

The case that A is asymptotically stable is a trivial case where doing nothing (all protocols set to zero) will already achieve state synchronization since in that case all states will converge to zero. As such, the requirement that the graph of a multi-agent system has a directed spanning tree is essentially necessary.

Proof of Theorem 2.5 Note that L has eigenvalue 0 with associated right eigenvector $\mathbf{1}$. Let

$$L = TJT^{-1} \tag{2.10}$$

where J is the Jordan canonical form of the Laplacian matrix L such that $J(1,1) = 0$ and the first column of T equals $\mathbf{1}$. Let

$$\eta := (T^{-1} \otimes I_n)x = \begin{pmatrix} \eta_1 \\ \vdots \\ \eta_N \end{pmatrix},$$

where $\eta_i \in \mathbb{C}^n$. In the new coordinates, the dynamics of η can be written as:

$$\dot{\eta} = (I_N \otimes A + J_L \otimes BF)\eta.$$

If (2.9) is globally asymptotically stable for $i = 2, \ldots, N$, we see from the above that $\eta_i(t) \to 0$ for $i = 2, \ldots, N$. This implies that

$$x(t) - (T \otimes I_n) \begin{pmatrix} \eta_1(t) \\ 0 \\ \vdots \\ 0 \end{pmatrix} \to 0.$$

Note that the first column of T is equal to the vector $\mathbf{1}$ and therefore

$$x_i(t) - \eta_1(t) \to 0$$

for $i = 1, \ldots, N$. This implies that we achieve state synchronization.

Conversely, suppose that the network (2.8) reaches state synchronization. In this case, we shall have

$$x(t) - \mathbf{1} \otimes x_1(t) \to 0$$

for all initial conditions. Then $\eta(t) - (T^{-1}\mathbf{1}) \otimes x_1(t) \to 0$. Since $\mathbf{1}$ is the first column of T, we have

$$T^{-1}\mathbf{1} = \begin{pmatrix} 1 \\ 0 \\ \vdots \\ 0 \end{pmatrix}.$$

Therefore, $\eta(t) - (T^{-1}\mathbf{1}) \otimes x_1(t) \to 0$ implies that $\eta_1(t) - x_1(t) \to 0$ and $\eta_i(t) \to 0$ for $i = 2, \ldots, N$ for all initial conditions. This implies that $A + \lambda_i BF$ is Hurwitz stable for $i = 2, \ldots, N$. ■

Remark 2.7 The function η_1, as defined in the proof of Theorem 2.5, satisfies

$$\dot{\eta}_1 = A\eta_1, \quad \eta_1(0) = (w \otimes I_n)x(0), \tag{2.11}$$

which can be shown by exploiting that 0 is a simple eigenvalue of the Laplacian. In the above, w is the first row of T^{-1}, i.e., the normalized left eigenvector with row sum equal to 1 associated with the zero eigenvalue.

The proof of Theorem 2.5 then yields that the *synchronized trajectory* x_s of the network is given by $x_s(t) = \eta_1(t)$. This shows that the modes of the synchronized trajectory are determined by the eigenvalues of A and the complete dynamics depends on both A and a weighted average of the initial conditions of agents.

Regarding the weighted average of the initial conditions of agents, we have the following lemma from [164, Lemma 2.29].

Lemma 2.8 *Consider a directed graph that satisfies Assumption 2.2 with its corresponding Laplacian matrix L. Let*

$$w = \begin{pmatrix} w_1 & \cdots & w_N \end{pmatrix}$$

be the normalized left eigenvector associated with the zero eigenvalue of L. Then, $w_i > 0$ *for root agents, and otherwise* $w_i = 0$.

In light of the above lemma, one can conclude that $\eta_1(0)$ is only a linear combination of initial conditions of root agents. Recall that the set of all root agents for a graph $\mathcal{G}$ is denoted by $\Pi_{\mathcal{G}}$. Therefore, the synchronized trajectory given by (2.11) yields that the synchronized trajectory is given by

$$x_s(t) = e^{At} \sum_{i \in \Pi_{\mathcal{G}}} w_i x_i(0), \tag{2.12}$$

which is the weighted average of the trajectories of root agents. As such, these root agents are called leaders.

Remark 2.9 If A is neutrally stable, i.e., all the eigenvalues of A are in the closed left half plane and those eigenvalues on the imaginary axis, if any, are semi-simple, then the synchronized trajectories are bounded. Otherwise, for at least some initial conditions, the synchronized trajectory will be unbounded.

Note that the above reduces the state synchronization problem for a *known* graph to a *simultaneous* stabilization problem. We need to find one state feedback that stabilizes $N-1$ different systems. It is notoriously difficult to find conditions for simulteneous stabilization problems with more than *two* different systems (see, for instance, [7]).

In light of the definition of Problem 2.3 where synchronization is formulated for a set of graphs $\mathbf{G}$, we basically obtain a *robust* stabilization problem, i.e., the stabilization of the system

$$\dot{x} = Ax + \lambda Bu, \tag{2.13}$$

via a protocol

$$u = Fx, \tag{2.14}$$

for any λ which is a nonzero eigenvalue of a Laplacian matrix associated with a graph in the set $\mathbf{G}$. There are more techniques available for robust stabilization.

Next we will show that the state synchronization problem for full-state coupling is always solvable for a set of graphs $\mathbf{G} = \mathbb{G}_\beta^N$, where β is a lower bound for the real part of all nonzero eigenvalues of any Laplacian matrices associated with a graph in the set $\mathbb{G}_\beta^N$. We will present below two protocol design methods based on robust stabilization. One relies on an algebraic Riccati equation (ARE), and the other is based on an asymptotic time-scale and eigenstructure assignment (ATEA) method.

2.4.2 ARE-Based Method

Protocol design 2.1 *Consider a MAS described by (2.1) and (2.5). We consider the protocol*

$$u_i = \rho F \zeta_i, \tag{2.15}$$

where $\rho \geq 1$ and $F = -B'P$ with $P > 0$ being the unique solution of the continuous-time algebraic Riccati equation

$$A'P + PA - 2\beta PBB'P + Q = 0, \tag{2.16}$$

where $Q > 0$ and β is a lower bound for the real part of the nonzero eigenvalues of all Laplacian matrices associated with a graph in the set $\mathbb{G}_\beta^N$.

Remark 2.10 The protocol (2.15) can be generalized to the following protocol:

$$u_i = \rho F \zeta_i, \tag{2.17}$$

where $\rho \geq 1$ and $F = -R^{-1}B'P$ with $P > 0$ being the unique solution of the continuous-time algebraic Riccati equation

$$A'P + PA - 2\beta PBR^{-1}B'P + Q = 0, \tag{2.18}$$

with $Q > 0$ and $R > 0$. Note that the protocol (2.15) is a special case of the protocol (2.17) with $R = I$.

Remark 2.11 We can equivalently use the following protocol:

$$u_i = \rho F \zeta_i, \tag{2.19}$$

where $F = -B'P$ with $P > 0$ being the unique solution of the continuous-time algebraic Riccati equation

$$A'P + PA - PBB'P + Q = 0, \tag{2.20}$$

with $Q > 0$, and $\rho \geq \frac{1}{2\beta}$.

Using an algebraic Riccati equation, we can design a suitable protocol provided (A, B) is stabilizable. The main result based on Protocol Design 2.1 is as follows.

Theorem 2.12 *Consider a MAS described by* (2.1) *and* (2.5). *Let any* $\beta > 0$ *be given, and consider the set of network graphs* $\mathbb{G}_\beta^N$ *as defined in Definition* 1.6.

If (A, B) *is stabilizable, then the state synchronization problem stated in Problem* 2.3 *with* $\mathbf{G} = \mathbb{G}_\beta^N$ *is solvable. In particular, the protocol* (2.15) *solves the state synchronization problem for any graph* $\mathcal{G} \in \mathbb{G}_\beta^N$. *Moreover, the synchronized trajectory is given by* (2.12).

Note that $\rho \geq 1$ and $Q > 0$ and, possibly, R are design parameters in the protocol design which affect issues such as convergence rate of the synchronization or overshoot effects.

Proof Note that $\mathcal{G} \in \mathbb{G}_\beta^N$ implies that the graph has a directed spanning tree and all nonzero eigenvalues of the Laplacian have a real part larger than or equal to β. Using Theorem 2.5, we only need to prove that the system

$$\dot{x} = (A - \lambda\rho BB'P)x \tag{2.21}$$

is asymptotically stable for any λ that satisfies $\mathrm{Re}(\lambda) \geq \beta$.

We observe that

$$\begin{aligned}
&(A - \lambda\rho BB'P)^*P + P(A - \lambda\rho BB'P) \\
&= -Q - 2(\rho\,\mathrm{Re}(\lambda) - \beta)PBB'P \\
&\leq -Q.
\end{aligned}$$

Therefore, $(A - \lambda\rho BB'P)$ is Hurwitz stable for any $\mathrm{Re}(\lambda) \geq \beta$, which proves the result. ■

2.4.3 ATEA-Based Method

If the graph $\mathcal{G}$ is an arbitrary element of $\mathbb{G}_\beta^N$, then using Theorem 2.5, we need to find a state feedback F such that (2.13) and (2.14) result in a stable system for all λ with $\mathrm{Re}\,\lambda \geq \beta$.

The ATEA-based design is basically a method of time-scale structure assignment in linear multivariable systems by high-gain feedback [96]. In the current case, we do not need the full structure presented in the above method. It is sufficient to note that there exists a nonsingular state transformation T_x such that

$$\bar{x} = \begin{pmatrix} \bar{x}_1 \\ \bar{x}_2 \end{pmatrix} = T_x x, \tag{2.22}$$

and the dynamics of $\bar{x}$ is represented as

$$\begin{aligned} \dot{\bar{x}}_1 &= \bar{A}_{11}\bar{x}_1 + \bar{A}_{12}\bar{x}_2 \\ \dot{\bar{x}}_2 &= \bar{A}_{21}\bar{x}_1 + \bar{A}_{22}\bar{x}_2 + \lambda \bar{B}u \end{aligned} \tag{2.23}$$

with $\bar{B}$ invertible, and, since (A, B) is stabilizable, we have that $(\bar{A}_{11}, \bar{A}_{12})$ is stabilizable. Using the above structure, we can design a suitable protocol for our MAS.

Protocol design 2.2 *Consider a MAS described by (2.1) and (2.5). Let T_x be such that the basis transformation (2.22) results in a system of the form (2.23).*

We choose F_1 such that $\bar{A}_{11} + \bar{A}_{12}F_1$ is asymptotically stable. In that case, a suitable controller for (2.23) achieving the required robust stability is

$$u = \frac{1}{\varepsilon}\bar{B}^{-1}(F_1\bar{x}_1 - \bar{x}_2). \tag{2.24}$$

for ε sufficiently small. This results in a protocol (in the original coordinates) $u_i = F_\varepsilon \zeta_i$ where

$$F_\varepsilon := \frac{1}{\varepsilon}\bar{B}^{-1}\begin{pmatrix} F_1 & -I \end{pmatrix} T_x. \tag{2.25}$$

The formal result is as follows.

Theorem 2.13 *Consider a MAS described by (2.1) and (2.5). Let any $\beta > 0$ be given and hence a set of network graphs $\mathbb{G}_\beta^N$ be defined.*

If (A, B) is stabilizable, then the state synchronization problem stated in Problem 2.3 with $\mathbf{G} = \mathbb{G}_\beta^N$ is solvable. In particular, there exists an ε^* such that for any $\varepsilon \in (0, \varepsilon^*]$, the protocol $u_i = F_\varepsilon \zeta_i$, where F_ε is defined by (2.24), solves the state synchronization problem for any graph $\mathcal{G} \in \mathbb{G}_\beta^N$. Moreover, the synchronized trajectory is given by (2.12).

Proof Consider the interconnection of (2.23) and (2.24). Then we can rewrite this system in the form:

$$\begin{aligned} \dot{\tilde{x}}_1 &= \tilde{A}_{11}\tilde{x}_1 + \tilde{A}_{12}\tilde{x}_2, \\ \dot{\tilde{x}}_2 &= \tilde{A}_{21}\tilde{x}_1 - \tfrac{\lambda}{\varepsilon}\tilde{x}_2, \end{aligned}$$

where:

$$\tilde{x}_1 = \bar{x}_1, \qquad \tilde{x}_2 = \bar{x}_2 - F_1\bar{x}_1$$

and $\tilde{A}_{11} = \bar{A}_{11} + \bar{A}_{12}F_1$ is asymptotically stable. Therefore, there exists a matrix $P_1 > 0$ such that:

$$P_1\tilde{A}_{11} + \tilde{A}_{11}'P_1 = -I.$$

Define a candidate Lyapunov function:

$$V = \tilde{x}_1^* P_1 \tilde{x}_1 + \tilde{x}_2^* \tilde{x}_2.$$

We find that:

$$\begin{aligned}
\dot{V} &= -\|\tilde{x}_1\|^2 - \tfrac{2}{\varepsilon}\operatorname{Re}(\lambda)\|\tilde{x}_2\|^2 + 2\operatorname{Re}\tilde{x}_1^*\left[P_1\tilde{A}_{12} + \tilde{A}_{21}^*\right]\tilde{x}_2 \\
&\leq -\|\tilde{x}_1\|^2 - \tfrac{2\beta}{\varepsilon}\|\tilde{x}_2\|^2 + r\|\tilde{x}_1\|\,\|\tilde{x}_2\| \\
&\leq -\tfrac{1}{2}\|\tilde{x}_1\|^2 - \tfrac{1}{2}\|\tilde{x}_2\|^2
\end{aligned}$$

since $\operatorname{Re}\lambda \geq \beta$ where:

$$r = 2\|P_1\tilde{A}_{12} + \tilde{A}_{21}^*\|$$

provided $\varepsilon < \varepsilon^*$ with:

$$\varepsilon^* = \frac{4\beta}{r^2+1}.$$

Using Theorem 2.5, this robust stabilization implies that the protocol $u_i = F_\varepsilon \zeta_i$ achieves synchronization. ■

2.4.4 Example: Comparison ATEA- and ARE-Based Methods

The ATEA-based method is superior to the ARE-based method for high performance (i.e., fast convergence). Namely, via ATEA-based method, one can achieve the same level of performance with a smaller gain. This yields a smaller bandwidth than via ARE-based methods. This feature has been documented in a different context (see [102]). The following example illustrates this feature.

This example illustrates ARE-based and ATEA-based methods for designing protocols. The network considered is comprised of *three* agents in the form of (2.1) with:

$$A = \begin{pmatrix} 0 & 1 & 0 & 1 \\ 0 & 0 & 1 & 0 \\ 0 & 0 & 0 & 1 \\ -1 & -1 & -1 & -1 \end{pmatrix}, \quad B = \begin{pmatrix} 1 & 0 \\ 0 & 0 \\ 0 & 0 \\ 0 & 1 \end{pmatrix}.$$

We consider a set of network graphs $\mathbb{G}_\beta^N$ with $\beta = 0.8$. We choose a network that belongs to this graph set that has the Laplacian matrix:

$$L = \begin{pmatrix} 0 & 0 & 0 \\ -1 & 1 & 0 \\ 0 & -2 & 2 \end{pmatrix}.$$

By using the Riccati-based method with $Q = I$, we get:

$$F = \begin{pmatrix} -1.1224 & -0.5994 & 0.4404 & 0.1397 \\ 0.1397 & -0.6851 & -1.4336 & -1.1374 \end{pmatrix}.$$

Choosing the parameter $\rho = 15$, then the state feedback gain $\|F\| = 24.7331$. The synchronized trajectories are shown in Fig. 2.1.

When using the direct eigenstructure assignment method with $\varepsilon = 0.4$, we get:

$$F = \frac{1}{\varepsilon} \begin{pmatrix} -1 & 0 & 0 & 0 \\ 0 & -1.5 & -2 & -1 \end{pmatrix}$$

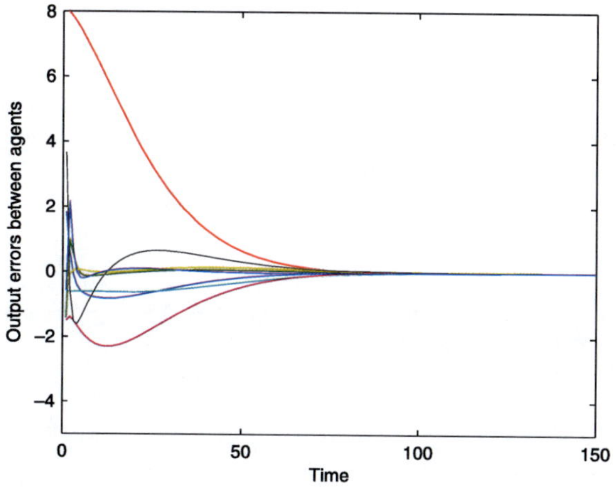

Fig. 2.1 Riccati-based method

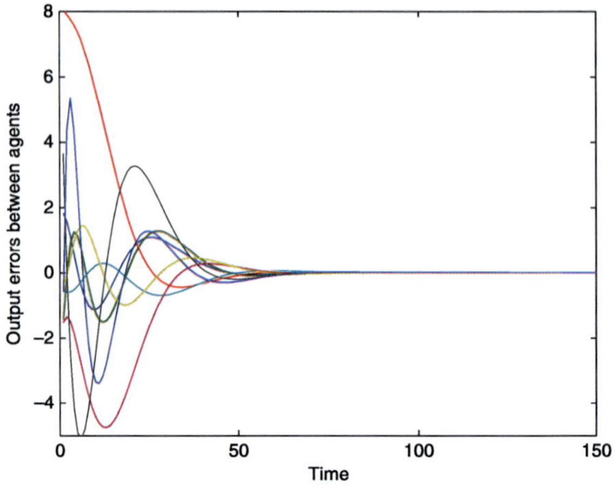

Fig. 2.2 Direct eigenstructure assignment method

with the state feedback gain $\|F\| = 6.7315$, and the synchronized trajectories are shown in Fig. 2.2. We see that with the similar settling time of 100 sec and overshoot equal to 8, ATEA-based method has a smaller gain than that of ARE-based method.

2.4.5 Neutrally Stable Agents

We next consider a special case where the agent dynamics are neutrally stable, that is, the eigenvalues of A on the imaginary axis, if any, are semi-simple. This implies that A is stable and there exists a positive definite matrix $P > 0$ such that:

$$A'P + PA \leq 0.$$

In this case, we show here that the consensus controller design no longer requires the knowledge of β and hence allows us to deal with a larger set of unknown network graphs $\mathbb{G}^N$ for any agent number N.

We design protocols for MAS with neutrally stable agents as follows.

Protocol design 2.3 *Consider a MAS described by* (2.1) *and* (2.5). *We consider a protocol:*

$$u_i = \rho F \zeta_i, \tag{2.26}$$

(continued)

Protocol design 2.3 (continued)
where $\rho > 0$ and $F = -B'P$ with $P > 0$ being a solution of the Lyapunov function:

$$A'P + PA \leq 0.$$

The main result based on the above design is as follows.

Theorem 2.14 *Consider a MAS described by (2.1) and (2.5). Let a set of network graphs $\mathbb{G}^N$ be defined.*

Suppose that the agent dynamics are neutrally stable. If (A, B) is stabilizable, then the state synchronization problem stated in Problem 2.3 with $\mathbf{G} = \mathbb{G}^N$ is solvable for any number N. In particular, the protocol (2.26) solves the state synchronization problem for any graph $\mathcal{G}$ which is in $\mathbb{G}^N$ for any N. Moreover, the synchronized trajectory is given by (2.12).

Proof We find that $\mathcal{G} \in \mathbb{G}^N$ implies that the graph has a directed spanning tree and all nonzero eigenvalues of the Laplacian have a real part larger than 0. Using Theorem 2.5, we only need to prove that the system:

$$\dot{x} = (A - \lambda\rho BB'P)x \tag{2.27}$$

is asymptotically stable for any λ that satisfies $\operatorname{Re}\lambda > 0$.

We have:

$$(A - \lambda\rho BB'P)^*P + P(A - \lambda\rho BB'P) \leq -2\rho\operatorname{Re}(\lambda)PBB'P.$$

Since (A, B) is stabilizable and $\operatorname{Re}\lambda > 0$, it follows from LaSalle's invariance principle that $A - \lambda\rho BB'P$ is Hurwitz stable, which proves the result. ■

2.5 Protocol Design for MAS with Partial-State Coupling

In this section, similar to the approach of the previous section, we show first that the state synchronization among agents in the network with partial-state coupling can be solved by equivalently solving a robust stabilization problem, and then we design a controller for such a robust stabilization problem.

2.5.1 Connection to a Robust Stabilization Problem

The MAS system described by (2.1) and (2.3) after implementing the linear static protocol (2.7) is described by:

$$\begin{cases} \dot{\bar{x}}_i = \begin{pmatrix} A & BC_c \\ 0 & A_c \end{pmatrix} \bar{x}_i + \begin{pmatrix} BD_c \\ B_c \end{pmatrix} \zeta_i, \\ y_i = \begin{pmatrix} C & 0 \end{pmatrix} \bar{x}_i, \\ \zeta_i = \sum_{j=1}^{N} a_{ij}(y_i - y_j), \end{cases} \tag{2.28}$$

for $i = 1, \ldots, N$, where:

$$\bar{x}_i = \begin{pmatrix} x_i \\ \chi_i \end{pmatrix}.$$

Define:

$$\bar{x} = \begin{pmatrix} \bar{x}_1 \\ \vdots \\ \bar{x}_N \end{pmatrix}, \quad \bar{A} = \begin{pmatrix} A & BC_c \\ 0 & A_c \end{pmatrix}, \quad \bar{B} = \begin{pmatrix} BD_c \\ B_c \end{pmatrix}, \quad \text{and} \quad \bar{C} = \begin{pmatrix} C & 0 \end{pmatrix}.$$

Then, the overall dynamics of the N agents can be written as:

$$\dot{\bar{x}} = (I_N \otimes \bar{A} + L \otimes \bar{B}\bar{C})\bar{x}. \tag{2.29}$$

A key observation, which we already used in the case of full-state coupling, is that synchronization for the system (2.29) is equivalent to the asymptotic stability of the following $N - 1$ subsystems:

$$\dot{\bar{\eta}}_i = (\bar{A} + \lambda_i \bar{B}\bar{C})\bar{\eta}_i, \quad i = 2, \ldots, N, \tag{2.30}$$

where λ_i for $i = 2, \ldots, N$ are the nonzero eigenvalues of the Laplacian matrix L. This is formalized in the following theorem.

Theorem 2.15 *The MAS* (2.29) *achieves state synchronization if and only if the system* (2.30) *is globally asymptotically stable for* $i = 2, \ldots, N$.

Proof As we did in the proof of Theorem 2.5, we define the Jordan form (2.10) of the Laplacian. Let:

$$\eta := \left(T^{-1} \otimes I_{n+n_c}\right) \bar{x} = \begin{pmatrix} \eta_1 \\ \vdots \\ \eta_N \end{pmatrix},$$

where $\eta_i \in \mathbb{C}^{n+n_c}$. In the new coordinates, the dynamics of η can be written as:

$$\dot{\eta} = \left[I_N \otimes \bar{A} + J_L \otimes \bar{B}\bar{C}\right] \eta.$$

If (2.30) is globally asymptotically stable for $i = 2, \ldots, N$, we see from the above that: $\eta_i(t) \to 0$ for $i = 2, \ldots, N$. This implies that

$$\bar{x}(t) - (T \otimes I_{n+n_c}) \begin{pmatrix} \eta_1(t) \\ 0 \\ \vdots \\ 0 \end{pmatrix} \to 0.$$

Note that the first column of T is equal to the vector $\mathbf{1}$ and therefore:

$$\bar{x}_i(t) - \eta_1(t) \to 0$$

for $i = 1, \ldots, N$. This implies that we achieve state synchronization.

Conversely, suppose that the network (2.29) reaches synchronization. In this case, we shall have:

$$\bar{x}(t) - \mathbf{1} \otimes x_1(t) \to 0$$

for all initial conditions. Then, $\eta(t) - (T^{-1}\mathbf{1}) \otimes x_1(t) \to 0$. Since $\mathbf{1}$ is the first column of T, we have:

$$T^{-1}\mathbf{1} = \begin{pmatrix} 1 \\ 0 \\ \vdots \\ 0 \end{pmatrix}.$$

Therefore, $\eta(t) - (T^{-1}\mathbf{1}) \otimes x_1(t) \to 0$ implies that $\eta_1(t) - x_1(t) \to 0$ and $\eta_i(t) \to 0$ for $i = 2, \ldots, N$ and for all initial conditions. This implies that $\bar{A} + \lambda_i \bar{B}\bar{C}$ is Hurwitz stable for $i = 2, \ldots, N$. ∎

Corollary 2.16 *Consider a MAS described by (2.1) and (2.3). If there exists a protocol of the form (2.7) that achieves state synchronization, then the associated graph must have a directed spanning tree, or A must be asymptotically stable.*

Since the case where A is asymptotically stable is a trivial case, the requirement that the graph of a multi-agent system has a directed spanning tree is essentially necessary.

Remark 2.17 Similar to what we did in Theorem 2.7, we consider η_1 defined in the proof of Theorem 2.15 and show that it satisfies:

$$\dot{\eta}_1 = \bar{A}\eta_1, \quad \eta_1(0) = (w \otimes I_{n+n_c})\bar{x}(0), \tag{2.31}$$

where w is the first row of T^{-1}, i.e., the normalized left eigenvector associated with the zero eigenvalue.

It can then be shown that the synchronized trajectory (i.e., the asymptotic behavior of the state of each agent) is given by the first component, consistent with the partitioning of $\bar{x}_i$, of the steady state behavior of $\eta_1(t)$. This shows that the modes of the synchronized trajectory are determined by the eigenvalues of $\bar{A}$, which is the union of the eigenvalues of A and A_c. The complete dynamics depends on both $\bar{A}$ and a weighted average of the initial conditions of the agents and their protocols.

Again, in light of Lemma 2.8, one can conclude that $\eta_1(0)$ is only a linear combination of initial conditions of root agents and their protocol. As such, the synchronized trajectory given by (2.31) can be written explicitly as:

$$x_s(t) = \begin{pmatrix} I & 0 \end{pmatrix} e^{\bar{A}t} \sum_{i \in \Pi_{\mathcal{G}}} w_i \bar{x}_i(0), \tag{2.32}$$

which is the weighted average of the trajectories of root agents and their protocol. In the case of state synchronization via a stable protocol, the synchronized trajectory is given by:

$$x_s(t) = e^{At} \sum_{i \in \Pi_{\mathcal{G}}} w_i x_i(0). \tag{2.33}$$

Remark 2.18 From the above, the modes of the synchronized trajectory always contain the eigenvalues of the matrix A. However, additional modes are determined through the protocol and are given by the eigenvalues of the matrix A_c. In other words, the unstable modes of the protocol (2.7), if any, can be present in the synchronized trajectories. Thus, it is reasonable that one might strengthen the requirements on the protocol by imposing that the protocol is stable.

Note that (2.30) can be viewed as the closed-loop of the system:

$$\begin{cases} \dot{x} = Ax + Bu, \\ y = \lambda_i Cx, \end{cases} \tag{2.34}$$

and a dynamic controller:

$$\begin{cases} \dot{\chi} = A_c\chi + B_c y, \\ u = C_c\chi + D_c y, \end{cases} \tag{2.35}$$

with λ_i ($i = 2, \ldots, N$) being the nonzero eigenvalues of the Laplacian matrix associated with a given graph $\mathcal{G}$. It is easy to see that owing to the linearity, (2.35) stabilizes (2.34) if it stabilizes:

$$\begin{cases} \dot{x} = Ax + \lambda_i Bu, \\ y = Cx. \end{cases} \tag{2.36}$$

For a given graph, this is a simultaneous stabilization problem. However, in light of the definition of Problem 2.4 where synchronization is formulated for a set of graphs $\mathbf{G}$, we basically obtain a *robust* stabilization problem, i.e., the stabilization of the system:

$$\begin{cases} \dot{x} = Ax + \lambda Bu, \\ y = Cx, \end{cases} \tag{2.37}$$

via a controller (2.35) for any λ which is a nonzero eigenvalue of a Laplacian matrix associated with a graph in the given set of graphs $\mathbf{G}$.

2.5.2 *Solvability Conditions*

We have seen in the previous subsection that synchronization of MAS for a set of graphs reduces to a robust stabilization problem. In this subsection, we will determine the solvability conditions for controllers that solve this robust stabilization problem. These conditions will be expressed in terms of two matrices which we will first define.

- Let $P_0 \geq 0$ be the unique solution (see [160]) of the Riccati equation:

$$A'P_0 + P_0A - P_0BB'P_0 = 0 \tag{2.38}$$

  such that $A - BB'P_0$ has all eigenvalues in the closed left half plane.

- Moreover, let $Q_0 \geq 0$ be the solution of the linear matrix inequality (see [116]):

$$\begin{pmatrix} AQ_0 + Q_0A' + BB' & Q_0C' \\ CQ_0 & 0 \end{pmatrix} \geq 0, \tag{2.39}$$

  such that:

$$\operatorname{rank} \begin{pmatrix} AQ_0 + Q_0A' + BB' & Q_0C' \\ CQ_0 & 0 \end{pmatrix} = \gamma$$

  and:

$$\operatorname{rank} \begin{pmatrix} sI - A & AQ_0 + Q_0A' + BB' & Q_0C' \\ -C & CQ_0 & 0 \end{pmatrix} = n + \gamma$$

  for all s in the open right half plane where

$$\gamma = \operatorname{normrank} C(sI - A)^{-1}B.$$

  Here normrank is the rank of the transfer matrix for all but finitely many values of s. In the single-input case, we have $\gamma = 1$ (excluding the trivial case where the transfer function is equal to zero).

Note that these two matrices are always well-defined provided (A, B) is stabilizable and (C, A) is detectable.

2.5.2.1 Single-Input Agents

In this chapter, we mostly consider synchronization problems where the associated graph is in the set $\mathbb{G}^N_{\alpha,\beta}$. It is convenient to use a different class as well. $\mathbb{G}^N_\rho$ denotes the set of graphs that contain a directed spanning tree and where the nonzero eigenvalues of the Laplacian are in the set:

$$\Lambda_\rho := \left\{ \lambda \in \mathbb{C}^+ \mid \lambda = \tfrac{\rho}{1+\rho} + \mu \text{ where } \mu \in \mathbb{C} \text{ such that } |1 - 2\rho\mu| \leq 1 \right\}, \tag{2.40}$$

for any $0 < \rho < 1$. On the other hand, $\mathbb{G}^N_{\rho,M}$ is the set of graphs $\mathbb{G}^N_\rho$ with the additional property that the associated Laplacian satisfies $\|L\| < M$.

Lemma 2.19 *For any $\alpha > \beta > 0$, there exists a $\rho \in (0, 1)$ such that:*

$$\mathbb{G}^N_{\alpha,\beta} \subset \mathbb{G}^N_\rho. \tag{2.41}$$

Conversely, for any $\rho \in (0, 1)$ and $M > 0$, there exists an $\alpha > \beta > 0$ such that:

$$\mathbb{G}^N_{\rho,M} \subset \mathbb{G}^N_{\alpha,\beta}. \tag{2.42}$$

Proof Given ρ and M, we choose:

$$\alpha = \max\left\{\frac{\rho}{1+\rho}, M\right\},$$
$$\beta = \frac{\rho}{1+\rho} + \frac{1}{\rho}.$$

Then it easy to verify that (2.42) is satisfied. Conversely, let α and β be given; then we choose ρ by:

$$\rho = \min\left\{\frac{\beta}{3(\alpha^2+1-2\beta)}, \frac{1}{2}\right\}. \tag{2.43}$$

We will show that (2.41) is satisfied. Consider a λ with $\operatorname{Re}\lambda \geq \beta$ and $|\lambda| \leq \alpha$. We set:

$$\lambda = \frac{\rho}{1+\rho} + \frac{1}{2\rho} + v.$$

It is not hard to verify that $\lambda \in \Lambda_\rho$ if and only if v satisfies:

$$|v| \leq \frac{1}{2\rho}. \tag{2.44}$$

We have for $\lambda = \lambda_1 + i\lambda_2$ that:

$$\begin{aligned}
\|v\|^2 &= \left(\lambda_1 - \frac{\rho}{1+\rho} - \frac{1}{2\rho}\right)^2 + \lambda_2^2 \\
&= |\lambda|^2 - 2\lambda_1\left(\frac{\rho}{1+\rho} + \frac{1}{2\rho}\right) + \left(\frac{\rho}{1+\rho} + \frac{1}{2\rho}\right)^2 \\
&\leq \alpha^2 - 2\beta\left(\frac{\rho}{1+\rho} + \frac{1}{2\rho}\right) + \left(\frac{\rho}{1+\rho} + \frac{1}{2\rho}\right)^2.
\end{aligned}$$

We find that:

$$\begin{aligned}
\|v\|^2 &\leq \alpha^2 + 1 - 2\beta\left(\frac{\rho}{1+\rho} + \frac{1}{2\rho}\right) + \frac{1}{4\rho^2} \\
&\leq \alpha^2 + 1 - 2\beta - \beta\frac{1-\rho}{\rho(1+\rho)} + \frac{1}{4\rho^2} \\
&\leq \alpha^2 + 1 - 2\beta - \beta\frac{1}{3\rho} + \frac{1}{4\rho^2}
\end{aligned}$$

where in the last step we exploited that $\rho \le \frac{1}{2}$. It is then easily verified that for ρ given by (2.43) we have (2.44). ■

The following theorem characterizes when our synchronization problem as formulated in Problem 2.4 is solvable.

Theorem 2.20 *Consider a MAS described by (2.1) and (2.3) with a scalar input with (A, B) stabilizable and (C, A) detectable. Let $\rho \in (0, 1)$. Then, the problem of synchronization with partial-state coupling as defined in Problem 2.4 is solvable for $\mathbf{G} = \mathbb{G}_\rho^N$ if and only if:*

$$\lambda_{\max}(P_0 Q_0) < 4\rho^2 \left[1 - \frac{\rho}{(\rho+1)^2} \right] \tag{2.45}$$

where $\lambda_{\max}$ denotes the spectral radius.

Proof of Theorem 2.20 Using Theorem 2.15, we need to verify whether for any $\beta > 0$ there exists a controller $\mathcal{H}$ of the form:

$$\mathcal{H}: \begin{cases} \dot{\chi} = A_c \chi + B_c y \\ u = C_c \chi + D_c y \end{cases} \tag{2.46}$$

for the system:

$$\mathcal{S}: \begin{cases} \dot{x} = Ax + \lambda B u \\ y = Cx \end{cases} \tag{2.47}$$

such that the interconnection is stable for all λ in Λ_ρ given by (2.40).

The interconnection of $\mathcal{S}$ and $\mathcal{H}$ can be written as:

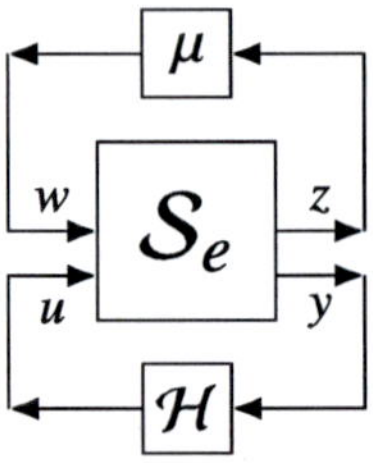

where:

$$\mathcal{S}_e: \begin{cases} \dot{x} = Ax + \frac{\rho}{1+\rho} Bu + Bw \\ y = Cx \\ z = u. \end{cases} \tag{2.48}$$

Denote the transfer function of the interconnection of $\mathcal{S}_e$ and $\mathcal{H}$ (with input w and output z) by S_{cl}. We find that the interconnection of $\mathcal{S}$ and $\mathcal{H}$ is stable if the interconnection of $\mathcal{S}_e$ and $\mathcal{H}$ is internally stable and:

$$1 - \mu S_{\text{cl}}(s) \neq 0 \tag{2.49}$$

for all μ such that $|1 - 2\rho\mu| \leq 1$ and for all $s \in \mathbb{C}^+$. We note that:

$$\begin{aligned}(1-2\rho\mu)(1-2\rho\mu^*) \leq 1 &\Longleftrightarrow 2\rho\mu\mu^* - \mu - \mu^* \leq 0\\ &\Longleftrightarrow \tfrac{1}{\mu} + \tfrac{1}{\mu^*} \geq 2\rho.\end{aligned} \tag{2.50}$$

This implies that the interconnection satisfies (2.49) if and only if:

$$S_{\text{cl}}(s) + S_{\text{cl}}^*(s) < 2\rho,$$

which is equivalent to:

$$(\rho - S_{\text{cl}}(s) - 1)^* \, (\rho - S_{\text{cl}}(s) - 1) < (\rho - S_{\text{cl}}(s) + 1)^* \, (\rho - S_{\text{cl}}(s) + 1)$$

for all $s \in \mathbb{C}^+$ which yields the requirement that:

$$\left\| (\rho - S_{\text{cl}} + 1)^{-1} \, (\rho - S_{\text{cl}} - 1) \right\|_\infty < 1.$$

Next, note that:

$$q = -v + (\rho - S_{\text{cl}}(s)) \, (v - q)$$

yields that the transfer function from v to q is equal to:

$$(\rho - S_{\text{cl}}(s) + 1)^{-1} \, (\rho - S_{\text{cl}}(s) - 1) \, .$$

Therefore, we can rewrite our original question as trying to find a stabilizing controller of the form (2.46) for the system:

$$\bar{\mathcal{S}}_e : \begin{cases} \dot{x} = Ax + \frac{\rho}{1+\rho} Bu + \frac{1+\rho}{2} B(v-q) \\ y = Cx \\ q = -\frac{2}{1+\rho} u + \rho(v-q) - v, \end{cases}$$

such that the closed-loop transfer matrix from v to q has an H_∞ norm less than 1. This system $\bar{\mathcal{S}}_e$ can be rewritten as:

$$\bar{\mathcal{S}}_e : \begin{cases} \dot{x} = Ax + Bu + Bv \\ y = Cx \\ q = -\frac{2}{(\rho+1)^2} u + \frac{\rho-1}{\rho+1} v. \end{cases} \tag{2.51}$$

In order to check whether we can make the H_∞ norm less than 1, we use the approach presented in [125]. In this reference, it is shown that we can make the H_∞ norm less than 1 if and only if four conditions are satisfied.

Firstly, there must exist a solution of the state feedback Riccati equation:

$$A'P + PA - \nu PBB'P = 0 \tag{2.52}$$

such that $A - \nu BB'P$ has all its eigenvalues in the closed left half plane. Here:

$$\nu = \rho(\rho^2 + \rho + 1).$$

We note:

$$P = \frac{1}{\nu} P_0$$

where P_0 is given in Theorem 2.20. Hence, this condition is always satisfied.

We have a singular problem, and hence, the second condition is the existence of a solution of a quadratic matrix inequality (which reduces to a linear matrix inequality in this special case):

$$\begin{pmatrix} AQ + QA' + \eta BB' & QC' \\ CQ & 0 \end{pmatrix} \geq 0, \tag{2.53}$$

such that:

$$\operatorname{rank} \begin{pmatrix} sI - A & AQ + QA' + \eta BB' & QC' \\ -C & CQ & 0 \end{pmatrix} = n + 1, \tag{2.54}$$

for all s in the open right half plane where:

$$\eta = \frac{(\rho + 1)^2}{4\rho}.$$

We find that:

$$Q = \eta Q_0$$

where Q_0 is given in Theorem 2.20. Hence, this second condition is also always satisfied.

The third condition is:

$$\lambda_{\max}(PQ) < 1. \tag{2.55}$$

We have:

$$PQ = \frac{\eta}{\nu} P_0 Q_0,$$

which immediately yields that (2.55) is equivalent to the condition (2.45).

The final condition from [125] is that for any $s_0 \in \mathbb{C}^0$ which is an invariant zero of either $(A, B, C, 0)$ or $(A, B, 0, I)$, there should exist a K such that:

$$\left\| -\tfrac{2}{(\rho+1)^2} KC(s_0 I - A - BKC)^{-1} B + \tfrac{\rho-1}{\rho+1} \right\| < 1.$$

It is easy to check that it is sufficient to prove that we can find a K such that:

$$\|KC(s_0 I - A - BKC)^{-1} B\| \leq 1.$$

If $s_0 I - A$ is invertible, we can choose $K = 0$. Otherwise, we choose S and T such that:

$$s_0 I - A = S \begin{pmatrix} I & 0 \\ 0 & 0 \end{pmatrix} T, \quad B = S \begin{pmatrix} B_1 \\ B_2 \end{pmatrix}, \quad C = \begin{pmatrix} C_1 & C_2 \end{pmatrix} T.$$

Stabilizability of (A, B) guarantees that B_2 is surjective, while detectability of (C, A) guarantees that C_2 is injective. We denote by B_2^r and C_2^ℓ right and left inverses of B_2 and C_2, respectively, such that $B_2^r B_2$ is an orthogonal projection. We then choose $K = -B_2^r C_2^\ell$. We obtain:

$$\begin{aligned} &KC(s_0 I - A - BKC)^{-1} B = \\ &\quad = -B_2^r \begin{pmatrix} C_2^\ell C_1 & I \end{pmatrix} \begin{pmatrix} I + B_1 B_2^r C_2^\ell C_1 & B_1 B_2^r \\ C_2^\ell C_1 & I \end{pmatrix}^{-1} \begin{pmatrix} B_1 \\ B_2 \end{pmatrix} \\ &\quad = -B_2^r \begin{pmatrix} C_2^\ell C_1 & I \end{pmatrix} \begin{pmatrix} I & -B_1 B_2^r \\ -C_2^\ell C_1 & I + C_2^\ell C_1 B_1 B_2^r \end{pmatrix} \begin{pmatrix} B_1 \\ B_2 \end{pmatrix} \\ &\quad = -B_2^r \begin{pmatrix} 0 & I \end{pmatrix} \begin{pmatrix} B_1 \\ B_2 \end{pmatrix} = -B_2^r B_2, \end{aligned}$$

and since the norm of an orthogonal projection is less than or equal to 1, we note that this final condition from [125] is always satisfied. In other words, the given synchronization problem is solvable if and only if (2.45) is satisfied. ■

Corollary 2.21 *Consider a MAS described by* (2.1) *and* (2.3) *with a scalar input with* (A, B) *stabilizable and* (C, A) *detectable. Then, the problem of synchronization with Partial-state coupling as defined in Problem* 2.4 *is solvable with* $\mathbb{G} = \mathbb{G}^N_{\alpha,\beta}$

for any $\alpha > \beta > 0$ *if and only if:*

$$P_0 Q_0 = 0.$$

Proof Given Lemma 2.19, the problem of synchronization with partial-state coupling as defined in Problem 2.4 is solvable with $\mathbb{G} = \mathbb{G}^N_{\alpha,\beta}$ for any $\alpha > \beta > 0$ if and only if the problem of synchronization with partial-state coupling as defined in Problem 2.4 is solvable with $\mathbb{G} = \mathbb{G}^N_\rho$ for any $\rho \in (0, 1)$. By Theorem 2.20, this is possible if and only if (2.45) is satisfied for any $\rho \in (0, 1)$ which is clearly equivalent to $P_0 Q_0 = 0$. ■

Remark 2.22 Note that for two important cases, we have $P_0 Q_0 = 0$:

- We have $P_0 = 0$ if and only if A is at most weakly unstable.
- We have $Q_0 = 0$ if and only if (A, B, C) is at most weakly non-minimum-phase.

Clearly, there might be other cases where $P_0 Q_0 = 0$, but the above two cases are the important cases and will be explicitly addressed in later sections.

2.5.2.2 Multi-input Agents

In the general case, we will not have necessary and sufficient conditions. However, we do have the following sufficient condition for partial-state synchronization.

Theorem 2.23 *Consider a MAS described by* (2.1) *and* (2.3) *with multiple inputs, and with* (A, B) *stabilizable and* (C, A) *detectable. Then, the problem of synchronization with partial-state coupling as defined in Problem* 2.4 *is solvable for* $\mathbf{G} = \mathbb{G}^N_{\alpha,\beta}$ *for any* $\alpha > \beta > 0$ *if* $P_0 Q_0 = 0$.

Proof We rely on arguments from the proof of Theorem 2.20 with some modifications to make the argument valid for multi-input systems.

We first choose ρ such that any λ with $\operatorname{Re}\lambda \geq \beta$ and $|\lambda| \leq \alpha$ is contained in Λ_ρ which is defined in (2.40). Our objective is to design a controller $\mathcal{H}$ of the form (2.46) which stabilizes (2.47) for any $\lambda \in \Lambda_\rho$. We again define the system $\mathcal{S}_e$ by (2.48), and we need to ensure that the interconnection of $\mathcal{S}_e$ and $\mathcal{H}$ is internally stable and:

$$\det\left(I - \mu S_{\text{cl}}(s)\right) \neq 0 \tag{2.56}$$

for all μ such that $|1 - 2\rho\mu| \leq 1$ and for all $s \in \mathbb{C}^+ \cup \mathbb{C}^0$. It is sufficient if we achieve:

$$S_{\text{cl}}(s) + S^*_{\text{cl}}(s) < 2\rho I. \tag{2.57}$$

After all if (2.56) is not satisfied, then there exists a vector $v \in \mathbb{C}^n$ such that $\mu S_{\text{cl}}(s)v = v$ which implies that:

$$\mu v^* S_{\text{cl}}(s)v = v^* v,$$

and hence:

$$v^* S_{\text{cl}}(s)v + v^* S_{\text{cl}}^*(s)v = \left(\tfrac{1}{\mu} + \tfrac{1}{\mu^*}\right) v^* v \geq 2\rho v^* v$$

using (2.50) which yields a contradiction with (2.57).

Note that in Theorem 2.20 we developed necessary conditions and used that (2.56) and (2.57) are identical for scalar input systems. Here we want sufficient conditions for solvability, and we can rely on the fact that for multi-input systems (2.57) implies (2.56). Using arguments similar to as before, it is sufficient that we find a controller $\mathcal{H}$ which stabilizes the system $\bar{\mathcal{S}}_e$ given by (2.51) and yields an H_∞ norm from v to q less than 1. The existence of such a controller follows arguments similar to those in the proof of Theorem 2.20. ■

The combination of Theorem 2.20 and Theorem 2.23 results in a necessary and sufficient condition for the single-input case.

Corollary 2.24 *Consider a MAS described by (2.1) and (2.3) with a single-input with (A, B) stabilizable and (C, A) detectable. Then, the problem of synchronization with partial-state coupling as defined in Problem 2.4 is solvable for $\mathbf{G} = \mathbb{G}_{\alpha,\beta}^N$ for any $\alpha > \beta > 0$ if only if $P_0 Q_0 = 0$.*

In the above, we have obtained conditions for the existence of protocols to achieve synchronization by reducing the problem to an H_∞ control problem. In principle, we can use the results from [23, 124] to design such a controller.

Note that this specific design is not very transparent regarding the structure of the protocol. Also in many cases, the standard H_∞ controller results in a protocol which is not stable, while for many of these special cases we can find stable or even static protocols. The advantage of static or stable protocols has been argued earlier in Remark 2.18.

However, the class of MAS that satisfies our sufficient condition contains two important cases:

At most weakly unstable agents: In this case, the eigenvalues of A are in the closed left half plane. This set contains stable agents, neutrally stable agents, and weakly unstable agents.

At most weakly non-minimum-phase and left-invertible agents: In this case, the invariant zeros of (A, B, C) are in the closed left half plane. This set contains minimum-phase agents, weakly minimum-phase agents, and weakly non-minimum-phase agents.

Note that a precise definition of weakly minimum-phase and weakly non-minimum-phase is given in Definition 23.22.

Our design will concentrate on the two important cases identified in the above: at most weakly unstable agents and at most weakly non-minimum-phase agents. The first case guarantees that $P_0 = 0$. The second case yields $Q_0 = 0$ when the system is left-invertible, but we will see that the condition regarding left invertibility can be removed via a pre-compensator. We will also strengthen the latter assumption to minimum-phase and show that this enables the design of stable protocols. Finally, we briefly discuss the general case when the agents are neither at most weakly unstable nor at most weakly non-minimum-phase.

There exists a lot of research on synchronization where the agents are assumed to be passive or passifiable (see, for instance, [4, 35, 175]). Passive systems have their poles in the closed left half plane and are at most weakly non-minimum-phase among other conditions. Therefore, this class of agents is in the intersection of the cases mentioned above. However, passive and passifiable agents are an important class since it allows for static protocols as we will see later on and therefore will be addressed separately.

The different cases mentioned above will be addressed in the following subsections.

2.5.3 *MAS with at Most Weakly Unstable Agents*

We present here two protocol designs for a MAS with at most weakly unstable agents. One is based on the algebraic Riccati equation (ARE), and another is based on the direct eigenstructure assignment method. In each case, we are able to obtain a stable protocol which results in a synchronized trajectory which is completely determined by the dynamics of the agents.

2.5.3.1 ARE-Based Method

In this subsection, we design a protocol for MAS with at most weakly unstable agents. Our design is based on a low-gain feedback with a CSS observer.

Protocol design 2.4 *Consider a MAS described by at most weakly unstable agents* (2.1) *and* (2.3).

We choose an observer gain K such that $A + KC$ is Hurwitz stable. Next, we consider a feedback gain $F_\delta = -B'P_\delta$ where $P_\delta > 0$ is the unique solution of the continuous-time algebraic Riccati equation:

$$A'P_\delta + P_\delta A - \beta P_\delta BB'P_\delta + \delta I = 0, \tag{2.58}$$

(continued)

Protocol design 2.4 (continued)
where $\delta > 0$ is a low-gain parameter and β is the lower bound of the real part of the nonzero eigenvalues of the Laplacian. This results in the protocol:

$$\begin{cases} \dot{\chi}_i = (A + KC)\chi_i - K\zeta_i, \\ u_i = F_\delta \chi_i. \end{cases} \tag{2.59}$$

Remark 2.25 The protocol (2.59) is equivalent to the following protocol:

$$\begin{cases} \dot{\chi}_i = (A + KC)\chi_i - K\zeta_i, \\ u_i = \frac{1}{\beta}\bar{F}_\delta \chi_i, \end{cases} \tag{2.60}$$

where $\bar{F}_\delta = -B'\bar{P}_\delta$ with $\bar{P}_\delta > 0$ being the unique solution of the continuous-time algebraic Riccati equation:

$$A'\bar{P}_\delta + \bar{P}_\delta A - \bar{P}_\delta BB'\bar{P}_\delta + \delta I = 0. \tag{2.61}$$

Remark 2.26 The intrinsic difference with full-state coupling is that in the above protocol design we use a low-gain feedback in the partial-state feedback. A minor change needed in the analysis is the factor 2β which becomes β in (2.58). This feedback is then coupled with a standard observer.

The main result regarding this design can be stated as follows.

Theorem 2.27 *Consider a MAS described by at most weakly unstable agents* (2.1) *and* (2.3). *Let any $\alpha \geq \beta > 0$ be given, and hence a set of network graphs $\mathbb{G}^N_{\alpha,\beta}$ be defined.*

If (A, B) is stabilizable and (C, A) is detectable, then the state synchronization problem via a stable protocol stated in Problem 2.4 *with $\mathbf{G} = \mathbb{G}^N_{\alpha,\beta}$ is solvable. In particular, there exists a $\delta^* > 0$ such that for any $\delta \in (0, \delta^*]$, protocol* (2.59) *solves the state synchronization problem for any graph $\mathcal{G}$ which is in $\mathbb{G}^N_{\alpha,\beta}$ for some N. Moreover, the synchronized trajectory is given by* (2.33).

We will rely on the following well-known result which can, for instance, be found in [52, Lemma 2.2.6].

Lemma 2.28 *Assume (A, B) is stabilizable and $\beta > 0$. In that case, Eq.* (2.58) *has a unique positive definite solution P_δ for any $\delta > 0$. Moreover, when A is at most weakly Unstable, then we obtain:*

$$\lim_{\delta \downarrow 0} P_\delta = 0. \tag{2.62}$$

***Proof of Theorem* 2.27** Using Theorem 2.15, we only need to prove that the system:

$$\begin{cases} \dot{x} = Ax - \lambda BB'P_\delta\chi, \\ \dot{\chi} = (A + KC)\chi - KCx, \end{cases} \tag{2.63}$$

is asymptotically stable for any λ that satisfies $\operatorname{Re}(\lambda) \geq \beta$ and $|\lambda| < \alpha$.

Define $e = x - \chi$. The system (2.63) can be rewritten in terms of x and e as:

$$\begin{cases} \dot{x} = (A - \lambda BB'P_\delta)x + \lambda BB'P_\delta e \\ \dot{e} = (A + KC + \lambda BB'P_\delta)e - \lambda BB'P_\delta x. \end{cases} \tag{2.64}$$

Since $\operatorname{Re}(\lambda) \geq \beta$, we have:

$$(A - \lambda BB'P_\delta)^* P_\delta + P_\delta(A - \lambda BB'P_\delta) \leq -\delta I - \beta P_\delta BB'P_\delta.$$

Define $V_1 = x^* P_\delta x$ and $u = -\sqrt{\beta}B'P_\delta x$. We can derive that:

$$\dot{V}_1 \leq -\delta\|x\|^2 - \|u\|^2 + \theta(\delta)\|e\|\|u\|,$$

where:

$$\theta(\delta) = \frac{\alpha}{\sqrt{\beta}} \left\| B'P_\delta \right\|.$$

Clearly, $\theta(\delta) \to 0$ as $\delta \to 0$.

Let Q be the positive definite solution of the Lyapunov equation:

$$(A + KC)'Q + Q(A + KC) = -2I.$$

Since $P_\delta \to 0$ and λ is bounded (we have $|\lambda| < \alpha$), there exists a δ_1 such that for all $\delta \in (0, \delta_1]$:

$$(A + KC + \lambda BB'P_\delta)'Q + Q(A + KC + \lambda BB'P_\delta) \leq -I.$$

Define $V_2 = e^* Qe$. We get:

$$\dot{V}_2 \leq -\|e\|^2 + M\|e\|\|u\|$$

where:

$$M = 2\frac{\alpha}{\sqrt{\beta}}\|QB\|.$$

Define $V = 4M^2 V_1 + 2V_2$. Then:

$$\dot{V} \leq -4M^2\delta\|x\|^2 - 2\|e\|^2 - 4M^2\|u\|^2 + (4M^2\theta(\delta) + 2M)\|e\|\,\|u\|.$$

There exists a $\delta^* \leq \delta_1$ such that $4M^2\theta(\delta) \leq 2M$ for all $\delta \in (0, \delta^*]$. Hence, for a $\delta \in (0, \delta^*]$:

$$\dot{V} \leq -4M^2\delta\|x\|^2 - \|e\|^2 - (\|e\| - 2M\|u\|)^2.$$

We conclude that the system (2.63) is asymptotically stable for any λ that satisfies $\text{Re}(\lambda) \geq \beta$ and $|\lambda| < \alpha$. ∎

The above protocol uses a sequence F_δ which is an H_2 low-gain sequence as defined in [106, Definition 4.73]. In the current context, it is important that this property is robust with respect to the uncertainty in the eigenvalues of the Laplacian. This result is obtained in the following lemma and will be crucial in Chap. 4.

Lemma 2.29 *Suppose (A, B) is stabilizable and all the eigenvalues of A are in the closed left half plane. Let F_δ be designed in Protocol Design 2.4. Then, we have the following properties:*

1. *The closed-loop system matrix $A + \lambda B F_\delta$ is Hurwitz stable for all $\delta > 0$ and for all λ with $\text{Re}\,\lambda > \beta$.*
2. *For any $\beta > 0$, there exists a $\delta^* > 0$ such that for all $\delta \in (0, \delta^*]$ there exist $\zeta_\delta > 0$ and $\rho_\delta > 0$ with $\zeta_\delta \to 0$ as $\delta \to 0$ such that:*

$$\|F_\delta e^{(A+\lambda B F_\delta)t}\| \leq \zeta_\delta e^{-\rho_\delta t}, \tag{2.65}$$

for all $t \geq 0$ and for all $\lambda \in \mathbb{C}$ with $\text{Re}\,\lambda > \beta$.

Remark 2.30 Note that $F_\delta \to 0$ as $\delta \to 0$ is not sufficient for (2.65). We need that the input signal is small for all $t \geq 0$. In [106, Definition 4.73], we basically established this property for one fixed value of λ.

Proof The first property is obvious from the previous theorem. Consider:

$$\dot{x} = (A + \lambda B F_\delta)x.$$

It is easily seen that:

$$\begin{aligned}\frac{d}{dt}x'(t)P_\delta x(t) &\leq -\delta x'(t)x(t)\\ &\leq -2\rho_\delta x'(t)P_\delta x(t),\end{aligned}$$

using that $\operatorname{Re}\lambda \geq \beta$ where $\rho_\delta = \delta\|P_1\|^{-1}$. Hence:

$$\|P_\delta^{1/2}x(t)\| \leq e^{-\rho_\delta t}\|P_\delta^{1/2}x(0)\|. \tag{2.66}$$

Finally:

$$\begin{aligned}\|F_\delta e^{(A+\lambda BF_\delta)t}x(0)\| &= \|B'P_\delta x(t)\| \\ &\leq \|B\|\|P_\delta^{1/2}\|e^{-\rho_\delta t}\|P_\delta^{1/2}x(0)\|.\end{aligned} \tag{2.67}$$

Since (2.67) is true for all $x(0) \in \mathbb{R}^n$, it follows trivially that:

$$\|F_\delta e^{(A+\lambda BF_\delta)t}\| \leq \|B\|\|P_\delta^{1/2}\|^2 e^{-\rho_\delta t} = \|B\|\|P_\delta\|e^{-\rho_\delta t}. \tag{2.68}$$

The proof is then completed by taking $\zeta_\delta = \|B\|\|P_\delta\|$. ■

We consider next a special case where the agent dynamics are neutrally stable, that is, those eigenvalues of A on the imaginary axis, if any, are semi-simple. We shall show that in this case the protocol design no longer requires the knowledge of β as long as the argument of the eigenvalues is bounded. This allows us to deal with a larger set of network graphs $\mathbb{G}^N_{\alpha,0,\theta}$.

As argued in Remark 2.9 for neutrally stable agents, we also have the nice property that the synchronized trajectory is bounded for stable protocols. The design below will result in stable protocols.

Protocol design 2.5 *Consider a MAS described by neutrally stable agents (2.1) and (2.3).*

We choose an observer gain K such that $A + KC$ is Hurwitz stable. Next, we consider a feedback gain $F_\delta = -\delta B'P$ where $P > 0$ satisfies the Lyapunov inequality $A'P + PA \leq 0$. This results in the protocol:

$$\begin{cases} \dot{\chi}_i = (A + KC)\chi_i - K\zeta_i, \\ u_i = F_\delta \chi_i, \end{cases} \tag{2.69}$$

for a sufficiently small δ.

The result is stated in the following theorem.

Theorem 2.31 *Consider a MAS described by neutrally stable agents (2.1) and (2.3). Let any $\alpha > 0$ and $\theta \in (0, \frac{\pi}{2})$ be given, and consider the set of network graphs $\mathbb{G}^N_{\alpha,0,\theta}$.*

If (A, B) is stabilizable and (C, A) is detectable, then the state synchronization problem stated in Problem 2.4 with $\mathbf{G} = \mathbb{G}^N_{\alpha,0,\theta}$ is solvable. In particular, there exists *a $\delta^* > 0$ such that for any $\delta \in (0, \delta^*]$, the stable protocol (2.69) solves the state synchronization problem for any graph $\mathcal{G} \in \mathbb{G}^N_{\alpha,0,\theta}$. Moreover, the synchronized trajectory is given by (2.33).*

Proof Based on Theorem 2.15, we need to prove only that the system

$$\begin{cases} \dot{x} = Ax - \delta\lambda BB'P\chi, \\ \dot{\chi} = (A+KC)\chi - KCx, \end{cases} \tag{2.70}$$

is asymptotically stable for any λ that satisfies $|\lambda| < \alpha$ and $|\arg(\lambda)| \leq \theta$. Note that

$$|\lambda| \leq \frac{1}{\cos\theta}\operatorname{Re}(\lambda).$$

Define $e = x - \chi$. The system (2.70) can be rewritten in terms of x and e as:

$$\begin{cases} \dot{x} = (A - \delta\lambda BB'P)x + \delta\lambda BB'Pe \\ \dot{e} = (A+KC+\delta\lambda BB'P)e - \delta\lambda BB'Px. \end{cases} \tag{2.71}$$

Define $V_1 = x^*Px$ and $u = -B'Px$. We can derive that:

$$\dot{V}_1 \leq -\delta\operatorname{Re}(\lambda)\|u\|^2 + \frac{\delta r_1}{\cos\theta}\operatorname{Re}(\lambda)\|e\|\|u\|,$$

where $r_1 = 2\|PB\|$.

Let Q be the positive definite solution of the Lyapunov equation:

$$(A+KC)'Q + Q(A+KC) = -2I.$$

Since λ is bounded in $|\lambda| < \alpha$, there exists a δ_1 such that for all $\delta \in (0, \delta_1]$:

$$(A+KC+\delta\lambda BB'P)'Q + Q(A+KC+\delta\lambda BB'P) \leq -I.$$

Define $V_2 = e^*Qe$. We get:

$$\dot{V}_2 \leq -\|e\|^2 + \frac{\delta r_2}{\cos\theta}\|e\|\|u\|$$

where $r_2 = 2\|QB\|$. Define $V = V_1 + V_2$. Then with:

$$r_3 = \frac{r_1 + r_2}{\cos\theta},$$

we can derive that:

$$\begin{aligned}\dot{V} &\leq -\frac{\delta}{2}\operatorname{Re}(\lambda)\|u\|^2 - \frac{\delta}{2}\operatorname{Re}(\lambda)(\|u\| - r_3\|e\|)^2 + \frac{\delta}{2}r_3^2\operatorname{Re}(\lambda)\|e\|^2 - \|e\|^2 \\ &\leq -\frac{\delta}{2}\operatorname{Re}(\lambda)\|u\|^2 - \frac{1}{2}\|e\|^2\end{aligned}$$

provided δ is chosen such that:

$$\frac{\delta}{2}r_3^2\operatorname{Re}(\lambda) \leq \frac{\delta}{2}r_3^2\alpha \leq \frac{1}{2}.$$

Since $\operatorname{Re}(\lambda) > 0$, this proves that the system is stable as required. ■

2.5.3.2 Direct Eigenstructure Assignment Method

In this section, we will design the protocol via direct eigenstructure assignment. We assume the agents are at most weakly unstable (i.e., the eigenvalues of A are in the closed left half plane). We use a canonical form used before in [163, Section 1.4] and [52, Section 2.2.1]. If (A, B) is controllable, then there exist a nonsingular state transformation T_x and an input transformation T_u, such that:

$$\tilde{A} = T_x^{-1}AT_x = \begin{pmatrix} A_1 & 0 & \cdots & 0 \\ 0 & A_2 & \ddots & \vdots \\ \vdots & \ddots & \ddots & 0 \\ 0 & \cdots & 0 & A_q \end{pmatrix},$$

$$\tilde{B} = T_x^{-1}BT_u = \begin{pmatrix} B_1 & B_{1,2} & \cdots & B_{1,q} & B_{1,q+1} \\ 0 & B_2 & \ddots & \vdots & \vdots \\ \vdots & \ddots & \ddots & B_{q-1,q} & \vdots \\ 0 & \cdots & 0 & B_q & B_{q,q+1} \end{pmatrix},$$

and for $i = 1, 2, \ldots, q$:

$$A_i = \begin{pmatrix} 0 & 1 & 0 & \cdots & 0 \\ \vdots & \ddots & \ddots & \ddots & \vdots \\ \vdots & & \ddots & 1 & 0 \\ 0 & \cdots & \cdots & 0 & 1 \\ -a_{n_i}^i & \cdots & -a_{n_3}^i & -a_{n_2}^i & -a_{n_1}^i \end{pmatrix}, \quad B_i = \begin{pmatrix} 0 \\ \vdots \\ \vdots \\ 0 \\ 1 \end{pmatrix}.$$

Clearly, (A_i, B_i) is controllable. Note that in [106, Section 4.3] and [55] a different canonical form was used where A has an upper triangular structure and B has a diagonal structure. However, as pointed out in [106, Remark 4.79], the resulting low-gain feedback can then exhibit peaking which, in our current context, makes the structure above more suitable.

Protocol design 2.6 *Consider a MAS described by at most weakly unstable agents (2.1) and (2.3). Choose T_x and T_u such that the system is in the canonical form described above.*

For each (A_i, B_i), let $F_{\delta,i} \in \mathbb{R}^{1\times n_i}$ be the unique state feedback gain such that the eigenvalues of $A_i + B_i F_{\delta,i}$ can be obtained from the eigenvalues of A_i by moving any eigenvalue λ_i on the imaginary axis to $\lambda_i - 2\delta$ while all the eigenvalues in the open left half complex plane remain at the same location. Note that $F_{\delta,i}$ can be obtained explicitly in terms of δ. Now define F_δ as:

$$F_\delta = T_u \begin{pmatrix} F_{\delta,1} & 0 & \cdots & 0 \\ 0 & F_{\delta,2} & \ddots & \vdots \\ \vdots & \ddots & \ddots & 0 \\ \vdots & & \ddots & F_{\delta,q} \\ 0 & \cdots & \cdots & 0 \end{pmatrix} T_x^{-1}. \tag{2.72}$$

Next, choose K such that $A + KC$ is Hurwitz stable. We then construct the dynamic protocol:

$$\begin{cases} \dot{\chi}_i = (A + KC)\chi_i - K\zeta_i, \\ u_i = \frac{2}{\beta} F_\delta \chi_i. \end{cases} \tag{2.73}$$

The following theorem guarantees that the above design has the required properties.

Theorem 2.32 *Consider a MAS described by at most weakly unstable agents (2.1) and (2.3). Let any $\alpha \geq \beta > 0$ be given and hence a set of network graphs $\mathbb{G}^N_{\alpha,\beta}$ be defined.*

If (A, B) is controllable and (C, A) is detectable, then the state synchronization problem stated in Problem 2.4 with $\mathbf{G} = \mathbb{G}^N_{\alpha,\beta}$ is solvable. In particular, there exists a $\delta^* > 0$ such that for any $\delta \in (0, \delta^*]$, the stable protocol (2.73) solves the state synchronization problem for any graph $\mathcal{G} \in \mathbb{G}^N_{\alpha,\beta}$. Moreover, the synchronized trajectory is given by (2.33).

Proof We will use an argument similar to that in the proof of Theorem 2.20. Using Theorem 2.15, we need to verify whether the controller $\mathcal{H}$ of the form:

$$\mathcal{H}: \begin{cases} \dot{\chi} = (A + KC)\chi - Ky, \\ u = \frac{2}{\beta} F_\delta \chi, \end{cases} \tag{2.74}$$

for the system:

$$\mathcal{S}: \begin{cases} \dot{x} = Ax + \lambda Bu \\ y = Cx, \end{cases}$$

is such that the interconnection is stable for all λ with $\mathrm{Re}(\lambda) \geq \beta$ and $|\lambda| < \alpha$. We write:

$$\lambda = \frac{\beta}{2} + \mu$$

and using our properties for λ, we get that $\mathrm{Re}\,\mu > \beta/2$ and $|\,\mathrm{Im}\,\mu| < \alpha$. Important for our purposes is that:

$$-\theta < \arg(\mu) < \theta \tag{2.75}$$

with:

$$\theta = \arcsin(\tfrac{2\alpha}{\beta}).$$

The interconnection of $\mathcal{S}$ and $\mathcal{H}$ can be written as:

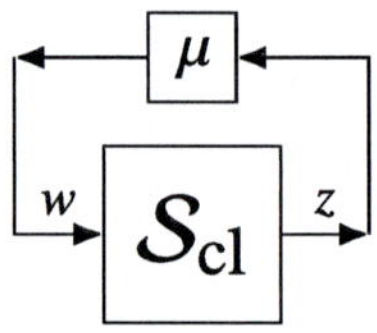

where:

$$S_{\text{cl}}: \begin{cases} \dot{x} = (A + BF_\delta)x + BF_\delta e + Bw, \\ \dot{e} = (A + KC - BF_\delta)e - BF_\delta x - Bw, \\ z = \frac{2}{\beta} F_\delta x + \frac{2}{\beta} F_\delta e. \end{cases}$$

S_{cl} denotes the transfer matrix of this system. By the small-gain theorem, we find that this closed-loop system is stable provided S_{cl} is asymptotically stable and for all μ and for all $s \in \mathbb{C}^+$ we have:

$$\det\left(I - \mu S_{\text{cl}}(s)\right) \neq 0. \tag{2.76}$$

We can easily verify that:

$$S_{\rm cl}(s) = \tfrac{2}{\beta}\begin{pmatrix} I & I\end{pmatrix}\begin{pmatrix} I & -G_{f,\delta}(s) \\ G_{c,\delta}(s) & I \end{pmatrix}^{-1}\begin{pmatrix} G_{f,\delta}(s) \\ -G_{c,\delta}(s)\end{pmatrix}$$

where:

$$\begin{aligned} G_{f,\delta}(s) &= F_\delta(sI - A - BF_\delta)^{-1}B \\ G_{c,\delta}(s) &= F_\delta(sI - A - KC + BF_\delta)^{-1}B. \end{aligned}$$

Note that $\|G_{c,\delta}\|_\infty \to 0$ as $\delta \to 0$ since $A + KC$ is Hurwitz and $F_\delta \to 0$. This implies that:

$$\|S_{\rm cl} - \tfrac{2}{\beta}G_{f,\delta}\|_\infty \to 0$$

as $\delta \to 0$. Lemma 2.33 can be used to show that $\|G_{f,\delta}\|_\infty$ is bounded which then yields that $\|S_{\rm cl}\|_\infty$ is bounded. Moreover, since $T_u^{-1}G_{f,\delta}T_u$ is upper triangular, we note that the lower triangular terms of $T_u^{-1}S_{\rm cl}T_u$ converge to zero. This implies that the determinant of (2.76) converges to:

$$\prod_{i=1}^{q}\left(1 - \tfrac{2\mu}{\beta}G_{f,\delta,i}(s)\right)$$

where

$$G_{f,\delta,i}(s) = F_{\delta,i}(sI - A_i - B_iF_{\delta,i})^{-1}B_i. \tag{2.77}$$

Therefore, to prove (2.76), it suffices to prove that:

$$1 - \tfrac{2\mu}{\beta}G_{f,\delta,i}(s)$$

is bounded away from zero for $i = 1, \ldots, q$. Since we have (2.75), the above follows directly if we establish that

$$G_{f,\delta,i}(s) + G_{f,\delta,i}(s)^* \leq 0 \tag{2.78}$$

for $i = 1, \ldots, q$. Define:

$$p_{0,i}(s) = \det(sI - A_i).$$

In that case, we have:

$$\det(sI - A_i - B_iF_{\delta,i}) = p_{0,i}(s - 2\delta)$$

and using the structure of the matrices A_i and B_i, we find that:

$$G_{f,\delta,i}(s) = -1 + \frac{p_{0,i}(s)}{p_{0,i}(s-2\delta)}.$$

Note that:

$$\frac{p_{0,i}(s)}{p_{0,i}(s-2\delta)}$$

is a product of terms of the form:

$$\frac{s-\alpha_i}{s-\alpha_i-2\delta}$$

with α_i on the imaginary axis, $s \in \mathbb{C}^+$ and $\delta > 0$. This shows that this term is less than 1 in absolute value. But then:

$$\left| \frac{p_{0,i}(s)}{p_{0,i}(s-2\delta)} \right| \leq 1$$

for $s \in \mathbb{C}^+$ which yields (2.78). ■

In Lemma 2.29, we showed an additional property for our sequence of state feedback gains that will be used in Chap. 4. Below, we establish that this result also holds in case the sequence of state feedback gains is designed based on a direct method.

Lemma 2.33 *Suppose (A, B) is stabilizable and all the eigenvalues of A are in the closed left half plane. Let:*

$$\tilde{F}_\delta = \tfrac{2}{\beta} F_\delta$$

be designed in Protocol Design 2.6. Then, we have the following properties:

1. *The closed-loop system matrix $A + \lambda B \tilde{F}_\delta$ is Hurwitz stable for all $\delta > 0$ and for all λ with $\operatorname{Re}\lambda > \beta$ and $|\lambda| < \alpha$.*
2. *For any $\beta > 0$, there exists a $\delta^* > 0$ such that for all $\delta \in (0, \delta^*]$ we have:*

$$\| \tilde{F}_\delta e^{(A+\lambda B\tilde{F}_\delta)t} \| \leq N\delta e^{-\delta t/4} \tag{2.79}$$

for all $t \geq 0$ and for all $\lambda \in \mathbb{C}$ with $\operatorname{Re}\lambda > \beta$ and $|\lambda| < \alpha$.

Proof The first property is obvious from the proof of Theorem 2.32. In order to show the second property, we first consider the single-input case. We write:

$$\lambda = \tfrac{\beta}{2} + \mu$$

and using our properties for λ, we get that $\operatorname{Re}\mu > \frac{\beta}{2}$ and $|\operatorname{Im}\mu| < \alpha$. Next, consider the system:

$$\begin{cases} \dot{x}_i = (A_i + \lambda B_i \tilde{F}_{\delta,i})x_i \\ u_i = \tilde{F}_{\delta,i}x_i, \end{cases}$$

where:

$$\tilde{F}_{\delta,i} = \frac{2}{\beta}F_{\delta,i}.$$

This can be rewritten as:

$$\begin{cases} \dot{x}_i = (A_i + B_i F_{\delta,i})x_i + B_i w_i \\ u_i = \frac{2}{\beta}F_{\delta,i}x_i \end{cases}$$

with $w_i = \mu u_i$. Defining $G_{f,\delta,i}$ by (2.77), we obtained (2.78) which implies that:

$$1 - \frac{2\mu}{\beta}G_{f,\delta,i}(s)$$

is bounded away from zero on the imaginary axis (uniformly with respect to μ).

From [52, Lemma 2.2.1], we obtain that there exists a N_i such that $\|G_{f,\delta,i}\|_\infty < N_i$, while $\|G_{f,\delta,i}\|_2 \to 0$ as $\delta \to 0$.

A slight modification of the results that have been established before in [52, Lemma 2.2.1] yields that for any $0 < \gamma_i < 1$, there exist constants $M_i > 0$ and $\delta^* > 0$ such that for all $\delta \in (0, \delta^*]$ and $t \geq 0$, we have:

$$\|F_{\delta,i}e^{(A_i+B_iF_{\delta,i})t}\| \leq M_i\delta e^{-\gamma_i\delta t} \tag{2.80}$$

for all $t \geq 0$ and for $i = 1, \ldots, q$.

The multi-input version of (2.80) can then be obtained through a standard recursive argument. Choose:

$$0.5 < \gamma_1 < \gamma_2 < \cdots < \gamma_q < 1.$$

We can then establish (2.79) through a recursive argument. We have:

$$\|F_{\delta,q}x_q(t)\| \leq M_q\delta e^{-\gamma_q\delta t}\|x_q(0)\| \tag{2.81}$$

using (2.80). Next:

$$\dot{x}_{q-1}(t) = (A_{q-1} + B_{q-1}F_{\delta,q})x_q + B_{q-1,q}F_qx_q(t)$$

yields:

$$\|F_{\delta,q-1}x_{q-1}(t)\| \leq M_{1,q-1\delta}e^{-\gamma_{q-1\delta}t}\|x_{q-1}(0)\|$$
$$+\int_0^t M_{1,q-1\delta}e^{-\gamma_{q-1\delta}(t-\tau)}M_{2,q\delta}e^{-\gamma_{q\delta}\tau}\|x_q(0)\|\mathrm{d}\tau,$$

which, in turn, yields:

$$\|F_{\delta,q-1}x_{q-1}(t)\| \leq \tilde{M}_{1,q-1\delta}e^{-\gamma_{q-1\delta}t}\|x(0)\|$$

for some $\tilde{M}_{1,q-1} > 0$ using that $\gamma_q > \gamma_{q-1}$. Completing this recursive argument, we end up with the bound:

$$\|F_\delta e^{(A+BF_\delta)t}\| \leq M\delta e^{-\delta t/2} \tag{2.82}$$

for some $\tilde{M} > 0$. This implies that there exists a N_g such that:

$$\tilde{G}_{f,\delta}(s) = F_\delta(sI - A - BF_\delta)^{-1}$$

satisfies $\|\tilde{G}_{f,\delta}(s-\delta/4)\|_\infty < N_g$, while $\|\tilde{G}_{f,\delta}(s-\delta/4)\|_2 \to 0$ as $\delta \to 0$.

Next, consider the system:

$$\begin{cases} \dot{x} = (A + \lambda B\tilde{F}_\delta)x \\ u = \tilde{F}_\delta x. \end{cases}$$

This can be rewritten as:

$$\begin{cases} \dot{x} = (A + BF_\delta)x + Bw \\ u = \frac{2}{\beta}F_\delta x \end{cases}$$

with $w = \mu u$. Note that the Laplace transform of:

$$\tilde{F}_\delta e^{(A+\lambda B\tilde{F}_\delta)t}$$

can be written as:

$$\tfrac{2}{\beta}\left(I - \tfrac{2\mu}{\beta}G_{f,\delta}\right)^{-1}\tilde{G}_{f,\delta}. \tag{2.83}$$

We know that:

$$I - \tfrac{2\mu}{\beta}G_{f,\delta}(s-\delta/4)$$

is upper triangular and has a bounded H_∞ norm. Moreover, its determinant:

$$\prod_{i=1}^{N}\left(1-\tfrac{2\mu}{\beta}G_{f,\delta,i}(s-\delta/4)\right)$$

is bounded away from zero. This implies that there exists a $\tilde{N}$ such that

$$\left\|\tfrac{2}{\beta}\left(I-\tfrac{2\mu}{\beta}G_{f,\delta}(s-\delta/4)\right)^{-1}\right\|_\infty<\tilde{N}. \tag{2.84}$$

We have by (2.82) that the L_∞-norm of the function r given by:

$$r(t)=e^{\delta t/4}F_\delta e^{(A+BF_\delta)t}$$

satisfies:

$$\|r\|_\infty<N_r$$

for some $N_r>0$. But then the function y, given by:

$$y(t)=e^{\delta t/4}\tilde{F}_\delta e^{(A+\lambda B\tilde{F}_\delta)t},$$

and r are related (in the frequency domain) by:

$$Y(s)=\tfrac{2}{\beta}\left(I-\tfrac{2\mu}{\beta}G_{f,\delta}(s-\delta/4)\right)^{-1}R(s).$$

The transfer matrix relating r and y has a bounded H_∞ norm by (2.84). But then by the continuous-time version of [9, Theorem 1], also the L_∞-induced operator norm of this system is bounded which implies that if $\|r\|_\infty\le N_r\delta$ then also $\|y\|_\infty\le N\delta$ for some suitable constant $N>0$. This implies (2.79) which completes the proof. ■

2.5.4 MAS with at Most Weakly Non-minimum-Phase Agents

As indicated before, we will study the problem of synchronization for MAS with weakly non-minimum-phase agents.

Theorem 2.15 shows that achieving synchronization can be reduced to a simultaneous stabilization problem. If the graph is unknown and only restrictions are available on the eigenvalues of the associated Laplacian, then achieving stabilization becomes a robust stabilization problem. We have seen earlier that if the system is weakly non-minimum-phase agents and left-invertible, then this robust stabilization problem is solvable.

Protocol design 2.7 *Consider a MAS described by agents (2.1) and (2.3). Assume the agents are at most weakly non-minimum phase and left invertible. For $0 < \rho < 1$, we construct $P \geq 0$ such that:*

$$A'P + PA - \nu PBB'P + C'C = 0 \tag{2.85}$$

where:

$$\nu = \frac{2\rho + \rho^2}{(1+\rho)^2}.$$

For any $\delta > 0$, let Q_δ be the unique solution of the algebraic Riccati equation:

$$AQ_\delta + Q_\delta A' + (1 - \delta^{-2})Q_\delta C'CQ_\delta + \tfrac{1}{(1+\rho)^2}BB' = 0. \tag{2.86}$$

Since(A, B, C) is weakly minimum-phase and left-invertible, we find that $Q_\delta \to 0$ as $\delta \downarrow 0$. Therefore, there exists a δ small enough such that the spectral radius of PQ_δ is less than 1. We then construct the dynamic protocol:

$$\begin{cases} \dot{\chi}_i = A\chi_i + \nu Bu_i + L_{p,q}(C\chi_i - \zeta_i) \\ u_i = -B'P\chi_i, \end{cases} \tag{2.87}$$

where:

$$L_{p,q} = -\delta^{-2}(I - Q_\delta P)^{-1} Q_\delta C'. \tag{2.88}$$

The main result is as follows.

Theorem 2.34 *Consider a MAS described by at most weakly non-minimum-phase, left-invertible agents (2.1) and (2.3). Let any $\alpha \geq \beta > 0$ be given, and hence a set of network graphs $\mathbb{G}^N_{\alpha,\beta}$ be defined.*

If (A, B) is stabilizable and (A, C) is detectable, then the state synchronization problem stated in Problem 2.4 with $\mathbf{G} = \mathbb{G}^N_{\alpha,\beta}$ is solvable.

More specifically, choose a ρ such that any $\lambda \in \mathbb{C}$ with $\operatorname{Re}\lambda > \beta$ and $|\lambda| < \alpha$ is in the set Λ_ρ defined in (2.40). Choose a δ such that the spectral radius of PQ_δ is less than 1 and then the protocol (2.87) solves the state synchronization problem for any graph $\mathcal{G}$ which is in $\mathbb{G}^N_{\alpha,\beta}$ for some N. Moreover, the synchronized trajectory is given by (2.32).

Proof of Theorem 2.34 We have seen earlier that we basically need to find a controller for the system (2.51) such that the H_∞ norm from v to q is less than 1. It is easily seen that this is achieved if we have a controller for the system:

$$\mathcal{S}: \begin{cases} \dot{x} = Ax + Bu + Bv \\ y = Cx \\ q = u, \end{cases} \tag{2.89}$$

such that the H_∞ norm from v to q is less than $\gamma = 1 + \rho$ where ρ is sufficiently small such that all λ with $\operatorname{Re}\lambda \geq \beta$ and $|\lambda| < \alpha$ are in the set Λ_ρ defined by (2.40).

We can then use standard H_∞ control to design a controller. However, this is a singular problem so we have to be careful. We basically design a controller such that if applied to the system:

$$\mathcal{S}: \begin{cases} \dot{x} = Ax + Bu + \begin{pmatrix} \frac{1}{1+\rho}B & 0 \end{pmatrix}\tilde{v} \\ y = Cx + \begin{pmatrix} 0 & \delta I \end{pmatrix}\tilde{v} \\ \tilde{q} = \begin{pmatrix} 0 \\ C \end{pmatrix} x + \begin{pmatrix} I \\ 0 \end{pmatrix} u, \end{cases}$$

the resulting H_∞ norm is less than 1. Then it is easily seen that the same controller applied to the system (2.89) achieves an H_∞ norm less than γ.

First the algebraic Riccati equation (2.85) is a standard H_2 algebraic Riccati equation and therefore has a solution since (A, B) is stabilizable. Since (A, B, C) is weakly minimum-phase and left-invertible, we have that (2.86) has a solution for δ small and $Q_\delta \to 0$ as $\delta \downarrow 0$. This implies that the spectral radius of PQ_δ is strictly less than 1 for sufficiently small δ.

We can then use the results from [124] to find that:

$$\begin{cases} \dot{\chi}_i = A\chi_i + \nu B u_i + L_{p,q}(C\chi_i - \zeta_i) \\ u_i = -B'P\chi_i, \end{cases}$$

with $L_{p,q}$ given by (2.88) achieves an H_∞ norm strictly less that $1+\delta$ when applied to the system (2.89). As noted earlier, this implies that the protocol (2.87) achieves synchronization. ∎

The above can be generalized since the agents can always be made left-invertible through a pre-compensator.

Theorem 2.35 *Consider an agent of the form* (2.1) *which is stabilizable and detectable. In that case, there exists an asymptotically stable pre-compensator:*

$$\begin{aligned} \dot{p}_i &= Kp_i + Lv_i \\ u_i &= Mp_i + Nv_i, \end{aligned}$$

such that the interconnection of (2.1) *and this pre-compensator which is given by:*

$$\begin{aligned}\dot{\bar{x}}_i &= \begin{pmatrix} A & BM \\ 0 & K \end{pmatrix}\bar{x}_i + \begin{pmatrix} BN \\ L \end{pmatrix} v_i \\ y_i &= \begin{pmatrix} C & 0 \end{pmatrix}\bar{x}_i,\end{aligned}$$

has the following properties:

1. *It is stabilizable and detectable.*
2. *It is left-invertible.*
3. *Its poles are the poles of the agent* (2.1) *plus the poles of the pre-compensator (i.e., the eigenvalues of K).*
4. *Its infinite zero structure is the same as the infinite zero structure of the agent* (2.1).
5. *Its invariant zeros are the invariant zeros of the agent* (2.1) *plus the eigenvalues of K and any additional invariant zeros that can be arbitrarily placed in the open left half complex plane.*

Remark 2.36 Note that if the system (2.1) is (weakly) minimum-phase, then we can guarantee that the interconnection of (2.1) and the pre-compensator is also (weakly) minimum-phase.

Proof The design of this pre-compensator is discussed in detail in Chap. 24. ■

Note that since this pre-compensator is asymptotically stable, it does not effect the synchronized trajectory. On the other hand, since the pre-compensator is minimum-phase, it preserves the property that the system is at most weakly non-minimum-phase. Therefore, we can, without loss of generality, present our design only for left-invertible systems.

An additional property that would be desirable is to have a static pre-compensator $u_i = Nv_i$. However, as argued in Chap. 24, this is not always possible if we want to preserve that the system is at most weakly non-minimum-phase.

If the original agents are weakly non-minimum-phase, then using the above pre-compensator, we can guarantee that the agents are weakly non-minimum-phase. If the original agents are weakly non-minimum-phase and left-invertible, we can then apply Theorem 2.34 as well.

As argued before, achieving synchronization with stable protocols is very desirable. However, the above design is in most cases not stable. In the next subsection, we will show that if the system is minimum-phase instead of only weakly non-minimum-phase, then we can actually find a stable protocol with the Protocol Design 2.7 as outlined in Theorem 2.34.

2.5.5 Minimum-Phase Agents

Compared to the weakly non-minimum-phase systems, for minimum-phase agents, we can always find stable protocols. Another advantage is that we have much better design methods. As before, we distinguish between ARE-based and direct methods.

In both design methods, we assume that the system is left-invertible. As argued in Theorem 2.35, this can always be achieved through a pre-compensator while preserving the property that the system is minimum-phase.

2.5.5.1 ARE-Based Method

Recall from the proof of Theorem 2.34 that we basically need to find a controller for the system:

$$\mathcal{S}: \begin{cases} \dot{x} = Ax + Bu + Bv \\ y = Cx \\ q = u, \end{cases} \tag{2.90}$$

such that the H_∞ norm from v to q is less than $\gamma = 1 + \rho$ where ρ is sufficiently small such that all λ with $\operatorname{Re}\lambda \geq \beta$ and $|\lambda| < \alpha$ are in the set Λ_ρ defined by (2.40).

Protocol design 2.8 *Consider a MAS described by agents (2.1) and (2.3). Assume that the agents are minimum phase and left invertible. For $0 < \rho < 1$, we construct $P_\rho \geq 0$ such that:*

$$A'P_\rho + P_\rho A - \tilde{\nu} P_\rho BB'P_\rho + I = 0 \tag{2.91}$$

where:

$$\tilde{\nu} = \frac{4\rho + \rho^2}{(2+\rho)^2}.$$

Define:

$$F_\rho = -B'P_\rho.$$

For any $\varepsilon > 0$, choose a δ sufficiently small such that there exists a unique solution Q_ε of the algebraic Riccati equation:

$$AQ_\varepsilon + Q_\varepsilon A' + \varepsilon^{-2} Q_\varepsilon F_\rho' F_\rho Q_\varepsilon - \delta^{-2} Q_\varepsilon C'CQ_\varepsilon + BB' = 0. \tag{2.92}$$

(continued)

Protocol design 2.8 (continued)
We then construct the stable dynamic protocol:

$$\begin{cases} \dot{\chi}_i = A\chi_i - \delta^{-2} Q_\varepsilon C'(C\chi_i - \zeta_i), \\ u_i = F_\rho \chi_i. \end{cases} \tag{2.93}$$

The main result is as follows.

Theorem 2.37 *Consider a MAS described by left-invertible, minimum-phase agents (2.1) and (2.3). Let any $\alpha \geq \beta > 0$ be given and hence a set of network graphs $\mathbb{G}^N_{\alpha,\beta}$ be defined.*

If (A, B) is stabilizable and (A, C) is detectable, then the state synchronization problem stated in Problem 2.4 with $\mathbf{G} = \mathbb{G}^N_{\alpha,\beta}$ is solvable.

More specifically, choose a ρ such that any $\lambda \in \mathbb{C}$ with $\operatorname{Re}\lambda > \beta$ and $|\lambda| < \alpha$ is in the set Λ_ρ defined in (2.40). Then there exists an ε^* such that the protocol (2.93) for any $\varepsilon < \varepsilon^*$ solves the state synchronization problem by stable protocol for any graph $\mathcal{G}$ in $\mathbb{G}^N_{\alpha,\beta}$ for some N. Moreover, the synchronized trajectory is given by (2.33).

Proof of Theorem 2.37 The existence of Q_ε for any $\delta > 0$ such that (2.92) follows from the fact that (A, B, C) is minimum-phase and left-invertible.

Note that the stability of the protocol follows from the stability of the matrix $A - \delta^{-2} Q_\varepsilon C'C$. This is a direct consequence of the properties of the Riccati equation (2.92). After all, we have:

$$\left[A - \delta^{-2} Q_\varepsilon C'C\right] Q_\varepsilon + Q_\varepsilon \left[A - \delta^{-2} Q_\varepsilon C'C\right]' \\ = -\varepsilon^{-2} Q_\varepsilon F_\rho' F_\rho Q_\varepsilon - \delta^{-2} Q_\varepsilon C'C Q_\varepsilon - BB'.$$

We have seen earlier that we basically need to find a controller for the system (2.90) such that the H_∞ norm from v to q is less than $1 + \rho$. If we look at the interconnection of (2.90) and the controller:

$$\begin{cases} \dot{\chi} = A\chi - \delta^{-2} Q_\varepsilon C'(C\chi - y), \\ u = F_\rho \chi, \end{cases}$$

then we note that the transfer function from v to q is equal to:

$$\begin{pmatrix} I & I \end{pmatrix} \begin{pmatrix} I & -G_f \\ G_c & I \end{pmatrix}^{-1} \begin{pmatrix} G_f & -G_c \end{pmatrix},$$

where:

$$G_f(s) = F_\rho(sI - A - BF_\rho)^{-1}B$$
$$G_c(s) = F_\rho(sI - A + BF_\rho - K_\varepsilon C)^{-1}B.$$

From standard H_∞ control theory, we find that:

$$\|G_f\|_\infty < 1 + \tfrac{\rho}{2}.$$

It is then obvious that the H_∞ norm from v to q is less than $1+\rho$ provided $\|G_c\|_\infty$ is sufficiently small. Note that:

$$G_c = (I - \tilde{G}_c)^{-1}\tilde{G}_c,$$

where:

$$\tilde{G}_c(s) = F_\rho(sI - A - K_\varepsilon C)^{-1}B.$$

Again from standard H_∞ theory, we have

$$\|\tilde{G}_c\|_\infty < \varepsilon.$$

Therefore, by choosing ε small, we can make $\|\tilde{G}_c\|_\infty$ sufficiently small such that the H_∞ norm from v to q is less than $1+\delta$. ■

2.5.5.2 Direct Method

If we look at the arguments in the proof of Theorem 2.37, we basically see that we need to design F_δ and K_ε such that

$$\|F_\delta(sI - A - BF_\delta)^{-1}B\| < 1 + \tfrac{\rho}{2}$$

for sufficiently small δ and

$$\|F_\delta(sI - A - K_\varepsilon C)^{-1}B\| \to 0$$

as $\varepsilon \to 0$.

We will see that the design for F_δ can be obtained along the lines of Protocol Design 2.2. There exists a nonsingular state transformation T_x for the system (2.90) such that:

$$\bar{x} = \begin{pmatrix}\bar{x}_1\\ \bar{x}_2\end{pmatrix} = T_x x, \tag{2.94}$$

and the dynamics of $\bar{x}$ is represented as:

$$\begin{aligned} \dot{\bar{x}}_1 &= \bar{A}_{11}\bar{x}_1 + \bar{A}_{12}\bar{x}_2 \\ \dot{\bar{x}}_2 &= \bar{A}_{21}\bar{x}_1 + \bar{A}_{22}\bar{x}_2 + \bar{B}u + \bar{B}v \end{aligned} \tag{2.95}$$

with $\bar{B}$ invertible, and, since (A, B) is stabilizable, we have that $(\bar{A}_{11}, \bar{A}_{12})$ is stabilizable. We choose F_1 such that $\bar{A}_{11} + \bar{A}_{12}F_1$ is asymptotically stable. In that case, a suitable F_δ is given by:

$$F_\delta := \frac{1}{\delta}\bar{B}^{-1}\left(F_1 \; -I\right)T_x, \tag{2.96}$$

for a δ sufficiently small.

For the design of K_ε, we use a different state space transformation. We first transform the system:

$$\begin{aligned} \dot{x} &= Ax + Bu \\ y &= Cx \end{aligned}$$

into the SCB form. We know this system is minimum-phase and left-invertible. Theorem 23.3 then tells us that there exist nonsingular matrices Γ_x, Γ_u, and Γ_y such that the system is in SCB form which, in the compact form presented in (23.30)–(23.32), yields:

$$\begin{aligned} \dot{x}_a &= A_{aa}x_a + L_{ad}y_d + L_{ab}y_b, \\ \dot{x}_b &= A_{bb}x_b + L_{bd}y_d, \\ \dot{x}_d &= A_{dd}x_d + B_d(u_d + E_{da}x_a + E_{db}x_b + E_{dd}x_d), \\ y_b &= C_bx_b, \\ y_d &= C_dx_d, \end{aligned}$$

with:

$$y = \Gamma_y \begin{pmatrix} y_b \\ y_d \end{pmatrix}, \qquad x = \Gamma_x \begin{pmatrix} x_a \\ x_b \\ x_d \end{pmatrix}, \qquad u = \Gamma_u u_d$$

where:

$$A_d = \begin{pmatrix} A_{q_1} & 0 & \cdots & 0 \\ 0 & A_{q_2} & \ddots & \vdots \\ \vdots & \ddots & \ddots & 0 \\ 0 & \cdots & 0 & A_{q_d} \end{pmatrix}, \qquad B_d = \begin{pmatrix} B_{q_1} & 0 & \cdots & 0 \\ 0 & B_{q_2} & \ddots & \vdots \\ \vdots & \ddots & \ddots & 0 \\ 0 & \cdots & 0 & B_{q_d} \end{pmatrix},$$

$$C_d = \begin{pmatrix} C_{q_1} & 0 & \cdots & 0 \\ 0 & C_{q_2} & \ddots & \vdots \\ \vdots & \ddots & \ddots & 0 \\ 0 & \cdots & 0 & C_{q_d} \end{pmatrix},$$

where integers $q_1, \ldots, q_d$ describe the infinite zero structure and $A_q \in \mathbb{R}^{q \times q}$, $B_q \in \mathbb{R}^{q \times 1}$, and $C_q \in \mathbb{R}^{1 \times q}$ are given by:

$$A_q = \begin{pmatrix} 0 & I_{q-1} \\ 0 & 0 \end{pmatrix}, \quad B_q = \begin{pmatrix} 0 \\ 1 \end{pmatrix}, \quad C_q = \begin{pmatrix} 1 & 0 \end{pmatrix}.$$

Define S_ε and $S_{q,\varepsilon} \in \mathbb{R}^{q \times q}$ as:

$$S_\varepsilon = \begin{pmatrix} S_{q_1,\varepsilon} & 0 & \cdots & 0 \\ 0 & S_{q_2,\varepsilon} & \ddots & \vdots \\ \vdots & \ddots & \ddots & 0 \\ 0 & \cdots & 0 & S_{q_d,\varepsilon} \end{pmatrix}, \qquad S_{q,\varepsilon} = \begin{pmatrix} 1 & 0 & \cdots & 0 \\ 0 & \varepsilon & \ddots & \vdots \\ \vdots & \ddots & \ddots & 0 \\ 0 & \cdots & 0 & \varepsilon^{q-1} \end{pmatrix}.$$

We choose K_1 such that $A_{bb} + K_1 C_b$ is asymptotically stable and K_2 equals:

$$K_2 = \begin{pmatrix} K_{2,q_1} & 0 & \cdots & 0 \\ 0 & K_{2,q_2} & \ddots & \vdots \\ \vdots & \ddots & \ddots & 0 \\ 0 & \cdots & 0 & K_{2,q_d} \end{pmatrix},$$

where $A_{q_i} + K_{2,q_i} C_{q_i}$ is asymptotically stable for $i = 1, \ldots, d$. We set:

$$K_\varepsilon = \Gamma_x \begin{pmatrix} -L_{ab} & -L_{ad} \\ K_1 & -L_{bd} \\ 0 & \varepsilon^{-1} S_\varepsilon^{-1} K_2 \end{pmatrix} \Gamma_y^{-1}. \tag{2.97}$$

Protocol design 2.9 *Consider a MAS described by agents (2.1) and (2.3). Assume that the agents are minimum-phase and left-invertible.*

We then construct the dynamic protocol:

$$\begin{cases} \dot{\chi}_i = A\chi_i - K_\varepsilon(C\chi_i - \zeta_i), \\ u_i = F_\delta \chi_i, \end{cases} \tag{2.98}$$

where F_δ and K_ε are chosen according to (2.96) and (2.97), respectively.

We obtain the following result.

Theorem 2.38 *Consider a MAS described by left-invertible, minimum-phase agents (2.1) and (2.3). Let any $\alpha \geq \beta > 0$ be given and hence a set of network graphs $\mathbb{G}^N_{\alpha,\beta}$ be defined.*

If (A, B) is stabilizable and (A, C) is detectable, then the state synchronization problem stated in Problem 2.4 with $\mathbf{G} = \mathbb{G}^N_{\alpha,\beta}$ is solvable.

More specifically, there exist ε^* and δ^* such that the protocol (2.98) solves the state synchronization problem by stable protocol for any graph $\mathcal{G} \in \mathbb{G}^N_{\alpha,\beta}$ for all $\varepsilon < \varepsilon^*$ and all $\delta < \delta^*$. Moreover, the synchronized trajectory is given by (2.33).

Proof of Theorem 2.38 We use a similar argument as in the proof of Theorem 2.37. We basically need to find a controller for the system (2.90) such that the H_∞ norm from v to q is less than $1 + \rho$. If we look at the interconnection of (2.90) and the controller:

$$\begin{cases} \dot{\chi} = A\chi - K_\varepsilon(C\chi - \zeta), \\ u = F_\delta \chi, \end{cases}$$

then we note that the transfer function from v to q is equal to:

$$\begin{pmatrix} I & I \end{pmatrix} \begin{pmatrix} I & -G_f \\ G_c & I \end{pmatrix}^{-1} \begin{pmatrix} G_f & -G_c \end{pmatrix},$$

where:

$$\begin{aligned} G_f(s) &= F_\delta(sI - A - BF_\delta)^{-1}B \\ G_c(s) &= F_\delta(sI - A + BF_\delta - K_\varepsilon C)^{-1}B. \end{aligned}$$

In our case, we find that:

$$G_f(s) = \begin{pmatrix} 0 & \frac{1}{\delta}\bar{B}^{-1} \end{pmatrix} \begin{pmatrix} sI - \tilde{A}_{11} & -A_{12} \\ -\tilde{A}_{21} & sI - \tilde{A}_{22} + \frac{1}{\delta}I \end{pmatrix} \begin{pmatrix} 0 \\ \bar{B} \end{pmatrix}$$

with $\tilde{A}_{11} = A_{11} + A_{12}F_1$ asymptotically stable. We get:

$$G_f(s) = (1 + \delta s)^{-1} \left(1 - (1 + \delta s)^{-1}\delta K_c(s)\right)^{-1},$$

with:

$$K_c(s) = \bar{B}^{-1} \left[\tilde{A}_{21}(sI - \tilde{A}_{11})^{-1}A_{12} + \tilde{A}_{22}\right] \bar{B}.$$

Since K_c is asymptotically stable, we see that for δ small enough we have:

$$\|G_f\|_\infty < 1 + \tfrac{\rho}{2}.$$

It is then obvious that the H_∞ norm from v to q is less than $1+\rho$ provided $\|G_c\|_\infty$ is sufficiently small. Note that:

$$G_c = (I - \tilde{G}_c)^{-1}\tilde{G}_c$$

where:

$$\tilde{G}_c(s) = F_\delta(sI - A - K_\varepsilon C)^{-1}B.$$

Therefore, we clearly obtain our objective when we guarantee that the H_∞ norm of:

$$G_k(s) = (sI - A - K_\varepsilon C)^{-1}B$$

is sufficiently small. We obtain:

$$G_k(s) = \Gamma_x \begin{pmatrix} 0 \\ 0 \\ I \end{pmatrix} \tilde{G}_k(s)\Gamma_u^{-1}$$

where:

$$\tilde{G}_k(s) = (sI - A_d - B_d E_{dd} - \varepsilon^{-1}S_\varepsilon^{-1}K_2C_d)^{-1}B_d.$$

We get:

$$\tilde{G}_k(s) = (I + \breve{G}_k(s)E_{dd})^{-1}\breve{G}_k(s)$$

with:

$$\breve{G}_k(s) = \begin{pmatrix} \varepsilon^{q_1}S_{q_1,\varepsilon}^{-1} & 0 & \cdots & 0 \\ 0 & \varepsilon^{q_2}S_{q_2,\varepsilon}^{-1} & \ddots & \vdots \\ \vdots & \ddots & \ddots & 0 \\ 0 & \cdots & 0 & \varepsilon^{q_d}S_{q_d,\varepsilon}^{-1} \end{pmatrix} (sI - A_d - K_2C_d)^{-1}B_d$$

and since:

$$\begin{pmatrix} \varepsilon^{q_1} S_{q_1,\varepsilon}^{-1} & 0 & \cdots & 0 \\ 0 & \varepsilon^{q_2} S_{q_2,\varepsilon}^{-1} & \ddots & \vdots \\ \vdots & \ddots & \ddots & 0 \\ 0 & \cdots & 0 & \varepsilon^{q_d} S_{q_d,\varepsilon}^{-1} \end{pmatrix} \leq \varepsilon I$$

we see that the H_∞ norm of $\check{G}_k$ can be made arbitrarily small by choosing ε small enough. This implies that also the H_∞ norm of $\tilde{G}_k$ and therefore G_k can be made arbitrarily small by choosing ε small enough. This completes the proof except for one detail. We need to establish that the protocol is stable. It is easily seen that we need to establish that:

$$A_d + B_d E_{dd} + \varepsilon^{-1} S_\varepsilon^{-1} K_2 C_d \tag{2.99}$$

is asymptotically stable given that A_{aa} and $A_{bb} + K_1 C_b$ are asymptoptically stable. We find that:

$$sI - A_d - B_d E_{dd} - \varepsilon^{-1} S_\varepsilon^{-1} K_2 C_d = \left[sI - A_d - \varepsilon^{-1} S_\varepsilon^{-1} K_2 C_d \right]^{-1} \left[I - \tilde{G}_k(s) E_d \right].$$

Clearly this matrix is invertible for all s in the closed right half plane for ε small enough, given that:

$$A_d + \varepsilon^{-1} S_\varepsilon^{-1} K_2 C_d$$

is asymptotically stable and the H_∞ norm of $\tilde{G}_k$ converges to zero as $\varepsilon \to 0$. This guarantees that the the matrix (2.99) is asymptotically stable and hence the protocol is stable. ∎

2.5.6 Unstable and Strictly Non-minimum-Phase Agents

In this section, we discuss protocol design for unstable and strictly non-minimum-phase agents, that is, poles and zeros are in the open right half plane.

In Theorem 2.20, we established for single-input systems a necessary condition for solvability for any α, β is that $P_0 Q_0 = 0$.

This enabled us to find a protocol by designing a controller to make the H_∞ from v to q smaller than $1 + \rho$ for the system (2.90). If we know that the network graph is in the set $\mathbf{G} \subset \mathbb{G}_{\alpha,\beta}^N$, then ρ should be such that any $\lambda \in \mathbb{C}$ with $\operatorname{Re} \lambda > \beta$ and $|\lambda| < \alpha$ is in the set Λ_ρ defined in (2.40).

This condition is sufficient for both single-input and multi-input systems. Hence, in this section, we consider the general case where the system can be both single-input or multi-input. This condition can actually be weakened slightly by finding a controller to make the H_∞ from v to q smaller than $1+\rho$ for the system:

$$\mathcal{S}: \begin{cases} \dot{x} = Ax + BS\bar{u} + BSv \\ y = Cx \\ q = \bar{u}. \end{cases} \tag{2.100}$$

This S is basically the effect of a preliminary static pre-compensator $u = S\bar{u}$.

The role of S has different interpretations. The following example illustrates the interpretation of a static, squaring-down pre-compensator.

Example 2.39 For the case:

$$A = \begin{pmatrix} 2 & 0 \\ 1 & 1 \end{pmatrix}, \qquad B = \begin{pmatrix} 1 & 0 \\ 0 & 1 \end{pmatrix}, \qquad C = \begin{pmatrix} 0 & 1 \end{pmatrix},$$

we see that for the system (2.90) we cannot make the H_∞ norm arbitrarily close to 1. Note that in this case the system is not left-invertible and, for instance, the proof of Theorem 2.34 no longer applies since the solution of (2.86) no longer converges to zero as $\delta \downarrow 0$. However, if we choose:

$$S = \begin{pmatrix} 3 \\ 1 \end{pmatrix},$$

then for the system (2.100), we can make the H_∞ norm arbitrarily close to 1.

Also when obtaining our solvability conditions, we basically have a model uncertainty Δ which was equal to δI_m. For multi-input systems, this Δ has a specific structure being a multiple of the identity matrix. In robust control, a classical technique to exploit this to make results less conservative is to use an arbitrary scaling matrix D such that $D\Delta = \Delta D$. In case S is a square invertible matrix, it has the interpretation of this scaling matrix used in robust control.

In the above, it does not help to consider a matrix S which is a wide matrix. On the other hand, in case of a tall matrix S, such as in Example 2.39, we can, without loss of generality, consider a square matrix S by adding zero columns. So we can restrict attention to square matrices S, but they need not be invertible.

We have the following design.

Protocol design 2.10 *Consider a MAS described by agents (2.1) and (2.3). Let a square matrix S and parameter ρ be given. Assume that there exists a $P_s \geq 0$ such that:*

$$A'P_s + P_sA - \nu P_sBSS'B'P_s + C'C = 0 \tag{2.101}$$

where:

$$\nu = \frac{2\rho + \rho^2}{(1+\rho)^2}.$$

Also let $Q_s > 0$ be the unique solution of the algebraic Riccati equation:

$$AQ_s + Q_sA' + (1-\delta^{-2})Q_sC'CQ_s + \tfrac{1}{(1+\rho)^2}BSS'B' = 0 \tag{2.102}$$

for some $\delta > 0$. Finally, assume that the spectral radius of P_sQ_s is less than 1. We then construct the dynamic protocol:

$$\begin{cases} \dot{\chi}_i = A\chi_i + \nu Bu_i + L_{p,q}(C\chi_i - \zeta_i) \\ u_i = -SS'B'P\chi_i, \end{cases} \tag{2.103}$$

where:

$$L_{p,q} = -\delta^{-2}(I - Q_sP_s)^{-1}Q_sC'.$$

Remark 2.40 Note that if (A, BS) is stabilizable, then P_s always exists. The main issue is to design S such that (A, BS) is stabilizable and the spectral radius of P_sQ_s is less than 1 for some $\delta > 0$. Note that P_s will always be invertible. It can then be shown that the existence of S, P_s, and Q_s satisfying the spectral radius condition is then equivalent to the existence of $R > 0$, $Q > 0$ and $W \geq 0$ such that:

$$\begin{pmatrix} RA' + AR - \nu BWB' & RC' \\ RC & -I \end{pmatrix} < 0,$$

$$\begin{pmatrix} AQ + QA' + \tfrac{1}{(1+\rho)^2}BWB' & QC' \\ CQ & 0 \end{pmatrix} \leq 0,$$

and:

$$Q < R.$$

Note that the above is a set of linear matrix inequalities which can be solved by standard numerical tools. Here, $W = SS'$, while R is a lower bound for P_s^{-1} and Q is an upper bound for $\lim_{\delta \downarrow 0} Q_s$.

The main result is stated in the following theorem.

Theorem 2.41 *Consider a MAS described by agents (2.1) with measurements (2.3). Assume that (A, B) is stabilizable and (C, A) is detectable. Let any matrix S of appropriate dimension be given.*

Let any $\alpha \geq \beta > 0$ be given, and hence a set of network graphs $\mathbb{G}^N_{\alpha,\beta}$ be defined. Choose a ρ such that for any $\lambda \in \mathbb{C}$ with $\operatorname{Re} \lambda > \beta$ and $|\lambda| < \alpha$ is in the set Λ_ρ defined in (2.40).

Assume that there exists a matrix S such that stabilizing solutions P_s and Q_s exist for algebraic Riccati equations (2.101) and (2.102), respectively, such that the spectral radius of $P_s Q_s$ is less than 1. In that case, the protocol (2.103) achieves state synchronization for any graph $\mathcal{G}$ in $\mathbb{G}^N_{\alpha,\beta}$ for some N.

Proof of Theorem 2.41 Consider Λ_ρ as defined in (2.40). We first choose a ρ such that any λ with $\operatorname{Re} \lambda \geq \beta$ and $|\lambda| \leq \alpha$ is contained in Λ_ρ.

We then follow the arguments of the proof of Theorem 2.34 which establishes that we achieve synchronization if we can find a controller which stabilizes the system $\bar{S}_e$ given by (2.51) and achieves an H_∞ norm from v to q less than 1. It is easy to verify that it is sufficient if we find a controller for (2.100) such that the said H_∞ norm is less than $1 + \rho$. The results from [124] immediately yields an appropriate controller via the same arguments as in the proof of Theorem 2.34. Combining this controller with $u = S\bar{u}$ then yields the desired protocol. ∎

2.5.7 Static Protocol Design

In most cases, we design dynamic protocols for a MAS with partial-state coupling. As we have seen, in many cases, we can actually design stable dynamic protocols which are desirable since they do not introduce additional synchronized trajectories. However, even more desirable are static protocols. In this section, we investigate the static protocol design for a MAS with partial-state coupling. As before, we will see that this problem reduces to a robust stabilization problem.

We then identify four classes of agents, for which static protocols can be designed:

- Squared-down passive
- Squared-down passifiable via static output feedback
- Squared-down passifiable via static input feedforward
- Squared-down minimum-phase with relative degree 1

Each of these cases will be studied next.

2.5.7.1 Connection to a Robust Stabilization Problem

We consider static protocols of the form (2.6) with ζ_i given by (2.3). In that case, the closed-loop system of the agent (2.1) and the protocol (2.6) can be described by:

$$\begin{cases} \dot{x}_i = Ax_i + BF\zeta_i, \\ y_i = Cx_i, \\ \zeta_i = \sum_{j=1}^{N} \ell_{ij} y_j, \end{cases} \tag{2.104}$$

for $i = 1, \ldots, N$. Define:

$$x = \begin{pmatrix} x_1 \\ \vdots \\ x_N \end{pmatrix}.$$

Then, the overall dynamics of the N agents can be written as:

$$\dot{x} = (I_N \otimes A + L \otimes BFC)x. \tag{2.105}$$

Using similar arguments as in Sect. 2.4.1, we obtain the following theorem and corollary.

Theorem 2.42 *The MAS* (2.105) *achieves state synchronization if and only if the system:*

$$\dot{\eta}_i = (A + \lambda_i BFC)\eta_i \tag{2.106}$$

is globally asymptotically stable for $i = 2, \ldots, N$. *Moreover, the synchronization trajectory is given by* (2.12).

Corollary 2.43 *Consider a MAS described by agents* (2.1) *and* (2.3). *If there exists a protocol of the form* (2.6) *that achieves state synchronization, then the associated graph must have a directed spanning tree or A must be asymptotically stable.*

In light of the definition of Problem 2.4 where synchronization is formulated for a set of graphs **G**, we basically obtain a robust stabilization problem, i.e., the stabilization of the system:

$$\begin{cases} \dot{x} = Ax + \lambda Bu, \\ y = Cx, \end{cases} \tag{2.107}$$

via a static controller:

$$u = Fy, \tag{2.108}$$

for any eigenvalue $\lambda \neq 0$ of a Laplacian matrix associated with a graph in the set **G**.

2.5.7.2 Squared-Down Passive Agents

In Sect. 1.3, we have defined the concept of squared-down passivity. In the current subsection, we will show that if the agents are squared-down passive, then we can actually find static protocols in the case of partial-state coupling. Let us first present the associated design.

Protocol design 2.11 *Consider a MAS described by squared-down passive agents (2.1) with respect to G_1 and G_2 with communication via (2.3). The static protocol is designed as:*

$$u_i = -\rho G_1 K G_2 \zeta_i, \tag{2.109}$$

where $\rho > 0$ and K is any positive definite matrix.

The main result regarding this design can be stated as follows.

Theorem 2.44 *Consider a MAS described by agents (2.1) and (2.3). Let $\mathbb{G}^N$ be defined in Definition 1.4. Assume that (2.1) is squared-down passive with respect to G_1 and G_2 such that (A, BG_1) is controllable and (A, G_2C) is observable.*

In that case, the state synchronization problem stated in Problem 2.4 is solvable with a static protocol. In particular, the protocol (2.109) for any $\rho > 0$ and $K > 0$ solves the state synchronization problem for any graph $\mathcal{G}$ which is in $\mathbb{G}^N$ for some N. Moreover, the synchronized trajectory is given by (2.12).

Proof Since:

$$\begin{aligned}
P(A - \lambda\rho BG_1KG_2C)+&(A - \lambda\rho BG_1KG_2C)'P \\
&= PA + A'P - 2\rho\operatorname{Re}(\lambda)PBG_1KG_2C \\
&\leq -2\rho\operatorname{Re}(\lambda)C'G_2'KG_2C,
\end{aligned}$$

we have that $A - \lambda\rho BG_1KG_2C$ is Hurwitz stable for any $\lambda \in \mathbb{C}^+$, which proves the result. ■

Remark 2.45 When the agents are square, $G_1 = G_2 = I$, the above result is presented in [168].

Remark 2.46 It is important to point out that for squared-down passive agents, protocol (2.109) requires no knowledge of the communication graph, and synchronization under this protocol is achieved for any number of agents and any graph as long as it contains a directed spanning tree. As such in this case we have scalability property.

2.5.7.3 Squared-Down Passifiable via Static Output Feedback

Next we will show that the static protocol (2.6) still works for the MAS where agents are squared-down passifiable via static output feedback, but some of the knowledge of network graphs (i.e., parameter β) is required. In other words, the problem can be solvable for a set of graphs $\mathbb{G}_\beta^N$ via the static protocol (2.6). We design the static protocol as follows.

Protocol design 2.12 *Consider a MAS described by agents (2.1) with communication via (2.3). Assume that the agents are squared-down passifiable via static output feedback with respect to G_1, G_2, and H. The static protocol is designed as,*

$$u_i = -\rho G_1 K G_2 \zeta_i, \tag{2.110}$$

where K is any positive definite matrix and ρ is a positive parameter to be designed later.

The main result based on the above design can be stated as follows.

Theorem 2.47 *Consider a MAS described by agents (2.1) and communication via (2.3). Assume the agents are squared-down passifiable via static output feedback with respect to G_1, G_2, and H, while (A, BG_1) is controllable and (A, G_2C) is observable with BG_1 and G_2C full-column and full-row rank, respectively. Let a $\beta > 0$ be given, and define $\mathbb{G}_\beta^N$ according to Definition 1.6.*

The state synchronization problem stated in Problem 2.4 but with a static protocol is solvable. In particular, there exists a ρ^* such that for any $\rho > \rho^*$, protocol (2.110) solves the state synchronization problem for any graph $\mathcal{G} \in \mathbb{G}_\beta^N$. Moreover, the synchronized trajectory is given by (2.12).

Proof We only need to prove that:

$$A - \lambda\rho B G_1 K G_2 C$$

is Hurwitz stable for any λ that satisfies $\operatorname{Re}(\lambda) > \beta$. Since the system is squared-down passifiable via static output feedback with respect to G_1, G_2, and H (A, BG_1) is controllable and (A, G_2C) is observable with BG_1 and G_2C full-column and full-row rank, respectively. In that case, there exists a positive definite matrix P such that (1.11) is satisfied. Moreover, for a fixed K, there exists a real number $b > 0$ such that:

$$H + H' \leq 2bK.$$

Then, we find that:

$$
\begin{aligned}
P(A-\lambda\rho BG_1KG_2C)+&(A-\lambda\rho BG_1KG_2C)'P \\
&\leq PBG_1HG_2C+C'G_2'H'G_1'B'P \\
&\quad -2\,\mathrm{Re}(\lambda)\rho PBG_1KG_2C \\
&= C'G_2'(H+H')G_2C-2\,\mathrm{Re}(\lambda)\rho C'G_2'KG_2C \\
&\leq 2bC'G_2'KG_2C-2\,\mathrm{Re}(\lambda)\rho C'G_2'KG_2C \\
&= -2(\mathrm{Re}(\lambda)\rho-b)C'G_2'KG_2C.
\end{aligned}
$$

Choosing $\rho^* = \frac{b}{\beta}$ immediately ensures that $A-\lambda\rho BG_1KG_2C$ is Hurwitz stable for $\rho > \rho^*$ since (A, G_2C) is observable, which proves the result. ∎

2.5.7.4 Squared-Down Passifiable Via Static Input feedforward

Next, we consider agents which are squared-down passifiable via static input feedforward. The crucial difference with the previous subsection is that squared-down passifiability via static output feedback in general requires a high-gain feedback, while squared-down passifiability via static input feedforward in general requires a low-gain feedback.

We design the static protocol based on the low-gain methodology as follows.

Protocol design 2.13 *Consider a MAS described by agents (2.1) with communication via (2.3). Assume the agents are squared-down passifiable via static input feedforward with respect to G_1, G_2, and R. The static protocol is designed as:*

$$
u_i = -\delta G_1KG_2\zeta_i, \tag{2.111}
$$

where $K > 0$ and the low-gain parameter $\delta > 0$ needs to be designed.

The main result based on the above design can be stated as follows.

Theorem 2.48 *Consider a MAS described by agents (2.1) and (2.3). Assume the agents are squared-down passifiable via static input feedforward with respect to G_1, G_2, and R such that (A, BG_1) is controllable and (A, G_2C) is observable with BG_1 and G_2C full-column and full-row rank, respectively. Let any $\alpha > 0$ be given, and define $\mathbb{G}_\alpha^N$ according to Definition 1.6.*

The state synchronization problem stated in Problem 2.4 but with a static protocol is solvable. In particular, there exists a δ^* such that for any $\delta < \delta^*$, the *protocol* (2.111) *solves the state synchronization problem for any graph* $\mathcal{G} \in \mathbb{G}_{\alpha}^{N}$. *Moreover, the synchronized trajectory is given by* (2.12).

In order to prove the above theorem, we need the following technical lemma.

Lemma 2.49 *Consider a Laplacian matrix* L *associated with a graph in* $\mathbb{G}_{\alpha}^{N}$. *In that case, any eigenvalue of* L *satisfies:*

$$|\lambda|^2 \le 2\alpha \operatorname{Re}(\lambda).$$

Proof Let $L = [\ell_{ij}]$. According to Gershgorin circle theorem, we know an eigenvalue λ is in a Gershgorin circle, that is:

$$|\lambda - \ell_{ii}|^2 \le \ell_{ii}^2$$

for some i. See, for instance, [28]. Here we use the fact that the Laplacian property implies:

$$\sum_{j=1, j\neq i}^{N} \ell_{ij} = -\ell_{ii}.$$

If $\operatorname{Re}(\lambda) = x$ and $\operatorname{Im}(\lambda) = y$, then the Gershgorin circle yields:

$$(x - \ell_{ii})^2 + y^2 \le \ell_{ii}^2,$$

or:

$$x^2 + y^2 \le 2\ell_{ii} x.$$

This yields:

$$|\lambda|^2 \le 2\ell_{ii} \operatorname{Re}(\lambda)$$

for some i. The proof is complete by noting that: $\ell_{ii} \le \alpha$. ■

Proof It is clear that we only need to prove that

$$\dot{\eta} = (A - \delta\lambda B G_1 K G_2 C)\eta \tag{2.112}$$

is asymptotically stable for all λ with $|\lambda| < \alpha$.

Next, by choosing a Lyapunov function $V = \eta' P \eta$, we obtain:

$$\begin{aligned}\dot{V} &= \eta' P(A - \delta\lambda B G_1 K G_2 C)\eta + \eta'(A - \delta\lambda B G_1 K G_2 C)' P\eta \\ &= \eta'(PA + A'P)\eta - \delta\lambda\eta'(PBG_1 - C'G_2')K G_2 C\eta\end{aligned}$$

$$
\begin{aligned}
&- \delta\lambda^* \eta' C' G_2' K (PBG_1 - C'G_2')'\eta - 2\delta \operatorname{Re}(\lambda)\eta' C' G_2' K G_2 C\eta \\
&= \begin{pmatrix} \eta \\ -\delta\lambda K G_2 C\eta \end{pmatrix}' G(P) \begin{pmatrix} \eta \\ -\delta\lambda K G_2 C\eta \end{pmatrix} + \delta^2 |\lambda|^2 \eta' C' G_2' K (R + R') K G_2 C\eta \\
&\qquad\qquad\qquad\qquad - 2\delta \operatorname{Re}(\lambda)\eta' C' G_2' K G_2 C\eta \\
&\leq \eta' C' G_2' \left[\delta^2 |\lambda|^2 K (R + R') K - 2\delta \operatorname{Re}(\lambda) K \right] G_2 C\eta \\
&\leq 2 \operatorname{Re}(\lambda)\eta' C' G_2' \left[\alpha\delta^2 K (R + R') K - \delta K \right] G_2 C\eta,
\end{aligned}
$$

where $G(P)$ is given by (1.12). We find that:

$$
\dot{V} < -\gamma \eta' C' G_2' K G_2 C\eta, \tag{2.113}
$$

for some $\gamma > 0$ provided δ is such that:

$$
\delta\alpha K (R + R') K < K,
$$

which is clearly satisfied for sufficiently small δ. Since (A, G_2C) is observable, we find that (2.113) guarantees the required asymptotic stability. ∎

2.5.7.5 Squared-Down Minimum-Phase Agent with Relative Degree 1

A non-square system (1.4) is called *squared-down minimum-phase with relative degree* 1 with a pre-compensator $G_1 \in \mathbb{R}^{m\times q}$ and a post-compensator $G_2 \in \mathbb{R}^{q\times p}$ if the square system (A, BG_1, G_2C) is minimum-phase with relative degree 1 ($\det(G_2CBG_1) \neq 0$). Note that for such a system (A, BG_1, G_2C), with input $\hat{u}$ where $u = G_1\hat{u}$, and output $\hat{y} = G_2 y$, there exist nonsingular state transformation matrices T_x and T_u with:

$$
\bar{x} = \begin{pmatrix} \bar{x}_1 \\ \bar{x}_2 \end{pmatrix} = T_x x, \qquad \bar{u} = T_u \hat{u}
$$

and the dynamics of $\bar{x}$ is represented as

$$
\begin{aligned}
\dot{\bar{x}}_1 &= A_{11}\bar{x}_1 + A_{12}\bar{x}_2, \\
\dot{\bar{x}}_2 &= A_{21}\bar{x}_1 + A_{22}\bar{x}_2 + \bar{u}, \\
\hat{y} &= \bar{x}_2,
\end{aligned} \tag{2.114}
$$

where $\bar{x}_1 \in \mathbb{R}^{n-m}$ and $\bar{x}_2 \in \mathbb{R}^m$. Moreover, A_{11} is Hurwitz stable.

Next, we can design a protocol for a MAS which is squared-down minimum-phase with relative degree 1.

Protocol design 2.14 *Consider a MAS described by squared-down minimum-phase with relative degree* 1 *agents* (2.1) *with communication via* (2.3). *The static protocol is designed as:*

$$u_i = -\rho G_1 T_u^{-1} G_2 \zeta_i, \tag{2.115}$$

where ρ is a parameter to be designed.

The main result based on the above design can be stated as follows.

Theorem 2.50 *Consider a MAS described by agents* (2.1) *and* (2.3). *Let any $\beta > 0$ be given, and define $\mathbb{G}_\beta^N$ according to Definition 1.6.*

If (A, B, C) is squared-down minimum-phase with relative degree 1, *then the state synchronization problem stated in Problem 2.4 is solvable. In particular, there exists a $\rho^* > 0$ such that for any $\rho > \rho^*$, protocol* (2.115) *solves the state synchronization problem for any graph $\mathcal{G} \in \mathbb{G}_\beta^N$. Moreover, the synchronized trajectory is given by* (2.12).

Proof We only need to prove that:

$$A - \lambda\rho B G_1 T_u^{-1} G_2 C \tag{2.116}$$

is asymptotically stable for all λ with $\mathrm{Re}(\lambda) > \beta$. Since the agent is squared-down minimum-phase, the stability of (2.116) is equivalent to the stability of the interconnection of the system:

$$\begin{aligned} \dot{\tilde{x}}_1 &= A_{11}\tilde{x}_1 + A_{12}\tilde{x}_2, \\ \dot{\tilde{x}}_2 &= A_{21}\tilde{x}_1 + A_{22}\tilde{x}_2 + \lambda\tilde{u}, \\ \hat{y} &= \tilde{x}_2, \end{aligned} \tag{2.117}$$

with a controller:

$$\tilde{u} = -\rho\hat{y} \tag{2.118}$$

for all λ with $\mathrm{Re}(\lambda) > \beta$. Here we used that:

$$G_2 C T_x^{-1} = \begin{pmatrix} 0 & I \end{pmatrix}.$$

The closed-loop system of (2.117) and (2.118) is written as:

$$\begin{aligned} \dot{\bar{x}}_1 &= A_{11}\bar{x}_1 + A_{12}\bar{x}_2, \\ \dot{\bar{x}}_2 &= A_{21}\bar{x}_1 + (A_{22} - \lambda\rho I)\bar{x}_2. \end{aligned} \tag{2.119}$$

Since A_{11} is Hurwitz stable, there exists a $P_1 > 0$ such that:

$$P_1 A_{11} + A_{11}' P_1 = -I.$$

Now choose $\rho^* > 0$ such that:

$$A_{22} - \beta\rho^* I$$

is Hurwitz stable. Therefore, there exists a solution $P_2 > 0$ such that:

$$P_2(A_{22} - \beta\rho^* I) + (A_{22} - \beta\rho^* I)' P_2 = -I.$$

Then for any $\rho > \rho_*$, we have:

$$P_2(A_{22} - \lambda\rho I) + (A_{22} - \lambda\rho I)^* P_2 \le -\kappa I$$

for all λ with $\mathrm{Re}(\lambda) > \beta$ where κ is such that:

$$\kappa = 1 + 2\beta(\rho - \rho_*)\|P_2^{-1}\|^{-1}.$$

Note that κ is an increasing function of ρ. Define a Lyapunov function:

$$V = \bar{x}_1' P_1 \bar{x}_1 + \bar{x}_2' P_2 \bar{x}_2.$$

Then, the time derivative of V is obtained as:

$$\begin{aligned}
\dot{V} &= -\|\bar{x}_1\|^2 - \kappa\|\bar{x}_2\|^2 + 2\,\mathrm{Re}(\bar{x}_2' A_{12}' P_1 \bar{x}_1) + 2\,\mathrm{Re}(\bar{x}_1' A_{21}' P_2 \bar{x}_2) \\
&\le -\|\bar{x}_1\|^2 - \kappa\|\bar{x}_2\|^2 + 2r_1\|\bar{x}_1\|\|\bar{x}_2\| + 2r_2\|\bar{x}_1\|\|\bar{x}_2\| \\
&= \begin{pmatrix} \|\bar{x}_1\| & \|\bar{x}_2\| \end{pmatrix} \begin{pmatrix} -1 & r_1 + r_2 \\ r_1 + r_2 & -\kappa \end{pmatrix} \begin{pmatrix} \|\bar{x}_1\| \\ \|\bar{x}_2\| \end{pmatrix}
\end{aligned}$$

where $r_1 \ge \|A_{12}' P_1\|$ and $r_2 \ge \|A_{21}' P_2\|$. It is clear that by choosing ρ sufficiently large will guarantee that $\kappa > (r_1 + r_2)^2$; thus, the closed-loop system (2.119) is asymptotically stable. This proves the result. ■

Remark 2.51 Compared with squared-down passivity, squared-down minimum-phase with relative degree 1 does not require the agents to be passive.

2.5.8 *Partial-State Coupling and Additional Communication*

Problems 2.3 and 2.4 are the most natural ones. However, in the literature for partial-state coupling initially (see [49, 114]), a different problem was addressed.

In this case, an additional communication of protocol states via the network's communication infrastructure is allowed. This is mathematically appealing in the analysis as it allows for a relaxation in the solvability conditions.

Specifically, agent i ($i = 1, \dots, N$) is presumed to have access to the quantity:

$$\hat{\xi}_i = \sum_{j=1}^{N} a_{ij}(\xi_i - \xi_j) = \sum_{j=1}^{N} \ell_{ij}\xi_j, \tag{2.120}$$

where $\xi_j \in \mathbb{R}^p$ is a variable produced internally by agent j as part of the protocol.

Problem 2.52 (Partial-State Coupling with Extra Communication) Consider a MAS described by (2.1), (2.3), and (2.120). Let $\mathbf{G}$ be a given set of graphs such that $\mathbf{G} \subseteq \mathbb{G}^N$. The *state synchronization* problem with extra communication and a set of network graphs $\mathbf{G}$ is to find, if possible, a linear time-invariant dynamic protocol of the form:

$$\begin{cases} \dot{\chi}_i = A_c\chi_i + B_{c,1}\zeta_i + B_{c,2}\hat{\xi}_i, \\ u_i = C_c\chi_i + D_{c,1}\zeta_i + D_{c,2}\hat{\xi}_i, \end{cases} \tag{2.121}$$

for $i = 1, \dots, N$ where $\chi_i \in \mathbb{R}^{n_c}$, such that for $\xi_i = C\chi_i$ and for any graph $\mathcal{G} \in \mathbf{G}$ and for all initial conditions of the agents and their protocol, the state synchronization among agents is achieved.

Previously, Theorem 2.20 showed that protocol design for MAS with partial-state coupling, as formulated in Problem 2.4, basically requires the agents to be either at most weakly unstable or at most weakly non-minimum-phase. In this section, we will see that allowing an additional communication layer of communication (2.120), as formulated in Problem 2.52, allows us to remove this restriction. However, whether this extra communication is feasible in practice is questionable.

As before, we can easily describe our approach, we first show that state synchronization among agents with additional communication can be solved by equivalently solving a robust stabilization problem, and then we design a controller for such a robust stabilization problem.

Choosing $\xi_i = C\chi_i$ in (2.120), the closed loop of agent (2.1) and protocol (2.121) with the network described by (2.3) and (2.120) can then be written as:

$$\dot{\bar{x}} = (I_N \otimes \bar{A} + L \otimes \bar{B}\bar{C})\bar{x}, \tag{2.122}$$

where:

$$\bar{x} = \begin{pmatrix} \bar{x}_1 \\ \vdots \\ \bar{x}_N \end{pmatrix}, \quad \bar{A} = \begin{pmatrix} A & BC_c \\ 0 & A_c \end{pmatrix}, \quad \bar{B} = \begin{pmatrix} BD_c \\ B_c \end{pmatrix}, \quad \text{and} \quad \bar{C} = \begin{pmatrix} -C & C \end{pmatrix},$$

and where $B_{c,1} = -B_{c,2} = B_c$ and $D_{c,1} = -D_{c,2} = D_c$ while:

$$\bar{x}_i = \begin{pmatrix} x_i \\ \chi_i \end{pmatrix}.$$

A key observation, which we already used for Problems 2.3 and 2.4, is that synchronization for the system (2.122) is equivalent to the asymptotic stability of the following $N-1$ subsystems:

$$\dot{\tilde{\eta}}_i = (\bar{A} + \lambda_i \bar{B}\bar{C})\tilde{\eta}_i, \quad i = 2, \dots, N, \tag{2.123}$$

where λ_i, $i = 2, \dots, N$ are the nonzero eigenvalues of the Laplacian matrix L.

Theorem 2.53 *The MAS (2.122) achieves state synchronization if and only if the system (2.123) is globally asymptotically stable for $i = 2, \dots, N$.*

Proof The proof is basically the same as the proof of Theorem 2.15. ■

Clearly, this implies that if the network is unknown and we only know certain properties of the eigenvalues of the associated Laplacian matrix L, then we need to guarantee that:

$$\dot{\tilde{\eta}} = (\bar{A} + \lambda \bar{B}\bar{C})\tilde{\eta}, \tag{2.124}$$

is asymptotically stable for all nonzero eigenvalues λ of the Laplacian matrix L associated with the graph set $\mathbf{G}$.

Similar to Corollary 2.16, we find that for Problem 2.52, it is basically necessary that the associated graph must have a directed spanning tree. Moreover, if we have a protocol (2.121) that achieves state synchronization, then the synchronized trajectory is given by (2.32) which reduces to (2.33) for stable protocols.

Consider a linear dynamic stable protocol in the form of:

$$\begin{cases} \dot{\chi}_i = (A + BF)\chi_i + K(\zeta_i - \hat{\zeta}_i), \\ u_i = F\chi_i, \end{cases} \tag{2.125}$$

where F is chosen such that $A + BF$ is Hurwitz stable and K is to be designed.

We have seen that in order to check whether the protocol (2.125) achieves state synchronization, we need to verify (2.124) is asymptotically stable for all nonzero eigenvalues λ of the Laplacian matrix L associated with the graph set $\mathbf{G}$. Given our protocol, we have:

$$\bar{A} + \lambda \bar{B}\bar{C} = \begin{pmatrix} A & BF \\ -\lambda KC & A + BF + \lambda KC \end{pmatrix},$$

the stability of which is equivalent to that of the matrix

$$\begin{pmatrix} 1 & -1 \\ 0 & 1 \end{pmatrix} \begin{pmatrix} A & BF \\ -\lambda KC & A+BF+\lambda KC \end{pmatrix} \begin{pmatrix} 1 & 1 \\ 0 & 1 \end{pmatrix} = \begin{pmatrix} A+\lambda KC & 0 \\ -\lambda KC & A+BF \end{pmatrix}.$$

It shows that $A + BF$ is Hurwitz stable, which requires that (A, B) is stabilizable. Moreover, we need to design K such that $A+\lambda KC$ is Hurwitz stable for any nonzero eigenvalues λ of the Laplacian matrix L associated with the graph set $\mathbf{G}$. Apparently it is a dual version of state synchronization problem in full-state coupling, that is, designing F such that $A + \lambda BF$ is Hurwitz stable for any nonzero eigenvalues λ of the Laplacian matrix L associated with the graph set $\mathbf{G}$. Hence, (A, C) is required to be detectable.

Clearly we can use ARE-based method to design K based on a dual version of the ARE-based design in Sect. 2.4.2 or a direct method to design K based on a dual version of the ATEA design in Sect. 2.4.3. In the following, we will give the protocol design based on ARE method.

Protocol design 2.15 *Consider a MAS described by agents (2.1) with communication via (2.3) and (2.120). Let $\xi_i = C\chi_i$ in (2.120) and consider the protocol:*

$$\begin{cases} \dot{\chi}_i = (A+BF)\chi_i + K(\zeta_i - \hat{\xi}_i), \\ u_i = F\chi_i, \end{cases} \tag{2.126}$$

with F such that $A + BF$ is Hurwitz stable while $K = -\rho QC'$, where $\rho > 1$ and $Q > 0$ is the unique solution of the continuous-time algebraic Riccati equation:

$$AQ + QA' - 2\beta QC'CQ + W = 0, \tag{2.127}$$

where $W > 0$.

Theorem 2.54 *Consider a MAS described by agents (2.1) with communication via (2.3) and (2.120). Let any $\beta > 0$ be given, and define $\mathbb{G}_\beta^N$ according to Definition 1.6.*

If (A, B) is stabilizable and (C, A) is detectable, then the state synchronization problem via stable protocol stated in Problem 2.52 with $\mathbf{G} = \mathbb{G}_\beta^N$ is solvable. More specifically, the protocol (2.126) solves the state synchronization problem by stable protocol for any graph $\mathcal{G} \in \mathbb{G}_\beta^N$ for any $\rho > 1$ and $W > 0$. Moreover, the synchronized trajectory is given by (2.33).

Proof Since we have:

$$
\begin{aligned}
&(A - \lambda\rho QC'C)Q + Q(A - \lambda\rho QC'C)^* \\
&= -W - 2(\rho\,\mathrm{Re}(\lambda) - \beta)QC'CQ \\
&\leq -W,
\end{aligned}
$$

the system (2.124) is asymptotically stable for any λ that satisfies $\mathrm{Re}(\lambda) \geq \beta$, which proves the result. ■

2.5.9 *Partial-State Coupling for Introspective Agents*

In the previous subsection, we considered the partial-state coupling problem with an additional communication layer, and we noted that this extra communication allows us to solve the synchronization problem if the agents are neither at most weakly unstable nor at most weakly non-minimum-phase. This extra communication layer might not be very realistic in practice. A more realistic situation might be introspective agents where each agent has a measurement available of its own state. Specifically, agent i is presumed to have access to the quantity:

$$
z_{m_i} = C_m x_i, \tag{2.128}
$$

for $i = 1, \ldots, N$. This results in the following problem.

Problem 2.55 (Partial-State Coupling with Introspective Agents) Consider a MAS described by (2.1), (2.3), and (2.128). Let $\mathbf{G}$ be a given set of graphs such that $\mathbf{G} \subseteq \mathbb{G}^N$. The *state synchronization* problem with introspective agents and a set of network graphs $\mathbf{G}$ is to find, if possible, a linear time-invariant dynamic protocol of the form:

$$
\begin{cases}
\dot{\bar{\chi}}_i = \bar{A}_c\bar{\chi}_i + \bar{B}_c\,\mathrm{col}\{\zeta_i, z_{m,i}\}, \\
u_i = \bar{C}_c\bar{\chi}_i + \bar{D}_c\,\mathrm{col}\{\zeta_i, z_{m,i}\},
\end{cases} \tag{2.129}
$$

for $i = 1, \ldots, N$ where $\bar{\chi}_i \in \mathbb{R}^{n_c}$, such that for any graph $\mathcal{G} \in \mathbf{G}$ and for all the initial conditions for the agents and their protocol, the state synchronization among agents is achieved.

We can use a specific structure for the protocol (2.129) where:

$$
\begin{cases}
\dot{\chi}_i &= A_c\chi_i + B_c\zeta_i, \\
\dot{\chi}_{p,i} &= A_p\chi_i + B_p z_{m,i}, \\
u_i &= C_c\chi_i + C_p\chi_i + D_c\zeta_i
\end{cases} \tag{2.130}
$$

for $i = 1, \ldots, N$, where (A_p, B_p, C_p) describe a pre-feedback and (A_c, B_c, C_c, D_c) describe a protocol with a standard structure based on the network measurement (2.3). Obviously, this pre-feedback can stabilize each agent, and then even without using network information, we achieve state synchronization since all agent states converge to zero. However, the pre-feedback can be designed with different objectives in mind:

- For instance, in the work of [16], they considered agents that are introspective which enabled them to use a pre-feedback to create passive agents. Obviously a feedback cannot affect the zeros so one of the restrictions that had to be imposed is that the system is at most weakly non-minimum-phase and therefore the problem could have been solved through a non-introspective protocol using the techniques from Sect. 2.5.4.
- Another design is to use the pre-feedback to make sure that the system becomes at most weakly unstable which enables us to design a protocol based on the techniques from Sect. 2.5.3.
- Finally, we can use the pre-feedback to shape the synchronized trajectory. If the pre-feedback is such that all the poles of the system are stable, then the synchronized trajectory will always be equal to zero. But by moving the poles of the system to specific locations on the imaginary axis, then we can shape the synchronized trajectory by imposing certain dynamics.

2.6 Application to Formation

The formation considered here is restricted to the rigid formation. It is shown in the literature (see [45, 174] for example) that the formation problem is closely related to the synchronization problem. The protocol design procedure in previous sections can be easily modified to solve the formation problem. In this section, both the full-state and the partial-state coupling are considered.

2.6.1 Full-State Coupling

In this subsection, we consider the formation problem for a multi-vehicle network with full-state coupling. We first formulate the formation problem and then analyze the transformation of the formation problem to the synchronization problem.

Consider a multi-agent system (MAS) composed of N vehicles with the dynamics:

$$\dot{x}_i = Ax_i + Bu_i, \quad (i = 1, \ldots, N) \tag{2.131}$$

where $x_i \in \mathbb{R}^{2n}$ represents position-like variables and velocity-like variables, i.e.:

$$x_i = \begin{pmatrix} x_{i,p} \\ x_{i,v} \end{pmatrix}.$$

We assume that A and B have the form:

$$A = \begin{pmatrix} 0 & I_n \\ A_1 & A_2 \end{pmatrix}, \quad B = \begin{pmatrix} 0 \\ I_n \end{pmatrix}, \tag{2.132}$$

which immediately implies the derivative relationship $\dot{x}_{i,p} = x_{i,v}$. Moreover, we assume that (A, B) is controllable.

The concept of formation that we will use is based on the definition used by Lafferriere et al. [45].

Definition 2.56 Given an arbitrary formation vector $\{h_1, \ldots, h_N\}$ with $h_i \in \mathbb{R}^n$, vehicles (2.131) achieve formation if:

$$\lim_{t\to\infty} \big((x_{i,p}(t) - h_i) - (x_{j,p}(t) - h_j)\big) = 0, \tag{2.133}$$

and:

$$\lim_{t\to\infty} \big(x_{i,v}(t) - x_{j,v}(t)\big) = 0. \tag{2.134}$$

Now let:

$$\bar{h}_i = \begin{pmatrix} h_i \\ 0 \end{pmatrix}, \quad (i = 1, \ldots, N). \tag{2.135}$$

The communication network provides each vehicle with a linear combination of relative state information, that is:

$$\zeta_i = \sum_{j=1}^{N} a_{ij} \big((x_i - \bar{h}_i) - (x_j - \bar{h}_j)\big) = \sum_{j=1}^{N} \ell_{ij}(x_j - \bar{h}_j), \tag{2.136}$$

where a_{ij} are the weights associated with the edges of the graph $\mathcal{G}$, while ℓ_{ij} are the coefficients of the associated Laplacian matrix L. Assume that the network has a directed spanning tree. Then, the problem is formulated as follows.

Problem 2.57 (Full-State Coupling) Consider a multi-vehicle system described by (2.131) and (2.136). Let $\mathbf{G}$ be a given set of graphs such that $\mathbf{G} \subseteq \mathbb{G}^N$. The *formation* problem with a set of network graphs $\mathbf{G}$ is to find, if possible, a linear static protocol of the form:

$$u_i = F_1\zeta_i + F_2 h_i, \tag{2.137}$$

for $i = 1, \ldots, N$ such that for any graph $\mathcal{G} \in \mathbf{G}$ and for all the initial conditions of vehicles, formation among vehicles is achieved.

Let:

$$\bar{x}_i = x_i - \bar{h}_i.$$

Choosing $F_2 = -A_1$, the closed-loop system of the vehicle (2.131) and the linear static protocol (2.137) can be written as:

$$\begin{aligned}
\dot{\bar{x}}_i &= Ax_i - BA_1 h_i + BF_1 \sum_{j=1}^{N} \ell_{ij}\bar{x}_j,\\
&= Ax_i - A\bar{h}_i + BF_1 \sum_{j=1}^{N} \ell_{ij}\bar{x}_j,\\
&= A\bar{x}_i + BF_1 \sum_{j=1}^{N} \ell_{ij}\bar{x}_j.
\end{aligned}$$

Let:

$$\bar{x} = \begin{pmatrix} \bar{x}_1 \\ \vdots \\ \bar{x}_N \end{pmatrix}.$$

Then the overall dynamics of the N vehicles in terms of $\bar{x}$ can be written as:

$$\dot{\bar{x}} = (I_N \otimes A + L \otimes BF_1)\bar{x}. \tag{2.138}$$

Note that the above overall system (2.138) is exactly the same as the overall system given in (2.8). Thus, the design of F_1 can be based on the protocol design in Protocol Design 2.1 using an ARE-based method or can be based on the protocol design in Protocol Design 2.2 using an ATEA-based method. The result, based on an ARE-based design, is stated in the following theorem.

Theorem 2.58 *Consider a multi-vehicle system described by* (2.131) *and* (2.136). *Let any $\beta > 0$ be given, and consider the set of network graphs $\mathbb{G}_\beta^N$ as defined in Definition 1.6.*

The formation problem stated in Problem 2.57 with $\mathbf{G} = \mathbb{G}_\beta^N$ *is solvable. In particular, the protocol (2.137) with* $F_2 = -A_1$ *and* $F_1 = \rho F$ *with* F *as designed in Protocol Design 2.1 and* $\rho > 1$ *solves the formation problem for any graph* $\mathcal{G} \in \mathbb{G}_\beta^N$.

Proof The proof follows that of Theorem 2.12 or Theorem 2.13. ∎

Remark 2.59 An alternative is to use the protocol (2.137) with $F_2 = -A_1$ and $F_1 = F_\varepsilon$ with F_ε designed via Protocol Design 2.2 for ε sufficiently small.

2.6.2 *Partial-State Coupling*

In this subsection, we consider the formation problem for a multi-vehicle network with partial-state coupling. Similar to the approach of full-state coupling, we first formulate the formation problem and then analyze the transformation of the formation problem to the synchronization problem.

Consider a multi-agent system (MAS) composed of N vehicles with the dynamics:

$$\begin{cases} \dot{x}_i = Ax_i + Bu_i, \\ y_i = Cx_i \end{cases} \qquad (i = 1, \dots, N) \tag{2.139}$$

where x_i and matrices A, B are defined as in (2.131). Partial state here implies that only the position-like variables are communicated over the network, i.e., the matrix C has the form:

$$C = \begin{pmatrix} I_n & 0 \end{pmatrix}, \tag{2.140}$$

which, together with the definition of A and B, implies that (C, A) is observable and (A, B, C) is minimum-phase. Then the output $y_i = x_{i,p}$ is the position variable.

The communication network provides each vehicle with a linear combination of relative output information (instead of relative state information (2.136) that we used in the case of full-state coupling). In other words:

$$\zeta_i = \sum_{j=1}^{N} a_{ij} \left((y_i - h_i) - (y_j - h_j) \right) = \sum_{j=1}^{N} \ell_{ij} (y_j - h_j). \tag{2.141}$$

Then, the problem is formulated as follows.

Problem 2.60 (Partial-State Coupling) Consider a multi-vehicle system described by (2.139) and (2.141). Let $\mathbf{G}$ be a given set of graphs such that $\mathbf{G} \subseteq \mathbb{G}^N$. The *formation* problem with a set of network graphs $\mathbf{G}$ is to find, if possible, a linear time-invariant dynamic protocol of the form:

$$\begin{cases} \dot{\chi}_i = A_c\chi_i + B_c\zeta_i, \\ u_i = C_c\chi_i + D_{c,1}\zeta_i + D_{c,2}h_i, \end{cases} \tag{2.142}$$

for $i = 1, \ldots, N$ where $\chi_i \in \mathbb{R}^{n_c}$, such that for any graph $\mathcal{G} \in \mathbf{G}$ and for all the initial conditions for the vehicles and their protocol, formation among agents is achieved.

Let:

$$\bar{x}_i = x_i - \bar{h}_i, \quad \bar{y}_i = y_i - h_i,$$

where $\bar{h}_i$ is defined in (2.135).

Similar to the full-state coupling case, we choose $D_{c,2} = -A_1$. In terms of the dynamics of $\bar{x}_i$ and output $\bar{y}_i$, the closed-loop system of vehicle (2.139) and the linear dynamic protocol (2.142) can then be written as:

$$\begin{cases} \dot{\tilde{x}}_i = \begin{pmatrix} A & BC_c \\ 0 & A_c \end{pmatrix}\tilde{x}_i + \begin{pmatrix} BD_{c,1} \\ B_c \end{pmatrix}\zeta_i, \\ \bar{y}_i = \begin{pmatrix} C & 0 \end{pmatrix}\tilde{x}_i, \\ \zeta_i = \sum_{j=1}^N \ell_{ij}\bar{y}_j, \end{cases}$$

for $i = 1, \ldots, N$ where:

$$\tilde{x}_i = \begin{pmatrix} \bar{x}_i \\ \chi_i \end{pmatrix}.$$

Define:

$$\tilde{x} = \begin{pmatrix} \tilde{x}_1 \\ \vdots \\ \tilde{x}_N \end{pmatrix}, \quad \bar{A} = \begin{pmatrix} A & BC_c \\ 0 & A_c \end{pmatrix}, \quad \bar{B} = \begin{pmatrix} BD_{c,1} \\ B_c \end{pmatrix}, \quad \text{and} \quad \bar{C} = \begin{pmatrix} C & 0 \end{pmatrix}.$$

Then, the overall dynamics of the N vehicles in terms of $\tilde{x}$ can be written as:

$$\dot{\tilde{x}} = (I_N \otimes \bar{A} + L \otimes \bar{B}\bar{C})\tilde{x}. \tag{2.143}$$

Note that the above overall system (2.143) is exactly the same as the overall system given in (2.29). Due to the structure of the system, it is easily seen that (A, B, C) is minimum-phase and therefore the protocol design of $(A_c, B_c, C_c, D_{c,1})$ can be obtained from Sect. 2.5.5.

Then, the result can be stated in the following theorem.

Theorem 2.61 *Consider a multi-vehicle system described by* (2.139) *and* (2.141). *Let any* $\alpha \geq \beta > 0$ *be given and hence a set of network graphs* $\mathbb{G}^N_{\alpha,\beta}$ *be defined.*

The formation problem stated in Problem 2.60 *with* $\mathbf{G} = \mathbb{G}^N_{\alpha,\beta}$ *is always solvable by a stable protocol. In particular, let* ρ *be such that any* $\lambda \in \mathbb{C}$ *with* $\operatorname{Re}\lambda > \beta$ *and* $|\lambda| < \alpha$ *is in the set* Λ_ρ *defined in* (2.40). *Then there exists an* ε^* *such that the protocol:*

$$\begin{cases} \dot{\chi}_i = (A + K_\varepsilon C)\chi_i - K_\varepsilon \zeta_i \\ u_i = F_\rho \chi_i - A_1 h_i \end{cases} \tag{2.144}$$

solves the formation problem for any graph $\mathcal{G} \in \mathbb{G}^N_{\alpha,\beta}$ *provided* $\varepsilon < \varepsilon^*$ *where* F_ρ *and* K_ε *are constructed in Protocol Design* 2.8.

Remark 2.62 If A happens to have all its eigenvalues in the closed left half plane, i.e., the vehicle dynamic is at most weakly unstable, then the protocol design for (A_c, B_c, C_c) can also be obtained from Sect. 2.5.3.

Proof We have shown how to reduce the formation problem to a synchronization problem. Moreover, the structure guarantees that the system is minimum-phase. Therefore, we can obtain the above result from Theorem 2.34. ■

Chapter 3
Synchronization of Discrete-Time Linear MAS

3.1 Introduction

This chapter creates similar results to Chap. 2 but for homogeneous linear discrete-time MAS, that is, solvability conditions for state synchronization of MAS with both full-state and partial-state couplings, dynamic protocol designs for at most weakly unstable agents and strictly unstable agents, and static protocol design for MAS with partial-state coupling. This chapter also considers introspective and non-introspective agents and investigates the dynamic protocol when an additional communication is provided among agents.

For homogeneous MAS with discrete-time agents, earlier work can be found in [24, 34, 48, 79, 91, 144] for essentially first- and second-order agents and in [36, 47, 51, 155, 176, 193, 196] for higher-order agents, while [24, 34, 36, 79] deal with full-state coupling and [51, 155] with partial-state coupling. With the exception of [155], the literature on synchronization of MAS with partial-state coupling utilizes an additional communication layer which involves the exchange of part of the state of the protocol associated with each agent over the existing communication network. The ubiquitous condition on the agent model in the literature is at most weakly unstable (see [155, 180] for MAS with full-state coupling and [47, 51, 144, 154, 180] for MAS with partial-state coupling). For general, at most weakly unstable agents, the protocols given in the literature utilize a "nonstandard" discrete-time Riccati equation. Recently, [36, 51] and [176] have considered synchronization for MAS with unstable agents.

The write-up of this chapter is partially based on [131].

A. Saberi et al., *Synchronization of Multi-Agent Systems in the Presence of Disturbances and Delays*, Systems & Control: Foundations & Applications,
https://doi.org/10.1007/978-3-030-88148-1_3

3.2 Multi-agent Systems

Consider a multi-agent system (MAS) composed of N identical linear time-invariant discrete-time agents of the form:

$$\begin{aligned} x_i(k+1) &= Ax_i(k) + Bu_i(k), \\ y_i(k) &= Cx_i(k), \end{aligned} \qquad (i = 1, \ldots, N) \tag{3.1}$$

where $x_i \in \mathbb{R}^n$, $u_i \in \mathbb{R}^m$, and $y_i \in \mathbb{R}^p$ are, respectively, the state, input, and output vectors of agent i. We make the following standard and necessary assumption for the agent dynamics.

Assumption 3.1 *(A, B) is stabilizable, and (A, C) is detectable.*

The communication network provides each agent with a linear combination of its own output relative to that of other neighboring agents. In particular, each agent $i \in \{1, \ldots, N\}$ has access to the quantity:

$$\zeta_i(k) = \frac{1}{1 + \sum_{j=1}^N a_{ij}} \sum_{j=1}^N a_{ij}(y_i(k) - y_j(k)), \tag{3.2}$$

where $a_{ij} \geq 0$, $a_{ii} = 0$ for $i, j \in \{1, \ldots, N\}$. The topology of the network can be described by a graph $\mathcal{G}$ with nodes corresponding to the agents in the network and edges given by the nonzero coefficients a_{ij}. In particular, $a_{ij} > 0$ implies that an edge exists from agent j to i. The weight of the edge equals the magnitude of a_{ij}.

In this context, the matrix $A_a = [a_{ij}]$ is referred to as the adjacency matrix. Based on the adjacency matrix, we can associate a Laplacian matrix L with a graph:

$$L = D_{\text{in}} - A_a$$

where:

$$D_{\text{in}} = \text{diag}\{d_{\text{in}}(i)\}$$

while:

$$d_{\text{in}}(i) = \sum_{j=1}^N a_{ij}$$

denotes the weighted in degree of vertex i. We define:

$$D = I - (I + D_{\text{in}})^{-1} L. \tag{3.3}$$

It is easily verified that D is a row stochastic matrix in the sense that the sum of the elements in each row is equal to 1. In other words, $D\mathbf{1} = \mathbf{1}$.

Next we write ζ_i as:

$$\zeta_i(k) = \sum_{j=1}^{N} d_{ij}(y_i(k) - y_j(k)). \tag{3.4}$$

We will use the following assumption on the network graph.

Assumption 3.2 *The graph $\mathcal{G}$ describing the communication topology of the network contains a directed spanning tree.*

By Lemma 1.14, we note that Assumption 3.2 guarantees that the row stochastic matrix D has a simple eigenvalue at 1 with corresponding right eigenvector $\mathbf{1}$ and all other eigenvalues are strictly within the unit disc. Let $\lambda_1, \ldots, \lambda_N$ denote the eigenvalues of row stochastic matrix D such that $\lambda_1 = 1$ and $|\lambda_i| < 1, i = 2, \ldots, N$.

3.3 Problem Formulation

Since all agents in the network are identical, we can pursue state synchronization among agents. State synchronization is defined as follows.

Definition 3.1 Consider a homogeneous network described by (3.1) and (3.4). The agents in the network achieve state synchronization if:

$$\lim_{k\to\infty} (x_i(k) - x_j(k)) = 0, \qquad \forall i, j \in \{1, \ldots, N\}. \tag{3.5}$$

In case $C = I$, the network exhibits full-state coupling. This implies that the quantity ζ_i becomes:

$$\zeta_i(k) = \sum_{j=1}^{N} d_{ij}(x_i(k) - x_j(k)), \tag{3.6}$$

which means agents in the network have access to the relative information of the full state of their neighboring agents compared to their own state.

We formulate two state synchronization problems for a network with full-state coupling and partial-state coupling, respectively.

Problem 3.2 (Full-State Coupling) Consider a MAS described by (3.1) and (3.6). Let $\mathbf{G}$ be a given set of digraphs such that $\mathbf{G} \subseteq \bar{\mathbb{G}}^N$. The *state synchronization* problem with a set of network graphs $\mathbf{G}$ is to find, if possible, a linear static protocol of the form:

$$u_i(k) = F\zeta_i(k), \tag{3.7}$$

such that for any graph $\mathcal{G} \in \mathbf{G}$ and for all the initial conditions of agents, the state synchronization among agents can be achieved.

Problem 3.3 (Partial-State Coupling) Consider a MAS described by (3.1) and (3.4). Let $\mathbf{G}$ be a given set of digraphs such that $\mathbf{G} \subseteq \bar{\mathbb{G}}^N$. The *state synchronization* problem with a set of network graphs $\mathbf{G}$ is to find, if possible, a linear time-invariant dynamic protocol of the form:

$$\begin{cases} \chi_i(k+1) = A_c\chi_i(k) + B_c\zeta_i(k), \\ \quad u_i(k) \quad = C_c\chi_i(k) + D_c\zeta_i(k), \end{cases} \tag{3.8}$$

where $\chi_i \in \mathbb{R}^{n_c}$, such that for any graph $\mathcal{G} \in \mathbf{G}$ and for all the initial conditions of agents and their protocol, the state synchronization among agents can be achieved.

Additionally, whenever the protocol (3.8) is required to be stable (i.e., the matrix A_c must be asymptotically stable), it is referred to as a partial-state synchronization via a stable protocol.

3.4 Protocol Design for MAS with Full-State Coupling

In this section, as in the continuous-time case, we establish a connection between the problem of state synchronization among agents in the network and a robust stabilization problem. Then, we design a controller for such a robust stabilization problem.

3.4.1 Connection to a Robust Stabilization Problem

The closed-loop system of agent (3.1) and the linear static protocol (3.7) is described by:

$$x_i(k+1) = Ax_i(k) + BF\zeta_i(k), \tag{3.9}$$

for $i = 1, \ldots, N$. Let:

$$x = \begin{pmatrix} x_1 \\ \vdots \\ x_N \end{pmatrix}.$$

Then, the overall dynamics of the N agents can be written as:

$$x(k+1) = (I_N \otimes A + (I - D) \otimes BF)x(k). \tag{3.10}$$

As in the continuous-time case, synchronization for the system (3.10) is equivalent to the asymptotic stability of the following $N-1$ subsystems:

$$\tilde{\eta}_i(k+1) = (A + (1-\lambda_i)BF)\tilde{\eta}_i(k), \quad i = 2, \ldots, N, \tag{3.11}$$

where λ_i, $i = 2, \ldots, N$ are those eigenvalues of D inside the unit disc.

Theorem 3.4 *The MAS* (3.10) *achieves state synchronization if and only if the system* (3.11) *is asymptotically stable for* $i = 2, \ldots, N$.

Combining this theorem with Lemma 1.14 and the associated Remark 1.15 immediately yields the following necessary condition for the state synchronization.

Corollary 3.5 *Consider a MAS described by* (3.1) *and* (3.6). *If there exists a protocol of the form* (3.7) *that achieves state synchronization, then the associated graph must have a directed spanning tree or A must be asymptotically stable.*

Since the case that A is asymptotically stable is trivial, the requirement that the graph of a multi-agent system has a directed spanning tree is essentially necessary.

Proof of Theorem 3.4 Note that D has an eigenvalue at 1 with associated right eigenvector as $\mathbf{1}$. Let:

$$D = TJ_DT^{-1}, \tag{3.12}$$

where J_D is the Jordan canonical form of the row stochastic matrix D such that $J_D(1,1) = 1$ and the first column of T equals $\mathbf{1}$. Let:

$$\eta := (T^{-1} \otimes I_n)x = \begin{pmatrix} \eta_1 \\ \vdots \\ \eta_N \end{pmatrix},$$

where $\eta_i \in \mathbb{C}^n$. In the new coordinates, the dynamics of η can be written as:

$$\eta(k+1) = (I_N \otimes A + (I - J_D) \otimes BF)\eta(k).$$

If (3.11) is globally asymptotically stable for $i = 2, \ldots, N$, we see from the above that $\eta_i(k) \to 0$ for $i = 2, \ldots, N$. This implies that:

$$x(k) - (T \otimes I_n)\begin{pmatrix} \eta_1(k) \\ 0 \\ \vdots \\ 0 \end{pmatrix} \to 0.$$

Note that the first column of T is equal to the vector $\mathbf{1}$ and therefore:

$$x_i(k) - \eta_1(t) \to 0$$

for $i = 1, \ldots, N$. This implies that we achieve state synchronization.

Conversely, suppose that the network (3.10) reaches state synchronization. In this case, we shall have:

$$x(k) - \mathbf{1} \otimes x_1(k) \to 0.$$

Then $\eta(k) - (T^{-1}\mathbf{1}) \otimes x_1(k) \to 0$. Since $\mathbf{1}$ is the first column of T, we have:

$$T^{-1}\mathbf{1} = \begin{pmatrix} 1 \\ 0 \\ \vdots \\ 0 \end{pmatrix}.$$

Therefore, $\eta(k) \to (T^{-1}\mathbf{1}) \otimes x_1(k)$ implies that $\eta_1(k) - x_1(k) \to 0$ and $\eta_i(k) \to 0$ for $i = 2, \ldots, N$. Then, $A + (1 - \lambda_i)BF$ is Schur stable for $i = 2, \ldots, N$. ■

Remark 3.6 It also becomes clear from the proof of Theorem 3.4 that the *synchronized trajectory* is given by $x_s(k) = \eta_1(k)$ which is governed by:

$$\eta_1(k+1) = A\eta_1(k), \quad \eta_1(0) = (w \otimes I_n)x(0), \tag{3.13}$$

where w is the first row of T^{-1}, i.e., the normalized left eigenvector associated with the eigenvalue 1. This shows that the modes of the synchronized trajectory are determined by the eigenvalues of A and the complete dynamics depends on both A and a weighted average of the initial conditions of agents.

Lemma 3.7 *Consider a directed graph which contains a directed spanning tree with its corresponding row stochastic matrix D. Let:*

$$w = \begin{pmatrix} w_1 & \cdots & w_N \end{pmatrix}$$

be a left normalized (in the sense that $w\mathbf{1} = 1$*) eigenvector associated with the eigenvalue* 1 *of D. Then,* $w_i > 0$ *for root agents, and otherwise* $w_i = 0$.

Proof According to the relationship between the row stochastic matrix D and the Laplacian matrix L as shown in (3.3), w is also a normalized left eigenvector associated with the eigenvalue 0 of $(I + D_{in})^{-1}L$. Since the graphs associated with Laplacian matrices L and $(I + D_{in})^{-1}L$ are the same except for a scaling of the weights, we conclude, in light of Lemma 2.8, that $w_i > 0$ for root agents and otherwise $w_i = 0$. ■

Based on the above lemma, one can conclude that $\eta_1(0)$ is only a linear combination of initial conditions of root agents. As such, the synchronized trajectory given by (3.13) can be written explicitly as:

$$x_s(k) = A^k \sum_{i \in \Pi_{\mathcal{G}}} w_i x_i(0), \tag{3.14}$$

which is the weighted average of the trajectories of root agents. As such, these root agents are called *leaders*.

Remark 3.8 If A is neutrally stable, i.e., all the eigenvalues of A are in the closed unit disc and those eigenvalues of A on the unit circle, if any, are semi-simple, then the synchronized trajectories are bounded. Otherwise, the synchronized trajectories could be unbounded.

Note that the above reduces the state synchronization problem for a *known* graph to a *simultaneous* stabilization problem. We need to find one state feedback that stabilizes $N - 1$ different systems.

In light of the definition of Problem 3.2 that synchronization is formulated for a set of graphs **G**, we basically obtain a *robust* stabilization problem, i.e., we need to find a controller:

$$u(k) = Fx(k), \tag{3.15}$$

such that when applied to the system:

$$x(k+1) = Ax(k) + (1 - \lambda)Bu(k), \tag{3.16}$$

the interconnection is stable for any λ which is an eigenvalue inside the unit disc for some row stochastic matrix D associated with a graph in the given set **G**.

3.4.2 Solvability Conditions

In the continuous-time case, for the synchronization problem with full-state coupling to be solvable, the only assumption we needed was that the network has a directed spanning tree. In the discrete-time case, we need an additional condition.

For single-input agents, we provide a necessary and sufficient condition which is given in Theorem 3.9. For multi-input agents, we provide a necessary condition and also a sufficient condition.

Before giving the result, we first recall the discrete-time algebraic Riccati equation and its solution.

Assume (A, B) is stabilizable. Let $P_0 \geq 0$ be a unique solution of H_2 discrete-time algebraic Riccati equation:

$$P_0 = A' P_0 A - A' P_0 B \left(I + B' P_0 B\right)^{-1} B' P_0 A, \tag{3.17}$$

such that:

$$A - B \left(I + B' P_0 B\right)^{-1} B' P_0 A \tag{3.18}$$

has all its eigenvalues in the closed unit disc. Note that such a solution always exists.

3.4.2.1 Single-Input Agents

We have the following result.

Theorem 3.9 *Consider a MAS described by (3.1) and (3.6) with a scalar input. The problem of synchronization with full-state coupling as defined in Problem 3.2 is solvable for* $\mathbf{G} = \bar{\mathbb{G}}_\beta^N$ *for a given* $\beta \in (0, 1)$ *if and only if:*

$$\frac{\beta^2}{1 - \beta^2} B' P_0 B < 1. \tag{3.19}$$

Proof of Theorem 3.9 We established in the above that synchronization with full-state coupling as defined in Problem 3.2 is equivalent to finding a state feedback $u = Fx$ such that (3.16) is asymptotically stable for all $\lambda \in \mathbb{C}$ with $|\lambda| < \beta$. It is well-known that this is equivalent to finding a state feedback $u = Fx$ for the system:

$$\begin{aligned} x(k+1) &= Ax(k) + Bu(k) + \beta B w(k) \\ z(k) &= u(k), \end{aligned} \tag{3.20}$$

such that the H_∞ norm from w to z is less than 1. Using the result from [125], we find that such a controller exists if and only if three conditions are satisfied. First of all, there must exist a $P \geq 0$ such that:

$$P = A'PA - \begin{pmatrix} A'PB & \beta A'PB \end{pmatrix} \begin{pmatrix} 1 + B'PB & \beta B'PB \\ \beta B'PB & -1 + \beta^2 B'PB \end{pmatrix}^{-1} \begin{pmatrix} B'PA \\ \beta B'PA \end{pmatrix}$$

while:

$$A - \begin{pmatrix} B & \beta B \end{pmatrix} \begin{pmatrix} 1 + B'PB & \beta B'PB \\ \beta B'PB & -1 + \beta^2 B'PB \end{pmatrix}^{-1} \begin{pmatrix} B'PA \\ \beta B'PA \end{pmatrix}$$

has all its eigenvalues in the closed unit disc. Note that $P \geq 0$, if it exists, is uniquely determined by the above two conditions. Next, through some algebraic manipulations, it can be shown that:

$$P = \frac{1}{1-\beta^2} P_0 \tag{3.21}$$

satisfies the first two conditions on P where $P_0 \geq 0$ is uniquely determined by (3.17) and the condition that (3.18) has all its eigenvalues in the closed unit disc.

The second condition we need is that:

$$\beta^2 B' P B < I. \tag{3.22}$$

This condition (3.22) is equivalent to (3.19).

The final condition is that for any eigenvalue z_0 of A on the unit disc, there must exist a matrix F such that:

$$\|\beta F(z_0 I - A - BF)^{-1} B\| < 1. \tag{3.23}$$

Similar to the proof of Theorem 2.20, we choose S and T such that:

$$z_0 I - A = S \begin{pmatrix} I & 0 \\ 0 & 0 \end{pmatrix} T, \quad B = S \begin{pmatrix} B_1 \\ B_2 \end{pmatrix}.$$

Stabilizability of (A, B) guarantees that B_2 is right-invertible. We denote by B_2^r the right-inverse of B_2 such that $B_2^r B_2$ is an orthogonal projection. In that case, we choose:

$$F = \begin{pmatrix} 0 & B_2^r \end{pmatrix} T,$$

and we find that:

$$\begin{aligned} \beta F(z_0 I - A - BF)^{-1} B &= \beta \begin{pmatrix} 0 & B_2^r \end{pmatrix} \begin{pmatrix} I & -B_1 B_2^r \\ 0 & -I \end{pmatrix}^{-1} \begin{pmatrix} B_1 \\ B_2 \end{pmatrix} \\ &= -\beta B_2^r B_2, \end{aligned}$$

and therefore (3.23) is satisfied given that $\beta < 1$ and $B_2^r B_2$ is an orthogonal projection. Therefore, based on [125], we know the appropriate protocol exists if and only if (3.19) is satisfied. ■

Corollary 3.10 *Consider a MAS described by (3.1) and (3.6) with a scalar input. The problem of synchronization with full-state coupling as defined in Problem 3.2 is solvable for any* $\mathbf{G}$ *with the property that* $\mathbf{G} \subseteq \bar{\mathbb{G}}_\beta^N$ *for some* $\beta \in (0, 1)$ *only if the agents are at most weakly unstable.*

***Proof of Corollary* 3.10** If $P_0 \neq 0$, then there exists a β_0 such that for $\beta > \beta_0$ the synchronization with full-state coupling as defined in Problem 3.2 is not solvable for $\mathbf{G} = \bar{\mathbb{G}}_\beta^N$.

It is well-known (see, for instance, [99]) that $B'P_0B = 0$ if and only if A has all its eigenvalues in the closed unit disc, i.e., the agents are at most weakly unstable. In that case, the above yields that synchronization with full-state coupling as defined in Problem 3.2 is solvable for $\mathbf{G} = \bar{\mathbb{G}}_\beta^N$ for any $\beta \in (0, 1)$. ■

3.4.2.2 Multi-input Agents

The following theorem deals with multi-input agents. It gives necessary conditions as well as sufficient conditions which unfortunately are not the same.

Theorem 3.11 *Consider a MAS described by multi-input agents* (3.1) *and* (3.6).

The problem of synchronization with full-state coupling as defined in Problem 3.2 *is solvable with* $\mathbf{G} = \bar{\mathbb{G}}_\beta^N$ *for a given* $\beta \in (0, 1)$ *only if:*

$$\beta < \frac{1}{|\lambda_{\max}(A)|}. \tag{3.24}$$

On the other hand, the above problem is solvable with $\mathbf{G} = \bar{\mathbb{G}}_\beta^N$ *for a given* $\beta \in (0, 1)$ *if the condition* (3.19) *is satisfied.*

Corollary 3.12 *Consider a MAS described by* (3.1) *and* (3.6). *The problem of synchronization with full-state coupling as defined in Problem* 3.2 *is solvable for any* $\mathbf{G} \subseteq \bar{\mathbb{G}}^N$ *only if the agents are at most weakly unstable.*

***Proof of Theorem* 3.11** *(Necessity)* For a given β, we know that the problem of synchronization with full-state coupling can be solved if and only if there exists a matrix F such that $A + BF$ is asymptotically stable and:

$$\det\left(I - \lambda F(zI - A - BF)^{-1}B\right) \neq 0 \tag{3.25}$$

for all $\lambda \in \mathbb{C}$ with $|\lambda| < \beta$ and for all $z \in \mathbb{C}$ with $|z| \geq 1$. Assume that z_0 denotes the location of an eigenvalue of A with the largest magnitude.

The proof is by contradiction. Assume that the above holds for a given F and $\beta|z_0| > 1$. In that case, there exists an eigenvector x such that:

$$(z_0 I - A)x = 0,$$

and this implies that:

$$y = (z_0 I - A - BF)x = -BFx$$

results in:

$$x = (z_0 I - A - BF)^{-1} y = -(z_0 I - A - BF)^{-1} BFx, \tag{3.26}$$

and hence:

$$Fx = -F(z_0 I - A - BF)^{-1} B(Fx),$$

which yields that:

$$\det\left(I + F(z_0 I - A - BF)^{-1} B\right) = 0.$$

Note that $Fx \neq 0$ since otherwise (3.26) would imply that $x = 0$ which is not true since x is an eigenvector. Let k be a sufficiently large integer such that:

$$\beta^{k+1} > \frac{1}{|z_0|^k}, \tag{3.27}$$

and consider:

$$f_r(z) = \det\left(I + \frac{r z^k}{\bar{z}_0^k} F(z^{k+1} I - A - BF)^{-1} B\right)$$

where $\bar{z}_0$ satisfies $\bar{z}_0^{k+1} = z_0$. Choose v such that:

$$\left\| \frac{z^k}{\bar{z}_0^k} F(z^{k+1} I - A - BF)^{-1} B \right\| < 1$$

for all $z \in \mathbb{C}$ with $|z| \geq v$. Such a v obviously exists. Define the region:

$$C = \{ z \in \mathbb{C} \mid 1 \leq |z| \leq v \}.$$

We note that f_r is an analytic function on C for all $r \in [0, 1]$. Moreover, f_1 has a zero in $\bar{z}_0$, and f_0 has no zeros in C. A straightforward application of Rouché's theorem (see [93, Theorem 10.43]) then yields that there exists a $r_0 \in (0, 1)$ for which f_r has a zero z_1 on the boundary of C. By construction, the function f_{r_0} cannot have a zero with $|z_1| = v$, and therefore we have $|z_1| = 1$. The fact that z_1 is a zero yields that:

$$\det\left(I + \frac{r_0 z_1^k}{\bar{z}_0^k} F(z_1^{k+1} I - A - BF)^{-1} B\right) = 0,$$

but then:

$$\lambda = -\frac{r_0 z_1^k}{\bar{z}_0^k}$$

satisfies $|\lambda| < \beta$ using (3.27) and $|z_1| = 1$. We then obtain:

$$\det\left(I - \lambda F(zI - A - BF)^{-1}B\right) = 0$$

where $z = z_1^{k+1}$ satisfies $|z| = 1$. This contradicts (3.25) which completes the proof.

(Sufficiency) Looking at the proof of Theorem 3.9, we note that the only step where we used that the input is scalar is the fact that finding a state feedback $u = Fx$ such that (3.16) is asymptotically stable for all $\lambda \in \mathbb{C}$ with $|\lambda| < \beta$ is equivalent to finding a state feedback $u = Fx$ for the system (3.20) such that the H_∞ norm from w to z is less than 1. However, finding a state feedback $u = Fx$ for the system (3.20) such that the H_∞ norm from w to z is less than 1 is still a ***sufficient*** condition in the multi-input case. The result of this theorem then immediately follows from the arguments in the proof of Theorem 3.9. ■

The above two theorems present results for state synchronization with the network associated with a graph in the set $\bar{\mathbb{G}}_\beta^N$ for a **fixed** β. It is interesting to also consider the case when state synchronization is possible for arbitrary $\beta \in (0, 1)$. The following theorem presents necessary and sufficient conditions for both single- and multi-input agents.

Theorem 3.13 *Consider a MAS described by* (3.1) *and* (3.6). *The problem of state synchronization with full-state coupling as defined in Problem* 3.2 *is solvable for any* $\mathbb{G} = \bar{\mathbb{G}}_\beta^N$ *for all* $\beta \in (0, 1)$ *if and only if the agents are at most weakly unstable.*

Proof For single-input agents, this is a direct result of Theorem 3.9 if we note that the agents are at most weakly unstable if and only if $P_0 = 0$ (see, for instance, [99]). For multi-input agents, the necessary condition of Theorem 3.11 shows that the agents must be at most weakly unstable, and since this implies that $P_0 = 0$, we find that Theorem 3.11 guarantees this condition is also sufficient. ■

In the following, we will present the protocol design for both at most weakly unstable agents and strictly unstable agents. For at most weakly unstable agents, we give the protocol design based on discrete ARE method and direct eigenstructure assignment method.

3.4.3 At Most Weakly Unstable Agents

Given the above necessary conditions, we will present two designs for the synchronization with full-state coupling for systems which are at most weakly unstable.

First, we consider designs based on Riccati equations, and then we consider eigenstructure assignment methods.

3.4.3.1 ARE-Based Method

In the literature, a nonstandard Riccati equation is used which goes back to [40]. In the context of synchronization, it was first used in [51]. The following result is based on [40, 115, 121].

Lemma 3.14 *Consider the nonstandard Riccati equation:*

$$P = A'PA + Q - (1-\beta^2)A'PB(B'PB+I)^{-1}B'PA, \tag{3.28}$$

where $\beta \in (0,1)$*. Assume that* (A,B) *is controllable and* A *is at most weakly unstable. In that case, the above equation has a unique positive definite solution. Moreover, when we denote the solution* P *of* (3.28) *for* $Q = \delta I$ *by* P_δ*, then we obtain:*

$$\lim_{\delta \downarrow 0} P_\delta = 0. \tag{3.29}$$

Remark 3.15 This is a nonstandard algebraic Riccati equation, and therefore many of the standard techniques to find the solution of this equation do not apply. However, it is shown in [40] that:

$$P_{k+1} = Q + \beta^2 A'P_kA + (1-\beta^2)A'(P_k^{-1} + BB')^{-1}A$$

with $P_1 = Q \leq P$ is an increasing sequence where:

$$P = \lim_{k\to\infty} P_k$$

is a solution of the nonstandard algebraic Riccati equation (3.28).

Proof The lemma basically follows from [40, 115, 121] except for the final result (3.29). To prove the final result, we consider P_δ. From the construction of P as the limit of a Riccati difference equation, it is clear that $P_\delta > 0$ is a decreasing function of δ. Define:

$$P_0 = \lim_{\delta\downarrow 0} P_\delta.$$

We find that:

$$P_0 = A'P_0A - (1-\beta^2)A'P_0B(B'P_0B+I)^{-1}B'P_0A.$$

Clearly, we have the linear matrix inequality:

$$\begin{pmatrix} -P_0 + A' P_0 A & A' P_0 B \\ B' P_0 A & \frac{1}{1-\beta^2}(I + B' P_0 B) \end{pmatrix} \geq 0.$$

If $Ax = \lambda x$, then:

$$0 \leq x'(-P_0 + A' P_0 A)x = \left(1 - |\lambda|^2\right) x' P_0 x,$$

and this yields $P_0 x = 0$ or $|\lambda| = 1$. In the latter case, we find:

$$(-P_0 + A' P_0 A)x = 0$$

which yields $\lambda A' P_0 x = P_0 x$. Moreover, from the linear matrix inequality, we then get $B' P_0 x = 0$ which yields a contradiction with stabilizability unless $P_0 x = 0$. Therefore, for all eigenvalues x of A, we have that $P_0 x = 0$. Similarly, we can prove that all generalized eigenvalues of A are in the kernel of P_0 and therefore P_0 must be identically zero. ■

Using the above nonstandard algebraic Riccati equation (3.28), we can design a suitable protocol provided A is at most weakly unstable.

Protocol design 3.1 *Consider a MAS described by (3.1) and (3.6). We consider the protocol:*

$$u_i = F \zeta_i, \tag{3.30}$$

where $F = -(B' P B + I)^{-1} B' P A$ with $P > 0$ being the unique solution of the discrete-time algebraic Riccati equation:

$$P = A' P A + Q - (1 - \beta^2) A' P B (B' P B + I)^{-1} B' P A, \tag{3.31}$$

where $Q > 0$ and β is the upper bound for the amplitude of the eigenvalues unequal to 1 of the row stochastic matrix D associated with a graph in the set $\bar{\mathbb{G}}_\beta^N$.

Note that $Q > 0$ is a design parameter which affects issues such as convergence rate of the synchronization or overshoot effects.

The following result establishes that the above design works and is based on the result from [51].

Theorem 3.16 *Consider a MAS described by (3.1) and (3.6). Let any $\beta \in (0, 1)$ be given, and consider the set of network graphs $\bar{\mathbb{G}}_\beta^N$ as defined in Definition 1.16.*

If A is at most weakly unstable and (A, B) is stabilizable, then the state synchronization problem stated in Problem 3.2 with $\mathbf{G} = \bar{\mathbb{G}}_\beta^N$ is solvable. In particular, the Protocol Design 3.1 solves the state synchronization problem for any graph $\mathcal{G} \in \bar{\mathbb{G}}_\beta^N$. Moreover, the synchronized trajectory is given by (3.14).

Proof Note that $\mathcal{G} \in \bar{\mathbb{G}}_\beta^N$ implies that the graph has a directed spanning tree and all its eigenvalues inside the unit disc of the row stochastic matrix satisfy $|\lambda| < \beta$. Using Theorem 3.4, we only need to prove that the system:

$$x(k+1) = \left[A - (1-\lambda)B(B'PB + I)^{-1}B'PA\right]x(k)$$

is asymptotically stable for any λ that satisfies $|\lambda| < \beta$.

We observe that:

$$\begin{aligned}
&(A + (1-\lambda)BF)'P(A + (1-\lambda)BF) - P \\
&\quad = A'PA - 2\,\mathrm{Re}(1-\lambda)A'PB(B'PB + I)^{-1}B'PA - P \\
&\qquad\quad + |1-\lambda|^2 A'PB(B'PB + I)^{-1}B'PB(B'PB + I)^{-1}B'PA \\
&\quad = A'PA + \left[-2\,\mathrm{Re}(1-\lambda) + |1-\lambda|^2\right]A'PB(B'PB + I)^{-1}B'PA - P \\
&\qquad\quad - |1-\lambda|^2 A'PB(B'PB + I)^{-2}B'PA \\
&\quad = A'PA + (|\lambda|^2 - 1)A'PB(B'PB + I)^{-1}B'PA - P \\
&\qquad\quad - |1-\lambda|^2 A'PB(B'PB + I)^{-2}B'PA \\
&\quad \leq A'PA + (\beta^2 - 1)A'PB(B'PB + I)^{-1}B'PA - P \\
&\quad = -Q < 0.
\end{aligned}$$

Therefore, $(A + (1-\lambda)BF)$ is Schur stable for any $|\lambda| < \beta$, which proves the result. ■

Note that protocol (3.30) achieves state synchronization independent of our choice for $Q > 0$. However, the choice of the parameter is important for the performance, i.e., the convergence rate of the synchronization as well as possible overshoot effects.

The above is based on a nonstandard algebraic Riccati equation which was presented because it was used in the synchronization literature. However, we can also use a standard H_2 algebraic Riccati equation. This results in the following design.

Protocol design 3.2 *Consider a MAS described by (3.1) and (3.6). We consider the protocol:*

$$u_i = F_\delta \zeta_i, \tag{3.32}$$

where:

$$F_\delta = -\tfrac{1}{1-\beta_+^2}(B'P_\delta B + I)^{-1}B'P_\delta A$$

with $P_\delta > 0$ given by:

$$P_\delta = A'P_\delta A + \delta I - A'P_\delta B(B'P_\delta B + I)^{-1}B'P_\delta A, \tag{3.33}$$

while $\delta > 0$ is sufficiently small such that:

$$\beta_+^2 B'P_\delta B < (1 - \beta_+^2)I, \tag{3.34}$$

where $\beta_+ \in [\beta, 1)$ with β the upper bound for the amplitude of the eigenvalues unequal to 1 of the row stochastic matrix D associated with a graph in the set $\bar{\mathbb{G}}_\beta^N$, while P_δ is the unique solution of the above discrete-time algebraic Riccati equation.

Remark 3.17 In full-state coupling, we can simply choose $\beta_+ = \beta$. However, in several other cases such as partial-state coupling or for agents subject to input saturation that will be studied in Chap. 4, we need to require $\beta_+ > \beta$.

The following result establishes that the above design works.

Theorem 3.18 *Consider a MAS described by (3.1) and (3.6). Let any $\beta \in (0, 1)$ be given, and consider the set of network graphs $\bar{\mathbb{G}}_\beta^N$ as defined in Definition 1.16.*

If A is at most weakly unstable and (A, B) is stabilizable, then for any $\beta_+ \in [\beta, 1)$, there exists a unique $P_\delta > 0$ satisfying (3.33) and (3.34) for a sufficiently small $\delta > 0$, and the design 3.2 solves the state synchronization problem for any graph $\mathcal{G} \in \bar{\mathbb{G}}_\beta^N$. Moreover, the synchronized trajectory is given by (3.14).

Proof We need to prove that (3.16) is asymptotically stable for any λ which is an eigenvalue inside the unit disc for some row stochastic matrix D associated with a graph in the given set $\bar{\mathbb{G}}_\beta^N$. This implies that $|\lambda| \leq \beta$. The system (3.16) can be written as:

$$x(k+1) = Ax(k) + Bu(k) + \beta_+ Bw(k) \tag{3.35}$$

provided:

$$w(k) = \frac{\lambda}{\beta_+} u(k). \tag{3.36}$$

The existence of δ and P_δ satisfying (3.33) and (3.34) is well known. With a little algebra, it can then be verified that:

$$P = \frac{1}{1-\beta_+^2} P_\delta \tag{3.37}$$

is such that:

$$P = A'PA + \delta I - \begin{pmatrix} A'PB & \beta_+ A'PB \end{pmatrix} \begin{pmatrix} I + B'PB & \beta_+ B'PB \\ \beta_+ B'PB & -I + \beta_+^2 B'PB \end{pmatrix}^{-1} \begin{pmatrix} B'PA \\ \beta_+ B'PA \end{pmatrix}$$

while $\beta_+^2 B'PB < I$ and:

$$A - \begin{pmatrix} B & \beta_+ B \end{pmatrix} \begin{pmatrix} I + B'PB & \beta_+ B'PB \\ \beta_+ B'PB & -I + \beta_+^2 B'PB \end{pmatrix}^{-1} \begin{pmatrix} B'PA \\ \beta_+ B'PA \end{pmatrix}$$

has all its eigenvalues in the open unit disc. We have:

$$\begin{aligned} &- x(k)'Px(k) + x(k+1)'Px(k+1) + \delta x(k)'x(k) + \|u(k)\|^2 - \|w(k)\|^2 \\ &= \begin{pmatrix} x(k) \\ u(k) \\ w(k) \end{pmatrix}' \begin{pmatrix} -P + A'PA + \delta I & A'PB & \beta_+ A'PB \\ B'PA & B'PB + I & \beta_+ B'PB \\ \beta_+ B'PA & \beta_+ B'PB & \beta_+^2 B'PB - I \end{pmatrix} \begin{pmatrix} x(k) \\ u(k) \\ w(k) \end{pmatrix}. \end{aligned}$$

If we maximize the right-hand side with respect to $w(k)$, we get:

$$w(k) = -\beta_+ (\beta_+^2 B'PB - I)^{-1} B'P \left[Ax(k) + Bu(k) \right],$$

using that $\beta_+^2 B'PB < I$. If we then minimize the right-hand side with respect to $u(k)$, we get:

$$\begin{aligned} u(k) = &- \left[B'PB + I - \beta_+^2 B'PB(\beta_+^2 B'PB - I)^{-1} B'PB \right]^{-1} \\ &\times \left[B'PA - \beta_+^2 B'PB(\beta_+^2 B'PB - I)^{-1} B'PA \right] x(k), \end{aligned} \tag{3.38}$$

which yields that:

$$-x(k)'Px(k) + x(k+1)'Px(k+1) + \delta x(k)'x(k) + \|u(k)\|^2 - \|w(k)\|^2 \leq 0$$

for any $w(k)$ provided $u(k)$ satisfies (3.38). But then we get for $w(k)$ given by (3.36):

$$\begin{aligned} x(k+1)'Px(k+1) & \\ &\leq x(k)'Px(k) - \delta x(k)'x(k) - \left(1 - \tfrac{|\lambda|^2}{\beta_+^2}\right)\|u(k)\|^2 \\ &\leq x(k)'Px(k) - \delta x(k)'x(k). \end{aligned} \tag{3.39}$$

Since for this choice for $w(k)$ Eq. (3.35) reduces to (3.16), this clearly guarantees the asymptotic stability of (3.16). Finally, we note that (3.38) can be written as:

$$u(k) = -\tfrac{1}{1-\beta_+^2}(B'P_\delta B + I)^{-1}B'P_\delta Ax(k)$$

using (3.37). ■

Remark 3.19 Note that on page 461 of [127], an expression is given for the state feedback of the discrete-time H_∞ control problem which is not consistent with the expression (3.38) derived in the above proof. It appears that the formula in [127] is not correct.

The above protocol uses a sequence F_δ which is an H_2 low-gain sequence as defined in [106, Definition 4.73]. Instead of only requiring that $F_\delta \to 0$ as $\delta \to 0$, we guarantee that the input signal is small. However, in the current context, it is important that this property is robust with respect to the uncertainty in the eigenvalues of the Laplacian. This result obtained in the following lemma is the discrete-time version of Lemma 2.29 and will be crucial in Chap. 4.

Lemma 3.20 *Suppose that (A, B) is stabilizable and all the eigenvalues of A are within the closed unit disc. Let F_δ be designed as in Protocol Design 3.2. Then, we have the following properties:*

1. *The closed-loop system matrix $A + \lambda BF_\delta$ is Schur stable for all $\delta > 0$ and for all λ with $|\lambda| < \beta$.*
2. *For any $\beta_+ > \beta$, there exists a $\delta^* > 0$ such that for all $\delta \in (0, \delta^*]$ there exist $\zeta_\delta > 0$ and $0 < \eta_\delta < 1$ with $\zeta_\delta \to 0$ as $\delta \to 0$ such that:*

$$\|F_\delta(A + (1-\lambda)BF_\delta)^k\| \leq \zeta_\delta \eta_\delta^k, \tag{3.40}$$

for all $k \geq 0$ and for all $\lambda \in \mathbb{C}$ with $|\lambda| < \beta$.

Proof The first property is obvious from the previous theorem. Consider:

$$x(k+1) = (A + (1-\lambda)BF_\delta)x(k).$$

It is easily seen that:

$$\begin{aligned} x'(k+1)P_\delta x(k+1) &\leq x'(k)P_\delta x(k) - \delta x'(k)x(k) \\ &\leq \eta_\delta^2 x'(k)P_\delta x(k), \end{aligned}$$

using that $|\lambda| \leq \beta$ where $\eta_\delta^2 = 1 - \delta\|P_1\|^{-1}$. Hence:

$$\|P_\delta^{1/2}x(k)\| \leq \eta_\delta^k \|P_\delta^{1/2}x(0)\|. \tag{3.41}$$

Note that (3.39) implies that:

$$\left(1 - \frac{|\lambda|^2}{\beta_+^2}\right) F_\delta' F_\delta \leq P_\delta.$$

Finally:

$$\begin{aligned} \|F_\delta(A + (1-\lambda)BF_\delta)^k x(0)\| &= \|F_\delta x(k)\| \\ &\leq \left(1 - \frac{|\lambda|^2}{\beta_+^2}\right)^{-1} \eta_\delta^k \|P_\delta^{1/2}x(0)\| \\ &\leq \left(1 - \frac{\beta^2}{\beta_+^2}\right)^{-1} \eta_\delta^k \|P_\delta^{1/2}x(0)\|. \end{aligned} \tag{3.42}$$

Since (3.42) is true for all $x(0) \in \mathbb{R}^n$, it follows trivially that:

$$\|F_\delta(A + (1-\lambda)BF_\delta)^k\| \leq \left(1 - \frac{\beta^2}{\beta_+^2}\right)^{-1} \|P_\delta^{1/2}\|\eta_\delta^k. \tag{3.43}$$

The proof is then completed by taking:

$$\zeta_\delta = \left(1 - \frac{\beta^2}{\beta_+^2}\right)^{-1} \|P_\delta^{1/2}\|.$$

■

Finally, we consider a special case where the agent dynamics are neutrally stable, that is, those eigenvalues of A on the unit circle, if any, are semi-simple. This implies that A is Schur stable and that there exists a positive definite matrix P such that $A'PA \leq P$. In this case, we shall show that the protocol design no longer requires the knowledge of β and hence allows us to deal with a larger set of network graphs $\bar{\mathbb{G}}^N$ (effectively the case $\beta = 1$).

Protocol design 3.3 *Consider a MAS described by neutrally stable agents (3.1) and (3.6).*

We consider a feedback gain:

$$F_\delta = -\delta B' P A, \tag{3.44}$$

where $\delta > 0$ is a low-gain parameter and the positive definite matrix P is a solution of the Lyapunov inequality:

$$A' P A \leq P. \tag{3.45}$$

This results in the protocol:

$$u_i(k) = F_\delta \zeta_i(k). \tag{3.46}$$

The result for neutrally stable systems is stated as follows.

Theorem 3.21 *Consider a MAS described by neutrally stable agents (3.1) and (3.6). Let the set of network graphs $\bar{\mathbb{G}}^N$ with any number N be given.*

If (A, B) is stabilizable, then the state synchronization problem via a protocol stated in Problem 3.2 with $\mathbf{G} = \bar{\mathbb{G}}^N$ is solvable. In particular, there exists a $\delta^* > 0$, which is independent of any network graph, such that for any $\delta \in (0, \delta^*]$, controller (3.46) solves the state synchronization problem for any graph $\mathcal{G} \in \bar{\mathbb{G}}^N$ for any N. Moreover, the synchronized trajectory is given by (3.14).

Proof From Theorem 3.4, we only need to prove that the system:

$$x(k+1) = \big[A - (1-\lambda)\delta B B' P A\big] x(k), \tag{3.47}$$

is asymptotically stable for $|\lambda| < 1$. Note that:

$$\begin{aligned}[A - (1-\lambda)\delta B B' P A]' P [A - (1-\lambda)\delta B B' P A] - P \\ \leq -2\operatorname{Re}(1-\lambda)\delta A' P B B' P A + |1-\lambda|^2 \delta^2 A' P B B' P B B' P A.\end{aligned}$$

There exists a δ^* such that for a $\delta \in (0, \delta^*]$, we have $\delta B' P B \leq \frac{1}{2}$. Since $|1-\lambda|^2 \leq 2\operatorname{Re}(1-\lambda)$ for $|\lambda| < 1$, we get:

$$[A - (1-\lambda)\delta B B' P A]^* P [A - (1-\lambda)\delta B B' P A] - P \leq -\tfrac{1}{2}|1-\lambda|^2 \delta A' P B B' P A$$

for $\delta \in (0, \delta^*]$. Since (A, B) is controllable and $|\lambda| < 1$, it follows from LaSalle's invariance principle that the system (3.47) is globally asymptotically stable. Note that δ^* depends only on agent's model. Hence, the protocol (3.44) works for any graph $\mathcal{G} \in \bar{\mathbb{G}}^N$ with any number N. ■

3.4.3.2 Direct Eigenstructure Assignment Method

In this section, we will design the protocol via direct eigenstructure assignment. In this section, we restrict attention to agents which are at most weakly unstable.

As we did in the continuous time, we use a canonical form used before in [163, Section 1.4] and [52, Section 2.3.1]. There exist a nonsingular state transformation T_x and an input transformation T_u, such that:

$$\tilde{A} = T_x^{-1} A T_x = \begin{pmatrix} A_1 & 0 & \cdots & 0 \\ 0 & A_2 & \ddots & \vdots \\ \vdots & \ddots & \ddots & 0 \\ 0 & \cdots & 0 & A_q \end{pmatrix}, \tag{3.48}$$

$$\tilde{B} = T_x^{-1} B T_u = \begin{pmatrix} B_1 & B_{1,2} & \cdots & B_{1,q} & B_{1,q+1} \\ 0 & B_2 & \ddots & \vdots & \vdots \\ \vdots & \ddots & \ddots & B_{q-1,q} & \vdots \\ 0 & \cdots & 0 & B_q & B_{q,q+1} \end{pmatrix}, \tag{3.49}$$

and for $i = 1, 2, \ldots, q$:

$$A_i = \begin{pmatrix} 0 & 1 & 0 & \cdots & 0 \\ \vdots & \ddots & \ddots & \ddots & \vdots \\ \vdots & & \ddots & 1 & 0 \\ 0 & \cdots & \cdots & 0 & 1 \\ -a_{n_i}^i & \cdots & -a_{n_3}^i & -a_{n_2}^i & -a_{n_1}^i \end{pmatrix}, \quad B_i = \begin{pmatrix} 0 \\ \vdots \\ \vdots \\ 0 \\ 1 \end{pmatrix}.$$

Clearly, (A_i, B_i) is controllable. Note that the geometric multiplicity of all eigenvalues of A_i is equal to 1. Based on the above, we can present a direct design method for MAS with at most weakly unstable agents.

Protocol design 3.4 *Consider a MAS described by at most weakly unstable agents (3.1) and (3.6). Choose T_x and T_u such that the system is in the canonical form described above.*

For each (A_i, B_i), let $F_{\delta,i} \in \mathbb{R}^{1\times n_i}$ be the unique state feedback gain such that the eigenvalues of $A_i + B_i F_{\delta,i}$ can be obtained from the eigenvalues of A_i

(continued)

Protocol design 3.4 (continued)
by moving an eigenvalue $\lambda_{i,m}$ *with algebraic multiplicity* $r_{i,m}$ *to the locations,*

$$(1-\delta)\lambda_{i,m}, (1-\delta^2)\lambda_{i,m}, \ldots, (1-\delta^{r_{i,m}})\lambda_{i,m}.$$

Note that $F_{\delta,i}$ *can be obtained explicitly in terms of* δ. *Now define* F_δ *as:*

$$F_\delta = T_u \begin{pmatrix} F_{\delta,1} & 0 & \cdots & 0 \\ 0 & F_{\delta,2} & \ddots & \vdots \\ \vdots & \ddots & \ddots & 0 \\ \vdots & & \ddots & F_{\delta,q} \\ 0 & \cdots & \cdots & 0 \end{pmatrix} T_x^{-1}. \tag{3.50}$$

We then construct the protocol:

$$u_i = F_\delta x_i. \tag{3.51}$$

Remark 3.22 Note that if we would place the eigenvalues of multiplicity higher than 1 all from λ_m to $(1-\delta)\lambda_m$, which would be consistent with our continuous-time design in Protocol Design 2.6, then the above design is not guaranteed to work. In particular, for:

$$A = \begin{pmatrix} 0 & 1 \\ -1 & 2 \end{pmatrix}, \quad B = \begin{pmatrix} 0 \\ 1 \end{pmatrix}$$

which has two eigenvalues in 1, placing both poles in $(1-\delta)$ would yield:

$$F_\delta = \begin{pmatrix} 2\delta - \delta^2 & -2\delta \end{pmatrix}$$

and:

$$\lim_{\delta \downarrow 0} \left\| F_\delta (zI - A - BF_\delta)^{-1} B \right\|_\infty = \tfrac{4}{3}$$

which implies that the design only works for $\beta \leq \frac{3}{4}$.

For the purpose of this chapter, the following lemma is crucial which establishes a crucial bound needed to establish robust stability.

Lemma 3.23 *Suppose that (A, B) is stabilizable, and all the eigenvalues of A are within the closed unit disc. Then, we have the following properties:*

1. *The closed-loop system matrix $A + BF_\delta$ is Schur stable for all $\delta > 0$,*
2. *We have:*

$$\lim_{\delta \downarrow 0} \| F_{\delta,i}(zI - A_i - B_i F_{\delta,i})^{-1} B_i \|_\infty = 1 \tag{3.52}$$

for $i = 1, \ldots, q$.

Proof Stability of the closed-loop system matrix $A + BF_\delta$ is obvious. In order to compute the H_∞ norm, since the peak value of the H_∞ norm of the transfer function occurs on the unit circle, we need to check (3.52) on the unit circle only. Let:

$$\lambda_1 = e^{j\theta_1}, \lambda_2 = e^{j\theta_2}, \ldots, \lambda_q = e^{j\theta_q}$$

be the distinct eigenvalues of A_i on the upper half of the unit circle where $\theta_i \in [0, \pi]$ with algebraic multiplicities $r_1, \ldots, r_q$. Define:

$$\mathcal{E}_k := \left\{ z \in \mathbb{C} \mid z = e^{j\theta} \text{ with } \theta \in [\theta_k(1-\varepsilon), \theta_k(1+\varepsilon)] \cap [0, \pi] \right\}$$

where ε is small enough to guarantee that $\mathcal{E}_{k_1} \cap \mathcal{E}_{k_2} = \emptyset$ for $k_1 \neq k_2$. Define:

$$\Omega = \left\{ z \in \mathbb{C} \mid z \notin \mathcal{E}_k \text{ for } k = 1, \ldots, q \text{ and } z = e^{j\theta} \text{ with } \theta \in [0, \pi] \right\}.$$

We have:

$$F_{\delta,i}(zI - A_i - B_i F_{\delta,i})^{-1} B_i = \frac{n(z)}{p(z)}$$

for:

$$p_\delta(z) = \det(zI - A_i - B_i F_{\delta,i}) = \prod_{k=1}^{q} p_{k,\delta}(z)$$

with:

$$p_{k,\delta}(z) = \prod_{s=1}^{r_k} (z - (1 - \delta^s)\lambda_k)$$

and:

$$n(z) = \det(zI - A_i) - \det(zI - A_i - B_i F_{\delta,i}) = \prod_{k=1}^{q} p_{k,0}(z) - \prod_{k=1}^{q} p_{k,\delta}(z).$$

Note that:

$$\sup_{z\in\Omega} n(z) \leq M\delta, \qquad \inf_{z\in\Omega} p(z) \geq \rho$$

for suitable M and ρ, and hence:

$$\sup_{z\in\Omega} \left| F_{\delta,i}(zI - A_i - B_i F_{\delta,i})^{-1} B_i \right| < \frac{M}{\rho}\delta$$

which can be made arbitrarily small for a sufficiently small δ. Next consider $\mathcal{E}_k$. We have:

$$\frac{n(z)}{p(z)} = \frac{p_{k,0}(z) - p_{k,\delta}(z)}{p_{k,\delta}(z)} + \frac{p_{k,0}(z)}{p_{k,\delta}(z)} \frac{p_0^k(z) - p_\delta^k(z)}{p_\delta^k(z)} \tag{3.53}$$

where p_δ^k is the polynomial defined by:

$$p_\delta^k(z) = \frac{p_\delta(z)}{p_{k,\delta}(z)}.$$

First note that:

$$\sup_{z\in\mathcal{E}_k} \left| \frac{p_0^k(z) - p_\delta^k(z)}{p_\delta^k(z)} \right| < M_k\delta \tag{3.54}$$

for a suitably chosen M_k since p_δ^k has no eigenvalues in $\mathcal{E}_k$ and hence is bounded away from zero while $p_\delta^k \to p_0^k$. Next, we investigate:

$$\begin{aligned}\frac{p_{k,0}(z) - p_{k,\delta}(z)}{p_{k,\delta}(z)} &= -\frac{\delta^{r_k}}{(s-1+\delta)} \prod_{k=2}^{r_k} \frac{s-1}{s-1+\delta^k} \\ &\quad - \frac{\delta^{r_k-1}}{(s-1+\delta)} \prod_{k=2}^{r_{k-1}} \frac{s-1}{s-1+\delta^k} \\ &\quad - \cdots - \frac{\delta^2}{(s-1+\delta)} \frac{(s-1)}{(s-1+\delta^2)} - \frac{\delta}{s-1+\delta}\end{aligned}$$

where $s = \lambda_k^{-1} z$ which yields that:

$$s \in \left\{ x \in \mathbb{C} \mid x = e^{j\theta} \text{ with } \theta \in [-\varepsilon, \varepsilon] \right\}$$

since $z \in \mathcal{E}_k$. We note that for $k > 1$:

$$\left| \frac{\delta^k}{s - 1 + \delta} \right| \leq \delta^{k-1}$$

since s is on the unit circle while:

$$\left| \frac{\delta}{s - 1 + \delta} \right| \leq 1,$$

and

$$\left| \frac{s - 1}{s - 1 + \delta^k} \right| \leq \frac{1}{1 - \delta^k}.$$

Using the above bounds, it is easily seen that we can achieve:

$$\left| \frac{p_{k,0}(z) - p_{k,\delta}(z)}{p_{k,\delta}(z)} \right| < 1 + N_k \delta$$

for some $N_k > 0$ provided $\delta < 1$. Similarly:

$$\begin{aligned} \frac{p_{k,0}(z)}{p_{k,\delta}(z)} = &- \frac{\delta^{r_k}}{(s - 1 + \delta)} \prod_{k=2}^{r_k} \frac{s - 1}{s - 1 + \delta^k} \\ &- \frac{\delta^{r_k - 1}}{(s - 1 + \delta)} \prod_{k=2}^{r_{k-1}} \frac{s - 1}{s - 1 + \delta^k} \\ &- \cdots - \frac{\delta^2}{(s - 1 + \delta)} \frac{(s - 1)}{(s - 1 + \delta^2)} - \frac{s - 1}{s - 1 + \delta} \end{aligned}$$

which yields that:

$$\left| \frac{p_{k,0}(z)}{p_{k,\delta}(z)} \right| < 1 + P_k \delta \tag{3.55}$$

for some $P_k > 0$. Putting all bounds together we have:

$$\sup_{z\in\mathcal{E}_k} \left| F_{\delta,i}(zI - A_i - B_i F_{\delta,i})^{-1} B_i \right| = \sup_{z\in\mathcal{E}_k} \left| \frac{n(z)}{p(z)} \right| < 1 + Q_k\delta$$

for some $Q_k > 0$. Since this is true for $k = 1, \ldots, q$ we get (3.52). ∎

The following theorem guarantees that Protocol Design 3.4 has the required properties.

Theorem 3.24 *Consider a MAS described by* (3.1) *and* (3.6). *Let any* $1 > \beta > 0$ *be given and hence a set of network graphs* $\bar{\mathbb{G}}_\beta^N$ *be defined.*

If (A, B) *is stabilizable and* A *is at most weakly unstable, then the state synchronization problem stated in Problem* 3.2 *with* $\mathbf{G} = \bar{\mathbb{G}}_\beta^N$ *is solvable. In particular, there exists a* $\delta^* > 0$ *such that for any* $\delta \in (0, \delta^*]$, *controller* (3.51) *solves the state synchronization problem for any graph* $\mathcal{G} \in \bar{\mathbb{G}}_\beta^N$. *Moreover, the synchronized trajectory is given by* (3.14).

Proof We need to prove that:

$$A + (1 - \lambda)BF_\delta$$

is asymptotically stable for all λ with $|\lambda| < \beta$ provided δ is sufficiently small. This is equivalent to establishing that:

$$\tilde{A} + (1 - \lambda)\tilde{B}\tilde{F} \tag{3.56}$$

is asymptotically stable where $\tilde{A}$ and $\tilde{B}$ are given by (3.48) and (3.49), respectively, while:

$$\tilde{F}_\delta = \begin{pmatrix} F_{\delta,1} & 0 & \cdots & 0 \\ 0 & F_{\delta,2} & \ddots & \vdots \\ \vdots & \ddots & \ddots & 0 \\ \vdots & & \ddots & F_{\delta,q} \\ 0 & \cdots & \cdots & 0 \end{pmatrix}.$$

Note that (3.56) is a block upper triangular matrix with:

$$A_i + (1 - \lambda)B_i F_{\delta,i}$$

on the diagonal. Therefore, it is sufficient to establish that the latter matrices are asymptotically stable. By Lemma 3.23, we know that for sufficiently small δ we have:

$$\|F_{\delta,i}(zI - A_i - B_i F_{\delta,i})^{-1} B_i\|_\infty < \frac{1}{\beta}.$$

This implies by the small-gain theorem that $A_i + (1-\lambda)B_i F_{\delta,i}$ is asymptotically stable for λ with $|\lambda| < \beta$. ∎

In Lemma 3.20, we showed an additional property for our sequence of state feedback gains that will be used in Chap. 4. Below, we establish that this result also holds in case the sequence of state feedback gains is designed based on a direct method. This is similar to Lemma 2.33 where we showed a similar property for the direct method in continuous time.

Lemma 3.25 *Suppose that (A, B) is stabilizable and all the eigenvalues of A are within the closed unit disc. Then, we have the following properties:*

1. *The closed-loop system matrix $A + \lambda B F_\delta$ is Schur stable for all $\delta > 0$ and for all λ with $|\lambda| < \beta$.*
2. *There exist constants $M > 0$ and $\delta^* > 0$ such that for all $\delta \in (0, \delta^*]$ and $k \geq 0$, we have:*

$$\left\| F_\delta \left(A + (1-\lambda) B F_\delta\right)^k \right\| \leq M\delta \tag{3.57}$$

for all λ with $|\lambda| < \beta$.

Proof Like in the continuous time, we first consider the single-input case. We claim that there exist constants $M_i > 0$ and $\delta^* > 0$ such that for all $\delta \in (0, \delta^*]$ and $k \geq 0$, we have:

$$\|F_{\delta,i}(A_i + B_i F_{\delta,i})^k\| \leq M_i \delta (1 - \delta^{\bar{r}_i})^k \tag{3.58}$$

where $i = 1, \ldots, q$, while $\bar{r}_i$ is the largest multiplicity of the eigenvalues of A_i. This is a very similar result to [52, Lemma 2.3.1], but because of our different scaling, it requires an intrinsically different proof. Note that compared to the continuous-time bound (2.80), we do not need the γ_i since we choose all eigenvalues distinct and hence have no Jordan structures. Let $\lambda_1, \ldots, \lambda_q$ denote the eigenvalues of A_i with algebraic multiplicities $r_1, \ldots, r_q$, respectively. Then $A_i + B_i F_{\delta,i}$ has eigenvectors:

$$\begin{pmatrix} 1 \\ (1-\delta)\lambda_m \\ (1-\delta)^2\lambda_m^2 \\ \vdots \\ (1-\delta)^{n-1}\lambda_m^{n_i-1} \end{pmatrix}, \begin{pmatrix} 1 \\ (1-\delta^2)\lambda_m \\ (1-\delta^2)^2\lambda_m^2 \\ \vdots \\ (1-\delta^2)^{n-1}\lambda_m^{n_i-1} \end{pmatrix}, \quad \ldots \quad, \begin{pmatrix} 1 \\ (1-\delta^{r_m})\lambda_m \\ (1-\delta^{r_m})^2\lambda_m^2 \\ \vdots \\ (1-\delta^{r_m})^{n_i-1}\lambda_m^{n-1} \end{pmatrix}$$

for $m = 1, \dots q$ where n_i denotes the size of the matrix A_i. Let S_δ be a matrix whose columns contain the above eigenvectors for the matrix $A_i + B_i F_{\delta,i}$. Clearly S_δ is invertible for $\delta > 0$. However, its inverse is unbounded as a function of δ. Consider the υth element of the row vector:

$$F_{\delta,i}(A_i + B_i F_{\delta,i})^k S_\delta.$$

There exist m and s such that:

$$\upsilon = r_1 + r_2 + \cdots + r_{m-1} + s$$

with $s \leq r_m$. This element is of the form:

$$(1-\delta^s)^k \lambda_m^k F_{\delta,i} \begin{pmatrix} 1 \\ (1-\delta^s)\lambda_m \\ (1-\delta^s)^2\lambda_m^2 \\ \vdots \\ (1-\delta^s)^{n_i-1}\lambda_m^{n_i-1} \end{pmatrix} = (1-\delta^s)^k \delta^{sr_m} \lambda_k^{t+sr_m} p^k((1-\delta^s)\lambda_m),$$

where:

$$\det(zI - A_i) = (z - \lambda_m)^{r_m} p^m(z)$$

for some polynomial p^m. This implies that this element is bounded by:

$$M(1-\delta^{r_m})^k \delta^{sr_m}.$$

In order to show that:

$$F_{\delta,i}(A_i + B_i F_{\delta,i})^k = \left[F_{\delta,i}(A_i + B_i F_{\delta,i})^k S_\delta\right] S_\delta^{-1}$$

is sufficiently small, we need to verify whether the υth row of S_δ^{-1} is less than:

$$N_1 \frac{1}{\delta^{sr_m - 1}}. \tag{3.59}$$

The matrix S_δ has a very nice structure as a Vandermonde matrix. There exists an explicit formula for the inverse of a Vandermonde matrix in [66]. This formula yields that the υth row of S_δ is bounded by:

$$N_2 \frac{1}{\delta^p}$$

with $p = sr_m - s[1 + \frac{1}{2}(s-1)]$ which is smaller than (3.59) when $s > 1$ and equals (3.59) for $s = 1$. This yields (3.58). We also obtain that there exists a N_i such that $\|G_{f,\delta,i}\|_\infty < N_i$, while $\|G_{f,\delta,i}\|_2 \to 0$ as $\delta \to 0$ where:

$$G_{f,\delta,i}(z) = F_{\delta,i}(zI - A_i - B_i F_{\delta,i})^{-1} B_i. \tag{3.60}$$

Next, consider the system:

$$\begin{cases} x_i(k+1) = \left(A_i + (1-\lambda) B_i F_{\delta,i}\right) x_i(k) \\ \quad u_i(k) \quad = F_{\delta,i} x_i(k). \end{cases}$$

This can be rewritten as:

$$\begin{cases} x_i(k+1) = (A_i + B_i F_{\delta,i}) x_i(k) + B_i w_i(k) \\ \quad u_i(k) \quad = F_{\delta,i} x_i(k) \end{cases}$$

with $w_i = -\lambda u_i$. We obtained (3.52) in Lemma 3.23. This implies that $G_{f,\delta,i}$, defined by (3.60), is such that:

$$1 + \lambda G_{f,\delta,i}(z)$$

is bounded away from zero on the unit circle for a small δ (uniformly with respect to λ) since $|\lambda| < \beta$.

To find the result for the multi-input case, we first use a recursive argument similar to the argument used in the proof of Lemma 2.33 in the continuous time. This does not require any intrinsic changes except for the obvious ones like replacing integration by summation. We obtain:

$$\|F_\delta (A + BF_\delta)^k\| \leq \tilde{M} \delta (1 - \delta^r/2)^k \tag{3.61}$$

for some $\tilde{M} > 0$.

We know from Lemma 3.23 that for a small δ, we have:

$$\|G_{f,\delta,i}\|_\infty < 2.$$

In that case:

$$F_{\delta,i}(zI - A_i - B_i F_{\delta,i})^{-1} = \begin{pmatrix} F_{\delta,i} & G_{f,\delta,i}(z) \end{pmatrix} \begin{pmatrix} I & 0 \\ F_{\delta,i} & I \end{pmatrix} V_i(z),$$

where:

$$V_i(z) = \begin{pmatrix} I \\ -\bar{F}_i \end{pmatrix} (zI - A_i - B_i \bar{F}_i)^{-1}$$

satisfies:

$$\begin{pmatrix} zI - A_i & B_i \end{pmatrix} V_i(z) = I$$

with $\bar{F}_i$ such that $A_i + B_i\bar{F}_i$ Schur stable. This clearly implies that:

$$\|F_{\delta,i}(zI - A_i - B_i F_{\delta,i})^{-1}\|_\infty < N_g$$

for some N_g since V and $G_{f,\delta,i}$ are both bounded in H_∞ norm. Define:

$$\tilde{G}_{f,\delta}(z) = F_\delta(zI - A - BF_\delta)^{-1}.$$

Using the block diagonal structure of A and F_δ and the upper triangular structure of B with:

$$\tilde{G}_{f,\delta}(z) = \begin{pmatrix} X_1 \\ \vdots \\ X_m \end{pmatrix}$$

we get:

$$X_m(z) = F_{\delta,m}(zI - A_m - B_m F_{\delta,m})^{-1},$$

and obviously $\|X_m\|_\infty < N_g$. Next

$$\begin{aligned} X_{m-1}(z) &= F_{\delta,m-1}(zI - A_{m-1} - B_{m-1}F_{\delta,m-1})^{-1} \\ &+ F_{\delta,m-1}(zI - A_{m-1} - B_{m-1}F_{\delta,m-1})^{-1}B_{m-1,m}F_{\delta,m}(zI - A_m - B_m F_{\delta,m})^{-1}. \end{aligned}$$

Clearly:

$$\|X_{m-1}\|_\infty < N_g + N_g^2\|B_{m-1,m}\|.$$

Recursively, we can establish that $\|X_i\|_\infty$ is bounded for all $i = 1, \ldots, m$. This implies that there exists a N_f such that:

$$\|\tilde{G}_{f,\delta}\|_\infty < N_f. \tag{3.62}$$

Next, consider the system:

$$\begin{cases} x(k+1) = (A + (1-\lambda)BF_\delta)x(k) \\ \quad u(k) \quad = F_\delta x(k). \end{cases}$$

This can be rewritten as:

$$\begin{cases} x(k+1) = (A + BF_\delta)x(k) + Bw(k) \\ u(k) \quad = F_\delta x(k) \end{cases}$$

with $w = -\lambda u$. Note that the z-transform of:

$$F_\delta(A + (1-\lambda)BF_\delta)^k$$

can be written as:

$$\left(I + \lambda G_{f,\delta}\right)^{-1} \tilde{G}_{f,\delta}, \tag{3.63}$$

where:

$$G_{f,\delta}(z) = F_\delta(zI - A - BF_\delta)^{-1}B.$$

We know that:

$$I + \lambda G_{f,\delta}(z)$$

is upper triangular and has a bounded H_∞ norm for a small δ. Moreover, its determinant:

$$\prod_{i=1}^{N} \left(1 + \lambda G_{f,\delta,i}(z)\right)$$

is bounded away from zero. This implies that there exists a $\tilde{N}$ such that:

$$\left\| \left(I + \lambda G_{f,\delta}\right)^{-1} \right\|_\infty < \tilde{N}. \tag{3.64}$$

We have by (3.61) that the L_∞-norm of the function r given by:

$$r(k) = F_\delta(A + BF_\delta)^k$$

satisfies:

$$\|r\|_\infty < \tilde{M}\delta$$

for some $N_r > 0$. But then the function y, given by:

$$y(k) = F_\delta(A + (1-\lambda)BF_\delta)^k,$$

and r are related (in the frequency domain) by

$$Y(z) = \left(I + \lambda G_{f,\delta}(z)\right)^{-1} R(z).$$

The transfer matrix relating r and y has a bounded H_∞ norm by (3.64). But then by Boyd and Doyle [9, Theorem 1], also the L_∞-induced operator norm of this system is bounded which implies that if $\|r\|_\infty \leq \tilde{M}\delta$, then also $\|y\|_\infty \leq N\delta$ for some suitable constant $N > 0$. This implies (3.57) which completes the proof. ■

3.4.4 *Strictly Unstable Agents*

In this section, we will present the protocol design for MAS with full-state coupling and strictly unstable agents, i.e., poles are outside the unit disc. Consider a MAS associated with a graph in the set $\bar{\mathbb{G}}_\beta^N$ where $\beta \in (0, 1)$ is fixed.

For single-input systems, we know that a necessary and sufficient condition is given by (3.19) where P_0 is uniquely determined by (3.17) and the condition that (3.18) has all its eigenvalues in the closed unit disc. For multi-input systems, we have the same sufficient condition given by (3.19) but a different necessary condition given in (3.24).

Since agents are unstable, the nonstandard Riccati equation-based method and direct eigenstructure assignment method can not be used any more. However, the standard H_2 discrete-time Riccati equation-based method can be utilized immediately.

Protocol design 3.5 *Consider a MAS described by* (3.1) *and* (3.6). *We consider the protocol:*

$$u_i = F_\delta \zeta_i, \tag{3.65}$$

where:

$$F_\delta = -\tfrac{1}{1-\beta_+^2}(B'P_\delta B + I)^{-1}B'P_\delta A$$

with $P_\delta > 0$ given by:

$$P_\delta = A'P_\delta A + \delta I - A'P_\delta B(B'P_\delta B + I)^{-1}B'P_\delta A, \tag{3.66}$$

with $\delta > 0$ such that:

$$\beta_+^2 B'P_\delta B < (1 - \beta_+^2)I, \tag{3.67}$$

(continued)

Protocol design 3.5 (continued)
where $\beta_+ \in [\beta, 1)$ *with* β *the upper bound for the amplitude of the eigenvalues unequal to* 1 *of the row stochastic matrix* D *associated with a graph in the set* $\bar{\mathbb{G}}^N_\beta$, *while* P_δ *is the unique solution of the above discrete-time algebraic Riccati equation.*

The main result is stated in the following theorem.

Theorem 3.26 *Consider a MAS described by strictly unstable, stabilizable agents* (3.1) *and* (3.6). *Let* P_0 *be uniquely determined by* (3.17) *such that all eigenvalues of* (3.18) *are in the closed unit disc.*

Let β *satisfy* (3.19). *In that case, there exists a* $\delta > 0$ *such that* $P_\delta > 0$, *which is uniquely determined by* (3.66) *and satisfies* (3.67). *Then the protocol* (3.65) *solves the state synchronization problem for any graph* $\mathcal{G} \in \bar{\mathbb{G}}^N_\beta$. *Moreover, the synchronized trajectory is given by* (3.14).

Proof The existence of $\delta > 0$ for which there exists a P_δ satisfying (3.33) and (3.34) follows from the fact that P_δ is a continuous function of δ and the fact that P_0 satisfies (3.19).

The rest of the proof is basically identical to the proof of Theorem 3.18. ∎

3.5 Protocol Design for MAS with Partial-State Coupling

As in the previous section for full-state coupling, the problem of partial-state coupling is also approached by reducing the protocol design for a MAS to a robust controller design.

3.5.1 Connection to a Robust Stabilization Problem

The interconnection of (3.1) and the protocol (3.8) can be written as:

$$\begin{cases} \bar{x}_i(k+1) = \begin{pmatrix} A & BC_c \\ 0 & A_c \end{pmatrix} \bar{x}_i(k) + \begin{pmatrix} BD_c \\ B_c \end{pmatrix} \zeta_i(k), \\ y_i(k) \quad = \begin{pmatrix} C & 0 \end{pmatrix} \bar{x}_i(k), \\ \zeta_i(k) \quad = \sum_{j=1}^N d_{ij}(y_i(k) - y_j(k)), \end{cases} \tag{3.68}$$

for $i = 1, \ldots, N$ where:

$$\bar{x}_i = \begin{pmatrix} x_i \\ \chi_i \end{pmatrix}.$$

Define:

$$\bar{x} = \begin{pmatrix} \bar{x}_1 \\ \vdots \\ \bar{x}_N \end{pmatrix}, \quad \bar{A} = \begin{pmatrix} A & BC_c \\ 0 & A_c \end{pmatrix}, \quad \bar{B} = \begin{pmatrix} BD_c \\ B_c \end{pmatrix}, \quad \text{and} \quad \bar{C} = \begin{pmatrix} C & 0 \end{pmatrix}.$$

Then, the overall dynamics of the N agents can be written as:

$$\bar{x}(k+1) = \left(I_N \otimes \bar{A} + (I - D) \otimes \bar{B}\bar{C}\right) \bar{x}(k). \tag{3.69}$$

A key observation, which we already used in the case of full-state coupling, is that synchronization for the system (3.69) is equivalent to the asymptotic stability of the following $N - 1$ subsystems:

$$\tilde{\eta}_i(k+1) = (\bar{A} + (1 - \lambda_i)\bar{B}\bar{C})\tilde{\eta}_i(k), \quad i = 2, \ldots, N, \tag{3.70}$$

where λ_i, $i = 2, \ldots, N$ are those eigenvalues of the row stochastic matrix D inside the unit disc.

Theorem 3.27 *The MAS* (3.69) *achieves state synchronization if and only if the system* (3.70) *is asymptotically stable for* $i = 2, \ldots, N$.

Proof As we did in the proof of Theorem 3.4, we define the Jordan form (3.12) of the row stochastic matrix D where the first column of T equals $\mathbf{1}$. Let:

$$\eta := (T^{-1} \otimes I_{n+n_c})\bar{x} = \begin{pmatrix} \eta_1 \\ \vdots \\ \eta_N \end{pmatrix},$$

where $\eta_i \in \mathbb{C}^{n+n_c}$. In the new coordinates, the dynamics of η can be written as:

$$\eta(k+1) = (I_N \otimes \bar{A} + (I - J_D) \otimes \bar{B}\bar{C})\eta(k).$$

If (3.70) is globally asymptotically stable for $i = 2, \ldots, N$, we see from the above that $\eta_i(k) \to 0$ for $i = 2, \ldots, N$. This implies that:

$$\bar{x}(k) - (T \otimes I_{n+n_c}) \begin{pmatrix} \eta_1(k) \\ 0 \\ \vdots \\ 0 \end{pmatrix} \to 0.$$

Note that the first column of T is equal to the vector $\mathbf{1}$ and therefore:

$$\bar{x}_i(k) - \eta_1(t) \to 0$$

for $i = 1, \ldots, N$. This implies that we achieve state synchronization.

Conversely, suppose that the network (3.69) reaches state synchronization. In this case, we shall have:

$$\bar{x}(k) \to \mathbf{1} \otimes \bar{x}_1(k).$$

Then $\eta(k) - (T^{-1}\mathbf{1}) \otimes \bar{x}_1(k) \to 0$. Since $\mathbf{1}$ is the first column of T, we have:

$$T^{-1}\mathbf{1} = \begin{pmatrix} 1 \\ 0 \\ \vdots \\ 0 \end{pmatrix}.$$

Therefore, $\eta(k) - (T^{-1}\mathbf{1}) \otimes \bar{x}_1(k) \to 0$ implies that $\eta_1(k) - \bar{x}_1(k) \to 0$ and $\eta_i(k) \to 0$ for $i = 2, \ldots, N$. Then, $\bar{A} + (1 - \lambda_i)\bar{B}\bar{C}$ is Schur stable for $i = 2, \ldots, N$. ■

Corollary 3.28 *Consider a MAS described by* (3.1) *and* (3.4)*. If there exists a protocol of the form* (3.8) *that achieves state synchronization, then the associated graph must have a directed spanning tree or A must be asymptotically stable.*

Since the case where A is asymptotically stable is a trivial case, the requirement that the graph of a multi-agent system has a directed spanning tree is essentially necessary.

Remark 3.29 It also becomes clear from Theorem 3.27 that the synchronized trajectory is given by $\eta_1(k)$ which is governed by:

$$\eta_1(k+1) = \bar{A}\eta_1(k), \quad \eta_1(0) = (w \otimes I_{n+n_c})\bar{x}(0), \tag{3.71}$$

where w is the first row of T^{-1}, i.e., the normalized left eigenvector with row sum equal to 1 associated with the eigenvalue 1. This shows that the modes of the synchronized trajectories are determined by the eigenvalues of $\bar{A}$ and the complete dynamics depends on both $\bar{A}$ and a weighted average of the initial conditions of agents and their protocols.

Again, in light of Lemma 3.7, one can conclude that $\eta_1(0)$ is only a linear combination of initial conditions of root agents and their protocol. As such, the synchronized trajectory given by (3.71) can be written explicitly as:

$$x_s(k) = \begin{pmatrix} I & 0 \end{pmatrix} \bar{A}^k \sum_{i \in \Pi_{\mathcal{G}}} w_i \bar{x}_i(0), \tag{3.72}$$

which is the weighted average of the trajectories of root agents and their protocol. In the case of state synchronization via stable protocol, the synchronized trajectory is given by:

$$x_s(k) = A^k \sum_{i \in \Pi_{\mathcal{G}}} w_i x_i(0). \tag{3.73}$$

Remark 3.30 From (3.72), the modes of the synchronized trajectory are a subset of or equal to the union of eigenvalues of matrices A and A_c. In other words, the unstable modes of the protocol (3.8), if any, may be present in the synchronized trajectories. Thus, it is reasonable that one can strengthen the requirements on the synchronization problem by imposing the stability of the protocol dynamics.

Note that (3.70) in Theorem 3.27 can be viewed as the closed-loop of the system:

$$\begin{cases} x(k+1) = Ax(k) + Bu(k), \\ z(k) \quad = (1-\lambda_i)Cx(k), \end{cases} \tag{3.74}$$

and a dynamic controller:

$$\begin{cases} \chi(k+1) = A_c\chi(k) + B_c z(k), \\ u(k) \quad = C_c\chi(k) + D_c z(k) \end{cases} \tag{3.75}$$

with λ_i $(i = 2, \dots, N)$ representing the eigenvalues inside the unit disc of the row stochastic matrices D associated with a given graph $\mathcal{G}$. It is easy to see that owing to the linearity, (3.75) stabilizes (3.74) if it stabilizes:

$$\begin{cases} x(k+1) = Ax(k) + (1-\lambda_i)Bu(k), \\ z(k) \quad = Cx(k). \end{cases} \tag{3.76}$$

For a given graph, this is a simultaneous stabilization problem. However, in light of the definition of Problem 3.3 where synchronization is formulated for a set of graphs $\mathbf{G}$, we basically obtain a *robust* stabilization problem, i.e., the stabilization of the system:

$$\begin{cases} x(k+1) = Ax(k) + (1-\lambda)Bu(k), \\ z(k) \quad = Cx(k), \end{cases} \tag{3.77}$$

via a controller:

$$\begin{cases} \chi(k+1) = A_c\chi(k) + B_c z(k), \\ u(k) \quad = C_c\chi(k) + D_c z(k), \end{cases} \tag{3.78}$$

for any λ which is an eigenvalue inside the unit disc for some row stochastic matrices D associated with the given set of graphs $\mathbf{G}$.

3.5.2 Solvability Conditions

In Sect. 3.4.2, we gave solvability conditions for synchronization problems with full-state coupling. In this subsection, we will address solvability conditions for synchronization problems with partial-state coupling. We again obtain necessary and sufficient condition for the solvability of the state synchronization problem of a MAS with single-input agents with any network graph in $\mathbf{G}_\beta^N$ with a given β. For multi-input agents, Theorem 3.33 provides a sufficient condition, for the solvability of the state synchronization problem of a MAS with any network graph in $\mathbf{G}_\beta^N$ with a given β. Finally, for any network graph in $\mathbf{G}_\beta^N$ with all β, Theorem 3.34 provides a necessary and sufficient condition for the solvability of the state synchronization problem.

Theorem 3.31 *Consider a MAS described by single-input agents* (3.1) *and* (3.4) *where* (A, B) *is stabilizable and* (C, A) *detectable.*

Let $P_0 \geq 0$ *be such that* (3.17) *is satisfied while* (3.18) *has all its eigenvalues in the closed unit disc.*

Let Q_0 *be such that:*

$$Q_0 = AQ_0A' + BB' - AQ_0C'(CQ_0C')^{-1}CQ_0A' \tag{3.79}$$

while:

$$A - AQ_0C'(CQ_0C')^{-1}C \tag{3.80}$$

has all its eigenvalues in the closed unit disc.

Then, the problem of synchronization with partial-state coupling as defined in Problem 3.3 *is solvable for* $\mathbf{G} = \bar{\mathbb{G}}_\beta^N$ *for a given* $\beta \in (0, 1)$ *if and only if*

$$\frac{\beta^2}{1-\beta^2}\lambda_{\max}(P_0Q_0) < 1. \tag{3.81}$$

Remark 3.32 Note that in the case of full-state coupling, we have $C = I$. In that case, it is easy to verify that $Q_0 = BB'$ which implies that (3.81) reduces to the condition (3.19).

Proof As in Theorem 3.9, we note that synchronization is similar to finding a measurement feedback of the form (3.78) for the system:

$$\begin{aligned} x(k+1) &= Ax(k) + Bu(k) + \beta Bw(k) \\ y(k) &= Cx(k) \\ z(k) &= \qquad\qquad u(k) \end{aligned} \tag{3.82}$$

such that the H_∞ norm from w to z is less than 1. Using the result from [125], we find that such a controller exists if and only if four conditions are satisfied. The first condition is a semi-stabilizing solution of the state feedback Riccati equation and the existence of this solution we already established in the proof of Theorem 3.9 resulting in the explicit expression (3.21).

The second condition is a semi-stabilizing solution $Q \geq 0$ of the observer Riccati equation:

$$Q = AQA' + \beta^2 BB' - AQC'(CQC')^{-1}CQA'$$

while $A - AQC'(CQC')^{-1}C$ has all its eigenvalues in the closed unit disc. In this case, $Q = \beta^2 Q_0$ satisfies the required conditions, while the existence of Q_0 is guaranteed by the standard H_2 optimal control.

The third condition is the requirement that:

$$\lambda_{\max}(PQ) < 1$$

which in our cases reduces to the condition (3.81). Finally, we need to check that for all eigenvalues z_0 of A on the unit circle, there exists a matrix K such that $z_0 I - A - BKC$ is invertible and

$$\beta \| KC(z_0 I - A - BKC)^{-1} B \| < 1. \tag{3.83}$$

We choose S and T such that:

$$z_0 I - A = S \begin{pmatrix} I & 0 \\ 0 & 0 \end{pmatrix} T, \quad B = S \begin{pmatrix} B_1 \\ B_2 \end{pmatrix}, \quad C = \begin{pmatrix} C_1 & C_2 \end{pmatrix} T.$$

Since (A, B) and (C, A) are stabilizable and detectable, respectively, we must have that B_2 is surjective and C_2 injective. We denote by B_2^r and C_2^ℓ right and left inverses of B_2 and C_2, respectively, such that $B_2^r B_2$ is an orthogonal projection. We then choose $K = -B_2^r C_2^\ell$. Then it is easily verified that:

$$\begin{aligned}
KC(z_0 I - A - BKC)^{-1}B &= \\
&= K \begin{pmatrix} C_1 & C_2 \end{pmatrix} \left(\begin{pmatrix} I & 0 \\ 0 & 0 \end{pmatrix} - \begin{pmatrix} B_1 \\ B_2 \end{pmatrix} K \begin{pmatrix} C_1 & C_2 \end{pmatrix} \right)^{-1} \begin{pmatrix} B_1 \\ B_2 \end{pmatrix} \\
&= -B_2^r \begin{pmatrix} C_2^\ell C_1 & I \end{pmatrix} \begin{pmatrix} I + B_1 B_2^r C_2^\ell C_1 & B_1 B_2^r \\ C_2^\ell C_1 & I \end{pmatrix}^{-1} \begin{pmatrix} B_1 \\ B_2 \end{pmatrix} \\
&= -B_2^r \begin{pmatrix} C_2^\ell C_1 & I \end{pmatrix} \begin{pmatrix} I & -B_1 B_2^r \\ -C_2^\ell C_1 & I + C_2^\ell C_1 B_1 B_2^r \end{pmatrix} \begin{pmatrix} B_1 \\ B_2 \end{pmatrix} \\
&= -B_2^r \begin{pmatrix} 0 & I \end{pmatrix} \begin{pmatrix} B_1 \\ B_2 \end{pmatrix} = -B_2^r B_2
\end{aligned}$$

and therefore (3.83) is satisfied given that $\beta < 1$ and $B_2^r B_2$ is an orthogonal projection. Therefore, our original problem is equivalent to an H_∞ control problem which is solvable if and only if P_0 and Q_0 satisfy (3.81). This completes the proof. ∎

For multi-input agents, we can obtain a sufficient condition when synchronization is possible. However, in this case, the algebraic Riccati equation for Q_0 becomes:

$$Q_0 = AQ_0A' + BB' - AQ_0C'(CQ_0C')^{\dagger}CQ_0A'$$

such that:

$$\begin{pmatrix} zI - A & AQ_0C' \\ C & CQ_0C' \end{pmatrix} = n + \operatorname{normrank} C(zI - A)^{-1}B$$

for all $z \in \mathbb{C}$ with $|z| \geq 1$ where $\dagger$ denotes the generalized inverse. This reduces to the previous case when (A, B, C) is right-invertible since this guarantees the invertibility of CQ_0C'.

Theorem 3.33 *Consider a MAS described by multi-input agents* (3.1) *and* (3.4). *The problem of synchronization with partial-state coupling as defined in Problem 3.3 is solvable with* $\mathbf{G} = \bar{\mathbb{G}}_\beta^N$ *for a given* $\beta \in (0, 1)$ *if the condition* (3.81) *is satisfied.*

Proof Similar to the arguments in the proof of Theorem 3.11 for the full-state feedback case, the result of Theorem 3.31 still provides a sufficient condition for synchronization for a multi-input system which results in this theorem. ∎

The following theorem considers the case when state synchronization is possible for any graph in $\bar{\mathbb{G}}^N$, i.e., for arbitrary β. It presents necessary and sufficient conditions for both single- and multi-input agents.

Theorem 3.34 *Consider a MAS described by* (3.1) *and* (3.4). *The problem of state synchronization with partial-state coupling as defined in Problem 3.3 is solvable for* $\mathbf{G} = \bar{\mathbb{G}}_\beta^N$ *for all* $\beta \in (0, 1)$ *if and only if the agents are at most weakly unstable.*

Proof If synchronization via partial-state coupling as defined in Problem 3.3 is solvable, it is clear that also synchronization via full-state coupling as defined in Problem 3.3 is solvable. From Theorem 3.10, we know that a necessary condition for solvability is that the agents are at most weakly stable.

On the other hand, if the system is at most weakly stable, then the sufficient condition (3.81) established in Theorem 3.33 is always satisfied since in that case $P_0 = 0$ and therefore synchronization with partial-state coupling as defined in Problem 3.3 is solvable. ∎

In the following, we will present the protocol design for both at most weakly unstable agents and strictly unstable agents. For at most weakly unstable agents, we give the protocol design based on discrete ARE method and direct eigenstructure assignment method.

3.5.3 At Most Weakly Unstable Agents

In this section, we design protocols for MAS with at most weakly unstable agents. We have three different designs. The first two designs are utilizing Riccati equations as discussed before in Sect. 3.4.3.1. The third design exploits the eigenstructure assignment method presented in Sect. 3.4.3.2. In each case, an observer is used with a CSS architecture which requires a low-gain based state feedback used in conjunction with the observer. We conclude this subsection by presenting a design for neutrally stable systems.

The first design is based on the nonstandard discrete-time algebraic Riccati equation we encountered before.

Protocol design 3.6 *Consider a MAS described by at most weakly unstable agents (3.1) and (3.2). We will use a CSS observer-based protocol for the agents.*

We choose an observer gain K such that $A+KC$ is Schur stable. Next, we consider a feedback gain:

$$F_\delta = -(B'P_\delta B + I)^{-1}B'P_\delta A, \tag{3.84}$$

where the positive definite matrix P_δ is the solution of the nonstandard discrete-time algebraic Riccati equation:

$$P_\delta = A'P_\delta A + \delta I - (1-\beta^2)A'P_\delta B(B'P_\delta B + I)^{-1}B'P_\delta A, \tag{3.85}$$

where $\delta > 0$ is a low-gain parameter and β is the upper bound of magnitude of the eigenvalues unequal to 1 of the row stochastic matrix D. This results in the protocol:

$$\begin{cases} \chi_i(k+1) = (A+KC)\chi_i(k) - K\zeta_i(k), \\ \quad u_i(k) \quad = F_\delta \chi_i(k). \end{cases} \tag{3.86}$$

The main result of this section is then stated as follows.

Theorem 3.35 *Consider a MAS described by at most weakly unstable agents (3.1) and (3.4). Let any $\beta \in (0,1)$ be given and hence a set of network graphs $\bar{\mathbb{G}}_\beta^N$ be defined.*

If (A, B) is stabilizable and (C, A) is detectable, then the state synchronization problem via a stable protocol stated in Problem 3.3 with $\mathbf{G} = \bar{\mathbb{G}}_\beta^N$ is solvable. In particular, there exists a $\delta^* > 0$ such that for any $\delta \in (0, \delta^*]$, protocol (3.86) solves the state synchronization problem for any graph $\mathcal{G} \in \bar{\mathbb{G}}_\beta^N$. Moreover, the synchronized trajectory is given by (3.73).

Proof From Theorem 3.27, we only need to prove that the system:

$$\begin{cases} x(k+1) = Ax(k) + (1-\lambda)BF_\delta\chi(k), \\ \chi(k+1) = (A+KC)\chi(k) - KCx(k), \end{cases} \tag{3.87}$$

is asymptotically stable for any λ that satisfies $|\lambda| < \beta$.

Define $e = x - \chi$. Then, (3.87) can be rewritten in terms of x and e as:

$$\begin{cases} x(k+1) = [A + (1-\lambda)BF_\delta]x(k) - (1-\lambda)BF_\delta e(k) \\ e(k+1) = [A + KC - (1-\lambda)BF_\delta]e(k) + (1-\lambda)BF_\delta x(k). \end{cases}$$

Let Q be the positive definite solution of the Lyapunov equation:

$$(A+KC)'Q(A+KC) - Q + 4I = 0.$$

Since $F_\delta \to 0$ as $\delta \to 0$, there exists a δ_1 such that for a $\delta \in (0, \delta_1]$:

$$(A + KC - (1-\lambda)BF_\delta)^*Q(A + KC - (1-\lambda)BF_\delta) - Q + 3I \leq 0.$$

Consider $V_1(k) = e(k)^*Qe(k)$. Let $\mu(k) = F_\delta x(k)$. To ease our presentation, we omit the time label (k) whenever this causes no confusion. We have:

$$\begin{aligned} V_1(k+1) &- V_1(k) \\ &\leq -3\|e\|^2 + 2\left|(1-\lambda)^*\mu^*B'Q[A+KC-(1-\lambda)BF_\delta]e\right| \\ &\qquad\qquad + |1-\lambda|^2\mu^*B'QB\mu \\ &\leq -3\|e\|^2 + M_1\|\mu\|\|e\| + M_2\|\mu\|^2, \end{aligned} \tag{3.88}$$

where:

$$M_1 = 4\|B'Q\|\|A+KC\| + 8\|B'Q\| \max_{\delta\in[0,1]}\{\|BF_\delta\|)\}, \quad M_2 = 4\|B'QB\|.$$

It should be noted that M_1 and M_2 are independent of δ and λ provided that $|\lambda| < \beta$.

Consider $V_2(k) = x(k)'P_\delta x(k)$. We have:

$$V_2(k+1) - V_2(k) \leq -\delta\|x\|^2 - (1-\beta)^2\|\mu\|^2$$
$$+ 2\,\mathrm{Re}\Big((1-\lambda)^* e^* F_\delta' B' P_\delta [A + (1-\lambda) B F_\delta] x\Big) + |1-\lambda|^2 e^* F_\delta' B P_\delta B F_\delta e$$

using arguments similar to the bounds in the proof of Theorem 3.16. Note that:

$$\begin{aligned} e^* F_\delta' B' P_\delta [A + (1-\lambda) B F_\delta] x &= e^* F_\delta' B' P_\delta A x + (1-\lambda) e^* F_\delta' B' P_\delta B \mu \\ &= -e^* F_\delta' (B' P_\delta B + I)\mu + (1-\lambda) e^* F_\delta' B' P_\delta B \mu \\ &= e^* [F_\delta' - \lambda F_\delta' B' P_\delta B]\mu, \end{aligned}$$

and hence:

$$V_2(k+1) - V_2(k) \leq -\delta\|x\|^2 - (1-\beta)^2\|\mu\|^2 + \theta_1(\delta)\|e\|\|\mu\| + \theta_2(\delta)\|e\|^2, \tag{3.89}$$

where:

$$\theta_1(\delta) = 4(\|F_\delta\| + \|F_\delta' B' P_\delta B\|), \quad \theta_2(\delta) = 4\|F_\delta' B P_\delta B F_\delta\|.$$

Consider a Lyapunov candidate $V(k) = V_1(k) + \kappa V_2(k)$ with:

$$\kappa = \frac{M_2 + M_1^2}{(1-\beta)^2}.$$

In view of (3.88) and (3.89), we get:

$$V(k+1) - V(k) \leq -\delta\kappa\|x\|^2 - M_1^2\|\mu\|^2 - [3 - \kappa\theta_2(\delta)]\|e\|^2 + [M_1 + \kappa\theta_1(\delta)]\|\mu\|\|e\|.$$

There exists a $\delta^* \leq \delta_1$ such that for a $\delta \in (0, \delta^*]$:

$$3 - \kappa\theta_2(\delta) \geq 2, \quad M_1 + \kappa\theta_1(\delta) \leq 2M_1.$$

This yields:

$$V(k+1) - V(k) \leq -\delta\kappa\|x\|^2 - \|e\|^2 - (\|e\| - M_1\|\mu\|)^2.$$

Therefore, for a $\delta \in (0, \delta^*]$, the system (3.87) is globally asymptotically stable for any λ that satisfies $|\lambda| < \beta$. ■

The second design is based on the standard H_2 discrete-time algebraic Riccati equation.

Protocol design 3.7 *Consider a MAS described by at most weakly unstable agents* (3.1) *and* (3.2). *We will use a CSS observer-based protocol for the agents.*

We choose an observer gain K such that $A + KC$ is Schur stable. Next, we consider a feedback gain:

$$F_\delta = -\tfrac{1}{1-\beta_+^2}(B'P_\delta B + I)^{-1}B'P_\delta A, \tag{3.90}$$

where the positive definite matrix P_δ is the solution of the discrete-time H_2 algebraic Riccati equation:

$$P_\delta = A'P_\delta A + \delta I - A'P_\delta B(B'P_\delta B + I)^{-1}B'P_\delta A, \tag{3.91}$$

where $\delta > 0$ is a low-gain parameter, while $1 > \beta_+ > \beta$ with β being the upper bound of magnitude of the eigenvalues unequal to 1 of the row stochastic matrix D. This results in the protocol:

$$\begin{cases} \chi_i(k+1) = (A + KC)\chi_i(k) - K\zeta_i(k), \\ u_i(k) \quad = F_\delta \chi_i(k). \end{cases} \tag{3.92}$$

The main result is then stated as follows.

Theorem 3.36 *Consider a MAS described by at most weakly unstable agents* (3.1) *and* (3.4). *Let any $\beta \in (0, 1)$ be given and hence a set of network graphs $\bar{\mathbb{G}}_\beta^N$ be defined.*

If (A, B) is stabilizable and (C, A) is detectable, then the state synchronization problem via a stable protocol stated in Problem 3.3 *with $\mathbf{G} = \bar{\mathbb{G}}_\beta^N$ is solvable. In particular, for any $\beta_+ \in [\beta, 1)$, there exists a $\delta^* > 0$ such that for any $\delta \in (0, \delta^*]$, protocol* (3.92) *solves the state synchronization problem for any graph $\mathcal{G} \in \bar{\mathbb{G}}_\beta^N$. Moreover, the synchronized trajectory is given by* (3.73).

Proof In the proof of Theorem 3.18, we have basically shown that:

$$G_{f,\delta}(z) = F_\delta(zI - A - BF_\delta)^{-1}B$$

has the property that:

$$\|G_{f,\delta}\|_\infty < \beta_+^{-1} \tag{3.93}$$

for a δ small enough such that (3.34) is satisfied. Next, define:

$$G_{k,\delta}(z) = F_\delta(zI - A - KC + BF_\delta)^{-1}B,$$

and we clearly have:

$$\lim_{\delta\to 0} \|G_{k,\delta}\|_\infty = 0. \tag{3.94}$$

Let us look at the interconnection of:

$$\begin{cases} \chi(k+1) = (A+KC)\chi(k) - Ky(k), \\ u(k) \quad = F_\delta \chi(k), \end{cases} \tag{3.95}$$

and:

$$\begin{cases} x(k+1) = Ax(k) + Bu(k) + \beta B w(k), \\ y(k) \quad = Cx(k), \\ z(k) \quad = u(k). \end{cases}$$

We basically need to prove that the closed-loop system has an H_∞ norm less than or equal to 1 to guarantee the robust stability which ensures that our protocol will achieve state synchronization.

Note that the closed-loop transfer function from w to z is given by:

$$G_{\text{cl}}(z) = \beta \begin{pmatrix} I & I \end{pmatrix} \begin{pmatrix} I & -G_{f,\delta} \\ G_{k,\delta} & I \end{pmatrix}^{-1} \begin{pmatrix} G_{f,\delta} \\ G_{k,\delta} \end{pmatrix}.$$

Using (3.93) and (3.94), we immediately find that the closed-loop transfer function satisfies:

$$\|G_{\text{cl}}\|_\infty \to \beta \|G_{f,\delta}\|_\infty < \frac{\beta}{\beta_+} < 1$$

as $\delta \to 0$. Hence for a sufficiently small δ we have a H_∞ norm less than 1 which, as noted before, guarantees that we achieve state synchronization. ∎

The final design is based on a direct method.

Protocol design 3.8 *Consider a MAS described by at most weakly unstable agents (3.1) and (3.2). We will use a CSS observer-based protocol for the agents.*

We choose an observer gain K such that $A+KC$ is Schur stable. Next, let F_δ be as defined in Protocol Design 3.4. This results in the protocol:

$$\begin{cases} \chi_i(k+1) = (A+KC)\chi_i(k) - K\zeta_i(k), \\ u_i(k) \quad = F_\delta \chi_i(k). \end{cases} \tag{3.96}$$

The main result is then stated as follows.

Theorem 3.37 *Consider a MAS described by at most weakly unstable agents* (3.1) *and* (3.4). *Let any* $\beta \in (0, 1)$ *be given and hence a set of network graphs* $\bar{\mathbb{G}}_\beta^N$ *be defined.*

If (A, B) *is stabilizable and* (C, A) *is detectable, then the state synchronization problem via a stable protocol stated in Problem* 3.3 *with* $\mathbf{G} = \bar{\mathbb{G}}_\beta^N$ *is solvable. In particular, there exists a* $\delta^* > 0$ *such that for any* $\delta \in (0, \delta^*]$, *protocol* (3.96) *solves the state synchronization problem for any graph* $\mathcal{G} \in \bar{\mathbb{G}}_\beta^N$. *Moreover, the synchronized trajectory is given by* (3.73).

Proof The proof is essentially the same as the proof of Theorem 3.36. ∎

Finally, we consider a special case where the agent dynamics are neutrally stable, that is, those eigenvalues of A matrix on the unit disc, if any, are semi-simple. This implies that A is stable and that there exists a positive definite matrix P such that $A'PA \leq P$. In this case, we shall show that the protocol design no longer requires the knowledge of β and hence allows us to deal with a larger set of network graphs $\bar{\mathbb{G}}^N$ (effectively the case $\beta = 1$).

Protocol design 3.9 *Consider a MAS described by at most weakly unstable agents* (3.1) *and* (3.2). *We will use a CSS observer-based protocol for the agents.*

We choose an observer gain K *such that* $A + KC$ *is Schur stable. Next, we consider a feedback gain:*

$$F_\delta = -\delta B'PA, \tag{3.97}$$

where $\delta > 0$ *is a low-gain parameter and the positive definite matrix* P *is a solution of the Lyapunov inequality:*

$$A'PA \leq P. \tag{3.98}$$

This results in the protocol:

$$\begin{cases} \chi_i(k+1) = (A + KC)\chi_i(k) - K\zeta_i(k), \\ \quad u_i(k) \quad = F_\delta \chi_i(k). \end{cases} \tag{3.99}$$

The result for neutrally stable systems is stated as follows.

Theorem 3.38 *Consider a MAS described by neutrally stable agents* (3.1) *and* (3.4). *Let the set of network graphs* $\bar{\mathbb{G}}^N$ *be given.*

If (A, B) *is stabilizable and* (C, A) *is detectable, then the state synchronization problem via a stable protocol stated in Problem* 3.3 *with* $\mathbf{G} = \bar{\mathbb{G}}^N$ *is solvable. In particular, there exists a* $\delta^* > 0$ *such that for any* $\delta \in (0, \delta^*]$, *controller* (3.99) *solves the state synchronization problem for any graph* $\mathcal{G} \in \bar{\mathbb{G}}^N$. *Moreover, the synchronized trajectory is given by* (3.73).

Proof From Theorem 3.27, we only need to prove that the system:

$$\begin{cases} x(k+1) = Ax(k) - (1-\lambda)\delta BB'PA\chi(k), \\ \chi(k+1) = (A+KC)\chi(k) - KCx(k), \end{cases} \tag{3.100}$$

is asymptotically stable for $|\lambda| < 1$.

Define $e(k) = x(k) - \chi(k)$. The system (3.100) can be rewritten in terms of x and e as:

$$\begin{cases} x(k+1) = (A - (1-\lambda)\delta BB'PA)x(k) + (1-\lambda)\delta BB'PAe(k), \\ e(k+1) = (A + KC + (1-\lambda)\delta BB'PA)e(k) - (1-\lambda)\delta BB'PAx(k). \end{cases} \tag{3.101}$$

Let Q be the positive definite solution of the Lyapunov equation:

$$(A+KC)'Q(A+KC) - Q + 4I = 0.$$

Since $|\lambda| < 1$, there exists a δ_1 such that for a $\delta \in (0, \delta_1]$:

$$(A + KC + (1-\lambda)\delta BB'PA)^* Q(A + KC + (1-\lambda)\delta BB'PA) - Q + 3I \le 0.$$

Consider $V_1(k) = e(k)^* Qe(k)$. Let $\mu(k) = \delta B'PAx(k)$. To ease our presentation, we omit again the time label (k) whenever this causes no confusion. We have:

$$\begin{aligned} &V_1(k+1) - V_1(k) \\ &\qquad \le -3\|e\|^2 + 2\left|((1-\lambda)^*\mu^* B'Q[A + KC + (1-\lambda)BB'PA]e)\right| \\ &\qquad\qquad + |1-\lambda|^2 \mu^* B'QB\mu \\ &\qquad \le -3\|e\|^2 + |1-\lambda| M_1 \|\mu\| \|e\| + |1-\lambda|^2 M_2 \|\mu\|^2, \end{aligned}$$

where:

$$M_1 = 2\|B'Q\| \|A + KC\| + 4\|B'Q\| \|BB'PA\|, \quad M_2 = \|B'QB\|.$$

It should be noted that M_1 and M_2 are independent of δ and λ. Consider $V_2(k) = x(k)'Px(k)$. Note that:

$$\begin{aligned} &[A - (1-\lambda)\delta BB'PA]'P[A - (1-\lambda)\delta BB'PA] - P \\ &\qquad \le -2\,\mathrm{Re}(1-\lambda)\delta A'PBB'PA + |1-\lambda|^2\delta^2 A'PBB'PBB'PA. \end{aligned}$$

There exists a δ_2 such that for a $\delta \in (0, \delta_2]$, we have $\delta B'PB \leq \frac{1}{2}$. Since $|1-\lambda|^2 \leq 2\operatorname{Re}(1-\lambda)$ for $|\lambda| < 1$, we get for a $\delta \in (0, \delta_2]$:

$$[A-(1-\lambda)\delta BB'PA]^* P[A-(1-\lambda)\delta BB'PA] - P \leq -\tfrac{1}{2}|1-\lambda|^2\delta A'PBB'PA.$$

Hence:

$$\begin{aligned}
&V_2(k+1) - V_2(k) \\
&\quad \leq -\tfrac{1}{2\delta}|1-\lambda|^2\|\mu\|^2 + 2\left|(1-\lambda^*)e^*A'PB\mu - |1-\lambda|^2\delta e^*A'PBB'PB\mu\right| \\
&\qquad\qquad + |1-\lambda|^2\delta^2 e^*A'PBB'PBB'PAe \\
&\quad \leq -\tfrac{1}{2\delta}|1-\lambda|^2\|\mu\|^2 + \theta_1|1-\lambda|\|e\|\|\mu\| + \theta_3|1-\lambda|^2\|e\|\|\mu\| + \theta_2\|e\|^2,
\end{aligned}$$

where:

$$\theta_1 = 2\|A'PB\|, \quad \theta_3 = 2\|A'PBB'PB\|, \quad \theta_2 = 4\|A'PBB'PB'PA\|.$$

Define a Lyapunov candidate $V(k) = V_1(k) + \delta\kappa V_2(k)$ with:

$$\kappa = 4 + 2M_2 + 2M_1^2.$$

We get that:

$$\begin{aligned}
V(k+1) - V(k) \leq &-(3-\delta\theta_2\kappa)\|e\|^2 - (2+M_1^2)|1-\lambda|^2\|\mu\|^2 \\
&+ (M_1 + \delta\theta_1\kappa)|1-\lambda|\|\mu\|\|e\| + \delta\theta_3\kappa|1-\lambda|^2\|\mu\|\|e\|.
\end{aligned}$$

There exists a δ_3 such that for a $\delta \in (0, \delta_3]$:

$$3 - \delta\theta_2\kappa \geq 2.5, \quad M_1 + \delta\theta_1\kappa \leq 2M_1 \text{ and } \delta\theta_3\kappa \leq 1.$$

This yields:

$$\begin{aligned}
&V(k+1) - V(k) \\
&\qquad \leq -2.5\|e\|^2 - (2+M_1^2)|1-\lambda|^2\|\mu\|^2 \\
&\qquad\qquad + (2M_1|1-\lambda| + |1-\lambda|^2)\|\mu\|\|e\| \\
&\qquad \leq -0.5\|e\|^2 - |1-\lambda|^2\|\mu\|^2 - (\|e\| - M_1|1-\lambda|\|\mu\|)^2 \\
&\qquad\qquad - |1-\lambda|^2(\tfrac{1}{2}\|e\| - \|\mu\|)^2 \\
&\qquad \leq -0.5\|e\|^2 - |1-\lambda|^2\|\mu\|^2.
\end{aligned}$$

Since (A, B) is controllable, it follows from LaSalle's invariance principle that the system (3.101) is globally asymptotically stable. ■

3.5.4 Strictly Unstable Agents

In this section, we will present the protocol design for MAS with partial-state coupling and strictly unstable agents, i.e., poles are outside the unit disc. Consider a MAS associated with a graph in the set $\bar{\mathbb{G}}_{\beta}^{N}$ where $\beta \in (0, 1)$ is fixed.

Since the agents are unstable, the nonstandard Riccati equation-based method and direct eigenstructure assignment method can not be used any more. However, the standard H_2 discrete-time Riccati equation-based method can be utilized immediately.

Protocol design 3.10 *Consider a MAS described by at most weakly unstable agents (3.1) and (3.2). We will use a CSS observer-based protocol for the agents.*

We choose an observer gain K such that $A + KC$ is Schur stable. Next, we consider a feedback gain:

$$F_{\delta} = -\tfrac{1}{1-\beta_{+}^{2}}(B' P_{\delta} B + I)^{-1} B' P_{\delta} A, \tag{3.102}$$

where the positive definite matrix P_δ is the solution of the discrete-time H_2 algebraic Riccati equation:

$$P_{\delta} = A' P_{\delta} A + \delta I - A' P_{\delta} B (B' P_{\delta} B + I)^{-1} B' P_{\delta} A, \tag{3.103}$$

where $\delta > 0$ is a low-gain parameter, while $1 > \beta_+ > \beta$ with β is the upper bound of magnitude of the eigenvalues unequal to 1 of the row stochastic matrix D. This results in the protocol:

$$\begin{cases} \chi_i(k+1) = (A + KC)\chi_i(k) - K\zeta_i(k), \\ u_i(k) \quad\;\; = F_{\delta}\chi_i(k). \end{cases} \tag{3.104}$$

The main result is stated in the following theorem.

Theorem 3.39 *Consider a MAS described by strictly unstable but stabilizable agents (3.1) and (3.4). Let P_0 be uniquely determined by (3.17) such that (3.18) has all its eigenvalues in the closed unit disc and Q_0 is uniquely determined by (3.79) such that (3.80) has all its eigenvalues in the closed unit disc.*

Let β satisfy (3.81). *In that case, the state synchronization problem via a stable protocol stated in Problem 3.3 with* $\mathbf{G} = \bar{\mathbb{G}}_\beta^N$ *is solvable. In particular, Protocol Design 3.10 with* $\beta_+ > \beta$ *and a* δ *sufficiently small results in a protocol* (3.104) *that solves the state synchronization problem for any graph* $\mathcal{G} \in \bar{\mathbb{G}}_\beta^N$. *Moreover, the synchronized trajectory is given by* (3.72).

Proof It is similar to the proof of Theorem 3.36. ■

3.5.5 *Static Protocol Design*

In this section, we investigate the static protocol design for a MAS with partial-state coupling.

We consider MAS with strictly proper agents. That means that these systems are never passive. We also note that the pre-compensator G_1, post-compensator G_2, or the feedback H results in a system which is still strictly proper and therefore our strictly proper agents are never squared-down passive or squared-down passifiable via static output feedback either.

The only concept related to passivity which is of potential use for us is therefore the case of systems which are squared-down passifiable via static input feedforward. After all, the input feedforward changes a strictly proper system into a proper system.

Finally, consider a G-minimum-phase systems with relative degree 1 system. In principle, this can be trivially defined for discrete-time systems, and we obtain a structure of the form (2.114). However, the associated protocol design inherently relies on a high-gain feedback. In discrete time, the eigenvalues of a stable system need to be in a bounded region (instead of a half plane). That makes these types of designs impossible.

As before, we first establish that state synchronization among agents in the network with partial-state coupling via static protocols can be solved by equivalently solving a robust stabilization problem.

3.5.5.1 Connection to a Robust Stabilization Problem

The static protocol (3.7) with ζ_i given in (3.4) will be used. Then, the closed-loop system of the agent (3.1) and the protocol (3.7) can be described by:

$$\begin{cases} x_i(k+1) &= Ax_i(k) + BF\zeta_i(k), \\ y_i(k) &= Cx_i(k), \\ \zeta_i(k) &= \sum_{j=1}^N d_{ij} y_j(k), \end{cases} \tag{3.105}$$

for $i = 1, \ldots, N$. Define:

$$x = \begin{pmatrix} x_1 \\ x_2 \\ \vdots \\ x_N \end{pmatrix}.$$

Then the overall dynamics of the N agents can be written as:

$$x(k+1) = (I_N \otimes A + (I - D) \otimes BFC)x(k). \tag{3.106}$$

With the similar arguments in Sect. 3.4.1, we can have the following theorem and corollary.

Theorem 3.40 *The MAS (3.106) achieves state synchronization if and only if the system:*

$$\eta_i(k+1) = (A + (1 - \lambda_i)BFC)\eta_i(k+1) \tag{3.107}$$

is globally asymptotically stable for $i = 2, \ldots, N$. Moreover, the synchronization trajectory is given by (3.14).

Corollary 3.41 *Consider a MAS described by agents (3.1) and (3.4). If there exists a protocol of the form (3.7) that achieves state synchronization, then the associated graph must have a directed spanning tree or A must be asymptotically stable.*

In light of the definition of Problem 3.2 that synchronization is formulated for a set of graphs **G**, we basically obtain a *robust* stabilization problem, i.e., we need to find a controller:

$$u(k) = Fy(k), \tag{3.108}$$

such that when applied to the system:

$$\begin{aligned} x(k+1) &= Ax(k) + (1 - \lambda)Bu(k), \\ y(k) &= Cx(k), \end{aligned} \tag{3.109}$$

the interconnection is stable for any λ which is an eigenvalue inside the unit disc for some row stochastic matrix D associated with a graph in the given set **G**.

3.5.5.2 Squared-Down Passifiable via Static Input Feedforward

As argued before, for discrete-time systems, we find that for only one class of agents, static protocols can be designed. We therefore consider systems which are squared-down passifiable via static input feedforward. We design the static protocol based on the low-gain methodology as follows.

Protocol design 3.11 *Consider a MAS described by agents (3.1) with communication via (3.4). Assume the agents are squared-down passifiable via static input feedforward with respect to G_1, G_2, and R. The static protocol is designed as:*

$$u_i(k) = -\delta G_1 K G_2 \zeta_i(k), \tag{3.110}$$

where $K > 0$ while $\delta > 0$ is a low-gain parameter.

The main result based on the above design can be stated as follows.

Theorem 3.42 *Consider a MAS described by agents (3.1) and communication via (3.4). Assume the agents are squared-down passifiable via static input feedforward with respect to G_1, G_2, and R such that (A, BG_1) is controllable and (A, G_2C) is observable with BG_1 and G_2C full-column and full-row rank, respectively. Let a set of network graphs $\bar{\mathbb{G}}^N$ be defined.*

In that case, the state synchronization problem stated in Problem 3.3 but with a static protocol is solvable for a set of graphs $\mathbf{G} = \bar{\mathbb{G}}^N$. In particular, there exists a δ^* such that for any $\delta < \delta^*$, protocol (3.110) solves the state synchronization problem for any graph $\mathcal{G} \in \bar{\mathbb{G}}^N$. Moreover, the synchronized trajectory is given by (3.73).

In order to prove the above theorem, we need the following technical lemma.

Lemma 3.43 *Consider a Laplacian matrix $I - D$ associated with a graph in $\bar{\mathbb{G}}^N$. In that case, any eigenvalue of D satisfies:*

$$|1-\lambda|^2 \leq 2\,\mathrm{Re}(1-\lambda).$$

Proof Let $I - D = [\bar{d}_{ij}]$, where $\bar{d}_{ii} = 1 - d_{ii}$ and $\bar{d}_{ij} = -d_{ij}$. Note that we have $\bar{d}_{ii} \leq 1$ with $d_{ii} \geq 0$. According to Gershgorin circle theorem, we know that an eigenvalue $1 - \lambda$ is in a Gershgorin circle, that is:

$$|1-\lambda-\bar{d}_{ii}|^2 \leq \bar{d}_{ii}^2$$

for some i (see[28] for instance). The Laplacian property implies that:

$$\sum_{j=1, j\neq i}^{N} \bar{d}_{ij} = -\bar{d}_{ii}.$$

If $\operatorname{Re}(1-\lambda) = x$ and $\operatorname{Im}(1-\lambda) = y$, then the Gershgorin circle yields:

$$(x - \bar{d}_{ii})^2 + y^2 \leq \bar{d}_{ii}^2$$

or

$$x^2 + y^2 \leq 2\bar{d}_{ii} x.$$

This yields:

$$|1-\lambda|^2 \leq 2\bar{d}_{ii} \operatorname{Re}(1-\lambda)$$

for some i. Since $\bar{d}_{ii} = 1 - d_{ii} \leq 1$, the result is proved. ■

Proof From Theorem 3.40, it is clear that we only need to prove that:

$$\eta(k+1) = (A - \delta(1-\lambda)BG_1KG_2C)\eta(k) \tag{3.111}$$

is asymptotically stable for any λ that satisfies $|\lambda| < 1$. Let R and $P > 0$ be such that (1.16) is satisfied.

By choosing a Lyapunov function $V(k) = \eta' P \eta$, we obtain:

$$\begin{aligned} V(k+1) - V(k) &\\ &= \eta'(k+1)P\eta(k+1) - \eta'(k)P\eta(k) \\ &= \eta'(k)\,[A - \delta(1-\lambda)BG_1KG_2C]'\,P\,[A - \delta(1-\lambda)BG_1KG_2C]\,\eta(k) \\ &\qquad\qquad - \eta'(k)P\eta(k) \end{aligned}$$

which yields:

$$\begin{aligned} V(k+1) - V(k) &\\ &= \eta'(k)(A'PA - P)\eta(k) - \delta(1-\lambda)\eta'(k)(A'PBG_1 - C'G_2')KG_2C\eta(k) \\ &\quad - \delta(1-\lambda)'\eta'(k)C'G_2'K(A'PBG_1 - C'G_2')'\eta(k) \\ &\quad + \delta^2(1-\lambda)'(1-\lambda)\eta'(k)C'G_2'KG_1'B'PBG_1KG_2C\eta(k) \\ &\quad - 2\delta\operatorname{Re}(1-\lambda)\eta'(k)C'G_2'G_2C\eta(k) \\ &= \begin{pmatrix} \eta(k) \\ -\delta(1-\lambda)KG_2C\eta(k) \end{pmatrix}' G(P) \begin{pmatrix} \eta(k) \\ -\delta(1-\lambda)KG_2C\eta(k) \end{pmatrix} \\ &\quad + \delta^2|1-\lambda|^2\eta'(k)C'G_2'K(R+R')KG_2C\eta(k) \\ &\quad - 2\delta\operatorname{Re}(1-\lambda)\eta'(k)C'G_2'KG_2C\eta(k) \end{aligned}$$

$$\leq \eta'(k)C'G_2'\Big[\delta^2|1-\lambda|^2K(R+R')K - 2\delta\,\mathrm{Re}(1-\lambda)K\Big]G_2C\eta(k),$$

$$\leq 2\,\mathrm{Re}(1-\lambda)\eta'(k)C'G_2'\Big[\delta^2K(R+R')K - \delta K\Big]G_2C\eta(k),$$

where $G(P)$ is given by (1.16). Choose a large enough $r > 0$ such that $K(R+R')K < rK$ and:

$$\delta^* = \frac{1}{r}.$$

Then, for all $\delta < \delta^*$, we have:

$$V(k+1) - V(k) < -\nu\eta'(k)C'G_2'G_2C\eta(k),$$

for some $\nu > 0$, which proves the required stability since (A, G_2C) is detectable. ■

3.5.6 Partial-State Coupling and Additional Communication

In continuous time, we have seen in Sect. 2.5.8 that the use of an additional communication channel between the protocols in which they share information about their state allows us to solve the synchronization problem without restrictions on poles and zeros of the agents.

In discrete time, the situation is a bit different. We have seen that for full-state coupling we already have to impose a restriction that the agents are at most weakly unstable. However, the sufficient condition of the solvability of the state synchronization problem for full-state coupling is given in (3.19), while for partial-state coupling is in (3.81). We will show that the use of an additional communication layer in the partial-state coupling will make use of the sufficient condition (3.19) in a dual version, rather than (3.81).

Similarly, we assume agent i for $i = 1, \ldots, N$ is presumed to have access to the quantity:

$$\hat{\zeta}_i(k) = \sum_{j=1}^{N} d_{ij}(\xi_i(k) - \xi_j(k)), \tag{3.112}$$

where $\xi_j \in \mathbb{R}^p$ is a variable produced internally by agent j as a part of the protocol.

Problem 3.44 (Partial-State Coupling with Extra Communication) Consider a MAS described by (3.1), (3.4), and (3.112). Let $\mathbf{G}$ be a given set of graphs such that $\mathbf{G} \subseteq \bar{\mathbb{G}}^N$. The *state synchronization* problem with extra communication and a set of network graphs $\mathbf{G}$ is to find, if possible, a linear time-invariant dynamic protocol of the form:

$$\begin{cases} \chi_i(k+1) = A_c\chi_i(k) + B_{c,1}\zeta_i(k) + B_{c,2}\hat{\zeta}_i(k), \\ u_i(k) \quad = C_c\chi_i(k) + D_{c,1}\zeta_i(k) + D_{c,2}\hat{\zeta}_i(k), \end{cases} \tag{3.113}$$

for $i = 1, \ldots, N$ where $\chi_i \in \mathbb{R}^{n_c}$, such that for $\xi_i = C\chi_i$ and for any graph $\mathcal{G} \in \mathbf{G}$ and for all initial conditions of the agents and their protocol, the state synchronization among agents is achieved.

Again we choose $\xi_i(k) = C\chi_i(k)$ in (3.112), the closed loop of agent (3.1), and protocol (3.113) with the network described by (3.4) and (3.112) can then be written as:

$$\bar{x}(k+1) = (I_N \otimes \bar{A} + (I - D) \otimes \bar{B}\bar{C})\bar{x}(k), \tag{3.114}$$

where:

$$\bar{x} = \begin{pmatrix} \bar{x}_1 \\ \vdots \\ \bar{x}_N \end{pmatrix}, \quad \bar{A} = \begin{pmatrix} A & BC_c \\ 0 & A_c \end{pmatrix}, \quad \bar{B} = \begin{pmatrix} BD_c \\ B_c \end{pmatrix}, \quad \text{and} \quad \bar{C} = \begin{pmatrix} -C & C \end{pmatrix},$$

and where $B_{c,1} = -B_{c,2} = B_c$ and $D_{c,1} = -D_{c,2} = D_c$ while:

$$\bar{x}_i = \begin{pmatrix} x_i \\ \chi_i \end{pmatrix}.$$

As for Problems 3.2 and 3.3, the synchronization for the system (3.114) is equivalent to the asymptotic stability of the following $N - 1$ subsystems:

$$\tilde{\eta}_i(k+1) = (\bar{A} + (1 - \lambda_i)\bar{B}\bar{C})\tilde{\eta}_i(k), \quad i = 2, \ldots, N, \tag{3.115}$$

where $\lambda_i, i = 2, \ldots, N$ are those eigenvalues of D inside the unit disc.

Theorem 3.45 *The MAS* (3.114) *achieves state synchronization if and only if the system* (3.115) *is asymptotically stable for* $i = 2, \ldots, N$.

Proof The proof is basically the same as the proof of Theorem 3.27. ■

Since Problem 3.44 is defined for a set of graphs $\mathbf{G}$, we need to guarantee that:

$$\tilde{\eta}(k+1) = (\bar{A} + (1 - \lambda)\bar{B}\bar{C})\tilde{\eta}(k), \tag{3.116}$$

is asymptotically stable for any λ that is an eigenvalue of matrix D inside the unit disc associated with a given graph set $\mathbf{G}$.

Similar to Corollary 3.28, we find that for Problem 3.44, it is basically necessary that the associated graph must have a directed spanning tree. Moreover, if we have a protocol (3.113) that achieves state synchronization, then the synchronized trajectory is given by (3.71) which reduces to (3.73) for stable protocols.

Consider a linear dynamic stable protocol in the form of:

$$\begin{cases} \chi_i(k+1) = (A+BF)\chi_i(k) + K(\zeta_i(k) - \hat{\zeta}_i(k)), \\ \quad u_i(k) \quad = F\chi_i(k), \end{cases} \tag{3.117}$$

where F is chosen such that $A + BF$ is Schur stable and K is to be designed.

Note that in order to check whether the protocol (3.117) achieves state synchronization, we need to verify (3.116) is asymptotically stable for any λ that is an eigenvalue of matrix D inside the unit disc associated with a given graph set $\mathbf{G}$. Given the protocol (3.117), we have:

$$\bar{A} + (1-\lambda)\bar{B}\bar{C} = \begin{pmatrix} A & BF \\ -(1-\lambda)KC & A+BF+(1-\lambda)KC \end{pmatrix},$$

the stability of which is equivalent to that of the matrix:

$$\begin{aligned} &\begin{pmatrix} 1 & -1 \\ 0 & 1 \end{pmatrix} \begin{pmatrix} A & BF \\ -(1-\lambda)KC & A+BF+(1-\lambda)KC \end{pmatrix} \begin{pmatrix} 1 & 1 \\ 0 & 1 \end{pmatrix} \\ &\qquad = \begin{pmatrix} A+(1-\lambda)KC & 0 \\ -(1-\lambda)KC & A+BF \end{pmatrix}. \end{aligned}$$

It shows that $A + BF$ is Schur stable, which requires that (A, B) is stabilizable. Moreover, we need to design K such that $A + (1-\lambda)KC$ is Schur stable for any λ that is an eigenvalue of matrix D inside the unit disc associated with a given graph set $\mathbf{G}$. Note that it is a dual version of state synchronization for full-state coupling, that is, designing F such that $A + (1-\lambda)BF$ is Schur stable for any λ that is an eigenvalue of matrix D inside the unit disc associated with a given graph set $\mathbf{G}$.

Following Theorems 3.9 and 3.11 in the full-state coupling, we immediately obtain the solvability conditions.

Theorem 3.46 *Consider a MAS described by (3.1), (3.4), and (3.112) with a scalar input. The problem of synchronization with partial-state coupling as defined in Problem 3.3 is solvable under a linear dynamic protocol in the form of (3.117) for $\mathbf{G} = \bar{\mathbb{G}}_\beta^N$ for a given $\beta \in (0, 1)$ if and only if:*

$$\frac{\beta^2}{1-\beta^2} C Q_0 C' < 1, \tag{3.118}$$

where $Q_0 \geq 0$ is the unique solution of H_2 discrete-time algebraic Riccati equation:

$$Q_0 = AQ_0A' - AQ_0C'\left(I + CQ_0C'\right)^{-1} CQ_0A', \tag{3.119}$$

such that:

$$A - A'Q_0C'\left(I + CQ_0C'\right)^{-1}C \tag{3.120}$$

has all its eigenvalues in the closed unit disc.

Theorem 3.47 *Consider a MAS described by* (3.1), (3.4), *and* (3.112).

(Necessity) The problem of synchronization with partial-state coupling as defined in Problem 3.2 *is solvable under a linear dynamic protocol in the form of* (3.117) *for* $\mathbf{G} = \bar{\mathbb{G}}_\beta^N$ *for a given* $\beta \in (0, 1)$ *only if:*

$$\beta < \frac{1}{|\lambda_{\max}(A)|}. \tag{3.121}$$

(Sufficiency) The above problem is solvable under a linear dynamic protocol in the form of (3.117) *for* $\mathbf{G} = \bar{\mathbb{G}}_\beta^N$ *for a given* $\beta \in (0, 1)$ *if the condition* (3.118) *is satisfied.*

The design of K can be obtained from a dual version of the nonstandard or standard ARE-based method or the direct eigenstructure assignment method for at most weakly unstable agents while a dual version of the standard ARE-based method for strictly unstable agents. In the following, we give a protocol design based on a standard ARE-based method for at most weakly unstable agents.

Protocol design 3.12 *Consider a MAS described by agents* (3.1) *with communication via* (3.4) *and* (3.112). *Let* $\xi_i = C\chi_i$ *in* (3.112), *and consider the protocol:*

$$\begin{cases} \chi_i(k+1) = (A + BF)\chi_i(k) + K_\delta(\zeta_i(k) - \hat{\xi}_i(k)), \\ \quad u_i(k) \quad = F\chi_i(k), \end{cases} \tag{3.122}$$

with F *such that* $A + BF$ *is Schur stable while*

$$K_\delta = -\tfrac{1}{1-\beta_+^2} A'Q_\delta C'(CQ_\delta C' + I)^{-1}$$

with $Q_\delta > 0$ *given by:*

$$Q_\delta = AQ_\delta A' + \delta I - AQ_\delta C'(CQ_\delta C' + I)^{-1}CQ_\delta A', \tag{3.123}$$

while $\delta > 0$ *is sufficiently small such that:*

$$\beta_+^2 CQ_\delta C' < (1 - \beta_+^2)I, \tag{3.124}$$

(continued)

Protocol design 3.12 (continued)
where $\beta_+ \in [\beta, 1)$ with β the upper bound for the amplitude of the eigenvalues unequal to 1 of the row stochastic matrix D associated with a graph in the set $\bar{\mathbb{G}}^N_\beta$, while Q_δ is the unique solution of the above discrete-time algebraic Riccati equation.

Theorem 3.48 *Consider a MAS described by agents* (3.1) *with communication via* (3.4) *and* (3.112)*. Let any $\beta > 0$ be given and hence a set of network graphs $\bar{\mathbb{G}}^N_\beta$ be defined.*

If A is at most weakly unstable, (A, B) is stabilizable, and (C, A) is detectable, then for any $\beta_+ \in [\beta, 1)$, there exists a unique Q_δ satisfying (3.123) *and* (3.124) *for a sufficiently small δ, and the design 3.12 solves the state synchronization problem for any graph $\mathcal{G} \in \bar{\mathbb{G}}^N_\beta$. Moreover, the synchronized trajectory is given by* (3.73)*.*

Proof Follow the proof of Theorem 3.18. ■

3.5.7 *Partial-State Coupling for Introspective Agents*

Extra communication does not help us to remove the restriction that the agents must be at most weakly unstable. However, introspective agents, where each agent has a measurement available of its own state, can remove this requirement. Specifically, we assume that agent i ($i = 1, \dots, N$) has access to the quantity:

$$z_{m_i} = C_m x_i. \tag{3.125}$$

This results in the following problem.

Problem 3.49 (Partial-State Coupling with Introspective Agents) Consider a MAS described by (3.1), (3.4), and (3.125). Let $\mathbf{G}$ be a given set of graphs such that $\mathbf{G} \subseteq \bar{\mathbb{G}}^N$. The *state synchronization* problem with introspective agents and a set of network graphs $\mathbf{G}$ is to find, if possible, a linear time-invariant dynamic protocol of the form:

$$\begin{cases} \bar{\chi}_i(k+1) = \bar{A}_c \bar{\chi}_i(k) + \bar{B}_{c,1}\zeta_i(k) + \bar{B}_{c,2} z_{m,i}(k)\}, \\ \quad u_i(k) \quad = \bar{C}_c \bar{\chi}_i(k) + \bar{D}_{c,1}\zeta_i(k) + \bar{D}_{c,2} z_{m,i}(k), \end{cases} \tag{3.126}$$

for $i = 1, \dots, N$ where $\bar{\chi}_i \in \mathbb{R}^{n_c}$, such that for any graph $\mathcal{G} \in \mathbf{G}$ and for all the initial conditions for the agents and their protocol, state synchronization among agents is achieved.

We can use a specific structure for the protocol (3.126) where:

$$\begin{cases} \chi_i(k+1) &= A_c\chi_i(k) + B_c\zeta_i(k), \\ \chi_{p,i}(k+1) &= A_p\chi_i(k) + B_p z_{m,i}(k), \\ u_i(k) &= C_c\chi_i(k) + C_p\chi_i(k) + D_c\zeta_i(k), \end{cases} \tag{3.127}$$

for $i = 1, \ldots, N$ where (A_p, B_p, C_p) describe a pre-feedback and (A_c, B_c, C_c, D_c) describe a protocol with a standard structure based on the network measurement (3.2). Obviously, this pre-feedback can stabilize each agent, and then, even without using network information, we achieve state synchronization since all agent states converge to zero. However, the pre-feedback can be designed with different objectives in mind:

For instance, in the work of [16], they considered agents that are introspective which enabled them to use a pre-feedback to create passive agents.

Another design is to use the pre-feedback to make sure that the system becomes at most weakly unstable which enables us to design a protocol based on the techniques from Sect. 3.5.

Finally, we can use the pre-feedback to shape the synchronized trajectory. If the pre-feedback is such that all poles of the system are stable, then the synchronized trajectory will always be equal to zero. But by moving poles of the system to specific locations on the unit circle, we can then shape the synchronized trajectory by imposing certain dynamics.

Chapter 4
Synchronization of Linear MAS Subject to Actuator Saturation

4.1 Introduction

This chapter considers the synchronization problems for homogeneous MAS when the input for each agent is subject to saturation.

One can draw the following conclusions from the existing literature for a linear system subject to actuator saturation (see, for example, [106]):

1. Synchronization in the presence of actuator saturation requires eigenvalues of agents to be in the closed left half plane for continuous-time systems and in the closed unit disc for discrete-time systems, that is, the agents are at most weakly unstable.
2. Synchronization in the global sense (i.e., when initial conditions of agents are anywhere) in general requires a nonlinear protocol.
3. A linear protocol can be used if we consider synchronization in the semi-global framework (i.e., initial conditions of agents are in a priori given compact set).

Global synchronization for full-state coupling has been studied by Meng et al. [68] (continuous) and Yang et al. [173] (discrete) for neutrally stable agents. The global framework has only been studied for static protocols under the assumption that the agents are neutrally stable and the network is detailed balanced. Partial-state coupling has been studied in [19] using an adaptive approach, but the observer requires extra communication and is introspective.

Semi-global synchronization has been studied in [136] and [135] in the case of full-state coupling. For partial-state coupling, we have [133, 137] and [156], but they all require extra communication and/or introspective. The paper [184] is non-introspective but still requires extra communication. The only result which is non-introspective and does not require extra communication is [138], but this paper requires solution of a nonconvex optimization problem to find a dynamic protocol. Moreover, an underlying assumption basically requires the agents to be passifiable

A. Saberi et al., *Synchronization of Multi-Agent Systems in the Presence of Disturbances and Delays*, Systems & Control: Foundations & Applications,
https://doi.org/10.1007/978-3-030-88148-1_4

via static input feedforward. For the latter case, this chapter will show that we can actually find static controllers.

In all of the above papers, the network is either undirected or is so-called detailed balanced (a slightly weaker condition than undirected). The only papers dealing with networks that are not detailed balanced are based on [50] which intrinsically requires the agents to be single integrators. This chapter shows that for semi-global stabilization, we can work with more general networks which contain a directed spanning tree but can be directed and need not be detailed balanced.

When the communication network provides partial-state coupling, the linear protocol is in general dynamic. However, for some classes of agents, (i.e., G-passive or G-passifiable via static input feedforward in continuous-time or G-passifiable via static input feedforward in discrete-time MAS), a static linear protocol can be utilized.

The write-up of this chapter is partially based on [62] and [186].

4.2 Semi-global Synchronization for Continuous-Time MAS

Consider a multi-agent system (MAS) composed of N identical linear time-invariant continuous-time agents subject to actuator saturation:

$$\begin{aligned} \dot{x}_i &= Ax_i + B\sigma(u_i), \quad x_i(0) = x_{i0} \\ y_i &= Cx_i, \end{aligned} \tag{4.1}$$

for $i = 1, \ldots, N$, where $x_i \in \mathbb{R}^n$, $u_i \in \mathbb{R}^m$, and $y_i \in \mathbb{R}^p$ are, respectively, the state, input, and output vectors of agent i, and:

$$\sigma(u_i) = \begin{pmatrix} \text{sat}(u_{i,1}) \\ \vdots \\ \text{sat}(u_{i,m}) \end{pmatrix} \quad \text{with} \quad u_i = \begin{pmatrix} u_{i,1} \\ \vdots \\ u_{i,m} \end{pmatrix} \tag{4.2}$$

where $\text{sat}(u)$ is the standard saturation function:

$$\text{sat}(u) = \text{sgn}(u) \min\{1, |u|\}.$$

We make the following standard assumption for the agent dynamics.

Assumption 4.1 *For each agent $i \in \{1, \ldots, N\}$, we assume that:*

- *(A, B) is stabilizable, and (A, C) is detectable.*
- *A is at most weakly unstable.*

The communication network among agents is exactly the same as that in Chap. 2 and provides each agent with the quantity:

$$\zeta_i = \sum_{j=1}^{N} a_{ij}(y_i - y_j) = \sum_{j=1}^{N} \ell_{ij} y_j \tag{4.3}$$

for $i = 1, \ldots, N$. The graph $\mathcal{G}$ describing the communication topology of the network is always assumed to have a directed spanning tree. When we consider full-state coupling, i.e., $C = I$, the quantity ζ_i becomes:

$$\zeta_i = \sum_{j=1}^{N} a_{ij}(x_i - x_j) = \sum_{j=1}^{N} \ell_{ij} x_j \tag{4.4}$$

for $i = 1, \ldots, N$.

4.2.1 Problem Formulation

We formulate below two state synchronization problems, one for a network with full-state coupling and the other for partial-state coupling.

Problem 4.1 (Full-State Coupling) Consider a MAS described by (4.1) and (4.4). Let $\mathbf{G}$ be a given set of graphs such that $\mathbf{G} \subseteq \mathbb{G}^N$. The *semi-global state synchronization* problem with a set of network graphs $\mathbf{G}$ is to find, if possible, a parameterized family of linear protocols of the form:

$$u_i = F_\delta \zeta_i, \quad (i = 1, \ldots, N) \tag{4.5}$$

where for any, arbitrarily large, compact set $\mathcal{X}$ there exists a δ^* such that for all $\delta < \delta^*$, state synchronization among agents is achieved for any graph $\mathcal{G} \in \mathbf{G}$ as long as $x_{i0} \in \mathcal{X}$ for $i = 1, \ldots, N$.

Problem 4.2 (Partial-State Coupling) Consider a MAS described by (4.1) and (4.3). Let $\mathbf{G}$ be a given set of graphs such that $\mathbf{G} \subseteq \mathbb{G}^N$. The *semi-global state synchronization* problem with a set of network graphs $\mathbf{G}$ is to find, if possible, an integer q and a parameterized family of linear dynamic protocols of the form:

$$\begin{cases} \dot{\chi}_i = A_{c,\delta}\chi_i + B_{c,\delta}\zeta_i, & \chi_i(0) = \chi_{i0} \\ u_i = C_{c,\delta}\chi_i + D_{c,\delta}\zeta_i, \end{cases} \tag{4.6}$$

for $i = 1, \ldots, N$ with $\chi_i \in \mathbb{R}^q$, where for any, arbitrarily large, compact set $\mathcal{V} \subset \mathbb{R}^{n+q}$ there exists a δ^* such that for all $\delta < \delta^*$, state synchronization among agents is achieved for any graph $\mathcal{G} \in \mathbf{G}$ as long as $(x_{i0}, \chi_{i0}) \in \mathcal{V}$ $(i = 1, \ldots, N)$.

4.2.2 Protocol Design for MAS with Full-State Coupling

We present here two protocol designs for a MAS with full-state coupling. The first design is based on an algebraic Riccati equation, and the other design is based on a direct eigenstructure assignment method.

4.2.2.1 ARE-Based Method

In this section, we design a protocol for MAS with input saturation. Our design is based on the low-gain method given in Sect. 2.5.3.1, where agents are at most weakly unstable.

Protocol design 4.1 *Consider a MAS described by* (4.1) *and* (4.4). *We consider the following parameterized family of protocols:*

$$u_i = F_\delta \zeta_i, \tag{4.7}$$

where $F_\delta = -B'P_\delta$ with $P_\delta > 0$ being the unique solution of the continuous-time algebraic Riccati equation (2.58).

Remark 4.3 The protocol (4.7) is equivalent to the following parameterized family of protocols:

$$u_i = \frac{1}{\beta}\bar{F}_\delta \zeta_i, \tag{4.8}$$

where $\bar{F} = -B'\bar{P}_\delta$ with $\bar{P}_\delta > 0$ being the unique solution of the continuous-time algebraic Riccati equation (2.61).

The main result in this subsection is stated as follows.

Theorem 4.4 *Consider a MAS described by* (4.1) *and* (4.4) *with input saturation. Let any $\beta > 0$ be given and hence a set of network graphs $\mathbb{G}_\beta^N$ be defined.*

If (A, B) is stabilizable and the agents are at most weakly unstable, then the semi-global state synchronization problem stated in Problem 4.1 with $\mathbf{G} = \mathbb{G}_\beta^N$ is solvable. In particular, for any given compact set $\mathcal{X} \subset \mathbb{R}^n$, there exists a δ^*, such that for any $\delta \in (0, \delta^*]$, the protocol (4.7) *achieves state synchronization for any graph $\mathcal{G} \in \mathbb{G}_\beta^N$ and with any initial condition $x_i(0) \in \mathcal{X}$ for $i = 1, \ldots, N$.*

In order to present the proof of the above result, we first need a technical lemma.

Lemma 4.5 *Assume that (A, B) is controllable and $\beta > 0$ is given. Let $F_\delta = -B'P_\delta$ with $P_\delta > 0$ being the unique solution of the continuous-time algebraic*

Riccati equation (2.58). *In that case:*

$$\| F_\delta (sI - A - \lambda B F_\delta)^{-1} B \|_\infty \leq \frac{2}{\operatorname{Re} \lambda} \tag{4.9}$$

for all $\delta > 0$ *and for all* λ *with* $\operatorname{Re} \lambda > \beta$. *Moreover, if* λ *is real, we have:*

$$\| I + \lambda F_\delta (sI - A - \lambda B F_\delta)^{-1} B \|_\infty \leq 1. \tag{4.10}$$

On the other hand, we have:

$$\| F_\delta (sI - A - \lambda B F_\delta)^{-1} \|_2^2 \leq \frac{1}{\operatorname{Re} \lambda} \operatorname{trace} P_\delta. \tag{4.11}$$

Proof Using the Riccati equation, we have:

$$(sI - A - \lambda B F_\delta)^* P_\delta + P_\delta (sI - A - \lambda B F_\delta) - \alpha P B B' P - \delta I \leq 0,$$

where $\alpha = \operatorname{Re} \lambda$. Postmultiplying with $(sI - A - \lambda B F_\delta)^{-1} B$ and premultiplying with its complex conjugate, we get:

$$\begin{aligned} G(s) &+ G(s)^* + \alpha G(s)^* G(s) \\ &= -\delta \left[(sI - A - \lambda B F_\delta)^{-1} B \right]^* \left[(sI - A - \lambda B F_\delta)^{-1} B \right] \leq 0, \end{aligned}$$

where:

$$G(s) = F_\delta (sI - A - \lambda B F_\delta)^{-1} B.$$

This yields:

$$\left[G(s) + \tfrac{1}{\alpha} I \right]^* \left[G(s) + \tfrac{1}{\alpha} I \right] \leq \tfrac{1}{\alpha^2} I,$$

which leads to:

$$\| \alpha G + I \|_\infty \leq 1,$$

which yields (4.10) if λ is real. On the other hand, we get:

$$\| \alpha G \|_\infty \leq 2$$

which yields (4.9) even if λ is complex. Finally, we have:

$$(A + \lambda B F_\delta)^* P_\delta + P_\delta (A + \lambda B F_\delta) + \alpha F_\delta' F_\delta \leq 0$$

which yields (4.11). ∎

Proof of Theorem 4.4 Consider the multi-agent system without saturation with the protocol (4.7). Define a unitary matrix U such that $U^{-1}LU = S$ where S is the Schur form of L with $S_{11} = 0$. The first column of U is an eigenvector of L associated with eigenvalue 0 with length 1, i.e., it is equal to $\mathbf{1}/\sqrt{N}$. We have $\|S\| < \alpha$.

Next, we define new coordinates:

$$\eta := (U^{-1} \otimes I_n)x = \begin{pmatrix} \eta_1 \\ \vdots \\ \eta_N \end{pmatrix},$$

then the dynamics of η can be written as:

$$\dot{\eta} = (I_N \otimes A + S \otimes BF)\eta.$$

Note that $x(0) \in \mathcal{X}$ for some compact set $\mathcal{X}$ guarantees that $\eta(0) \in \mathcal{X}_\eta$ for some compact set $\mathcal{X}_\eta$ because the transformation matrix U is unitary. We first analyze the behavior with respect to the new coordinates and then use this to conclude synchronization in terms of the original coordinates.

Given $U = [u_{ij}]$ with $u_{ij} = 0$ for $i < j$ and $u_{ii} = \lambda_i$, we find that:

$$\dot{\eta}_i = (A + \lambda_i BF_\delta)\eta_i - \sum_{j=i+1}^{N} BF_\delta \eta_j,$$

while for $i = N$ we get:

$$\dot{\eta}_N = (A + \lambda_N BF_\delta)\eta_N.$$

We know that $A + \lambda_i BF$ is asymptotically stable for $i = 2, \ldots, N$, according to Lemma 2.29. We find that $\eta_i(t) \to 0$ for $i = 2, \ldots, N$. This immediately implies that we achieve synchronization if the saturation does not get activated. The rest of the proof will establish that for all δ sufficiently small, the saturation indeed does not get activated.

Using Lemma 2.29, we note that there exists a $\zeta_{\delta,N}$ such that:

$$\|F_\delta \eta_N(t)\| \leq \zeta_{\delta,N} \tag{4.12}$$

with $\zeta_{\delta,N} \to 0$ as $\delta \downarrow 0$ where we used that $\eta_N(0)$ is in a compact set and hence bounded. Assume that we have shown:

$$\|F_\delta \eta_i(t)\| \leq \zeta_{\delta,i} \tag{4.13}$$

with $\zeta_{\delta,i} \to 0$ as $\delta \downarrow 0$ for $i = j+1, \dots, N$. Next, consider $i = j$. We have:

$$\dot{\eta}_j = (A + \lambda_j B F_\delta)\eta_j - \sum_{i=j+1}^{N} u_{ji} B F_\delta \eta_i.$$

Note that $|u_{ij}| < \alpha$ given that $\|L\| < \alpha$. We have:

$$F_\delta \eta_j(t) = F_\delta e^{(A+\lambda_j B F_\delta)t} \eta_j(0) - \sum_{i=j+1}^{N} \int_0^t u_{ji} F_\delta e^{(A+\lambda_j B F_\delta)(t-\tau)} B F_\delta \eta_i(\tau)\, d\tau. \tag{4.14}$$

We have (4.13) for $i > j$. Moreover, Lemma 2.29 implies that:

$$\|F_\delta e^{(A+\lambda_j B F_\delta)t}\| \le \zeta_\delta,$$

while Lemma 4.5 implies that:

$$\|F_\delta (sI - A - \lambda_j B F_\delta)^{-1} B\|_\infty < \frac{2}{\beta}. \tag{4.15}$$

But then by Doyle and Chu [22, page 12], the L_∞-induced operator norm of this system is bounded by $2n/\beta$. We get:

$$\|F_\delta \eta_j(t)\| \le \zeta_\delta \|\eta_j(0)\| + \tfrac{2n\alpha}{\beta} \sum_{i=j+1}^{N} u_{ji} \zeta_{\delta,i}.$$

Using that $\eta_i(0)$ is contained in compact sets for $i = j, \dots, N$ and hence bounded, we note that there exists a $\zeta_{\delta,j}$ such that:

$$\|F_\delta \eta_j(t)\| \le \zeta_{\delta,j}$$

with $\zeta_{\delta,j} \to 0$ as $\delta \downarrow 0$. In this way, we can recursively establish the existence of $\zeta_{\delta,i}$ such that:

$$\|F_\delta \eta_i(t)\| \le \zeta_{\delta,i} \tag{4.16}$$

for $i = 2, \dots N$. We have:

$$(L \otimes F_\delta) \begin{pmatrix} x_1 \\ x_2 \\ \vdots \\ x_N \end{pmatrix} = (US \otimes F_\delta) \begin{pmatrix} \eta_1 \\ \eta_2 \\ \vdots \\ \eta_N \end{pmatrix} = (US \otimes F_\delta) \begin{pmatrix} 0 \\ \eta_2 \\ \vdots \\ \eta_N \end{pmatrix}$$

where we use that $S_{11} = 0$. Since U^{-1} is unitary and hence bounded while (4.16) is satisfied for $i = 2, \dots N$, we find that there exists a δ^* such that for a $\delta < \delta^*$

$$\|(L \otimes F_\delta)u_i(t)\| < 1$$

for all $t \geq 0$ and all $i = 1, \dots N$, which implies that the saturation never gets activated. ■

We next consider a special case where the agent dynamics are neutrally stable, that is, the eigenvalues of A on the imaginary axis, if any, are semi-simple. This implies that there exists a positive definite matrix $P > 0$ such that:

$$A'P + PA \leq 0.$$

We design protocols for MAS with neutrally stable agents as follows.

Protocol design 4.2 *Consider a MAS described by* (4.1) *and* (4.3) *with input saturation. We consider a protocol:*

$$u_i = \delta F \zeta_i, \tag{4.17}$$

where $\rho > 0$ *and* $F = -B'P$ *with* $P > 0$ *being a solution of the Lyapunov function:*

$$A'P + PA \leq 0.$$

The main result based on the above design is as follows.

Theorem 4.6 *Consider a MAS described by* (4.1) *and* (4.3) *with input saturation. Let a set of network graphs* $\mathbb{G}^N_{\alpha,\beta}$ *be defined.*

Suppose that the agent dynamics are neutrally stable. If (A, B) *is stabilizable, then the state synchronization problem as formulated in Problem* 2.3 *with* $\mathbf{G} = \mathbb{G}^N_{\alpha,\beta}$ *is solvable for any number* N. *In particular, there exists a* δ^*, *such that for any* $\delta \in (0, \delta^*]$, *the protocol* (2.26) *solves the state synchronization problem for any graph* $\mathcal{G}$ *which is in* $\mathbb{G}^N$ *for some* N.

Remark 4.7 For the special case of neutrally stable agents with a graph which is detailed balanced, it has been shown in [68] that the above design 4.2 actually achieves global synchronization where δ need not be chosen small.

Proof We use the same arguments as in the proof of Theorem 4.4. The only thing we need is the bounds established in Lemmas 2.29 and 4.5 for the Riccati-based design.

Similar to the argument used in the proof of Lemma 4.5 but relying on the Lyapunov equation instead of the Riccati equation, we get:

$$\|G + \tfrac{1}{\beta} I\|_\infty \leq \tfrac{1}{\beta},$$

where:

$$G(s) = \delta F(sI - A - \lambda \delta BF)^{-1} B,$$

which leads to:

$$\|G\|_\infty \leq \tfrac{2}{\beta}.$$

Moreover, our design easily shows that $x(t)'Px(t)$ is decreasing over time. Hence, there trivially exists for any $\xi > 0$ a δ^*, such that for any $\delta \in (0, \delta^*]$, we have:

$$\|\delta B' P x(t)\| < \xi. \qquad \blacksquare$$

4.2.2.2 Direct Eigenstructure Assignment Method

In this section, we will design the protocol via direct eigenstructure assignment, which is already given in Sect. 2.5.3.2.

Protocol design 4.3 *Consider a MAS described by (4.1) and (4.4) with saturated inputs. We consider the protocol:*

$$u_i = \frac{2}{\beta} F_\delta \zeta_i, \tag{4.18}$$

where F_δ is given in (2.72). Note that it is the state feedback gain used in the Protocol Design (2.73).

The main result is stated in the following theorem.

Theorem 4.8 *Consider a MAS described by (4.1) and (4.4) with saturated inputs. Let any $\beta > 0$ be given and hence a set of network graphs $\mathbb{G}_\beta^N$ be defined.*

If (A, B) is stabilizable and A is at most weakly unstable, then the semi-global state synchronization problem stated in Problem 4.1 with $\mathbf{G} = \mathbb{G}_\beta^N$ is solvable for any graph $\mathcal{G} \in \mathbb{G}_\beta^N$. In particular, for any given bounded set $\mathcal{X} \subset \mathbb{R}^n$, there exists a δ^*, such that for any $\delta \in (0, \delta^*]$, the protocol (4.18) solves the state synchronization problem for any initial condition $x_i(0) \in \mathcal{X}$ $(i = 1, \ldots, N)$.

Proof We can basically use the same proof as for Theorem 4.4 except replacing the bound from Lemma 2.29 by the bound from Lemma 2.33 and replacing the inequality (4.15) in Lemma 4.5 by:

$$\|\tilde{F}_\delta(sI - A - \lambda B\tilde{F}_\delta)B\|_\infty \leq 4N\|B\| \tag{4.19}$$

which follows directly from (2.79) by noting that the L_1 norm is an upper bound for the H_∞ norm. ∎

4.2.3 *Protocol Design for MAS with Partial-State Coupling*

We present here two protocol designs for MAS with partial-state coupling. One is based on an algebraic Riccati equation (ARE) and the other on direct eigenstructure assignment.

4.2.3.1 ARE-Based Method

The low-gain based Protocol Design 2.4 for MAS with at most weakly unstable agents in Sect. 2.5.3.1 works as well for MAS with input saturation which is being considered in this section; this is because we can choose the low-gain δ sufficiently small so that the saturation nonlinearity is not activated. The main result in this subsection is stated as follows. We first repeat the Protocol Design 2.4 here.

Protocol design 4.4 *Consider a MAS described by at most weakly unstable agents with input saturation* (4.1) *and measurements* (4.3). *We will use a CSS observer-based protocol for the agents.*

We choose an observer gain K such that $A + KC$ is Hurwitz stable. Next, we consider a feedback gain $F = -B'P_\delta$ where $P_\delta > 0$ is the unique solution of the continuous-time algebraic Riccati equation (2.58). *This results in the protocol:*

$$\begin{cases} \dot{\chi}_i = (A + KC)\chi_i - K\zeta_i, \\ u_i = F_\delta \chi_i. \end{cases} \tag{4.20}$$

Theorem 4.9 *Consider a MAS described by* (4.1) *and* (4.3) *with input saturation. Let any $\alpha \geq \beta > 0$ be given and hence a set of network graphs $\mathbb{G}^N_{\alpha,\beta}$ be defined.*

If (A, B) is stabilizable, (C, A) is detectable, and A is at most weakly unstable, then the semi-global state synchronization problem stated in Problem 4.2 with $\mathbf{G} = \mathbb{G}^N_{\alpha,\beta}$ is solvable. In particular, there exists an integer n, such that for any given *compact set $\mathcal{V} \subset \mathbb{R}^{2n}$, there exists a δ^*, such that for any $\delta \in (0, \delta^*]$, the protocol (4.20) solves the semi-global state synchronization problem for any graph $\mathcal{G} \in \mathbb{G}^N_{\alpha,\beta}$ and for any initial conditions $(x_{i0}, \chi_{i0}) \in \mathcal{V}$ with $i = 1, \dots, N$.*

Remark 4.10 If the agents are neutrally stable, then we can replace the Riccati equation-based state feedback in the above design by the Lyapunov equation-based Protocol Design 4.2. The proof of the above theorem only requires minor modifications in that case.

In order to prove this theorem, we need a technical lemma which will be used to show that the saturation will not get activated.

Lemma 4.11 *Consider:*

$$A_{e,\delta} = \begin{pmatrix} A & BF_\delta \\ -\lambda KC & A + KC \end{pmatrix}, \qquad B_e = \begin{pmatrix} B \\ 0 \end{pmatrix} \qquad F_{e,\delta} = \begin{pmatrix} 0 & F_\delta \end{pmatrix}.$$

For any $\alpha > \beta > 0$, there exists a $\delta^* > 0$ such that we have:

1. *The closed-loop system matrix $A_{e,\delta}$ is Hurwitz stable for all $\delta \in (0, \delta^*]$ and for all λ with $\operatorname{Re}\lambda > \beta$ and $|\lambda| < \alpha$.*
2. *There exists a $\zeta_{e,\delta} > 0$ with $\zeta_{e,\delta} \to 0$ as $\delta \to 0$ such that:*

$$\|F_{e,\delta} e^{A_{e,\delta} t}\| \leq \zeta_{e,\delta} \tag{4.21}$$

for all $\delta \in (0, \delta^*]$, for all $t \geq 0$, and for all $\lambda \in \mathbb{C}$ with $\operatorname{Re}\lambda > \beta$ and $|\lambda| < \alpha$.

3. *For any $\mu > 0$, there exists a $\delta^* > 0$ such that:*

$$\|F_{e,\delta}(sI - A_{e,\delta})^{-1} B_e\|_\infty \leq \frac{2|\lambda|}{\operatorname{Re}\lambda} + \mu \tag{4.22}$$

for all $\delta \in (0, \delta^*]$ and for all $\lambda \in \mathbb{C}$ with $\operatorname{Re}\lambda > \beta$. Moreover, if λ is real:

$$\|I + F_{e,\delta}(sI - A_{e,\delta})^{-1} B_e\|_\infty \leq 1 + \mu \tag{4.23}$$

for all $\delta \in (0, \delta^*]$ and for all $\lambda \in \mathbb{C}$ with $\operatorname{Re}\lambda > \beta$.

Proof We have:

$$\tilde{A}_{e,\delta} = \begin{pmatrix} A + \lambda BF_\delta & \lambda BF_\delta \\ -\lambda BF_\delta & A + KC - \lambda BF_\delta \end{pmatrix} = \begin{pmatrix} I & 0 \\ -I & \lambda^{-1} I \end{pmatrix} A_{e,\delta} \begin{pmatrix} I & 0 \\ \lambda I & \lambda I \end{pmatrix},$$

$$\tilde{B}_e = \begin{pmatrix} I & 0 \\ -I & \lambda^{-1} I \end{pmatrix} B_e,$$

and:

$$\tilde{F}_{e,\delta} = \begin{pmatrix} \lambda F_\delta & \lambda F_\delta \end{pmatrix} = F_{e,\delta} \begin{pmatrix} I & 0 \\ \lambda I & \lambda I \end{pmatrix}.$$

Since λ is bounded, we can prove the properties of the lemma for $\tilde{A}_{e,\delta}$, $\tilde{B}_{e,\delta}$, and $\tilde{F}_{e,\delta}$, and the original result immediately follows.

Consider:

$$\begin{aligned} \dot{x}_1 &= (A + \lambda B F_\delta)x_1 + \lambda B F_\delta x_2 \\ \dot{x}_2 &= -\lambda B F_\delta x_1 + (A + KC - \lambda B F_\delta)x_2 \\ z &= \lambda F_\delta x_1 + \lambda F_\delta x_2. \end{aligned}$$

In the frequency domain, this yields:

$$F_\delta X_1(s) = \lambda G_{f,\delta}(s) F_\delta X_2(s) + F_\delta (sI - A - \lambda B F_\delta)^{-1} x_{1,0}$$

$$F_\delta X_1(s) = \lambda G_{c,\delta}(s) F_\delta X_1(s) + F_\delta (sI - A - KC + \lambda B F_\delta)^{-1} x_{2,0}$$

where $x_1(0) = x_{1,0}$ and $x_2(0) = x_{2,0}$ and:

$$G_{f,\delta}(s) = F_\delta (sI - A - \lambda B F_\delta)^{-1} B$$

$$G_{c,\delta}(s) = F_\delta (sI - A - KC + \lambda B F_\delta)^{-1} B.$$

We find that the Laplace transform of $z(t)$ equals:

$$Z(s) = \lambda \begin{pmatrix} I & I \end{pmatrix} \begin{pmatrix} I & -\lambda G_{f,\delta} \\ -\lambda G_{c,\delta} & I \end{pmatrix}^{-1} \begin{pmatrix} R_1(s) \\ R_2(s) \end{pmatrix}$$

where R_1 and R_2 are the Laplace transform of:

$$r_1(t) = F_\delta e^{(A + \lambda B F_\delta)t} x_{1,0}$$

$$r_2(t) = F_\delta e^{(A + KC - \lambda B F_\delta)t} x_{2,0}$$

respectively. From Lemma 2.29, we know that there exist $\zeta_{1,\delta}$ and $\zeta_{2,\delta}$ such that:

$$\|r_1(t)\| < \zeta_{1,\delta} \|x_1(0)\|, \qquad \|r_2(t)\| < \zeta_{2,\delta} \|x_2(0)\|$$

for all $t > 0$ and for all λ with $\operatorname{Re} \lambda > \beta$ and $|\lambda| < \alpha$. On the other hand:

$$\left\| \lambda \begin{pmatrix} I & I \end{pmatrix} \begin{pmatrix} I & -\lambda G_{f,\delta} \\ -\lambda G_{c,\delta} & I \end{pmatrix}^{-1} \right\|_\infty < \tilde{M}_e$$

for all λ with $\operatorname{Re}\lambda > \beta$ and $|\lambda| < \alpha$ because of Lemma 4.5, and $F_\delta \to 0$ and $\delta \to 0$ imply that:

$$\|G_{f,\delta}\|_\infty < M, \quad \text{and} \quad \|G_{c,\delta}\|_\infty \to 0 \tag{4.24}$$

for $\delta \to 0$. Since the H_∞ norm is bounded by $\tilde{M}_e$, then by Doyle and Chu [22, page 12], the L_∞-induced operator norm is bounded by $4n\tilde{M}_e$, and we obtain:

$$\|z(t)\| \leq 4n\tilde{M}_e \max\left\{ \zeta_{1,\delta}\|x_1(0)\|, \zeta_{2,\delta}\|x_2(0)\| \right\}$$

for all $t > 0$ and for all λ with $\operatorname{Re}\lambda > \beta$ and $|\lambda| < \alpha$. The first and second properties of our lemma then follow immediately. For the final property, we use:

$$\|\tilde{F}_{e,\delta}(sI - \tilde{A}_{e,\delta})^{-1}\tilde{B}_e - \lambda G_{c,f}\|_\infty \leq \mu$$

for all δ sufficiently small. The results then follow from the bounds in Lemma 4.5. ■

Proof of Theorem 4.9 As noted in Sect. 2.5.1, if we ignore the saturation elements, we can write the overall dynamics of the agents in the form (2.29) which, in our case, becomes:

$$\dot{\bar{x}} = \left[I_N \otimes \begin{pmatrix} A & BF_\delta \\ 0 & A + KC \end{pmatrix} + L \otimes \begin{pmatrix} 0 & 0 \\ -KC & 0 \end{pmatrix} \right] \bar{x}.$$

We can now use the same recursive argument as we did in the proof of Theorem 4.4 with the bounds we obtained in Lemma 4.11 to establish that the saturation elements never get activated. But then Theorem 2.27 establishes that our design achieves state synchronization. ■

4.2.3.2 Direct Eigenstructure Assignment Method

The direct eigenstructure assignment protocol design for MAS with at most weakly unstable agents as given in Sect. 2.5.3.2 works for MAS with input saturation as well, under the same idea that the low-gain δ is chosen sufficiently small so that the saturation nonlinearity is not activated. We rewrite the design here.

Protocol design 4.5 *Consider a MAS described by at most weakly unstable agents with input saturation (4.1) and measurements (4.3).*

As in Sect. 2.5.3.2, for each (A_i, B_i), let $F_{\delta_i,i} \in \mathbb{R}^{1\times n_i}$ be the unique state feedback gain such that the eigenvalues of $A_i + B_i F_{\delta_i,i}$ can be obtained from

(continued)

Protocol design 4.5 (continued)
the eigenvalues of A_i by moving any eigenvalue λ_i on the imaginary axis to $\lambda_i - 2\delta_i$, while all the eigenvalues in the open left half complex plane remain at the same location. Note that $F_{\delta_i,i}$ can be obtained explicitly in terms of δ_i, and define F_δ as in (2.72).

Next choose K such that $A + KC$ is Hurwitz stable. We then construct the dynamic protocol:

$$\begin{cases} \dot{\chi}_i = (A + KC)\chi_i - K\zeta_i, \\ u_i = \frac{2}{\beta} F_\delta \chi_i. \end{cases} \tag{4.25}$$

The main result in this subsection is stated as follows.

Theorem 4.12 *Consider a MAS described by (4.1) and (4.3) with input saturation. Let any $\alpha \geq \beta > 0$ be given and hence a set of network graphs $\mathbb{G}^N_{\alpha,\beta}$ be defined.*

If (A, B) is stabilizable, (C, A) is detectable, and A is at most weakly unstable, then the semi-global state synchronization problem stated in Problem 4.2 with $\mathbf{G} = \mathbb{G}^N_{\alpha,\beta}$ is solvable for any graph $\mathcal{G} \in \mathbb{G}^N_{\alpha,\beta}$. In particular, there exists an integer n, such that for any given compact set $\mathcal{V} \subset \mathbb{R}^{2n}$, there exists a δ^*, such that for any $\delta \in (0, \delta^*]$, the protocol (4.25) solves the semi-global state synchronization problem for any initial conditions $(x_{i0}, \chi_{i0}) \in \mathcal{V}$ with $i = 1, \ldots, N$.

Proof We use the same arguments as in the proof of Theorem 4.9. As noted in Sect. 2.5.1, if we ignore the saturation elements, we can write the overall dynamics of the agents in the form (2.29) which, in our case, becomes:

$$\dot{\bar{x}} = \left[I_N \otimes \begin{pmatrix} A & BF_\delta \\ 0 & A + KC \end{pmatrix} + L \otimes \begin{pmatrix} 0 & 0 \\ -KC & 0 \end{pmatrix} \right] \bar{x}.$$

We can now use the same recursive argument as we did in the proof of Theorem 4.4. The only thing we need is the bounds established in Lemma 4.11 for the Riccati-based design. However, these bounds can be similarly obtained for the state feedbacks designed via the direct method using the bounds in Lemma 2.33 and the bound (4.19) obtained in the proof of Theorem 4.8. This yields that for a δ small enough, we can guarantee that the saturation elements never get activated.

Since the system behavior remains completely linear, Theorem 2.32 then establishes that our design achieves state synchronization. ∎

4.2.4 Static Protocol Design

In this section, we consider static protocol design for agents which are either squared-down passive or squared-down passifiable via static input feedforward. For systems with input saturation, we intrinsically rely on a low-gain feedback design. Systems which are squared-down passifiable via static output feedback require a minimal size of the gain unless they are already squared-down passive. Even systems which are squared-down passifiable via static output feedback and are already neutrally stable do not work in conjunction with a low-gain static feedback design. Therefore, this class of systems will not be considered.

4.2.4.1 Squared-Down Passive Agents

In the input saturation case, we choose $G_1 = I$ and $G_2 = G$. The main reason not to consider a pre-compensator is that in case of saturation the pre-compensator would get affected by the saturation element.

In this case, the protocol design in Chap. 2 for squared-down passive agents also works for agents with input saturation when the linear system described by (A, B, C) is squared-down passive. We rewrite the protocol design as follows.

Protocol design 4.6 *Consider a MAS described by agents (4.1) with input saturation and communication via (4.3). Moreover, assume the linear system described by (A, B, C) is squared-down passive with respect to $G_1 = I$ and $G_2 = I$. The static protocol is designed as:*

$$u_i = -\delta G\zeta_i, \tag{4.26}$$

where $\delta > 0$ is the parameter to be designed.

The result based on the above design is stated in the following theorem.

Theorem 4.13 *Consider a MAS described by agents (4.1) and (4.3) with input saturation. Let a set of network graphs $\mathbb{G}^N$ with any number N be defined. Assume (4.1) is squared-down passive with respect to $G_1 = I$ and $G_2 = G$ while (A, B) is controllable and (A, GC) is observable with B and GC full-column and full-row rank, respectively.*

In that case, the semi-global state synchronization problem stated in Problem 4.2 is solvable with a static protocol for a set of graphs $\mathbf{G} = \mathbb{G}^N_{\alpha,\beta}$. In particular, for any compact set $\mathcal{X} \subset \mathbb{R}^{Nn}$, there exists a $\delta^* > 0$ such that for any $\delta \in (0, \ \delta^*]$, the static protocol (4.26) solves the semi-global state synchronization problem for any initial condition of the agents in the set $\mathcal{X}$ and for any graph $\mathcal{G} \in \mathbb{G}^N_{\alpha,\beta}$.

The proof will make use of the following lemma.

Lemma 4.14 *Assume that the continuous-time linear system with minimal realization* (A, B, C) *is squared-down passive with* $G_1 = I$ *and* $G_2 = G$. *In that case, we have:*

1. *The closed-loop system matrix* $A - \delta\lambda BGC$ *is Hurwitz stable for all* $\delta > 0$ *and for all* λ *with* $\operatorname{Re}(\lambda) > \beta$.
2. *For any* $\beta > 0$, *there exists a* $\xi > 0$ *such that for all* $\delta > 0$ *we have:*
$$\|GCe^{(A-\delta\lambda BGC)t}\| \leq \xi \tag{4.27}$$
for all $t \geq 0$ *and for all* $\lambda \in \mathbb{C}$ *with* $\operatorname{Re}(\lambda) > \beta$.
3. *We have:*
$$\|\delta GC(sI - A + \delta\lambda BGC)^{-1}B\|_\infty < \tfrac{1}{\beta} \tag{4.28}$$
for all $\delta > 0$ *and for all* λ *with* $\operatorname{Re}(\lambda) > \beta$.

Proof For Property 1, follow the proof of Theorem 2.44 in Chap. 2 by setting $\rho = 1$, $G_1 = I$, $G_2 = G$, and $K = \delta I$.

Next, for Property 2, we consider:

$$\dot{x} = (A - \delta\lambda BGC)x.$$

It is easily seen that:

$$\frac{d}{dt}x'(t)Px(t) \leq -2\delta\beta x'(t)C'G'GCx(t) \leq 0$$

using that $\operatorname{Re}(\lambda) \geq \beta$. This implies that:

$$\|GCe^{(A-\delta\lambda BGC)t}x_0\| \leq \xi\|x_0\|$$

which immediately implies (4.27) for:

$$\xi = \|GC\|\sqrt{\frac{\lambda_{\max}(P)}{\lambda_{\min}(P)}}, \tag{4.29}$$

where $\lambda_{\min}$ and $\lambda_{\max}$ denote the smallest and largest eigenvalues, respectively.

For Property 3, we know $A - \delta\lambda BGC$ is stable. Thus, we have:

$$(sI - A + \delta\lambda BGC)^*P + P(sI - A + \delta\lambda BGC) \geq 2\delta\beta C'G'GC,$$

for $\operatorname{Re}(s) = 0$ using (1.9) and $\operatorname{Re}(\lambda) \geq \beta$. Postmultiplying with:

$$(sI - A + \delta\lambda BGC)^{-1}B$$

and premultiplying with its complex conjugate, we get:

$$G(s) + G(s)^* + 2\beta G(s)^* G(s) \leq 0$$

where:

$$G(s) = -\delta GC(sI - A + \delta\lambda BGC)^{-1} B.$$

This yields:

$$\left[G(s) + \tfrac{1}{2\beta} I\right]^* \left[G(s) + \tfrac{1}{2\beta} I\right] \leq \tfrac{1}{4\beta^2} I,$$

which leads to:

$$\left\| G(s) + \tfrac{1}{2\beta} I \right\|_\infty \leq \tfrac{1}{2\beta}.$$

This immediately yields (4.28). ■

Proof of Theorem 4.13 We consider the MAS without saturation under protocol (4.26). Define a unitary matrix U such that $U^{-1}LU = S$ where S is the Schur form of L with $S(1, 1) = 0$ (see, for example, [29]). The first column of U is an eigenvector of L associated with eigenvalue 0 with length 1, i.e., it is equal to $\mathbf{1}/\sqrt{N}$. We have $\|S\| < \alpha$.

Next, we define new coordinates:

$$\eta := (U^{-1} \otimes I_n)x = \begin{pmatrix} \eta_1 \\ \vdots \\ \eta_N \end{pmatrix},$$

then the dynamics of η can be written as:

$$\dot{\eta} = (I_N \otimes A + \delta S \otimes BGC)\eta.$$

Note that $x(0) \in \mathcal{X}$ for some compact set $\mathcal{X}$ guarantees that $\eta(0) \in \mathcal{X}_\eta$ for some compact set $\mathcal{X}_\eta$ because the transformation matrix U is unitary. We first analyze the behavior with respect to the new coordinates and then use this to conclude synchronization in terms of the original coordinates.

Given $S = [s_{ij}]$ with $s_{ij} = 0$ for $i < j$ and $s_{ii} = \lambda_i$, we find that:

$$\dot{\eta}_i = (A - \delta\lambda_i BGC)\eta_i - \delta \sum_{j=i+1}^{N} s_{ij} BGC\eta_j$$

while for $i = N$ we get:

$$\dot{\eta}_N = (A - \delta\lambda_N BGC)\eta_N.$$

Using Lemma 4.14, we note that there exists a $\xi_{\delta,N}$ such that:

$$\|\delta GC\eta_N(t)\| \leq \xi_{\delta,N} \tag{4.30}$$

with $\xi_{\delta,N} \to 0$ as $\delta \to 0$ where we used that $\eta_N(0)$ is in a compact set and hence bounded. Assume that we have shown:

$$\|\delta GC\eta_N(t)\| \leq \xi_{\delta,i} \tag{4.31}$$

with $\xi_{\delta,i} \to 0$ as $\delta \to 0$ for $i = j+1, \ldots, N$. Next, consider $i = j$. We have:

$$\dot{\eta}_j = (A - \delta\lambda_j BGC)\eta_j - \delta \sum_{i=j+1}^{N} s_{ji} BGC\eta_i.$$

Note that $|s_{ij}| < \alpha$ given that $\|L\| < \alpha$. We have:

$$\begin{aligned} GC\eta_j(t) &= GCe^{(A-\delta\lambda_j BGC)t}\eta_j(0) \\ &\quad - \sum_{i=j+1}^{N} \int_0^t s_{ji}\delta GCe^{(A-\delta\lambda_i BGC)(t-\tau)} BGC\eta_i(\tau)\, d\tau. \end{aligned} \tag{4.32}$$

We have (4.31) for $i > j$. Moreover, Lemma 4.14 implies that:

$$\|\delta GCe^{(A-\delta\lambda_j BGC)t}\| \leq \xi_\delta,$$

while:

$$\|\delta GC(sI - A + \delta\lambda_j BGC)^{-1}B\|_\infty < \tfrac{1}{\beta}. \tag{4.33}$$

But then by Doyle and Chu [22, page 12], the L_∞-induced operator norm of this system is bounded by $2n/\beta$. We get:

$$\|\delta GC\eta_j(t)\| \leq \xi_\delta\|\eta_j(0)\| + \sum_{i=j+1}^{N} \tfrac{2n\alpha}{\beta}\|\eta_i(0)\|$$

using that $|s_{ij}| < \alpha$. Since $\eta_i(0)$ is contained in compact sets for $i = j, \ldots, N$ and hence bounded, we note that there exists a $\xi_{\delta,j}$ such that:

$$\|\delta GC\eta_j(t)\| \leq \xi_{\delta,j}$$

with $\xi_{\delta,j} \to 0$ as $\delta \to 0$. In this way, we can recursively establish the existence of $\xi_{\delta,j}$ such that:

$$\|\delta GC\eta_i(t)\| \leq \xi_{\delta,i} \tag{4.34}$$

for $i = 2, \ldots N$. We have:

$$(L \otimes \delta GC)\begin{pmatrix} x_1 \\ x_2 \\ \vdots \\ x_N \end{pmatrix} = (US \otimes \delta GC)\begin{pmatrix} \eta_1 \\ \eta_2 \\ \vdots \\ \eta_N \end{pmatrix} = (US \otimes \delta GC)\begin{pmatrix} 0 \\ \eta_2 \\ \vdots \\ \eta_N \end{pmatrix}$$

(recall that the first column of S equals zero). Therefore:

$$\left\| \begin{pmatrix} u_1(t) \\ u_2(t) \\ \vdots \\ u_N(t) \end{pmatrix} \right\|^2 \leq \delta^2\alpha^2 \sum_{i=2}^{N} \xi_{\delta,i}^2,$$

since U is unitary, $\|S\| < \alpha$, and the obtained bounds for $GC\eta_i(t)$. Hence, for:

$$\delta < \delta_* = \frac{1}{\alpha\sqrt{\sum_{i=2}^{N} \xi_{\delta,i}^2}}, \tag{4.35}$$

we have:

$$\|u_i(t)\| < 1$$

for all $t \geq 0$ and all $i = 1, \ldots N$ which implies that the saturation never gets activated.

Since the saturation never gets activated, the behavior of the multi-agent system with saturation and the protocol (4.26) is the same as the behavior of the multi-agent system without saturation and the same protocol.

Using (1.9) and Property 1 in Lemma 4.14, it is clear that the multi-agent system without saturation and the protocol (4.26) achieves synchronization, but then also the multi-agent system with saturation and the same protocol achieves semi-global synchronization which completes the proof. ■

4.2.4.2 Squared-Down Passifiable via Static Input Feedforward

Again, we choose $G_1 = I$ and $G_2 = G$. We will show that previous protocol design also works for agents which are squared-down passifiable via static input feedforward agents.

The result is stated as follows.

Theorem 4.15 *Consider a MAS described by agents (4.1) and (4.3) with input saturation. Let any $\alpha, \beta > 0$ be given and hence a set of network graphs $\mathbb{G}^N_{\alpha,\beta}$ be defined. Assume that the system described by (A, B, C) is squared-down passifiable via static input feedforward with respect to $G_1 = I$, $G_2 = G$, and R. Assume that (A, B) is controllable and (A, GC) is observable with B and GC full-column and full-row rank, respectively.*

The semi-global state synchronization problem stated in Problem 4.2 is solvable with a static protocol for a set of graphs $\mathbf{G} = \mathbb{G}^N_{\alpha,\beta}$. In particular, for any compact set $\mathcal{X} \subset \mathbb{R}^{Nn}$, there exists a $\delta^* > 0$ such that for any $\delta \in (0, \delta^*]$ the static protocol (4.26) solves the semi-global state synchronization problem for any initial condition of the agents in the set $\mathcal{X}$ and for any graph $\mathcal{G} \in \mathbb{G}^N_{\alpha,\beta}$.

Lemma 4.16 *In this case, the three properties of Lemma 4.14 are still satisfied; however, for property 3, we have to obtain a new bound:*

$$\|\delta GC(sI - A + \delta\lambda BGC)^{-1} B\|_\infty < \tfrac{2}{\beta} \tag{4.36}$$

and choose a δ sufficiently small.

Proof Since the continuous-time linear system with minimal realization (A, B, C) is G-passifiable via static input feedforward, we find there exists an R and $P > 0$ satisfying (1.12). We obtain that:

$$\begin{aligned} P(A-\delta\lambda BGC) &+ (A - \delta\lambda BGC)^* P \\ &= \begin{pmatrix} I \\ -\delta\lambda GC \end{pmatrix}^* G(P) \begin{pmatrix} I \\ -\delta\lambda GC \end{pmatrix} \\ &\quad + \delta^2 |\lambda|^2 C'G'(R + R')GC \\ &\quad - 2\delta \operatorname{Re}(\lambda) C'G'GC \\ &\leq -\delta\beta C'G'GC \end{aligned}$$

with $\delta > 0$, where $G(P)$ is given by (1.12) provided δ is such that:

$$\delta\alpha^2(R + R') < \beta I,$$

which is clearly satisfied for all sufficiently small δ. Since (A, GC) is detectable, this implies that $A - \delta\lambda BGC$ is Hurwitz stable for any λ with $\operatorname{Re}\lambda > \beta$. In other words, Property 1 is satisfied. Then, Property 2 also follows similarly as in the proof of Lemma 4.14.

For Property 3, we start with (1.12). We obtain:

$$\begin{pmatrix} I & -\delta\lambda^* C' G' \\ 0 & I \end{pmatrix} G(P) \begin{pmatrix} I & 0 \\ -\delta\lambda GC & I \end{pmatrix} \leq 0.$$

If we work this out, we get:

$$\begin{pmatrix} \bar{A}^* P + P\bar{A} + cC'G'GC & PB \\ B'P & -dI \end{pmatrix} + V_\delta \leq 0$$

where $\bar{A} = A - \delta\lambda BGC$ and V_δ is given by:

$$V_\delta = \begin{pmatrix} C'G' & 0 \\ 0 & I \end{pmatrix} \begin{pmatrix} w_{11} & w_{12} \\ w_{21} & w_{22} \end{pmatrix} \begin{pmatrix} GC & 0 \\ 0 & I \end{pmatrix}$$

with $c = \delta\beta$ and $d = 4\delta^{-1}\beta^{-1}$, while:

$$\begin{aligned} w_{11} &= 2\delta \operatorname{Re}(\lambda) I - \delta^2 |\lambda|^2 (R + R') - cI \\ w_{12} &= -I + \delta\lambda^* (R + R') \\ w_{21} &= -I + \delta\lambda (R + R') \\ w_{22} &= -R - R' + dI. \end{aligned}$$

If we can show that $V_\delta \geq 0$, then we obtain:

$$\begin{pmatrix} \bar{A}^* P + P\bar{A} + cC'G'GC & PB \\ B'P & -dI \end{pmatrix} \leq 0,$$

and from the continuous-time bounded-real lemma, we find that:

$$\| GC(sI - A + \delta\lambda BGC)^{-1} B \|_\infty < \sqrt{\tfrac{d}{c}} = \tfrac{2}{\delta\beta} \tag{4.37}$$

which immediately yields (4.36). In order to show that $V_\delta \geq 0$, we first note that:

$$\begin{aligned} w_{11} &\geq \delta\beta I - \delta^2 \alpha^2 (R + R') \geq \tfrac{1}{2}\delta\beta I > 0 \\ w_{22} &> -R - R' + \tfrac{4}{\delta\beta} I > \tfrac{3}{\delta\beta} I \end{aligned}$$

for all δ sufficiently small. Finally:

$$w_{11} - w_{12} w_{22}^{-1} w_{21} \geq \tfrac{1}{2}\delta\beta I - \tfrac{1}{3}\delta\beta w_{12} w_{21} \geq 0$$

for all δ sufficiently small. Combining these bounds, we note that $V_\delta \geq 0$ and therefore (4.37) is satisfied when:

$$\delta < \frac{1}{\|R + R'\|} \min\left\{ \frac{\beta}{2\alpha^2}, \frac{1}{\beta}, \frac{1}{5\alpha} \right\}. \tag{4.38}$$

■

Proof of Theorem 4.15 The proof is similar to that of Theorem 4.13. We only need to use Lemma 4.16 instead of Lemma 4.14 to show that the saturation never gets activated and hence semi-global synchronization of MAS will be achieved. ■

4.3 Semi-global Synchronization for Discrete-Time MAS

Consider a multi-agent system (MAS) composed of N identical linear time-invariant agents subject to actuator saturation:

$$\begin{aligned} x_i(k+1) &= Ax_i(k) + B\sigma(u_i(k)), \quad x_i(0) = x_{i0} \\ y_i(k) &= Cx_i(k), \end{aligned} \tag{4.39}$$

for $i = 1, \ldots, N$, where $x_i \in \mathbb{R}^n$, $u_i \in \mathbb{R}^m$, and $y_i \in \mathbb{R}^p$ are, respectively, the state, input, and output vectors of agent i and $\sigma(\cdot)$ as defined in (4.2). We make the following standard assumption for the agent dynamics.

Assumption 4.2 *For each agent $i \in \{1, \ldots, N\}$, we assume that:*

- *(A, B) is stabilizable, and (A, C) is detectable.*
- *A is at most weakly unstable.*

As mentioned before, if agents are weakly unstable, then the synchronized trajectory is in general unbounded and the saturation gets activated. A linear design is then no longer possible. An alternative Lyapunov function-based argument appears to only work for graphs which are detailed balanced.

The communication network among agents is exactly the same as that in Chap. 3, which satisfies Assumption 3.2 and provides each agent with the quantity:

$$\zeta_i(k) = \sum_{j=1}^{N} d_{ij}(y_i(k) - y_j(k)), \quad i = 1, \ldots, N. \tag{4.40}$$

When the communication network is with full-state coupling, i.e., C has full-column rank, without loss of generality $C = I$, the quantity ζ_i becomes:

$$\zeta_i(k) = \sum_{j=1}^{N} d_{ij}(x_i(k) - x_j(k)). \tag{4.41}$$

4.3.1 Problem Formulation

As usual, we formulate two state synchronization problems, one for a network with full-state coupling and the other for partial-state coupling.

Problem 4.17 (Full-State Coupling) Consider a MAS described by (4.39) and (4.41). Let $\mathbf{G}$ be a given set of graphs such that $\mathbf{G} \subseteq \bar{\mathbb{G}}^N$. The *semi-global state synchronization* problem with a set of network graphs $\mathbf{G}$ is to find, if possible, for any arbitrarily large compact set $\mathcal{X} \in \mathbb{R}^n$, a linear protocol of the form:

$$u_i(k) = F\zeta_i(k), \quad (i = 1, \ldots, N) \tag{4.42}$$

such that for any graph $\mathcal{G} \in \mathbf{G}$, the state synchronization among agents can be achieved as long as $x_{i0} \in \mathcal{X}$ for $i = 1, \ldots, N$.

Problem 4.18 (Partial-State Coupling) Consider a MAS described by (4.39) and (4.40). Let $\mathbf{G}$ be a given set of graphs such that $\mathbf{G} \subseteq \bar{\mathbb{G}}^N$. The *semi-global state synchronization* problem with a set of network graphs $\mathbf{G}$ is to find, if possible, an integer q such that for all arbitrarily large compact sets $\mathcal{V} \in \mathbb{R}^{n+q}$, there exists a linear dynamic protocol of the form:

$$\begin{cases} \chi_i(k+1) = A_c\chi_i(k) + B_c\zeta_i(k), & \chi_i(0) = \chi_{i0} \\ u_i(k) \quad\;\; = C_c\chi_i(k), \end{cases} \tag{4.43}$$

for $i = 1, \ldots, N$, where $\chi_i \in \mathbb{R}^q$, such that for any graph $\mathcal{G} \in \mathbf{G}$, the state synchronization among agents can be achieved as long as $(x_{i0}, \chi_{i0}) \in \mathcal{V}$ for $i = 1, \ldots, N$.

4.3.2 Protocol Design for MAS with Full-State Coupling

We again present two protocol designs for a MAS with full-state coupling. As before, one is based on an algebraic Riccati equation (ARE) and the other on direct eigenstructure assignment.

4.3.2.1 ARE-Based Method

In principle, Protocol Design 3.1 based on the nonstandard Riccati equation can be used to achieve state synchronization for agents subject to input saturation. However, in this section, we will focus on Protocol Design 3.2 based on the standard H_2 Riccati equation. The following design is almost identical to Protocol Design 3.2 except that in this case we require $\beta_+ > \beta$ and exclude the case where $\beta_+ = \beta$.

Protocol design 4.7 *Consider a MAS described by (4.39) and (4.41) with input saturation. We consider the protocol:*

$$u_i = F_\delta \zeta_i, \tag{4.44}$$

where:

$$F_\delta = -\frac{1}{1-\beta_+^2}(B'P_0B + I)^{-1}B'P_0A \tag{4.45}$$

with $P_0 > 0$ being the unique solution of the discrete-time algebraic Riccati equation:

$$P_0 = A'P_0A + \delta I - A'P_0B(B'P_0B + I)^{-1}B'P_0A, \tag{4.46}$$

while δ is sufficiently small such that:

$$\beta_+^2 B'P_0B < (1-\beta_+^2)I,$$

where $\beta < \beta_+ < 1$ with β the upper bound of the eigenvalues inside the unit disc for some row stochastic matrix D associated with a graph in a set of graphs $\bar{\mathbb{G}}_\beta^N$.

The main result based on the above design is stated as follows.

Theorem 4.19 *Consider a MAS described by (4.39) and (4.41) with input saturation. Let any $1 > \beta > 0$ be given, and hence a set of network graphs $\bar{\mathbb{G}}_\beta^N$ be defined.*

If A is at most weakly unstable and (A, B) is stabilizable, then for any $\beta_+ \in [\beta, 1)$, there exists a unique $P_\delta > 0$ satisfying (3.33) and (3.34) for all sufficiently small $\delta > 0$. Moreover, Protocol Design 4.7 solves the semi-global state synchronization problem stated in Problem 4.17 with $\mathbf{G} = \bar{\mathbb{G}}_\beta^N$. In particular, for any given compact set $\mathcal{X} \subset \mathbb{R}^n$ and $\beta_+ > \beta$, there exists a δ^*, such that for any $\delta \in (0, \delta^*]$, the protocol (4.45) achieves state synchronization for any graph $\mathcal{G} \in \bar{\mathbb{G}}_\beta^N$ and for any initial conditions $x_{i0} \in \mathcal{X}$ with $i = 1, \ldots, N$.

In order to present the proof of the above result, we first need a technical lemma which is the discrete-time version of Lemma 4.5.

Lemma 4.20 *Assume that (A, B) is controllable and $\beta > 0$ is given. Let F_δ be given by (4.45) with $P_\delta > 0$ being the unique solution of the discrete-time algebraic Riccati equation (4.46) with $\beta < \beta_+ < 1$. In that case, there exists a $M > 0$ such that:*

$$\|F_\delta(zI - A + (1-\lambda)BF_\delta)^{-1}B\|_\infty < M \tag{4.47}$$

for all $\delta > 0$ and for all λ with $|\lambda| < \beta$.

Proof As noted earlier in (3.93), we have:

$$G_{f,\delta} = F_\delta(zI - A - BF_\delta)^{-1}B$$

satisfying:

$$\|G_{f,\delta}\|_\infty < \beta_+^{-1}.$$

We have:

$$F_\delta(zI - A - (1-\lambda)BF_\delta)^{-1}B = G_{f,\delta}(z)\left[I + \lambda G_{f,\delta}(z)\right]^{-1},$$

which yields:

$$\|F_\delta(zI - A + (1-\lambda)BF_\delta)^{-1}B\|_\infty < \frac{1}{\beta_+ - \beta}.$$

■

Proof of Theorem 4.19 The proof follows the same arguments as the proof of Theorem 4.4. Using the bounds obtained in Lemmas 3.20 and 4.20, we can show that the saturation elements do not get activated for a δ small enough. Then the result obtained in Theorem 3.18 guarantees that state synchronization is achieved. ■

4.3.2.2 Direct Eigenstructure Assignment Method

We will show that Protocol Design 3.4 also works in this section.

Protocol design 4.8 *Consider a MAS described by (4.39) and (4.41) with saturated inputs. We consider the protocol:*

$$u_i = F_\delta \zeta_i, \tag{4.48}$$

with:

$$F_\delta = T_u \begin{pmatrix} F_{\delta_1,1} & 0 & \dots & 0 \\ 0 & F_{\delta_2,2} & \ddots & \vdots \\ \vdots & \ddots & \ddots & 0 \\ \vdots & & \ddots & F_{\delta_q,q} \\ 0 & \dots & \dots & 0 \end{pmatrix} T_x^{-1},$$

where T_u, T_x, and $F_{\delta_i,i}$ for $i = 1, \dots, q$ are as defined in Sect. 3.4.3.2.

The main result is stated in the following theorem.

Theorem 4.21 *Consider a MAS described by (4.39) and (4.41) with saturated inputs. Let any $1 > \beta > 0$ be given and hence a set of network graphs $\bar{\mathbb{G}}_\beta^N$ be defined.*

If (A, B) is stabilizable and A is at most weakly unstable, then the semi-global state synchronization problem stated in Problem 4.17 with $\mathbf{G} = \bar{\mathbb{G}}_\beta^N$ is solvable. In particular, for any given bounded set $\mathcal{X} \subset \mathbb{R}^n$, there exists a δ^*, such that for any $\delta \in (0, \delta^*]$, the protocol (4.48) solves the state synchronization problem for any initial conditions $x_i(0) \in \mathcal{X}$ with $i = 1, \ldots, N$.

Proof We first note that we have (3.62) and (3.64) where:

$$G_{f,\delta} = F_\delta(zI - A - BF_\delta)^{-1}B$$
$$\tilde{G}_{f,\delta}(z) = F_\delta(zI - A - BF_\delta)^{-1}.$$

This immediately shows that there exists an N such that:

$$\left\| F_\delta\,(zI - A + (1-\lambda)BF_\delta)^{-1} \right\|_\infty < N \tag{4.49}$$

for all λ satisfying $|\lambda| < \beta$ since:

$$F_\delta\,(zI - A - (1-\lambda)BF_\delta)^{-1} = \left[I + \lambda G_{f,\delta}(z)\right]^{-1} \tilde{G}_{f,\delta}(z).$$

The proof then follows the same arguments as the proof of Theorem 4.4. Using the above bound and the bound obtained in Lemma 3.25, we can show that the saturation elements do not get activated for a δ small enough. Then the result obtained in Theorem 3.24 guarantees that state synchronization is achieved. ∎

4.3.3 *Protocol Design for MAS with Partial-State Coupling*

Again, both ARE-based method and direct eigenstructure assignment method are presented for a MAS with partial-state coupling.

4.3.3.1 ARE-Based Method

Protocol design 4.9 *Consider a MAS described by (4.39) and (4.40) with input saturation.*

We consider the protocol:

$$\begin{cases} \chi_i(k+1) = (A+KC)\chi_i(k) - K\zeta_i(k), \\ u_i(k) \quad = F_\delta \chi_i(k), \end{cases} \tag{4.50}$$

(continued)

Protocol design 4.9 (continued)
where K is such that $A + KC$ is Schur stable and F_δ is designed as in (4.45), that is:

$$F_\delta = -(B'P_\delta B + I)^{-1}B'P_\delta A$$

with $P_\delta > 0$ is the unique solution of the discrete-time Riccati equation (4.46).

The main result in this subsection is stated as follows.

Theorem 4.22 *Consider a MAS described by (4.39) and (4.40) with input saturation. Let any $\beta > 0$ be given and hence a set of network graphs $\bar{\mathbb{G}}_\beta^N$ be defined.*

If (A, B) is stabilizable and (C, A) is detectable, then the semi-global state synchronization problem stated in Problem 4.18 with $\mathbf{G} = \bar{\mathbb{G}}_\beta^N$ is solvable via a stable protocol. In particular, there exists an integer q, such that for any given compact set $\mathcal{V} \subset \mathbb{R}^{n+q}$, there exists a δ^*, such that for any $\delta \in (0, \delta^*]$, the stable protocol (4.50) solves the semi-global state synchronization problem for any graph $\mathcal{G} \in \bar{\mathbb{G}}_\beta^N$ and for any initial conditions $(x_{i0}, \chi_{i0}) \in \mathcal{V}$ with $i = 1, \ldots, N$.

Lemma 4.23 *Consider:*

$$A_{e,\delta} = \begin{pmatrix} A & BF_\delta \\ (1-\lambda)KC & A+KC \end{pmatrix}, \qquad B_e = \begin{pmatrix} B \\ 0 \end{pmatrix}, \qquad F_{e,\delta} = \begin{pmatrix} 0 & F_\delta \end{pmatrix}.$$

For any $\beta > 0$, there exists a $\delta^* > 0$ such that we have:

1. *The closed-loop system matrix $A_{e,\delta}$ is Schur stable for all $\delta \in (0, \delta^*]$ and for all λ with $|\lambda| < \beta$.*
2. *There exists a $\zeta_{e,\delta} > 0$ with $\zeta_{e,\delta} \to 0$ as $\delta \to 0$ such that:*

$$\|F_{e,\delta}A_{e,\delta}^k\| \leq \zeta_{e,\delta} \tag{4.51}$$

for all $\delta \in (0, \delta^*]$, for all $k \geq 0$ and for all $\lambda \in \mathbb{C}$ with $|\lambda| < \beta$.

3. *There exists a $M_e > 0$ such that:*

$$\|F_{e,\delta}(zI - A_{e,\delta})^{-1}B_e\|_\infty \leq M_e, \tag{4.52}$$

for all $\delta \in (0, \delta^*]$ and for all $\lambda \in \mathbb{C}$ with $|\lambda| < \beta$.

Proof We can basically use the proof of Lemma 4.11 for continuous-time systems; instead of the bound from Lemma 2.29, we use the bound from Lemma 3.20. Moreover, the bound from Lemma 4.5 is replaced by the bound from Lemma 4.20. ∎

Proof of Theorem 4.22 As noted in Sect. 2.5.1, if we ignore the saturation elements, we can write the overall dynamics of the agents in the form (3.69) which, in our case, becomes:

$$\bar{x}(k+1) = \left[I_N \otimes \begin{pmatrix} A & BF_\delta \\ 0 & A+KC \end{pmatrix} + L \otimes \begin{pmatrix} 0 & 0 \\ -KC & 0 \end{pmatrix} \right] \bar{x}(k).$$

We can now use the same recursive argument as we did in the proof of Theorem 4.4 with the bounds we obtained in Lemma 4.23 to establish that the saturation elements never get activated. But then Theorem 3.36 establishes that our design achieves state synchronization. ■

4.3.3.2 Direct Eigenstructure Assignment Method

Protocol design 4.10 *Consider the following protocol:*

$$\begin{cases} \chi_i(k+1) = (A+KC)\chi_i(k) - K\zeta_i(k), \\ \quad u_i(k) \quad = F_\delta \chi_i(k), \end{cases} \tag{4.53}$$

where K is such that $A+KC$ is Schur stable and F_δ is given in (3.50).

The main result in this subsection is stated as follows.

Theorem 4.24 *Consider a MAS described by* (4.39) *and* (4.40) *with input saturation. Let any $\beta > 0$ be given and hence a set of network graphs $\bar{\mathbb{G}}_\beta^N$ be defined.*

Under Assumptions 4.2 *and* 3.2, *the semi-global state synchronization problem stated in Problem* 4.18 *with $\mathbf{G} = \bar{\mathbb{G}}_\beta^N$ is solvable via a stable protocol. In particular, there exists an integer q, such that for any given compact set $\mathcal{V} \subset \mathbb{R}^{n+q}$, there exists a δ^*, such that for any $\delta \in (0, \delta^*]$, the protocol* (4.53) *solves the semi-global state synchronization problem for any graph $\mathcal{G} \in \bar{\mathbb{G}}_\beta^N$ and for all initial conditions $(x_{i0}, \chi_{i0}) \in \mathcal{V}$ with $i = 1, \ldots, N$.*

Lemma 4.25 *Consider:*

$$A_{e,\delta} = \begin{pmatrix} A & BF_\delta \\ (1-\lambda)KC & A+KC \end{pmatrix}, \qquad B_e \begin{pmatrix} B \\ 0 \end{pmatrix}, \qquad F_{e,\delta} = \begin{pmatrix} 0 & F_\delta \end{pmatrix}.$$

For any $\beta > 0$, there exists a $\delta^* > 0$ such that we have:

1. *The closed-loop system matrix $A_{e,\delta}$ is Schur stable for all $\delta \in (0, \delta^*]$ and for all λ with $|\lambda| < \beta$.*

2. *There exists a* $\zeta_{e,\delta} > 0$ *with:* $\zeta_{e,\delta} \to 0$ *as* $\delta \to 0$ *such that:*

$$\|F_{e,\delta}A_{e,\delta}^{k}\| \leq \zeta_{e,\delta} \tag{4.54}$$

for all $\delta \in (0, \delta^*]$, *for all* $k \geq 0$, *and for all* $\lambda \in \mathbb{C}$ *with* $|\lambda| < \beta$.

3. *There exists a* $M_e > 0$ *such that:*

$$\|F_{e,\delta}(zI - A_{e,\delta})^{-1}B_e\|_\infty \leq M_e, \tag{4.55}$$

for all $\delta \in (0, \delta^*]$ *and for all* $\lambda \in \mathbb{C}$ *with* $|\lambda| < \beta$.

Proof We can basically use the proof of Lemma 4.11 for continuous-time systems. Instead of the bound from Lemma 2.29, we use the bound from Lemma 3.25. Moreover, the bound from Lemma 4.5 is replaced by the bound (4.49). ■

Proof of Theorem 4.24 As noted in Sect. 2.5.1, if we ignore the saturation elements, we can write the overall dynamics of the agents in the form (3.69) which, in our case, becomes:

$$\bar{x}(k+1) = \left[I_N \otimes \begin{pmatrix} A & BF_\delta \\ 0 & A+KC \end{pmatrix} + L \otimes \begin{pmatrix} 0 & 0 \\ -KC & 0 \end{pmatrix}\right] \bar{x})k).$$

We can now use the same recursive argument as we did in the proof of Theorem 4.4 with the bounds we obtained in Lemma 4.25 to establish that the saturation elements never get activated. But then Theorem 3.37 establishes that our design achieves state synchronization. ■

4.3.4 Static Protocol Design

In this section, we also consider the static protocol design for squared-down passifiable via static input feedforward agents. As noted before, in discrete time, strictly proper agents are never squared-down passive or passifiable via output feedback.

4.3.4.1 Squared-Down Passifiable via Static Input Feedforward

Similar to the continuous-time case, we choose $G_1 = I$ and $G_2 = G$. Then, the protocol design in Chap. 3 also works for with the input saturation as long as the linear system (A, B, C) is squared-down passifiable via static input feedforward. The protocol design is rewritten as follows.

Protocol design 4.11 *Consider a MAS described by agents (4.39) with input saturation and communication via (4.40). The static protocol is designed as:*

$$u_i(k) = -\delta G\zeta_i(k), \tag{4.56}$$

where $\delta > 0$ is the low-gain parameter to be designed.

The result based on the above design is stated as follows.

Theorem 4.26 *Consider a MAS described by agents (4.39) with input saturation and communication via (4.40). Assume that the linear system (A, B, C) is squared-down passifiable via static input feedforward with respect to $G_1 = I$, $G_2 = G$, and R, while (A, B) is controllable and (A, GC) is observable with B and GC full-column and full-row rank, respectively. Let any $\delta \in (0, 1)$ be given.*

In that case, the semi-global state synchronization problem stated in Problem 4.18 but with a static protocol is solvable for a set of graphs $\mathbf{G} = \bar{\mathbb{G}}_\beta^N$. In particular, for any compact set, $\mathcal{X} \subset \mathbb{R}^{Nn}$, there exists a δ^* such that for any $\delta < \delta^*$, protocol (4.56) solves the state synchronization problem for any initial condition of the agents in the set $\mathcal{X}$ and for any graph $\mathcal{G} \in \bar{\mathbb{G}}_\beta^N$.

We need the following lemma, which is the discrete-time version of Lemma 4.16.

Lemma 4.27 *Assume that the discrete-time linear system with minimal realization (A, B, C) is squared-down passifiable via static input feedforward with respect to $G_1 = I$, $G_2 = G$, and R, while (A, B) is controllable and (A, GC) is observable with B and GC full-column and full-row rank, respectively. We have the following properties:*

1. *There exists a $\delta^* > 0$ such that the closed-loop system matrix $A - \delta(1-\lambda)BGC$ is Schur stable for all $\delta \in (0, \delta^*]$ and for all λ with $|\lambda| < \beta$.*
2. *For any $\beta > 0$, there exists a $\delta^* > 0$ and $\xi > 0$ such that for all $\delta \in (0, \delta^*]$ we have:*

$$\|G_2C(A - \delta(1-\lambda)BGC)^k\| \leq \xi, \tag{4.57}$$

for all $k \geq 0$ and for all $\lambda \in \mathbb{C}$ with $|\lambda| < \beta$.

3. *We have:*

$$\|\delta G_2C(zI - A + \delta(1-\lambda)BGC)^{-1}B\|_\infty < \frac{2}{1-\beta} \tag{4.58}$$

for all $\delta > 0$ and for all λ with $|\lambda| < \beta$.

Proof Since the system is squared-down passifiable via static input feedforward with respect to $G_1 = I$, $G_2 = G$, and R, there exists a P such that $\bar{G}(P)$ as given by (1.16) is negative semidefinite. For the first property, we obtain:

$$\begin{aligned}
(A-\delta(1-\lambda)BGC)^*P(A-\delta(1-\lambda)BGC)-P&\\
=\begin{pmatrix} I \\ -\delta(1-\lambda)GC\end{pmatrix}^* \bar{G}(P)\begin{pmatrix} I \\ -\delta(1-\lambda)GC\end{pmatrix}&\\
-2\delta\,\mathrm{Re}(1-\lambda)C'G'GC&\\
+\delta^2|1-\lambda|^2C'G'(R+R')GC&\\
\leq -2\delta(1-\beta)C'G'GC&\\
+\delta^2(1+\beta)^2C'G'(R+R')GC&\\
\leq -\delta(1-\beta)C'G'GC&
\end{aligned} \tag{4.59}$$

provided δ is such that:

$$\delta(1+\beta)^2(R+R') \leq (1-\beta)I.$$

Since (A, GC) is observable, we find that (4.59) guarantees the required asymptotic stability.

Next, for Property 2, we consider:

$$x(k+1) = (A-\delta(1-\lambda)BGC)x(k).$$

Using (4.59), we find:

$$x'(k+1)Px(k+1) - x'(k)Px(k) \leq -\delta(1-\beta)x'(t)C'G'GCx(t) \leq 0$$

which yields Property 2 for ξ given by (4.29).

For Property 3, we start with (1.16). We obtain:

$$\begin{pmatrix} I & -\delta(1-\lambda)^*C'G' \\ 0 & I \end{pmatrix} \bar{G}(P) \begin{pmatrix} I & 0 \\ -\delta(1-\lambda)GC & I\end{pmatrix} \leq 0.$$

If we work this out, we get:

$$\begin{pmatrix} \bar{A}^*P\bar{A} - P + cC'G'GC & \bar{A}^*PB \\ B'P\bar{A} & B'PB - dI\end{pmatrix} + V_\delta \leq 0,$$

where $\bar{A} = A - \delta(1-\lambda)BGC$ and V_δ is given by:

$$V_\delta = \begin{pmatrix} C'G' & 0 \\ 0 & I\end{pmatrix}\begin{pmatrix} w_{11} & w_{12} \\ w_{21} & w_{22}\end{pmatrix}\begin{pmatrix} GC & 0 \\ 0 & I\end{pmatrix}$$

with $c = \delta(1-\beta)$ and $d = 4\delta^{-1}(1-\beta)^{-1}$ while:

$$\begin{aligned}
w_{11} &= \delta(1-\lambda)^* I + \delta(1-\lambda)I - \delta^2|1-\lambda|^2(R+R') - cI \\
w_{12} &= -I + \delta(1-\lambda)^*(R+R') \\
w_{21} &= -I + \delta(1-\lambda)(R+R') \\
w_{22} &= -R - R' + dI.
\end{aligned}$$

If we can show that $V_\delta \geq 0$, then we obtain:

$$\begin{pmatrix} \bar{A}^* P \bar{A} - P + cC'G'GC & \bar{A}^* PB \\ B'P\bar{A} & B'PB - dI, \end{pmatrix} \leq 0$$

and from the discrete-time bounded real lemma, we find that:

$$\|GC(zI - A + \delta(1-\lambda)BGC)^{-1}B\|_\infty < \sqrt{\tfrac{d}{c}} = \tfrac{2}{\delta(1-\beta)}$$

which immediately yields (4.58). In order to show that $V_\delta \geq 0$, we first note that:

$$w_{11} \geq \delta(1-\beta)I - \delta^2(1+\beta)^2(R+R') \geq \tfrac{1}{2}\delta(1-\beta)I > 0$$

for all δ sufficiently small and:

$$w_{22} > \tfrac{3}{(1-\beta)\delta} I$$

for a δ sufficiently small. Finally:

$$w_{11} - w_{12}w_{22}^{-1}w_{21} \geq \tfrac{1}{2}\delta(1-\beta)I - \tfrac{1}{3}\delta(1-\beta)w_{12}w_{21} \geq 0$$

for all δ sufficiently small. Combining these bounds, we note that $V_\delta \geq 0$ when:

$$\delta < \frac{1}{\|R+R'\|} \min\left\{ \frac{1-\beta}{2(1+\beta)^2}, \frac{1}{1-\beta}, \frac{1}{5(1+\beta)} \right\}. \tag{4.60}$$

■

Proof of Theorem 4.26 The proof is similar to that of Theorem 4.13. Using the bounds from Lemma 4.27, we show that the saturation never gets activated and then it is easily seen that the semi-global synchronization of MAS is achieved since the saturation never gets activated. ■

4.4 Global Synchronization for MAS with Partial-State Coupling via Static Protocol

In this section, we will consider global synchronization for MAS with partial-state coupling via static protocols. The network graphs are assumed to be detailed balanced. The agents in the continuous-time MAS can be G-passive or G-passifiable via static input feedforward. While in a discrete-time MAS, the agents are G-passifiable via static input feedforward.

4.4.1 Continuous-Time MAS

Before we give our result, we have to define the class of network graphs we are going to consider.

Definition 4.28 For any given real numbers $\beta > 0$ and a positive integer N, the set $\mathbb{G}^{b,N}$ contains all weighted, directed graphs composed of N nodes satisfying the following properties:

- The graph has a directed spanning tree.
- The graph is detailed balanced, i.e., there exists a diagonal matrix $R > 0$ such that the corresponding Laplacian matrix satisfies:

$$RL + L'R \geq 0.$$

The class of Laplacian matrices for which such an R exists have been studied in [88, Theorem 4.31].

We formulate the global synchronization problem for a continuous-time MAS as follows.

Problem 4.29 Consider a MAS described by (4.1) and (4.3). Let $\mathbf{G}$ be a given set of graphs such that $\mathbf{G} \subseteq \mathbb{G}^N$. The *global state synchronization* problem with a set of network graphs $\mathbf{G}$ is to find, if possible, a linear static protocol of the form:

$$u_i = F\zeta_i \tag{4.61}$$

for $i = 1, \ldots, N$, such that for any graph $\mathcal{G} \in \mathbf{G}$ and for initial conditions of the agents, state synchronization among agents is achieved.

We will show that the protocol design for the semi-global state synchronization problem also solves the global state synchronization problem. The result for MAS with squared-down passive agents is stated in the following theorem.

Theorem 4.30 *Consider a MAS described by agents* (4.1) *with input saturation and comminication via* (4.3). *Let a set of strongly connected and detailed balanced network graphs* $\mathbb{G}^{b,N}$ *be defined. Assume that the linear system* (A, B, C) *is* *squared-down passive with respect to* $G_1 = I$ *and* $G_2 = G$ *such that* (A, B) *is controllable and* (A, GC) *is observable while* B *and* GC *full-column and full-row rank, respectively*

In that case, the global state synchronization problem stated in Problem 4.29 is solvable for a set of graphs $\mathbf{G} = \mathbb{G}^{b,N}$*. In particular, for any* $\delta > 0$*, the static protocol (4.26) solves the global state synchronization problem for any graph* $\mathcal{G} \in \mathbb{G}^{b,N}$.

We will use the following lemma to prove the above theorem.

Lemma 4.31 ([151]) *For two vectors* $v, w \in \mathbb{R}^m$*, we have the following inequality:*

$$\|v - \sigma(v)\| \leq v'\sigma(v).$$

Proof of Theorem 4.30 The overall dynamics of the N agents can be written as:

$$\begin{cases} \dot{x} = (I_N \otimes A)x + (L \otimes B)\sigma(u), \\ y = (I_N \otimes C)x, \end{cases} \tag{4.62}$$

where $x = \text{col}\{x_1, \ldots, x_N\}$.

Since the graph is strongly connected and detailed balanced, there exists a diagonal matrix Z such that ZL is symmetric. Since our agents are G-passive, there exists a $P > 0$ satisfying (1.9). We choose the following Lyapunov function for the system (4.62):

$$V_2 = x'(ZL \otimes P)x \tag{4.63}$$

where Z is a positive definite, diagonal matrix such that:

$$ZL \geq 0$$

which is possible since the graph is detailed balanced. Calculating the time derivative along the trajectories of system (4.62), we have:

$$\dot{V} = x'(ZL \otimes (PA + A'P))x - 2x'(ZL \otimes PB)\sigma(\delta(L \otimes GC)x).$$

Using (1.9), we have:

$$\dot{V} \leq -2\,((L \otimes GC)x)'\,(Z \otimes I)\,\sigma(\delta(L \otimes GC)x) \leq 0$$

since $Z > 0$. Defining:

$$v = \delta(L \otimes GC)x,$$

we find that:

$$\dot{V} \leq -2\delta^{-1} v' (Z \otimes I)\, \sigma(v) \leq 0 \tag{4.64}$$

which yields:

$$\int_0^\infty v'\sigma(v) \mathrm{d}t < \infty \tag{4.65}$$

by integrating (4.64) and using that Z is positive definite and diagonal. Next, we have:

$$\dot{x} = (I_N \otimes A)x - \delta(L \otimes BGC)x + (I_N \otimes B)w, \tag{4.66}$$

where:

$$w = v - \sigma(v).$$

By Lemma 4.31 and (4.65), we have:

$$\int_0^\infty \|w\| \mathrm{d}t \leq \int_0^\infty v'\sigma(v) \mathrm{d}t < \infty.$$

Let H be such that:

$$HLH^{-1} = \begin{pmatrix} \bar{L} & 0 \\ 0 & 0 \end{pmatrix}$$

with $\bar{L} \in \mathbb{R}^{(N-1)\times(N-1)}$ nonsingular, and the eigenvalues of $\bar{L}$ will then be the nonzero eigenvalues of L. Decompose:

$$H = \begin{pmatrix} H_1 \\ h_2 \end{pmatrix}$$

with h_2 a row vector. We have:

$$\ker H_1 L = \ker L, \qquad H_1 L = \bar{L} H_1,$$

while $h_2 L = 0$. From (4.66), we obtain that:

$$(H_1 L \otimes I)\dot{x} = (I_N \otimes A)(H_1 L \otimes I)x - \delta(\bar{L} \otimes BGC)(H_1 L \otimes I)x - (H_1 L \otimes B)w.$$

Since $I \otimes A - \delta(\bar{L} \otimes BGC)$ is asymptotically stable and $w \in L_1$, we find that:

$$(H_1 L \otimes I)x \to 0 \qquad (H_1 L \otimes I)x \in L_1$$

and since $\ker H_1 L = \ker L$, we obtain:

$$(L \otimes I)x \to 0 \qquad (L \otimes I)x \in L_1 \tag{4.67}$$

which clearly guarantees that synchronization is achieved. ∎

The result for MAS with squared-down passifiable via static input feedforward agents is stated in the following theorem.

Theorem 4.32 *Consider a MAS described by agents (4.1) and (4.3) with input saturation. Let a set of strongly connected and detailed balanced network graphs $\mathbb{G}_\alpha^{b,N}$ be defined. Assume that the linear system described by (A, B, C) is squared-down passifiable via static input feedforward with respect to G_1, G_2 and R while (A, B) is controllable and (A, GC) is observable with B and GC full-column and full-row rank, respectively.*

In that case, the global state synchronization problem stated in Problem 4.29 is solvable for a set of graphs $\mathbf{G} = \mathbb{G}_\alpha^{b,N}$. In particular, there exists a δ^* such that for any $\delta \in (0, \delta^*]$, the static protocol (4.26) solves the global state synchronization problem for any graph $\mathcal{G} \in \mathbb{G}_\alpha^{b,N}$.

Proof of Theorem 4.32 The closed-loop system of (4.62) and (4.26) is written as:

$$\dot{x} =(I_N \otimes A)x - I_N \otimes B\sigma(\delta(L \otimes GC)x). \tag{4.68}$$

Since the linear system described by (A, B, C) is squared-down passifiable via static input feedforward with respect to $G_1 = I$, $G_2 = G$, and R, there exists a $P > 0$ satisfying (1.12). We choose the Lyapunov function (4.63) for subsystem (4.68). Calculating the time derivative along the trajectories of system (4.68), we have:

$$\begin{aligned}\dot{V} =& \begin{pmatrix} x \\ -\sigma(v) \end{pmatrix}' (ZL \otimes G(P)) \begin{pmatrix} x \\ -\sigma(v) \end{pmatrix} \\ &- 2\delta^{-1} v' (Z \otimes I)\, \sigma(v) \\ &+ \sigma(v)' \left(ZL \otimes (R + R')\right) \sigma(v)\end{aligned}$$

where $v = \delta(L \otimes GC)x$. By using $G(P) \leq 0$ and:

$$-v'\sigma(v) \leq -\sigma'(v)\sigma(v),$$

we have:

$$\begin{aligned}\dot{V} \leq& - \delta^{-1} v' (Z \otimes I)\, \sigma(v) - \delta^{-1} \sigma'(v) (Z \otimes I)\, \sigma(v) \\ &+ \sigma(v)' \left(ZL \otimes (R + R')\right) \sigma(v) \\ \leq& - \delta^{-1} v' (Z \otimes I)\, \sigma(v) \leq 0\end{aligned}$$

for all sufficiently small δ such that:

$$\delta^{-1} Z \otimes I \geq ZL \otimes (R + R') \tag{4.69}$$

which is possible since $\|L\| < \alpha$. Thus, by integrating the above inequality, we find that:

$$\int_0^\infty v'\sigma(v)\mathrm{d}t < \infty$$

since diagonal matrix Z is positive definite. The rest of the proof is identical to the proof of Theorem 4.30. ■

4.4.2 Discrete-Time MAS

Problem 4.33 Consider a MAS described by (4.39) and (4.40). Let $\mathbf{G}$ be a given set of graphs such that $\mathbf{G} \subseteq \bar{\mathbb{G}}^N$. The *global state synchronization* problem with a set of network graphs $\mathbf{G}$ is to find, if possible, a linear static protocol of the form:

$$u_i(k) = F\zeta_i(k) \tag{4.70}$$

for $i = 1, \ldots, N$, such that for any graph $\mathcal{G} \in \mathbf{G}$ and for all initial conditions of the agents, state synchronization among agents is achieved.

In the discrete-time case, the protocol design for the semi-global state synchronization problem also achieves the global state synchronization problem for a MAS where the linear system described by (A, B, C) is squared-down passifiable via static input feedforward when the graph set $\mathbf{G} = \bar{\mathbb{G}}^{b,N}$. The result is stated in the following theorem.

Theorem 4.34 *Consider a MAS described by agents* (4.39) *and* (4.40) *with input saturation. Let a set of strongly connected and detailed balanced network graphs* $\bar{\mathbb{G}}^{b,N}$ *be defined. Assume that the linear system described by* (A, B, C) *is squared-down passifiable via static input feedforward with respect to* $G_1 = I$, $G_2 = G$, *and* R *while* (A, B) *is controllable and* (A, GC) *is observable with* B *and* GC *full-column and full-row rank, respectively.*

In that case, the global state synchronization problem stated in Problem 4.33 is solvable for a set of graphs $\bar{\mathbb{G}}^{b,N}$. *In particular, there exists a* δ^* *such that for any* $\delta \in (0, \delta^*]$, *the static protocol* (4.56) *solves the global state synchronization problem for any graph* $\mathcal{G} \in \bar{\mathbb{G}}^{b,N}$.

Proof of Theorem 4.34 The overall dynamics of the N agents can be written as:

$$\begin{cases} x(k+1) = (I_N \otimes A)x(k) + ((I - D) \otimes B)\sigma(u(k)), \\ y(k) = (I_N \otimes C)x(k), \end{cases} \tag{4.71}$$

where $x = \text{col}\{x_1, \ldots, x_N\}$.

The closed-loop system of (4.71) and the protocol (4.56) is written as:

$$x(k+1) = (I_N \otimes A)x(k) - (I_N \otimes B)\sigma\left(\delta((I - D) \otimes GC)x(k)\right). \tag{4.72}$$

By employing the same idea as in the proof of Theorem 4.32, we can prove global state synchronization problem of discrete-time MAS with input saturation for a $\delta \in (0, \delta^*]$ with a δ^* such that:

$$\delta^{-1} Z \otimes I \geq Z(I - D) \otimes (R + R') \tag{4.73}$$

is satisfied. The detailed proof is omitted. ■

Chapter 5
Synchronization of Continuous-Time MAS with Nonlinear Time-Varying Agents

5.1 Introduction

This chapter considers synchronization problems for homogeneous continuous-time MAS with nonlinear time-varying agents. As in the chapter of continuous-time MAS with linear agents, two kinds of communication networks are considered, one pertaining to full-state coupling and the other to partial-state coupling. The solvability of both synchronization problems is restricted to a class of agent models, which can be transformed to a certain canonical form and satisfy some properties.

Some authors have also studied synchronization in networks with nonlinear agent dynamics (e.g., [3, 16, 85, 169, 191, 192]). Explicit protocol design for nonlinear networks has, to a large degree, centered on the relatively strict assumption of passivity. Passivity can in some cases be ensured by first applying local pre-feedbacks to the system; however, this requires the system to be introspective. Grip et al. [33] addresses the issue of state synchronization for homogeneous networks consisting of SISO, non-introspective agents. However, all the above nonlinear work consider networks with partial-state coupling and are not applicable to the case of full-state coupling.

The write-up of this chapter is partially based on [33].

5.2 MAS with Full-State Coupling

In this section, we first formulate the synchronization problems for a class of nonlinear time-varying agents that are already in a required canonical form and then develop protocol designs. Subsequent to that we present a method of transforming general nonlinear time-varying agents to the canonical form.

A. Saberi et al., *Synchronization of Multi-Agent Systems in the Presence of Disturbances and Delays*, Systems & Control: Foundations & Applications,
https://doi.org/10.1007/978-3-030-88148-1_5

5.2.1 Problem Formulation

Consider a multi-agent system (MAS) composed of N identical nonlinear time-varying continuous-time agents of the form

$$\dot{x}_i = A_d x_i + \phi(t, x_i) + B_d(u_i + E x_i), \tag{5.1}$$

where $x_i \in \mathbb{R}^n$ and $u_i \in \mathbb{R}^m$ are the state and the input of agent i, respectively. Here the system A_d, B_d, and ϕ are in the so-called strict feedback form (see, for instance, [41, Section 14.3]). In particular, we assume that all controllability indices of this system to be equal to ρ. In that case, we can choose A_d and B_d to have the special form

$$A_d = \begin{pmatrix} 0 & I_{(\rho-1)m} \\ 0 & 0 \end{pmatrix}, \quad B_d = \begin{pmatrix} 0 \\ I_m \end{pmatrix}. \tag{5.2}$$

Furthermore, x_i and ϕ have the following structure:

$$x_i = \begin{pmatrix} x_{i1} \\ \vdots \\ x_{i\rho} \end{pmatrix}, \quad x_{ij} = \begin{pmatrix} x_{ij1} \\ \vdots \\ x_{ijm} \end{pmatrix}, \quad \phi = \begin{pmatrix} \phi_1 \\ \vdots \\ \phi_\rho \end{pmatrix}, \quad \phi_j = \begin{pmatrix} \phi_{j1} \\ \vdots \\ \phi_{jm} \end{pmatrix},$$

where x_i, x_{ij}, and x_{ijk} are in $\mathbb{R}^{\rho m}$, $\mathbb{R}^\rho$, and $\mathbb{R}$, respectively. Moreover ϕ, ϕ_j, and ϕ_{jk} are functions mapping $\mathbb{R}^{\rho m+1}$ to $\mathbb{R}^{\rho m}$, $\mathbb{R}^\rho$, and $\mathbb{R}$, respectively.

Then, we impose the following assumption on the time-varying nonlinearity $\phi(t, x_i)$.

Assumption 5.1 *Assume that $\phi(t, x)$ is continuously differentiable and globally Lipschitz continuous with respect to x, uniformly in t and piecewise continuous with respect to t. This implies that there exists a M (independent of t and x) such that*

$$\|\phi(t, x_1) - \phi(t, x_2)\| \le M\|x_1 - x_2\|$$

for all $t \in \mathbb{R}$ and all $x_1, x_2 \in \mathbb{R}^{\rho m}$.

Moreover, the nonlinearity has a lower-triangular structure in the sense that

$$\frac{\partial \phi_j(t, x_i)}{\partial x_{ik}} = 0, \quad \forall k > j. \tag{5.3}$$

The communication network among agents is exactly the same as that in Chap. 2, which satisfies Assumption 2.2 and provides each agent with the quantity

$$\zeta_i = \sum_{j=1}^{N} a_{ij}(x_i - x_j) = \sum_{j=1}^{N} \ell_{ij} x_j, \tag{5.4}$$

for $i = 1, \ldots, N$. We formulate next state synchronization problem for a network with full-state coupling as follows.

Problem 5.1 (Full-State Coupling) Consider a MAS described by (5.1) and (5.4). Let $\mathbf{G}$ be a given set of graphs such that $\mathbf{G} \subseteq \mathbb{G}^N$. The *state synchronization* problem with a set of network graphs $\mathbf{G}$ is to find, if possible, a linear static protocol of the form

$$u_i = F\zeta_i, \tag{5.5}$$

for $i = 1, \ldots, N$ such that, for any graph $\mathcal{G} \in \mathbf{G}$ and for all the initial conditions of agents, the state synchronization among agents can be achieved.

5.2.2 *Protocol Design*

In this section, we will present a two-step protocol design. We can find two designs, one relying on a Riccati equation and the other relying on a direct method.

Protocol design 5.1 *Consider a MAS described by* (5.1) *and* (5.4). *Let* $\alpha > \beta > 0$ *and assume the graph associated with the MAS is in the set* $\mathbb{G}^N_{\alpha,\beta}$.

Choose $F = -B'P$ *with* $P > 0$ *being the unique solution of the continuous-time algebraic Riccati equation*

$$A'P + PA - 2\beta PBB'P + Q = 0, \tag{5.6}$$

where $Q > 0$. *Let* $\varepsilon > 0$ *be a high-gain parameter. Define a time-scaling matrix*

$$S_\varepsilon = \begin{pmatrix} I_m & 0 & \cdots & 0 \\ 0 & \varepsilon I_m & \ddots & \vdots \\ \vdots & \ddots & \ddots & 0 \\ 0 & \cdots & 0 & \varepsilon^{\rho-1} I_m \end{pmatrix}. \tag{5.7}$$

(continued)

Protocol design 5.1 (continued)
The protocol is designed as

$$u_i = \varepsilon^{-\rho} F S_\varepsilon \zeta_i. \tag{5.8}$$

The main result based on the above design is stated in the following theorem.

Theorem 5.2 *Consider a MAS described by (5.1) and (5.4). Let any $\alpha \geq \beta > 0$ be given, and hence a set of network graphs $\mathbb{G}^N_{\alpha,\beta}$ be defined.*

Under Assumption 5.1, the state synchronization problem stated in Problem 5.1 with $\mathbf{G} = \mathbb{G}^N_{\alpha,\beta}$ is solvable. In particular, there exists an $\varepsilon^* \in (0, 1]$ such that for all $\varepsilon \in (0, \varepsilon^*]$, protocol (5.8) solves the state synchronization problem for any graph $\mathcal{G} \in \mathbb{G}^N_{\alpha,\beta}$.

The proof of the theorem relies on the following two lemmas.

Lemma 5.3 *Let a Laplacian matrix $L \in \mathbb{R}^{N\times N}$ be given associated with a graph that contains a directed spanning tree. We define $\bar{L} \in \mathbb{R}^{(N-1)\times(N-1)}$ as the matrix $\bar{L} = [\bar{\ell}_{ij}]$ with*

$$\bar{\ell}_{ij} = \ell_{ij} - \ell_{Nj}.$$

Then the eigenvalues of $\bar{L}$ are equal to the nonzero eigenvalues of L.

Proof We have

$$\bar{L} = \begin{pmatrix} I & -\mathbf{1} \end{pmatrix} L \begin{pmatrix} I \\ 0 \end{pmatrix}$$

Assume that λ is a nonzero eigenvalue of L with eigenvector x. Then

$$\bar{x} = \begin{pmatrix} I & -\mathbf{1} \end{pmatrix} x$$

where $\mathbf{1}$ is a vector with all 1's, satisfies,

$$\begin{pmatrix} I & -\mathbf{1} \end{pmatrix} Lx = \begin{pmatrix} I & -\mathbf{1} \end{pmatrix} \lambda x = \lambda \bar{x}$$

and since $\bar{L}\mathbf{1} = 0$ we find that

$$\bar{L}\bar{x} = \begin{pmatrix} I & -\mathbf{1} \end{pmatrix} L = \lambda \bar{x}.$$

This shows that λ is an eigenvector of $\bar{L}$ if $\bar{x} \neq 0$. It is easily seen that $\bar{x} = 0$ if and only if $\lambda = 0$. Conversely if $\bar{x}$ is an eigenvector of $\bar{L}$ with eigenvalue λ, then it is easily verified that

$$x = L \begin{pmatrix} I \\ 0 \end{pmatrix} \bar{x}$$

is an eigenvector of L with eigenvalue λ. ∎

Lemma 5.4 *The matrix*

$$\hat{A} = (I_{N-1} \otimes A_d) + (S \otimes B_d F), \tag{5.9}$$

is asymptotically stable for any upper-triangular matrix S whose eigenvalues satisfy $\operatorname{Re}\lambda_i > \beta$ *for all* $i = 1, \ldots, N-1$ *and* $\|S\| < \alpha$.

Moreover, there exists a $\hat{P} > 0$ *such that*

$$\hat{A}'\hat{P} + \hat{P}\hat{A} \leq -I \tag{5.10}$$

is satisfied for all possible upper-triangular matrices S.

Proof Define

$$\hat{A}_i = A_d + \lambda_i B_d F$$

and

$$\hat{B} = B_d F.$$

Note that the proof of Theorem 2.12 shows that $\hat{A}_i$ is asymptotically stable. Then

$$\hat{A} = \begin{pmatrix} \hat{A}_1 & s_{1,2}\hat{B} & \cdots & s_{1,N-1}\hat{B} \\ 0 & \hat{A}_2 & \ddots & \vdots \\ \vdots & \ddots & \ddots & s_{N-2,N-1}\hat{B} \\ 0 & \cdots & 0 & \hat{A}_{N-1} \end{pmatrix},$$

where λ_i are the eigenvalues of S and $s_{i,j}$ with $|s_{ij}| < \alpha$ are the elements of the matrix S. We have

$$\hat{A}_i'P + P\hat{A}_i \leq -I$$

for $i = 1, \ldots, N-1$ where

$$P = \|Q^{-1}\|P.$$

Via Schur complement, it is easy to verify that if matrices $A_{11} < -kI$, $A_{22} < 0$, and A_{12} are given, then there exists a μ sufficiently large such that the matrix

$$\begin{pmatrix} A_{11} & A_{12} \\ A_{12}' & \mu A_{22} \end{pmatrix} < -(k-1)I.$$

Now define the matrix

$$\hat{P} = \begin{pmatrix} \alpha_1 P & 0 & \cdots & 0 \\ 0 & \ddots & \ddots & \vdots \\ \vdots & \ddots & \ddots & 0 \\ 0 & \cdots & 0 & \alpha_{N-1} P \end{pmatrix}.$$

Then we have

$$\hat{A}'\hat{P} + \hat{P}\hat{A} \le \begin{pmatrix} -\alpha_1 I & \alpha_1 s_{1,2} P\hat{B} & \cdots & \alpha_1 s_{1,N-1} P\hat{B} \\ \alpha_1 s_{1,2}\hat{B}'P & -\alpha_2 I & \ddots & \vdots \\ \vdots & \ddots & \ddots & \alpha_{N-2} s_{N-2,N-1} P\hat{B} \\ \alpha_1 s_{1,N-1}\hat{B}'P & \cdots & \alpha_{N-2} s_{N-2,N-1}\hat{B}'P & -\alpha_{N-1} I \end{pmatrix}.$$

Using the above Schur argument, it is not hard to show that if

$$\begin{pmatrix} -\alpha_1 I & \alpha_1 s_{1,2} P\hat{B} & \cdots & \alpha_1 s_{1,N-2} P\hat{B} \\ \alpha_1 s_{1,2}\hat{B}'P & -\delta\alpha_2 I & \ddots & \vdots \\ \vdots & \ddots & \ddots & \alpha_{N-3} s_{N-3,N-2} P\hat{B} \\ \alpha_1 s_{1,N-2}\hat{B}'P & \cdots & \alpha_{N-3} s_{N-3,N-2}\hat{B}'P & -\delta\alpha_{N-2} I \end{pmatrix} \le -2\delta I,$$

then there exists an α_{N-1} such that

$$\begin{pmatrix} -\delta\alpha_1 I & \alpha_1 s_{1,2} P\hat{B} & \cdots & \alpha_1 s_{1,N-1} P\hat{B} \\ \alpha_1 s_{1,2}\hat{B}'P & -\alpha_2 I & \ddots & \vdots \\ \vdots & \ddots & \ddots & \alpha_{N-2} s_{N-2,N-1} P\hat{B} \\ \alpha_1 s_{1,N-1}\hat{B}'P & \cdots & \alpha_{N-2} s_{N-2,N-1}\hat{B}'P & -\alpha_{N-1} I \end{pmatrix} \le -I.$$

Using a recursive argument, we can then prove that there exist $\alpha_1, \ldots, \alpha_{N-1}$ such that

$$\hat{A}'\hat{P} + \hat{P}\hat{A} \le -I.$$

■

Proof of Theorem 5.2 Let $\bar{x}_i = x_i - x_N$ for $i = 1, \dots N$. Then, state synchronization among agents is achieved if $\bar{x}_i \to 0$ for $i = 1, \dots, N-1$.

By Taylor's theorem (see, e.g., [77, Theorem 11.1]), we can write

$$\phi(t, x_N) - \phi(t, x_i) = \Phi_i(t)\bar{x}_i,$$

where

$$\Phi_i(t) = \int_0^1 \frac{\partial \phi}{\partial x_i}(t, x_i + p\bar{x}_i)\mathrm{d}p. \tag{5.11}$$

Due to the Lipschitz property of the nonlinearity, $\Phi_i(t)$ is uniformly bounded, and the lower-triangular structure of the nonlinearity implies that $\Phi_i(t)$ is lower-triangular. Now, the dynamics of $\bar{x}_i$ can be written as

$$\dot{\bar{x}}_i = A_d\bar{x}_i + B_d\varepsilon^{-\rho}FS_\varepsilon\sum_{j=1}^{N-1}\bar{\ell}_{ij}\bar{x}_j + B_dE\bar{x}_i - \Phi_i(t)\bar{x}_i,$$

where $\bar{\ell}_{ij} = \ell_{ij} - \ell_{Nj}$. Defining $\xi_i = S_\varepsilon\bar{x}_i$, we have

$$\varepsilon\dot{\xi}_i = A_d\xi_i + B_dF\sum_{j=1}^{N-1}\bar{\ell}_{ij}\xi_j + W_{i\varepsilon}\xi_i,$$

where

$$W_{i\varepsilon} = \varepsilon^\rho B_dES_\varepsilon^{-1} - \varepsilon S_\varepsilon\Phi_i(t)S_\varepsilon^{-1}.$$

The first term of $W_{i\varepsilon}$ is obviously $O(\varepsilon)$. The second term is $O(\varepsilon)$ uniformly in t since $\Phi_i(t)$ is lower-triangular and bounded. Let

$$\xi = \begin{pmatrix}\xi_1\\ \vdots \\ \xi_{N-1}\end{pmatrix}, \quad W_\varepsilon = \begin{pmatrix} W_{1\varepsilon} & 0 & \cdots & 0\\ 0 & W_{2\varepsilon} & \ddots & \vdots \\ \vdots & \ddots & \ddots & 0 \\ 0 & \cdots & 0 & W_{(N-1)\varepsilon}\end{pmatrix}, \quad \text{and} \quad \bar{L} = [\bar{\ell}_{ij}].$$

We get

$$\varepsilon\dot{\xi} = \left[(I_{N-1}\otimes A_d) + (\bar{L}\otimes B_dF)\right]\xi + W_\varepsilon\xi.$$

Define a unitary matrix U such that $U^{-1}\bar{L}U = S$ where S is the Schur form of $\bar{L}$. According to Lemma 5.3 we have that the eigenvalues of S satisfy $\mathrm{Re}\,\lambda \geq \beta$ and

$\|S\| < \alpha$. Then

$$\varepsilon\dot{\nu} = \hat{A}\nu + \hat{W}_\varepsilon \nu,$$

where $\hat{A}$ as defined in (5.9) while

$$\begin{aligned}\nu &= (U \otimes I_{p\rho})\xi,\\ \hat{W}_\varepsilon &= (U \otimes I_{p\rho})W_\varepsilon(U^{-1} \otimes I_{p\rho}).\end{aligned}$$

According to Lemma 5.4, there exists a $\hat{P} > 0$ which does not depend on the Laplacian matrix but only on the bounds α and β for the Laplacian such that

$$\hat{P}\hat{A} + \hat{A}'\hat{P} \leq -I.$$

Consider the Lyapunov function $V = \varepsilon\nu'\hat{P}\nu$, for which we have

$$\begin{aligned}\dot{V} &= -\|\nu\|^2 + 2\,\mathrm{Re}(\nu'\hat{W}_\varepsilon\hat{P}\nu)\\ &\leq -\|\nu\|^2 + \varepsilon r_1\|\nu\|^2\end{aligned}$$

where r_1 is such that $\varepsilon r_1 \geq \|\hat{W}_\varepsilon\hat{P}\|$ for any $\varepsilon \in [0, 1]$. By choosing an ε small enough, we get $\dot{V} \leq 0$. It follows that $\nu \to 0$, which implies that $\bar{x}_i \to 0$ for $i = 1, \ldots, N-1$. ∎

We proceed now to another protocol design.

Protocol design 5.2 *Consider a MAS described by (5.1) and (5.4). Let β be a lower bound for the real part of the nonzero eigenvalues of all Laplacian matrices associated with a graph in the set $\mathbb{G}^N_{\alpha,\beta}$.*

Note that

$$A_d = \begin{pmatrix} A_{11} & A_{12}\\ A_{21} & 0\end{pmatrix}, \quad B_d = \begin{pmatrix}0\\1\end{pmatrix}.$$

Let $\varepsilon > 0$ be a high-gain parameter. Choose F_ε by

$$F_\varepsilon = \tfrac{1}{\varepsilon}\left(F_1 - I\right)$$

with F_1 such that $A_{11} + A_{12}F_1$ is asymptotically stable. Define a time-scaling matrix by (5.7). The protocol is designed as follows.

$$u_i = \varepsilon^{-\rho}F_\varepsilon S_\varepsilon \zeta_i. \tag{5.12}$$

Theorem 5.5 *Consider a MAS described by* (5.1) *and* (5.4). *Let any* $\alpha \geq \beta > 0$ *be given, and hence a set of network graphs* $\mathbb{G}^N_{\alpha,\beta}$ *be defined.*

Under Assumption 5.1, *the state synchronization problem stated in Problem* 5.1 *with* $\mathbf{G} = \mathbb{G}^N_{\alpha,\beta}$ *is solvable. In particular, there exists an* $\varepsilon^* \in (0, 1]$ *such that for all* $\varepsilon \in (0, \varepsilon^*]$, *protocol* (5.12) *solves the state synchronization problem for any graph* $\mathcal{G} \in \mathbb{G}^N_{\alpha,\beta}$.

Proof Similar to the proof of Theorem 5.2, we can rewrite the closed loop system in the form

$$\varepsilon\dot{\xi} = \left[(I_{N-1} \otimes A_d) + (\bar{L} \otimes B_d F_\varepsilon)\right]\xi + W_\varepsilon \xi, \tag{5.13}$$

with $\|W_\varepsilon\| \leq r\varepsilon$ for some $r > 0$. We have

$$\begin{pmatrix} I & 0 \\ -F_1 & I \end{pmatrix} (A_d + \lambda B_d F_\varepsilon) \begin{pmatrix} I & 0 \\ F_1 & I \end{pmatrix} = \begin{pmatrix} A_{11} + A_{12}F_1 & A_{12} \\ \tilde{A}_{21} & -\frac{\lambda}{\varepsilon} I \end{pmatrix} \tag{5.14}$$

with $\tilde{A}_{21}$ independent of ε and λ. Let P_1 be such that

$$(A_{11} + A_{12}F_1)'P_1 + P_1(A_{11} + A_{12}F_1) < -2I.$$

In that case, for $\varepsilon < \varepsilon^*$ with ε^* sufficiently small we have

$$\begin{pmatrix} A_{11} + A_{12}F_1 & A_{12} \\ \tilde{A}_{21} & -\frac{\lambda}{\varepsilon} I \end{pmatrix}' \begin{pmatrix} P_1 & 0 \\ 0 & I \end{pmatrix} + \begin{pmatrix} P_1 & 0 \\ 0 & I \end{pmatrix} \begin{pmatrix} A_{11} + A_{12}F_1 & A_{12} \\ \tilde{A}_{21} & -\frac{\lambda}{\varepsilon} I \end{pmatrix} < -I$$

for all λ with $\operatorname{Re}\lambda > \beta$. This implies that

$$P = \begin{pmatrix} I & -F_1 \\ 0 & I \end{pmatrix} \begin{pmatrix} P_1 & 0 \\ 0 & I \end{pmatrix} \begin{pmatrix} I & 0 \\ -F_1 & I \end{pmatrix}$$

satisfies

$$P(A_d + \lambda B_d F_\varepsilon) + (A_d + \lambda_i B_d F_\varepsilon)'P \leq -\gamma I$$

for some γ independent of ε and λ. But then, similar to the proof of Lemma 5.4, we can show that

$$\hat{P} = \begin{pmatrix} \alpha_1 P & 0 & \cdots & 0 \\ 0 & \ddots & \ddots & \vdots \\ \vdots & \ddots & \ddots & 0 \\ 0 & \cdots & 0 & \alpha_{N-1}P \end{pmatrix}$$

for suitable $\alpha_1, \ldots, \alpha_{N-1}$ independent of $\varepsilon < \varepsilon^*$ but yields

$$\left[(I_{N-1} \otimes A_d) + (\bar{L} \otimes B_d F_\varepsilon)\right]' \hat{P} + \hat{P}\left[(I_{N-1} \otimes A_d) + (\bar{L} \otimes B_d F_\varepsilon)\right] < -I$$

for any $\varepsilon < \varepsilon^*$. But then $\xi' \hat{P} \xi$ is a Lyapunov function for the system (5.13) for $\varepsilon < \varepsilon^*$ which shows the required stability. ∎

5.2.3 *Transforming Nonlinear Time-Varying Systems to the Canonical Form*

We discuss next the transformation of general nonlinear time-varying systems to the canonical form (5.1).

Consider a nonlinear time-varying system of the form

$$\dot{\tilde{x}}_i = \tilde{A}\tilde{x}_i + \tilde{B}\tilde{u}_i + \tilde{\phi}(t, \tilde{x}_i), \tag{5.15}$$

where $\tilde{x}_i \in \mathbb{R}^{\tilde{n}}$, $\tilde{u}_i \in \mathbb{R}^m$.

Assume the controllability indices of $\tilde{A}$, $\tilde{B}$ are not all the same. In that case, a linear pre-compensator can be designed to be of the form

$$\begin{aligned} \dot{x}_{i,c} &= A_c x_{i,c} + B_c \bar{u}_i, \\ \tilde{u}_i &= C_c x_{i,c} \end{aligned} \tag{5.16}$$

such that the interconnection of the nonlinear system (5.15) and the pre-compensator (5.16) is such that all controllability indices are the same and equal to some integer ρ. The interconnection can then be written as

$$\dot{\bar{x}}_i = \bar{A}\bar{x}_i + \bar{B}\bar{u}_i + \bar{\phi}(t, \bar{x}_i), \tag{5.17}$$

where $\bar{x}_i \in \mathbb{R}^n$, $\bar{u}_i \in \mathbb{R}^m$.

Theorem 5.6 *Consider the nonlinear time-varying system (5.17) with $\bar{x} \in \mathbb{R}^n$ and $\bar{u}_i \in \mathbb{R}^m$. Assume that all controllability indices of $(\bar{A}, \bar{B})$ are equal to ρ and $\bar{\phi}(t, x)$ is continuously differentiable and globally Lipschitz continuous with respect to x, uniformly in t, and piecewise continuous with respect to t. This implies that there exists a M (independent of t and x) such that*

$$\|\bar{\phi}(t, x_1) - \bar{\phi}(t, x_2)\| \leq M\|x_1 - x_2\|$$

for all $t \in \mathbb{R}$ and all $x_1, x_2 \in \mathbb{R}^n$. Let $\Gamma_x \in \mathbb{R}^{n\times n}$ and $\Gamma_u \in \mathbb{R}^m$ be nonsingular state and input transformations such that there exists a matrix E such that

$$\Gamma_x^{-1}\bar{A}\Gamma_x = A_d + B_d E, \qquad \Gamma_x^{-1}\bar{B}\Gamma_u = B_d,$$

with A_d and B_d given by (5.2) and define $\bar{x}_i = \Gamma_x x_i$ and $\bar{u}_i = \Gamma_u u_i$. Then either

- *the system with state x_i, input u_i satisfies the canonical form (5.1); or*
- *there exists no set of linear, nonsingular state and input transformations that take the system to the canonical form.*

Proof First, note that the linear portion of (5.1) has the required form. Thus, all we have to show is that all transformations that preserve the appropriate form of the linear portion of the system are equivalent with respect to satisfying Assumption 5.1. Consider therefore the system (5.17) and let $\breve{\Gamma}_x \in \mathbb{R}^{n\times n}$ and $\breve{\Gamma}_u \in \mathbb{R}^m$ denote state and input transformations such that

$$\breve{\Gamma}_x^{-1}(A_d + B_d E)\breve{\Gamma}_x = A_d + B_d\breve{E}, \qquad \breve{\Gamma}_x^{-1}B_d\breve{\Gamma}_u = B_d$$

for some suitably chosen matrix $\breve{E}$. Define $x_i = \breve{\Gamma}_x\breve{x}_i$ and $u_i = \breve{\Gamma}_u\breve{u}_i$. Then we can write

$$\dot{\breve{x}}_i = A_d\breve{x}_i + \breve{\phi}(t, \breve{x}_i) + B_d(\breve{u}_i + \breve{E}\breve{x}_i),$$

with

$$\breve{\phi}(t, \breve{x}_i) = \breve{\Gamma}_x^{-1}\phi\left(t, \breve{\Gamma}_x\breve{x}_i\right),$$

and we need to show that $\breve{\phi}$ satisfies (5.3). Let

$$\breve{\Gamma}_x = \begin{pmatrix} T_{1,1} & \cdots & T_{1,\rho} \\ \vdots & \ddots & \vdots \\ T_{\rho,1} & \cdots & T_{\rho,\rho} \end{pmatrix},$$

where $T_{i,j} \in \mathbb{R}^{m\times m}$. Note that

$$\breve{\Gamma}_x B_d = B_d\breve{\Gamma}_u,$$

which implies that

$$T_{1,\rho} = \cdots = T_{\rho-1,\rho} = 0. \tag{5.18}$$

Furthermore

$$\breve{\Gamma}_x(A_d + B_d\breve{E}) = (A_d + B_d E)\breve{\Gamma}_x.$$

Since only the last m rows of $B_d\breve{E}$ and B_dE contain possibly nonzero terms, this implies that the first $(\rho-1)m$ rows of $\breve{\Gamma}_x A_d$ and $A_d\breve{\Gamma}_x$ are equal. It then follows that

$$T_{2,1} = \cdots = T_{\rho,1} = 0 \text{ and } T_{i,j} = T_{i+1,j+1}$$

for $i, j \in \{1, \rho-1\}$. Together with (5.18) we find that $T_{i,j} = 0$ for $i \neq j$ and $T_{i,i} = T_{j,j}$ for $i, j \in \{1, \ldots, N\}$. Therefore, $\breve{\Gamma}_x = \gamma I$ for some $\gamma \in \mathbb{R}$, which means that $\breve{\phi}$ satisfies Assumption 5.1 if and only if ϕ satisfies Assumption 5.1. ■

5.3 MAS with Partial-State Coupling

5.3.1 Problem Formulation

Consider a multi-agent system (MAS) composed of N identical nonlinear time-varying continuous-time agents that can be represented in the canonical form:

$$\begin{aligned}
\dot{x}_{ia} &= A_a x_{ia} + L_{ad} y_i, \\
\dot{x}_{id} &= A_d x_{id} + \phi_d(t, x_{ia}, x_{id}) + B_d(u_i + E_{da}x_{ia} + E_{dd}x_{id}), \\
y_i &= C_d x_{id},
\end{aligned} \tag{5.19}$$

where

$$x_i := \begin{pmatrix} x_{ia} \\ x_{id} \end{pmatrix} \in \mathbb{R}^n, \qquad u_i \in \mathbb{R}^m, \qquad y_i \in \mathbb{R}^m$$

are, respectively, states, inputs, and outputs of agent i. The linear part of the agents has a uniform relative degree denoted by ρ. A_d and B_d have the special form of (5.2), while C_d has the special form

$$C_d = \begin{pmatrix} I_m & 0 \end{pmatrix}. \tag{5.20}$$

It is easy to see that (A_d, B_d, C_d) is controllable and observable. Note that x_{id} and ϕ_d have the structure

$$x_{id} = \begin{pmatrix} x_{id1} \\ \vdots \\ x_{id\rho} \end{pmatrix}, \quad x_{idj} = \begin{pmatrix} x_{idj1} \\ \vdots \\ x_{idjm} \end{pmatrix}, \quad \phi_d = \begin{pmatrix} \phi_{d1} \\ \vdots \\ \phi_{d\rho} \end{pmatrix}, \quad \phi_{dj} = \begin{pmatrix} \phi_{dj1} \\ \vdots \\ \phi_{djm} \end{pmatrix},$$

where x_{id}, x_{idj}, and x_{idjk} are in $\mathbb{R}^{\rho m}$, $\mathbb{R}^{\rho}$, and $\mathbb{R}$, respectively. Moreover ϕ_d, ϕ_{dj}, and ϕ_{djk} are functions mapping $\mathbb{R}^{n+1}$ to $\mathbb{R}^{\rho m}$, $\mathbb{R}^{\rho}$, and $\mathbb{R}$, respectively.

We impose the following assumption on the time-varying nonlinearity ϕ_d.

Assumption 5.2 *Assume that $\phi_d(t, x_{ia}, x_{id})$ is continuously differentiable and globally Lipschitz continuous with respect to (x_{ia}, x_{id}), uniformly in t, and piecewise continuous with respect to t. This implies that there exists a M (independent of t, x_a, and x_d) such that*

$$\|\phi_d(t, x_{a1}, x_{d1}) - \phi_d(t, x_{a2}, x_{d2})\| \leq M\|x_{a1} - x_{a2}\| + M\|x_{d1} - x_{d2}\|$$

for all $t \in \mathbb{R}$ and all $x_{a1}, x_{a2}, x_{d1}, x_{d2}$.

Moreover, the nonlinearity has the lower-triangular structure

$$\frac{\partial \phi_{dj}(t, x_{ia}, x_{id})}{\partial x_{idk}} = 0, \quad \forall k > j. \tag{5.21}$$

The communication network among agents is again the same as that in Chap. 2 and provides each agent with the quantity

$$\zeta_i = \sum_{j=1}^{N} a_{ij}(y_i - y_j) = \sum_{j=1}^{N} \ell_{ij} y_j, \tag{5.22}$$

for $i = 1, \ldots, N$ where the graph $\mathcal{G}$ describing the communication topology of the network contains a directed spanning tree.

The state synchronization problem for a network with partial-state coupling is formulated as follows.

Problem 5.7 (Partial-State Coupling) Consider a MAS described by (5.19) and (5.22). Let $\mathbf{G}$ be a given set of graphs such that $\mathbf{G} \subseteq \mathbb{G}^N$. The *state synchronization* problem with a set of network graphs $\mathbf{G}$ is to find, if possible, a linear dynamic protocol of the form

$$\begin{cases} \dot{\chi}_i = A_c \chi_i + B_c \zeta_i, \\ u_i = C_c \chi_i, \end{cases} \tag{5.23}$$

for $i = 1, \ldots, N$ where $\chi_i \in \mathbb{R}^{n_c}$, such that, for any graph $\mathcal{G} \in \mathbf{G}$ and for all the initial conditions of agents and their protocol, the state synchronization among agents can be achieved.

5.3.2 *Minimum-Phase Agents*

In this section, we consider MAS with minimum-phase agents. That means A_a in the agent model (5.19) is Hurwitz stable. We design here a protocol based on a low-and-high-gain Riccati-based method.

Protocol design 5.3 *Consider a MAS described by* (5.19) *and* (5.22)*. We will use a CSS observer-based protocol for the agents.*

We choose an observer gain K such that $A_d - KC_d$ is Hurwitz stable. Next, we consider a feedback gain

$$F_\delta = -B_d' P_\delta$$

with $P_\delta > 0$ being the unique solution of the continuous-time algebraic Riccati equation

$$P_\delta A_d + A_d' P_\delta - \beta P_\delta B_d B_d' P_\delta + \delta I = 0, \tag{5.24}$$

where $\delta > 0$ is a low-gain parameter and $\beta > 0$ is a lower bound on the real part of the nonzero eigenvalues of the Laplacian matrix.

Then, we define

$$F_{\delta\varepsilon} = \varepsilon^{-\rho} F_\delta S_\varepsilon, \quad K_\varepsilon = \varepsilon^{-1} S_\varepsilon^{-1} K,$$

where S_ε is the high-gain scaling matrix defined in (5.7) *with $\varepsilon > 0$ being a high-gain parameter. The protocol is designed as*

$$\begin{aligned}
\dot{\hat{x}}_{ia} &= A_a \hat{x}_{ia} + L_{ad} C_d \hat{x}_{id}, \\
\dot{\hat{x}}_{id} &= A_d \hat{x}_{id} + \phi_d(t, \hat{x}_{ia}, \hat{x}_{id}) + K_\varepsilon(\zeta_i - C_d \hat{x}_{id}) + B_d(E_{da}\hat{x}_{ia} + E_{dd}\hat{x}_{id}), \\
\tilde{u}_i &= F_{\delta\varepsilon} \hat{x}_{id}.
\end{aligned} \tag{5.25}$$

Remark 5.8 Note that the estimate

$$\hat{x}_i := \begin{pmatrix} \hat{x}_{ia} \\ \hat{x}_{id} \end{pmatrix}$$

can be interpreted as an estimate of $\sum_{j=1}^N \ell_{ij} x_j$.

Our result regarding this design can be stated as follows.

Theorem 5.9 *Consider a MAS described by* (5.19) *and* (5.22)*. Let any $\alpha \geq \beta > 0$ be given, and hence a set of network graphs $\mathbb{G}_{\alpha,\beta}^N$ be defined.*

Under Assumption 5.2, the state synchronization problem via a stable protocol stated in Problem 5.7 with the set of graphs $\mathbf{G} = \mathbb{G}_{\alpha,\beta}^N$ is solvable. In particular, there exists a $\delta^* \in (0, 1]$, such that for each $\delta \in (0, \delta^*]$, there exists an $\varepsilon^*(\delta) \in (0, 1]$ such that for all $\varepsilon \in (0, \varepsilon^*(\delta)]$, protocol (5.25) *solves the state synchronization problem for any graph $\mathcal{G} \in \mathbb{G}_{\alpha,\beta}^N$.*

Proof For each $i \in \{1, \ldots, N-1\}$, let

$$\bar{x}_i = \begin{pmatrix} \bar{x}_{ia} \\ \bar{x}_{id} \end{pmatrix} := x_i - x_N, \qquad \bar{\hat{x}}_i := \begin{pmatrix} \bar{\hat{x}}_{ia} \\ \bar{\hat{x}}_{id} \end{pmatrix} = \hat{x}_i - \hat{x}_N, \qquad \text{where} \quad \hat{x}_i = \begin{pmatrix} \hat{x}_{ia} \\ \hat{x}_{id} \end{pmatrix}.$$

Then, state synchronization among agents is achieved if $\bar{x}_i \to 0$ for $i = 1, \ldots, N-1$.

Similar to the proof of Theorem 5.2, we can write

$$\phi_d(t, x_{Na}, x_{Nd}) - \phi_d(t, x_{ia}, x_{id}) = \Phi_{ia}(t)\bar{x}_{ia} + \Phi_{id}(t)\bar{x}_{id},$$

using Taylor's theorem where $\Phi_{ia}(t)$ and $\Phi_{id}(t)$ are given by

$$\begin{aligned}
\Phi_{ia}(t) &= \int_0^1 \frac{\partial \phi_d}{\partial x_{ia}}(t, x_{ia} + p\bar{x}_{ia}, x_{id} + p\bar{x}_{id})\mathrm{d}p, \\
\Phi_{id}(t) &= \int_0^1 \frac{\partial \phi_d}{\partial x_{id}}(t, x_{ia} + p\bar{x}_{ia}, x_{id} + p\bar{x}_{id})\mathrm{d}p.
\end{aligned}$$

Due to the Lipschitz property of the nonlinearity, the elements of $\Phi_{ia}(t)$ and $\Phi_{id}(t)$ are uniformly bounded, and the lower-triangular structure of the nonlinearity implies that $\Phi_{id}(t)$ is lower-triangular. Moreover, we have

$$\phi_d(t, \hat{x}_{Na}, \hat{x}_{Nd}) - \phi_d(t, \hat{x}_{ia}, \hat{x}_{id}) = \hat{\Phi}_{ia}(t)\bar{\hat{x}}_{ia} + \hat{\Phi}_{id}(t)\bar{\hat{x}}_{id},$$

for matrices $\hat{\Phi}_{ia}(t)$ and $\hat{\Phi}_{id}(t)$ with the same properties. We can now write

$$\begin{aligned}
\dot{\bar{x}}_{ia} &= A_a \bar{x}_{ia} + L_{ad} C_d \bar{x}_{id}, \\
\dot{\bar{\hat{x}}}_{ia} &= A_a \bar{\hat{x}}_{ia} + L_{ad} C_d \bar{\hat{x}}_{id}, \\
\dot{\bar{x}}_{id} &= A_d \bar{x}_{id} + \Phi_{ia}(t)\bar{x}_{ia} + \Phi_{id}(t)\bar{x}_{id} + B_d(F_{\delta\varepsilon}\bar{\hat{x}}_{id} + E_{da}\bar{x}_{ia} + E_{dd}\bar{x}_{id}), \\
\dot{\bar{\hat{x}}}_{id} &= A_d \bar{\hat{x}}_{id} - \hat{\Phi}_{ia}(t)\bar{\hat{x}}_{ia} - \hat{\Phi}_{id}(t)\bar{\hat{x}}_{id} + B_d(E_{da}\bar{\hat{x}}_{ia} + E_{dd}\bar{\hat{x}}_{id}) \\
&\qquad + \textstyle\sum_{j=1}^{N-1} \bar{\ell}_{ij} K_\varepsilon C_d \bar{x}_{id} - K_\varepsilon C_d \bar{\hat{x}}_{id}.
\end{aligned} \tag{5.26}$$

where $\bar{\ell}_{ij} = \ell_{ij} - \ell_{Nj}$. Next, define

$$\xi_{ia} = \bar{x}_{ia}, \quad \hat{\xi}_{ia} = \bar{\hat{x}}_{ia}, \quad \xi_{id} = S_\varepsilon \bar{x}_{id}, \quad \text{and} \quad \hat{\xi}_{id} = S_\varepsilon \bar{\hat{x}}_{id}.$$

Then, using the identities

$$S_\varepsilon A_d S_\varepsilon^{-1} = \varepsilon^{-1} A_d, \qquad S_\varepsilon B_d = \varepsilon^{\rho-1} B_d, \quad \text{and} \quad C_d S_\varepsilon^{-1} = C_d,$$

we have

$$
\begin{aligned}
\dot{\xi}_{ia} &= A_a \xi_{ia} + V_{iad}\xi_{id}, \\
\dot{\hat{\xi}}_{ia} &= A_a \hat{\xi}_{ia} + \hat{V}_{iad}\hat{\xi}_{id}, \\
\varepsilon\dot{\xi}_{id} &= A_d \xi_{id} + B_d F_\delta \hat{\xi}_{id} + V^\varepsilon_{ida}\xi_{ia} + V^\varepsilon_{idd}\xi_{id}, \\
\varepsilon\dot{\hat{\xi}}_{id} &= A_d \hat{\xi}_{id} + \hat{V}^\varepsilon_{ida}\hat{\xi}_{ia} + \hat{V}^\varepsilon_{idd}\hat{\xi}_{id} + \sum_{j=1}^{N-1} \bar{\ell}_{ij} K C_d \xi_{jd} - K C_d \hat{\xi}_{id},
\end{aligned}
$$

where

$$
\begin{aligned}
V_{iad} &= \hat{V}_{iad} = L_{ad} C_d, \\
V^\varepsilon_{ida} &= \varepsilon^\rho B_d E_{da} - \varepsilon S_\varepsilon \Phi_{ia}(t), \\
\hat{V}^\varepsilon_{ida} &= \varepsilon^\rho B_d E_{da} - \varepsilon S_\varepsilon \hat{\Phi}_{ia}(t), \\
V^\varepsilon_{idd} &= \varepsilon^\rho B_d E_{dd} S_\varepsilon^{-1} - \varepsilon S_\varepsilon \Phi_{id}(t) S_\varepsilon^{-1}, \\
\hat{V}^\varepsilon_{idd} &= \varepsilon^\rho B_d E_{dd} S_\varepsilon^{-1} - \varepsilon S_\varepsilon \hat{\Phi}_{id}(t) S_\varepsilon^{-1}.
\end{aligned}
$$

Clearly $\|V_{iad}\|$ and $\|\hat{V}_{iad}\|$ are ε-independent, while $\|V^\varepsilon_{ida}\|$ and $\|\hat{V}^\varepsilon_{ida}\|$ are $O(\varepsilon)$. Since $\Phi_{id}(t)$ and $\hat{\Phi}_{id}(t)$ have the lower-triangular structure,

$$
\|\varepsilon S_\varepsilon \Phi_{id}(t) S_\varepsilon^{-1}\| \le \varepsilon \|\Phi_{id}(t)\| \quad \text{and} \quad \|\varepsilon S_\varepsilon \hat{\Phi}_{id}(t) S_\varepsilon^{-1}\| \le \varepsilon \|\hat{\Phi}_{id}(t)\|.
$$

Moreover, $\|\varepsilon^\rho B_d E_{dd} S_\varepsilon^{-1}\| \le \varepsilon \|B_d E_{dd}\|$. Hence $\|V^\varepsilon_{idd}\|$ and $\|\hat{V}^\varepsilon_{idd}\|$ are $O(\varepsilon)$.

Define $\bar{L} = [\bar{\ell}_{ij}]$, $i, j \in \{1, \dots, N-1\}$. By Lemma 5.3 the eigenvalues of $\bar{L}$ are the nonzero eigenvalues of L. Let

$$
\xi_a = \begin{pmatrix} \xi_{1a} \\ \vdots \\ \xi_{(N-1)a} \end{pmatrix}, \quad \hat{\xi}_a = \begin{pmatrix} \hat{\xi}_{1a} \\ \vdots \\ \hat{\xi}_{(N-1)a} \end{pmatrix},
$$

$$
\xi_d = \begin{pmatrix} \xi_{1d} \\ \vdots \\ \xi_{(N-1)d} \end{pmatrix}, \quad \text{and} \quad \hat{\xi}_d = \begin{pmatrix} \hat{\xi}_{1d} \\ \vdots \\ \hat{\xi}_{(N-1)d} \end{pmatrix}.
$$

Then

$$\begin{aligned}
\dot{\xi}_a &= (I_{N-1} \otimes A_a)\xi_a + V_{ad}\xi_d, \\
\dot{\hat{\xi}}_a &= (I_{N-1} \otimes A_a)\hat{\xi}_a + \hat{V}_{ad}\hat{\xi}_d, \\
\varepsilon\dot{\xi}_d &= (I_{N-1} \otimes A_d)\xi_d + (I_{N-1} \otimes B_d F_\delta)\hat{\xi}_d + V_{da}^\varepsilon\xi_a + V_{dd}^\varepsilon\xi_d, \\
\varepsilon\dot{\hat{\xi}}_d &= (I_{N-1} \otimes A_d)\hat{\xi}_d + \hat{V}_{da}^\varepsilon\hat{\xi}_a + \hat{V}_{dd}^\varepsilon\hat{\xi}_d + (\bar{L} \otimes KC_d)\xi_d - (I_{N-1} \otimes KC_d)\hat{\xi}_d,
\end{aligned}$$

where $V_{ad} = \text{diag}(V_{1ad}, \dots, V_{(N-1)ad})$, and $\hat{V}_{ad}, V_{da}^\varepsilon, \hat{V}_{da}^\varepsilon, V_{dd}^\varepsilon$, and $\hat{V}_{dd}^\varepsilon$ are similarly defined. Define a unitary U such that $U^{-1}\bar{L}U = S$, where S is the Schur form of $\bar{L}$, and let

$$\begin{aligned}
\nu_a &= (SU^{-1} \otimes I_{n-p\rho})\xi_a, \\
\tilde{\nu}_a &= \nu_a - (SU^{-1} \otimes I_{n-p\rho})\hat{\xi}_a, \\
\nu_d &= (SU^{-1} \otimes I_{p\rho})\xi_d, \\
\tilde{\nu}_d &= \nu_d - (U^{-1} \otimes I_{p\rho})\hat{\xi}_d.
\end{aligned}$$

Then

$$\begin{aligned}
\dot{\nu}_a &= (I_{N-1} \otimes A_a)\nu_a + W_{ad}\nu_d, \\
\dot{\tilde{\nu}}_a &= (I_{N-1} \otimes A_a)\tilde{\nu}_a + W_{ad}\nu_d - \hat{W}_{ad}(\nu_d - \tilde{\nu}_d), \\
\varepsilon\dot{\nu}_d &= (I_{N-1} \otimes A_d)\nu_d + (S \otimes B_d F_\delta)(\nu_d - \tilde{\nu}_d) + W_{da}^\varepsilon\nu_a + W_{dd}^\varepsilon\nu_d, \\
\varepsilon\dot{\tilde{\nu}}_d &= (I_{N-1} \otimes A_d)\tilde{\nu}_d + (S \otimes B_d F_\delta)(\nu_d - \tilde{\nu}_d) + W_{da}^\varepsilon\nu_a - \hat{W}_{da}^\varepsilon(\nu_a - \tilde{\nu}_a) \\
&\quad + W_{dd}^\varepsilon\nu_d - \hat{W}_{dd}^\varepsilon(\nu_d - \tilde{\nu}_d) - (I_{N-1} \otimes KC_d)\tilde{\nu}_d,
\end{aligned}$$

where

$$\begin{aligned}
W_{ad} &= (SU^{-1} \otimes I_{n-m\rho})V_{ad}(US^{-1} \otimes I_{m\rho}), \\
\hat{W}_{ad} &= (SU^{-1} \otimes I_{n-m\rho})\hat{V}_{ad}(U \otimes I_{m\rho}), \\
W_{da}^\varepsilon &= (SU^{-1} \otimes I_{m\rho})V_{da}^\varepsilon(US^{-1} \otimes I_{n-m\rho}), \\
W_{dd}^\varepsilon &= (SU^{-1} \otimes I_{m\rho})V_{dd}^\varepsilon(US^{-1} \otimes I_{m\rho}), \\
\hat{W}_{da}^\varepsilon &= (U^{-1} \otimes I_{m\rho})\hat{V}_{da}^\varepsilon(US^{-1} \otimes I_{n-m\rho}), \\
\hat{W}_{dd}^\varepsilon &= (U^{-1} \otimes I_{m\rho})\hat{V}_{dd}^\varepsilon(U \otimes I_{m\rho}).
\end{aligned}$$

Finally, let N_a and N_d be defined such that

$$\eta_a := N_a \begin{pmatrix} \nu_a \\ \tilde{\nu}_a \end{pmatrix} = \begin{pmatrix} \nu_{1a} \\ \tilde{\nu}_{1a} \\ \vdots \\ \nu_{(N-1)a} \\ \tilde{\nu}_{(N-1)a} \end{pmatrix}, \qquad \eta_d := N_d \begin{pmatrix} \nu_d \\ \tilde{\nu}_d \end{pmatrix} = \begin{pmatrix} \nu_{1d} \\ \tilde{\nu}_{1d} \\ \vdots \\ \nu_{(N-1)d} \\ \tilde{\nu}_{(N-1)d} \end{pmatrix}.$$

Then

$$\begin{aligned} \dot{\eta}_a &= \tilde{A}_a \eta_a + \tilde{W}_{ad} \eta_d, \\ \varepsilon \dot{\eta}_d &= \tilde{A}_\delta \eta_d + \tilde{W}^\varepsilon_{da} \eta_a + \tilde{W}^\varepsilon_{dd} \eta_d, \end{aligned} \tag{5.27}$$

where $\tilde{A}_a = (I_{2(N-1)} \otimes A_a)$ and

$$\tilde{A}_\delta = I_{N-1} \otimes \begin{pmatrix} A_d & 0 \\ 0 & A_d - KC_d \end{pmatrix} + S \otimes \begin{pmatrix} B_d F_\delta & -B_d F_\delta \\ B_d F_\delta & -B_d F_\delta \end{pmatrix}.$$

Moreover

$$\begin{aligned} \tilde{W}_{ad} &= N_a \begin{pmatrix} W_{ad} & 0 \\ W_{ad} - \hat{W}_{ad} & \hat{W}_{ad} \end{pmatrix} N_d^{-1}, \\ \tilde{W}^\varepsilon_{da} &= N_d \begin{pmatrix} W^\varepsilon_{da} & 0 \\ W^\varepsilon_{da} - \hat{W}^\varepsilon_{da} & \hat{W}^\varepsilon_{da} \end{pmatrix} N_a^{-1}, \\ \tilde{W}^\varepsilon_{dd} &= N_d \begin{pmatrix} W^\varepsilon_{dd} & 0 \\ W^\varepsilon_{dd} - \hat{W}^\varepsilon_{dd} & \hat{W}^\varepsilon_{dd} \end{pmatrix} N_d^{-1}. \end{aligned}$$

Due to the upper-triangular structure of S, the eigenvalues of $\tilde{A}_\delta$ are the eigenvalues of the matrices

$$\bar{A}_\delta := \begin{pmatrix} A_d + \lambda B_d F_\delta & -\lambda B_d F_\delta \\ \lambda B_d F_\delta & A_d - KC_d - \lambda B_d F_\delta \end{pmatrix}, \tag{5.28}$$

for each eigenvalue λ of $\bar{L}$ along the diagonal of S. Following along the lines of [117], we show that $\bar{A}_\delta$ is Hurwitz stable for all sufficiently small δ. Let $Q > 0$ be the solution of the Lyapunov equation

$$Q(A_d - KC_d) + (A_d - KC_d)'Q = -I,$$

and define

$$\bar{P}_\delta = \begin{pmatrix} P_\delta & 0 \\ 0 & \sqrt{\|P_\delta\|}Q \end{pmatrix} \text{ and } \bar{X}_\delta = \begin{pmatrix} X_{11} & X_{12} \\ X_{21} & X_{22} \end{pmatrix} = \bar{P}_\delta \bar{A}_\delta + \bar{A}_\delta' \bar{P}_\delta.$$

We define

$$\begin{aligned} X_{11} &= P_\delta A_d + A_d' P_\delta - 2\,\mathrm{Re}(\lambda) F_\delta' F_\delta, \\ X_{12} &= \lambda F_\delta' F_\delta + \lambda^* \sqrt{\|P_\delta\|} F_\delta' B_d' Q, \\ X_{21} &= X_{12}^*, \\ X_{22} &= \sqrt{\|P_\delta\|}\left[Q(A_d - KC_d - \lambda B_d F_\delta) + (A_d - KC_d - \lambda B_d F_\delta)^* Q\right] \end{aligned}$$

which together form the matrix $\bar{X}_\delta$. Using (5.24), we know that since $\mathrm{Re}\,\lambda \geq \beta$,

$$X_{11} = -\delta I - (2\,\mathrm{Re}\,\lambda - \beta) F_\delta' F_\delta \leq -\delta I - \beta F_\delta' F_\delta,$$

and we also have

$$\begin{aligned} X_{22} &= -\sqrt{\|P_\delta\|}(I + \lambda Q B_d F_\delta + \lambda^* F_\delta' B_d' Q) \\ &= -\tfrac{1}{2}\sqrt{\|P_\delta\|} I - \sqrt{\|P_\delta\|}\left[\tfrac{1}{2} I + \lambda Q B_d F_\delta + \lambda^* F_\delta' B_d' Q\right]. \end{aligned}$$

It follows that

$$\bar{X}_\delta \leq -\begin{pmatrix} \delta I & 0 \\ 0 & \frac{1}{2}\sqrt{\|P_\delta\|} I \end{pmatrix} - \begin{pmatrix} F_\delta' & 0 \\ 0 & I \end{pmatrix} W \begin{pmatrix} F_\delta & 0 \\ 0 & I \end{pmatrix},$$

where the blocks of W are given by

$$\begin{aligned} W_{11} &= \beta, \\ W_{12} &= -\lambda F_\delta - \lambda^* \sqrt{\|P_\delta\|} B_d' Q, \\ W_{21} &= W_{21}^*, \\ W_{22} &= \sqrt{\|P_\delta\|}\left[\tfrac{1}{2} I + \lambda Q B_d F_\delta + \lambda^* F_\delta' B_d' Q\right]. \end{aligned}$$

We only need to show that W is positive semidefinite. To this end, let

$$x = \begin{pmatrix} x_1 \\ x_2 \end{pmatrix}, \qquad \text{where } x_1 \in \mathbb{C},\ x_2 \in \mathbb{C}^{m\rho},$$

be an arbitrary vector. Then we have

$$x'Wx \geq \begin{pmatrix} |x_1| & \|x_2\| \end{pmatrix} \begin{pmatrix} \beta & -|\lambda|(\|F_\delta\| + \sqrt{\|P_\delta\|}\|QB_d\|) \\ \star & \sqrt{\|P_\delta\|}\left[\frac{1}{2} - 2|\lambda|\|QB_d\|\|F_\delta\|\right] \end{pmatrix} \begin{pmatrix} |x_1| \\ \|x_2\| \end{pmatrix},$$

where $\star$ denotes a symmetric element. The first-order leading principal minor of the above is $\beta > 0$. The second-order leading principal minor is

$$\tfrac{1}{2}\beta\sqrt{\|P_\delta\|} - 2\beta\sqrt{\|P_\delta\|}|\lambda|\|QB_d\|\|F_\delta\| - |\lambda|^2(\|F_\delta\| + \sqrt{\|P_\delta\|}\|QB_d\|)^2.$$

Since all the eigenvalues of A_d are in the closed left half complex plane, we know that $P_\delta \to 0$ as $\delta \to 0$ (see Lemma 2.28). Noting that $\|F_\delta\|$ is $O(\|P_\delta\|)$, we see that the second and third terms of the above expression are $O(\|P_\delta\|)$, and thus they are dominated by the first term for all sufficiently small δ. It follows that W is positive definite for all sufficiently small δ and $\bar{X}_\delta$ is therefore negative definite. Letting δ be small enough that this holds for all λ with $|\lambda| < \alpha$, we can therefore conclude that $\tilde{A}_\delta$ is Hurwitz stable.

Similar to the proof of Lemma 5.4, we can find $\alpha_1, \ldots, \alpha_{N-1}$ such that

$$\tilde{P}_\delta = \begin{pmatrix} \alpha_1 \bar{P}_\delta & 0 & \cdots & 0 \\ 0 & \ddots & \ddots & \vdots \\ \vdots & \ddots & \ddots & 0 \\ 0 & \cdots & 0 & \alpha_{N-1}\bar{P}_\delta \end{pmatrix}$$

satisfies the Lyapunov inequality

$$\tilde{P}_\delta \tilde{A}_\delta + \tilde{A}_\delta' \tilde{P}_\delta < -I,$$

where $\alpha_1, \ldots, \alpha_{N-1}$, and hence $\tilde{P}_\delta$, are independent of the Laplacian matrix but only depend on the bounds α and β. Next, let $\tilde{P}_a > 0$ be the solution of the Lyapunov equation

$$\tilde{P}_a \tilde{A}_a + \tilde{A}_a' \tilde{P}_a = -I.$$

$\tilde{P}_a > 0$ exists because $\tilde{A}_a$ is Hurwitz stable. Consider the Lyapunov function

$$V = \varepsilon \eta_d' \tilde{P}_\delta \eta_d + \varepsilon \eta_a' \tilde{P}_a \eta_a.$$

We then have

$$\begin{aligned} \dot{V} = -\|\eta_d\|^2 &+ 2\,\mathrm{Re}(\eta_d' \tilde{P}_\delta \tilde{W}_{da}^{\varepsilon} \eta_a) + 2\,\mathrm{Re}(\eta_d' \tilde{P}_\delta \tilde{W}_{dd}^{\varepsilon} \eta_d) \\ &- \varepsilon\|\eta_a\|^2 + 2\varepsilon\,\mathrm{Re}(\eta_a' \tilde{P}_a \tilde{W}_{ad} \eta_d) \\ \leq -(1 - 2\varepsilon\gamma_1)\|\eta_d\|^2 &- \varepsilon\|\eta_a\|^2 + 2\varepsilon\gamma_2\|\eta_d\|\|\eta_a\|, \end{aligned}$$

where

$$\|\tilde{P}_\delta \tilde{W}_{dd}^\varepsilon\| \leq \varepsilon\gamma_1,$$
$$\|\tilde{P}_\delta \tilde{W}_{da}^\varepsilon\| + \varepsilon\|\tilde{P}_a \tilde{W}_{ad}\| \leq \varepsilon\gamma_2.$$

Note that because U is unitary while S and S^{-1} are bounded, given α and β we can choose γ_1 and γ_2 independent of the specific graph in our set.

Let ε be chosen small enough that $2 - 4\varepsilon\gamma_1 \geq 1$. Then

$$\dot{V} \leq -\tfrac{1}{2}\begin{pmatrix}\|\eta_d\| & \|\eta_a\|\end{pmatrix}\begin{pmatrix}1 & -2\varepsilon\gamma_2 \\ -2\varepsilon\gamma_2 & 2\varepsilon\end{pmatrix}\begin{pmatrix}\|\eta_d\| \\ \|\eta_a\|\end{pmatrix}.$$

It is easy to verify that the above upper bound is negative for $2\varepsilon\gamma_2^2 < 1$. It follows that $\eta_a \to 0$ and $\eta_d \to 0$ for ε sufficiently small, which implies that $\bar{x}_i \to 0$ for $i = 1, \ldots, N-1$. ∎

5.3.3 Transforming Nonlinear Time-Varying Systems to the Canonical Form

The protocol design for nonlinear time-varying agents requires the system to be given in the canonical form (5.19). Given any arbitrarily nonlinear time-varying system, one would therefore like to know (i) whether it is possible to transform the given system to this canonical form and (ii) how the appropriate transformation can be constructed. If we limit ourselves to linear state and input transformation, then both questions are simultaneously answered in this section.

Consider a nonlinear time-varying system of the form

$$\begin{aligned}\dot{\tilde{x}}_i &= \tilde{A}\tilde{x}_i + \tilde{B}\tilde{u}_i + \tilde{\phi}(t, \tilde{x}_i),\\ y_i &= \tilde{C}\tilde{x}_i,\end{aligned} \tag{5.29}$$

where $\tilde{x}_i \in \mathbb{R}^{\tilde{n}}$, $\tilde{u}_i \in \mathbb{R}^m$, $y_i \in \mathbb{R}^p$ and $(\tilde{A}, \tilde{B}, \tilde{C})$, is right-invertible. From Chap. 24, there exists a pre-compensator in the form of

$$\begin{aligned}\dot{x}_{i,c} &= A_c x_{i,c} + B_c \bar{u}_i,\\ \tilde{u}_i &= C_c x_{i,c},\end{aligned} \tag{5.30}$$

such that the interconnection of the nonlinear system (5.29) and the pre-compensator (5.30) is invertible and has uniform relative degree. Let the interconnection be given by

$$\begin{aligned}\dot{\bar{x}}_i &= \bar{A}\bar{x}_i + \bar{B}\bar{u}_i + \bar{\phi}(t, \bar{x}_i),\\ y_i &= \bar{C}\bar{x}_i,\end{aligned} \tag{5.31}$$

where $\bar{x}_i \in \mathbb{R}^n$, $\bar{u}_i \in \mathbb{R}^m$, $y_i \in \mathbb{R}^m$.

Theorem 5.10 *Consider the nonlinear time-varying system (5.31). Assume that $(\bar{A}, \bar{B}, \bar{C})$ is minimum-phase and has uniform relative degree $\rho \geq 1$. Moreover $\bar{\phi}(t, \bar{x})$ is continuously differentiable and globally Lipschitz continuous with respect to $\bar{x}$, and it is also Lipschitz continuous uniformly in t, and piecewise continuous with respect to t. This implies that there exists a M (independent of t and $\bar{x}$) such that*

$$\|\bar{\phi}(t, \bar{x}_1) - \bar{\phi}(t, \bar{x}_2)\| \leq M\|\bar{x}_1 - \bar{x}_2\|$$

for all $t \in \mathbb{R}$ and all $\bar{x}_1, \bar{x}_2 \in \mathbb{R}^{\rho m}$.

Let $\Gamma_x \in \mathbb{R}^{n\times n}$ and $\Gamma_u \in \mathbb{R}^p$ be nonsingular state and input transformations such that the triple

$$(A, B, C) = (\Gamma_x^{-1}\bar{A}\Gamma_x,\ \Gamma_x^{-1}\bar{B}\Gamma_u,\ \bar{C}\Gamma_x)$$

is in the SCB form as presented in (23.27), i.e., it has the form

$$A = \begin{pmatrix} A_a & L_{ad}C_d \\ B_d E_{da} & A_d + B_d E_{dd} \end{pmatrix}, \quad B = \begin{pmatrix} 0 \\ B_d \end{pmatrix}, \quad C = \begin{pmatrix} 0 & C_d \end{pmatrix}$$

with A_d, B_d, C_d given by (5.2) and (5.20). Moreover, define $\bar{x}_i = \Gamma_x x_i$ and $\bar{u}_i = \Gamma_u u_i$. Then either

- *the system with state x_i, input u_i, and output y_i satisfies the canonical form (5.19); or*
- *there exists no set of linear, nonsingular state and input transformations that take the system to the canonical form.*

Proof The linear portion of (5.19) has the required form. Thus, similar to the proof of Theorem 5.6, all we have to show is that all transformations that preserve the appropriate form of the linear portion of the system are equivalent with respect to satisfying Assumption 5.1. Let $\check{\Gamma}_x \in \mathbb{R}^{n\times n}$ and $\check{\Gamma}_u \in \mathbb{R}^p$ denote the state and input transformations such that

$$\check{\Gamma}_x^{-1} A \check{\Gamma}_x = \begin{pmatrix} A_a & \check{L}_{ad}C_d \\ B_d \check{E}_{da} & A_d + B_d \check{E}_{dd} \end{pmatrix} = \check{A},$$

for suitable matrices $\check{L}_{ad}$, $\check{E}_{da}$, and $\check{E}_{dd}$, while

$$\check{\Gamma}_x^{-1} B \check{\Gamma}_u = \begin{pmatrix} 0 \\ B_d \end{pmatrix} = \check{B}, \quad C\check{\Gamma}_x = \begin{pmatrix} 0 & C_d \end{pmatrix} = \check{C}.$$

Define $x_i = \check{\Gamma}_x \check{x}_i$ and $u_i = \check{\Gamma}_u \check{u}_i$, and partition

$$\check{x}_i = \begin{pmatrix} \check{x}_{ia} \\ \check{x}_{id} \end{pmatrix}, \text{ where } \check{x}_{ia} \in \mathbb{R}^{n-p\rho} \text{ and } \check{x}_{id} \in \mathbb{R}^{p\rho}.$$

Then we can write

$$
\begin{aligned}
\dot{\breve{x}}_{ia} &= \breve{A}_a \breve{x}_{ia} + \breve{L}_{ad} y_i + \breve{\phi}_a(t, \breve{x}_{ia}, \breve{x}_{id}),\\
\dot{\breve{x}}_{id} &= A_d \breve{x}_{id} + \breve{\phi}_d(t, \breve{x}_{ia}, \breve{x}_{id}) + B_d(\breve{u}_i + \breve{E}_{da}\breve{x}_{ia} + \breve{E}_{dd}\breve{x}_{id}),\\
y_i &= C_d \breve{x}_{id},
\end{aligned}
$$

and we need to show that $\breve{\phi}_a(t, \breve{x}_{ia}, \breve{x}_{id}) = 0$ while $\breve{\phi}_d(t, \breve{x}_{ia}, \breve{x}_{id})$ satisfies (5.21). Let

$$
\breve{\Gamma}_x = \begin{pmatrix} \Gamma_{xaa} & \Gamma_{xad} \\ \Gamma_{xda} & \Gamma_{xdd} \end{pmatrix}
$$

be partitioned according to the dimensions of x_{ia} and x_{id}, and define

$$
O_\rho(A, C) = \begin{pmatrix} C \\ \vdots \\ A^{\rho-1}C \end{pmatrix},
$$

for matrices A and C of compatible dimensions. Note that

$$
O_\rho(\breve{A}, \breve{C}) = O_\rho(A, C) = \begin{pmatrix} 0 & O_\rho(A_d, C_d) \end{pmatrix} = \begin{pmatrix} 0 & I_{p\rho} \end{pmatrix}.
$$

On the other hand,

$$
O_\rho(\breve{A}, \breve{C}) = O_\rho(A, C)\breve{\Gamma}_x = \begin{pmatrix} 0 & I_{p\rho} \end{pmatrix} \breve{\Gamma}_x = \begin{pmatrix} \Gamma_{xda} & \Gamma_{xdd} \end{pmatrix}.
$$

It follows that $\Gamma_{xda} = 0$ and $\Gamma_{xdd} = I_{p\rho}$, which implies that $\breve{x}_{id} = x_{id}$. We have

$$
\Gamma_{xad} = \begin{pmatrix} \Gamma_1 & \cdots & \Gamma_\rho \end{pmatrix}.
$$

We will show that $\Gamma_{xad} = 0$ by induction. We have $\breve{\Gamma}_x \breve{B} = B\breve{\Gamma}_u$, which implies that

$$
\Gamma_m = \Gamma_{xad} B_d = 0.
$$

Next assume $\Gamma_k = 0$ for some $1 < k < m$. We have $\breve{\Gamma}_x \breve{A} = A\breve{\Gamma}_x$, which implies that

$$
\Gamma_{xaa}\breve{L}_{ad}C_d + \Gamma_{xad}(A_d + B_d\breve{E}_{dd}) = A_a\Gamma_{xad} + L_{ad}C_d.
$$

It follows that

$$
(\Gamma_{xaa}\breve{L}_{ad} - L_{ad})C_d = A_a\Gamma_{xad} - \Gamma_{xad}A_d.
$$

This yields

$$0 = A_a \Gamma_{k-1} - \Gamma_k$$

and therefore $A_a \Gamma_{k-1} = 0$. Since A_a is Hurwitz stable and hence invertible, we find that $\Gamma_{k-1} = 0$. By induction, $\Gamma_{xad} = 0$. In conclusion, we find that

$$\breve{\Gamma}_x = \begin{pmatrix} \Gamma_{xaa} & 0 \\ 0 & I \end{pmatrix},$$

which implies that $\breve{\phi}_a = 0$ while $\breve{\phi}_d$ satisfies (5.21). ■

Appendix

Let a system

$$\dot{\tilde{x}} = \tilde{A}\tilde{x} + \tilde{B}\tilde{u} \tag{5.32}$$

be given with $\tilde{x} \in \mathbb{R}^{\tilde{n}}$ and $\tilde{u} \in \mathbb{R}^m$. Let the controllability indices of this system be equal to $\rho_1, \ldots, \rho_m$. Assume that $(\tilde{A}, \tilde{B})$ is controllable and that $\tilde{B}$ has full-column rank. According to [94, Theorem 1], there exist nonsingular transformation matrices T_x and T_u such that the transformed system, with $\breve{x} = T_x \tilde{x}$ and $\breve{u} = T_u \tilde{u}$, is in the canonical form

$$\dot{\breve{x}}_j = A_j \breve{x}_j + B_j \left[\breve{u}_j + \sum_{q=1}^{m} A_{jq} \breve{x}_q \right], \tag{5.33}$$

for $j = 1, \ldots, m$ where $\breve{x}_j \in \mathbb{R}^{\rho_j}$, $\breve{u}_j \in \mathbb{R}$, and

$$\breve{x} = \begin{pmatrix} \breve{x}_1 \\ \vdots \\ \breve{x}_m \end{pmatrix}, \quad \breve{u} = \begin{pmatrix} \breve{u}_1 \\ \vdots \\ \breve{u}_m \end{pmatrix}, \quad A_j = \begin{pmatrix} 0 & I_{\rho_j - 1} \\ 0 & 0 \end{pmatrix}, \quad B_j = \begin{pmatrix} 0 \\ 1 \end{pmatrix}.$$

Note that $\sum_{j=1}^{m} \rho_j = \tilde{n}$. Next, we will construct a pre-compensator such that all the controllability indices are the same. Let $\rho = \max\{\rho_1, \ldots, \rho_m\}$. For each subsystem of dimension ρ_j, we will add $\rho - \rho_j$ integrators before the input $\breve{u}_j$. Thus, the pre-compensator for the subsystem j with $\rho_j < \rho$ is

$$\begin{aligned} \dot{x}_{j,c} &= A_{j,c} x_{j,c} + B_{j,c} u_j, \\ \breve{u}_j &= C_{j,c} x \end{aligned} \tag{5.34}$$

where u_j is the new input, and

$$A_{j,c} = \begin{pmatrix} 0 & I_{\rho-\rho_j-1} \\ 0 & 0 \end{pmatrix}, \quad B_{j,c} = \begin{pmatrix} 0 \\ 1 \end{pmatrix}, \quad C_{j,c} = \begin{pmatrix} 1 & 0 \end{pmatrix}.$$

If $\rho_j = \rho$ we simply set $\breve{u}_j = u_j$. Next, use the pre-compensators and cascade them with the system (5.32). The interconnection will be a system where all controllability indices are equal to ρ and hence can be represented in the form (after a basis transformation for state and input)

$$\dot{x} = A_d x + B_d(u + Ex),$$

with A_d and B_d of the form (5.2).

Chapter 6
Synchronization of Continuous-Time Linear MAS with Unknown Input Delay

6.1 Introduction

Recently, synchronization in a network with time delay has attracted a great deal of interest. In many applications, it is immediately clear that delays play a crucial role in the process of achieving synchronization. As clarified in [10], we can identify two kinds of delay. Firstly there is *communication delay*, which results from limitations on the communication between agents. Secondly we have *input delay* which is due to computational limitations of an individual agent.

This chapter focuses on input delay for continuous-time systems, while Chap. 7 considers discrete-time systems. Then, Chap. 8 considers communication delay.

Many works have focused on dealing with input delay, progressing from single- and double-integrator agent dynamics (see, e.g., [6, 11, 58, 79, 139, 140] to more general agent dynamics (see, e.g., [126, 150, 153, 180]). Its objective is to derive an upper bound on the input delay such that agents can still achieve synchronization. Clearly, such an upper bound always depends on the agent dynamics and the network properties.

Specifically, it is shown in [79] that a network of single-integrator agents subject to uniform constant input delay can achieve consensus with a particular linear local control protocol if and only if the delay is bounded by a maximum that is inversely proportional to the largest eigenvalue of the graph Laplacian associated with the network. This result was later on generalized in [6] to nonuniform constant or time-varying delays. Sufficient conditions for consensus among agents with first-order dynamics were also obtained in [139]. The results in [79] were extended in [11, 58] to double integrator dynamics.

The papers dealing with general agent dynamics show that in order to tolerate a time delay, a low-gain theory can be used in the controller design. This technique clearly requires that the agents in the MAS are at most weakly unstable, i.e., all

A. Saberi et al., *Synchronization of Multi-Agent Systems in the Presence of Disturbances and Delays*, Systems & Control: Foundations & Applications,
https://doi.org/10.1007/978-3-030-88148-1_6

the eigenvalues of the agents are in the closed left half plane for continuous-time systems.

As said above, this chapter considers synchronization problems for homogeneous continuous-time MAS in the presence of input delay. We first assume that the input delay is uniform, i.e., each agent has the same input delay. Next, we consider nonuniform input delay where each agent can have a different input delay. In all cases, we consider both full-state and partial-state coupling.

The write-up of this chapter is partially based on [153] and [188].

6.2 MAS in the Presence of Uniform Input Delay

We consider synchronization for MAS in the presence of uniform input delay. As in the previous chapters, for both full- and partial-state coupling networks, the synchronization problem can be solved by equivalently solving a robust stabilization problem.

Consider a MAS composed of N identical linear time-invariant continuous-time agents of the form

$$\begin{aligned} \dot{x}_i(t) &= Ax_i(t) + Bu_i(t-\tau), \\ y_i(t) &= Cx_i(t), \end{aligned} \tag{6.1}$$

for $i = 1, \ldots, N$, where τ is an unknown constant satisfying $\tau \in [0, \bar{\tau}]$ with $\bar{\tau}$ an upper bound for the input delay, and $x_i \in \mathbb{R}^n$, $u_i \in \mathbb{R}^m$, and $y_i \in \mathbb{R}^p$ are, respectively, the state, input, and output vectors of agent i. We make the following assumptions for the agent dynamics.

Assumption 6.1 *We have*

- *(A, B) is stabilizable and (A, C) is detectable;*
- *The agents are at most weakly unstable, that is, A has all its eigenvalues in the closed left half plane.*

The communication network among agents is exactly the same as that in Chap. 2 and provides each agent with the quantity

$$\zeta_i = \sum_{j=1}^{N} a_{ij}(y_i - y_j) = \sum_{j=1}^{N} \ell_{ij} y_j, \quad i = 1, \ldots, N \tag{6.2}$$

for partial-state coupling, and

$$\zeta_i = \sum_{j=1}^{N} a_{ij}(x_i - x_j) = \sum_{j=1}^{N} \ell_{ij} x_j, \quad i = 1, \ldots, N \tag{6.3}$$

for full-state coupling. As usual, we assume that the graph $\mathcal{G}$ describing the communication topology of the network contains a directed spanning tree.

6.2.1 Problem Formulation

We formulate here two state synchronization problems, one for a network with full-state coupling and the other for partial-state coupling.

Problem 6.1 (Full-State Coupling) Consider a MAS described by (6.1) and (6.3). Let $\mathbf{G}$ be a given set of graphs such that $\mathbf{G} \subseteq \mathbb{G}^N$. The *state synchronization* problem with a set of network graphs $\mathbf{G}$ is to find, if possible, a linear protocol of the form

$$u_i = F\zeta_i, \tag{6.4}$$

for $i = 1, \ldots, N$ such that, for any graph $\mathcal{G} \in \mathbf{G}$, for any $\tau \leq \bar{\tau}$, and for any initial conditions of the agents, state synchronization among agents is achieved.

Problem 6.2 (Partial-State Coupling) Consider a MAS described by (6.1) and (6.2). Let $\mathbf{G}$ be a given set of graphs such that $\mathbf{G} \subseteq \mathbb{G}^N$. The *state synchronization* problem with a set of network graphs $\mathbf{G}$ is to find, if possible, a linear dynamic protocol of the form

$$\begin{cases} \dot{\chi}_i = A_c\chi_i + B_c\zeta_i, \\ u_i = C_c\chi_i, \end{cases} \tag{6.5}$$

for $i = 1, \ldots, N$ where $\chi_i \in \mathbb{R}^{n_c}$, such that, for any graph $\mathcal{G} \in \mathbf{G}$, for any $\tau \leq \bar{\tau}$, and for any initial conditions of the agents and the protocol, state synchronization among agents is achieved.

6.2.2 Stability of Linear Time-Delay Systems

We need some preliminary results from the literature. The following result, which can, for instance, be found in [177], is a crucial result and forms the foundation of many of the proofs in this chapter.

Lemma 6.3 *Consider a linear time-delay system*

$$\dot{x}(t) = Ax(t) + A_d x(t - \tau). \tag{6.6}$$

Assume $A + A_d$ is Hurwitz. Then, (6.6) *is globally asymptotically stable for all $\tau \in [0, \bar{\tau}]$ if and only if*

$$\det\left[j\omega I - A - e^{-j\omega\tau} A_d\right] \neq 0, \quad \forall \omega \in \mathbb{R}, \quad \text{and } \forall \tau \in [0, \bar{\tau}].$$

Next, we recall some classical robustness properties of low-gain compensators. Consider an uncertain system

$$\begin{cases} \dot{x} = Ax + \mu Bu \\ y = Cx, \end{cases} \tag{6.7}$$

where (A, B) is stabilizable, (A, C) is detectable, and A has all its eigenvalues in the closed left half plane. Here $\mu \in \mathbb{C}$ is input uncertainty. We will consider both the ARE-based and the direct method for low-gain design.

We first consider the ARE-based approach. For any $\delta > 0$, let P_δ be the positive definite solution of ARE

$$A' P_\delta + A P_\delta - P_\delta B' B P_\delta + \delta I = 0. \tag{6.8}$$

The robustness of a low-gain state feedback $u = -B' P_\delta x$ is inherited from that of a classical LQR. From the proof of Theorem 2.12 (choosing $\beta = 1$, $\rho = 1/2$, and $\lambda = 2\mu$), we can state the following.

Lemma 6.4 *Let (A, B) be a given stabilizable pair with the eigenvalues of A in the closed left half plane. Let P_δ be given by* (6.8)*. In that case $A - \mu B B' P_\delta$ is Hurwitz stable for any $\mu \in \mathbb{C}$ with $2 \operatorname{Re} \mu > 1$.*

Next consider the low-gain feedback F_δ given by (2.72) which was a design based on a direct method. We have a similar result as for the ARE-based design. Note that it does not directly follow from an earlier result such as Lemma 2.33 because we have no upper bound for the uncertain parameter.

Lemma 6.5 *Let (A, B) be a given stabilizable pair with the eigenvalues of A in the closed left half plane while F_δ is given by* (2.72)*. In that case, we have that $A + \mu B F_\delta$ is Hurwitz stable for any $\mu \in \mathbb{C}$ with $\operatorname{Re} \mu > 1$.*

Proof We define

$$G_{f,\delta}(s) = F_\delta (sI - A - BF_\delta)^{-1} B.$$

In the proof of Theorem 2.32, it has been shown that $G_{f,\delta}$ is asymptotically stable. Moreover, there exists a T_u such that $T_u^{-1} G_{f,\delta} T_u$ is upper triangular with asymptotically stable diagonal elements $G_{f,\delta,i}$ satisfying

$$G_{f,\delta,i}(s) + G_{f,\delta,i}^*(s) \leq 0. \tag{6.9}$$

Let $\mu = 1 + \mu_0$ with $\operatorname{Re}\mu_0 > 0$. In that case

$$sI - A - \mu BF_\delta = (sI - A - BF_\delta)^{-1}\Big[I - \mu_0(sI - A - BF_\delta)^{-1}BF_\delta\Big].$$

Since $A + BF_\delta$ is asymptotically stable, we find that $A + \mu BF_\delta$ is asymptotically stable if

$$\det\Big[I - \mu_0(sI - A - BF_\delta)^{-1}BF_\delta\Big] = \prod_{i=1}^{q}\big(1 - \mu_0 G_{f,\delta,i}(s)\big)$$

is nonzero for all $s \in \mathbb{C}^+$. This follows from (6.9) and the fact that $\operatorname{Re}\mu_0 > 0$. The result then follows immediately. ■

The next two lemmas prove similar properties for low-gain observer-based compensators. Again, we consider designs based on Riccati equations and the direct method separately.

Consider first the ARE-based design. From the proof of Theorem 2.27 (choosing $\beta = 1$ and $\lambda = \mu$), we can state the following.

Lemma 6.6 *Let K be such that $A+KC$ is asymptotically stable and let P_δ be given by (6.8). Let $\mathcal{W}$ be a bounded set such that*

$$\mathcal{W} \subseteq \{s \in \mathbb{C} \mid \operatorname{Re} s \geq 1\}. \tag{6.10}$$

There exists a δ^* such that for any $\delta \in (0, \delta^*]$, the closed-loop system comprising of (6.7) and the low-gain compensator

$$\begin{cases} \dot{\chi} = (A + KC)\chi - Ky, \\ u = -B'P_\delta\chi, \end{cases} \tag{6.11}$$

is asymptotically stable for any $\mu \in \mathcal{W}$.

Next, we consider the direct method. We can state the following which is a direct consequence of the proof of Theorem 2.32 for $\beta = 1$ and $\lambda = \mu$.

Lemma 6.7 *Let K be such that $A+KC$ is asymptotically stable and let F_δ be given by (2.72). For any bounded set $\mathcal{W}$ satisfying (6.10), there exists a δ^* such that for all $\delta \in (0, \delta^*]$, the closed-loop system of (6.7) and the low-gain compensator*

$$\begin{cases} \dot{\chi} = (A + KC)\chi - Ky, \\ u = 2F_\delta\chi, \end{cases} \tag{6.12}$$

is asymptotically stable for any $\mu \in \mathcal{W}$.

Finally, for neutrally stable systems, we can extend the above results using a Lyapunov-based low-gain controller. This is directly a consequence of the proof of Theorem 2.31 for $\mu = \lambda$.

Lemma 6.8 *Consider the system (6.7). Suppose that the agent dynamics are neutrally stable, i.e., there exists a $P > 0$ such that $A'P + PA \leq 0$. For any a priori given $\theta \in (0, \frac{\pi}{2})$, we define a bounded $\mathcal{W}$ satisfying*

$$\mathcal{W} \subseteq \{s \in \mathbb{C} \mid \operatorname{Re} s > 0, \quad \arg(s) \in [-\theta, \theta]\}.$$

Given $\mathcal{W}$, there exists a δ^* such that for any $\delta \in (0, \delta^*]$, the closed-loop system of (6.7) and the low-gain compensator

$$\begin{cases} \dot{\chi} = (A + KC)\chi - Ky, \\ u = -\delta B'P\chi, \end{cases} \tag{6.13}$$

is asymptotically stable for any $\mu \in \mathcal{W}$.

6.2.3 *Protocol Design for MAS with Full-State Coupling*

In this section, we establish again a connection between the problem of state synchronization among agents in the network in the presence of uniform input delay and a robust stabilization problem. Then, we obtain a protocol for our MAS achieving synchronization by designing a controller for the associated robust stabilization problem.

6.2.3.1 Connection to a Robust Stabilization Problem

After implementing the linear static protocol (6.4), the MAS system described by (6.1) and (6.3) is described by

$$\dot{x}_i(t) = Ax_i(t) + BF\zeta_i(t - \tau)$$

for $i = 1, \ldots, N$. Let

$$x = \begin{pmatrix} x_1 \\ \vdots \\ x_N \end{pmatrix}.$$

Then, the overall dynamics of the N agents can be written as

$$\dot{x}(t) = (I_N \otimes A)x(t) + (L \otimes BF)x(t - \tau). \tag{6.14}$$

Recall the observation from Sect. 2.4.1 that synchronization for the system (6.14) is equivalent to the asymptotic stability of the following $N-1$ subsystems:

$$\dot{\tilde{\eta}}_i(t) = A\tilde{\eta}_i(t) + \lambda_i BF\tilde{\eta}_i(t-\tau), \quad i = 2, \dots, N, \tag{6.15}$$

for any $\tau \in [0, \bar{\tau}]$, where λ_i, $i = 2, \dots, N$ are the $N-1$ nonzero eigenvalues of the Laplacian matrix L. This equivalence is formalized in the following theorem.

Theorem 6.9 *The MAS (6.14) achieves state synchronization if and only if the system (6.15) is globally asymptotically stable for $i = 2, \dots, N$ and for any $\tau \in [0, \bar{\tau}]$.*

Proof Note that L has eigenvalue 0 with associated right eigenvector $\mathbf{1}$. Let

$$L = TJT^{-1} \tag{6.16}$$

where J is the Jordan canonical form of the Laplacian matrix L such that $J(1, 1) = 0$ and the first column of T equals $\mathbf{1}$. Let

$$\eta := (T^{-1} \otimes I_n)x = \begin{pmatrix} \eta_1 \\ \vdots \\ \eta_N \end{pmatrix},$$

where $\eta_i \in \mathbb{C}^n$. In the new coordinates, the dynamics of η can be written as

$$\dot{\eta}(t) = (I_N \otimes A)\eta(t) + (J_L \otimes BF)\eta(t-\tau). \tag{6.17}$$

The poles of system (6.17) are given by the roots of its characteristic equation

$$H(s) = \det\left\{sI - (I_N \otimes A) - e^{-s\tau}(J_L \otimes BF)\right\} = 0,$$

which, due to the diagonal structure of $I_N \otimes A$ and the upper triangular structure of $J_L \otimes BF$, are the union of the poles of the $N-1$ subsystems given in (6.15) and the eigenvalues of A.

Now if (6.15) is globally asymptotically stable for $i = 2, \dots, N$ and for any $\tau \in [0, \bar{\tau}]$, we see from the above that $\eta_i(t) \to 0$ for $i = 2, \dots, N$ and for any $\tau \in [0, \bar{\tau}]$. This implies that

$$x(t) - (T \otimes I_n)\begin{pmatrix} \eta_1(t) \\ 0 \\ \vdots \\ 0 \end{pmatrix} \to 0.$$

Note that the first column of T is equal to the vector $\mathbf{1}$ and therefore

$$x_i(t) - \eta_1(t) \to 0$$

for $i = 1, \ldots, N$. This implies that we achieve state synchronization.

Conversely, suppose the network (6.14) reaches state synchronization. In this case, we shall have

$$x(t) - \mathbf{1} \otimes x_1(t) \to 0.$$

Then $\eta(t) - (T^{-1}\mathbf{1}) \otimes x_1(t) \to 0$. Since $\mathbf{1}$ is the first column of T, we have

$$T^{-1}\mathbf{1} = \begin{pmatrix} 1 \\ 0 \\ \vdots \\ 0 \end{pmatrix}.$$

Therefore, $\eta(t) - (T^{-1}\mathbf{1}) \otimes x_1(t) \to 0$ implies that $\eta_1(t) - x_1(t) \to 0$ and $\eta_i(t) \to 0$ for $i = 2, \ldots, N$ for all the initial conditions. This implies that (6.15) is globally asymptotically stable for $i = 2, \ldots, N$. ■

We immediately have the following corollary.

Corollary 6.10 *Consider a MAS described by* (6.1) *and* (6.3). *If there exists a protocol of the form* (6.4) *that achieves state synchronization, then the associated graph must have a directed spanning tree or A must be asymptotically stable.*

Since A being asymptotically stable is a trivial case, the requirement that the graph of a multi-agent system has a directed spanning tree is essentially necessary in this section as well.

Remark 6.11 It also becomes clear from the proof of Theorem 6.9 that the *synchronized trajectory* is given by $x_s(t) = \eta_1(t)$, which is governed by

$$\dot{\eta}_1 = A\eta_1, \quad \eta_1(0) = (w \otimes I_n)x(0), \tag{6.18}$$

where w is the first row of T^{-1}, i.e., the normalized left eigenvector associated with the zero eigenvalue. This shows that the modes of the synchronized trajectory are determined by the eigenvalues of A, and the complete dynamics depends on both A and a weighted average of the initial conditions of agents at time 0.

In light of Lemma 2.8, one can conclude that $\eta_1(0)$ is only a linear combination of the initial conditions of root agents. Therefore, the synchronized trajectory given by (6.18) can be written explicitly as

$$\eta_1(t) = e^{At} \sum_{i \in \Pi_{\mathcal{G}}} w_i x_i(0), \tag{6.19}$$

which is the weighted average of the trajectories of root agents. Recall that $\Pi_{\mathcal{G}}$ denotes the set of all root agents for a graph $\mathcal{G}$.

Remark 6.12 If A is neutrally stable, i.e., all the eigenvalues of A are in the closed left half plane and those eigenvalues on the imaginary axis, if any, are semi-simple, then the synchronized trajectories are bounded. Otherwise, for at least some initial conditions the synchronized trajectory will be unbounded.

In light of the definition of Problem 6.1 that synchronization is formulated for a set of graphs, we basically obtain a robust stabilization problem, i.e., the stabilization of the system

$$\dot{x}(t) = Ax(t) + \lambda Bu(t-\tau), \tag{6.20}$$

via a protocol

$$u = Fx,$$

for any λ that are nonzero eigenvalues of possible Laplacian matrices associated with the given set of graphs $\mathbf{G}$ and for any $\tau \in [0, \bar{\tau}]$.

6.2.3.2 ARE-Based Method

Using a Riccati-based low-gain method, we design the following protocol.

Protocol design 6.1 *Consider a MAS described by (6.1) and (6.3). We consider the protocol*

$$u_i = \rho F_\delta \zeta_i, \tag{6.21}$$

where $\rho > 0$ is a design parameter, and $F_\delta = -B'P_\delta$ with $P_\delta > 0$ being the unique solution of the continuous-time algebraic Riccati equation

$$A'P_\delta + P_\delta A - P_\delta BB'P_\delta + \delta I = 0, \tag{6.22}$$

where $\delta \in (0, 1]$ is a low-gain parameter to be designed later.

The main result based on the above design is stated as follows.

Theorem 6.13 *Consider a MAS described by (6.1) and (6.3). Let any $\alpha \geq \beta > 0$ and $\frac{\pi}{2} > \theta \geq 0$ be given, and hence a set of network graphs $\mathbb{G}^N_{\alpha,\beta,\theta}$ be defined. Let*

$$\omega_{\max} = \begin{cases} 0, & A \text{ is Hurwitz,} \\ \max\{\omega \in \mathbb{R} \mid \det(j\omega I - A) = 0\}, & \text{otherwise.} \end{cases}$$

If A is at most weakly unstable and (A, B) is stabilizable, then the state synchronization problem stated in Problem 6.1 with $\mathbf{G} = \mathbb{G}^N_{\alpha,\beta,\theta}$ is solvable if

$$\bar{\tau}\omega_{\max} < \frac{\pi}{2} - \theta. \tag{6.23}$$

In particular, for any given $\bar{\tau}$ satisfying (6.23), there exist a $\rho > 0$ and a $\delta^* > 0$ such that for this ρ and any $\delta \in (0, \delta^*]$, controller (6.21) solves the state synchronization problem for any graph $\mathcal{G} \in \mathbb{G}^N_{\alpha,\beta,\theta}$ and any $\tau \in [0, \bar{\tau}]$. Moreover, the synchronized trajectory is given by (6.19).

Remark 6.14 It is transparent that in condition (6.23), $\bar{\tau}\omega_{\max}$ characterizes the delay effect and θ represents the uncertainty of the network graph. Together the uncertainties $\bar{\tau}\omega_{\max}$ and θ should not exceed $\pi/2$. We show that the low-gain feedback can accommodate these uncertainties if (6.23) is satisfied, by providing an infinite gain margin and a phase margin that can be made arbitrarily close to $\pi/2$.

Proof Let $\lambda = |\lambda|e^{j\psi}$, where $|\lambda| \in (\beta, \alpha)$ and $\psi \in [-\theta, \theta]$. According to Theorem 6.9, we only need to prove that

$$\dot{x} = Ax - \rho|\lambda|e^{j\psi}BB'P_\delta x(t-\tau) \tag{6.24}$$

is asymptotically stable for any $\tau \in [0, \bar{\tau}]$, $|\lambda| \in (\beta, \alpha)$, and $\psi \in [-\theta, \theta]$.

Since $\bar{\tau}$ and θ satisfy the condition (6.23), we have that $\bar{\tau}\omega_{\max} + \theta < \frac{\pi}{2}$. Choose ρ such that

$$\rho > \frac{1}{\beta\cos(\bar{\tau}\omega_{\max} + \theta)}. \tag{6.25}$$

Let this ρ be fixed. We first consider the system without delay (i.e., $\tau = 0$). We know (6.25) implies that $\rho|\lambda|\cos(\theta) > 1$, and hence, by Lemma 6.4, we have that

$$A - \rho|\lambda|e^{j\psi}BB'P_\delta$$

is Hurwitz stable since

$$\operatorname{Re} \rho|\lambda|e^{j\psi} \geq 1.$$

Since the system without delay is asymptotically stable, it follows from Lemma 6.3 that system (6.24) is asymptotically stable if

$$\det\Big[j\omega I - A + \rho|\lambda|e^{j(\psi-\omega\tau)}BB'P_\delta\Big] \neq 0, \tag{6.26}$$

for all $\omega \in \mathbb{R}$, $\tau \in [0, \bar{\tau}]$ and $\psi \in [-\theta, \theta]$.

Step 2. We need to prove (6.26). We note that given (6.25), there exists a $\phi > 0$ such that

$$\rho > \frac{1}{\beta \cos(\bar{\tau}\omega_{\max} + \theta)}, \qquad \forall|\omega| < \omega_{\max} + \phi. \tag{6.27}$$

Next we split the proof of (6.26) into two cases where $|\omega| < \omega_{\max} + \phi$ and $|\omega| \geq \omega_{\max} + \phi$, respectively.

If $|\omega| \geq \omega_{\max}+\phi$, we have $\det(j\omega I - A) \neq 0$, which yields $\sigma_{\min}(j\omega I - A) > 0$. Hence, there exists a $\mu > 0$ such that

$$\sigma_{\min}(j\omega I - A) > \mu, \quad \forall\omega, \quad \text{such that } |\omega| \geq \omega_{\max} + \phi.$$

To see this, note that for ω satisfying $|\omega| > \bar{\omega} := \max\{\|A\| + 1, \omega_{\max} + \phi\}$, we have $\sigma_{\min}(j\omega I - A) > |\omega| - \|A\| > 1$. But for ω with $|\omega| \in [\omega_{\max} + \phi, \bar{\omega}]$, there exists a $\mu \in (0, 1]$ such that $\sigma_{\min}(j\omega I - A) \geq \mu$, which is due to the fact that $\sigma_{\min}(j\omega I - A)$ depends continuously on ω.

Given ρ and $|\lambda| \in (\beta, \alpha)$, there exists a $\delta^* > 0$ such that $\rho|\lambda|\|BB'P_\delta\| \leq \mu/2$ for $\delta < \delta^*$. Then

$$\sigma_{\min}(j\omega I - A - \rho|\lambda|e^{j(\psi-\omega\tau)}BB'P_\delta) \geq \mu - \tfrac{\mu}{2} \geq \tfrac{\mu}{2}.$$

Therefore, the condition (6.26) holds for $|\omega| \geq \omega_{\max} + \phi$.

It remains to verify (6.26) with $|\omega| < \omega_{\max} + \phi$. By the definition of ϕ, we find that

$$\rho|\lambda|\cos(\psi - \omega\tau) > \rho\beta\cos(\theta + |\omega|\bar{\tau}) > 1,$$

and hence by Lemma 6.4

$$A - \rho|\lambda|e^{j(\psi-\omega\tau)}BB'P_\delta$$

is Hurwitz stable, for

$$\omega \in (-\omega_{\max} - \phi, \omega_{\max} + \phi), \quad |\lambda| \in (\alpha, \beta), \quad \psi \in [-\theta, \theta] \text{ and } \tau \in [0, \bar{\tau}].$$

(See Fig. 6.1.) Therefore, (6.26) also holds with $|\omega| < \omega_{\max} + \phi$ which completes the proof. ■

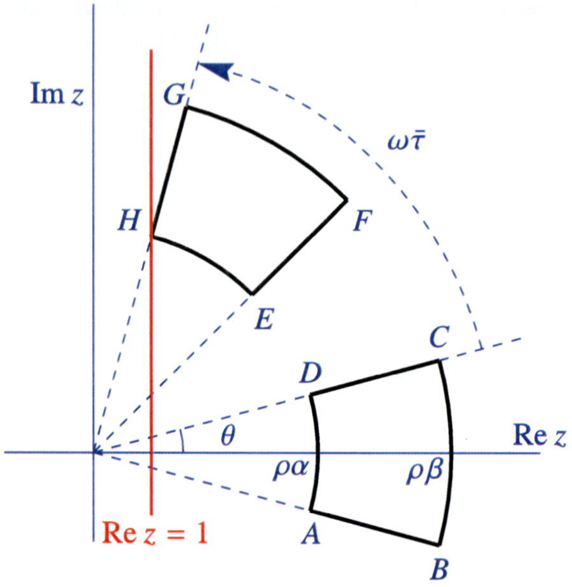

Fig. 6.1 Note that $\rho|\lambda|e^{j\psi}$ is originally located in $ABCD$ and $\rho|\lambda|e^{j(\psi-\omega\bar{\tau})}$ will be located in $EFGH$. The area $EFGH$ will not cross the vertical line Re $z = 1$ for $|\omega| < \omega_{\max} + \phi$

Remark 6.15 The consensus controller design depends only on the agent model and on parameters $\bar{\tau}$, α, β, and θ and is independent of specific network graph $\mathcal{G}$ provided that it belongs to the graph set $\mathbb{G}^N_{\alpha,\beta,\theta}$.

In the special case where $\omega_{\max} = 0$, i.e., the eigenvalues of A are either zero or in the open left half plane, then arbitrary, bounded, input delay can be tolerated as formulated in the following corollary:

Corollary 6.16 *Consider a MAS described by* (6.1) *and* (6.3)*. Let any* $\alpha \geq \beta > 0$ *and* $\frac{\pi}{2} > \theta \geq 0$ *be given, and hence a set of network graphs* $\mathbb{G}^N_{\alpha,\beta,\theta}$ *be defined.*

Suppose the eigenvalues of matrix A are either zero or in the open left half plane. If (A, B) *is stabilizable, then the state synchronization problem stated in Problem* 6.4 *with* $\mathbf{G} = \mathbb{G}^N_{\alpha,\beta,\theta}$ *is always solvable. In particular, for any given* $\bar{\tau} > 0$*, there exist a* $\rho > 0$ *and a* $\delta^* > 0$ *such that for this* ρ *and any* $\delta \in (0, \delta^*]$*, controller* (6.21) *solves the state synchronization problem for any graph* $\mathcal{G} \in \mathbb{G}^N_{\alpha,\beta,\theta}$ *and any* $\tau \in [0, \bar{\tau}]$*.*

We next consider a special case where the agent dynamics are neutrally stable, that is, the eigenvalues of A on the imaginary axis, if any, are semi-simple. This implies that A is stable. Then there exists a positive definite matrix $P > 0$ such that $A'P + PA \leq 0$. In this case, we show here that the consensus controller design no longer requires the knowledge of β and hence allows us to deal with a larger set of unknown network graphs that can be denoted as $\mathcal{G}^N_{\alpha,0,\theta}$.

Protocol design 6.2 *Consider a MAS described by* (6.1) *and* (6.3). *We consider a feedback gain*

$$F = -B'P,$$

where $P > 0$ is a solution of the Lyapunov function

$$A'P + PA \leq 0,$$

and $\delta > 0$ is a low-gain parameter. Then, the protocol is designed as

$$u_i = \delta F \zeta_i. \tag{6.28}$$

Based on the above design, we have the following theorem.

Theorem 6.17 *Consider a MAS described by* (6.1) *and* (6.3). *Let any $\alpha > 0$ and and $\frac{\pi}{2} > \theta \geq 0$ be given, and hence a set of network graphs $\mathbb{G}^N_{\alpha,0,\theta}$ be defined.*

Suppose that the agent dynamics are neutrally stable. If (A, B) is stabilizable, then the state synchronization problem stated in Problem 6.4 *with $\mathbf{G} = \mathbb{G}^N_{\alpha,0,\theta}$ is solvable if the condition* (6.23) *holds. In particular, for any given $\bar{\tau}$ satisfying* (6.23), *there exists a $\delta^* > 0$ such that for any $\delta \in (0, \delta^*]$, controller* (6.28) *solves the state synchronization problem for any graph $\mathcal{G} \in \mathbb{G}^N_{\alpha,0,\theta}$ and any $\tau \in [0, \bar{\tau}]$. Moreover, the synchronized trajectory is given by* (6.19).

Proof From Theorem 6.9, we only need to prove that the system

$$\dot{x}(t) = Ax(t) - |\lambda|\delta e^{j\psi} BB'Px(t-\tau), \tag{6.29}$$

is asymptotically stable for $|\lambda| \in (0, \alpha)$, $\psi \in [-\theta, \theta]$, and $\tau \in [0, \bar{\tau}]$. We first consider the system without delay (i.e. $\tau = 0$). We have

$$[A - \delta|\lambda|e^{j\psi}BB'P]^*P + P[A - \delta|\lambda|e^{j\psi}BB'P] \leq -2|\lambda|\delta\cos(\psi)PBB'P.$$

Since (A, B) is stabilizable, we conclude that (6.29) is asymptotically stable.

The system without delay is asymptotically stable, and therefore by Lemma 6.3, the system (6.29) is asymptotically stable if and only if

$$\det\left[j\omega I - A + \delta|\lambda|e^{j(\psi-\omega\tau)}BB'P\right] \neq 0,$$
$$\forall \omega \in \mathbb{R},\ |\lambda| \in (0, \alpha),\ \psi \in [-\theta, \theta],\ \tau \in [0, \bar{\tau}]. \tag{6.30}$$

There exists a $\phi > 0$ such that $\omega\bar{\tau} + \theta < \frac{\pi}{2}$ for all $\omega \in \mathbb{R}$ such that $|\omega| < \omega_{\max} + \phi$.

For a given $|\lambda| \in (0, \alpha)$, we can show with a similar argument as in the proof of Theorem 6.13 that there exist a $\mu > 0$ and a δ_1 such that for $\delta \in (0, \delta_1]$ and $|\lambda| \in (0, \alpha)$,

$$\sigma_{\min}(j\omega I - A + \delta|\lambda|e^{j\psi - j\omega\tau}BB'P) > \tfrac{\mu}{2}, \qquad \forall\omega \text{ such that } |\omega| \geq \omega_{\max} + \phi.$$

Hence, (6.30) is satisfied for $|\omega| \geq \omega_{\max} + \phi$. It remains to show (6.30) for $|\omega| < \omega_{\max} + \phi$. Note that $\psi - \omega\tau \in (-\frac{\pi}{2}, \frac{\pi}{2})$ by the definition of ϕ and hence $\cos(\psi - \omega\tau) > 0$. Then

$$\begin{aligned}[A - \delta|\lambda|e^{j\psi - j\omega\tau}BB'P]^*P + P[A - \delta|\lambda|e^{j\psi - j\omega\tau}BB'P]&\\ \leq -2|\lambda|\delta\cos(\psi - \omega\tau)PBB'P.&\end{aligned}$$

Since (A, B) is stabilizable, we conclude that $A - \delta|\lambda|e^{j\psi - j\omega\tau}BB'P$ is Hurwitz stable, and hence (6.30) also holds with $|\omega| < \omega_{\max} + \phi$. ■

6.2.3.3 Direct Eigenstructure Assignment Method

In this section, we design the protocol via direct eigenstructure assignment, which is already given in Sect. 2.5.3.2.

Protocol design 6.3 *Consider a MAS described by (6.1) and (6.3). We consider the protocol*

$$u_i = \rho F_\delta \zeta_i, \tag{6.31}$$

where $\rho > 0$ and $\delta > 0$ are design parameters with F_δ given by (2.72).

Note that in the above the same state feedback gain is used as in the Protocol Design 2.6. The main result based on the above design is stated as follows.

Theorem 6.18 *Consider a MAS described by (6.1) and (6.3). Let any $\alpha \geq \beta > 0$ and $\frac{\pi}{2} > \theta \geq 0$ be given, and hence a set of network graphs $\mathbb{G}^N_{\alpha,\beta,\theta}$ be defined.*

If A is at most weakly unstable and (A, B) is stabilizable, then the state synchronization problem stated in Problem 6.1 with $\mathbf{G} = \mathbb{G}^N_{\alpha,\beta,\theta}$ is solvable if the condition (6.23) holds. In particular, for any given $\bar{\tau}$ satisfying (6.23), there exist a $\rho > 0$ and a $\delta^* > 0$ such that for this ρ and any $\delta \in (0, \delta^*]$, controller (6.31) solves the state synchronization problem for any graph $\mathcal{G} \in \mathbb{G}^N_{\alpha,\beta,\theta}$ and any $\tau \in [0, \bar{\tau}]$. Moreover, the synchronized trajectory is given by (6.19).

Proof Given that (6.23) is satisfied, we can find $\phi > 0$ such that

$$\theta + \bar{\tau}(\omega_{\max} + \phi) < \frac{\pi}{2}.$$

This implies that for a sufficiently large ρ

$$\operatorname{Re} \rho\lambda e^{-j\omega\tau} \geq \rho|\lambda| \cos(\psi - \omega\tau) \geq \rho\beta \cos(\theta - \omega_{\max}\bar{\tau}) > 1 \tag{6.32}$$

for all ω satisfying $|\omega| \leq \omega_{\max} + \phi$. From Theorem 6.9, we only need to prove that the system

$$\dot{x}(t) = Ax(t) - \rho\lambda B F_\delta x(t - \tau), \tag{6.33}$$

is asymptotically stable for all λ and τ such that $|\lambda| < \alpha$, $\operatorname{Re}\lambda \geq \beta$, $|\arg\lambda| < \theta$, and $\tau \in [0, \bar{\tau}]$. We first consider the case without delay, i.e., $\tau = 0$, which follows directly from Lemma 6.5 since (6.32) implies that

$$\operatorname{Re} \rho\lambda > 1.$$

Given that the system is asymptotically stable without delays, we can use Lemma 6.3, to show that (6.33) is asymptotically stable for any $\tau \in [0, \bar{\tau}]$ if and only if

$$\det\left[j\omega I - A - \rho\lambda e^{-j\omega\tau} B F_\delta\right] \neq 0,$$
$$\forall \omega \in \mathbb{R}, \ \operatorname{Re}\lambda \geq \beta, \ |\arg\lambda| \leq \theta, \ \tau \in [0, \bar{\tau}]. \tag{6.34}$$

Similar to the proof of Theorem 6.13, we can show that for δ sufficiently small, (6.34) is satisfied for $|\omega| > \omega_{\max} + \phi$. For $|\omega| \leq \omega_{\max} + \phi$, we use (6.32). From Lemma 6.5, we find that

$$A + \rho\lambda e^{-j\omega\tau} B F_\delta$$

is asymptotically stable for all $\omega \in \mathbb{R}$, for all λ such that $\operatorname{Re}\lambda \geq \beta$, $|\arg\lambda| \leq \theta$ and for all $\tau \in [0, \bar{\tau}]$ and therefore (6.34) is satisfied which guarantees that we achieve synchronization. ■

6.2.4 *Protocol Design for MAS with Partial-State Coupling*

In this section, we consider protocol design for MAS with partial-state coupling. We will see that the basic approach is very similar to the previous section except that an observer will be needed. As before, we show first that the state synchronization among agents in a network with partial-state coupling and in the presence of uniform input delay can be solved by equivalently solving a robust stabilization problem.

6.2.4.1 Connection to a Robust Stabilization Problem

The MAS system described by agent (6.1) and (6.2) after implementing the linear dynamical protocol (6.5) is described by

$$\begin{aligned} \dot{\bar{x}}_i(t) &= \begin{pmatrix} A & BC_c \\ 0 & A_c \end{pmatrix}\bar{x}_i(t) + \begin{pmatrix} 0 \\ B_c \end{pmatrix}\zeta_i(t-\tau), \\ y_i(t) &= \begin{pmatrix} C & 0 \end{pmatrix}\bar{x}_i(t), \\ \zeta_i(t) &= \textstyle\sum_{j=1}^N \ell_{ij} y_j(t), \end{aligned} \tag{6.35}$$

for $i = 1, \dots, N$ where

$$\bar{x}_i = \begin{pmatrix} x_i \\ \chi_j \end{pmatrix}.$$

Define

$$\bar{x} = \begin{pmatrix} \bar{x}_1 \\ \vdots \\ \bar{x}_N \end{pmatrix}, \quad \bar{A} = \begin{pmatrix} A & BC_c \\ 0 & A_c \end{pmatrix}, \quad \bar{B} = \begin{pmatrix} 0 \\ B_c \end{pmatrix}, \quad \text{and} \quad \bar{C} = \begin{pmatrix} C & 0 \end{pmatrix}.$$

Then, the overall dynamics of the N agents can be written as

$$\dot{\bar{x}}(t) = (I_N \otimes \bar{A})\bar{x}(t) + (L \otimes \bar{B}\bar{C})\bar{x}(t-\tau). \tag{6.36}$$

With a similar observation as in Sect. 2.5.1, the synchronization for the system (6.36) is equivalent to the asymptotic stability of the following $N-1$ subsystems:

$$\dot{\tilde{\eta}}_i(t) = \bar{A}\tilde{\eta}_i(t) + \lambda_i \bar{B}\bar{C}\tilde{\eta}_i(t-\tau), \quad i = 2, \dots, N, \tag{6.37}$$

for any $\tau \in [0, \bar{\tau}]$, where $\lambda_i, i = 2, \dots, N$ are the $N-1$ nonzero eigenvalues of the Laplacian matrix L.

Theorem 6.19 *The MAS (6.36) achieves state synchronization if and only if the system (6.37) is globally asymptotically stable for $i = 2, \dots, N$ and all $\tau \in [0, \bar{\tau}]$.*

Proof Note that L has eigenvalue 0 with associated right eigenvector $\mathbf{1}$. Let

$$L = TJT^{-1} \tag{6.38}$$

where J is the Jordan canonical form of the Laplacian matrix L such that $J(1,1) = 0$ and the first column of T equals $\mathbf{1}$. Let

$$\eta := \left(T^{-1} \otimes I_{n+n_c}\right)\bar{x} = \begin{pmatrix} \eta_1 \\ \vdots \\ \eta_N \end{pmatrix}$$

where $\eta_i \in \mathbb{C}^{n+n_c}$. In the new coordinates, the dynamics of η can be written as

$$\dot{\eta}(t) = (I_N \otimes \bar{A})\eta(t) + (J_L \otimes \bar{B}\bar{C})\eta(t-\tau). \tag{6.39}$$

The eigenvalues of system (6.39) are given by the roots of its characteristic equation

$$H(s) = \det\left\{sI - (I_N \otimes \bar{A}) - e^{-s\tau}(J_L \otimes \bar{B}\bar{C})\right\} = 0,$$

which, due to the diagonal structure of $I_N \otimes \bar{A}$ and the upper triangular structure of $J_L \otimes \bar{B}\bar{C}$, are the union of the eigenvalues of the $N-1$ subsystems given in (6.37) and the eigenvalues of $\bar{A}$.

Now if (6.37) is globally asymptotically stable for $i = 2, \ldots, N$, we see from the above that $\eta_i(t) \to 0$ for $i = 2, \ldots, N$. This implies that

$$\bar{x}(t) - (T \otimes I_{n+n_c})\begin{pmatrix} \eta_1(t) \\ 0 \\ \vdots \\ 0 \end{pmatrix} \to 0.$$

Note that the first column of T is equal to the vector $\mathbf{1}$ and therefore

$$\bar{x}_i(t) - \eta_1(t) \to 0$$

for $i = 1, \ldots, N$. This implies that we achieve state synchronization.

Conversely, suppose the network (6.36) reaches synchronization. In this case, we shall have

$$\bar{x}(t) - \mathbf{1} \otimes x_1(t) \to 0.$$

Then $\eta(t) \to (T^{-1}\mathbf{1}) \otimes x_1(t)$. Since $\mathbf{1}$ is the first column of T, we have

$$T^{-1}\mathbf{1} = \begin{pmatrix} 1 \\ 0 \\ \vdots \\ 0 \end{pmatrix}.$$

Therefore, $\eta(t) \to (T^{-1}\mathbf{1}) \otimes x_1(t)$ implies that $\eta_1(t) \to x_1(t)$ and $\eta_i(t) \to 0$ for $i = 2, \ldots, N$ for any $\tau \in [0, \bar{\tau}]$ and for all the initial conditions. ■

We immediately have the following result.

Corollary 6.20 *Consider a MAS described by (6.1) and (6.2). If there exists a protocol of the form (6.5) that achieves state synchronization, then the associated graph must have a directed spanning tree or A must be asymptotically stable.*

Since A being asymptotically stable is a trivial case, the requirement that the graph of a multi-agent system has a directed spanning tree is essentially necessary.

Remark 6.21 Again, from Theorem 6.19 it is clear that the synchronized trajectory is given by $x_s(t) = \eta_1(t)$, which is governed by

$$\dot{\eta}_1 = \bar{A}\eta_1, \quad \eta_1(0) = (w \otimes I_{n+n_c})\bar{x}(0), \tag{6.40}$$

where w is the first row of T^{-1}, i.e., the normalized left eigenvector associated with the zero eigenvalue. This shows that the modes of the synchronized trajectory are determined by the eigenvalues of $\bar{A}$ and the complete dynamics depends on both $\bar{A}$ and a weighted average of the initial conditions of the agents and their protocol.

In light of Lemma 2.8, one can conclude that $\eta_1(0)$ is only a linear combination of the initial conditions of root agents and their protocol. As such, the synchronized trajectory given by (6.40) can be written explicitly as

$$\eta_1(t) = e^{\bar{A}t} \sum_{i \in \Pi_{\mathcal{G}}} w_i \bar{x}_i(0), \tag{6.41}$$

which is the weighted average of the trajectories of root agents and their protocol. In the case of state synchronization via a stable protocol, the synchronized trajectory is given by

$$\eta_1(t) = e^{At} \sum_{i \in \Pi_{\mathcal{G}}} w_i x_i(0). \tag{6.42}$$

In light of the definition of Problem 6.2 that synchronization is formulated for a set of graphs, from the simultaneous system (6.37) we basically obtain a *robust* stabilization problem, i.e., the stabilization of the system

$$\begin{cases} \dot{x}(t) = Ax(t) + Bu(t - \tau), \\ z(t) = \lambda C x(t), \end{cases} \tag{6.43}$$

via a protocol

$$\begin{cases} \dot{\chi}(t) = A_c \chi(t) + B_c z(t), \\ u(t) = C_c \chi(t), \end{cases} \tag{6.44}$$

for any $\tau \in [0, \bar{\tau}]$ and for all λ that are the non-zero eigenvalues of all possible Laplacian matrices associated with the given set of graphs $\mathbf{G}$.

Define an auxiliary system

$$\begin{aligned} \dot{x}(t) &= Ax(t) + \lambda Bu(t-\tau), \\ z(t) &= Cx(t), \end{aligned} \tag{6.45}$$

and a protocol

$$\begin{aligned} \dot{\chi}(t) &= A_c\chi(t) + B_c z(t), \\ u(t) &= C_c\chi(t), \end{aligned} \tag{6.46}$$

where A_c, B_c, and C_c are given in (6.5). We need the result of the following lemma.

Lemma 6.22 *Problem 6.2 is solvable via the controller (6.5) for any graph $\mathcal{G} \in \mathbf{G}$ and for any $\tau \in [0, \bar{\tau}]$ if the closed-loop system of (6.45) and (6.46) is globally asymptotically stable for any nonzero eigenvalue λ of a Laplacian matrix associated with a graph $\mathcal{G} \in \mathbf{G}$ and for any $\tau \in [0, \bar{\tau}]$.*

Proof The closed-loop system of (6.43) and (6.44) has a set of eigenvalues determined by

$$\det \begin{pmatrix} sI - A & -BC_c \\ -\lambda e^{s\tau} B_c C & sI - A_c \end{pmatrix} = 0.$$

On the other hand, the eigenvalues of closed-loop system of (6.45) and (6.46) are given by

$$\det \begin{pmatrix} sI - A & -\lambda e^{s\tau} BC_c \\ -B_c C & sI - A_c \end{pmatrix} = 0.$$

Note that

$$\begin{pmatrix} \frac{1}{\lambda} e^{-s\tau} I & 0 \\ 0 & I \end{pmatrix} \begin{pmatrix} sI - A & -\lambda e^{s\tau} BC_c \\ -B_c C & sI - A_c \end{pmatrix} \begin{pmatrix} \lambda e^{s\tau} I & 0 \\ 0 & I \end{pmatrix} = \begin{pmatrix} sI - A & -BC_c \\ -\lambda e^{s\tau} B_c C & sI - A_c \end{pmatrix}.$$

We find that both the above two closed-loop systems have the same set of eigenvalues. Therefore, the result is proved. ∎

6.2.4.2 ARE-Based Method

We design the following protocol based on a low-gain method utilizing an algebraic Riccati equation.

Protocol design 6.4 *Consider a MAS described by (6.1) and (6.2). We choose an observer gain K such that $A + KC$ is Hurwitz stable, and consider a feedback gain $F_\delta = -B'P_\delta$ where P_δ is the unique solution of the ARE (6.22).*

Then, the linear dynamic protocol can be constructed as

$$\begin{aligned}\dot{\chi}_i &= (A + KC)\chi_i - K\zeta_i,\\ u_i &= \rho F_\delta \chi_i,\end{aligned} \tag{6.47}$$

where $\rho > 0$ and δ are design parameters to be chosen later.

The main result based on the above design is stated as follows.

Theorem 6.23 *Consider a MAS described by (6.1) and (6.2). Let any $\alpha \geq \beta > 0$ and $\frac{\pi}{2} > \theta \geq 0$ be given, and hence a set of network graphs $\mathbb{G}^N_{\alpha,\beta,\theta}$ be defined.*

Assume A is at most weakly unstable and (A, B, C) is stabilizable and detectable. The state synchronization problem stated in Problem 6.2 with $\mathbf{G} = \mathbb{G}^N_{\alpha,\beta,\theta}$ is solvable if the condition (6.23) holds. In particular, for any given $\bar{\tau}$ satisfying (6.23), there exist a $\rho > 0$ and a $\delta^* > 0$ such that for this ρ and any $\delta \in (0, \delta^*]$, the stable dynamic controller (6.47) solves the state synchronization problem for any graph $\mathcal{G} \in \mathbb{G}^N_{\alpha,\beta,\theta}$ and any $\tau \in [0, \bar{\tau}]$. Moreover, the synchronized trajectory is given by (6.42).

Proof According to Lemma 6.22, we only need to prove that the system

$$\begin{cases}\dot{x}(t) = Ax(t) - \rho|\lambda|e^{j\psi}BB'P_\delta\chi(t-\tau),\\ \dot{\chi}(t) = (A + KC)\chi(t) - KCx(t),\end{cases} \tag{6.48}$$

is globally asymptotically stable for any λ that satisfies $|\lambda| \in (\beta, \alpha)$ and $\psi \in [-\theta, \theta]$, and for any $\tau \in [0, \bar{\tau}]$.

Define

$$\bar{\mathcal{A}} = \begin{pmatrix} A & 0 \\ -KC & A + KC \end{pmatrix}, \quad \bar{\mathcal{B}} = \begin{pmatrix} B \\ 0 \end{pmatrix}, \quad \bar{\mathcal{F}}_\delta = \begin{pmatrix} 0 & -B'P_\delta \end{pmatrix}.$$

First of all, for the given α, β, and $\psi \in (-\theta, \theta)$, there exists a ρ such that

$$\rho > \frac{1}{\beta\cos(\theta + \omega_{\max}\bar{\tau})}. \tag{6.49}$$

Let this ρ be fixed. By Lemma 6.6, there exists a δ_1 such that for a $\delta \in (0, \delta_1]$, $\bar{\mathcal{A}} + \rho|\lambda|e^{j\psi}\bar{\mathcal{B}}\bar{\mathcal{F}}_\delta$ is Hurwitz stable for λ satisfying $|\lambda| \in (\beta, \alpha)$ and $\psi \in (-\theta, \theta)$.

It follows from Lemma 6.3 that (6.48) is asymptotically stable if

$$\det\left[j\omega I - \bar{\mathcal{A}} - \rho|\lambda|e^{j(\psi-\omega\tau)}\bar{\mathcal{B}}\bar{\mathcal{F}}_\delta\right] \neq 0,\ \forall\omega\in\mathbb{R},$$

$$\forall|\lambda|\in(\beta,\alpha),\ \forall\psi\in(-\theta,\theta),\ \forall\tau\in[0,\bar{\tau}]. \tag{6.50}$$

Given (6.49), there exists a $\phi > 0$ such that

$$\rho|\lambda|\cos(\theta+\omega\bar{\tau}) > 1, \quad \forall|\omega| < \omega_{\max}+\phi. \tag{6.51}$$

We can show, as in the proof of Theorem 6.13, that there exists a $\delta_2 \leq \delta_1$ such that for $\delta \in (0, \delta_2]$, the condition (6.50) holds for $|\omega| \geq \omega_{\max}+\phi$.

For $|\omega| < \omega_{\max}+\phi$, it follows from Lemma 6.6 that

$$\bar{\mathcal{A}} + \rho|\lambda|e^{j(\psi-\omega\tau)}\bar{\mathcal{B}}\bar{\mathcal{F}}_\delta$$

is Hurwitz stable. Therefore, the condition (6.50) also holds with $|\omega| < \omega_{\max}+\phi$. ∎

As discussed in the full-state coupling, any arbitrarily bounded communication delay can be tolerated in the special case where $\omega_{\max} = 0$, i.e., the eigenvalues of A are either zero or in the open left half plane. Then we immediately have the following corollary.

Corollary 6.24 *Consider a MAS described by (6.1) and (6.2). Let any $\alpha \geq \beta > 0$ and $\frac{\pi}{2} > \theta \geq 0$ be given, and hence a set of network graphs $\mathbb{G}^N_{\alpha,\beta,\theta}$ be defined.*

Suppose the eigenvalues of matrix A are either zero or in the open left half plane. If (A, B) is stabilizable and (C, A) is detectable, then the state synchronization problem stated in Problem 6.5 with $\mathbf{G} = \mathbb{G}^N_{\alpha,\beta,\theta}$ is always solvable. In particular, for any given $\bar{\tau} > 0$, there exist a $\rho > 0$ and a $\delta^* > 0$ such that for this ρ and any $\delta \in (0, \delta^*]$, the stable dynamic controller (6.47) solves the state synchronization problem for any graph $\mathcal{G} \in \mathbb{G}^N_{\alpha,\beta,\theta}$ and any $\tau \in [0, \bar{\tau}]$. Moreover, the synchronized trajectory is given by (6.42).

We consider next a special case where the agent dynamics are neutrally stable.

Protocol design 6.5 *Consider a MAS described by (6.1) and (6.2). We will use a CSS observer-based protocol for the agents.*

We choose an observer gain K such that $A + KC$ is Hurwitz stable and consider a feedback gain

$$F = -B'P \tag{6.52}$$

(continued)

Protocol design 6.5 (continued)
where $P > 0$ is a solution of the Lyapunov inequality

$$A'P + PA \leq 0. \tag{6.53}$$

Then, the linear dynamic protocol can be designed as

$$\begin{cases} \dot{\chi}_i = (A + KC)\chi_i - K\zeta_i \\ u_i = \delta F \chi_i, \end{cases} \tag{6.54}$$

where $\delta > 0$ is a low-gain parameter.

In this case, we show here that the consensus controller design no longer requires the knowledge of β and hence allows us to deal with a larger set of unknown network graphs that can be denoted as $\mathcal{G}^N_{\alpha,0,\theta}$. The result based on the above design is stated in the following theorem.

Theorem 6.25 *Consider a MAS described by (6.1) and (6.2). Let any $\alpha > 0$ and $\frac{\pi}{2} > \theta \geq 0$ be given, and hence a set of network graphs $\mathbb{G}^N_{\alpha,0,\theta}$ be defined.*

Suppose that the agent dynamics are neutrally stable. If (A, B) is stabilizable and (C, A) is detectable, then the state synchronization problem stated in Problem 6.5 with $\mathbf{G} = \mathbb{G}^N_{\alpha,0,\theta}$ is solvable for any graph $\mathcal{G} \in \mathbb{G}^N_{\alpha,0,\theta}$ if the condition (6.23) holds. In particular, for any given $\bar{\tau}$ satisfying (6.23), there exists a $\delta^* > 0$ such that for any $\delta \in (0, \delta^*]$, the stable dynamic controller (6.54) solves the state synchronization problem for any graph $\mathcal{G} \in \mathbb{G}^N_{\alpha,0,\theta}$ and any $\tau \in [0, \bar{\tau}]$. Moreover, the synchronized trajectory is given by (6.42).

Proof From Lemma 6.22 it follows that Theorem 6.25 holds if there exists a $\delta^* > 0$ such that for $\delta \in (0, \delta^*]$, the system

$$\begin{cases} \dot{x}(t) = Ax(t) - \delta|\lambda|e^{j\psi}BB'P\chi(t-\tau) \\ \dot{\chi}(t) = (A + KC)\chi(t) - KCx(t), \end{cases} \tag{6.55}$$

is asymptotically stable for any $|\lambda| \in (0, \alpha)$, $\psi \in [-\theta, \theta]$, and $\tau \in [0, \bar{\tau}]$.

Define

$$\bar{\mathcal{A}} = \begin{pmatrix} A & 0 \\ -KC & A + KC \end{pmatrix}, \quad \bar{\mathcal{B}} = \begin{pmatrix} B \\ 0 \end{pmatrix}, \quad \bar{\mathcal{F}}_\delta = \begin{pmatrix} 0 & -\delta B'P \end{pmatrix}.$$

By Lemma 6.8, there exists a δ_1 such that for $\delta \in (0, \delta_1]$, $\bar{\mathcal{A}} + |\lambda|e^{j\psi}\bar{\mathcal{B}}\bar{\mathcal{F}}_\delta$ is Hurwitz stable. In other words, the system without delay is asymptotically stable. It then follows from Lemma 6.3 that (6.55) is asymptotically stable if

$$\det\left[j\omega I - \bar{\mathcal{A}} - |\lambda|e^{j(\psi-\omega\tau)}\bar{\mathcal{B}}\bar{\mathcal{F}}_\delta\right] \neq 0, \ \forall\omega\in\mathbb{R},$$
$$\forall|\lambda|\in(0,\alpha), \ \forall\psi\in(-\theta,\theta), \ \forall\tau\in[0,\bar{\tau}]. \qquad (6.56)$$

As before there exist a $\phi > 0$ and a $\delta_2 \leq \delta_1$ such that for $\delta \in (0, \delta_2]$, the condition (6.56) holds for $|\omega| \geq \omega_{\max} + \phi$. On the other hand, for $|\omega| < \omega_{\max} + \phi$, it follows from Lemma 6.8 that $\bar{\mathcal{A}} + |\lambda|e^{j(\psi-\omega\tau)}\bar{\mathcal{F}}_\delta$ is Hurwitz stable. Therefore, the condition (6.56) also holds with $|\omega| < \omega_{\max} + \phi$. ■

6.2.4.3 Direct Eigenstructure Assignment Method

We present here a protocol via direct eigenstructure assignment, which is already given in Sect. 2.5.3.2.

Protocol design 6.6 *Consider a MAS described by (6.1) and (6.2). We choose an observer gain K such that $A + KC$ is Hurwitz stable and consider a feedback gain F_δ given in (2.72). Then, the linear dynamic protocol can be constructed as*

$$\begin{aligned}\dot{\chi}_i &= (A + KC)\chi_i - K\zeta_i,\\ u_i &= \rho F_\delta \chi_i,\end{aligned} \qquad (6.57)$$

where $\rho > 0$ and δ are design parameters.

The main result based on the above design is stated as follows.

Theorem 6.26 *Consider a MAS described by (6.1) and (6.2). Let any $\alpha \geq \beta > 0$ and $\frac{\pi}{2} > \theta \geq 0$ be given, and hence a set of network graphs $\mathbb{G}^N_{\alpha,\beta,\theta}$ be defined.*

Assume A is at most weakly unstable and (A, B, C) is stabilizable and detectable. The state synchronization problem stated in Problem 6.2 with $\mathbf{G} = \mathbb{G}^N_{\alpha,\beta,\theta}$ is solvable if the condition (6.23) holds. In particular, for any given $\bar{\tau}$ satisfying (6.23), there exist a $\rho > 0$ and a $\delta^* > 0$ such that for this ρ and any $\delta \in (0, \delta^*]$, controller (6.57) solves the state synchronization problem for any graph $\mathcal{G} \in \mathbb{G}^N_{\alpha,\beta,\theta}$ and any $\tau \in [0, \bar{\tau}]$. Moreover, the synchronized trajectory is given by (6.42).

Proof According to Lemma 6.22, we only need to prove that the system

$$\begin{cases}\dot{x}(t) = Ax(t) + \rho|\lambda|e^{j\psi}BF_\delta\chi(t-\tau),\\ \dot{\chi}(t) = (A + KC)\chi(t) - KCx(t),\end{cases} \qquad (6.58)$$

is globally asymptotically stable for any λ that satisfies $|\lambda| \in (\beta, \alpha)$ and $\psi \in [-\theta, \theta]$ and for any $\tau \in [0, \bar{\tau}]$.

Define

$$\bar{\mathcal{A}} = \begin{pmatrix} A & 0 \\ -KC & A + KC \end{pmatrix}, \quad \bar{\mathcal{B}} = \begin{pmatrix} B \\ 0 \end{pmatrix}, \quad \bar{\mathcal{F}}_\delta = \begin{pmatrix} 0 & F_\delta \end{pmatrix}.$$

First of all, for the given α, β, and $\psi \in (-\theta, \theta)$, there exists a ρ such that

$$\rho > \frac{2}{\beta \cos(\theta + \omega_{\max} \bar{\tau})}. \tag{6.59}$$

Let this ρ be fixed. By Lemma 6.7, there exists a δ_1 such that for a $\delta \in (0, \delta_1]$, $\bar{\mathcal{A}} + \rho|\lambda|e^{j\psi}\bar{\mathcal{B}}\bar{\mathcal{F}}_\delta$ is Hurwitz stable for λ satisfying $|\lambda| \in (\beta, \alpha)$ and $\psi \in (-\theta, \theta)$. It follows from Lemma 6.3 that (6.58) is asymptotically stable if

$$\det\left[j\omega I - \bar{\mathcal{A}} - \rho|\lambda|e^{j(\psi - \omega\tau)}\bar{\mathcal{B}}\bar{\mathcal{F}}_\delta\right] \neq 0, \ \forall \omega \in \mathbb{R},$$
$$\forall |\lambda| \in (\beta, \alpha), \ \forall \psi \in (-\theta, \theta), \ \forall \tau \in [0, \bar{\tau}]. \tag{6.60}$$

Given (6.59), there exists a $\phi > 0$ such that

$$\rho|\lambda| \cos(\theta + \omega\bar{\tau}) > 2, \quad \forall |\omega| < \omega_{\max} + \phi. \tag{6.61}$$

We can show, as in the proof of Theorem 6.13, that there exists a $\delta_2 \leq \delta_1$ such that for $\delta \in (0, \delta_2]$, the condition (6.50) holds for $|\omega| \geq \omega_{\max} + \phi$.

For $|\omega| < \omega_{\max} + \phi$, it follows from Lemma 6.7 that

$$\bar{\mathcal{A}} + \rho|\lambda|e^{j(\psi - \omega\tau)}\bar{\mathcal{B}}\bar{\mathcal{F}}_\delta$$

is Hurwitz stable. Therefore, the condition (6.60) also holds with $|\omega| < \omega_{\max} + \phi$. ∎

6.2.5 *Static Protocol Design with Partial-State Coupling*

In this section, we consider the static protocol design for two classes of agents identified in Sect. 2.5.7. These are squared-down passive agents and squared-down passifiable via static input feedforward agents. With delay we intrinsically rely on a low-gain feedback design. Systems which are squared-down passifiable via static output feedback require a minimal size of the gain unless they are already squared-down passive. Even systems which are squared-down passifiable via static output feedback and are already neutrally stable do not work in conjunction with a low-gain static feedback design. Therefore, this class of systems will not be considered.

We show first that the state synchronization among agents in the network with partial-state coupling and input delay via static protocols can be solved by equivalently solving a robust stabilization problem.

6.2.5.1 Connection to a Robust Stabilization Problem

The static protocol (6.4) with ζ_i given in (6.2) will be used. Then, the closed-loop system of the agent (6.1) and the protocol (6.4) can be described by

$$\begin{cases} \dot{x}_i(t) = Ax_i(t) + BF\zeta_i(t-\tau), \\ y_i(t) = Cx_i(t), \\ \zeta_i(t) = \sum_{j=1}^{N} \ell_{ij} y_j(t), \end{cases} \tag{6.62}$$

for $i = 1, \ldots, N$. Define

$$x = \begin{pmatrix} x_1 \\ \vdots \\ x_N \end{pmatrix}.$$

The overall dynamics of the N agents can be written as

$$\dot{x}(t) = (I_N \otimes A)x(t) + (L \otimes BFC)x(t-\tau). \tag{6.63}$$

Along the lines of the proof of Theorem 6.9, we immediately have the following result.

Theorem 6.27 *The MAS (6.63) achieves state synchronization if and only if the $N-1$ subsystems*

$$\dot{\tilde{\eta}}_i(t) = A\tilde{\eta}_i(t) + \lambda_i BFC\tilde{\eta}_i(t-\tau), \quad i = 2, \ldots, N, \tag{6.64}$$

are globally asymptotically stable for $i = 2, \ldots, N$ and for any $\tau \in [0, \bar{\tau}]$, where λ_i, $i = 2, \ldots, N$ are the $N-1$ nonzero eigenvalues of the Laplacian matrix L.

Similarly, we immediately have the following corollary and the synchronized trajectory.

Corollary 6.28 *Consider a MAS described by (6.1) and (6.2). If there exists a protocol of the form (6.4) that achieves state synchronization, then the associated graph must have a directed spanning tree or A must be asymptotically stable.*

The synchronized trajectory is given explicitly as

$$\eta_1(t) = e^{At} \sum_{i \in \Pi_{\mathcal{G}}} w_i x_i(0), \tag{6.65}$$

which is the weighted average of the trajectories of root agents.

In light of the definition of Problem 6.2 that synchronization is formulated for a set of graphs, we basically obtain a robust stabilization problem, i.e., the stabilization of the system

$$\begin{aligned} \dot{x}(t) &= Ax(t) + \lambda Bu(t-\tau), \\ y(t) &= Cx(t), \end{aligned} \tag{6.66}$$

via a protocol

$$u(t) = Fy(t),$$

for any λ that are nonzero eigenvalues of possible Laplacian matrices associated with the given set of graphs **G** and for any $\tau \in [0, \bar{\tau}]$.

6.2.5.2 Squared-Down Passive Agents

Next, we can design a protocol provided the system is squared-down passive.

Protocol design 6.7 *Consider a MAS described by agents (6.1) with communication via (6.2). Assume (6.1) is squared-down passive with respect to G_1 and G_2 such that (A, BG_1) is controllable and (A, G_2C) is observable. The static protocol is designed as*

$$u_i = -\delta G_1 K G_2 \zeta_i, \tag{6.67}$$

where $\delta > 0$ is a low-gain parameter and K is any positive definite matrix.

The main result regarding this design can be stated as follows.

Theorem 6.29 *Consider a MAS described by agents (6.1) and (6.2). Let any $\alpha > 0$ and $\theta \in (0, \frac{\pi}{2})$ be given, and consider the set of network graphs $\mathbb{G}^N_{\alpha,0,\theta}$.*

If the agents are squared-down passive with respect to G_1 and G_2 such that (A, BG_1) is controllable and (A, G_2C) is observable, then the state synchronization problem stated in Problem 6.2 is solvable with a static protocol if the condition (6.23) holds. In particular, for any given $\bar{\tau}$ satisfying (6.23), there exist for any $K > 0$, a $\delta^* > 0$ such that for any $\delta \in (0, \delta^*]$, controller (6.67) solves the state synchronization problem for any graph $\mathcal{G} \in \mathbb{G}^N_{\alpha,0,\theta}$ and any $\tau \in [0, \bar{\tau}]$. Moreover, the synchronized trajectory is given by (6.65).

Proof Let $\lambda = |\lambda|e^{j\psi}$, where $|\lambda| < \alpha$ and $\psi \in [-\theta, \theta]$. According to Theorem 6.27, we only need to prove that

$$\dot{x}(t) = Ax(t) - \delta|\lambda|e^{j\psi}BG_1KG_2Cx(t-\tau) \tag{6.68}$$

is asymptotically stable for any $\tau \in [0, \bar{\tau}]$ and $\psi \in [-\theta, \theta]$.

From Theorem 2.44, we have $A - \delta|\lambda|e^{j\psi}BG_1KG_2C$ Hurwitz stable for any $\lambda \in \mathbb{C}^+$. Hence the system without delay is asymptotically stable. Then it follows from Lemma 6.3 that the system (6.68) is asymptotically stable if

$$\det\left[j\omega I - A + \delta|\lambda|e^{j(\psi-\omega\tau)}BG_1KG_2C\right] \neq 0, \tag{6.69}$$

for all $\omega \in \mathbb{R}$, $\tau \in [0, \bar{\tau}]$ and $\psi \in [-\theta, \theta]$.

We split the proof of (6.69) into two cases where $|\omega| < \omega_{\max} + \phi$ and $|\omega| \geq \omega_{\max} + \phi$, respectively.

If $|\omega| \geq \omega_{\max} + \phi$, there exists a $\mu > 0$ such that

$$\sigma_{\min}(j\omega I - A) > \mu, \quad \forall \omega, \text{ such that } |\omega| \geq \omega_{\max} + \phi.$$

Then, with the same arguments as in the proof of Theorem 6.13, we obtain

$$\sigma_{\min}(j\omega I - A - \delta|\lambda|e^{j(\psi-\omega\tau)}BG_1KG_2C) \geq \mu - \tfrac{\mu}{2} \geq \tfrac{\mu}{2},$$

for small enough δ. Therefore, the condition (6.69) holds for $|\omega| \geq \omega_{\max} + \phi$.

If $|\omega| < \omega_{\max} + \phi$, we find that

$$|\psi - \omega\tau| < \theta + |\omega|\bar{\tau} < \tfrac{\pi}{2},$$

and hence $\text{Re}(|\lambda|e^{j(\psi-\omega\tau)}) > 0$. Therefore by Theorem 2.44, we have that

$$A - \delta|\lambda|e^{j(\psi-\omega\tau)}BG_1KG_2C$$

is Hurwitz stable. Therefore (6.69) also holds with $|\omega| < \omega_{\max} + \phi$ which completes the proof. ∎

6.2.5.3 Squared-Down Passifiable via Static Input Feedforward

Protocol Design 2.13 for a MAS with agents which are squared-down passifiable via static input feedforward can tolerate input delay. We reproduce the protocol design here.

Protocol design 6.8 *Consider a MAS described by agents (6.1) with communication via (6.2). Assume the agents are squared-down passifiable via static input feedforward with respect to G_1, G_2, and R. The static protocol is designed as*

$$u_i = -\delta G_1 K G_2 \zeta_i, \tag{6.70}$$

with $K > 0$ and $\delta > 0$ is a low-gain parameter to be designed.

The main result based on the above design can be stated as follows.

Theorem 6.30 *Consider a MAS described by agents (6.1) and (6.2). Let any $\alpha > 0$ and $\theta \in (0, \frac{\pi}{2})$ be given, and consider the set of network graphs $\mathbb{G}^N_{\alpha,0,\theta}$.*

Assume the agents are squared-down passifiable via static input feedforward with respect to G_1, G_2, and R such that (A, BG_1) is controllable and (A, G_2C) is observable with BG_1 and G_2C full-column and full-row rank, respectively. The state synchronization problem stated in Problem 6.2 but with a static protocol is solvable if the condition (6.23) holds. In particular, for any given $\bar{\tau}$ satisfying (6.23), there exist a $\delta^* > 0$ such that for any $\delta \in (0, \delta^*]$, controller (6.70) solves the state synchronization problem for any graph $\mathcal{G} \in \mathbb{G}^N_{\alpha,0,\theta}$ and any $\tau \in [0, \bar{\tau}]$. Moreover, the synchronized trajectory is given by (6.65).

Proof Let $\lambda = |\lambda|e^{j\psi}$, where $|\lambda| < \alpha$ and $\psi \in [-\theta, \theta]$. Similarly to the passive case in Theorem 6.29, we only need to prove that

$$\dot{x} = Ax - \delta|\lambda|e^{j\psi} BG_1 K G_2 C x(t - \tau) \tag{6.71}$$

is asymptotically stable for any $\tau \in [0, \bar{\tau}]$ and $\psi \in [-\theta, \theta]$.

When the system (6.71) is without delay, i.e., $\tau = 0$, from Theorem 2.48, we have $A - \delta|\lambda|e^{j\psi} BG_1 K G_2 C$ Hurwitz stable. Then it follows from Lemma 6.3 that the system (6.71) is asymptotically stable if

$$\det\left[j\omega I - A + \delta|\lambda|e^{j(\psi-\omega\tau)} BG_1 K G_2 C\right] \neq 0, \tag{6.72}$$

for all $\omega \in \mathbb{R}$, $\tau \in [0, \bar{\tau}]$ and $\psi \in [-\theta, \theta]$.

Next we split the proof of (6.72) into two cases where $|\omega| < \omega_{\max} + \phi$ and $|\omega| \geq \omega_{\max} + \phi$, respectively.

If $|\omega| \geq \omega_{\max} + \phi$, there exists a $\mu > 0$ such that

$$\sigma_{\min}(j\omega I - A) > \mu, \quad \forall \omega, \text{ such that } |\omega| \geq \omega_{\max} + \phi.$$

Then, with the exact arguments used in the proof of Theorem 6.13 and the boundedness of $|\lambda|$, we obtain

$$\sigma_{\min}(j\omega I - A - \delta|\lambda|e^{j(\psi-\omega\tau)}BG_1KG_2C) \geq \mu - \tfrac{\mu}{2} \geq \tfrac{\mu}{2},$$

for small enough δ. Therefore, the condition (6.72) holds for $|\omega| \geq \omega_{\max} + \phi$.

If $|\omega| < \omega_{\max} + \phi$, we find that $\tilde{\lambda} = |\lambda|e^{j(\psi-\omega\tau)}$ satisfies $|\tilde{\lambda}| < \alpha$ and

$$\operatorname{Re}\tilde{\lambda} = |\lambda|\cos(\psi - \omega\tau) > 0$$

since

$$|\psi - \omega\tau| < \theta + (\omega_{\max} + \phi)\bar{\tau} < \frac{\pi}{2}.$$

Following Theorem 2.48, we have that

$$A - \delta\tilde{\lambda}BG_1KG_2C$$

is Hurwitz stable which immediately yields that (6.72) also holds for $|\omega| < \omega_{\max}+\phi$ which completes the proof. ∎

6.2.6 *Protocol Design for a Known Communication Topology*

It is clear from Theorem 6.13 that in the case where the network is directed and its associated Laplacian matrix has complex eigenvalues, we have to yield some delay tolerance to accommodate the uncertainty of complex Laplacian eigenvalues. However, it will be shown in this section that if the communication topology is exactly known, we can achieve a larger delay tolerance. In this section, we shall study the synchronization problem in a known network which can be either directed or undirected.

We will use the following lemma (see [25, Theorem 1]).

Lemma 6.31 *Given a Laplacian matrix L and any positive real number β, there exists a diagonal matrix*

$$D = \operatorname{diag}\{d_i\} > 0, \tag{6.73}$$

such that the eigenvalues of DL, denoted by $\lambda_i(DL)$ ($i = 1, \ldots, N$), are real and satisfy

$$\lambda_1(DL) = 0,$$

and

$$\lambda_i(DL) > \beta,$$

for $i = 2, \ldots, N$.

Note that D depends on L and hence the network topology. Let $\alpha > 0$ be such that $\lambda_i(DL) < \alpha$ for $i = 2, \ldots, N$.

6.2.6.1 MAS with Full-State Coupling

The protocol design for MAS with full-state coupling is as follows.

Protocol design 6.9 *Consider a MAS described by* (6.1) *and* (6.3). *We consider the protocol*

$$u_i = \rho d_i F_\delta \zeta_i, \tag{6.74}$$

where d_i *is the* i^{th} *element of matrix* D *defined in* (6.73), *and* ρ *and* F_δ *are given in Protocol Design* 6.1.

The result is stated in the following theorem.

Theorem 6.32 *Consider a MAS described by* (6.1) *and* (6.3). *Suppose the network graph* $\mathcal{G}$ *is known.*

If (A, B) *is stabilizable, then the state synchronization problem stated in Problem* 6.1 *with this graph* $\mathcal{G}$ *is solvable if*

$$\bar{\tau}\omega_{\max} < \frac{\pi}{2}. \tag{6.75}$$

In particular, for any given $\bar{\tau}$ *satisfying* (6.75), *there exist a* $\rho > 0$ *and a* $\delta^* > 0$ *such that for this* ρ *and any* $\delta \in (0, \delta^*]$, *controller* (6.74) *solves the state synchronization problem for the given graph* $\mathcal{G}$ *and any* $\tau \in [0, \bar{\tau}]$. *Moreover, the synchronized trajectory is given by* (6.19).

Proof Using the same arguments as in the preceding sections, we can write the overall dynamics of N agents as

$$\dot{x}(t) = (I_N \otimes A)x(t) + \rho(DL \otimes BF_\delta)x(t - \tau).$$

Since all the eigenvalues of DL are real and $\lambda_i(DL) \in (\beta, \alpha)$ $(i = 2, \ldots, N)$, the rest follows from the proof of Theorem 6.13. ∎

Next we consider the case where agents are neutrally stable.

Protocol design 6.10 *Consider a MAS described by (6.1) and (6.3). We design a feedback gain*

$$F_i = -d_i B' P, \tag{6.76}$$

where $P > 0$ is a solution of the Lyapunov function

$$A'P + PA \leq 0, \tag{6.77}$$

while $\delta > 0$ is a low-gain parameter. Then, the protocol is designed as

$$u_i = \delta F_i \zeta_i. \tag{6.78}$$

Based on the above design, we have the following result.

Theorem 6.33 *Consider a MAS described by (6.1) and (6.3). Suppose the network graph $\mathcal{G}$ is known.*

Suppose that the agent dynamics are neutrally stable. If (A, B) is stabilizable, then the state synchronization problem stated in Problem 6.4 with this graph $\mathcal{G}$ is solvable if the condition (6.75) holds. In particular, for any given $\bar{\tau}$ satisfying (6.75), there exists a $\delta^* > 0$ such that for any $\delta \in (0, \delta^*]$, controller (6.78) solves the state synchronization problem for this given graph $\mathcal{G}$ and any $\tau \in [0, \bar{\tau}]$.

6.2.6.2 MAS with Partial-State Coupling

The protocol design for MAS with partial-state coupling is as follows.

Protocol design 6.11 *Consider a MAS described by (6.1) and (6.2). The linear dynamic protocol can be constructed as*

$$\begin{aligned} \dot{\chi}_i &= (A + KC)\chi_i - K\zeta_i, \\ u_i &= \rho d_i F_\delta \chi_i, \end{aligned} \tag{6.79}$$

where d_i is the ith element of matrix D defined in Lemma 6.31 and $\rho > 0$, K, and F_δ are given in Protocol Design 6.4.

The result based on the above design is as follows.

Theorem 6.34 *Consider a MAS described by* (6.1) *and* (6.2). *Suppose the network graph $\mathcal{G}$ is known.*

If (A, B) is stabilizable and (C, A) is detectable, then the state synchronization problem stated in Problem 6.2 with this graph $\mathcal{G}$ is solvable if the condition (6.75) *holds. In particular, for any given $\bar{\tau}$ satisfying* (6.75)*, there exist a $\rho > 0$ and a $\delta^* > 0$ such that for this ρ and any $\delta \in (0, \delta^*]$, the stable dynamic controller* (6.79) *solves the state synchronization problem for the given graph $\mathcal{G}$ and any $\tau \in [0, \bar{\tau}]$. Moreover, the synchronized trajectory is given by* (6.42).

Proof It follows the proofs of Theorems 6.23 and 6.32. ∎

Next we consider a special case where agents are neutrally stable.

Protocol design 6.12 *Consider a MAS described by* (6.1) *and* (6.2). *We choose an observer gain K such that $A + KC$ is Hurwitz stable and consider a feedback gain*

$$F = -\delta B'P \tag{6.80}$$

where $P > 0$ is a solution of the Lyapunov inequality

$$A'P + PA \leq 0. \tag{6.81}$$

Then, the linear dynamic protocol can be designed as

$$\begin{cases} \dot{\chi}_i = (A + KC)\chi_i - K\zeta_i \\ u_i = \delta d_i F \chi_i, \end{cases} \tag{6.82}$$

where $\delta > 0$ is a low-gain parameter.

The result based on the above design is stated as follows.

Theorem 6.35 *Consider a MAS described by* (6.1) *and* (6.2). *Suppose the network graph $\mathcal{G}$ is known.*

Suppose that the agent dynamics are neutrally stable. If (A, B) is stabilizable and (C, A) is detectable, then the state synchronization problem stated in Problem 6.5 with this graph $\mathcal{G}$ is solvable if the condition (6.75) *holds. In particular, for any given $\bar{\tau}$ satisfying* (6.75)*, there exists a $\delta^* > 0$ such that for any $\delta \in (0, \delta^*]$, the stable dynamic controller* (6.82) *solves the state synchronization problem for this given graph $\mathcal{G}$ and any $\tau \in [0, \bar{\tau}]$. Moreover, the synchronized trajectory is given by* (6.42).

6.3 MAS in the Presence of Nonuniform Input Delay

In this section, we consider synchronization for MAS in the presence of nonuniform input delay. Both full-state and partial-state coupling networks are considered.

Consider a MAS composed of N identical linear time-invariant continuous-time agents of the form

$$\begin{aligned} \dot{x}_i(t) &= Ax_i(t) + Bu_i(t-\tau_i), \\ y_i(t) &= Cx_i(t), \end{aligned} \qquad (i = 1, \ldots, N) \tag{6.83}$$

where $\tau_1, \ldots, \tau_N$ are unknown constants satisfying $\tau_i \in [0, \bar{\tau}]$ for $i = 1, \ldots, N$, and $x_i \in \mathbb{R}^n$, $u_i \in \mathbb{R}^m$, and $y_i \in \mathbb{R}^p$ are, respectively, the state, input, and output vectors of agent i. Assumption 6.1 on the agent dynamics holds in this section as well.

The communication network among agents, which is undirected in this section, is similar to that in Chap. 2, which provides each agent with the quantity

$$\zeta_i = \sum_{j=1}^{N} a_{ij}(y_i - y_j) = \sum_{j=1}^{N} \ell_{ij} y_j, \quad i = 1, \ldots, N, \tag{6.84}$$

for partial-state coupling, and

$$\zeta_i = \sum_{j=1}^{N} a_{ij}(x_i - x_j) = \sum_{j=1}^{N} \ell_{ij} x_j, \quad i = 1, \ldots, N, \tag{6.85}$$

for full-state coupling. Consider the set $\mathbb{G}_{\alpha,\beta}^{u,N}$ as defined in Definition 1.12. Note that the main restriction compared to the case of uniform delay is that we assume that the network is undirected.

6.3.1 Problem Formulation

As always, we formulate two state synchronization problems for a network, one for the full-state coupling and the other for partial-state coupling.

Problem 6.36 (Full-State Coupling) Consider a MAS described by (6.83) and (6.85). Let $\mathbf{G}$ be a given set of graphs such that $\mathbf{G} \subseteq \mathbb{G}^N$. The *state synchronization* problem with a set of network graphs $\mathbf{G}$ is to find, if possible, a linear static protocol of the form

$$u_i = F\zeta_i, \tag{6.86}$$

for $i = 1, \ldots, N$ such that, for any graph $\mathcal{G} \in \mathbf{G}$, for any $\tau_1, \ldots, \tau_N \leq \bar{\tau}$, and for all the initial conditions of agents, state synchronization among agents is achieved.

Problem 6.37 (Partial-State Coupling) Consider a MAS described by (6.83) and (6.84). Let $\mathbf{G}$ be a given set of graphs such that $\mathbf{G} \subseteq \mathbb{G}^N$. The *state synchronization* problem with a set of network graphs $\mathbf{G}$ is to find, if possible, a linear dynamic protocol of the form

$$\begin{cases} \dot{\chi}_i = A_c \chi_i + B_c \zeta_i, \\ u_i = C_c \chi_i, \end{cases} \tag{6.87}$$

for $i = 1, \ldots, N$ where $\chi_i \in \mathbb{R}^{n_c}$, such that, for any graph $\mathcal{G} \in \mathbf{G}$, for any $\tau_1, \ldots, \tau_N \leq \bar{\tau}$, and for all the initial conditions of agents and the protocol, state synchronization among agents is achieved.

6.3.2 Stability of Linear Time-Delay System

We need some preliminary results from the literature. The following lemma is a generalization of (6.3) and can also be found in [177].

Lemma 6.38 *Consider a linear time-delay system*

$$\dot{x}(t) = Ax(t) + \sum_{i=1}^{m} A_i x(t - \tau_i), \tag{6.88}$$

where $x(t) \in \mathbb{R}^n$ and $\tau_i \in \mathbb{R}$. Assume that $A + \sum_{i=1}^{m} A_i$ is Hurwitz stable. Then, (6.88) is asymptotically stable for all $\tau_1, \ldots, \tau_N \in [0, \bar{\tau}]$ if

$$\det[j\omega I - A - \sum_{i=1}^{m} e^{-j\omega\tau_i} A_i] \neq 0,$$

for all $\omega \in \mathbb{R}$, and for all $\tau_1, \ldots, \tau_N \in [0, \bar{\tau}]$.

6.3.3 Protocol Design for MAS with Full-State Coupling

In this section, we will show how to achieve state synchronization for case of nonuniform delays. The main restriction that we impose is that the network must be undirected.

Protocol design 6.13 *Consider a MAS described by (6.83) and (6.85). We consider the protocol*

$$u_i = \rho F_\delta \zeta_i, \tag{6.89}$$

where $F_\delta = -B'P_\delta$ with P_δ the positive definite solution of the algebraic Riccati equation (6.22), while ρ and δ are design parameters.

Let

$$\omega_{\max} = \begin{cases} 0, & A \text{ is Hurwitz.} \\ \max\{\omega \in \mathbb{R} \mid \det(j\omega I - A) = 0\}, & \text{otherwise.} \end{cases}$$

The main result based on the above design is stated as follows.

Theorem 6.39 *Consider a MAS described by (6.83) and (6.85). Let any $\alpha \geq \beta > 0$ be given, and hence a set of network graphs $\mathbb{G}^{u,N}_{\alpha,\beta}$ be defined.*

If A is at most weakly unstable and (A, B) is stabilizable, then the state synchronization problem stated in Problem 6.36 with $\mathbf{G} = \mathbb{G}^{u,N}_{\alpha,\beta}$ is solvable if

$$\bar{\tau}\omega_{\max} < \frac{\pi}{2}. \tag{6.90}$$

In particular, for any given $\bar{\tau}$ satisfying (6.90), there exist a $\rho > 0$ and a $\delta^* > 0$ such that for this ρ and any $\delta \in (0, \delta^*]$, controller (6.89) solves the state synchronization problem for any graph $\mathcal{G} \in \mathbb{G}^{u,N}_{\alpha,\beta}$ and any $\tau_1, \ldots, \tau_N \in [0, \bar{\tau}]$.

Proof Define $T_1 \in \mathbb{R}^{(N-1)\times N}$ and $T_2 \in \mathbb{R}^{N\times(N-1)}$

$$T_1 = \begin{pmatrix} I & -\mathbf{1} \end{pmatrix}, \qquad T_2 = \begin{pmatrix} I \\ 0 \end{pmatrix}. \tag{6.91}$$

We define $\bar{x}_i := x_i - x_N$ as the state synchronization error for agent $i = 1, \ldots, N-1$ and

$$\bar{x} = \begin{pmatrix} \bar{x}_1 \\ \bar{x}_2 \\ \vdots \\ \bar{x}_{N-1} \end{pmatrix}.$$

Using that

$$LT_2T_1x = LT_2\bar{x} = L\begin{pmatrix} x_1 - x_N \\ x_2 - x_N \\ \vdots \\ x_{N-1} - x_N \\ x_N - x_N \end{pmatrix} = Lx, \tag{6.92}$$

we can write the full closed-loop system as

$$\dot{\bar{x}} = (I \otimes A)\bar{x} - \rho(T_1DLT_2 \otimes BB'P_\delta)\bar{x}, \tag{6.93}$$

where D represents the delays

$$D = \operatorname{diag}\{D_1, \ldots, D_N\} \quad \text{where} \quad (D_iz_i)(t) = z_i(t - \tau_i). \tag{6.94}$$

We first show the system is stable without the delays, i.e., when $D = I$

$$\dot{\bar{x}} = (I \otimes A)\bar{x} - \rho(T_1LT_2 \otimes BB'P_\delta)\bar{x}.$$

Note that from (6.92), i.e., $LT_2T_1 = L$, it is easy to see that the eigenvalues of T_1LT_2 are exactly the nonzero eigenvalue of L. Similar to the proof of Theorem 6.9 this system is Hurwitz stable if the $N - 1$ systems

$$\dot{\xi}_i = (A - \rho\lambda_iBB'P_\delta)\xi_i,$$

are Hurwitz stable where $\lambda_2, \ldots, \lambda_N$ are the nonzero eigenvalues of L which satisfy $\lambda_i \geq \beta$. The required stability for sufficiently large ρ follows immediately from Lemma 6.4.

Next, we consider the case with delays. We define

$$\tilde{D}(\omega) = \operatorname{diag}\{e^{-j\omega\tau_1}, \ldots, e^{-j\omega\tau_N}\}. \tag{6.95}$$

It follows from Lemma 6.38 that the system (6.93) is asymptotically stable if

$$\det\left[j\omega I - (I \otimes A) + \rho(T_1\tilde{D}(\omega)LT_2 \otimes BB'P_\delta)\right] \neq 0, \tag{6.96}$$

for all $\omega \in \mathbb{R}$, for all $\tau_1, \ldots, \tau_N \in [0, \bar{\tau}]$ and all possible L associated to a network graph in $\mathbb{G}_{\alpha,\beta}^{u,N}$.

Given (6.90), we note that for ρ sufficiently large there exists a $\phi > 0$ such that

$$\beta\rho\cos(\omega\bar{\tau}) > 1, \quad \forall|\omega| < \omega_{\max} + \phi. \tag{6.97}$$

Next we split the proof of (6.96) into two cases where $|\omega| < \omega_{\max} + \phi$ and $|\omega| \geq \omega_{\max} + \phi$, respectively.

As noted before, if $|\omega| \geq \omega_{\max} + \phi$, there exists a $\mu > 0$ such that

$$\sigma_{\min}(j\omega I - A) > \mu, \quad \forall \omega, \quad \text{such that } |\omega| \geq \omega_{\max} + \phi.$$

Given ρ and L bounded, there exists a $\delta^* > 0$ such that

$$\rho \|(T_1 \tilde{D}(\omega) L T_2 \otimes B B' P_\delta)\| \leq \tfrac{\mu}{2}$$

for $\delta < \delta^*$. It is then easy to see that the condition (6.96) holds for $|\omega| \geq \omega_{\max} + \phi$.

Next, for $|\omega| < \omega_{\max} + \phi$, we use a Lyapunov argument. Define

$$P_{L,\delta} = T_2' L T_2 \otimes P_\delta. \tag{6.98}$$

We already noted earlier that $L T_2 T_1 = L$ and the eigenvalues of $T_1 L T_2$ are the nonzero eigenvalues of L. Hence $L T_2$ is injective and since $L \geq 0$ we find $T_2' L T_2 > 0$ and therefore $P_{L,\delta} > 0$. We get

$$\begin{aligned}
P_{L,\delta} & \left[(I \otimes A) - \rho(T_1 \tilde{D}(\omega) L T_2 \otimes B B' P_\delta)\right] \\
&\qquad + \left[(I \otimes A) - \rho(T_1 \tilde{D}(\omega) L T_2 \otimes B B' P_\delta)\right]^* P_{L,\delta} \\
&= T_2' L T_2 \otimes (A P_\delta + P_\delta A') \\
&\qquad - \rho \left[T_2' L (T_2 T_1 \tilde{D}(\omega) + \tilde{D}(\omega)^* T_1' T_2') L T_2 \otimes P_\delta B B' P_\delta\right] \\
&= -\delta T_2' L T_2 \otimes I + T_2' L T_2 \otimes P_\delta B B' P_\delta \\
&\qquad - \rho T_2' L \left[\tilde{D}(\omega) + \tilde{D}(\omega)^*\right] L T_2 \otimes P_\delta B B' P_\delta.
\end{aligned}$$

For $|\omega| < \omega_{\max} + \phi$ we have (6.97) which yields

$$\rho L \left[\tilde{D}(\omega) + \tilde{D}(\omega)^*\right] L > 2\beta^{-1} L^2 > 2L,$$

and hence

$$\begin{aligned}
P_{L,\delta} & \left[(I \otimes A) - \rho(T_1 \tilde{D}(\omega) L T_2 \otimes B B' P_\delta)\right] \\
&\qquad + \left[(I \otimes A) - \rho(T_1 \tilde{D}(\omega) L T_2 \otimes B B' P_\delta)\right]^* P_{L,\delta} \\
&\qquad\qquad \leq -\delta T_2' L T_2 \otimes I - T_2' L T_2 \otimes P_\delta B B' P_\delta.
\end{aligned} \tag{6.99}$$

We find that

$$(I \otimes A) - \rho(T_1 \tilde{D}(\omega) L T_2 \otimes BB' P_\delta)$$

is asymptotically stable and therefore (6.96) is satisfied. ■

Remark 6.40 The consensus controller design depends only on the agent model and parameters $\bar{\tau}$, β, and α and is independent of specific network topology.

6.3.4 Protocol Design for MAS with Partial-State Coupling

In this section, we consider the state synchronization problem for homogeneous MAS with partial-state coupling and with nonuniform unknown constant input delay. As before, we use a protocol consisting of the static feedback used for the case of full-state coupling combined with an observer.

Protocol design 6.14 *Consider a MAS described by (6.83) and (6.84). We choose an observer gain K such that $A + KC$ is Hurwitz stable, and consider a feedback gain $F_\delta = -B' P_\delta$ where P_δ is the solution of the algebraic Riccati equation (6.22). We design the linear dynamical protocol*

$$\begin{aligned} \dot{\chi}_i &= (A + KC)\chi_i - K\zeta_i, \\ u_i &= \rho F_\delta \chi_i, \end{aligned} \tag{6.100}$$

for $i \in \{1, \ldots, N\}$ where ρ and δ are design parameters to be chosen later.

The main result based on the above design is stated as follows.

Theorem 6.41 *Consider a MAS described by (6.83) and (6.84). Let any $\alpha \geq \beta > 0$ be given, and hence a set of network graphs $\mathbb{G}^{u,N}_{\alpha,\beta}$ be defined.*

If (A, B) is stabilizable and (C, A) is detectable, then the state synchronization problem stated in Problem 6.37 with $\mathbf{G} = \mathbb{G}^{u,N}_{\alpha,\beta}$ is solvable if the condition (6.90) holds. In particular, for any given $\bar{\tau}$ satisfying (6.90), there exist a $\rho > 0$ and a $\delta^* > 0$ such that for this ρ and any $\delta \in (0, \delta^*]$, the stable dynamic controller (6.100) solves the state synchronization problem for any graph $\mathcal{G} \in \mathbb{G}^{u,N}_{\alpha,\beta}$ and any $\tau_1, \ldots, \tau_N \in [0, \bar{\tau}]$.

Proof We will use the notation $\chi = \mathrm{col}\{\chi_i\}$ and $x = \mathrm{col}\{x_i\}$. We note that

$$\begin{aligned} \dot{x}(t) &= (I \otimes A)x - \rho D(I \otimes BB' P_\delta)\chi \\ \dot{\chi}(t) &= -(L \otimes KC)x + (I \otimes (A + KC))\chi. \end{aligned}$$

We can define

$$\chi = L\bar{\chi} + \mathbf{1}r$$

where

$$\dot{\bar{\chi}}(t) = -(I \otimes KC)x + (I \otimes (A + KC))\bar{\chi}$$

By analyzing $\mathbf{1}'\chi$ and recalling that $\mathbf{1}'L = 0$, we can derive

$$\dot{r} = (A + KC)r$$

which is asymptotic stable. Since we consider the asymptotic behavior of the system, we can ignore this asymptotically stable signal. Therefore we analyze the following system:

$$\begin{pmatrix} \dot{x} \\ \dot{\tilde{\chi}} \end{pmatrix} = \left[\begin{pmatrix} \tilde{A} & 0 \\ -\tilde{K}\tilde{C} & \tilde{A} + \tilde{K}\tilde{C} \end{pmatrix} + \begin{pmatrix} \tilde{B} \\ 0 \end{pmatrix} (DL \otimes I) \begin{pmatrix} 0 & -\rho \tilde{B}' \tilde{P}_\delta \end{pmatrix} \right] \begin{pmatrix} x \\ \tilde{\chi} \end{pmatrix},$$

where

$$\tilde{A} = I_N \otimes A, \qquad \tilde{B} = I_N \otimes B, \qquad \tilde{C} = I_N \otimes C,$$
$$\tilde{K} = I_N \otimes K, \qquad \tilde{P}_\delta = I_N \otimes P_\delta,$$

with the operator D defined in (6.94). We introduce

$$e = \begin{pmatrix} T_1 & 0 \\ 0 & T_2 \end{pmatrix} \begin{pmatrix} x \\ \tilde{\chi} \end{pmatrix},$$

with T_1 and T_2 as defined in (6.91). Clearly, if we show that $e(t) \to 0$, then we immediately obtain that the system achieves synchronization. We obtain, using that $LT_2T_1 = L$, that

$$\dot{e} = [\mathcal{A} + \mathcal{B}(T_1 DLT_2 \otimes I)\mathcal{F}_\delta]\, e, \tag{6.101}$$

where

$$\mathcal{A} = \begin{pmatrix} \bar{A} & 0 \\ -\bar{K}\bar{C} & \bar{A} + \bar{K}\bar{C} \end{pmatrix}, \quad \mathcal{B} = \begin{pmatrix} \bar{B} \\ 0 \end{pmatrix}, \quad \mathcal{F}_\delta = \begin{pmatrix} 0 & -\rho \bar{B}' \bar{P}_\delta \end{pmatrix},$$

while

$$\bar{A} = I_{N-1} \otimes A, \qquad \bar{B} = I_{N-1} \otimes B, \qquad \bar{C} = I_{N-1} \otimes C,$$
$$\bar{K} = I_{N-1} \otimes K, \qquad \bar{P}_\delta = I_{N-1} \otimes P_\delta.$$

We first establish that the system without delay is asymptotically stable. As before

$$\dot{\tilde{e}} = \left[\tilde{\mathcal{A}} + \tilde{\mathcal{B}}(T_1 L T_2 \otimes I)\tilde{\mathcal{F}}_\varepsilon\right]\tilde{e}$$

is asymptotically stable if the $N-1$ systems

$$\dot{\xi}_i = \begin{pmatrix} A & -\rho\lambda_i BB'P_\delta \\ -KC & A+KC \end{pmatrix}\xi_i,$$

are asymptotically stable where $\lambda_1, \ldots, \lambda_{N-1}$ are the eigenvalues of $T_1 L T_2$ which are the nonzero eigenvalues of L and therefore satisfy $\operatorname{Re}\lambda_i \geq \beta$. Lemma 6.6 establishes then the required stability for ρ sufficiently large.

Choose ρ such that additionally we have

$$\rho\beta\cos(\bar{\tau}\omega_{\max}) > 1, \tag{6.102}$$

which is possible since $\bar{\tau}\omega_{\max} < \frac{\pi}{2}$.

We will next show that the system (6.101) is stable in the presence of delays. The system (6.101) can also be expressed as

$$\dot{\tilde{e}} = \left[\tilde{\mathcal{A}} + \tilde{\mathcal{B}}(T_1 D L T_2 \otimes I)\tilde{\mathcal{F}}_\varepsilon\right]\tilde{e} \tag{6.103}$$

using

$$\tilde{e} = \begin{pmatrix} \tilde{e}_1 \\ \tilde{e}_2 \end{pmatrix} = \begin{pmatrix} I & 0 \\ -I & I \end{pmatrix} e,$$

where

$$\tilde{\mathcal{A}} = \begin{pmatrix} \bar{A} & 0 \\ 0 & \bar{A}+\bar{K}\bar{C} \end{pmatrix}, \quad \tilde{\mathcal{B}} = \begin{pmatrix} \bar{B} \\ -\bar{B} \end{pmatrix}, \quad \tilde{\mathcal{F}}_\delta = -\rho\begin{pmatrix} \bar{B}'\bar{P}_\delta & \bar{B}'\bar{P}_\delta \end{pmatrix}.$$

Let $\tilde{D}(\omega)$ be defined by (6.95). Based on Lemma 6.38, we only need to check whether

$$\det\left[j\omega I - \tilde{\mathcal{A}} - \tilde{\mathcal{B}}(T_1\tilde{D}(\omega)LT_2 \otimes I)\tilde{\mathcal{F}}\right] \neq 0, \tag{6.104}$$

is satisfied for all $\omega \in \mathbb{R}$, for all $\tau_1, \ldots, \tau_N \in [0, \bar{\tau}]$ and for all possible L associated to a network graph in $\mathbb{G}^{u,N}_{\beta,\alpha}$.

As in the full-state case, we note that given (6.102), there exists a $\phi > 0$ such that

$$\rho\beta\cos(\bar{\tau}(\omega_{\max}+\phi)) > 1. \tag{6.105}$$

Next, as before, we split the proof of (6.104) into two cases where $|\omega| < \tilde{\omega}_{\max} + \phi$ and $|\omega| \geq \tilde{\omega}_{\max} + \phi$, respectively.

If $|\omega| \geq \tilde{\omega}_{\max} + \phi$, we have $\det(j\omega I - \tilde{\mathcal{A}}) \neq 0$, which yields $\underline{\sigma}(j\omega I - \tilde{\mathcal{A}}) > 0$. Hence, there exists a $\mu > 0$ such that

$$\underline{\sigma}(j\omega I - \tilde{\mathcal{A}}) > \mu, \quad \forall\omega, \text{ such that } |\omega| \geq \omega_{\max} + \phi. \tag{6.106}$$

Given ρ, there exists a $\delta^* > 0$ such that

$$\|\tilde{\mathcal{B}}(T_1\tilde{D}(\omega)LT_2 \otimes I)\tilde{\mathcal{F}}\| \leq \frac{\mu}{2} \tag{6.107}$$

for $\delta < \delta^*$. Combining (6.106) and (6.107), we obtain

$$\underline{\sigma}\left[j\omega I - \tilde{\mathcal{A}} - \tilde{\mathcal{B}}(T_1\tilde{D}(\omega)LT_2 \otimes I)\tilde{\mathcal{F}}\right] \geq \frac{\mu}{2}.$$

Therefore, (6.104) holds for $|\omega| \geq \tilde{\omega}_{\max} + \phi$.

It remains to verify (6.104) with $|\omega| < \tilde{\omega}_{\max} + \phi$. Through a Lyapunov argument we will show that

$$\tilde{\mathcal{A}} - \tilde{\mathcal{B}}(T_1\tilde{D}(\omega)LT_2 \otimes I)\tilde{\mathcal{F}} \tag{6.108}$$

is Hurwitz for any fixed ω satisfying $|\omega| < \tilde{\omega}_{\max} + \phi$. This then clearly implies (6.104). Consider

$$V(\tilde{e}) = \tilde{e}' \begin{pmatrix} P_{L,\delta} & 0 \\ 0 & \mu Q \end{pmatrix} \tilde{e},$$

with $P_{L,\delta}$ as defined in (6.98) while Q is such that

$$\left(\bar{A} + \bar{K}\bar{C}\right)' Q + Q\left(\bar{A} + \bar{K}\bar{C}\right) \leq -2I. \tag{6.109}$$

The existence of Q is obvious since $\bar{A} + \bar{K}\bar{C}$ is Hurwitz stable. This implies that

$$\begin{aligned}\left[\tilde{A} + \tilde{K}\tilde{C} + \rho T_1\tilde{D}(\omega)LT_2 \otimes BB'P_\delta\right]^* \tilde{Q} \\ + \tilde{Q}\left[\tilde{A} + \tilde{K}\tilde{C} + \rho T_1\tilde{D}(\omega)LT_2 \otimes BB'P_\delta\right] \leq -I\end{aligned} \tag{6.110}$$

for a δ small enough since

$$\rho T_1\tilde{D}(\omega)LT_2 \otimes BB'P_\delta$$

can be made arbitrarily small for small δ as L and $\tilde{D}(\omega)$ are bounded.

Using (6.99) and the above inequality together with some algebra, we obtain

$$\begin{aligned}\dot{V} &\leq -\delta \tilde{e}_1'(T_2'LT_2 \otimes I)\tilde{e}_1 - \tilde{v}_1'\tilde{v}_1 - \mu\tilde{e}_2'\tilde{e}_2 \\ &\quad - 2\rho\tilde{v}_1'(L^{1/2}\tilde{D}(\omega)LT_2 \otimes B'P_\delta)\tilde{e}_2 + 2\mu\rho\tilde{e}_2'Q(T_1\tilde{D}(\omega)L^{1/2} \otimes B)\tilde{v}_1\end{aligned}$$

where

$$\tilde{v}_1 = \left[L^{1/2}T_2 \otimes B'P_\delta\right]\tilde{e}_1. \tag{6.111}$$

Let $\tilde{M}$ be such that

$$2\rho\|Q(T_1\tilde{D}(\omega)L^{1/2} \otimes B)\| \leq \tilde{M}$$

for all possible $\tilde{D}(\omega)$ and choose μ such that $4\tilde{M}^2\mu = 1$. Next choose δ small enough such that

$$2\rho\|L^{1/2}\tilde{D}(\omega)LT_2 \otimes B'P_\delta\| < \frac{1}{4\tilde{M}}$$

The above yields that

$$\begin{aligned}\dot{V} &\leq -\delta\|\tilde{e}_1\|^2 - \|\tilde{v}_1\|^2 - \mu\|\tilde{e}_2\|^2 + \sqrt{\mu}\|\tilde{v}_1\|\|\tilde{e}_2\| \\ &\leq -\delta\|\tilde{e}_1\|^2 - \tfrac{1}{2}\|\tilde{v}_1\|^2 - \tfrac{1}{2}\mu\|\tilde{e}_2\|^2.\end{aligned}$$

This shows (6.108) is Hurwitz for any fixed ω satisfying $|\omega| < \tilde{\omega}_{\max} + \phi$. Therefore (6.104) is satisfied which completes the proof. ■

6.3.5 *Static Protocol Design*

In this section, we also consider the static protocol design for two classes of agents: squared-down passive agents and squared-down passifiable via static input feedforward agents with nonuniform delay.

As argued before with delay, we intrinsically rely on a low-gain feedback design. Systems which are squared-down passifiable via static output feedback require a minimal size of the gain unless they are already squared-down passive. Therefore, this class of systems will not be considered.

We show first that the state synchronization among agents in the network with partial-state coupling and input delay via static protocols can be solved by equivalently solving a robust stabilization problem.

6.3.5.1 Squared-Down Passive Agents

It will be shown that the protocol design for uniform delay also works for nonuniform delay.

Protocol design 6.15 *Consider a MAS described by squared-down passive agents (6.83) with communication via (6.84). The static protocol is designed as*

$$u_i = -\delta G_1 K G_2 \zeta_i, \tag{6.112}$$

where $\delta > 0$ is a low-gain parameter to be designed, K is any symmetric and positive definite matrix, and G_1, G_2 are given in (1.9).

The main result regarding this design can be stated as follows.

Theorem 6.42 *Consider a MAS described by agents (6.83) and (6.84). Let any $\alpha > 0$ be given and hence a set of undirected network graphs $\mathbb{G}_{\alpha,0,}^{u,N}$.*

If (A, B, C) is squared-down passive with G_1, G_2 given in (1.9), then the state synchronization problem stated in Problem 6.37 but with a static protocol is solvable if the condition (6.90) holds. In particular, for any given $\bar{\tau}$ satisfying (6.90), there exist a $\delta^* > 0$ such that for any $\delta \in (0, \delta^*]$, controller (6.112) solves the state synchronization problem for any graph $\mathcal{G} \in \mathbb{G}_{\alpha,0}^{u,N}$ and any $\tau_1, \ldots, \tau_N \in [0, \bar{\tau}]$.

Proof Define $\bar{x}_i = x_i - x_N$ as the state synchronization error for agent $i = 1, \ldots, N-1$ and $\bar{x} = \operatorname{col}\{\bar{x}_i\}$. Then, following the steps in the proof of Theorem 6.39, we obtain the closed-loop system as

$$\dot{\bar{x}} = (I \otimes A)\bar{x} - \delta(T_1 D L T_2 \otimes B G_1 K G_2 C)\bar{x}, \tag{6.113}$$

where T_1, T_2, and D are defined as before. Since the eigenvalues of $T_1 L T_2$ are the nonzero eigenvalues of L, the system is stable without the delay according to Theorem 2.44. Then, from Lemma 6.3 that system (6.113) is asymptotically stable if

$$\det\Big[j\omega I - (I \otimes A) + \delta(T_1 \tilde{D}(\omega) L T_2 \otimes B G_1 K G_2 C)\Big] \neq 0, \tag{6.114}$$

for all $\omega \in \mathbb{R}$, for all $\tau_1, \ldots, \tau_N \in [0, \bar{\tau}]$ and all possible L associated to a network graph in $\mathbb{G}_{\alpha,0}^{u,N}$.

As before, we split the proof of (6.114) into two cases where $|\omega| < \omega_{\max} + \phi$ and $|\omega| \geq \omega_{\max} + \phi$ respectively. Here ϕ is such that

$$(\omega_{\max} + \phi)\bar{\tau} < \tfrac{\pi}{2}. \tag{6.115}$$

If $|\omega| \geq \omega_{\max} + \phi$, there exists a $\mu > 0$ such that

$$\sigma_{\min}(j\omega I - A) > \mu, \quad \forall \omega \text{ such that } |\omega| \geq \omega_{\max} + \phi.$$

Then, using the same arguments as in the proof of Theorem 6.39 and the boundedness of L, there exists a $\delta^* > 0$ such that

$$\|\delta(T_1 \tilde{D}(\omega) L T_2 \otimes B G_1 K G_2 C)\| \geq \tfrac{\mu}{2},$$

for $\delta < \delta^*$. Therefore, the condition (6.114) holds for $|\omega| \geq \omega_{\max} + \phi$.

Next, for $|\omega| < \omega_{\max} + \phi$, we use a Lyapunov argument. Define

$$P_L = T_2' L T_2 \otimes P, \tag{6.116}$$

where P is given in (1.9). As noted before, we have $T_2' L T_2 > 0$ which implies that $P_L > 0$. Then we get

$$\begin{aligned}
P_L &\left[(I \otimes A) - \delta(T_1 \tilde{D}(\omega) L T_2 \otimes B G_1 K G_2 C)\right] \\
&\qquad + \left[(I \otimes A) - \delta(T_1 \tilde{D}(\omega) L T_2 \otimes B G_1 K G_2 C)\right]^* P_L \\
&= T_2' L T_2 \otimes (AP + PA') \\
&\qquad - \delta\left[T_2' L (T_2 T_1 \tilde{D}(\omega) + \tilde{D}(\omega)^* T_1' T_2') L T_2 \otimes P B G_1 K G_2 C\right] \\
&< -\delta T_2' L \left[\tilde{D}(\omega) + \tilde{D}(\omega)^*\right] L T_2 \otimes C' G_2' K G_2 C.
\end{aligned}$$

The last inequality holds because $L T_2 T_1 = L$, $PA + A'P \leq 0$ and $P B G_1 = C' G_2'$. Since

$$\tilde{D}(\omega) + \tilde{D}(\omega)^* = 2\cos(\omega\tau) \geq 2\cos((\omega_{\max} + \phi)\bar{\tau}) > 0,$$

we obtain that

$$(I \otimes A) - \delta(T_1 \tilde{D}(\omega) L T_2 \otimes B G_1 K G_2 C)$$

is asymptotically stable and therefore the condition (6.114) holds for $|\omega| < \omega_{\max} + \phi$. ∎

6.3.5.2 Squared-Down Passifiable via Static Input Feedforward

The protocol design for uniform delay can also be applied directly for nonuniform delay.

Protocol design 6.16 *Consider a MAS described by agents (6.83) with communication via (6.84). Assume the agents are squared-down passifiable via static input feedforward with respect to G_1, G_2, and R. The static protocol is designed as*

$$u_i = -\delta G_1 K G_2 \zeta_i, \tag{6.117}$$

where $K > 0$ while $\delta > 0$ is the low-gain parameter to be designed.

The main result based on the above design can be stated as follows.

Theorem 6.43 *Consider a MAS described by agents (6.83) and (6.84). Let any $\alpha, \beta > 0$ be given, and consider the set of network graphs $\mathbb{G}^{u,N}_{\alpha,0}$.*

Assume the agents are squared-down passifiable via static input feedforward with respect to G_1, G_2, and R such that (A, BG_1) is controllable and (A, G_2C) is observable with BG_1 and G_2C full-column and full-row rank, respectively. The state synchronization problem stated in Problem 6.37 but with a static protocol is solvable if the condition (6.90) holds. In particular, for any given $\bar{\tau}$ satisfying (6.90), there exists a $\delta^* > 0$ such that for any $\delta \in (0, \delta^*]$, controller (6.117) solves the state synchronization problem for any graph $\mathcal{G} \in \mathbb{G}^{u,N}_{\alpha,0}$ and any $\tau_1, \ldots, \tau_N \in [0, \bar{\tau}]$.

Proof According to the proof of Theorem 6.42, we obtain the following closed-loop system:

$$\dot{\bar{x}} = (I \otimes A)\bar{x} - \delta(T_1 D L T_2 \otimes B G_1 K G_2 C)\bar{x}, \tag{6.118}$$

where $\bar{x} = \operatorname{col}\{\bar{x}_1, \ldots \bar{x}_N\}$ with $\bar{x}_i = x_i - x_N$ for $i = 1, \ldots, N-1$.

According to Theorem 2.48, the system (6.118) without delay is Hurwitz stable. Then, from Lemma 6.3 the system (6.118) is asymptotically stable if

$$\det\Big[j\omega I - (I \otimes A) + \delta(T_1 \tilde{D}(\omega) L T_2 \otimes B G_1 K G_2 C)\Big] \neq 0, \tag{6.119}$$

for all $\omega \in \mathbb{R}$, for all $\tau_1, \ldots, \tau_N \in [0, \bar{\tau}]$ and all possible L associated to a network graph in $\mathbb{G}^{u,N}_{\alpha,0}$.

Next, we split the proof of (6.119) into two cases where $|\omega| < \omega_{\max} + \phi$ and $|\omega| \geq \omega_{\max} + \phi$, respectively, where ϕ is such that (6.115) is satisfied.

If $|\omega| \geq \omega_{\max} + \phi$, there exists a $\mu > 0$ such that

$$\sigma_{\min}(j\omega I - A) > \mu, \quad \forall \omega, \text{ such that } |\omega| \geq \omega_{\max} + \phi.$$

Then, since L is bounded, there exists a $\delta^* > 0$ such that

$$\|\delta(T_1 \tilde{D}(\omega) L T_2 \otimes B G_1 K G_2 C)\| \leq \tfrac{\mu}{2},$$

for any $\delta < \delta^*$. Therefore, the condition (6.119) holds for $|\omega| \geq \omega_{\max} + \phi$.

Next, for $|\omega| < \omega_{\max} + \phi$, we use a Lyapunov argument. Define

$$P_L = T_2' L T_2 \otimes P, \tag{6.120}$$

where P is given in (1.12). We already noted earlier that $LT_2T_1 = L$ and the eigenvalues of T_1LT_2 are the nonzero eigenvalues of L. We also noted earlier that $T_2'LT_2 > 0$ and hence $P_L > 0$. We get

$$\begin{aligned}
&P_L[(I \otimes A) - \delta T_1 \tilde{D}(\omega) L T_2 \otimes B G_1 K G_2 C)] \\
&\qquad + [(I \otimes A) - \delta T_1 \tilde{D}(\omega) L T_2 \otimes B G_1 K G_2 C)]^* P_L \\
&= \begin{pmatrix} I \\ -\delta T_1 \tilde{D}(\omega) L T_2 \otimes K G_2 C \end{pmatrix}' T_2' L T_2 \otimes G(P) \begin{pmatrix} I \\ -\delta T_1 \tilde{D}(\omega) L T_2 \otimes K G_2 C \end{pmatrix} \\
&\qquad - \delta T_2' L T_2 T_1 \tilde{D}(\omega) L T_2 \otimes C' G_2' K G_2 C \\
&\qquad - \delta T_2' L \tilde{D}(\omega)^* T_1' T_2' L T_2 \otimes C' G_2' K G_2 C \\
&\qquad + \delta^2 T_2' L \tilde{D}(\omega)^* T_1' T_2' L T_2 T_1 \tilde{D}(\omega) L T_2 \otimes C' G_2' K (R + R') K G_2 C \\
&\leq -\delta T_2' L [\tilde{D}(\omega) + \tilde{D}(\omega)^*] L T_2 \otimes C' G_2' K G_2 C \\
&\qquad + \delta^2 T_2' L \tilde{D}(\omega)^* L \tilde{D}(\omega) L T_2 \otimes C' G_2' K (R + R') K G_2 C
\end{aligned}$$

using that $LT_1T_2 = L$ where $G(P)$ is given by (1.12). For $|\omega| < \omega_{\max} + \phi$ we have (6.115) which yields

$$\tilde{D}(\omega) + \tilde{D}(\omega)' \geq 2\cos((\omega_{\max} + \phi)\bar{\tau}) I.$$

Clearly for a δ sufficiently small, we have

$$\delta \tilde{D}(\omega)' L \tilde{D}(\omega) \otimes K(R + R')K \leq \cos((\omega_{\max} + \phi)\bar{\tau}) I \otimes K,$$

and hence

$$\begin{aligned}
P_{L,\delta} &\left[(I \otimes A) - \delta(T_1 \tilde{D}(\omega) L T_2 \otimes B G_1 K G_2 C)\right] \\
&\qquad + \left[(I \otimes A) - \delta(T_1 \tilde{D}(\omega) L T_2 \otimes B G_1 K G_2 C)\right]^* P_{L,\delta} \\
&\qquad\qquad \leq -\delta \cos((\omega_{\max} + \phi)\bar{\tau}) T_2' L^2 T_2 \otimes C' G_2' K G_2 C.
\end{aligned} \tag{6.121}$$

We noted that $T_2'LT_2$ and T_1LT_2 are invertible and hence

$$(T_2' L T_2)(T_1 L T_2) = T_2' L^2 T_2$$

is invertible. Since clearly $T_2' L^2 T_2 \geq 0$, we find that $T_2' L^2 T_2 > 0$. We obtain that

$$(I \otimes A) - \delta(T_1 \tilde{D}(\omega) L T_2 \otimes B G_1 K G_2 C)$$

is asymptotically stable and therefore the condition (6.119) holds for $|\omega| < \omega_{\max} + \phi$. ∎

6.4 Application to Formation

We again consider the application of protocol design for synchronization problems to a rigid formation problem. As mentioned in Sect. 2.6, the protocol design procedure in previous sections can be easily modified to solve the formation problem. We address the two cases of uniform and nonuniform delay separately.

6.4.1 *Uniform Input Delay*

In this subsection both full-state coupling and partial-state coupling are considered for multi-agent systems subject to uniform input delay.

6.4.1.1 Full-State Coupling

In this section, we consider the formation problem for a multi-vehicle network with full-state coupling in the presence of uniform input delay. We first formulate the formation problem and then analyze the transformation of the formation problem to the synchronization problem.

Consider a multi-agent system (MAS) composed of N vehicles with the dynamics

$$\dot{x}_i(t) = A x_i(t) + B u_i(t - \tau), \quad (i = 1, \ldots, N) \tag{6.122}$$

where A and B are defined in (2.132), that is

$$x_i = \begin{pmatrix} x_{i,p} \\ x_{i,v} \end{pmatrix}, \quad A = \begin{pmatrix} 0 & I_n \\ A_1 & A_2 \end{pmatrix}, \quad B = \begin{pmatrix} 0 \\ I_n \end{pmatrix},$$

and τ is an unknown constant satisfying $\tau \in [0, \bar{\tau}]$. Note that (A, B) is controllable.

Given an arbitrary formation vector $\{h_1, \ldots, h_N\}$ with $h_i \in \mathbb{R}^n$, our goal is to achieve formation among the vehicles, i.e.,

$$\lim_{t \to \infty} \big((x_{i,p}(t) - h_i) - (x_{j,p}(t) - h_j)\big) = 0,$$

and

$$\lim_{t\to\infty} \left(x_{i,v}(t) - x_{j,v}(t)\right) = 0.$$

Let

$$\bar{h}_i = \begin{pmatrix} h_i \\ 0 \end{pmatrix}, \quad (i = 1, \ldots, N). \tag{6.123}$$

The communication network provides each vehicle with a linear combination of relative state information, that is

$$\zeta_i = \sum_{j=1}^{N} a_{ij} \left((x_i - \bar{h}_i) - (x_j - \bar{h}_j)\right) = \sum_{j=1}^{N} \ell_{ij}(x_j - \bar{h}_j), \tag{6.124}$$

where a_{ij} are the weights associated with the edges of the graph $\mathcal{G}$, while ℓ_{ij} are the coefficients of the associated Laplacian matrix L. The network satisfies Assumption 2.2. Then, the problem is formulated as follows.

Problem 6.44 (Full-State Coupling) Consider a multi-vehicle system described by (6.122) and (6.124). Let $\mathbf{G}$ be a given set of graphs such that $\mathbf{G} \subseteq \mathbb{G}^N$. The *formation* problem with a set of network graphs $\mathbf{G}$ is to find, if possible, a linear static protocol of the form

$$u_i = F_1\zeta_i + F_2 h_i, \tag{6.125}$$

for $i = 1, \ldots, N$ such that, for any graph $\mathcal{G} \in \mathbf{G}$ and for all the initial conditions of the vehicles, formation among the vehicles is achieved.

Let

$$\bar{x}_i(t) = x_i(t) - \bar{h}_i.$$

Choosing $F_2 = -A_1$, in terms of the dynamics of $\bar{x}_i$, the closed-loop system of vehicle (6.122) and the linear static protocol (6.125) can be written as

$$\begin{aligned}
\dot{\bar{x}}_i(t) &= Ax_i(t) - BA_1h_i + BF_1\sum_{j=1}^{N} \ell_{ij}\bar{x}_j(t-\tau), \\
&= Ax_i(t) - A\bar{h}_i + BF_1\sum_{j=1}^{N} \ell_{ij}\bar{x}_j(t-\tau), \\
&= A\bar{x}_i(t) + BF_1\sum_{j=1}^{N} \ell_{ij}\bar{x}_j(t-\tau).
\end{aligned}$$

Let

$$\bar{x} = \begin{pmatrix} \bar{x}_1 \\ \vdots \\ \bar{x}_N \end{pmatrix}.$$

Then the overall dynamics of the N vehicles in terms of $\bar{x}$ can be written as

$$\dot{\bar{x}}(t) = (I_N \otimes A)\bar{x}(t) + (L \otimes BF_1)\bar{x}(t-\tau). \tag{6.126}$$

Note that the above overall system (6.126) is exactly the same as the overall system given in (6.14). Thus, the design of F_1 can be based on the protocol design in Design 6.1 using an ARE-based method or can be based on the protocol design in Design 6.3 using a direct eigenstructure assignment method. The result, based on an ARE-based design, is stated in the following theorem.

Theorem 6.45 *Consider a multi-vehicle system described by* (6.122) *and* (6.124). *Let any* $\alpha \geq \beta > 0$ *and* $\frac{\pi}{2} > \theta \geq 0$ *be given, and hence a set of network graphs* $\mathbb{G}^N_{\alpha,\beta,\theta}$ *be defined.*

The formation problem stated in Problem 6.44 with $\mathbf{G} = \mathbb{G}^N_{\alpha,\beta,\theta}$ *is solvable if the condition* (6.23) *holds. In particular, for any given* $\bar{\tau}$ *satisfying* (6.23), *there exist a* $\rho > 0$ *and a* $\delta^* > 0$ *such that for this* ρ *and any* $\delta \in (0, \delta^*]$, *the protocol*

$$u_i = \rho F_\delta \zeta_i - A_1 h_i, \tag{6.127}$$

solves the formation problem for any graph $\mathcal{G} \in \mathbb{G}^N_{\alpha,\beta,\theta}$ *and any* $\tau \in [0, \bar{\tau}]$, *where* F_δ *and* ρ *are constructed in Protocol Design 6.1.*

Proof The proof follows along the steps of the proof of Theorem 6.13. ■

6.4.1.2 Partial-State Coupling

In this subsection, we consider the formation problem for a multi-vehicle network with partial-state coupling in the presence of uniform input delay. Similar to the approach of full-state coupling, we first formulate the formation problem and then analyze the transformation of the formation problem to the synchronization problem.

Consider a multi-agent system (MAS) composed of N vehicles with the dynamics

$$\begin{cases} \dot{x}_i(t) = Ax_i(t) + Bu_i(t-\tau), \\ y_i(t) = Cx_i(t), \end{cases} \quad (i = 1, \ldots, N) \tag{6.128}$$

where A, B, and C are defined in (2.132) and (2.140), that is

$$x_i = \begin{pmatrix} x_{i,p} \\ x_{i,v} \end{pmatrix}, \quad A = \begin{pmatrix} 0 & I_n \\ A_1 & A_2 \end{pmatrix}, \quad B = \begin{pmatrix} 0 \\ I_n \end{pmatrix}, \quad C = \begin{pmatrix} I_n & 0 \end{pmatrix},$$

which immediately implies that (A, B) is controllable, (C, A) is observable and (A, B, C) is minimum-phase, and τ is an unknown constant satisfying $\tau \in [0, \bar{\tau}]$. Remember that the output $y_i = x_{i,p}$ is the position variable.

The communication network provides each vehicle with a linear combination of relative output information (instead of relative state information (6.124) that we used in the case of full-state coupling). In other words

$$\zeta_i = \sum_{j=1}^{N} a_{ij} \big((y_i - h_i) - (y_j - h_j)\big) = \sum_{j=1}^{N} \ell_{ij}(y_j - h_j). \tag{6.129}$$

Then, the problem is formulated as follows.

Problem 6.46 (Partial-State Coupling) Consider a multi-vehicle system described by (6.128) and (6.129). Let $\mathbf{G}$ be a given set of graphs such that $\mathbf{G} \subseteq \mathbb{G}^N$. The *formation* problem with a set of network graphs $\mathbf{G}$ is to find, if possible, a linear time-invariant dynamic protocol of the form

$$\begin{cases} \dot{\chi}_i = A_c \chi_i + B_c \zeta_i, \\ u_i = C_c \chi_i + D_{c,2} h_i, \end{cases} \tag{6.130}$$

for $i = 1, \ldots, N$ where $\chi_i \in \mathbb{R}^{n_c}$, such that, for any graph $\mathcal{G} \in \mathbf{G}$ and for all the initial conditions for the vehicles and their protocol, formation among agents is achieved.

Again, let

$$\bar{x}_i = x_i - \bar{h}_i, \quad \bar{y}_i = y_i - h_i,$$

where $\bar{h}_i$ is defined in (6.123).

Similar to the full-state coupling case, we choose $D_{c,2} = -A_1$. In terms of the dynamics of $\bar{x}_i$ and output $\bar{y}_i$, the closed-loop system of vehicle (6.128) and the linear dynamic protocol (6.130) can then be written as

$$\begin{cases} \dot{\tilde{x}}_i(t) = \begin{pmatrix} A & BC_c \\ 0 & A_c \end{pmatrix} \tilde{x}_i(t) + \begin{pmatrix} 0 \\ B_c \end{pmatrix} \zeta_i(t - \tau), \\ \bar{y}_i(t) = \begin{pmatrix} C & 0 \end{pmatrix} \tilde{x}_i(t), \\ \zeta_i(t) = \sum_{j=1}^{N} \ell_{ij} \bar{y}_j(t), \end{cases}$$

for $i = 1, \ldots, N$ where

$$\tilde{x}_i = \begin{pmatrix} \bar{x}_i \\ \chi_i \end{pmatrix}.$$

Define

$$\tilde{x} = \begin{pmatrix} \tilde{x}_1 \\ \vdots \\ \tilde{x}_N \end{pmatrix}, \quad \bar{A} = \begin{pmatrix} A & BC_c \\ 0 & A_c \end{pmatrix}, \quad \bar{B} = \begin{pmatrix} 0 \\ B_c \end{pmatrix}, \quad \text{and} \quad \bar{C} = \begin{pmatrix} C & 0 \end{pmatrix}.$$

Then, the overall dynamics of the N vehicles in terms of $\tilde{x}$ can be written as

$$\dot{\tilde{x}}(t) = (I_N \otimes \bar{A})\tilde{x}(t) + (L \otimes \bar{B}\bar{C})\tilde{x}(t - \tau). \tag{6.131}$$

Note that the above overall system (6.131) is exactly the same as the overall system given in (6.36). Thus, the protocol design of (A_c, B_c, C_c) can be obtained from Design 6.4 using an ARE-based method or from Design 6.6 using a direct eigenstructure assignment method. If A happens to be neutrally stable, then the protocol design for (A_c, B_c, C_c) can be obtained from Design 6.5.

The result, utilizing an ARE-based design, is stated in the following theorem.

Theorem 6.47 *Consider a multi-vehicle system described by* (6.128) *and* (6.129). *Let any* $\alpha \geq \beta > 0$ *and* $\frac{\pi}{2} > \theta \geq 0$ *be given, and hence a set of network graphs* $\mathbb{G}^N_{\alpha,\beta,\theta}$ *be defined.*

The formation problem stated in Problem 6.46 *with* $\mathbf{G} = \mathbb{G}^N_{\alpha,\beta,\theta}$ *is always solvable by a stable protocol if the condition* (6.23) *holds. In particular, for any given* $\bar{\tau}$ *satisfying* (6.23), *there exist a* $\rho > 0$ *and a* $\delta^* > 0$ *such that for this* ρ *and any* $\delta \in (0, \delta^*]$, *the stable dynamic protocol*

$$\begin{aligned} \dot{\chi}_i &= (A + KC)\chi_i - K\zeta_i, \\ u_i &= \rho F_\delta \chi_i - A_1 h_i, \end{aligned} \tag{6.132}$$

solves the formation problem for any graph $\mathcal{G} \in \mathbb{G}^N_{\alpha,\beta,\theta}$ *and any* $\tau \in [0, \bar{\tau}]$, *where* F_δ *and* ρ *are constructed in Design* 6.4.

Proof The proof follows that of Theorem 6.23. ■

6.4.2 Nonuniform Input Delay

The protocol design procedure in previous sections can be again easily modified to solve the rigid formation problem in this section. Both full-state coupling and partial-state coupling are considered.

6.4.2.1 Full-State Coupling

In this subsection, we consider the formation problem for a multi-vehicle network with full-state coupling in the presence of nonuniform input delay. We first formulate the formation problem and then analyze the transformation of the formation problem to the synchronization problem.

Consider a multi-agent system (MAS) composed of N vehicles with the dynamics

$$\dot{x}_i(t) = Ax_i(t) + Bu_i(t-\tau_i), \quad (i = 1, \dots, N) \tag{6.133}$$

where

$$x_i = \begin{pmatrix} x_{i,p} \\ x_{i,v} \end{pmatrix} \in \mathbb{R}^{2n},$$

and A and B are defined in Sect. 2.6, and $\tau_1, \dots, \tau_N$ are unknown constants satisfying $\tau_i \in [0, \bar{\tau}]$ for $i = 1, \dots, N$. We know that (A, B) is controllable.

Given an arbitrary formation vector be $\{h_1, \dots, h_N\}$ with $h_i \in \mathbb{R}^n$, we achieve formation among the vehicles in such a way that

$$\lim_{t\to\infty} \big((x_{i,p}(t) - h_i) - (x_{j,p}(t) - h_j)\big) = 0,$$

and

$$\lim_{t\to\infty} \big(x_{i,v}(t) - x_{j,v}(t)\big) = 0,$$

for $i, j \in \{1, \dots, N\}$.

The communication network provides each vehicle with a linear combination of relative state information, that is

$$\zeta_i = \sum_{j=1}^{N} a_{ij}\big((x_i - \bar{h}_i) - (x_j - \bar{h}_j)\big) = \sum_{j=1}^{N} \ell_{ij}(x_j - \bar{h}_j), \tag{6.134}$$

where a_{ij} are the weights associated with the edges of the graph $\mathcal{G}$, while ℓ_{ij} are the coefficients of the associated Laplacian matrix L. Moreover, $\bar{h}_j$ is given in (6.123). Then, the problem is formulated as follows.

Problem 6.48 (Full-State Coupling) Consider a multi-vehicle system described by (6.133) and (6.134). Let $\mathbf{G}$ be a given set of graphs such that $\mathbf{G} \subseteq \mathbb{G}^N$. The *formation* problem with a set of network graphs $\mathbf{G}$ is to find, if possible, a linear static protocol of the form

$$u_i = F_1\zeta_i + F_2h_i, \tag{6.135}$$

for $i = 1, \ldots, N$ such that, for any graph $\mathcal{G} \in \mathbf{G}$ and for all the initial conditions of the vehicles, formation among the vehicles is achieved.

Let

$$\bar{x}_i = x_i - \bar{h}_i.$$

Next we choose $F_2 = -A_1$. Similar to the case of uniform delays, the closed-loop system of vehicle (6.133) and the linear static protocol (6.135) can be written as

$$\dot{\bar{x}}_i(t) = A\bar{x}_i(t) + BF_1 \sum_{j=1}^{N} \ell_{ij}\bar{x}_j(t - \tau_i). \tag{6.136}$$

Note that the above overall system is exactly the same as the overall system given in (6.93). Thus, the design of F_1 can be based on the protocol design in Protocol Design 6.13 using an ARE-based method. The result is stated in the following theorem.

Theorem 6.49 *Consider a multi-vehicle system described by (6.133) and (6.134). Let any $\alpha \geq \beta > 0$ be given, and hence a set of network graphs $\mathbb{G}_{\alpha,\beta}^{u,N}$ be defined.*

The formation problem stated in Problem 6.48 with $\mathbf{G} = \mathbb{G}_{\alpha,\beta}^{u,N}$ is solvable if the condition (6.90) holds. In particular, for any given $\bar{\tau}$ satisfying (6.90), there exist a $\rho > 0$ and a $\delta^* > 0$ such that for this ρ and any $\delta \in (0, \delta^*]$, the protocol

$$u_i = \rho F_\delta \zeta_i - A_1 h_i$$

for $i = 1, \ldots, N$ solves the formation problem for any graph $\mathcal{G} \in \mathbb{G}_{\alpha,\beta}^{u,N}$ and any $\tau_i \in [0, \bar{\tau}]$, where F_δ and ρ are constructed in Protocol Design 6.13.

Proof The proof follows that of Theorem 6.39. ■

6.4.2.2 Partial-State Coupling

In this subsection, we consider the formation problem for a multi-vehicle network with partial-state coupling in the presence of nonuniform input delay. As before, we first formulate the formation problem and then analyze the transformation of the formation problem to the synchronization problem.

Consider a multi-agent system (MAS) composed of N vehicles with the dynamics

$$\begin{cases} \dot{x}_i(t) = Ax_i(t) + Bu_i(t - \tau_i), \\ y_i(t) = Cx_i(t), \end{cases} \quad (i = 1, \ldots, N) \tag{6.137}$$

where x_i and A, B, and C are defined as in Sect. 2.6, which immediately implies that (A, B) is controllable, (C, A) is observable and (A, B, C) is minimum-phase, and $\tau_1, \ldots, \tau_N$ are unknown constants satisfying $\tau_i \in [0, \bar{\tau}]$. Note that the output $y_i = x_{i,p}$ is the position variable.

The communication network provides each vehicle with a linear combination of relative output information, that is

$$\zeta_i = \sum_{j=1}^{N} a_{ij}\big((y_i - h_i) - (y_j - h_j)\big) = \sum_{j=1}^{N} \ell_{ij}(y_j - h_j). \tag{6.138}$$

Then, the problem is formulated as follows.

Problem 6.50 (Partial-State Coupling) Consider a multi-vehicle system described by (6.137) and (6.138). Let $\mathbf{G}$ be a given set of graphs such that $\mathbf{G} \subseteq \mathbb{G}^N$. The *formation* problem with a set of network graphs $\mathbf{G}$ is to find, if possible, a linear time-invariant dynamic protocol of the form

$$\begin{cases} \dot{\chi}_i = A_c\chi_i + B_c\zeta_i, \\ u_i = C_c\chi_i + D_{c,2}h_i, \end{cases} \tag{6.139}$$

for $i = 1, \ldots, N$ where $\chi_i \in \mathbb{R}^{n_c}$, such that, for any graph $\mathcal{G} \in \mathbf{G}$ and for all the initial conditions for the vehicles and their protocol, formation among agents is achieved.

Again, let

$$\bar{x}_i = x_i - \bar{h}_i, \quad \bar{y}_i = y_i - h_i$$

where $\bar{h}_i$ is defined in (6.123).

Using Design 6.14, we choose

$$A_c = A + KC, \qquad B_c = -K, \qquad C_c = \rho F_\delta$$

in (6.139). Then we select $D_{c,2} = -A_1$. Using the same arguments as in the case of uniform delays, the closed-loop system of dynamic $\bar{x}_i$ and the linear dynamic protocol (6.139) can then be written as

$$\begin{cases} \dot{\bar{x}}_i(t) = A\bar{x}_i(t) - \rho BB'P_\delta\chi_i(t - \tau_i), \\ \dot{\chi}_i(t) = (A + KC)\chi_i(t) - KC\sum_{j=1}^{N}\ell_{ij}\bar{x}_j, \end{cases}$$

for $i = 1, \ldots, N$. It is easy to see that the above overall system is exactly the same as the interconnection of (6.83) and (6.84) and (6.100). Thus, we can use the same arguments as in the proof of Theorem 6.41 to show that we achieve formation.

The result is stated in the following theorem.

Theorem 6.51 *Consider a multi-vehicle system described by* (6.137) *and* (6.138). *Let any* $\alpha \geq \beta > 0$ *be given, and hence a set of network graphs* $\mathbb{G}_{\alpha,\beta}^{u,N}$ *be defined.*

The formation problem stated in Problem 6.50 *with* $\mathbf{G} = \mathbb{G}_{\alpha,\beta}^{u,N}$ *is always solvable by a stable protocol if the condition* (6.90) *holds. In particular, for any given* $\bar{\tau}$ *satisfying* (6.90), *there exist a* $\rho > 0$ *and a* $\delta^* > 0$ *such that for this* ρ *and any* $\delta \in (0, \delta^*]$, *the stable dynamic protocol*

$$\begin{aligned} \dot{\chi}_i &= (A + KC)\chi_i - K\zeta_i, \\ u_i &= \rho F_\delta \chi_i - A_1 h_i, \end{aligned} \tag{6.140}$$

for $i = 1, \ldots, N$ *solves the formation problem for any graph* $\mathcal{G} \in \mathbb{G}_{\alpha,\beta}^{u,N}$ *and any* $\tau_i \in [0, \bar{\tau}]$ *for* $i = 1, \ldots, N$, *where* F_δ *and* ρ *are constructed in Protocol Design* 6.14.

Proof The proof follows that of Theorem 6.41. ■

Chapter 7
Synchronization of Discrete-Time Linear MAS with Unknown Input Delay

7.1 Introduction

For discrete-time MAS, this chapter is a counterpart of Chap. 6 which deals with continuous-time systems. As in Chap. 6 as well as in reference [10], we can identify two kinds of delay. Firstly there is *communication delay*, which results from limitations on the communication between agents. Secondly we have *input delay* which is due to computational limitations of an individual agent. We consider both uniform and nonuniform input delay. In each case, both full- and partial-state coupling networks are considered.

To the best of our knowledge, the only relevant results in the context of synchronization of discrete-time agents are reported in [139] where first-order agents are considered and the results are obtained using a frequency-domain approach.

The write-up of this chapter is partially based on [155] and [188].

7.2 MAS in the Presence of Uniform Input Delay

We consider synchronization problems for MAS in the presence of uniform input delay. As before, each synchronization problem can be solved by equivalently solving a robust stabilization problem.

Consider a MAS composed of N identical linear time-invariant discrete-time agents of the form

$$\begin{aligned} x_i(k+1) &= Ax_i(k) + Bu_i(k-\kappa), \\ y_i(k) &= Cx_i(k), \end{aligned} \tag{7.1}$$

A. Saberi et al., *Synchronization of Multi-Agent Systems in the Presence of Disturbances and Delays*, Systems & Control: Foundations & Applications,
https://doi.org/10.1007/978-3-030-88148-1_7

for $i = 1, \ldots, N$ where κ is an integer satisfying $\kappa \in [0, \bar{\kappa}]$ with $\bar{\kappa}$ the upper bound of the input delay and $x_i \in \mathbb{R}^n$, $u_i \in \mathbb{R}^m$, and $y_i \in \mathbb{R}^p$ are respectively the state, input, and output vectors of agent i.

The communication network among agents is exactly the same as that in Chap. 3 which provides each agent with the quantity

$$\zeta_i(k) = \sum_{j=1}^{N} d_{ij}(y_i(k) - y_j(k)), \quad i = 1, \ldots, N \tag{7.2}$$

in case of partial-state coupling, and

$$\zeta_i(k) = \sum_{j=1}^{N} d_{ij}(x_i(k) - x_j(k)), \quad i = 1, \ldots, N \tag{7.3}$$

in case of full-state coupling.

Similar to the continuous-time case, we formulate here two state synchronization problems, one for a network with full-state coupling and the other for partial-state coupling.

Problem 7.1 (Full-State Coupling) Consider a MAS described by (7.1) and (7.3). Let $\mathbf{G}$ be a given set of graphs such that $\mathbf{G} \subseteq \bar{\mathbb{G}}^N$. The *state synchronization* problem with a set of network graphs $\mathbf{G}$ is to find, if possible, a linear protocol of the form

$$u_i = F\zeta_i, \tag{7.4}$$

for $i = 1, \ldots, N$ such that, for any graph $\mathcal{G} \in \mathbf{G}$, for any integer $\kappa \leq \bar{\kappa}$, and for all the initial conditions of the agents, state synchronization among agents is achieved.

Problem 7.2 (Partial-State Coupling) Consider a MAS described by (7.1) and (7.2). Let $\mathbf{G}$ be a given set of graphs such that $\mathbf{G} \subseteq \bar{\mathbb{G}}^N$. The *state synchronization* problem with a set of network graphs $\mathbf{G}$ is to find, if possible, a linear dynamic protocol of the form

$$\begin{cases} \chi_i(k+1) = A_c\chi_i(k) + B_c\zeta_i(k), \\ u_i(k) \quad = C_c\chi_i(k), \end{cases} \tag{7.5}$$

for $i = 1, \ldots, N$ where $\chi_i \in \mathbb{R}^{n_c}$, such that, for any graph $\mathcal{G} \in \mathbf{G}$, for any integer $\kappa \leq \bar{\kappa}$, and for all the initial conditions of the agents and the protocol, state synchronization among agents is achieved.

As motivated in Sect. 3.4.1, we restrict here our attention to agents which are at most weakly unstable, that is, A has all its eigenvalues in the closed unit disc.

7.2.1 Stability of Linear Time-Delay System

We need a preliminary result which is a minor modification of a result found in [152]. This is similar to Lemma 6.3 in continuous-time.

Lemma 7.3 *Consider a linear time-delay system*

$$x(k+1) = Ax(k) + A_1 x(k-\kappa), \tag{7.6}$$

where $x(k) \in \mathbb{R}^n$ and $\kappa \in \mathbb{N}$. Suppose $A + A_1$ is Schur stable. Then, (7.6) is asymptotically stable if and only if

$$\det[e^{j\omega}I - A - e^{-j\omega\kappa_r}A_1] \neq 0,$$

for all $\omega \in [-\pi, \pi]$, and for all $\kappa_r \in \mathbb{R}$ with $0 < \kappa_r < \kappa$.

Another result which plays a key role in this chapter can be found in, e.g., [47]. This is similar to Lemma 6.4 in the continuous-time and shows robustness of the ARE-based design.

Lemma 7.4 *Consider a linear uncertain system*

$$x(k+1) = Ax(k) + \lambda Bu(k), \qquad x(0) = x_0, \tag{7.7}$$

where $\lambda \in \mathbb{C}$ is unknown. Assume that (A, B) is stabilizable and A has all its eigenvalues in the closed unit disc. A low-gain state feedback $u = F_\delta x$ is constructed, where

$$F_\delta = -(B'P_\delta B + I)^{-1}B'P_\delta A, \tag{7.8}$$

with P_δ being the unique positive definite solution of the H_2 algebraic Riccati equation

$$P_\delta = A'P_\delta A + \delta I - A'P_\delta B(B'P_\delta B + I)^{-1}B'P_\delta A. \tag{7.9}$$

Then, $A + \lambda BF_\delta$ is Schur stable for any $\lambda \in \mathbb{C}$ satisfying

$$\lambda \in \Omega_\delta := \left\{ z \in \mathbb{C} : \left| z - \left(1 + \tfrac{1}{\gamma_\delta}\right) \right| < \tfrac{\sqrt{1+\gamma_\delta}}{\gamma_\delta} \right\}, \tag{7.10}$$

where $\gamma_\delta = \lambda_{\max}(B'P_\delta B)$. As $\delta \to 0$, Ω_δ approaches the set

$$H_1 := \{z \in \mathbb{C} : \operatorname{Re} z > \tfrac{1}{2}\}$$

in the sense that any compact subset of H_1 is contained in Ω_δ for a δ small enough.

The following lemma considers the robustness properties of low-gain feedbacks designed via the direct method. Consider the low-gain feedback F_δ given by (3.50) which was designed based on a direct method.

Lemma 7.5 *Consider a linear uncertain system (7.7) where $\lambda \in \mathbb{C}$ is unknown. Assume that (A, B) is stabilizable and A has all its eigenvalues in the closed unit disc. A low-gain state feedback $u = F_\delta x$ is constructed, where F_δ is given by (3.50). In that case, for any given $\beta \in (0, 1)$ and a sufficiently small $\delta > 0$, we have $A + \lambda B F_\delta$ Schur stable for any $|1 - \lambda| < \beta$.*

Proof This is an immediate consequence of the proof of Theorem 3.24. ∎

We have similar robustness results in the case of partial-state coupling. We first consider the ARE-based design.

Lemma 7.6 *Consider a linear uncertain system*

$$\begin{cases} x(k+1) = Ax(k) + \lambda Bu(k), & x(0) = x_0 \\ y(k) \quad = Cx(k), \end{cases} \tag{7.11}$$

where $\lambda \in \mathbb{C}$ is unknown. Assume that (A, B) is stabilizable, (A, C) is detectable, and A has all its eigenvalues in the closed unit disc. Design a low-gain dynamic measurement feedback controller as

$$\begin{cases} \chi(k+1) = A\chi(k) - K\left[y(k) - C\chi(k)\right], & \chi(0) = \chi_0 \\ u(k) \quad = F_\delta \chi(k), \end{cases} \tag{7.12}$$

where K is selected such that $A + KC$ is Schur stable, while F_δ is given by (7.8).

Then, for any compact set

$$\mathcal{S} \subset H_2 := \{z \in \mathbb{C} : \operatorname{Re} z > 1\},$$

there exists a δ^* such that for all $\delta \in (0, \delta^*]$, the closed-loop system of (7.11) and (7.12) is asymptotically stable for any $\lambda \in \mathcal{S}$.

Proof Define $e = x - \chi$. The closed-loop system of (7.11) and (7.12) in terms of x and e can be written as

$$\begin{cases} x(k+1) = (A + \lambda B F_\delta)x(k) - \lambda B F_\delta e(k) \\ e(k+1) = (A + KC - \lambda B F_\delta)e(k) + \lambda B F_\delta x(k). \end{cases} \tag{7.13}$$

First of all, by (7.9) we have

$$\begin{aligned} (A + \lambda B F_\delta)^* P_\delta (A + \lambda B F_\delta) - P_\delta \\ &= -\delta I + F_\delta' \left[|1 - \lambda|^2 (B' P_\delta B + I) - |\lambda|^2 I\right] F_\delta \\ &\leq -\delta I + \left[(1 + \gamma_\delta)|1 - \lambda|^2 - |\lambda|^2\right] F_\delta' F_\delta, \end{aligned} \tag{7.14}$$

where γ_δ is defined in Lemma 7.4. Define a set

$$\tilde{\Omega}_\delta := \left\{ z \in \mathbb{C} : \left| z - \left(1 + \tfrac{1}{\gamma_\delta}\right) \right| \leq \tfrac{1}{\gamma_\delta} \right\}. \tag{7.15}$$

It is easy to see that for any compact set $\mathcal{S} \subset H_2$, there exists a δ_1 such that $\mathcal{S} \subset \tilde{\Omega}_\delta$ for all $\delta \in (0, \delta_1]$. Moreover, $\lambda \in \tilde{\Omega}_\delta$ is equivalent to

$$(1 + \gamma_\delta)|1 - \lambda|^2 - |\lambda|^2 \leq -1,$$

and hence (7.14) yields

$$(A + \lambda B F_\delta)^* P_\delta (A + \lambda B F_\delta) - P_\delta \leq -\delta I - F_\delta' F_\delta. \tag{7.16}$$

Next, let Q be the positive definite solution of the Lyapunov equation

$$(A + KC)' Q (A + KC) - Q + 4I = 0.$$

Since $F_\delta \to 0$ as $\delta \to 0$, for any compact set $\mathcal{W}$, there exists a $\delta_2 \leq \delta_1$ such that for all $\delta \in (0, \delta_2]$, we have

$$(A + KC - \lambda B F_\delta)' Q (A + KC - \lambda B F_\delta) - Q + 3I \leq 0.$$

Then, consider the function $V_1(k) = e(k)^* Q e(k)$. Let $\mu(k) = F_\delta x(k)$. To ease our presentation, we shall omit the time label (k) whenever this does not cause any confusion. Then

$$\begin{aligned} V_1(k+1) - V_1(k) & \\ &\leq -3\|e\|^2 + 2\,\mathrm{Re}\Big(\lambda^* \mu^* B' Q [A + KC - \lambda B F_\delta] e\Big) + |\lambda|^2 \mu^* B' Q B \mu \\ &\leq -3\|e\|^2 + M_1 \|\mu\| \|e\| + M_2 \|\mu\|^2, \end{aligned} \tag{7.17}$$

where

$$M_1 = 2\|B'Q\| \max_{\substack{\lambda \in \mathcal{S} \\ \delta \in [0,1]}} \{|\lambda| \|A + KC - \lambda B F_\delta\|\}, \quad M_2 = \|B'QB\| \max_{\lambda \in \mathcal{S}} |\lambda|^2.$$

It should be noted that M_1 and M_2 are independent of specific δ and λ. Next, consider $V_2(k) = x(k)' P_\delta x(k)$. We have

$$\begin{aligned} V_2(k+1) - V_2(k) \leq -\delta\|x\|^2 - \|\mu\|^2 + 2\,\mathrm{Re}\Big(\lambda^* e^* F_\delta' B' P_\delta [A + \lambda B F_\delta] x\Big) \\ + |\lambda|^2 e^* F_\delta' B P_\delta B F_\delta e. \end{aligned}$$

Note that

$$\begin{aligned}
2\left|\lambda^* e^* F_\delta' B' P_\delta (A+\lambda B F_\delta)x\right| &= 2\left|\lambda^* e^* F_\delta' B' P_\delta A x + |\lambda|^2 e^* F_\delta' B' P_\delta B\mu\right| \\
&= 2\left|-\lambda^* e^* F_\delta' (B' P_\delta B + I)\mu + |\lambda|^2 e^* F_\delta' B' P_\delta B\mu\right| \\
&\leq \theta_1(\delta)\|e\|\|\mu\|,
\end{aligned}$$

where

$$\theta_1(\delta) = 2(1+\gamma_\delta)\|F_\delta\| \max_{\lambda\in\mathcal{S}}\{|\lambda|\} + 2\gamma_\delta\|F_\delta\| \max_{\lambda\in\mathcal{S}}\{|\lambda|^2\}.$$

Then

$$V_2(k+1) - V_2(k) \leq -\delta\|x\|^2 - \|\mu\|^2 + \theta_1(\delta)\|e\|\|\mu\| + \theta_2(\delta)\|e\|^2, \tag{7.18}$$

where

$$\theta_2(\delta) = \gamma_\delta\|F_\delta\|^2 \max_{\lambda\in\mathcal{S}}\{|\lambda|^2\}.$$

Finally, consider a Lyapunov candidate $V(k) = V_1(k) + cV_2(k)$ with

$$c = M_2 + M_1^2.$$

In view of (7.17) and (7.18), we get

$$V(k+1)-V(k) \leq -\delta c\|x\|^2 - M_1^2\|\mu\|^2 - [3-c\theta_2(\delta)]\|e\|^2 + [M_1 + c\theta_1(\delta)]\|\mu\|\|e\|.$$

There exists a $\delta^* \leq \delta_1$ such that for all $\delta \in (0, \delta^*]$

$$3 - c\theta_2(\delta) \geq 2, \quad M_1 + c\theta_1(\delta) \leq 2M_1.$$

This yields

$$V(k+1) - V(k) \leq -\delta c\|x\|^2 - \|e\|^2 - (\|e\| - M_1\|\mu\|)^2.$$

Therefore, for all $\delta \in (0, \delta^*]$, the system (7.13) is globally asymptotically stable. ■

Next we obtain a similar result using an observer-based controller and a state feedback designed via the direct method.

Lemma 7.7 *Consider a linear uncertain system (7.11) where $\lambda \in \mathbb{C}$ is unknown. Assume that (A, B) is stabilizable, (A, C) is detectable, and A has all its eigenvalues in the closed unit disc. Design a low-gain dynamic measurement feedback controller (7.12) where K is selected such that $A + KC$ is Schur stable and F_δ is given by (3.50).*

Then, for any $\beta < 1$ there exists a δ^* such that for all $\delta \in (0, \delta^*]$, the interconnection of (7.11) *and* (7.12) *is asymptotically stable for any λ with $|1-\lambda| < \beta$.*

Proof This is a direct consequence of the proof of Theorem 3.36. ■

7.2.2 *Protocol Design for MAS with Full-State Coupling*

As in the previous chapters, we establish first a connection between the problem of state synchronization among agents in the network in the presence of uniform input delay and a robust stabilization problem, and then we design a controller for such a robust stabilization problem.

7.2.2.1 Connection to a Robust Stabilization Problem

The MAS system described by (7.1) and (7.3) after implementing the linear static protocol (7.4) is written as

$$x_i(k+1) = Ax_i(k) + BF\zeta_i(k-\kappa) \tag{7.19}$$

for $i = 1, \ldots, N$. Let

$$x = \begin{pmatrix} x_1 \\ \vdots \\ x_N \end{pmatrix}.$$

Then, the overall dynamics of the N agents can be written as

$$x(k+1) = (I_N \otimes A)x(k) + ((I - D) \otimes BF)x(k-\kappa), \tag{7.20}$$

where D is the row stochastic matrix associated with the graph. As in Sect. 3.4.1, the synchronization for the system (7.20) is equivalent to the asymptotic stability of the following $N-1$ subsystems:

$$\tilde{\eta}_i(k+1) = A\tilde{\eta}_i(k) + (1-\lambda_i)BF\tilde{\eta}_i(k-\kappa), \quad i = 2, \ldots, N, \tag{7.21}$$

for any integer $\kappa \in [0, \bar{\kappa}]$, where λ_i, $i = 2, \ldots, N$ are those eigenvalues of D inside the unit disc which is formally established in the following theorem.

Theorem 7.8 *The MAS* (7.20) *achieves state synchronization if and only if the system* (7.21) *is asymptotically stable for $i = 2, \ldots, N$ and for any integer $\kappa \in [0, \bar{\kappa}]$.*

Proof As in the proof of Theorem 3.4, let

$$D = TJ_DT^{-1}, \tag{7.22}$$

where J_D is the Jordan canonical form of the row stochastic matrix D such that $J_D(1,1) = 1$ and the first column of T equals $\mathbf{1}$. Let

$$\eta := (T^{-1} \otimes I_n)x = \begin{pmatrix} \eta_1 \\ \vdots \\ \eta_N \end{pmatrix},$$

where $\eta_i \in \mathbb{C}^n$. In the new coordinates, the dynamics of η can be written as

$$\eta(k+1) = (I_N \otimes A)\eta(k) + ((I - J_D) \otimes BF)\eta(k-\kappa). \tag{7.23}$$

The eigenvalues of system (7.23) are given by the roots of its characteristic equation

$$H(z) = \det\left\{zI - (I_N \otimes A) - z^{-\kappa}((I - J_D) \otimes BF)\right\} = 0,$$

which, due to the diagonal structure of $I_N \otimes A$ and the upper-triangular structure of $J_D \otimes BF$, are the union of the eigenvalues of the $N-1$ subsystems given in (7.21) and the eigenvalues of A.

Now if (7.21) is globally asymptotically stable for $i = 2, \ldots, N$, we see from the above that $\eta_i(k) \to 0$ for $i = 2, \ldots, N$. This implies that

$$x(k) - (T \otimes I_n)\begin{pmatrix} \eta_1(k) \\ 0 \\ \vdots \\ 0 \end{pmatrix} \to 0.$$

Note that the first column of T is equal to the vector $\mathbf{1}$ and therefore

$$x_i(k) - \eta_1(k) \to 0$$

for $i = 1, \ldots, N$. This implies that we achieve state synchronization.

Conversely, suppose the network (7.20) reaches state synchronization. In this case, we shall have

$$x(k) - \mathbf{1} \otimes x_1(k) \to 0.$$

Then $\eta(k) - (T^{-1}\mathbf{1}) \otimes x_1(k) \to 0$. Since $\mathbf{1}$ is the first column of T, we have

$$T^{-1}\mathbf{1} = \begin{pmatrix} 1 \\ 0 \\ \vdots \\ 0 \end{pmatrix}.$$

Therefore, $\eta(k) - (T^{-1}\mathbf{1}) \otimes x_1(k) \to 0$ implies that $\eta_1(k) - x_1(k) \to 0$ and $\eta_i(k) \to 0$ for $i = 2, \ldots, N$. ∎

Immediately, we have the following corollary.

Corollary 7.9 *Consider a MAS described by* (7.1) *and* (7.3). *If there exists a protocol of the form* (7.4) *that achieves state synchronization, then the associated graph must have a directed spanning tree, or A must be asymptotically stable.*

Since the case of A being asymptotically stable is trivial, the requirement that the graph of a MAS has a directed spanning tree is essentially necessary.

Remark 7.10 It also becomes clear from the proof of Theorem 7.8 that the *synchronized trajectory* is given by $x_s(k) = \eta_1(k)$ which is governed by

$$\eta_1(k+1) = A\eta_1(k), \quad \eta_1(0) = (w \otimes I_n)x(0), \tag{7.24}$$

where w is the first row of T^{-1}, i.e., the normalized left eigenvector associated with the eigenvalue 1. This shows that the modes of the synchronized trajectory are determined by the eigenvalues of A and the complete dynamics depends on both A and a weighted average of the initial conditions of agents.

In light of Lemma 3.7, one can conclude that $\eta_1(0)$ is only a linear combination of the initial conditions of root agents. As such, the synchronized trajectory given by (7.24) can be written explicitly as

$$\eta_1(k) = A^k \sum_{i \in \Pi_{\mathcal{G}}} w_i x_i(0), \tag{7.25}$$

which is the weighted average of the trajectories of root agents.

Remark 7.11 If A is neutrally stable, i.e., all the eigenvalues of A are in the closed unit disc, and those eigenvalues of A on the unit disc, if any, are semi-simple, then the synchronized trajectories are bounded. Otherwise, the synchronized trajectories could be unbounded.

In light of the definition of Problem 7.1 that synchronization is formulated for a set of graphs, we obtain a *robust* stabilization problem, i.e., the stabilization of the system

$$x(k+1) = Ax(k) + (1-\lambda)Bu(k-\kappa), \tag{7.26}$$

via a protocol

$$u(k) = Fx(k),$$

for any λ that represents all the eigenvalues inside the unit disc of any possible stochastic row matrix D associated with a graph in the given set $\bar{\mathbb{G}}_\beta$.

7.2.2.2 ARE-Based Method

Based on a low-gain method, together with the standard H_2 discrete-time algebraic Riccati equation, we design the following protocol.

Protocol design 7.1 *Consider a MAS described by* (7.1) *and* (7.3). *We consider the protocol*

$$u_i = \rho F_\delta \zeta_i, \tag{7.27}$$

where $\rho > 0$ *is a design parameter, and* $F_\delta = -(B'P_\delta B + I)^{-1}B'P_\delta A$ *with* $P_\delta > 0$ *being the unique positive definite solution of the discrete-time algebraic Riccati equation*

$$P_\delta = A'P_\delta A + \delta I - A'P_\delta B(B'P_\delta B + I)^{-1}B'P_\delta A, \tag{7.28}$$

where $\delta \in (0, 1]$ *is the low-gain parameter to be chosen later.*

Define

$$\omega_{\max} = \begin{cases} 0, & A \text{ is Schur stable.} \\ \max\{\omega \in [0, \pi] \mid \det\left(e^{j\omega}I - A\right) = 0\}, & \text{otherwise} \end{cases} \tag{7.29}$$

The main result of this subsection is stated as follows.

Theorem 7.12 *Consider a MAS described by* (7.1) *and* (7.3). *Let any* $\beta \in (0, 1]$ *be given, and hence a set of network graphs* $\bar{\mathbb{G}}_\beta^N$ *be defined.*

If A *is at most weakly unstable and* (A, B) *is stabilizable, then the state synchronization problem stated in Problem* 7.1 *is solvable with* $\mathbf{G} = \bar{\mathbb{G}}_\beta^N$ *if*

$$\bar{\kappa}\omega_{\max} < \arccos(\beta). \tag{7.30}$$

In particular, for any given $\bar{\kappa}$ *satisfying* (7.30), *there exist a* $\rho > 0$ *and a* $\delta^* > 0$ *such that for this* ρ *and any* $\delta \in (0, \delta^*]$, *controller* (7.27) *solves the state synchronization* *problem for any graph $\mathcal{G} \in \bar{\mathbb{G}}_{\beta}^{N}$ and for any integer $\kappa \in [0, \bar{\kappa}]$. Moreover, the synchronized trajectory is given by (7.25).*

Proof According to Theorem 7.8, we only need to prove that the system

$$x(k+1) = Ax(k) + (1-\lambda)\rho B F_{\delta} x(k-\kappa), \tag{7.31}$$

is asymptotically stable for all $|\lambda| < \beta$ and $\kappa \in [0, \bar{\kappa}]$.

Let $\tilde{\lambda} = 1 - \lambda$. We find that $\tilde{\lambda} \in \mathcal{D}(1, \beta)$. Here $\mathcal{D}(z_0, r)$ denotes an open disc centered at z_0 with radius r. Since $\bar{\kappa}$ satisfies condition (7.30), there exists a $\rho > \frac{1}{1-\beta}$ such that

$$\bar{\kappa}\omega_{\max} < \arccos(\beta + \tfrac{1}{\rho}). \tag{7.32}$$

Note that ρ is independent of δ. Let this ρ be fixed.

For $\tilde{\lambda} \in \mathcal{D}(1, \beta)$ and for the previously selected ρ, we have

$$\tilde{\lambda}\rho \in \mathcal{D}(\rho, \rho\beta) \subset H_2 = \{z \in \mathbb{C} : \mathrm{Re}(z) > 1\}$$

since $\rho(1-\beta) > 1$. Since $\mathcal{D}(\rho, \rho\beta)$ is contained in a compact subset of $H_2 \subset H_1$, by Lemma 7.4, there exists a δ_1 such that for a $\delta \in (0, \delta_1]$, $\mathcal{D}(\rho, \rho\beta) \subset \Omega_{\delta}$, and hence, $A + \tilde{\lambda}\rho B F_{\delta}$ is Schur stable, where Ω_{δ} is the disc margin defined in (7.10). According to Lemma 7.3, the system (7.31) is asymptotically stable if

$$\det\left\{e^{j\omega}I - A - \tilde{\lambda}\rho e^{-j\omega\kappa_r} B F_{\varepsilon}\right\} \neq 0,$$
$$\forall \omega \in [-\pi, \pi],\ \forall \tilde{\lambda} \in \mathcal{D}(1, \beta),\ \forall \kappa_r \in [0, \bar{\kappa}]. \tag{7.33}$$

By (7.32), there exists a $\phi > 0$ independent of δ such that

$$\bar{\kappa}\omega < \arccos\left(\beta + \tfrac{1}{\rho}\right), \quad \text{for } |\omega| < \omega_{\max} + \phi.$$

We first show that there exists a $\delta_2 \in (0, \delta_1]$ such that for all $\delta \in (0, \delta_2]$, (7.33) holds for $\pi \geq |\omega| \geq \omega_{\max} + \phi$. To see this, note that

$$\det(e^{j\omega}I - A) \neq 0 \text{ for all } \omega_{\max} + \phi \leq |\omega| \leq \pi,$$

which implies that

$$\sigma_{\min}(e^{j\omega}I - A) > 0 \text{ for all } \omega_{\max} + \phi \leq |\omega| \leq \pi.$$

Since $\sigma_{\min}(e^{j\omega}I - A)$ depends continuously on ω and the set $\{\omega_{\max} + \phi \leq |\omega| \leq \pi\}$ is compact, there exists a μ such that

$$\sigma_{\min}(e^{j\omega}I - A) > \mu, \quad \text{for all } \omega_{\max} + \phi \leq |\omega| \leq \pi.$$

Let $\bar{\lambda} = \tilde{\lambda}\rho e^{-j\omega\kappa_r}$. We have

$$|\bar{\lambda}| \leq 2\rho, \quad \forall \tilde{\lambda} \in \mathcal{D}(1, \beta), \ \forall \omega, \ \forall \kappa_r.$$

Choose a $\delta_3 \leq \delta_2$ such that $\|F_\delta\| \leq \frac{\mu}{2\rho}\|B\|^{-1}$ for all $\delta \leq \delta_3$. In that case,

$$\sigma_{\min}(e^{j\omega}I - A - \bar{\lambda}BF_\delta) > \mu - |\bar{\lambda}|\|B\|\|F_\delta\| > 0, \qquad \text{for all } \omega_{\max} + \phi \leq |\omega| \leq \pi,$$

and hence, (7.33) holds for $\omega_{\max} + \phi \leq |\omega| \leq \pi$.

It remains to verify the condition (7.33) for $|\omega| < \omega_{\max} + \phi$. Let us consider the gain $\tilde{\lambda}\rho e^{-j\omega\kappa_r}$. We have $\tilde{\lambda}\rho \in \mathcal{D}(\rho, \rho\beta)$. It is evident from Fig. 7.1 that, since

$$|\omega|\kappa_r \leq |\omega|\bar{\kappa} < \arccos\left(\beta + \tfrac{1}{\rho}\right),$$

we have

$$\tilde{\lambda}\rho e^{-j\omega\kappa_r} \in \mathcal{D}(\rho e^{-j\omega\kappa_r}, \rho\beta) \subset H_2, \quad \forall \kappa_r \in [0, \bar{\kappa}].$$

Since β, ρ, ϕ, $\bar{\kappa}$, and $\omega_{\max}$ are independent of δ, by Lemma 7.4, there exists a $\delta_4 \leq \delta_3$ such that for all $\delta \in (0, \delta_4]$

$$\mathcal{D}(\rho e^{-j\omega\kappa_r}, \rho\beta) \subset \Omega_\delta, \quad \forall |\omega| < \omega_{\max} + \phi, \ \forall \kappa_r \in [0, \bar{\kappa}].$$

This is visualized in Fig. 7.1. This implies that

$$\tilde{\lambda}e^{j\omega\kappa_r} \in \Omega_\delta, \ \forall |\omega| < \omega_{\max} + \phi, \ \kappa_r \in [0, \bar{\kappa}] \text{ and } \tilde{\lambda} \in \mathcal{D}(1, \beta).$$

In this case, by Lemma 7.4, $A + \tilde{\lambda}\rho e^{-j\omega\kappa_r}BF_\delta$ is Schur stable. Hence, the condition (7.33) holds. ∎

From Theorem 7.12 and its proof, we immediately have the following result for the case when A is either Schur stable or has all its unstable eigenvalues at 1. This results from the fact that in this case we have $\omega_{\max} = 0$.

Corollary 7.13 *Consider a MAS described by* (7.1) *and* (7.3). *Let any* $\beta \in (0, 1]$ *be given, and hence a set of network graphs* $\bar{\mathbb{G}}_\beta^N$ *be defined.*

Assume A is either Schur stable or has all its unstable eigenvalues at 1. *If* (A, B) *is stabilizable, then the state synchronization problem stated in Problem* 7.1 *with* $\mathbf{G} = \bar{\mathbb{G}}_\beta^N$ *is always solvable via the controller* (7.27). *In particular, for any given*

Fig. 7.1 $\mathcal{D}(\rho e^{-j\omega\kappa_r}, \rho\beta)$ is circle with center A

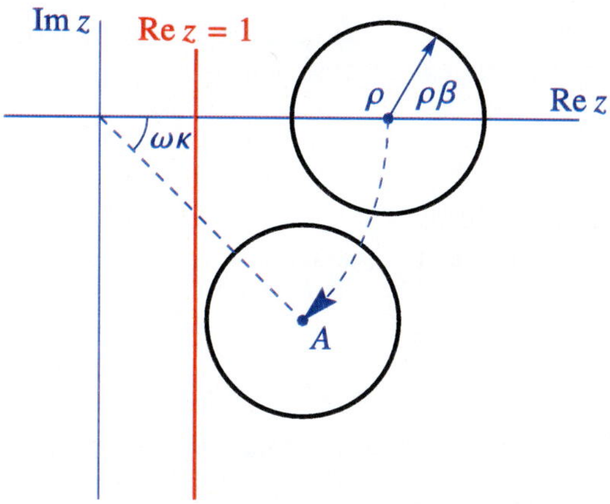

$\bar{\kappa} > 0$, *there exist a* $\rho > 0$ *and a* $\delta^* > 0$ *such that for this* ρ *and any* $\delta \in (0, \delta^*]$, *controller* (7.27) *solves the state synchronization problem for any graph* $\mathcal{G} \in \bar{\mathbb{G}}_\beta^N$ *and any integer* $\kappa \in [0, \bar{\kappa}]$. *Moreover, the synchronized trajectory is given by* (7.25).

If the communication topology is undirected, D is symmetric and has only real eigenvalues. Of course, the result in Theorem 7.12 still holds, but in fact a stronger result can be proved. It turns out that the delay tolerance is independent of network topology.

Corollary 7.14 *Consider a MAS described by* (7.1) *and* (7.3). *Let any* $\beta \in (0, 1]$ *be given, and hence a set of undirected network graphs* $\bar{\mathbb{G}}_\beta^{u,N}$ *be defined.*

If (A, B) *is stabilizable, then the state synchronization problem stated in Problem 7.1 with* $\mathbf{G} = \bar{\mathbb{G}}_\beta^{u,N}$ *is solvable if*

$$\bar{\kappa}\omega_{\max} < \tfrac{\pi}{2}. \tag{7.34}$$

In particular, for any given $\bar{\kappa}$ *satisfying* (7.34), *there exist a* $\rho > 0$ *and a* $\delta^* > 0$ *such that for this* ρ *and any* $\delta \in (0, \delta^*]$, *controller* (7.27) *solves the state synchronization problem for any graph* $\mathcal{G} \in \bar{\mathbb{G}}_\beta^{u,N}$ *and any integer* $\kappa \in [0, \bar{\kappa}]$. *Moreover, the synchronized trajectory is given by* (7.25).

Proof According to Theorem 7.8, we only need to show that for any $\beta \in (0, 1]$ and $\bar{\kappa} \in \mathbb{Z}^+$ satisfying (7.34), there exist a $\rho > 0$ and a $\delta^* \in (0, 1]$ such that for a $\delta \in (0, \delta^*]$, the system

$$x(k+1) = Ax(k) + (1-\lambda)\rho B F_\delta x(k-\kappa), \tag{7.35}$$

is asymptotically stable for all $\lambda \in (-\beta, \beta)$ and $\kappa \in [0, \bar{\kappa}]$.

With $\tilde{\lambda} = 1 - \lambda$, we have $\tilde{\lambda} \in (1 - \beta, 1 + \beta)$. For any $\beta \in (0, 1)$ and $\bar{\kappa}$ satisfying (7.34), there exists a ρ such that

$$\rho(1 - \beta)\cos(\bar{\kappa}\omega_{\max}) > 1. \tag{7.36}$$

Let this ρ be fixed.

We note that, since $\tilde{\lambda}\rho > 1$, by Lemma 7.4, there exists a $\delta_1 \in (0, 1]$ such that for all $\delta \in (0, \delta_1]$

$$\tilde{\lambda}\rho \in \Omega_\delta.$$

Then Lemma 7.3 says that the system (7.35) is asymptotically stable if

$$\det\left\{e^{j\omega}I - A - \tilde{\lambda}\rho e^{-j\omega\kappa_r} BF_\delta\right\} \neq 0,$$
$$\forall \omega \in [-\pi, \pi],\ \forall\tilde{\lambda} \in (1 - \beta, 1 + \beta),\ \kappa_r \in [0, \bar{\kappa}]. \tag{7.37}$$

By (7.36), there exists a $\phi > 0$ independent of δ such that

$$\tilde{\lambda}\rho\cos(\omega\bar{\kappa}) > \tfrac{1}{2}, \quad \forall\, |\omega| < \omega_{\max} + \phi,\ \forall\, \tilde{\lambda} \in (1 - \beta, 1 + \beta).$$

With the same arguments as in the proof of Theorem 7.12, we can show that there exists a $\delta_2 \in (0, \delta_1]$ such that for all $\delta \in (0, \delta_2]$, the condition (7.37) holds for $|\omega| \in [\omega_{\max} + \phi, \pi]$.

For $|\omega| < \omega_{\max} + \phi$, by the definition of ϕ and Lemma 7.4, there exists a $\delta_3 \in (0, \delta_2]$ such that for all $\delta \in (0, \delta_3]$

$$\tilde{\lambda}\rho e^{-j\omega\kappa_r} \in \Omega_\delta, \quad \forall\tilde{\lambda} \in (1 - \beta, 1 + \beta),\ |\omega| < \omega_{\max} + \phi,\ \kappa_r \in [0, \bar{\kappa}].$$

Therefore, $A + \tilde{\lambda}\rho e^{-j\omega\kappa_r} BF_\delta$ is Schur stable for any

$$|\omega| < \omega_{\max} + \phi, \quad \tilde{\lambda} \in (1 - \beta, 1 + \beta) \text{ and } \kappa_r \in [0, \bar{\kappa}].$$

In conclusion, the condition (7.37) holds. ■

7.2.2.3 Direct Eigenstructure Assignment Method

The feedback gain (3.50) in Sect. 3.4.3.2, which is based on the direct eigenstructure assignment method, also works for the delay case in this section, that is

$$u_i = \rho F_\delta \zeta_i, \tag{7.38}$$

where

$$F_\delta = T_u \begin{pmatrix} F_{\delta_1,1} & 0 & \dots & \dots & 0 \\ 0 & F_{\delta_2,2} & \ddots & & \vdots \\ \vdots & \ddots & \ddots & \ddots & \vdots \\ \vdots & & \ddots & F_{\delta_q,q} & 0 \\ 0 & \dots & \dots & 0 & 0 \end{pmatrix} T_x^{-1},$$

with T_u, T_x, and $F_{\delta_i,i}$ $(i = 1, \dots, q)$ are defined in Sect. 3.4.3.2. Parameters ρ and δ are chosen according to β and $\bar{\kappa}$.

The main result of this subsection is stated as follows.

Theorem 7.15 *Consider a MAS described by (7.1) and (7.3) with (A, B) controllable and A being at most weakly unstable. Let any $\beta \in (0, 1]$ be given, and hence a set of network graphs $\bar{\mathbb{G}}_\beta^N$ be defined.*

The state synchronization problem stated in Problem 7.1 is solvable if the condition (7.30) holds. In particular, for any given $\bar{\kappa}$ satisfying (7.30), there exist a $\rho > 0$ and a $\delta^* > 0$ such that for this ρ and any $\delta \in (0, \delta^*]$, controller (7.38) solves the state synchronization problem for any graph $\mathcal{G} \in \bar{\mathbb{G}}_\beta^N$ and for any integer $\kappa \in [0, \bar{\kappa}]$. Moreover, the synchronized trajectory is given by (7.25).

Proof The proof is similar to the proof of Theorem 7.12. We only need to prove that the system

$$x(k+1) = Ax(k) + (1-\lambda)\rho B F_\delta x(k-\kappa), \tag{7.39}$$

is asymptotically stable for all $|\lambda| < \beta$ and $\kappa \in [0, \bar{\kappa}]$.

Let $\tilde{\lambda} = 1 - \lambda$. We find that $\tilde{\lambda} \in \mathcal{D}(1, \beta)$. Since $\bar{\kappa}$ satisfies the condition (7.30), there exists a $\rho > 0$ sufficiently small such that

$$\omega_{\max}\bar{\kappa} < \arccos(\beta + 4\rho). \tag{7.40}$$

We choose $\tilde{\beta}$ such that

$$1 - 4\rho^2 < \tilde{\beta} < 1.$$

Then

$$\rho\tilde{\lambda}e^{-j\omega\kappa_r} \in \mathcal{D}(1, \tilde{\beta}). \tag{7.41}$$

After all $\tilde{\lambda} \in \mathcal{D}(1, \beta)$ implies that

$$\tilde{\lambda}e^{-j\omega\kappa_r} = v + wj \in \mathcal{D}(e^{-j\omega\kappa_r}, \beta)$$

with v, $w \in \mathbb{R}$. Clearly, (7.40) implies that

$$v > \cos(\omega\kappa_r) - \beta > 4\rho.$$

Moreover, we obviously have

$$v^2 + w^2 \leq 4.$$

But then

$$(1-\rho v)^2 + \rho^2 w^2 \leq 1 - 2\rho v + \rho^2(v^2+w^2) \leq 1 - 4\rho^2 \leq \tilde{\beta},$$

which yields (7.41). To prove that the system is asymptotically stable without delay, we note that this is a direct consequence of Lemma 7.5 by noting that $\rho\tilde{\lambda} \in \mathcal{D}(1, \tilde{\beta})$. In that case, according to Lemma 7.3, the system (7.39) is asymptotically stable if

$$\det\left\{e^{j\omega}I - A - \tilde{\lambda}\rho e^{-j\omega\kappa_r} BF_\varepsilon\right\} \neq 0,$$
$$\forall\omega \in [-\pi, \pi],\ \forall\tilde{\lambda} \in \mathcal{D}(1, \beta),\ \forall\kappa_r \in [0, \bar{\kappa}]. \qquad (7.42)$$

By (7.40), there exists a $\phi > 0$ independent of δ such that

$$\omega\bar{\kappa} < \arccos(\beta + 4\rho), \qquad \text{for } |\omega| < \omega_{\max} + \phi.$$

Similar to the proof of Theorem 7.12, we can show that for a sufficiently small δ we have (7.42) satisfied for $\omega_{\max} + \phi \leq |\omega| \leq \pi$. It remains to verify the condition (7.42) for $|\omega| < \omega_{\max} + \phi$. But using (7.41), it is an immediate consequence of Lemma 7.5 that $A + \tilde{\lambda}\rho e^{-j\omega\kappa_r} BF_\delta$ is Schur stable. Hence, the condition (7.42) holds. ∎

7.2.3 *Protocol Design for MAS with Partial-State Coupling*

In this section, by paralleling the approach of the previous section, we show first that the problem of state synchronization among agents in the network with partial-state coupling and in the presence of uniform input delay can be solved by equivalently solving a robust stabilization problem, and then we design a controller for such a robust stabilization problem.

7.2.3.1 Connection to a Robust Stabilization Problem

The MAS described by (7.1) and (7.2) after implementing the linear dynamic protocol (7.5) is written as

$$\begin{cases} \bar{x}_i(k+1) = \begin{pmatrix} A & BC_c \\ 0 & A_c \end{pmatrix} \bar{x}_i(k) + \begin{pmatrix} 0 \\ B_c \end{pmatrix} \zeta_i(k-\kappa), \\ y_i(k) \quad = \begin{pmatrix} C & 0 \end{pmatrix} \bar{x}_i(k), \\ \zeta_i(k) \quad = \sum_{j=1}^{N} d_{ij}(y_i(k) - y_j(k)), \end{cases}$$

for $i = 1, \ldots, N$ where

$$\bar{x}_i = \begin{pmatrix} x_i \\ \chi_i \end{pmatrix}.$$

Define

$$\bar{x} = \begin{pmatrix} \bar{x}_1 \\ \vdots \\ \bar{x}_N \end{pmatrix}, \quad \bar{A} = \begin{pmatrix} A & BC_c \\ 0 & A_c \end{pmatrix}, \quad \bar{B} = \begin{pmatrix} 0 \\ B_c \end{pmatrix} \text{ and } \bar{C} = \begin{pmatrix} C & 0 \end{pmatrix}. \tag{7.43}$$

The overall dynamics of the N agents can be written as

$$\bar{x}(k+1) = (I_N \otimes \bar{A})\bar{x}(k) + \left\{(I - D) \otimes \bar{B}\bar{C}\right\} \bar{x}(k-\kappa). \tag{7.44}$$

As observed in the case of full-state coupling, synchronization for the system (7.44) is equivalent to the asymptotic stability of the following $N-1$ subsystems:

$$\tilde{\eta}_i(k+1) = \bar{A}\eta_i(k) + (1-\lambda_i)\bar{B}\bar{C}\tilde{\eta}_i(k-\kappa), \quad i = 2, \ldots, N, \tag{7.45}$$

for any integer $\kappa \in [0, \bar{\kappa}]$, where λ_i for $i = 2, \ldots, N$ are the eigenvalues of the row stochastic matrix D inside the unit disc.

Theorem 7.16 *The MAS (7.44) achieves state synchronization if and only if the system (7.45) is asymptotically stable for $i = 2, \ldots, N$ and for any integer $\kappa \in [0, \bar{\kappa}]$.*

Proof As in the proof of Theorem 3.4, we define the Jordan form (3.12) of the stochastic row matrix D. Let

$$\eta := (T^{-1} \otimes I_{n+n_c})\bar{x} = \begin{pmatrix} \eta_1 \\ \vdots \\ \eta_N \end{pmatrix},$$

where $\eta_i \in \mathbb{C}^{n+n_c}$. In the new coordinates, the dynamics of η can be written as

$$\eta(k+1) = (I_N \otimes \bar{A})\eta(k) + ((I - J_D) \otimes \bar{B}\bar{C})\eta(k-\kappa). \tag{7.46}$$

The eigenvalues of system (7.46) are given by the roots of its characteristic equation

$$H(z) = \det\left\{zI - (I_N \otimes \bar{A}) - z^{-\kappa}((I - J_L) \otimes \bar{B}\bar{C})\right\} = 0,$$

which, due to the diagonal structure of $I_N \otimes \bar{A}$ and the upper triangular structure of $J_L \otimes \bar{B}\bar{C}$, are the union of the eigenvalues of the $N-1$ subsystems given in (7.45) and the eigenvalues of $\bar{A}$.

Now if (7.45) is globally asymptotically stable for $i = 2, \ldots, N$, we see from the above that $\eta_i(k) \to 0$ for $i = 2, \ldots, N$. This implies that

$$\bar{x}(k) - (T \otimes I_{n+n_c}) \begin{pmatrix} \eta_1(k) \\ 0 \\ \vdots \\ 0 \end{pmatrix} \to 0.$$

Note that the first column of T is equal to the vector $\mathbf{1}$ and therefore

$$\bar{x}_i(k) - \eta_1(k) \to 0$$

for $i = 1, \ldots, N$. This implies that we achieve state synchronization.

Conversely, suppose the network (7.44) reaches state synchronization. In this case, we shall have

$$\bar{x}(k) - \mathbf{1} \otimes \bar{x}_1(k) \to 0.$$

Then $\eta(k) - (T^{-1}\mathbf{1}) \otimes \bar{x}_1(k) \to 0$. Since $\mathbf{1}$ is the first column of T, we have

$$T^{-1}\mathbf{1} = \begin{pmatrix} 1 \\ 0 \\ \vdots \\ 0 \end{pmatrix}.$$

Therefore, $\eta(k) - (T^{-1}\mathbf{1}) \otimes \bar{x}_1(k) \to 0$ implies that $\eta_1(k) - \bar{x}_1(k) \to 0$ and $\eta_i(k) \to 0$ for $i = 2, \ldots, N$. ■

Corollary 7.17 *Consider a MAS described by* (7.1) *and* (7.2)*. If there exists a protocol of the form* (7.5) *that achieves state synchronization, then the associated graph must have a directed spanning tree or A must be asymptotically stable.*

Since A being asymptotically stable is trivial, the requirement that the graph of a MAS has a directed spanning tree is essentially necessary.

Remark 7.18 It also becomes clear from the proof of Theorem 7.16 that the *synchronized trajectory* is given by $x_s(k) = \eta_1(k)$ which is governed by

$$\eta_1(k+1) = \bar{A}\eta_1(k), \quad \eta_1(0) = (w \otimes I_{n+n_c})\bar{x}(0), \tag{7.47}$$

where w is the first row of T^{-1}, i.e., the normalized left eigenvector associated with the eigenvalue 1. This shows that the modes of the synchronized trajectory are determined by the eigenvalues of $\bar{A}$ and the complete dynamics depends on both $\bar{A}$ and a weighted average of the initial conditions of agents.

In light of Lemma 3.7, one can conclude that $\eta_1(0)$ is only a linear combination of the initial conditions of root agents. As such, the synchronized trajectory given by (7.47) can be written explicitly as

$$\eta_1(k) = \bar{A}^k \sum_{i \in \Pi_{\mathcal{G}}} w_i \bar{x}_i(0), \tag{7.48}$$

which is the weighted average of the trajectories of root agents.

In light of the definition of Problem 7.2 that synchronization is formulated for a set of graphs, and from the robust system (7.45), we basically obtain a *robust* stabilization problem, i.e., the stabilization of the system

$$\begin{cases} x(k+1) = Ax(k) + Bu(k-\kappa), \\ \quad z(k) \quad = (1-\lambda)Cx(k), \end{cases} \tag{7.49}$$

and a protocol

$$\begin{cases} \chi(k+1) = A_c\chi(k) + B_c z(k), \\ \quad u(k) \quad = C_c\chi(k), \end{cases} \tag{7.50}$$

for any integer $\kappa \in [0, \bar{\kappa}]$ and any λ that are the eigenvalues inside the unit disc of the row stochastic matrix D associated with a graph in the given set $\bar{\mathbb{G}}_{\beta}^{N}$.

Define an auxiliary system

$$\begin{aligned} x(k+1) &= Ax(k) + (1-\lambda)Bu(k-\kappa), \\ z(k) &= Cx(k). \end{aligned} \tag{7.51}$$

We need the result of the following lemma.

Lemma 7.19 *For any given $\beta \in (0, 1]$, Problem 7.2 is solvable via the controller (7.5) if the closed-loop system of (7.50) and (7.51) is globally asymptotically stable for any λ that satisfies $|\lambda| < \beta$ and for any integer $\kappa \in [0, \bar{\kappa}]$.*

Proof The closed-loop system of (7.49) and (7.50) can be written as

$$\bar{x}(k+1) = \bar{A}\bar{x}(k) + (1-\lambda)\bar{B}\bar{C}\bar{x}(k-\kappa), \tag{7.52}$$

where

$$\bar{x} = \begin{pmatrix} x(k) \\ \chi(k) \end{pmatrix},$$

while $\bar{A}$, $\bar{B}$, and $\bar{C}$ are given in (7.43).

Now define

$$\tilde{A} = \begin{pmatrix} A & 0 \\ B_c C & A_c \end{pmatrix}, \quad \tilde{B} = \begin{pmatrix} B \\ 0 \end{pmatrix}, \quad \tilde{C} = \begin{pmatrix} 0 & C_c \end{pmatrix}, \quad \tilde{x} = \begin{pmatrix} x(k) \\ \chi(k) \end{pmatrix}.$$

Then, the closed-loop system of (7.51) and (7.50) can be written as

$$\tilde{x}(k+1) = \tilde{A}\tilde{x}(k) + (1-\lambda)\tilde{B}\tilde{C}\tilde{x}(k-\kappa). \tag{7.53}$$

Let $\tilde{\lambda} = 1 - \lambda$. We find that

$$\begin{aligned}
&\begin{pmatrix} \frac{1}{\tilde{\lambda}} z^{\kappa} I & 0 \\ 0 & I \end{pmatrix} (zI - \tilde{A} - \tilde{\lambda} z^{-\kappa} \tilde{B}\tilde{C}) \begin{pmatrix} \tilde{\lambda} z^{-\kappa} I & 0 \\ 0 & I \end{pmatrix} \\
&\quad = \begin{pmatrix} \frac{1}{\tilde{\lambda}} z^{\kappa} I & 0 \\ 0 & I \end{pmatrix} \begin{pmatrix} zI - A & -\tilde{\lambda} z^{-\kappa} B C_c \\ -B_c C & zI - A_c \end{pmatrix} \begin{pmatrix} \tilde{\lambda} z^{-\kappa} I & 0 \\ 0 & I \end{pmatrix} \\
&\quad = \begin{pmatrix} zI - A & -BC_c \\ -\tilde{\lambda} z^{-\kappa} B_c C & zI - A_c \end{pmatrix} \\
&\quad = z - \bar{A} - \tilde{\lambda} z^{-\kappa} \bar{B}\bar{C}.
\end{aligned}$$

This implies that the characteristic function of each system has the same number of zeros outside the unit disc. Hence the stability of system (7.52) and (7.53) is equivalent. ∎

7.2.3.2 ARE-Based Method

We design the following protocol based on a low-gain method, together with the standard H_2 discrete-time algebraic Riccati equation.

Protocol design 7.2 *Consider a MAS described by (7.1) and (7.2). We choose an observer gain K such that $A + KC$ is Schur stable and a feedback gain $F_\delta = -(B'P_\delta B + I)^{-1}B'P_\delta A$ where P_δ is the solution of the ARE (7.28).*

Then, the linear dynamic protocol can be constructed as

$$\begin{aligned} \chi_i(k+1) &= (A+KC)\chi_i(k) - K\zeta_i(k), \\ u_i(k) &= \rho F_\delta \chi_i(k), \end{aligned} \tag{7.54}$$

where $\rho > 0$ and δ are design parameters to be chosen later.

The main result based on the above design is stated as follows.

Theorem 7.20 *Consider a MAS described by (7.1) and (7.2). Let any $\beta \in (0, 1]$ be given, and hence a set of network graphs $\bar{\mathbb{G}}_\beta^N$ be defined.*

If A is at most weakly unstable, (A, B) is stabilizable and (C, A) is detectable, then the state synchronization problem stated in Problem 7.2 is solvable with $\mathbf{G} \in \bar{\mathbb{G}}_\beta^N$ if the condition (7.30) holds. In particular, for any given $\bar{\kappa}$ satisfying (7.30), there exist a $\rho > 0$ and a $\delta^* > 0$ such that for this ρ and any $\delta \in (0, \delta^*]$, controller (7.54) solves the state synchronization problem for any graph $\mathcal{G} \in \bar{\mathbb{G}}_\beta^N$ and any $\kappa \in [0, \bar{\kappa}]$. Moreover, the synchronized trajectory is given by (7.48).

Proof Following Lemma 7.19, we only need to prove that there exist a ρ and a δ^* such that for any $\delta \in (0, \delta^*]$, the system

$$\begin{cases} x(k+1) = Ax(k) + (1-\lambda)\rho B F_\delta \chi(k-\kappa) \\ \chi(k+1) = (A+KC)\chi(k) - KCx(k), \end{cases} \tag{7.55}$$

is asymptotically stable for any $\lambda \in \mathcal{D}(0, \beta)$ and $\kappa \in [0, \bar{\kappa}]$.

Let $\tilde{\lambda} = 1 - \lambda$. Then, we have $\tilde{\lambda} \in \mathcal{D}(1, \beta)$. First of all, since $\bar{\kappa}$ satisfies the condition (7.30), there exists a $\rho > \frac{2}{1-\beta}$ such that

$$\omega_{\max}\bar{\kappa} < \arccos(\beta + \tfrac{2}{\rho}). \tag{7.56}$$

Note that ρ is independent of δ. Let this ρ be fixed. Since $(1-\beta)\rho > 2$, we have for $\tilde{\lambda} \in \mathcal{D}(1, \beta)$ that $\operatorname{Re}(\tilde{\lambda}\rho) > 2$. It follows from Lemma 7.6 that there exists a δ_1 such that for $\delta \in (0, \delta_1]$, the system (7.55) is asymptotically stable for $\kappa = 0$, that is, the matrix $\bar{\mathcal{A}} + \tilde{\lambda}\rho\bar{\mathcal{B}}\bar{C}$ is Schur stable, where

$$\bar{\mathcal{A}} = \begin{pmatrix} A & 0 \\ -KC & A+KC \end{pmatrix}, \quad \bar{\mathcal{B}} = \begin{pmatrix} B \\ 0 \end{pmatrix} \quad \text{and} \quad \bar{C} = \begin{pmatrix} 0 & \beta F_\delta \end{pmatrix}.$$

Then Lemma 7.3 tells that (7.55) with $\delta \in (0, \delta_1]$ is asymptotically stable for $\tilde{\lambda} \in \mathcal{D}(1, \beta)$ and $\kappa \in [0, \bar{\kappa}]$ if

$$\det\left\{e^{j\omega}I - \bar{\mathcal{A}} - \tilde{\lambda}\rho e^{-j\omega\kappa_r}\bar{\mathcal{B}}\bar{C}\right\} \neq 0,$$
$$\forall \omega \in [-\pi, \pi],\ \tilde{\lambda} \in \mathcal{D}(1, \beta),\ \kappa_r \in [0, \bar{\kappa}]. \tag{7.57}$$

The rest of the proof basically follows along the same lines as in the proof of Theorem 7.12. There exists a $\phi > 0$ independent of δ such that

$$\omega\bar{\kappa} < \arccos\left(\beta + \tfrac{2}{\rho}\right), \quad \text{for } |\omega| < \omega_{\max} + \phi.$$

For $|\omega| \geq \omega_{\max} + \phi$, we can show that there exists a $\delta_2 \leq \delta_1$ such that for all $\delta \in (0, \delta_2]$, the condition (7.57) holds using the same argument as in the proof of Theorem 7.12.

For $|\omega| < \omega_{\max} + \phi$, by the definition of ρ and ϕ

$$\operatorname{Re}(\tilde{\lambda}\rho e^{-j\omega\kappa_r}) > 2$$

for any $\tilde{\lambda} \in \mathcal{D}(1, \beta)$, $|\omega| < \omega_{\max} + \phi$ and $\kappa_r \in [0, \bar{\kappa}]$. Obviously, for $\tilde{\lambda} \in \mathcal{D}(1, \beta)$ and ρ given by (7.56), $\tilde{\lambda}\rho e^{-j\omega\kappa_r}$ can then be bounded in some compact set, say $\mathcal{S}$, which only depends on β and $\bar{\kappa}$ and is located inside $\{z \in \mathbb{C} : \operatorname{Re}(z) > 1\}$. Then, by Lemma 7.6, we can find $\delta^* \leq \delta_2$ such that for $\delta \in (0, \delta^*]$, the matrix

$$\bar{\mathcal{A}} + \tilde{\lambda}\rho e^{-j\omega\kappa_r}\bar{\mathcal{B}}\bar{C}$$

is Schur stable, and hence, the condition (7.57) holds for $|\omega| < \omega_{\max} + \phi$, $\tilde{\lambda} \in \mathcal{D}(1, \beta)$ and $\kappa_r \in [0, \bar{\kappa}]$. ∎

When the communication topology is undirected, the following result can be proved by using an argument similar to that in the proof of Corollary 7.14.

Corollary 7.21 *Consider a MAS described by (7.1) and (7.2). Let any $\beta \in (0, 1]$ be given, and hence a set of undirected network graphs $\bar{\mathbb{G}}_\beta^N$ be defined.*

If (A, B) is stabilizable and (C, A) is detectable, then the state synchronization problem stated in Problem 7.2 is solvable if

$$\bar{\kappa}\omega_{\max} < \frac{\pi}{2}. \tag{7.58}$$

In particular, for any given $\bar{\kappa}$ satisfying (7.58), there exist a $\rho > 0$ and a $\delta^* > 0$ such that for this ρ and any $\delta \in (0, \delta^*]$, controller (7.54) solves the state synchronization problem for any graph $\mathcal{G} \in \bar{\mathbb{G}}_\beta^N$ and for any integer $\kappa \in [0, \bar{\kappa}]$. Moreover, the synchronized trajectory is given by (7.48).

7.2.3.3 Direct Eigenstructure Assignment Method

The dynamic low-gain protocol (6.5) based on the direct eigenstructure assignment method can be constructed as

$$\begin{aligned} \chi_i(k+1) &= (A+KC)\chi_i(k) - K\zeta_i(k), \\ u_i(k) &= \rho F_\delta \chi_i(k), \end{aligned} \tag{7.59}$$

where F_δ is defined in (3.50) in Sect. 3.4.3.2 and K is chosen such that $A+KC$ is Schur stable. Here parameters ρ and δ are also chosen according to β and $\bar{\kappa}$.

The main result of this subsection is stated as follows.

Theorem 7.22 *Consider a MAS described by (7.1) and (7.2) with (A, B) controllable and A being at most weakly unstable. Let any $\beta \in (0, 1]$ be given, and hence a set of network graphs $\bar{\mathbb{G}}_\beta^N$ be defined.*

The state synchronization problem stated in Problem 7.2 is solvable for any graph $\mathcal{G} \in \bar{\mathbb{G}}_\beta^N$ if the condition (7.30) holds. In particular, for any given $\bar{\kappa}$ satisfying (7.30), there exist a $\rho > 0$ and a $\delta^* > 0$ such that for this ρ and any $\delta \in (0, \delta^*]$, controller (7.59) solves the state synchronization problem for any graph $\mathcal{G} \in \bar{\mathbb{G}}_\beta^N$ and for any integer $\kappa \in [0, \bar{\kappa}]$. Moreover, the synchronized trajectory is given by (7.48).

Proof The proof is similar to the proof of Theorem 7.20. Again, we only need to prove that there exist a ρ and a δ^* such that for any $\delta \in (0, \delta^*]$, the system (7.55) is asymptotically stable for any $\lambda \in \mathcal{D}(0, \beta)$ and $\kappa \in [0, \bar{\kappa}]$ but with F_δ this time given by (3.50).

Let $\tilde{\lambda} = 1 - \lambda$. Then, we have $\tilde{\lambda} \in \mathcal{D}(1, \beta)$. First of all, since $\bar{\kappa}$ satisfies the condition (7.30), for a sufficiently small ρ we have

$$\omega_{\max}\bar{\kappa} < \arccos(\beta + 4\rho). \tag{7.60}$$

Note that this choice for ρ is independent of δ. Let this ρ be fixed. We choose $\tilde{\beta}$ such that

$$1 - 4\rho^2 < \tilde{\beta} < 1,$$

in which case we have (7.41).

Using (7.41), it then follows from Lemma 7.7 that there exists a δ_1 such that for all $\delta \in (0, \delta_1]$, the system (7.55) is asymptotically stable for $\kappa = 0$, that is, the matrix $\bar{\mathcal{A}} + \tilde{\lambda}\rho\bar{\mathcal{B}}\bar{C}$ is Schur stable, where

$$\bar{\mathcal{A}} = \begin{pmatrix} A & 0 \\ -KC & A+KC \end{pmatrix}, \quad \bar{\mathcal{B}} = \begin{pmatrix} B \\ 0 \end{pmatrix} \quad \text{and } \bar{C} = \begin{pmatrix} 0 & \beta F_\delta \end{pmatrix}.$$

Then Lemma 7.3 yields that (7.55) with $\delta \in (0, \delta_1]$ is asymptotically stable for $\tilde{\lambda} \in \mathcal{D}(1, \beta)$ and $\kappa \in [0, \bar{\kappa}]$ if

$$\det\left\{e^{j\omega} I - \bar{\mathcal{A}} - \tilde{\lambda}\rho e^{-j\omega\kappa_r}\bar{\mathcal{B}}\bar{\mathcal{C}}\right\} \neq 0,$$
$$\forall \omega \in [-\pi, \pi],\ \tilde{\lambda} \in \mathcal{D}(1, \beta),\ \kappa_r \in [0, \bar{\kappa}]. \tag{7.61}$$

The rest of the proof basically follows along the same lines as in the proof of Theorem 7.15. There exists a $\phi > 0$ independent of δ such that

$$\omega\bar{\kappa} < \arccos(\beta + 4\rho), \quad \text{for } |\omega| < \omega_{\max} + \phi.$$

For $|\omega| \geq \omega_{\max} + \phi$, we can show that there exists a $\delta_2 \leq \delta_1$ such that for all $\delta \in (0, \delta_2]$, the condition (7.61) holds using the same argument as in the proof of Theorem 7.12.

For $|\omega| < \omega_{\max} + \phi$, by the definition of ρ and ϕ, we have (7.41) for any $\tilde{\lambda} \in \mathcal{D}(1, \beta)$, $|\omega| < \omega_{\max} + \phi$ and $\kappa_r \in [0, \bar{\kappa}]$. Then, by Lemma 7.7, we can find a $\delta^* \leq \delta_2$ such that for $\delta \in (0, \delta^*]$, the matrix

$$\bar{\mathcal{A}} + \tilde{\lambda}\rho e^{-j\omega\kappa_r}\bar{\mathcal{B}}\bar{\mathcal{C}}$$

is Schur stable, and hence, the condition (7.61) holds for $|\omega| < \omega_{\max} + \phi$, $\tilde{\lambda} \in \mathcal{D}(1, \beta)$ and $\kappa_r \in [0, \bar{\kappa}]$. ■

7.2.4 *Static Protocol Design with Partial-State Coupling*

In this section, we consider the static protocol design for squared-down passifiable via static input feedforward discrete-time agents. We show first that the state synchronization among agents in the network with partial-state coupling and input delay via static protocols can be solved by equivalently solving a robust stabilization problem.

7.2.4.1 Connection to a Robust Stabilization Problem

The static protocol (7.4) with ζ_i given in (7.2) is used. Then, the closed-loop system of the agent (7.1) and the protocol (7.4) can be described by

$$\begin{cases} x_i(k+1) = Ax_i(k+1) + BF\zeta_i(k-\kappa), \\ y_i(k) \quad = Cx_i(k), \\ \zeta_i(k) \quad = \sum_{j=1}^{N} d_{ij}(y_i(k) - y_j(k)), \end{cases} \tag{7.62}$$

for $i = 1, \ldots, N$. Define

$$x = \begin{pmatrix} x_1 \\ \vdots \\ x_N \end{pmatrix}.$$

Then, the overall dynamics of the N agents can be written as

$$x(k+1) = (I_N \otimes A)x(k) + ((I - D) \otimes BFC)x(k - \kappa). \tag{7.63}$$

Following the proof of Theorem 7.8, we immediately have the result.

Theorem 7.23 *The MAS (7.63) achieves state synchronization if and only if each of the $N - 1$ subsystems*

$$\tilde{\eta}_i(k+1) = A\tilde{\eta}_i(k) + (1 - \lambda_i)BFC\tilde{\eta}_i(k - \kappa), \quad i = 2, \ldots, N, \tag{7.64}$$

is globally asymptotically stable for $i = 2, \ldots, N$ and for any integer $\kappa \in [0, \bar{\kappa}]$, where λ_i, $i = 2, \ldots, N$ are those eigenvalues of D inside the unit disc.

Similarly, we immediately have the following corollary and the synchronized trajectory.

Corollary 7.24 *Consider a MAS described by (7.1) and (7.2). If there exists a protocol of the form (7.4) that achieves state synchronization, then the associated graph must have a directed spanning tree or A must be asymptotically stable.*

The synchronized trajectory is given explicitly as

$$\eta_1(k) = A^k \sum_{i \in \Pi_{\mathcal{G}}} w_i x_i(0), \tag{7.65}$$

which is the weighted average of the trajectories of root agents.

In light of the definition of Problem 7.2 that synchronization is formulated for a set of graphs, we basically obtain a robust stabilization problem, i.e., the stabilization of the system

$$\begin{aligned} x(k+1) &= Ax(k) + (1 - \lambda)Bu(k - \kappa), \\ y(k) &= Cx(k), \end{aligned} \tag{7.66}$$

via a controller

$$u(k) = Fy(k),$$

for any λ that represents all eigenvalues inside the unit disc of any possible stochastic matrix D associated with the given set of graphs $\mathbf{G}$ and for any $\kappa \in [0, \bar{\kappa}]$.

7.2.4.2 Squared-Down Passifiable via Static Input Feedforward

The Protocol Design 3.11 for a MAS with agents that are squared-down passifiable via static input feedforward can tolerate a certain level of input delay. We reproduce the protocol design here.

Protocol design 7.3 *Consider a MAS described by agents* (7.1) *with communication via* (7.2). *Assume the agents are squared-down passifiable via static input feedforward with respect to* G_1, G_2, *and* R. *The static protocol is designed as*

$$u_i(k) = -\delta G_1 K G_2 \zeta_i(k), \tag{7.67}$$

where $K > 0$ *and* $\delta > 0$ *is the low-gain parameter to be designed.*

The main result based on the above design can be stated as follows.

Theorem 7.25 *Consider a MAS described by agents* (7.1) *and* (7.2). *Let any* $1 > \beta > 0$ *be given, and hence a set of network graphs* $\bar{\mathbb{G}}_\beta^N$ *be defined.*

Assume the agents are squared-down passifiable via static input feedforward with respect to G_1, G_2, *and* R *such that* (A, BG_1) *is controllable and* (A, G_2C) *is observable with* BG_1 *and* G_2C *full-column and full-row rank, respectively, then the state synchronization problem stated in Problem* 7.2 *but with a static protocol is solvable if the condition* (7.30) *holds. In particular, for any given* $\bar{\kappa}$ *satisfying* (7.30), *there exists a* $\delta^* > 0$ *such that for any* $\delta \in (0, \delta^*]$, *controller* (7.67) *solves the state synchronization problem for any graph* $\mathcal{G} \in \bar{\mathbb{G}}_\beta^N$ *and any integer* $\kappa \in [0, \bar{\kappa}]$. *Moreover, the synchronized trajectory is given by* (7.65).

Proof According to Theorem 7.23, we only need to prove that

$$x(k+1) = Ax(k) - \delta(1-\lambda)BG_1KG_2Cx(k-\kappa) \tag{7.68}$$

is asymptotically stable for all $|\lambda| < \beta$ and for any $\kappa \in [0, \bar{\kappa}]$.

Since $\bar{\kappa}$ satisfies the condition (7.30), there exists a $\bar{\rho}$ such that

$$\omega_{\max}\bar{\kappa} < \arccos(\beta + \bar{\rho}).$$

When the system (7.68) is without delay, i.e., $\kappa = 0$, from Theorem 3.42, we have $A - \delta(1-\lambda)BG_1KG_2C$ Schur stable. Then it follows from Lemma 7.3 that the system (7.68) is asymptotically stable if

$$\det\left[e^{j\omega}I - A + \delta(1-\lambda)e^{-j\omega\kappa_r}BG_1KG_2C\right] \neq 0, \tag{7.69}$$

for all $\omega \in [-\pi, \pi]$, for all $\kappa^r \in \mathbb{R}$ with $0 < \kappa^r \leq \bar{\kappa}$ and for all λ with $|\lambda| < \beta$.

Again we split the proof of (7.69) into two cases where $|\omega| < \omega_{\max} + \phi$ and $|\omega| \geq \omega_{\max} + \phi$ respectively. Here ϕ is such that

$$\omega\bar{\kappa} < \arccos(\beta + \bar{\rho}), \qquad \text{for all } \omega \text{ such that } |\omega| < \omega_{\max} + \phi.$$

If $|\omega| \geq \omega_{\max} + \phi$, there exists a $\mu > 0$ such that

$$\sigma_{\min}(e^{j\omega}I - A) < \mu, \quad \forall\omega \text{ such that } |\omega| \geq \omega_{\max} + \phi.$$

Then, with the exact arguments as in the proof of Theorem 7.12 and the boundedness of $|\lambda|$, we obtain

$$\sigma_{\min}(e^{j\omega}I - A - \delta(1-\lambda)e^{-j\omega\kappa_r}BG_1KG_2C) \geq \mu - \tfrac{\mu}{2} \geq \tfrac{\mu}{2},$$

for a small enough δ. Therefore, the condition (7.69) holds for $|\omega| \geq \omega_{\max} + \phi$.

If $|\omega| < \omega_{\max} + \phi$, we find that

$$\operatorname{Re}(1-\lambda)e^{-j\omega\kappa_r} > \operatorname{Re}(1-\lambda)e^{-j\omega_{\max}\bar{\kappa}} > \cos(\omega_{\max}\bar{\kappa}) - \beta > \bar{\rho},$$

and

$$|(1-\lambda)e^{-j\omega\kappa_r}| < 1 + \beta.$$

With the same arguments as in Theorem 3.42, we choose

$$\delta^* = \frac{2\bar{\rho}}{r(1+\beta)^2},$$

such that for all $\delta < \delta^*$, we have

$$A - \delta(1-\lambda)e^{-j\omega\kappa_r}BG_1KG_2C$$

Schur stable for all $|\omega| < \omega_{\max} + \phi$. That proves the result. ■

7.3 MAS in the Presence of Nonuniform Input Delay

In this section, we consider synchronization for MAS in the presence of nonuniform input delay. Both full- and partial-state coupling networks are considered. However, we restrict attention to undirected and weighted communication networks.

7.3.1 Multi-agent Systems

Consider a MAS composed of N identical linear time-invariant discrete-time agents of the form

$$\begin{aligned} x_i(k+1) &= Ax_i(k) + Bu_i(k-\kappa_i), \\ y_i(k) &= Cx_i(k), \end{aligned} \qquad (i = 1, \ldots, N) \tag{7.70}$$

where $\kappa_1, \ldots, \kappa_N$ are unknown constants satisfying $\kappa_i \in [0, \bar{\kappa}]$ $(i = 1, \ldots, N)$ with $\bar{\kappa}$ the upper bound of the input delay, and $x_i \in \mathbb{R}^n$, $u_i \in \mathbb{R}^m$, and $y_i \in \mathbb{R}^p$ are, respectively, the state, input, and output vectors of agent i.

The communication network among agents is exactly the same as that in Chap. 3, which provides each agent with the quantity

$$\zeta_i(k) = \sum_{j=1}^{N} d_{ij}(y_i(k) - y_j(k)), \quad i = 1, \ldots, N \tag{7.71}$$

for the partial-state coupling, and

$$\zeta_i(k) = \sum_{j=1}^{N} d_{ij}(x_i(k) - x_j(k)), \quad i = 1, \ldots, N \tag{7.72}$$

for the full-state coupling.

Unlike the directed graph in the previous section, the graphs considered in this section are undirected.

7.3.2 Problem Formulation

We formulate here two state synchronization problems, one for the network with full-state coupling and the other for partial-state coupling.

Problem 7.26 (Full-State Coupling) Consider a MAS described by (7.70) and (7.72). Let $\mathbf{G}$ be a given set of graphs such that $\mathbf{G} \subseteq \bar{\mathbb{G}}^N$. The *state synchronization* problem with a set of network graphs $\mathbf{G}$ is to find, if possible, a linear protocol of the form

$$u_i = F\zeta_i, \tag{7.73}$$

for $i = 1, \ldots, N$ such that, for any graph $\mathcal{G} \in \mathbf{G}$, for any integers $\kappa_1, \ldots, \kappa_N \leq \bar{\kappa}$, and for all the initial conditions of agents, the state synchronization among agents is achieved.

Problem 7.27 (Partial-State Coupling) Consider a MAS described by (7.1) and (7.2). Let $\mathbf{G}$ be a given set of graphs such that $\mathbf{G} \subseteq \bar{\mathbb{G}}^N$. The *state synchronization* problem with a set of network graphs $\mathbf{G}$ is to find, if possible, a linear dynamic protocol of the form

$$\begin{cases} \chi_i(k+1) = A_c\chi_i(k) + B_c\zeta_i(k), \\ \quad u_i(k) \quad = C_c\chi_i(k), \end{cases} \tag{7.74}$$

for $i = 1, \ldots, N$ where $\chi_i \in \mathbb{R}^n$, such that, for any graph $\mathcal{G} \in \mathbf{G}$, for any integers $\kappa_1, \ldots, \kappa_N \leq \bar{\kappa}$, and for all the initial conditions of agents and the protocol, the state synchronization among agents is achieved.

We need the following preliminary result which is the equivalent of Lemma 7.3 expanded to cover nonuniform delays.

Lemma 7.28 *Consider a linear time-delay system*

$$x(k+1) = Ax(k) + \sum_{i=1}^{m} A_i x(k-\kappa_i), \tag{7.75}$$

where $x(k) \in \mathbb{R}^n$ and $\kappa_i \in \mathbb{N}$. Suppose $A + \sum_{i=1}^{m} A_i$ is Schur stable. Then, (7.75) is asymptotically stable if

$$\det[e^{j\omega}I - A - \sum_{i=1}^{m} e^{-j\omega\kappa_i^r} A_i] \neq 0,$$

for all $\omega \in [-\pi, \pi]$ and for all $\kappa_i^r \in \mathbb{R}$ with $0 < \kappa_i^r \leq \kappa_i$ $(i = 1, \ldots, N)$.

7.3.3 Protocol Design for MAS with Full-State Coupling

Essentially we use the same protocol as in the case of uniform delays.

Protocol design 7.4 *Consider a MAS described by (7.70) and (7.72). We consider the protocol for agent $i \in \{1, \ldots, N\}$*

$$u_i = \rho F_\delta \zeta_i, \tag{7.76}$$

where $F_\delta = -(B'P_\delta B+I)^{-1}B'P_\delta A$ with P_δ being the unique positive definite solution of the discrete-time algebraic Riccati equation (7.28), and ρ and δ are design parameters.

Recall the definition of $\omega_{\max}$ from (7.29). The main result based on the above design is stated as follows.

Theorem 7.29 *Consider a MAS described by (7.70) and (7.72). Let any $\beta \in (0, 1]$ be given, and hence a set of undirected network graphs $\bar{\mathbb{G}}_{\beta}^{u,N}$ be defined.*

If A is at most weakly unstable and (A, B) is stabilizable, then the state synchronization problem stated in Problem 7.26 with $\mathbf{G} = \bar{\mathbb{G}}_{\beta}^{u,N}$ is solvable if

$$\bar{\kappa}\omega_{\max} < \frac{\pi}{2}. \tag{7.77}$$

In particular, for any given $\bar{\kappa}$ satisfying (7.77), there exist a $\rho > 0$ and a $\delta^* > 0$ such that for this ρ and any $\delta \in (0, \delta^*]$, controller (7.76) solves the state synchronization problem for any graph $\mathcal{G} \in \bar{\mathbb{G}}_{\beta}^{u,N}$ and for any integers $\kappa_1, \ldots, \kappa_N \in [0, \bar{\kappa}]$.

Proof of Theorem 7.29 To clarify the controller design, we introduce here a delay operator D_i for the agent i such that $(D_i x_i)(k) = x_i(k - \kappa_i)$. In the frequency domain, $\tilde{D}_i(\omega) = z^{-\kappa_i} = e^{-j\omega\kappa_i}$.

We use arguments similar to those in the proof of Theorem 6.39 for the case of continuous-time. As before, we define $T_1 \in \mathbb{R}^{(N-1)\times N}$ and $T_2 \in \mathbb{R}^{N\times(N-1)}$

$$T_1 = \begin{pmatrix} I & -\mathbf{1} \end{pmatrix}, \qquad T_2 = \begin{pmatrix} I \\ 0 \end{pmatrix}. \tag{7.78}$$

Since $D\mathbf{1} = \mathbf{1}$ we get, similar to (6.92), that

$$(I - D) = (I - D)T_2T_1. \tag{7.79}$$

We define $\bar{x}_i := x_i - x_N$ as the state synchronization error for agent $i = 1, \ldots, N-1$, and

$$\bar{x} = \begin{pmatrix} \bar{x}_1 \\ \bar{x}_2 \\ \vdots \\ \bar{x}_{N-1} \end{pmatrix}.$$

Using

$$(I - D)T_2\bar{x} = (I - D)x, \tag{7.80}$$

we can write the full closed-loop system as

$$\bar{x}(k+1) = (I \otimes A)\bar{x}(k) - \rho(T_1 D_\kappa (I - D)T_2 \otimes BB'P_\delta)\bar{x}(k), \tag{7.81}$$

where D represents the row stochastic matrix and D_κ represents the delays

$$D_\kappa = \operatorname{diag}\{D_i\} \quad \text{where} \quad (D_i z_i)(k) = z_i(k - \kappa_i). \tag{7.82}$$

We choose ρ such that

$$\rho(1-\beta)\cos(\bar{\kappa}\omega_{\max}) > 2. \tag{7.83}$$

We first show the system is stable without the delays, i.e., when $D_\kappa = I$. We have

$$\bar{x}(k+1) = (I \otimes A)\bar{x}(k) - \rho(T_1(I-D)T_2 \otimes F_\delta)\bar{x}(k).$$

Note that from (7.80) it is easy to see that the eigenvalues of $T_1(I-D)T_2$ are exactly the nonzero eigenvalue of $(I-D)$. Similar to the proof of Theorem 6.9, we can show that this system is Schur stable if the $N-1$ systems

$$\xi_i(k+1) = (A - \rho(1-\lambda_i)F_\delta)\bar{x}(k),$$

are Schur stable where $\lambda_1, \ldots, \lambda_{N-1}$ are the eigenvalues unequal to 1 of D which satisfy $\lambda_i \le \beta$. In that case according to Lemma 7.4, there exists a δ_1 such that we have the required stability for the system without delays for all $\delta < \delta_1$.

Next, we consider the case with delays. We define

$$\tilde{D}_\kappa(\omega) = \operatorname{diag}\{e^{-j\omega\kappa_i^r}\}. \tag{7.84}$$

It follows from Lemma 7.28 that the system (7.81) is asymptotically stable if

$$\det\Big[[e^{j\omega}I - (I \otimes A) + \rho T_1\tilde{D}_\kappa(\omega)(I-D)T_2 \otimes BF_\delta\Big] \neq 0, \tag{7.85}$$

for all $\omega \in [-\pi, \pi]$, for all $\kappa_1^r, \ldots, \kappa_N^r \in \mathbb{R}$ with $0 < \kappa_i^r \le \bar{\kappa}$ $(i = 1, \ldots, N)$ and all possible D associated with a network graph in $\bar{\mathbb{G}}_\beta^{u,N}$.

There exists a $\phi > 0$ independent of δ such that

$$\rho(1-\beta)\cos(\omega\bar{\kappa}) > 2, \quad \text{for} \quad |\omega| < \omega_{\max} + \phi.$$

For $\omega_{\max} + \phi \le |\omega| \le \pi$, we can show that there exists a $\delta_2 \le \delta_1$ such that for all $\delta \in (0, \delta_2]$, the condition (7.85) holds using the same argument as in the proof of Theorem 7.12.

Next, we consider the case $|\omega| \le \omega_{\max} + \phi$. It is sufficient to prove that

$$A_\omega = (I \otimes A) + \rho T_1\tilde{D}_\kappa(\omega)(I-D)T_2 \otimes BF_\delta$$

is asymptotically stable. We use a Lyapunov equation approach. Define

$$P_{D,\delta} = T_2'(I - D)T_2 \otimes P_\delta. \tag{7.86}$$

Similar to the continuous-time, we can show that $T_2'(I - D)T_2 > 0$ and hence $P_{D,\delta} > 0$. We get

$$\begin{aligned}
P_{D,\delta} - A_\omega^* P_{D,\delta} A_\omega
&= P_{D,\delta} - (I \otimes A')P_{D,\delta}(I \otimes A) \\
&\quad - \rho T_2'(I - D)\tilde{D}_\kappa(\omega)^* T_1' T_2'(I - D)T_2 \otimes F_\delta' B' P_\delta A \\
&\quad - \rho T_2'(I - D)T_2 T_1 \tilde{D}_\kappa(\omega)(I - D)T_2 \otimes A' P_\delta B F_\delta \\
&\quad - \rho^2 T_2'(I - D)\tilde{D}_\kappa(\omega)^* T_1' T_2'(I - D)T_2 T_1 \tilde{D}_\kappa(\omega)(I - D)T_2 \otimes F_\delta' B' P_\delta B F_\delta.
\end{aligned}$$

Using that $(I - D) = (I - D)T_2T_1$ we get

$$\begin{aligned}
P_{D,\delta} - A_\omega^* P_{D,\delta} A_\omega
&= T_2'(I - D)T_2 \otimes P_\delta - T_2'(I - D)T_2 \otimes A' P_\delta A \\
&\quad - \rho T_2'(I - D)\tilde{D}_\kappa(\omega)^*(I - D)T_2 \otimes F_\delta' B' P_\delta A \\
&\quad - \rho T_2'(I - D)\tilde{D}_\kappa(\omega)(I - D)T_2 \otimes A' P_\delta B F_\delta \\
&\quad - \rho^2 T_2'(I - D)\tilde{D}_\kappa(\omega)^*(I - D)\tilde{D}_\kappa(\omega)(I - D)T_2 \otimes F_\delta' B' P_\delta B F_\delta.
\end{aligned}$$

Next, we note that

$$F_\delta' B' P_\delta A = A' P_\delta B F_\delta = -F_\delta(I + B' P_\delta B)F_\delta$$

$$\rho\left[\tilde{D}_\kappa(\omega) + \tilde{D}_\kappa(\omega)^*\right] > \tfrac{4}{1-\beta} I$$

and, finally, we choose γ such that

$$\rho^2 T_2'(I - D)\tilde{D}_\kappa(\omega)^*(I - D)\tilde{D}_\kappa(\omega)(I - D)T_2 < \gamma T_2'(I - D)T_2.$$

We find

$$\begin{aligned}
P_{D,\delta} - A_\omega^* P_{D,\delta} A_\omega
&\geq T_2'(I - D)T_2 \otimes P_\delta - T_2'(I - D)T_2 \otimes A' P_\delta A \\
&\quad + \tfrac{4}{1-\beta} T_2'(I - D)^2 T_2 \otimes F_\delta(I + B' P_\delta B)F_\delta \\
&\quad - \gamma T_2'(I - D)T_2 \otimes F_\delta' B' P_\delta B F_\delta.
\end{aligned} \tag{7.87}$$

We have

$$(I - D)^2 \geq (1 - \beta)(I - D).$$

Moreover, for a δ sufficiently small, we have

$$\gamma B' P_\delta B < I + B' P_\delta B.$$

Finally, we have the algebraic Riccati equation for P_δ. Combining these properties with the inequality (7.87), we get

$$P_{D,\delta} - A_\omega^* P_{D,\delta} A_\omega \geq \delta T_2'(I - D)T_2 \otimes I + 2T_2'(I - D)T_2 \otimes F_\delta' F_\delta \tag{7.88}$$

which, since $T_2'(I - D)T_2$ is positive definite, yields the required stability. ■

In the special case where $\omega_{\max} = 0$, i.e., the eigenvalue of A is either 1 or in the unit circle, arbitrarily bounded communication delay can be tolerated as formulated in the following corollary.

Corollary 7.30 *Consider a MAS described by* (7.70) *and* (7.72). *Let any* $\beta \in (0, 1]$ *be given, and hence a set of undirected network graphs* $\bar{\mathbb{G}}_\beta^{u,N}$ *be defined.*

Suppose $\omega_{\max} = 0$. *If* (A, B) *is stabilizable, then the state synchronization problem stated in Problem* 7.26 *is always solvable via the controller* (7.76). *In particular, for any given* $\bar{\kappa}$ *satisfying* (7.77), *there exist a* $\rho > 0$ *and a* $\delta^* > 0$ *such that for this* ρ *and any* $\delta \in (0, \delta^*]$, *controller* (7.76) *solves the state synchronization problem for any graph* $\mathcal{G} \in \bar{\mathbb{G}}_\beta^N$ *and for any integers* $\kappa_1, \ldots, \kappa_N \in [0, \bar{\kappa}]$.

7.3.4 Protocol Design for MAS with Partial-State Coupling

In this section, we consider the state synchronization problem for homogeneous linear discrete-time MAS with partial-state coupling and with nonuniform unknown input delay.

Protocol design 7.5 *Consider a MAS described by* (7.70) *and* (7.71). *We choose an observer gain* K *such that* $A + KC$ *is Schur stable and a feedback gain* $F_\delta = -(B' P_\delta B + I)^{-1} B' P_\delta A$ *where* P_δ *is the unique positive definite solution of the discrete-time algebraic Riccati equation* (7.28). *We design the protocol*

$$\begin{cases} \chi_i(k+1) = (A + KC)\chi_i(k) - K\zeta_i(k), \\ \quad u_i(k) \quad = \rho F_\delta \chi_i(k), \end{cases} \tag{7.89}$$

(continued)

Protocol design 7.5 (continued)
where ρ and low-gain δ are design parameters to be chosen later.

The result based on the above design is given in the following theorem.

Theorem 7.31 *Consider a MAS described by (7.70) and (7.71). Let any $\beta \in (0, 1]$ be given, and hence a set of undirected network graphs $\bar{\mathbb{G}}_{\beta}^{u,N}$ be defined.*

If A is at most weakly unstable, (A, B) is stabilizable, and (C, A) is detectable, then the state synchronization problem stated in Problem 7.27 with $\mathbf{G} = \bar{\mathbb{G}}_{\beta}^{u,N}$ is solvable if the condition (7.77) holds. In particular, for any given $\bar{\kappa}$ satisfying (7.77), there exist a $\rho > 0$ and a $\delta^* > 0$ such that for this ρ and any $\delta \in (0, \delta^*]$, controller (7.89) solves the state synchronization problem for any graph $\mathcal{G} \in \bar{\mathbb{G}}_{\beta}^{u,N}$ and for any integers $\kappa_1, \ldots, \kappa_N \in [0, \bar{\kappa}]$.

Proof We use the notation $\chi = \operatorname{col}\{\chi_i\}$ and $x = \operatorname{col}\{x_i\}$. We note that

$$\begin{aligned} x(k+1)) &= (I \otimes A)x(k) - \rho D_{\kappa}(I \otimes BF_{\delta})\chi(k) \\ \chi(k+1) &= -((I - D) \otimes KC)x(k) + (I \otimes (A + KC))\chi(k). \end{aligned}$$

We can define

$$\chi = L\bar{\chi} + \mathbf{1}r$$

where

$$\bar{\chi}(k+1) = -(I \otimes KC)x(k) + (I \otimes (A + KC))\bar{\chi}(k).$$

By analyzing $\mathbf{1}'\chi$ and recalling that $\mathbf{1}'(I - D) = 0$, we can derive

$$r(k+1) = (A + KC)r(k)$$

which is asymptotically stable. Since we consider the asymptotic behavior of the system, we can ignore this asymptotically stable signal. We obtain

$$\begin{pmatrix} x(k+1) \\ \bar{\chi}(k+1) \end{pmatrix} = \left[\begin{pmatrix} \tilde{A} & 0 \\ -\tilde{K}\tilde{C} & \tilde{A} + \tilde{K}\tilde{C} \end{pmatrix} + \begin{pmatrix} \tilde{B} \\ 0 \end{pmatrix} D_{\kappa}(I - D) \otimes \begin{pmatrix} 0 & \rho\tilde{F}_{\delta} \end{pmatrix} \right] \begin{pmatrix} x(k) \\ \bar{\chi}(k) \end{pmatrix},$$

where

$$\begin{aligned} &\tilde{A} = I_N \otimes A, \qquad \tilde{B} = I_N \otimes B, \qquad \tilde{C} = I_N \otimes C, \\ &\tilde{K} = I_N \otimes K, \qquad \tilde{F}_{\delta} = I_N \otimes F_{\delta}, \end{aligned}$$

with the operator D_κ defined in (7.82). We introduce

$$e = \begin{pmatrix} T_1 & 0 \\ 0 & T_1 \end{pmatrix} \begin{pmatrix} x \\ \bar{\chi} \end{pmatrix}$$

with T_1 and T_2 as defined in (7.78). Clearly if we show that $e(k) \to 0$, then we immediately obtain that the system achieves synchronization. We obtain

$$e(k+1) = [\mathcal{A} + \mathcal{B}(D_\kappa(I - D) \otimes I)\mathcal{F}_\delta]\, e, \tag{7.90}$$

where

$$\mathcal{A} = \begin{pmatrix} \bar{A} & 0 \\ -\bar{K}\bar{C} & \bar{A} + \bar{K}\bar{C} \end{pmatrix}, \quad \mathcal{B} = \begin{pmatrix} \bar{B} \\ 0 \end{pmatrix}, \quad \mathcal{F}_\delta = \begin{pmatrix} 0 & \rho\bar{F}_\delta \end{pmatrix},$$

while

$$\begin{aligned} \bar{A} &= I_{N-1} \otimes A, & \bar{B} &= I_{N-1} \otimes B, & \bar{C} &= I_{N-1} \otimes C, \\ \bar{K} &= I_{N-1} \otimes K, & \bar{F}_\delta &= I_{N-1} \otimes F_\delta. \end{aligned}$$

We first establish that the system without delay is asymptotically stable. As before

$$\tilde{e}(k+1) = \left[\tilde{\mathcal{A}} + \tilde{\mathcal{B}}(T_1(I - D)T_2 \otimes I)\tilde{\mathcal{F}}_\varepsilon\right] \tilde{e}(k)$$

is asymptotically stable if the $N-1$ systems

$$\xi_i(k+1) = \begin{pmatrix} A & \rho\lambda_i F_\delta \\ -KC & A + KC \end{pmatrix} \xi_i(k),$$

are asymptotically stable where $\lambda_1, \ldots, \lambda_{N-1}$ are the eigenvalues of $T_1(I - D)T_2$ which are the nonzero eigenvalues of the symmetric matrix $I - D$ and therefore satisfy $\lambda_i \geq \beta$. Lemma 7.5 then establishes the required stability for ρ sufficiently large.

Choose ρ such that additionally we have

$$\rho(1 - \beta)\cos(\bar{\kappa}\omega_{\max}) > 2. \tag{7.91}$$

We show next that the system (7.90) is stable in the presence of delays. The system (7.90) can also be expressed as

$$\tilde{e}(k+1) = \left[\tilde{\mathcal{A}} + \tilde{\mathcal{B}}(T_1 D_\kappa(I - D)T_2 \otimes I)\tilde{\mathcal{F}}_\varepsilon\right] \tilde{e}(k) \tag{7.92}$$

using

$$\tilde{e} = \begin{pmatrix} \tilde{e}_1 \\ \tilde{e}_2 \end{pmatrix} = \begin{pmatrix} I & 0 \\ -I & I \end{pmatrix} e,$$

where

$$\tilde{\mathcal{A}} = \begin{pmatrix} \bar{A} & 0 \\ 0 & \bar{A} + \bar{K}\bar{C} \end{pmatrix}, \quad \tilde{\mathcal{B}} = \begin{pmatrix} \bar{B} \\ -\bar{B} \end{pmatrix}, \quad \tilde{\mathcal{F}}_\delta = \rho \begin{pmatrix} \bar{F}_\delta & \bar{F}_\delta \end{pmatrix}.$$

Let $\tilde{D}_\kappa(\omega)$ be defined by (7.84). Based on Lemma 7.28, we only need to check whether

$$\det\left[e^{j\omega} I - \tilde{\mathcal{A}} - \tilde{\mathcal{B}}(T_1 \tilde{D}_\kappa(\omega)(I - D)T_2 \otimes I)\tilde{\mathcal{F}}\right] \neq 0, \tag{7.93}$$

is satisfied for all $\omega \in [-\pi, \pi]$, for all $\kappa_1, \ldots, \kappa_N \in [0, \bar{\kappa}]$, and for all possible D associated with a network graph in $\mathbb{G}^{u,N}_{\beta,\alpha}$.

As in the full-state case, we note that given (7.91), there exists a $\phi > 0$ such that

$$\rho(1 - \beta)\cos(\bar{\kappa}(\omega_{\max} + \phi)) > 2. \tag{7.94}$$

Next, as before, we split the proof of (7.93) into two cases where $|\omega| < \tilde{\omega}_{\max} + \phi$ and $|\omega| \geq \tilde{\omega}_{\max} + \phi$, respectively.

If $|\omega| \geq \tilde{\omega}_{\max} + \phi$, we can use similar arguments as in the proof of Theorem 7.12 to show that (7.93) holds for $|\omega| \geq \tilde{\omega}_{\max} + \phi$.

It remains to verify (7.93) with $|\omega| < \tilde{\omega}_{\max} + \phi$. Through a Lyapunov argument, we show here that

$$\tilde{\mathcal{A}} - \tilde{\mathcal{B}}(T_1 \tilde{D}_\kappa(\omega)(I - D)T_2 \otimes I)\tilde{\mathcal{F}} \tag{7.95}$$

is Schur stable for any fixed ω satisfying $|\omega| < \tilde{\omega}_{\max} + \phi$. This then clearly implies (7.93). Consider

$$V(k) = \tilde{e}'(k) \begin{pmatrix} P_{D,\delta} & 0 \\ 0 & \mu Q \end{pmatrix} \tilde{e}(k),$$

with $P_{D,\delta}$ as defined in (7.86) while Q is such that

$$Q - (\bar{A} + \bar{K}\bar{C})' Q (\bar{A} + \bar{K}\bar{C}) \geq 2I. \tag{7.96}$$

The existence of Q is obvious since $\bar{A} + \bar{K}\bar{C}$ is Schur stable. This implies that

$$Q - A(\omega)^* Q A(\omega) \geq I, \tag{7.97}$$

with

$$A(\omega) = \tilde{A} + \tilde{K}\tilde{C} - \rho T_1 \tilde{D}_\kappa(\omega)(I - D)T_2 \otimes BF_\delta$$

for all δ small enough since

$$\rho T_1 \tilde{D}_\kappa(\omega)(I - D)T_2 \otimes BF_\delta$$

can be made arbitrarily small by choosing δ small enough since D and $\tilde{D}_\kappa(\omega)$ are bounded.

Using (7.88) and the above inequality together with some algebra, we obtain

$$\begin{aligned} V(k) - V(k+1) \geq \delta \tilde{e}_1'(T_2'(I - D)T_2 \otimes I)\tilde{e}_1 + 2\tilde{v}_1'\tilde{v}_1 + \mu \tilde{e}_2'\tilde{e}_2 \\ + 2\tilde{e}_2' X_1 \tilde{v}_1 - 2\mu \tilde{e}_2' X_2 \tilde{v}_1, \end{aligned}$$

where

$$\tilde{v}_1 = \left[(I - D)^{1/2} T_2 \otimes F_\delta\right] \tilde{e}_1, \tag{7.98}$$

and

$$\begin{aligned} X_1 &= \rho T_2'(I - D)\tilde{D}_\kappa^*(\omega)(I - D)^{1/2} \otimes A' P_\delta B \\ &\quad + \rho^2 T_2'(I - D)\tilde{D}_\kappa^*(\omega)(I - D)\tilde{D}_\kappa(\omega)(I - D)^{1/2} \otimes F_\delta' B' P_\delta B \\ X_2 &= \rho A'(\omega) Q (T_1 \tilde{D}_\kappa(\omega)(I - D)^{1/2} \otimes B. \end{aligned}$$

Choose μ small enough such that

$$9\mu X_2' X_2 \leq I$$

for all possible $\tilde{D}_\kappa(\omega)$ and D. Next choose δ small enough such that

$$9 X_1' X_1 \leq \mu I$$

for all possible $\tilde{D}_\kappa(\omega)$ and D. The above yields that

$$V(k) - V(k+1) \geq \delta \tilde{e}_1'(T_2'(I - D)T_2 \otimes I)\tilde{e}_1 + \tfrac{1}{3}\tilde{v}_1'\tilde{v}_1 + \tfrac{1}{3}\mu \tilde{e}_2'\tilde{e}_2.$$

This shows that (7.95) is Schur stable for any fixed ω satisfying $|\omega| < \tilde{\omega}_{\max} + \phi$. Therefore (7.93) is satisfied which completes the proof. ∎

7.3.5 *Static Protocol Design with Partial-State Coupling*

In this section, we also consider the static protocol design for discrete-time agents which are squared-down passifiable via static input feedforward.

7.3.5.1 Squared-Down Passifiable via Static Input Feedforward

It is shown here that the protocol design for uniform delay also works for nonuniform delay.

Protocol design 7.6 *Consider a MAS described by agents (7.70) with communication via (7.71). Assume the agents are squared-down passifiable via static input feedforward with respect to G_1, G_2, and R. The static protocol is designed as*

$$u_i = -\delta G_1 K G_2 \zeta_i, \tag{7.99}$$

where $\delta > 0$ is a low-gain parameter to be designed.

The main result based on the above design can be stated as follows.

Theorem 7.32 *Consider a MAS described by agents (7.70) and (7.71). Let any $\beta \in (0, 1]$ be given, and hence a set of undirected network graphs $\bar{\mathbb{G}}_{\beta}^{u,N}$ be defined.*
Assume the agents are squared-down passifiable via static input feedforward with respect to G_1, G_2, and R such that (A, BG_1) is controllable and (A, G_2C) is observable with BG_1 and G_2C full-column and full-row rank, respectively. In that case, the state synchronization problem stated in Problem 7.27 but with a static protocol is solvable if the condition (7.77) holds. In particular, for any given $\bar{\kappa}$ satisfying (7.77), there exists a $\delta^* > 0$ such that for any $\delta \in (0, \delta^*]$, controller (7.99) solves the state synchronization problem for any graph $\mathcal{G} \in \bar{\mathbb{G}}_{\beta}^{u,N}$ and any integers $\kappa_1, \ldots, \kappa_N \in [0, \bar{\kappa}]$.

Proof Define $\bar{x}_i = x_i - x_N$ as the state synchronization error for agent $i = 1, \ldots, N-1$ and $\bar{x} = \operatorname{col}\{\bar{x}_i\}$. According to the proof of Theorem 7.29, we can obtain the closed-loop system as

$$\bar{x}(k+1) = (I \otimes A)\bar{x}(k) - \delta(T_1 D_{\kappa}(I - D)T_2 \otimes BG_1 K G_2 C)\bar{x}(k). \tag{7.100}$$

According to Theorem 3.42, the system (7.100) without delay is Schur stable. Then, from Lemma 7.3 that the system (7.100) is asymptotically stable if

$$\det\left[e^{j\omega} I - (I \otimes A) + \delta T_1 \tilde{D}_{\kappa}(\omega)(I - D)T_2 \otimes BG_1 G_2 C\right] \neq 0, \tag{7.101}$$

for all $\omega \in [-\pi, \pi]$, for all $\kappa_1^r, \dots, \kappa_N^r \in \mathbb{R}$ with $0 < \kappa_i^r \le \bar{\kappa}$ $(i = 1, \dots, N)$ and all possible D associated with a network graph in $\bar{\mathbb{G}}_\beta^{u,N}$. Choose γ such that

$$(1 - \beta)\cos(\bar{\kappa}\omega_{\max}) > \gamma.$$

Next, we split the proof of (7.101) into two cases where $|\omega| < \omega_{\max} + \phi$ and $|\omega| \ge \omega_{\max} + \phi$, respectively, where $\phi > 0$ is such that

$$(1 - \beta)\cos(\bar{\kappa}(\omega_{\max} + \phi)) > \gamma.$$

If $|\omega| \ge \omega_{\max} + \phi$, there exists a $\mu > 0$ such that

$$\sigma_{\min}(e^{j\omega} I - A) < \mu, \quad \forall \omega, \text{ such that } |\omega| \ge \omega_{\max} + \phi.$$

Then, with the boundedness of D, there exists a $\delta^* > 0$ such that

$$\|\delta(T_1 \tilde{D}_\kappa(\omega)(I - D)T_2 \otimes BG_1KG_2C)\| \le \tfrac{\mu}{2},$$

for any $\delta < \delta^*$. Therefore, the condition (7.101) holds for $|\omega| \ge \omega_{\max} + \phi$.

Next, for $|\omega| < \omega_{\max} + \phi$, we use a Lyapunov argument. Define

$$P_D = T_2'(I - D)T_2 \otimes P, \tag{7.102}$$

where P is a solution of equation (1.16). We already noted in the proof of Theorem 7.29 that $T_2(I - D)T_2 > 0$ and hence $P_D > 0$. We need to prove A_ω is Schur stable where

$$A_\omega = (I \otimes A) - \delta T_1 \tilde{D}_\kappa(\omega)(I - D)T_2 \otimes BG_1KG_2C).$$

We get

$$\begin{aligned}
&P_D - A_\omega^* P_D A_\omega \\
&\quad = -\begin{pmatrix} I \\ -\delta T_1 \tilde{D}_\kappa(\omega)(I - D)T_2 \otimes KG_2C \end{pmatrix}' \left[T_2'(I - D)T_2 \otimes G(P)\right] \\
&\qquad\qquad \times \begin{pmatrix} I \\ -\delta T_1 \tilde{D}_\kappa(\omega)(I - D)T_2 \otimes KG_2C \end{pmatrix} \\
&\qquad + \delta T_2'(I - D)\tilde{D}_\kappa(\omega)(I - D)T_2 \otimes C'G_2'KG_2C \\
&\qquad + \delta T_2'(I - D)\tilde{D}_\kappa(\omega)^*(I - D)T_2 \otimes C'G_2'KG_2C \\
&\qquad - \delta^2 T_2'(I - D)\tilde{D}_\kappa(\omega)^*(I - D)\tilde{D}_\kappa(\omega)(I - D)T_2 \otimes C'G_2'K(R + R')KG_2C \\
&\quad \ge \delta T_2'(I - D)[\tilde{D}_\kappa(\omega) + \tilde{D}_\kappa(\omega)^*](I - D)T_2 \otimes C'G_2'KG_2C \\
&\qquad - \delta^2 T_2'(I - D)\tilde{D}_\kappa(\omega)^*(I - D)\tilde{D}_\kappa(\omega)(I - D)T_2 \otimes C'G_2'K(R + R')KG_2C.
\end{aligned}$$

We have

$$(I - D)[\tilde{D}_\kappa(\omega) + \tilde{D}_\kappa(\omega)^*](I - D) \geq 2\gamma(I - D),$$

and we choose δ small enough such that

$$\delta(I - D)\tilde{D}_\kappa(\omega)^*(I - D)\tilde{D}_\kappa(\omega)(I - D) \otimes K(R + R')K \leq \gamma(I - D) \otimes K$$

and we find

$$P_D - A_\omega^* P_D A_\omega \geq \delta\gamma T_2'(I - D)T_2 \otimes C'G_2'KG_2C$$

which clearly results in the required asymptotic stability of A_ω. ■

Chapter 8
Synchronization of Continuous-Time Linear MAS with Unknown Communication Delay

8.1 Introduction

In the case of communication delay, only for a constant synchronization trajectory we can preserve the diffusive nature of the network. This diffusive nature is an intrinsic part of the currently available design techniques, and hence only this case has been studied. Tian and Liu [139] and Xiao and Wang [170] consider single-integrator dynamics in the network, and it is demonstrated that the communication delay does not affect the synchronizability of the network. Münz et al. [75] and [76] give the consensus conditions for networks with higher-order but SISO dynamics. In [53], second-order dynamics are investigated, but the communication delays are assumed known.

In this chapter, we deal with general higher-order agent dynamics, and we provide solutions for regulated output/state synchronization problem for directed or undirected, weighted networks composed of general higher-order agent dynamics and with unknown, nonuniform, and asymmetric communication delays.

The write-up of this chapter is partially based on [185].

8.2 Multi-Agent Systems

Consider a multi-agent system (MAS) composed of N identical linear time-invariant continuous-time agents of the form

$$\begin{aligned} \dot{x}_i(t) &= Ax_i(t) + Bu_i(t), \\ y_i(t) &= Cx_i(t), \end{aligned} \tag{8.1}$$

A. Saberi et al., *Synchronization of Multi-Agent Systems in the Presence of Disturbances and Delays*, Systems & Control: Foundations & Applications,
https://doi.org/10.1007/978-3-030-88148-1_8

for $i = 1, \ldots, N$ where $x_i \in \mathbb{R}^n$, $u_i \in \mathbb{R}^m$, and $y_i \in \mathbb{R}^p$ are the state, input, and output, respectively, of agent i. We make the following assumptions for the agent dynamics.

Assumption 8.1 *We assume that*

- (A, B) *is stabilizable and* (A, C) *is detectable;*
- *All eigenvalues of* A *are in the closed left half complex plane.*

The communication network provides each agent with a linear combination of its own output relative to that of other neighboring agents with communication delays. In particular, each agent $i \in \{1, \ldots, N\}$ has access to the quantity

$$\zeta_i(t) = \sum_{j=1}^{N} a_{ij}(y_i(t) - y_j(t - \tau_{ij})), \tag{8.2}$$

where $a_{ij} \geq 0$ and $a_{ii} = 0$. The communication topology of the network can be represented by a weighted graph $\mathcal{G}$ with each node indicating an agent in the network and the weight of an edge is given by the coefficient a_{ij}. Here $\tau_{ij} \in \mathbb{R}^+$ represents an unknown constant communication delay from agent j to agent i for $i \neq j$. This communication delay implies that it takes τ_{ij} seconds for agent j to transfer its state information to agent i.

In the previous chapters, an intrinsic feature used is the diffusive nature of the network. If synchronization is achieved, i.e., if we have $x_i(t) = x_j(t)$ for all i, j, then we have $\zeta_i(t) = 0$, and no information is transferred over the network. However, this feature fails in the case of communication delay.

The approach used in this chapter is to impose output regulation. In other words, we prescribe a synchronized trajectory, and we require in addition to state synchronization that the outputs of all agents follow this prescribed signal. In order to do so, we assume some of the agents receive this prescribed signal, and then the protocol has to be designed such that we achieve synchronization and, additionally, the synchronized trajectory is equal to the prescribed signal. For systems with communication delay, we impose constant trajectories for the synchronized trajectory since that will restore the diffusive nature of the network.

Therefore, in this chapter, our goal is to achieve state synchronization while the output converges to a, a priori given, constant trajectory, denoted by $y_r \in \mathbb{R}^p$. This is often called the reference trajectory. Some of the agents have access to relative information about y_r. If agent i has information available about y_r, then its measurement is modified as

$$\bar{\zeta}_i(t) = \zeta_i(t) + (y_i(t) - y_r).$$

On the other hand, if agent i has no direct information available about y_r, then the agent has the same information as before

$$\bar{\zeta}_i(t) = \zeta_i(t).$$

Let $\iota_i = 1$ if agent i has information about y_r and $\iota_i = 0$ otherwise. Then, the above can be combined and we have in case of partial-state synchronization

$$\bar{\zeta}_i(t) = \zeta_i(t) + \iota_i(y_i(t) - y_r). \tag{8.3}$$

for $i = 1, \dots, N$.

We defined the *expanded Laplacian* $\bar{L} = [\bar{\ell}_{ij}]$ as

$$\begin{aligned} \bar{\ell}_{ii} &= \ell_{ii} + \iota_i \\ \bar{\ell}_{ij} &= \ell_{ij} \qquad\qquad i \neq j. \end{aligned}$$

Clearly, in order for all the agents to follow this prescribed trajectory, it is clear that for any agent i there must exist an agent j which has access to the reference trajectory such that the associated network graph has a directed path from j to i. This property is directly connected to the invertibility of the expanded Laplacian.

Lemma 8.1 *The expanded Laplacian matrix $\bar{L}$ is invertible if and only if for each $i \in \{1, \dots, N\}$, there exists a $j \in \{1, \dots, N\}$ such that $\iota_j = 1$ and the associated graph has a directed path from j to i.*

Remark 8.2 If only one agent i has access to the reference trajectory, then the expanded Laplacian is invertible if and only if the associated network graph contains a directed spanning tree with agent i as its root.

In the case of full-state coupling, i.e., $C = I$, we have $y_i = x_i$, for $i = 1, \dots, N$ and $x_r \in \mathbb{R}^n$ is a, a priori given, constant trajectory. Then, we have

$$\zeta_i(t) = \sum_{j=1}^{N} a_{ij}(x_i(t) - x_j(t - \tau_{ij})), \tag{8.4}$$

and

$$\bar{\zeta}_i(t) = \zeta_i(t) + \iota_i(x_i(t) - x_r). \tag{8.5}$$

For full-state coupling the standard protocol, we used many times in this book

$$u_i = F\bar{\zeta}_i$$

will in most cases not work. It would actually require that, at the very least, $Ax_r = 0$. We will use a dynamic protocol in line with the partial-state coupling. In other words, for the full-state coupling case, we obtain Problem 8.4 with $C = I$. Therefore, this case will not be discussed seperately.

Definition 8.3 Given ι and a given real number $\beta > 0$, let $\mathbb{G}^N_{\iota,\beta}$ denote the set of directed network graphs with N nodes such that the eigenvalues of the expanded Laplacian matrix $\bar{L}$, denoted by $\lambda_1, \ldots, \lambda_N$ satisfy $\mathrm{Re}(\lambda_i) > \beta$.

The subset of $\mathbb{G}^N_{\iota,\beta}$ of undirected graphs is denoted by $\mathbb{G}^{u,N}_{\iota,\beta}$.

8.3 Protocol Design for MAS with Partial-State Coupling

We consider the following problem.

Problem 8.4 (Partial-State Coupling) Consider a MAS described by (8.1) and (8.3). Let a given constant trajectory $y_r \in \mathbb{R}^p$ be available to at least one agent. The problem of *state synchronization with output regulation* given a set of graphs $\mathbf{G} \subseteq \mathbb{G}^N$ in the presence of an unknown, nonuniform, and arbitrarily large communication delay is to find a distributed linear dynamic protocol of the form

$$\begin{cases} \dot{x}_{i,c} &= A_c x_{i,c} + B_c \bar{\zeta}_i, \\ u_i &= C_c x_{i,c}, \end{cases} \qquad (i = 1, \ldots, N) \tag{8.6}$$

for each agent such that

$$\lim_{t\to\infty} x_i(t) - x_j(t) = 0, \tag{8.7}$$

for all $i, j \in \{1, \ldots, N\}$, while the output of each agent converges to the given constant trajectory, i.e.

$$\lim_{t\to\infty} y_i(t) - y_r = 0, \tag{8.8}$$

for all $i \in \{1, \ldots, N\}$, for any graph $\mathcal{G} \in \mathbf{G}$ and for any communication delay $\tau_{ij} \in \mathbb{R}^+$.

We consider here state synchronization and output regulation for networks with partial-state coupling and unknown, nonuniform, and arbitrarily large communication delays. We study this problem for undirected graphs. The special case that the graph is known is studied in Sect. 8.4 and allows us to consider undirected networks.

In general, we have to restrict our choice of y_r. Let

$$\begin{aligned} \mathscr{C}_y &= \left\{ y \in \mathbb{R}^p \,\middle|\, \begin{pmatrix} 0 \\ y \end{pmatrix} \in \mathrm{Im} \begin{pmatrix} A & B \\ C & 0 \end{pmatrix} \right\} \\ &= \left\{ y \in \mathbb{R}^p \,\middle|\, \exists x \in \mathbb{R}^n, u \in \mathbb{R}^m : Ax + Bu = 0, Cx = y \right\}. \end{aligned} \tag{8.9}$$

It turns out that our problem is solvable if and only if $y_r \in \mathscr{C}_y$. Note that $\mathscr{C}_y = \mathbb{R}^p$ if (A, B, C) is right-invertible and has no invariant zeros in the origin.

Protocol design 8.1 *Consider a MAS described by* (8.1) *and* (8.3). *Let R be an injective matrix such that $\mathscr{C}_y = \operatorname{Im} R$. Choose Π and Γ such that*

$$\begin{pmatrix} 0 \\ R \end{pmatrix} = \begin{pmatrix} A & B \\ C & 0 \end{pmatrix} \begin{pmatrix} \Pi \\ \Gamma \end{pmatrix} \tag{8.10}$$

and

$$\operatorname{rank} \begin{pmatrix} A & B\Gamma \\ C & 0 \end{pmatrix} = n + \operatorname{rank} \Gamma. \tag{8.11}$$

Choose Γ_1 and Γ_2 such that $\operatorname{Im} \Gamma_1 = \operatorname{Im} \Gamma$ *and*

$$(\Gamma_1 \ \Gamma_2) \tag{8.12}$$

is square and invertible. Define

$$\tilde{A} = \begin{pmatrix} A & B\Gamma_1 \\ 0 & 0 \end{pmatrix}, \quad \tilde{B} = \begin{pmatrix} B\Gamma_2 & 0 \\ 0 & I \end{pmatrix}, \quad \tilde{C} = (C \ 0)\,. \tag{8.13}$$

Finally, choose K such that $\tilde{A} + K\tilde{C}$ is Hurwitz stable, and let P_δ be the unique solution of the algebraic Riccati equation

$$\tilde{A}'P_\delta + P_\delta\tilde{A} - P_\delta\tilde{B}\tilde{B}'P_\delta + \delta I = 0. \tag{8.14}$$

We consider the protocol

$$\begin{cases} \dot{x}_{i,c} = \begin{pmatrix} \tilde{A} + K\tilde{C} & 0 \\ -\rho B_1\tilde{B}'P_\delta & 0 \end{pmatrix} x_{i,c} + \begin{pmatrix} -K \\ 0 \end{pmatrix} \bar{\zeta}_i, \\ u_i \ = (-\rho D_1\tilde{B}'P_\delta \quad \Gamma_1) x_{i,c}, \end{cases} \tag{8.15}$$

for $i = 1, \ldots, N$ where ρ and δ are design parameters to be chosen later on, while B_1 and D_1 are given by

$$B_1 = (0 \ I)\,, \qquad D_1 = (\Gamma_2 \ 0)\,. \tag{8.16}$$

Theorem 8.5 *Consider a MAS described by* (8.1) *and* (8.5). *Let a given constant trajectory $y_r \in \mathbb{R}^p$ be available to some of the agents with $\iota_i = 1$ if agent i has access to y_r and $\iota_i = 0$, otherwise. Let any $\beta > 0$ be given, and hence a set of network graphs $\mathbb{G}^{u,N}_{\iota,\beta}$ be defined.*

Then, under Assumption 8.1, Problem 8.4 is solvable for the set of undirected graphs $\mathbb{G} = \mathbb{G}^{u,N}_{\iota,\beta}$ if and only if $y_r \in \mathscr{C}_y$.

More specifically, there exist ρ and $\delta > 0$ such that the linear protocol (8.15) achieves state synchronization and output regulation for any undirected graph $\mathcal{G} \in \mathbb{G}^{u,N}_{\iota,\beta}$, for any communication delay $\tau_{ij} \in \mathbb{R}^+$, and for any $y_r \in \mathscr{C}_y$.

Before, we can prove this theorem, we need a preliminary lemma.

Lemma 8.6 *Let β be a lower bound for the eigenvalues of $\bar{L}$. Then, for all communication delays $\tau_{ij} \in \mathbb{R}^+$ for $i, j = 1, \ldots, N$ and for all $\omega \in \mathbb{R}$, the real parts of all eigenvalues of $\bar{L}_{j\omega}(\tau)$ are greater than or equal to β where*

$$\bar{L}_s(\tau) = \begin{pmatrix} \bar{\ell}_{11} & \bar{\ell}_{12}e^{-\tau_{12}s} & \cdots & \bar{\ell}_{1N}e^{-\tau_{1N}s} \\ \vdots & \vdots & \ddots & \vdots \\ \bar{\ell}_{\varrho 1}e^{-\tau_{\varrho 1}s} & \bar{\ell}_{\varrho\varrho} & \cdots & \bar{\ell}_{\varrho N}e^{-\tau_{\varrho N}s} \\ \vdots & \vdots & \ddots & \vdots \\ \bar{\ell}_{N1}e^{-\tau_{N1}s} & \bar{\ell}_{N2}e^{-\tau_{N2}s} & \cdots & \bar{\ell}_{NN} \end{pmatrix} \tag{8.17}$$

where τ denotes a vector consisting of all τ_{ij} $(i \neq j)$ with $i, j \in \{1, \ldots, N\}$.

Proof All eigenvalues of $\bar{L}_{j\omega}(\tau)$ are in the set

$$\left\{ v'\bar{L}_{j\omega}(\tau)v \mid v \in \mathbb{C}^N, \|v\| = 1 \right\}.$$

Therefore it is sufficient to establish that all elements in this set have a real part greater than or equal to β.

Since $\bar{L}$ is symmetric and β is a lower bound for the eigenvalues of $\bar{L}$, we find that $v'\bar{L}v$ is real and greater than or equal to β, provided $\|v\| = 1$.

Next, consider an arbitrary vector $v \in \mathbb{C}^N$. We have

$$v'\bar{L}_{j\omega}(\tau)v = \sum_{i=1}^{N} |v_i|^2 \bar{\ell}_{ii} + \sum_{i=1}^{N} \sum_{k,k\neq i}^{N} v_i' v_k \bar{\ell}_{ik} e^{-\tau_{ik} j\omega}.$$

Since $\bar{\ell}_{ik}$ $(i \neq k)$ is negative or equal to zero, we get

$$\operatorname{Re}\left(v'\bar{D}_{j\omega}(\tau)v\right) \geq \sum_{i=1}^{N} |v_i|^2 \bar{\ell}_{ii} + \sum_{i=1}^{N} \sum_{k,k\neq i}^{N} |v_i' v_k| \bar{\ell}_{ik} = \begin{pmatrix} |v_1| \\ \vdots \\ |v_N| \end{pmatrix}' \bar{L} \begin{pmatrix} |v_1| \\ \vdots \\ |v_N| \end{pmatrix} \geq \beta,$$

which completes the proof. ∎

Proof of Theorem 8.5 For any individual agent to be able to track a constant reference signal y_r, there must exist a x_0 and a u_0 such that

$$\begin{pmatrix} A & B \\ C & 0 \end{pmatrix} \begin{pmatrix} x_0 \\ u_0 \end{pmatrix} = \begin{pmatrix} 0 \\ y_r \end{pmatrix}. \tag{8.18}$$

Clearly, such x_0 and u_0 exist only if y_r is in the set $\mathscr{C}_y$ which is therefore a necessary condition for the solvability of our problem.

Next, we show that the condition $y_r \in \mathscr{C}_y$ guarantees the existence of a controller which achieves output synchronization. Let R be an injective matrix such that $\mathscr{C}_y = \operatorname{Im} R$. In that case we can find Π and Γ such that (8.10) is satisfied. To see that we can impose the rank condition (8.11), we note that (C, A) detectable implies that the first n columns are linearly independent. If the rank condition (8.11) is not satisfied, then there exist x and v such that

$$\begin{pmatrix} A & B\Gamma \\ C & 0 \end{pmatrix} \begin{pmatrix} x \\ v \end{pmatrix} = 0$$

with $B\Gamma v \neq 0$, $v \perp \ker \Gamma$ and $v'v = 1$. But then

$$\begin{pmatrix} A & B \\ C & 0 \end{pmatrix} \begin{pmatrix} \Pi - xv' \\ \Gamma(I - vv') \end{pmatrix} = \begin{pmatrix} 0 \\ R \end{pmatrix}$$

which shows that $\bar{\Pi} = \Pi - xv'$ and $\bar{\Gamma} = \Gamma(I - vv')$ also satisfy the above equation but with $\operatorname{rank} \bar{\Gamma} < \operatorname{rank} \Gamma$. Recursively, we can find a solution of (8.10) which also satisfies the extra rank condition (8.11).

We design a pre-compensator

$$\begin{aligned} \dot{p}_i &= \begin{pmatrix} 0 & I \end{pmatrix} v_i, \qquad p_i(t) \in \mathbb{R}^{\nu} \\ u_i &= \Gamma_1 p_i + \begin{pmatrix} \Gamma_2 & 0 \end{pmatrix} v_i \end{aligned} \tag{8.19}$$

where Γ_1 and Γ_2 are chosen such that $\operatorname{Im} \Gamma_1 = \operatorname{Im} \Gamma$ and (8.12) is square and invertible while ν is such that $\Gamma_1 \in \mathbb{R}^{n \times \nu}$.

The interconnection of (8.1) and (8.19) is of the form

$$\begin{cases} \dot{\tilde{x}}_i(t) = \tilde{A}\tilde{x}_i(t) + \tilde{B}v_i(t), \\ y_i(t) = \tilde{C}\tilde{x}_i(t), \end{cases} \tag{8.20}$$

with

$$\tilde{x}_i = \begin{pmatrix} x_i \\ p_i \end{pmatrix}$$

where $\tilde{A}$, $\tilde{B}$, and $\tilde{C}$ are defined according to (8.13).

We need to verify that the system remains stabilizable and detectable if we use a pre-compensator. The stabilizability of this system follows immediately from (8.12) and the stabilizability of (A, B). To show the detectability of (8.20), we need to verify that

$$\operatorname{rank}\begin{pmatrix} sI - A & -B\Gamma_1 \\ 0 & sI \\ C & 0 \end{pmatrix} = n + \nu$$

for all s in the closed right half complex plane. For $s \neq 0$, this is achieved immediately from the detectability of (C, A). For $s = 0$, we have

$$\operatorname{rank}\begin{pmatrix} -A & -B\Gamma_1 \\ 0 & 0 \\ C & 0 \end{pmatrix} = \operatorname{rank}\begin{pmatrix} -A & -B\Gamma \\ C & 0 \end{pmatrix} = n + \operatorname{rank}\Gamma.$$

We obtain the required detectability when we note that $\operatorname{rank}\Gamma = \operatorname{rank}\Gamma_1$ and $\operatorname{rank}\Gamma_1 = \nu$ (since Γ_1 is injective).

Next, a protocol for MAS (8.20) is designed as

$$\begin{cases} \dot{\chi}_i = (\tilde{A} + K\tilde{C})\chi_i - K\bar{\zeta}_i, \\ v_i = -\rho\tilde{B}' P_\delta \chi_i, \end{cases} \tag{8.21}$$

where K is chosen such that $\tilde{A} + K\tilde{C}$ is Hurwitz stable and P_δ is the unique solution of the algebraic Riccati equation (8.14).

Next, we prove that with the above protocol, the states of each agent synchronize, while the output of each agent converges to the constant trajectory y_r. Let W be such that $\Gamma_1 W = V\Gamma$. We choose

$$\tilde{\Pi} = \begin{pmatrix} \Pi \\ W \end{pmatrix},$$

where W is such that $\Gamma_1 W = \Gamma$. It is then easily seen that $\tilde{A}\tilde{\Pi} = 0$ and $\tilde{C}\tilde{\Pi} = R$.

For $i = 1, \ldots, N$, define $\bar{x}_i = \tilde{x}_i - \tilde{\Pi} z$ where z is such that $y_r = Rz$ and the output synchronization error $e_i = y_i - y_r$. Then, we get the error dynamics

$$\begin{cases} \dot{\bar{x}}_i = \tilde{A}\bar{x}_i + \tilde{B}v_i, \\ e_i = \tilde{C}\bar{x}_i. \end{cases} \tag{8.22}$$

Moreover, $\bar{\zeta}_i(t)$ can be rewritten as

$$\bar{\zeta}_i(t) = \sum_{j=1}^{N} a_{ij}\tilde{C}(\bar{x}_j(t) - \bar{x}_j(t - \tau_{ij})) + \iota_i\tilde{C}\bar{x}_i(t).$$

Let

$$\bar{x} = \begin{pmatrix} \bar{x}_1 \\ \bar{x}_2 \\ \vdots \\ \bar{x}_N \end{pmatrix} \text{ and } \chi = \begin{pmatrix} \chi_1 \\ \chi_2 \\ \vdots \\ \chi_N \end{pmatrix}.$$

Then, the full closed-loop system can be written in the frequency domain as

$$\begin{pmatrix} s\bar{x} \\ s\chi \end{pmatrix} = \begin{pmatrix} I_N \otimes \tilde{A} & -\rho I_N \otimes \tilde{B}\tilde{B}' P_\delta \\ -\bar{L}_s(\tau) \otimes K\tilde{C} & I_N \otimes (\tilde{A} + K\tilde{C}) \end{pmatrix} \begin{pmatrix} \bar{x} \\ \chi \end{pmatrix}, \tag{8.23}$$

where $\bar{L}_s(\tau)$ is the expanded Laplacian matrix in the frequency domain as defined in (8.17).

Next, we prove (8.23) is asymptotically stable for all communication delays $\tau_{ij} \in \mathbb{R}^+$. We first prove stability without communication delay and then prove stability for the case that includes a communication delay.

When there is no communication delay in the network, similar to Sect. 2.5.1, the stability of system (8.23) is equivalent to asymptotic stability of the matrix

$$\begin{pmatrix} \tilde{A} & -\rho\tilde{B}\tilde{B}' P_\delta \\ -\lambda_i K\tilde{C} & \tilde{A} + K\tilde{C} \end{pmatrix}$$

for all $i \in \{1, \ldots, N\}$, where λ_i is the eigenvalue of the expanded Laplacian matrix $\bar{L}$ with its lower bound of β. Note that

$$\begin{pmatrix} \lambda_i & 0 \\ 0 & I \end{pmatrix} \begin{pmatrix} \tilde{A} & -\rho\tilde{B}\tilde{B}' P_\delta \\ -\lambda_i K\tilde{C} & \tilde{A} + K\tilde{C} \end{pmatrix} \begin{pmatrix} \lambda_i^{-1} & 0 \\ 0 & I \end{pmatrix} = \begin{pmatrix} \tilde{A} & -\rho\lambda_i\tilde{B}\tilde{B}' P_\delta \\ -K\tilde{C} & \tilde{A} + K\tilde{C} \end{pmatrix}.$$

By choosing

$$\rho > \frac{1}{\beta}, \tag{8.24}$$

and using the result of Lemma 6.6, we find that there exists a δ^* such that for any $\delta \in (0, \delta^*]$, the system (8.23) is asymptotically stable without any communication delay.

In the case when communication delay is present, according to Lemma 6.38, the closed-loop system (8.23) is asymptotically stable for all communication delays $\tau_{ij} \in \mathbb{R}^+$, if

$$\det\left[j\omega I - \begin{pmatrix} I_N \otimes \tilde{A} & -\rho I_N \otimes \tilde{B}\tilde{B}' P_\delta \\ -\bar{L}_{j\omega}(\tau) \otimes K\tilde{C} & I_N \otimes (\tilde{A} + K\tilde{C}) \end{pmatrix} \right] \neq 0 \tag{8.25}$$

for all $\omega \in \mathbb{R}$ and any communication delay $\tau_{ij} \in \mathbb{R}^+$. Condition (8.25) is satisfied if the matrix

$$\begin{pmatrix} I_N \otimes \tilde{A} & -\rho I_N \otimes \tilde{B}\tilde{B}' P_\delta \\ -\bar{L}_{j\omega}(\tau) \otimes K\tilde{C} & I_N \otimes (\tilde{A} + K\tilde{C}) \end{pmatrix} \tag{8.26}$$

has no eigenvalues on the imaginary axis for all $\omega \in \mathbb{R}$ and for all communication delays $\tau_{ij} \in \mathbb{R}^+$.

Lemma 8.6 implies that all the eigenvalues of $\bar{L}_{j\omega}(\tau)$ are greater than or equal to β. Hence, when ρ satisfies (8.24), by Lemma 6.6, there exists a δ^* such that for any $\delta \in (0, \delta^*]$, the matrix (8.26) has no eigenvalues on the imaginary axis. As noted before, this implies that the closed-loop system (8.23) is asymptotically stable for any communication delay $\tau_{ij} \in \mathbb{R}^+$.

Note that the asymptotic stability yields that $\bar{x}_i(t) \to 0$ and $e_i(t) \to 0$ as $t \to \infty$. This yields state synchronization and output regulation respectively. Finally, by combining the pre-compensator (8.19) and protocol (8.21), we get the linear dynamic protocol (8.15). ∎

8.4 MAS with a Known Communication Topology

We study here a variation of Problem 8.4 for a MAS with a known graph $\mathcal{G}$. The main advantage of knowing the graph is that we can also handle directed graphs.

Problem 8.7 (Partial-State Coupling and Known Graph) Consider a MAS described by (8.1) and (8.3) associated with a known graph $\mathcal{G}$. The problem of *state synchronization and output regulation* for networks with unknown, nonuniform, and arbitrarily large communication delay is to find a distributed linear dynamic protocol of the form (8.6) for each agent such that

$$\lim_{t \to \infty} x_i(t) - x_j(t) = 0, \tag{8.27}$$

for all $i, j \in \{1, \ldots, N\}$ and

$$\lim_{t \to \infty} y_i(t) - y_r = 0, \tag{8.28}$$

for all $i \in \{1, \ldots, N\}$, for the given graph $\mathcal{G}$ and for any communication delay $\tau_{ij} \in \mathbb{R}^+$.

We note that if our directed graph is balanced, then the results of Theorem 8.5 still hold if we define β as the smallest eigenvalue of $\bar{L} + \bar{L}'$. However, if the graph is not balanced, the derivation presented before might not be valid. In that case, we design a protocol for one individual graph (instead of for a set), i.e., we assume the directed graph $\mathcal{G}$ is given.

Again, the choice of the constant trajectory y_r needs to be restricted to the set given by (8.9).

Theorem 8.8 *Consider a MAS described by* (8.1) *and* (8.2) *with an associated directed graph $\mathcal{G}$. Let a given constant trajectory $y_r \in \mathbb{R}^p$ be available to some of the agents with $\iota_i = 1$ if agent i has access to y_r and $\iota_i = 0$, otherwise. Assume for any agent i there exists an agent j which has access to y_r such that the graph contains a directed path from agent j to agent i.*

Then, under Assumptions 8.1, Problem 8.7 is solvable if and only if $y_r \in \mathscr{C}_y$. Specifically, given the directed graph $\mathcal{G}$, there exist ρ and $\delta > 0$ such that the linear protocol (8.15) *achieves state synchronization and output regulation for any communication delay $\tau_{ij} \in \mathbb{R}^+$ and for any $y_r \in \mathscr{C}_y$.*

Remark 8.9 In the previous section for undirected graphs, we only used certain limited information about the network in order to design our distributed protocol. For directed graphs, we make explicit use of our knowledge of the network to design our protocol (more specifically, to find a lower bound for our design parameter ρ). If we have a finite set of possible graphs, then we can still find a protocol that works for every graph in this finite set (use as a lower bound for ρ, the maximum of the lower bounds for each individual graph in the set).

Proof of Theorem 8.8 The proof follows the proof of Theorem 8.5, except for the choice of the parameter ρ given in (8.24).

Here, the parameter ρ should be designed in such a way that the eigenvalues of

$$\rho \bar{L}_{j\omega}(\tau) \tag{8.29}$$

have real parts greater than 1 for any communication delay $\tau_{ij} \in \mathbb{R}^+$ and for all $\omega \in \mathbb{R}$.

By Lemma 8.1, we know the expanded Laplacian matrix $\bar{L}$ is invertible and has its eigenvalues in the open right half complex plane. Given that $\bar{L}$ is an invertible M-matrix, there exists a diagonal positive matrix $D = \operatorname{diag}\{d_i\}$ such that

$$D\bar{L} + \bar{L}'D > 0 \tag{8.30}$$

(in the case of an undirected or balanced graph, we can simply choose $D = I$). This follows from [5]. Since this matrix (8.30) is positive definite, we find that

$$\operatorname{Re}(v'D\bar{L}v) = \tfrac{1}{2}v'(D\bar{L} + \bar{L}'D)v$$

is greater than or equal to some positive constant β for all v with $\|v\| = 1$. Following the proof in Lemma 8.6, we obtain

$$\operatorname{Re}(v'D\bar{L}_{j\omega}(\tau)v) \geq \beta. \tag{8.31}$$

Let λ be an eigenvalue of $\bar{L}_{j\omega}(\tau)$ with eigenvector v. In other words, $\bar{L}_{j\omega}(\tau)v = \lambda v$. Combining with inequality (8.31), we get

$$\mathrm{Re}(\lambda(\max d_i)v'v) \geq \mathrm{Re}(\lambda v'Dv) = \mathrm{Re}\, v'D\bar{L}_{j\omega}(\tau)v \geq \beta,$$

and hence

$$\mathrm{Re}(\lambda) \geq \frac{\beta}{\max d_i}.$$

Thus, when choosing

$$\rho > \frac{\max d_i}{\beta},$$

there exists an δ^* such that for any $\delta \in (0, \delta^*]$, condition (8.29) is satisfied for any communication delay $\tau_{ij} \in \mathbb{R}^+$. ■

Chapter 9
Synchronization of Discrete-Time Linear MAS with Unknown Communication Delay

9.1 Introduction

This chapter considers synchronization problems for homogeneous discrete-time linear MAS in the presence of communication delay.

Similar to the continuous-time version in the previous chapter, our aim is state synchronization and output regulation to a reference trajectory. The communication delay in this chapter can also be arbitrarily large, unknown, and asymmetric. The network can be full- or partial-state coupled. The general result is for undirected graphs. However, if the graph is known, then we can also handle directed graphs.

9.2 Multi-Agent Systems

Consider a multi-agent system (MAS) composed of N identical linear time-invariant discrete-time agents of the form

$$\begin{aligned} x_i(k+1) &= Ax_i(k) + Bu_i(k), \\ y_i(k) &= Cx_i(k), \end{aligned} \qquad (i = 1, \ldots, N) \tag{9.1}$$

where $x_i \in \mathbb{R}^n, u_i \in \mathbb{R}^m, and y_i \in \mathbb{R}^p$ are the state, input, and output of agent i. We make the following assumptions for the agent dynamics.

A. Saberi et al., *Synchronization of Multi-Agent Systems in the Presence of Disturbances and Delays*, Systems & Control: Foundations & Applications,
https://doi.org/10.1007/978-3-030-88148-1_9

Assumption 9.1 *We assume that*

- *(A, B) is stabilizable and (A, C) is detectable;*
- *All eigenvalues of A are in the closed unit disc.*

The communication network provides each agent with a linear combination of its own output relative to that of other neighboring agents with communication delays. In particular, each agent $i \in \{1, \dots, N\}$ has access to the quantity

$$\zeta_i(k) = \sum_{j=1}^{N} a_{ij}(y_i(k) - y_j(k - \kappa_{ij})), \tag{9.2}$$

where $a_{ij} \geq 0$ while $a_{ii} = 0$ and $\kappa_{ij} \in \mathbb{N}$ represents an unknown constant communication delay from agent j to agent i for $i \neq j$. This communication delay implies that it takes κ_{ij} seconds for agent j to transfer its state information to agent i.

As described in the continuous-time chapter, we will seek state synchronization among agents, i.e.

$$\lim_{k\to\infty} x_i(k) - x_j(k) = 0,$$

and we prescribe a synchronized trajectory y_r for the output of each agent as the reference trajectory and the output of each agent should converge to the given constant trajectory, i.e.

$$\lim_{t\to\infty} y_i(t) - y_r = 0. \tag{9.3}$$

Some of the agents have access to relative information about y_r. If agent i has information available about y_r, then its measurement is modified as

$$\bar{\zeta}_i(k) = \zeta_i(k) + (y_i(k) - y_r).$$

On the other hand, if agent i has no direct information available about y_r, then the agent has the same information as before:

$$\bar{\zeta}_i(k) = \zeta_i(k).$$

Let $\iota_i = 1$ if agent i has information about y_r and $\iota_i = 0$ otherwise. Then, the above can be combined, and we have the case of partial-state synchronization:

$$\bar{\zeta}_i(k) = \zeta_i(k) + \iota_i(y_i(k) - y_r), \tag{9.4}$$

for $i = 1, \dots, N$.

In this context the matrix $A_a = [a_{ij}]$ is referred to as the adjacency matrix. Based on the adjacency matrix, we can associate a Laplacian matrix L to a graph:

$$L = D_{\text{in}} - A_a$$

where

$$D_{\text{in}} = \operatorname{diag}\{d_{\text{in}}(i)\}$$

while

$$d_{\text{in}}(i) = \sum_{j=1}^{N} a_{ij}$$

denotes the weighted in degree of vertex i. We defined the *expanded Laplacian* $\bar{L} = [\bar{\ell}_{ij}]$ as

$$\begin{aligned} \bar{\ell}_{ii} &= \ell_{ii} + \iota_i \\ \bar{\ell}_{ij} &= \ell_{ij} \qquad\qquad i \neq j. \end{aligned}$$

We define

$$\bar{D} = I - (2I + D_{\text{in}})^{-1} \bar{L}. \tag{9.5}$$

It is easily verified that $\bar{D}$ is a row stochastic matrix. Note that we use a factor 2 in (9.5) which was not used in (3.3) which is needed here because the expanded Laplacian matrix has larger diagonal elements compared to the original Laplacian.

Clearly, in order for all agents to follow this prescribed trajectory, it is clear that for any agent i there must exist an agent j which has access to the reference trajectory such that the associated network graph has a directed path from j to i. This property is directly connected to the invertibility of the expanded row stochastic matrix.

Lemma 9.1 *The expanded row stochastic matrix $\bar{D}$ has all eigenvalues in the open unit circle if and only if for each $i \in \{1, \ldots, N\}$ there exists a $j \in \{1, \ldots, N\}$ such that $\iota_j = 1$ and the associated graph has a directed path from j to i.*

Proof We know from Lemma 8.1 that $\bar{L}$ is invertible if and only if for each $i \in \{1, \ldots, N\}$ there exists a $j \in \{1, \ldots, N\}$ such that $\iota_j = 1$ and the associated graph has a directed path from j to i. If $\bar{L}v = 0$ then we immediately obtain that $\bar{D}v = v$ which implies that $\bar{D}$ has an eigenvalue on the unit circle. On the other hand, assume $\bar{L}$ is invertible. A row stochastic matrix has all eigenvalues in the closed unit disc. As noted in the proof of Lemma 1.14, the only eigenvalue on the unit disc can be in 1, but an eigenvalue in 1 immediately yields that $\bar{L}$ is singular which gives a contradiction. Hence all eigenvalues are inside the unit disc. ■

Remark 9.2 If only one agent i has access to the reference trajectory, then the expanded row stochastic matrix is invertible if and only if the associated network graph contains a directed spanning tree with agent i as its root.

In the case of full-state coupling, i.e., $C = I$, we have $y_i = x_i$, for $i = 1, \ldots, N$ and $x_r \in \mathbb{R}^n$ is a, a priori given, constant trajectory. Then, we have

$$\zeta_i(k) = \sum_{j=1}^{N} d_{ij}(x_i(k) - x_j(k - \kappa_{ij})), \tag{9.6}$$

and

$$\bar{\zeta}_i(k) = \zeta_i(k) + \iota_i(x_i(k) - x_r). \tag{9.7}$$

For full-state coupling the standard protocol, we used many times,

$$u_i(k) = F\bar{\zeta}_i(k),$$

will in most cases does not work. It would actually require that, at the very least, $Ax_r = 0$. We will use a dynamic protocol in line with the partial-state coupling. In other words, for the full-state coupling case, we obtain Problem 8.4 with $C = I$. Therefore, this case will not be discussed separately.

In the following section, we study state synchronization and output regulation. The general result requires that the graph is undirected. If the graph is known, then we can obtain a similar result even if the graph is directed.

Definition 9.3 Given ι and a given real number $\beta \in (0, 1)$, let $\bar{\mathbb{G}}^N_{\iota,\beta}$ denote the set of directed network graphs with N nodes for which the corresponding expanded row stochastic matrix $\bar{D}$ has the property that $|\lambda_i| < \beta$ for $i = 1, \ldots, N$.

The subset of $\bar{\mathbb{G}}^N_{\iota,\beta}$ of undirected graphs is denoted by $\bar{\mathbb{G}}^{u,N}_{\iota,\beta}$.

9.3 Protocol Design for MAS with Partial-State Coupling

We consider the problem.

Problem 9.4 (Partial-State Coupling) Consider a MAS described by (9.1) and (9.4). Let a given constant trajectory $y_r \in \mathbb{R}^p$ be available to at least one agent. The problem of *state synchronization with output regulation* given a set of graphs $\mathbf{G} \subseteq \bar{\mathbb{G}}^N$ in the presence of unknown, nonuniform, and arbitrarily large communication delay is to find a distributed linear dynamic protocol of the form

$$\begin{cases} x_{i,c}(k+1) = A_c x_{i,c}(k) + B_c \bar{\zeta}_i(k), \\ u_i(k) \quad\ = C_c x_{i,c}(k), \end{cases} \quad (i = 1, \ldots, N) \tag{9.8}$$

for each agent such that

$$\lim_{k\to\infty} x_i(k) - x_j(k) = 0, \tag{9.9}$$

for all $i, j \in \{1, \ldots, N\}$, while the output of each agent converges to the given constant trajectory, i.e.

$$\lim_{k\to\infty} y_i(k) - y_r = 0, \tag{9.10}$$

for all $i \in \{1, \ldots, N\}$, for any graph $\mathcal{G} \in \mathbf{G}$ and for any communication delay $\kappa_{ij} \in \mathbb{R}^+$.

We consider here the output synchronization problem for networks with partial-state coupling and unknown, nonuniform, and arbitrarily large communication delays. We study this problem for undirected graphs. The special case that the graph is known is studied in Sect. 8.4 and allows us to consider undirected networks.

In general, we have to restrict our choice of y_r. Let

$$\begin{aligned} \mathscr{C}_y &= \left\{ y \in \mathbb{R}^p \,\middle|\, \begin{pmatrix} 0 \\ y \end{pmatrix} \in \operatorname{Im} \begin{pmatrix} A-I & B \\ C & 0 \end{pmatrix} \right\} \\ &= \left\{ y \in \mathbb{R}^p \,|\, \exists x \in \mathbb{R}^n, u \in \mathbb{R}^m : Ax + Bu = x,\ Cx = y \right\}. \end{aligned} \tag{9.11}$$

It turns out that our problem is solvable if and only if $y_r \in \mathscr{C}_y$. Note that $\mathscr{C}_y = \mathbb{R}^p$ if (A, B, C) is right-invertible and has no invariant zeros in one.

The main result relies on the following design.

Protocol design 9.1 *Consider a MAS described by (9.1) and (9.4). Let R be an injective matrix such that $\mathscr{C}_y = \operatorname{Im} R$. Choose Π and Γ such that*

$$\begin{pmatrix} 0 \\ R \end{pmatrix} = \begin{pmatrix} A-I & B \\ C & 0 \end{pmatrix} \begin{pmatrix} \Pi \\ \Gamma \end{pmatrix} \tag{9.12}$$

and

$$\operatorname{rank} \begin{pmatrix} A-I & B\Gamma \\ C & 0 \end{pmatrix} = n + \operatorname{rank} \Gamma. \tag{9.13}$$

Choose Γ_1 and Γ_2 such that $\operatorname{Im} \Gamma_1 = \operatorname{Im} \Gamma$ *and*

$$(\Gamma_1\ \Gamma_2) \tag{9.14}$$

(continued)

Protocol design 9.1 (continued)
is square and invertible.

Define $\tilde{A}$, $\tilde{B}$, *and* $\tilde{C}$ *according to*

$$\tilde{A} = \begin{pmatrix} A & B\Gamma_1 \\ 0 & I \end{pmatrix}, \quad \tilde{B} = \begin{pmatrix} B\Gamma_2 & 0 \\ 0 & I \end{pmatrix}, \quad \tilde{C} = \begin{pmatrix} C & 0 \end{pmatrix}. \tag{9.15}$$

Moreover, choose K *such that* $\tilde{A} + K\tilde{C}$ *is Schur stable. Let* P_δ *be the unique solution of the discrete-time algebraic Riccati equation*

$$\tilde{A}' P_\delta \tilde{A} - P_\delta - \tilde{A}' P_\delta \tilde{B}(I + \tilde{B}' P_\delta \tilde{B})^{-1} \tilde{B}' P_\delta \tilde{A} + \delta I = 0, \tag{9.16}$$

where δ *is a design parameter to be chosen later on. Then we consider the protocol*

$$\begin{cases} x_{i,c}(k+1) = \begin{pmatrix} \tilde{A} + K\tilde{C} & 0 \\ B_1 F_\delta & 0 \end{pmatrix} x_{i,c}(k) + \begin{pmatrix} -K \\ 0 \end{pmatrix} \bar{\zeta}_i(k), \\ u_i(k) \quad = \begin{pmatrix} H_1 F_\delta & \Gamma_1 \end{pmatrix} x_{i,c}(k), \end{cases} \tag{9.17}$$

with

$$\begin{aligned} F_\delta &= -(I + \tilde{B}' P_\delta \tilde{B})^{-1} \tilde{B}' P_\delta \tilde{A}, \\ B_1 &= \begin{pmatrix} 0 & I \end{pmatrix}, \\ H_1 &= \begin{pmatrix} \Gamma_2 & 0 \end{pmatrix}. \end{aligned} \tag{9.18}$$

Theorem 9.5 *Consider a MAS described by* (9.1) *and* (9.2) *with an associated undirected graph. Let a given constant trajectory* $y_r \in \mathbb{R}^p$ *be available to some of the agents with* $\iota_i = 1$ *if agent* i *has access to* y_r *and* $\iota_i = 0$*, otherwise. Let any* $\beta \in (0, 1)$ *be given, and hence a set of network graphs* $\bar{\mathbb{G}}^{u,N}_{\iota,\beta}$ *be defined.*

Then, under Assumption 9.1, *Problem* 9.4 *is solvable for the set of undirected graphs* $\mathbf{G} = \bar{\mathbb{G}}^{u,N}_{\iota,\beta}$ *if and only if* $y_r \in \mathscr{C}_y$.

More specifically, there exists a $\delta > 0$ *such that the linear protocol* (9.17) *achieves state synchronization and output regulation for any undirected graph* $\mathcal{G} \in \bar{\mathbb{G}}^{u,N}_{\iota,\beta}$, *for any communication delay* $\kappa_{ij} \in \mathbb{N}$, *and for any* $y_r \in \mathscr{C}_y$.

Before we can prove these theorems, we need a preliminary lemma.

Lemma 9.6 *Let β be an upper bound for the eigenvalues of $\bar{D}$. Then, for all communication delays $\kappa_{ij} \in \mathbb{N}$ for $i, j = 1, \ldots, N$ and for all $\omega \in \mathbb{R}$, all eigenvalues of $\bar{D}_{j\omega}(\kappa)$ are less than or equal to β where*

$$\bar{D}_{j\omega}(\kappa) = \begin{pmatrix} \bar{d}_{11} & \bar{d}_{12}e^{-j\omega\kappa_{12}} & \ldots & \bar{d}_{1N}e^{-j\omega\kappa_{1N}} \\ \vdots & \vdots & \ddots & \vdots \\ \bar{d}_{\varrho 1}e^{-j\omega\kappa_{\varrho 1}} & \bar{d}_{\varrho\varrho} & \ldots & \bar{d}_{\varrho N}e^{-j\omega\kappa_{\varrho N}} \\ \vdots & \vdots & \ddots & \vdots \\ \bar{d}_{N1}e^{-j\omega\kappa_{N1}} & \bar{d}_{N2}e^{-j\omega\kappa_{N2}} & \ldots & \bar{d}_{NN} \end{pmatrix} \tag{9.19}$$

where κ denotes a vector consisting of all κ_{ij} ($i \neq j$) with $i, j \in \{1, \ldots, N\}$.

Proof All eigenvalues of $\bar{D}_{j\omega}(\kappa)$ are in the set

$$\left\{ v'\bar{D}_{j\omega}(\kappa)v \mid v \in \mathbb{C}^N, \|v\| = 1 \right\}.$$

Therefore, it is sufficient to establish that all elements in this set have amplitude less than or equal to β.

Since $\bar{D}$ is symmetric and β is an upper bound for the amplitude of eigenvalues of $\bar{L}$, we find that $v'\bar{D}v$ is less than or equal to β, provided $\|v\| = 1$.

Next, consider an arbitrary vector $v \in \mathbb{C}^N$. We have

$$v'\bar{D}_{j\omega}(\kappa)v = \sum_{i=1}^{N} |v_i|^2 \bar{d}_{ii} + \sum_{i=1}^{N} \sum_{m,m\neq i}^{N} v_i' v_m \bar{d}_{im} e^{-\kappa_{ik} j\omega}.$$

Since $\bar{d}_{im}$ are all nonnegative, we get

$$|v'\bar{D}_{j\omega}(\kappa)v| \leq \sum_{i=1}^{N} |v_i|^2 \bar{d}_{ii} + \sum_{i=1}^{N} \sum_{k,k\neq i}^{N} |v_i' v_k| \bar{d}_{ik} = \begin{pmatrix} |v_1| \\ \vdots \\ |v_N| \end{pmatrix}' \bar{D} \begin{pmatrix} |v_1| \\ \vdots \\ |v_N| \end{pmatrix} \leq \beta,$$

which completes the proof. ∎

Proof of Theorem 9.5 For any individual agent to be able to track a constant reference signal y_r, there must exist a x_0 and a u_0 such that

$$\begin{pmatrix} A - I & B \\ C & 0 \end{pmatrix} \begin{pmatrix} x_0 \\ u_0 \end{pmatrix} = \begin{pmatrix} 0 \\ y_r \end{pmatrix}. \tag{9.20}$$

Clearly, such x_0 and u_0 exist only if y_r is in the set $\mathscr{C}_y$ which is therefore a necessary condition for the solvability of our problem.

Next, we show that the condition $y_r \in \mathscr{C}_y$ guarantees the existence of a controller which achieves output synchronization. Let R be an injective matrix such that $\mathscr{C}_y = \operatorname{Im} R$. In that case we can find Π and Γ such that (9.12) and (9.13) are satisfied. To see that we can impose the rank condition (9.13), we note that (C, A) detectable implies that the first n columns are linearly independent. If the rank condition (9.13) is not satisfied, then there exist x and v such that

$$\begin{pmatrix} A-I & B\Gamma \\ C & 0 \end{pmatrix} \begin{pmatrix} x \\ v \end{pmatrix} = 0$$

with $B\Gamma v \neq 0$, $v \perp \ker \Gamma$ and $v'v = 1$. But then

$$\begin{pmatrix} A-I & B \\ C & 0 \end{pmatrix} \begin{pmatrix} \Pi - xv' \\ \Gamma(I - vv') \end{pmatrix} = 0$$

which shows that $\bar{\Pi} = \Pi - xv'$ and $\bar{\Gamma} = \Gamma(I - vv')$ also satisfy the above equation but with $\operatorname{rank} \bar{\Gamma} < \operatorname{rank} \Gamma$. Recursively, we can find a solution of (9.12) which also satisfies the extra rank condition (9.13). Choose an injective Γ_1 such that $\operatorname{Im} V\Gamma = \operatorname{Im} \Gamma_1$ and choose Γ_2 such that

$$\begin{pmatrix} \Gamma_1 & \Gamma_2 \end{pmatrix} \tag{9.21}$$

is square and invertible.

We design a pre-compensator in the form of (9.22):

$$\begin{aligned} p_i(k+1) &= p_i(k) + \begin{pmatrix} 0 & I \end{pmatrix} v_i(k), \qquad p_i(k) \in \mathbb{R}^\nu \\ u_i(k) &= \Gamma_1 p_i(k) + \begin{pmatrix} \Gamma_2 & 0 \end{pmatrix} v_i(k), \end{aligned} \tag{9.22}$$

where $\nu = \operatorname{rank} \Gamma$. The interconnection of (9.1) and (9.22) is of the form

$$\begin{cases} \tilde{x}_i(k+1) = \tilde{A}\tilde{x}_i(k) + \tilde{B}v_i(k), \\ y_i(k) = \tilde{C}\tilde{x}_i(k), \end{cases} \tag{9.23}$$

where $\tilde{A}$, $\tilde{B}$, and $\tilde{C}$ are defined according to (9.15). We need to verify that the system remains stabilizable and detectable if we use a pre-compensator. The stabilizability of this system follows immediately from (9.14) and the stabilizability of (A, B).

To show the detectability of (9.23), we need to verify that

$$\operatorname{rank} \begin{pmatrix} zI - A & -B\Gamma_1 \\ 0 & (z-1)I \\ C & 0 \end{pmatrix} = n + \nu$$

for all z outside or on the unit circle, where ν is such that $\Gamma_1 \in \mathbb{R}^{n\times\nu}$. For $z \neq 1$, this follows immediately from the detectability of (C, A). For $z = 1$, we have

$$\operatorname{rank}\begin{pmatrix} I-A & -B\Gamma_1 \\ 0 & 0 \\ C & 0 \end{pmatrix} = \operatorname{rank}\begin{pmatrix} I-A & -B\Gamma \\ C & 0 \end{pmatrix} = n + \operatorname{rank}\Gamma.$$

We obtain the required detectability when we note that $\operatorname{rank}\Gamma = \operatorname{rank}\Gamma_1$ and $\operatorname{rank}\Gamma_1 = \upsilon$ (since Γ_1 is injective).

Next, a protocol for the MAS (9.23) is designed as

$$\begin{cases} \chi_i(k+1) = (\tilde{A}+K\tilde{C})\chi_i(k) - K\bar{\zeta}_i(k), \\ v_i(k) \quad = -F_\delta \chi_i(k), \end{cases} \tag{9.24}$$

where K is chosen such that $\tilde{A}+K\tilde{C}$ is Schur stable and P_δ is the unique solution of the algebraic Riccati equation (9.16).

Next, we prove that with the above protocol, the output of each agent converges to the constant trajectory y_r. First we need to show that there exists a $\tilde{\Pi}$ such that $\tilde{A}\tilde{\Pi} = 0$ and $\tilde{C}\tilde{\Pi} = I$. It is easily verified that a suitable $\tilde{\Pi}$ is given by

$$\tilde{\Pi} = \begin{pmatrix} \Pi \\ W \end{pmatrix},$$

where W is such that $\Gamma_1 W = \Gamma$. For $i = 1, \ldots, N$, define $\bar{x}_i(k) = \tilde{x}_i(k) - \tilde{\Pi}s$ where s is such that $y_r = Rs$, and the output synchronization error $e_i(k) = y_i(k) - y_r$. Then, we get the error dynamics:

$$\begin{cases} \bar{x}_i(k) = \tilde{A}\bar{x}_i(k) + \tilde{B}v_i(k), \\ e_i(k) = \tilde{C}\bar{x}_i(k). \end{cases} \tag{9.25}$$

Moreover, $\bar{\zeta}_i(k)$ can be rewritten as

$$\bar{\zeta}_i(k) = \sum_{j=1}^{N} a_{ij}\tilde{C}(\bar{x}_j(k) - \bar{x}_j(k-\kappa_{ij})) + \iota_i\tilde{C}\bar{x}_i(t).$$

Let

$$\bar{x}(k) = \begin{pmatrix} \bar{x}_1(k) \\ \bar{x}_2(k) \\ \vdots \\ \bar{x}_N(k) \end{pmatrix} \text{ and } \chi(k) = \begin{pmatrix} \chi_1(k) \\ \chi_2(k) \\ \vdots \\ \chi_N(k) \end{pmatrix}.$$

Then, the full closed-loop system can be written in the frequency domain as

$$\begin{pmatrix} z\bar{x}(z) \\ z\chi(z) \end{pmatrix} = \begin{pmatrix} I_N \otimes \tilde{A} & -I_N \otimes \tilde{B}F_\delta \\ -(I - \bar{D}_z(\kappa)) \otimes K\tilde{C} & I_N \otimes (\tilde{A} + K\tilde{C}) \end{pmatrix} \begin{pmatrix} \bar{x}(z) \\ \chi(z) \end{pmatrix}, \tag{9.26}$$

where $\bar{D}_z(\kappa)$ is the expanded row stochastic matrix given by

$$\bar{D}_z(\kappa) = \left[\bar{d}_{ij} z^{-\kappa_{ij}}\right].$$

Next, we prove (9.26) is asymptotically stable for all communication delays $\kappa_{ij} \in \mathbb{N}$. We first prove stability without communication delay and then prove stability for the case that includes a communication delay.

When there is no communication delay in the network, the stability of system (9.26) is equivalent to asymptotic stability of the matrix

$$\begin{pmatrix} \tilde{A} & \tilde{B}F_\delta \\ -(1-\lambda_i)K\tilde{C} & \tilde{A} + K\tilde{C} \end{pmatrix}$$

for all $i \in \{1, \ldots, N\}$, where λ_i is the eigenvalue of the expanded row stochastic matrix $\bar{D}$ satisfying $|\lambda_i| \leq \beta$. Note that

$$\begin{pmatrix} 1-\lambda_i & 0 \\ 0 & I \end{pmatrix} \begin{pmatrix} \tilde{A} & \tilde{B}F_\delta \\ -(1-\lambda_i)K\tilde{C} & \tilde{A} + K\tilde{C} \end{pmatrix} \begin{pmatrix} (1-\lambda_i)^{-1} & 0 \\ 0 & I \end{pmatrix}$$
$$= \begin{pmatrix} \tilde{A} & (1-\lambda_i)\tilde{B}F_\delta \\ -K\tilde{C} & \tilde{A} + K\tilde{C} \end{pmatrix}.$$

From the proof of Theorem 3.35, we find that there exists a δ^* such that for any $\delta \in (0, \delta^*]$ system (9.26) is asymptotically stable without any communication delay.

In the case of communication delay, according to Lemma 7.28, the closed-loop system (9.26) is asymptotically stable for all communication delays $\kappa_{ij} \in \mathbb{N}$, if

$$\det\left[e^{j\omega} I - \begin{pmatrix} I_N \otimes \tilde{A} & I_N \otimes \tilde{B}F_\delta \\ -(I - \bar{D}_{j\omega}(\kappa^r)) \otimes K\tilde{C} & I_N \otimes (\tilde{A} + K\tilde{C}) \end{pmatrix} \right] \neq 0 \tag{9.27}$$

for all $\omega \in \mathbb{R}$ and any $\kappa_{ij}^r \in \mathbb{R}^+$ where κ^r denotes a vector consisting of all κ_{ij}^r $(i \neq j)$ with $i, j \in \{1, \ldots, N\}$, while $\bar{D}_{j\omega}$ is defined in (9.19). Condition (9.27) is satisfied if the matrix

$$\begin{pmatrix} I_N \otimes \tilde{A} & I_N \otimes \tilde{B}F_\delta \\ -(I - \bar{D}_{j\omega}(\kappa^r)) \otimes K\tilde{C} & I_N \otimes (\tilde{A} + K\tilde{C}) \end{pmatrix} \tag{9.28}$$

has no eigenvalues on the unit circle for all $\omega \in \mathbb{R}$ and for all $\kappa_{ij}^r \in \mathbb{R}^+$.

Lemma 9.6 implies that all the eigenvalues of $\bar{D}_{j\omega}(\kappa^r)$ have amplitude less than β. Hence, the proof of Theorem 3.35 implies that there exists a δ^* such that for any $\delta \in (0, \delta^*]$, the matrix (9.28) has no eigenvalues on the imaginary axis. As noted before, this implies that the closed-loop system (9.26) is asymptotically stable for any communication delay $\kappa_{ij} \in \mathbb{N}$.

Finally, by combining the pre-compensator (9.22) and protocol (9.24), we get the linear dynamic protocol (9.17). ∎

9.4 MAS with a Known Communication Topology

We study here a variation of Problem 9.4 for a MAS with a known graph $\mathcal{G}$. The main advantage of knowing the graph is that we can also handle directed graphs.

Problem 9.7 (Partial-State Coupling and Known Graph) Consider a MAS described by (9.1) and (9.4) associated with a given directed graph $\mathcal{G}$. The problem of *state synchronization and output regulation* for networks with unknown, nonuniform, and arbitrarily large communication delay is to find a distributed linear dynamic protocol of the form (9.8) for each agent such that

$$\lim_{k\to\infty} (x_i(k) - x_j(k)) = 0, \tag{9.29}$$

for all $i, j \in \{1, \ldots, N\}$ and

$$\lim_{k\to\infty} (y_i(k) - y_r) = 0, \tag{9.30}$$

for all $i \in \{1, \ldots, N\}$, for the given directed graph $\mathcal{G}$, and for any communication delay $\kappa_{ij} \in \mathbb{N}$, the output of each agent converges to the constant trajectory.

We note that if our directed graph is balanced, then the results of Theorem 9.5 still hold if we define $\beta < 1$ as an upper bound of $\frac{1}{2}\|\bar{D}+\bar{D}'\|$. However, if the graph is not balanced, the derivation presented before might not be valid. In that case, we design a protocol for one individual graph (instead of for a set), i.e., we assume the directed graph $\mathcal{G}$ is given.

Again, the choice of the constant trajectory y_r needs to be restricted to the set given by (8.9).

Theorem 9.8 *Consider a MAS described by (9.1) and (9.2) with an associated directed graph $\mathcal{G}$. Let a given constant trajectory $y_r \in \mathbb{R}^p$ be available to some of the agents with $\iota_i = 1$ if agent i has access to y_r and $\iota_i = 0$, otherwise. Assume for any agent i there exists an agent j which has access to y_r such that the graph contains a directed path from agent j to agent i.*

Then, under Assumption 9.1, Problem 9.7 is solvable if and only if $y_r \in \mathscr{C}_y$. Specifically, given the directed graph $\mathcal{G}$, there exists a $\delta > 0$ such that the linear protocol (9.17) achieves state synchronization and output regulation for any communication delay $\kappa_{ij} \in \mathbb{R}^+$ and for any $y_r \in \mathscr{C}_y$.

Remark 9.9 As in the continuous-time chapter, we will make explicit use of our knowledge of the network to design our protocol for directed graphs,. If we have a finite set of possible graphs, then we can still find a protocol that works for every graph in this finite set (use as an upper bound for δ, the maximum of the lower bounds for each individual graph in the set).

Proof of Theorem 9.8 The proof follows the proof of Theorem 9.5, except for the choice of the parameter δ.

We know the row stochastic matrix $\bar{D}$ has all its eigenvalues inside the unit circle.

By Qu [88, Theorem 4.36], if the associated graph is irreducible and all eigenvalues of $\bar{D}$ are strictly less than 1, there exists a positve diagonal matrix P and $\tilde{\beta} < 1$ such that

$$\bar{D}'P\bar{D} < \tilde{\beta}P.$$

This implies that

$$\left(P^{1/2}\bar{D}_{j\omega}(\tau)P^{-1/2}\right)'\left(P^{1/2}\bar{D}_{j\omega}(\tau)P^{-1/2}\right) < \tilde{\beta}I.$$

We find that

$$v'P^{1/2}\bar{D}P^{-1/2}v \le \|P^{1/2}\bar{D}P^{-1/2}\|v'v \le \tilde{\beta}^{1/2}v'v$$

for all $v \in \mathbb{R}^N$. We have

$$v'P^{1/2}\bar{D}_{j\omega}(\tau)P^{-1/2} = \sum_{i=1}^{N}|v_i|^2\bar{d}_{ii} + \sum_{i=1}^{N}\sum_{m,m\neq i}^{N} v_i'v_m p_i^{1/2}p_m^{-1/2}\bar{d}_{im}e^{-\tau_{ik}j\omega},$$

where $p_1,\ldots,p_N$ are the diagonal elements of P. Since $\bar{d}_{im}$ are all nonnegative, we get

$$\begin{aligned}|v'P^{1/2}\bar{D}_{j\omega}(\tau)P^{-1/2}v| &\le \sum_{i=1}^{N}|v_i|^2\bar{d}_{ii} + \sum_{i=1}^{N}\sum_{k,k\neq i}^{N}|v_i'v_k|p_i^{1/2}p_k^{-1/2}\bar{d}_{ik} \\ &= \begin{pmatrix}|v_1|\\ \vdots \\ |v_N|\end{pmatrix}' P^{1/2}\bar{D}P^{-1/2}\begin{pmatrix}|v_1|\\ \vdots \\ |v_N|\end{pmatrix} \le \tilde{\beta}^{1/2}.\end{aligned}$$

Hence all the eigenvalues of $\bar{D}_{j\omega}(\tau)$ which are equal to the eigenvalues of $P^{1/2}\bar{D}_{j\omega}(\tau)P^{-1/2}$ are less than $\tilde{\beta}$ in magnitude. If the graph is not irreducible, we can obtain the same result using the strongly connected components. After all if $\bar{D}$ has a block triangular structure, then $\bar{D}_{j\omega}(\tau)$ has the same block triangular structure and the eigenvalues of the whole matrix are the union of the eigenvalues of the blocks on the diagonal. We can guarantee that the blocks on the diagonal are irreducible, and hence the previous argument applies.

Except for using this new bound for the eigenvalues of $\bar{D}_{j\omega}(\tau)$), the rest of the proof is identical to the proof of Theorem 9.5. ■

Chapter 10
Synchronization of Linear MAS Subject to Actuator Saturation and Unknown Input Delay

10.1 Introduction

This chapter considers synchronization problems for a homogeneous multi-agent system (MAS) when the input for each agent is subject to saturation and unknown input delay. In earlier chapters we have addressed delays in Chap. 6 (continuous-time) and in Chap. 7 (discrete-time). We have also studied MAS with saturation in Chap. 4. The objective of this chapter is to combine these results to the case where we have both delays and saturation.

The write-up of this chapter is partially based on [190] and[189].

10.2 Semi-Global Synchronization for Continuous-Time MAS

Consider a MAS composed of N identical linear time-invariant continuous-time agents subject to actuator saturation and unknown nonuniform input delay

$$
\begin{aligned}
\dot{x}_i(t) &= Ax_i(t) + B\sigma(u_i(t-\tau_i)), \\
y_i(t) &= Cx_i(t), \\
x_i(\varsigma) &= \phi_i(\varsigma), \quad \varsigma \in [-\bar{\tau}, 0]
\end{aligned}
\tag{10.1}
$$

for $i = 1, \ldots, N$, where $x_i \in \mathbb{R}^n$, $u_i \in \mathbb{R}^m$, and $y_i \in \mathbb{R}^p$ are, respectively, the state, input, and output vectors of agent i, while $\tau_i \in [0, \bar{\tau}]$ is an unknown constant, $\bar{\tau}$ is a known upper bound, and $\phi_i \in \mathfrak{C}^n_{\bar{\tau}}$. Here $\mathfrak{C}^n_{\bar{\tau}} := \mathbb{C}([-\bar{\tau}, 0], \mathbb{R}^n)$ denotes the Banach space of all continuous functions from $[-\bar{\tau}, 0] \to \mathbb{R}^n$ with norm

A. Saberi et al., *Synchronization of Multi-Agent Systems in the Presence of Disturbances and Delays*, Systems & Control: Foundations & Applications,
https://doi.org/10.1007/978-3-030-88148-1_10

$$\|x\|_C = \sup_{t\in[-\bar{\tau},0]} \|x(t)\|.$$

Moreover, σ is defined as in (4.2). We make the following standard assumption for the agent dynamics.

Assumption 10.1 *For each agent $i \in \{1, \ldots, N\}$, we assume that*

- *(A, B) is stabilizable, and (A, C) is detectable;*
- *A is at most weakly unstable.*

The communication network among agents is exactly the same as that in Chap. 2 and provides each agent with the quantity

$$\zeta_i = \sum_{j=1}^{N} a_{ij}(y_i - y_j) = \sum_{j=1}^{N} \ell_{ij} y_j \tag{10.2}$$

for $i = 1, \ldots, N$. The graph $\mathcal{G}$ describing the communication topology of the network is assumed to be undirected, weighted, and strongly connected. When we consider full-state coupling, i.e., $C = I$, the quantity ζ_i becomes

$$\zeta_i = \sum_{j=1}^{N} a_{ij}(x_i - x_j) = \sum_{j=1}^{N} \ell_{ij} x_j \tag{10.3}$$

for $i = 1, \ldots, N$.

10.2.1 Problem Formulation

We formulate below two state synchronization problems, one for a network with full-state coupling and the other for partial-state coupling.

Problem 10.1 (Full-State Coupling) Consider a MAS described by (10.1) and (10.3) with a given upper bound $\bar{\tau}$ for the input delay. Let $\mathbf{G}$ be a given set of graphs such that $\mathbf{G} \subseteq \mathbb{G}^N$. The *semi-global state synchronization* problem with a set of network graphs $\mathbf{G}$ is to find, if possible, for any a priori given bounded set of initial conditions $\mathcal{W} \subset \mathfrak{C}^n_{\bar{\tau}}$, a parameterized family of linear protocols of the form

$$u_i = F_\delta \zeta_i, \quad (i = 1, \ldots, N) \tag{10.4}$$

where there exists a δ^* such that for all $\delta < \delta^*$, state synchronization among agents is achieved for any graph $\mathcal{G} \in \mathbf{G}$ and for any input delay $\tau_i \in [0, \bar{\tau}]$ and any initial conditions $\phi_i \in \mathcal{W}$ for $i = 1, \ldots, N$.

Problem 10.2 (Partial-State Coupling) Consider a MAS described by (10.1) and (10.2) with a given upper bound $\bar{\tau}$ for the input delay. Let $\mathbf{G}$ be a given set of graphs such that $\mathbf{G} \subseteq \mathbb{G}^N$. The *semi-global state synchronization* problem with a set of network graphs $\mathbf{G}$ is to find, if possible, a positive integer q and for any a priori given bounded set of initial conditions $\mathcal{W} \subset \mathfrak{C}^n_{\bar{\tau}} \times \mathbb{R}^q$, a parameterized family of linear dynamic protocols of the form

$$\begin{cases} \dot{\chi}_i(t) = A_{c,\delta}\chi_i(t) + B_{c,\delta}\zeta_i(t), \\ u_i(t) = C_{c,\delta}\chi_i(t) + D_{c,\delta}\zeta_i(t), \\ \chi_i(0) = \psi_i(0), \end{cases} \tag{10.5}$$

for $i = 1, \ldots, N$ with $\chi_i \in \mathbb{R}^q$, where there exists a δ^* such that for all $\delta < \delta^*$, state synchronization among agents is achieved for any graph $\mathcal{G} \in \mathbf{G}$, for any input delay $\tau_i \in [0, \bar{\tau}]$ and any initial conditions $(\phi_i, \psi_i) \in \mathcal{W}$ for $i = 1, \ldots, N$.

10.2.2 Protocol Design for MAS with Full-State Coupling

The low-gain-based Protocol Design 6.1 for MAS with unknown input delay in Sect. 6.2.3.2 works as well for MAS with both input saturation and unknown input delay, which is being considered in this section; this is because we can choose the low-gain δ sufficiently small so that the saturation nonlinearity is not activated.

Protocol design 10.1 *Consider a MAS described by* (10.1) *and* (10.3). *We consider the following parameterized family of protocols:*

$$u_i = \rho F_\delta \zeta_i, \tag{10.6}$$

where $\rho > 0$ is a design parameter and $F_\delta = -B'P_\delta$ with $P_\delta > 0$ being the unique solution of the continuous-time algebraic Riccati equation (6.22).

The main result in this subsection is stated as follows.

Theorem 10.3 *Consider a MAS described by* (10.1) *and* (10.3) *with an input delay upper bound $\bar{\tau}$ and input saturation. Let any $\alpha > \beta > 0$ be given.*

If (A, B) is stabilizable and the agents are at most weakly unstable, then the semi-global state synchronization problem stated in Problem 10.1 with $\mathbf{G} = \mathbb{G}^{u,N}_{\alpha,\beta}$ is solvable if

$$\bar{\tau}\omega_{\max} < \tfrac{\pi}{2}. \tag{10.7}$$

Moreover, for any a priori given compact set of initial conditions $\mathcal{W} \subset \mathfrak{C}^n_{\bar{\tau}}$, *there exist* $\rho > 0$ *and* $\delta^* > 0$ *such that for this* ρ *and any* $\delta \in (0, \delta^*]$, *the protocol* (10.6) *achieves state synchronization for any graph* $\mathcal{G} \in \mathbb{G}^{u,N}_{\alpha,\beta}$, *for any input delay* $\tau_i \in [0, \bar{\tau}]$, *and for any initial condition* $\phi_i \in \mathcal{W}$ *for* $i = 1, \ldots, N$.

In order to present the proof of the above result, we first need some technical lemmas.

Lemma 10.4 *Suppose* (A, B) *is stabilizable and all the eigenvalues of* A *are in the closed left half plane. Let* $F_\delta = -B'P_\delta$ *be designed with* P_δ *given in* (6.22). *Then, we have the following properties:*

1. *The closed-loop system matrix* $A + \nu\lambda B F_\delta$ *is Hurwitz stable for all* $\delta > 0$, $\nu = 1/\beta$, *and for all real* $\lambda > \beta$.
2. *For any* $\beta > 0$, *there exists a* $\delta^* > 0$ *such that for all* $\delta \in (0, \delta^*]$ *there exist* $r_{1,\delta} > 0$ *and* $\rho_\delta > 0$ *with* $r_{1,\delta} \to 0$ *as* $\delta \to 0$ *such that*

$$\left\| F_\delta e^{(A+\nu\lambda B F_\delta)t} \right\| \leq r_{1,\delta} e^{-\rho_\delta t}, \tag{10.8}$$

for all $t \geq 0$, $\nu = 1/\beta$ *and for all real* $\lambda > \beta$.

3. *We have*

$$\left\| I + \nu\lambda F_\delta (sI - A - \nu\lambda B F_\delta)^{-1} B \right\|_\infty < 1 \tag{10.9}$$

for all $\delta > 0$, $\nu = 1/\beta$ *and for all real* $\lambda > \beta$.

Proof The first two properties have been established in Lemma 2.29 where we should note the connection between algebraic Riccati equations (2.58) and (2.61) as established in Remark 2.25. Hence the F_δ in Lemma 2.29 is equal to νF_δ in the notation of Lemma 10.4.

The last property follows from Lemma 4.5 given that λ is real. ∎

Lemma 10.5 *Assume* L *is associated with an undirected graph. If* $L = R_e J_e R_e'$ *with* R_e *unitary and* $J_e = \operatorname{diag}\{J, 0\}$ *with* J *is diagonal, then we have*

$$T_1 L T_2 = R J R^{-1}, \tag{10.10}$$

with $R = T_1 R_e T_2$ *and* $R^{-1} = T_2' R_e' T_2$, *where* $T_1 \in \mathbb{R}^{(N-1)\times N}$ *and* $T_2 \in \mathbb{R}^{N\times(N-1)}$ *are given by*

$$T_1 = \begin{pmatrix} I & -\mathbf{1}_{N-1} \end{pmatrix}, \quad T_2 = \begin{pmatrix} I \\ 0 \end{pmatrix}.$$

Proof Since $L\mathbf{1}_N = 0$, we have

$$LT_2T_1 = L\begin{pmatrix} I & -\mathbf{1}_{N-1} \\ 0 & 0 \end{pmatrix} = L\begin{pmatrix} I & 0 \\ 0 & 1 \end{pmatrix} - L\begin{pmatrix} 0 & \mathbf{1}_N \end{pmatrix} = L.$$

Therefore, LT_2T_1 has $N-1$ nonzero eigenvalues, i.e., $\lambda_2, \ldots, \lambda_N$. Then T_1LT_2 has the same $N-1$ nonzero eigenvalues and hence T_1LT_2 is invertible. Define $R = T_1R_eT_2$.

We have

$$R_e = \begin{pmatrix} R_{11} & \frac{1}{\sqrt{N}}\mathbf{1} \\ R_{21} & \frac{1}{\sqrt{N}} \end{pmatrix},$$

given that $L\mathbf{1} = 0$. Then, it is found that

$$\begin{aligned} RT_2'R_e' &= T_1R_eT_2T_2'R_e' \\ &= T_1\begin{pmatrix} R_{11} \\ R_{21} \end{pmatrix}\begin{pmatrix} R_{11}' & R_{21}' \end{pmatrix} \\ &= T_1\begin{pmatrix} R_{11} & \frac{1}{\sqrt{N}}\mathbf{1}_{N-1} \\ R_{21} & \frac{1}{\sqrt{N}} \end{pmatrix}\begin{pmatrix} R_{11}' & R_{21}' \\ \frac{1}{\sqrt{N}}\mathbf{1}_{N-1}' & \frac{1}{\sqrt{N}} \end{pmatrix} \\ &= T_1R_eR_e' = T_1. \end{aligned}$$

The third equality holds because

$$T_1\begin{pmatrix} \mathbf{1}_{N-1} \\ 1 \end{pmatrix} = T_1\mathbf{1}_N = 0.$$

Therefore

$$R^{-1}T_1 = T_2'R_e',$$

which yields $R^{-1} = T_2'R_e'T_2$, and, moreover

$$T_1LT_2R = T_1LT_2T_1R_eT_2 = T_1LR_eT_2 = T_1R_eJ_eT_2 = T_1R_eT_2J = RJ.$$

Hence, (10.10) is satisfied. ■

Lemma 10.6 *Suppose (A, B) is stabilizable and all the eigenvalues of A are in the closed left half plane. Let $F_\delta = -B'P_\delta$ be designed with P_δ given in (6.22). Then, we have*

$$\left\| \nu(J \otimes F_\delta)e^{(I_{N-1}\otimes A + \nu J \otimes BF_\delta)t} \right\| \leq r_{2,\delta}, \tag{10.11}$$

where J is the Jordan form of matrix T_1LT_2 and L is the Laplacian matrix with its associated graph in $\mathbb{G}^{u,N}_{\alpha,\beta}$, and moreover, $r_{2,\delta} \to 0$ as $\delta \to 0$.

Proof Consider a system

$$\dot{x} = (I_{N-1} \otimes A + \nu T_1 L T_2 \otimes BF_\delta)x \tag{10.12}$$

and let

$$\eta := (R^{-1} \otimes I_n)x = \begin{pmatrix} \eta_2 \\ \vdots \\ \eta_N \end{pmatrix},$$

where R is as defined in Lemma 10.5. Then, the dynamics of η can be written as

$$\dot{\eta} = (I_{N-1} \otimes A + \nu J \otimes BF_\delta)\eta. \tag{10.13}$$

Note that, since we have an undirected network, J is diagonal. Hence (10.13) is equivalent to

$$\dot{\eta}_i = (A + \nu\lambda_i BF_\delta)\eta_i$$

for $i = 2, \ldots, N$. Then, following the results of the above Lemma 10.4, we can achieve

$$\|F_\delta \eta_i(t)\| \leq \tilde{r}_{i,\delta}$$

with $\tilde{r}_{i,\delta} \to 0$ as $\delta \to 0$ for $i = 2, \ldots, N$. Let

$$r_{2,\delta} = \frac{\alpha}{\beta} \max\{\tilde{r}_{2,\delta}, \ldots, \tilde{r}_{N,\delta}\}.$$

We have

$$\left\| \nu(J \otimes F_\delta) e^{(I_{N-1} \otimes A + \nu J \otimes BF_\delta)t} \right\| \leq r_{2,\delta},$$

which completes the proof. ∎

Proof of Theorem 10.3 Let $\mathbf{D}_i$ $(i = 1, \ldots, N)$ be a delay operator for agent i such that $(\mathbf{D}_i u_i)(t) = u_i(t - \tau_i)$. In the frequency domain, $\hat{\mathbf{D}}_i(\omega) = e^{-j\omega\tau_i}$. Define

$$x = \operatorname{col}\{x_1, \ldots, x_N\}, \qquad u = \operatorname{col}\{u_1, \ldots, u_N\}$$

and

$$\mathbf{D} = \operatorname{diag}\{\mathbf{D}_1, \ldots, \mathbf{D}_N\}, \qquad \hat{\mathbf{D}}(\omega) = \operatorname{diag}\{\hat{\mathbf{D}}_1(\omega), \ldots, \hat{\mathbf{D}}_N(\omega)\};$$

the overall dynamics of multi-agent system described by (10.1) and (10.3) can be represented by

$$\begin{cases} \dot{x}(t) = (I_N \otimes A)x(t) + (I_N \otimes B)\sigma[(\mathbf{D}u)(t)], \\ u(t) = (L \otimes \rho F_\delta)x(t). \end{cases} \tag{10.14}$$

If the input $u(t) = (L \otimes \rho F_\delta)x(t)$ can be squeezed small enough, i.e., the input can avoid triggering saturation, the overall dynamics (10.14) becomes a system without saturation:

$$\begin{cases} \dot{x}(t) = (I_N \otimes A)x(t) + (I_N \otimes B)(\mathbf{D}u)(t), \\ u(t) = (L \otimes \rho F_\delta)x(t). \end{cases} \tag{10.15}$$

The synchronization of (10.15) has been proven in Theorem 6.39, and we will show synchronization for the system (10.14) by establishing that the system does not saturate provided δ is small enough.

We define

$$\bar{x}_i = x_i - x_N, \qquad \bar{x} = \operatorname{col}\{\bar{x}_1, \ldots, \bar{x}_{N-1}\}.$$

Since

$$u_i = \rho F_\delta \sum \ell_{ij}(x_j - x_N) = \rho F_\delta \sum \ell_{ij}\bar{x}_j,$$

we have

$$u = (LT_2 \otimes \rho F_\delta)\bar{x} \tag{10.16}$$

and

$$\dot{\bar{x}} = (I_{N-1} \otimes A)\bar{x} + \rho(T_1 \mathbf{D} L T_2) \otimes BF_\delta)\bar{x}. \tag{10.17}$$

The following step is to show that we can avoid the saturation if $(T_1 L T_2 \otimes \rho F_\delta)\bar{x}(t)$ is sufficiently small. Applying Cauchy-Schwarz inequality, we can prove that for any $t \geq 0$

$$\begin{aligned} &\left| \|(T_1 L T_2 \otimes \rho F_\delta)\bar{x}(t)\|^2 - \|(T_1 L T_2 \otimes \rho F_\delta)\bar{x}(0)\|^2 \right| \\ &\qquad \leq 2\|(T_1 L T_2 \otimes \rho F_\delta)\dot{\bar{x}}\|_2 \|(T_1 L T_2 \otimes \rho F_\delta)\bar{x}\|_2, \end{aligned}$$

which means that it is sufficient to establish that

$$\|(T_1 L T_2 \otimes \rho F_\delta)\bar{x}\|_2 \leq r_{3,\delta} \tag{10.18}$$

and

$$\left\|(T_1 L T_2 \otimes \rho F_\delta)\dot{\bar{x}}\right\|_2 \leq r_{4,\delta} \tag{10.19}$$

where $r_{3,\delta} \to 0$ and $r_{4,\delta} \to 0$ as $\delta \to 0$. We use here that $(T_1 L T_2 \otimes \rho F_\delta)\bar{x}(0)$ is arbitrarily small for all sufficiently small δ since $\bar{x}(0)$ is in a bounded set of initial conditions while $F_\delta \to 0$ as $\delta \to 0$.

Now we define the linear time-invariant operator $g_\delta : v_\delta \to w_\delta$ with the state space representation:

$$\begin{cases} \dot{\xi} &= (I_{N-1} \otimes A + \nu(J \otimes BF_\delta))\xi + (I_{N-1} \otimes B)v_\delta, \\ w_\delta &= \nu(J \otimes F_\delta)\xi. \end{cases} \tag{10.20}$$

We also define another linear time-invariant operator ϑ by

$$g(t) = \vartheta(f)(t) = R^{-1}T_1 \begin{pmatrix} h_1 \\ \vdots \\ h_N \end{pmatrix}$$

where

$$h_i(t) = \begin{cases} \left[\left(\frac{\rho}{\nu}\mathbf{D}_i - I\right) e_i' R_e T_2 \otimes I_m\right] f(t) & \text{if } t > \bar{\tau} \\ \left[-e_i' R_e T_2 \otimes I_m\right] f(t) & \text{otherwise} \end{cases}$$

for $i = 1, \ldots, N$ with e_i denoting the ith unit vector in $\mathbb{R}^N$. We can see that the Laplace transform of these two operators are given by

$$\begin{aligned} G_\delta(j\omega) &= \nu(J \otimes F_\delta)(j\omega I - (I_{N-1} \otimes A) - \nu(J \otimes BF_\delta))^{-1}(I_{N-1} \otimes B), \\ \Delta(j\omega) &= R^{-1}T_1(\frac{\rho}{\nu}\hat{\mathbf{D}}(\omega) - I)R_e T_2 \otimes I_m \\ &= T_2' R_e'(\frac{\rho}{\nu}\hat{\mathbf{D}}(\omega) - I)R_e T_2 \otimes I_m. \end{aligned}$$

Given (10.16), we observe that u is small if and only if

$$(T_1 L T_2 \otimes \rho F_\delta)\bar{x}$$

is small, since

$$\operatorname{Im} L \perp \ker T_1.$$

Next, define $\tilde{x} = (R^{-1} \otimes I_n)\bar{x}$. Then the dynamics of $\tilde{x}$ can be written as

$$\dot{\tilde{x}} = (I_{N-1} \otimes A + \nu(J \otimes BF_\delta))\tilde{x} + (I_{N-1} \otimes B)\vartheta(\nu(J \otimes F_\delta)\tilde{x}) + (I_{N-1} \otimes B)v_\delta, \tag{10.21}$$

for $t \geq 0$, where

$$v_\delta(t) = R^{-1}T_1 \begin{pmatrix} \bar{h}_1 \\ \vdots \\ \bar{h}_N \end{pmatrix}$$

while

$$\bar{h}_i(t) = \begin{cases} \rho e_i' R_e T_2 J \otimes F_\delta \tilde{x}(t - \tau_i) & t < \tau_i, \\ 0 & t \geq \tau_i, \end{cases}$$

for $i = 1, \ldots, N$.

Note that v_δ vanishes for $t \geq \bar{\tau}$. Moreover, since $F_\delta \to 0$, we have $\|v_\delta\|_\infty \to 0$ and $\|v_\delta\|_2 \to 0$ as $\delta \to 0$ for all $\phi \in \mathcal{W}$.

Moreover, we have

$$(T_1 L T_2 \otimes \rho F_\delta)\bar{x} = (RJ \otimes \rho F_\delta)\tilde{x}$$

and from Lemma 10.5 we note that R and R^{-1} are both bounded in norm (recall that R_e is unitary). Hence u is small if $(J \otimes \rho F_\delta)\tilde{x}$ is small. From (10.21), we obtain

$$\begin{aligned} \nu(J \otimes F_\delta)\tilde{x}(t) &= \nu(J \otimes F_\delta)e^{(I_{N-1}\otimes A+\nu(J\otimes BF_\delta))t}\tilde{x}(0) \\ &\qquad + (g_\delta \circ \vartheta)(\nu(J \otimes F_\delta)\tilde{x})(t) + g_\delta(v_\delta)(t) \end{aligned}$$

and hence

$$\begin{aligned} &\nu(J \otimes F_\delta)\tilde{x}(t) \\ &= (1 - g_\delta \circ \vartheta)^{-1}\left[\nu(J \otimes F_\delta)e^{(I_{N-1}\otimes A+\nu(J\otimes BF_\delta))t}\tilde{x}(0) + g_\delta(v_\delta)(t)\right]. \end{aligned} \tag{10.22}$$

From (10.22), we get

$$\|\nu(J \otimes F_\delta)\tilde{x}\|_2 \leq \|(I - \Delta G_\delta)^{-1}\|_\infty \left\| \nu(J \otimes F_\delta)e^{(I_{N-1}\otimes A+\nu(J\otimes BF_\delta))t}\tilde{x}(0)\right\|_2 + \|(I - \Delta G_\delta)^{-1}\|_\infty \|w_\delta\|_2. \tag{10.23}$$

Taking the derivative of (10.22), we obtain

$$\nu(J \otimes F_\delta)\dot{\tilde{x}}(t) = (1 - g_\delta \circ \vartheta)^{-1}\left[\nu(J \otimes F_\delta)e^{(I_{N-1}\otimes A+\nu(J\otimes BF_\delta))t}\tilde{\tilde{x}}(0 + \dot{w}_\delta(t)\right]$$

where $\tilde{\tilde{x}}(0) = (I_{N-1} \otimes A + \nu(J \otimes BF_\delta))\tilde{x}(0)$. We obtain

$$\left\|\nu(J \otimes F_\delta)\dot{\tilde{x}}\right\|_2 \leq \|(I - \Delta G_\delta)^{-1}\|_\infty \left\| \nu(J \otimes F_\delta)e^{(I_{N-1}\otimes A+\nu(J\otimes BF_\delta))t}\tilde{\tilde{x}}(0)\right\|_2 + \|(I - \Delta G_\delta)^{-1}\|_\infty \|\dot{w}_\delta\|_2. \tag{10.24}$$

In order to show that the upper bound in (10.23) and (10.24) converges to 0 as $\delta \to 0$, we first note that (10.11) is satisfied. Next, we analyze $\|w_\delta\|_2$ and $\|\dot{w}_\delta\|_2$. According to the definition of operator g_δ, we have $w_\delta = g_\delta(v_\delta)(t)$. Then

$$\|w_\delta\|_2 \leq \|G_\delta\|_\infty \|v_\delta\|_2 \leq 2\|v_\delta\|_2.$$

Therefore, for any given initial condition $\phi_i \in \mathcal{W}$ $(i = 1, \ldots, N)$, $\|w_\delta\|_2 \to 0$ as $\delta \to 0$.

Next, we will prove $\|\dot{w}_\delta\|_2 \to 0$ as $\delta \to 0$. For $t \in [0, \bar{\tau}]$, the derivative of $w_\delta(t)$ is

$$\dot{w}_\delta(t) = \nu(J \otimes F_\delta)(I_{N-1} \otimes A + \nu(J \otimes BF_\delta))\xi(t) + \nu(J \otimes F_\delta)(I_{N-1} \otimes B)v_\delta(t)$$

with $\xi(0) = 0$. This yields

$$\int_0^{\bar{\tau}} \|\dot{w}_\delta(t)\|^2 \mathrm{d}t \to 0 \quad \text{as } \delta \to 0 \tag{10.25}$$

since $F_\delta \to 0$ and J is bounded. Then, for $t > \bar{\tau}$, v_δ vanishes and

$$\begin{aligned}\dot{w}_\delta(t) &= \nu(J \otimes F_\delta)(I_{N-1} \otimes A + \nu(J \otimes BF_\delta))e^{(I_{N-1}\otimes A+\nu(J\otimes BF_\delta))t}\xi(\bar{\tau}),\\ &= \nu(J \otimes F_\delta)e^{(I_{N-1}\otimes A+\nu(J\otimes BF_\delta))t}(I_{N-1} \otimes A + \nu(J \otimes BF_\delta))\xi(\bar{\tau}).\end{aligned}$$

Here $\xi(\bar{\tau})$ is bounded. By applying the final property of Lemma 4.5, we find that

$$\int_{\bar{\tau}}^{\infty} \|\dot{w}_\delta(t)\|^2 \mathrm{d}t \to 0 \quad \text{as } \delta \to 0. \tag{10.26}$$

Combining (10.25) and (10.26) gives

$$\|\dot{w}_\delta(t)\|_2 \to 0 \quad \text{as } \delta \to 0.$$

From (10.23) and (10.24), we can obtain (10.18) and (10.19) provided that we have $\|(I - \Delta G_\delta)^{-1}\|_\infty$ is bounded independent of δ, using that $(T_1 L T_2 \otimes \rho F_\delta)\bar{x}$ is small if and only if $(J \otimes \rho F_\delta)\tilde{x}$ is small.

Next, we show that $\|(I - \Delta G_\delta)^{-1}\|_\infty$ is indeed bounded. Since $\bar{\tau}\omega_{\max} < \frac{\pi}{2}$, we can choose ρ such that

$$\rho\beta \cos(\bar{\tau}\omega_{\max}) > 1. \tag{10.27}$$

Note that this ρ is independent of low-gain parameter δ and condition (10.27) implies that $\rho\beta > 1$. Let this ρ be fixed during the remaining proof.

Given (10.27), there exists a $\varpi > 0$ such that

$$\rho\beta \cos(\bar{\tau}(\omega_{\max} + \varpi)) > 1. \tag{10.28}$$

For $|\omega| < \omega_{\max} + \varpi$, we find that

$$\begin{aligned}
\Delta(j\omega) + \Delta(j\omega)' &= T_2' R_e' (\frac{\rho}{\nu}\hat{\mathbf{D}}(\omega) - I) R_e T_2 \otimes I_m + T_2' R_e' (\frac{\rho}{\nu}\hat{\mathbf{D}}(\omega)' - I) R_e T_2 \otimes I_m \\
&= T_2' R_e' (\frac{\rho}{\nu}\hat{\mathbf{D}}(\omega) + \frac{\rho}{\nu}\hat{\mathbf{D}}(\omega)' - 2I) R_e T_2 \otimes I_m \\
&\geq T_2' R_e' (2\rho\beta \cos(\omega\bar{\tau}) - 2) R_e T_2 \otimes I_m \\
&\geq 0,
\end{aligned}$$

because ρ is chosen to satisfy $\rho\beta \cos(\bar{\tau}(\omega_{\max} + \varpi)) > 1$ in (10.28). Furthermore, we obtain that

$$\begin{aligned}
\|\Delta\|_\infty &= \left\| T_2' R_e' (\frac{\rho}{\nu}\hat{\mathbf{D}} - I) R_e T_2 \otimes I_m \right\|_\infty \\
&= \left\| (\frac{\rho}{\nu}\hat{\mathbf{D}} - I) R_e T_2 T_2' R_e' \otimes I_m \right\|_\infty \\
&\leq \left\| (\frac{\rho}{\nu}\hat{\mathbf{D}} - I) \otimes I_m \right\|_\infty \\
&\leq 1 + \frac{\rho}{\nu},
\end{aligned}$$

since $\|\hat{\mathbf{D}}\|_\infty \leq 1$. Then, we have that, for $|\omega| < \omega_{\max} + \varpi$

$$\begin{aligned}\Delta(j\omega)'\Delta(j\omega) &\leq \Delta(j\omega)'\Delta(j\omega) + \Delta(j\omega) + \Delta(j\omega)' \\ &\quad + \left[I - (2+\frac{\rho}{\nu})^{-2}(I+\Delta(j\omega)')(I+\Delta(j\omega))\right] \\ &\leq \left[1 - (2+\frac{\rho}{\nu})^{-2}\right](I+\Delta(j\omega)')(I+\Delta(j\omega)),\end{aligned}$$

which leads to

$$(I+\Delta(j\omega)')^{-1}\Delta(j\omega)'\Delta(j\omega)(I+\Delta(j\omega))^{-1} \leq \left[1-(2+\frac{\rho}{\nu})^{-2}\right]I.$$

Hence, there exists a $\gamma > 0$ that is independent of parameter δ, such that

$$\|\Delta(j\omega)(I+\Delta(j\omega))^{-1}\| \leq 1-\gamma.$$

Moreover, from Property 3 in Lemma 10.4, we can immediately obtain that

$$\|G_\delta\|_\infty \leq 2 \text{ and } \|I+G_\delta\|_\infty < 1.$$

Hence we obtain, for $|\omega| < \omega_{\max} + \varpi$, that

$$\begin{aligned}&\det[I - \Delta(j\omega)G_\delta(j\omega)] \\ &\quad = \det[I + \Delta(j\omega) - \Delta(j\omega)(I+G_\delta(j\omega))] \\ &\quad = \det(I+\Delta(j\omega))\det\left[I - (I+\Delta(j\omega))^{-1}\Delta(j\omega)(I+G_\delta(j\omega))\right] \\ &\quad \geq 1-(1-\gamma) = \gamma,\end{aligned}$$

which means that, for some ϱ that is independent of parameter δ, the following holds:

$$\underline{\sigma}(I - \Delta(j\omega)G_\delta(j\omega)) > \varrho$$

for all $|\omega| < \omega_{\max} + \varpi$, for all $\tau_i \in [0, \bar{\tau}]$ and all possible L associated with a network graph in $\mathbb{G}^{u,N}_{\alpha,\beta}$. Therefore, we have that, for $|\omega| < \omega_{\max} + \varpi$

$$\left\|(I - \Delta(j\omega)G_\delta(j\omega))^{-1}\right\| \leq \frac{1}{\varrho}. \tag{10.29}$$

For $|\omega| \geq \omega_{\max} + \varpi$, we know that $\Delta(j\omega)G_\delta(j\omega) \to 0$ as $\delta \to 0$ uniformly in ω. Therefore, we can conclude that for a small enough δ, we have (10.29) for $|\omega| \geq \omega_{\max} + \varpi$. This completes the proof of Theorem 10.3. ■

10.2.3 Protocol Design for MAS with Partial-State Coupling

The low-gain-based Protocol Design 6.4 for MAS with unknown input delay in Sect. 6.2.4.2 works as well for MAS with both input saturation and unknown input delay, which is being considered in this section. This is also because we can choose the low-gain δ sufficiently small so that the saturation nonlinearity is not activated. The main result in this subsection is stated as follows.

Protocol design 10.2 *Consider a MAS described by at most weakly unstable agents with input saturation* (10.1) *and measurements* (10.2). *We will use a CSS observer-based protocol for the agents.*

We choose an observer gain K such that $A + KC$ is Hurwitz stable. Next, we consider a feedback gain $F = -B'P_\delta$ where $P_\delta > 0$ is the unique solution of the continuous-time algebraic Riccati equation (6.22). *This results in the protocol*

$$\begin{cases} \dot{\chi}_i = (A + KC)\chi_i - K\zeta_i, \\ u_i = \rho F_\delta \chi_i, \end{cases} \tag{10.30}$$

where $\rho > 0$ and δ are design parameters to be chosen later.

Theorem 10.7 *Consider a MAS described by* (10.1) *and* (10.2) *with an input delay upper bound $\bar{\tau}$ and input saturation. Let any $\alpha > \beta > 0$ be given and hence a set of network graphs $\mathbb{G}^{u,N}_{\alpha,\beta}$ be defined.*

If (A, B) is stabilizable, (A, C) is detectable, and A is at most weakly unstable, then the semi-global state synchronization problem stated in Problem 10.2 *with $\mathbf{G} = \mathbb{G}^{u,N}_{\alpha,\beta}$ is solvable if condition* (10.7) *is satisfied. In particular, there exists an integer n, and for a priori given compact set of initial conditions $\mathcal{W} \subset \mathfrak{C}^n_{\bar{\tau}} \times \mathbb{R}^n$, there exist $\rho > 0$ and $\delta^* > 0$, such that for this ρ any $\delta \in (0, \delta^*]$, the protocol* (10.30) *achieves state synchronization for any graph $\mathcal{G} \in \mathbb{G}^{u,N}_{\alpha,\beta}$, for any input delay $\tau_i \in [0, \bar{\tau}]$, and for any initial condition $(\phi_i, \psi) \in \mathcal{W}$ for $i = 1, \ldots, N$.*

In order to prove this theorem, we need a technical lemma which will be used to show that the saturation will not get activated.

Lemma 10.8 *Consider*

$$\hat{A} = \begin{pmatrix} A & 0 \\ -KC & A + KC \end{pmatrix}, \quad \hat{B} = \begin{pmatrix} B \\ 0 \end{pmatrix}, \quad \hat{F}_\delta = \begin{pmatrix} 0 & F_\delta \end{pmatrix}.$$

Then, for any $\alpha > \beta > 0$, there exists a δ^* such that we have the following.

1. *The closed-loop system matrix* $\hat{A} + \nu\lambda\hat{B}\hat{F}_\delta$ *is Hurwitz stable for all* $\delta \in [0, \delta^*]$ *and for* $\nu = 1/\beta$ *and all* $\beta < \lambda < \alpha$.
2. *There exists a* $\hat{r}_{1,\delta}$ *with* $\hat{r}_{1,\delta} \to 0$ *as* $\delta \to 0$ *such that*

$$\left\| \hat{F}_\delta e^{(\hat{A}+\nu\lambda\hat{B}\hat{F}_\delta)t} \right\| \leq \hat{r}_{1,\delta} \tag{10.31}$$

for all $\delta \in [0, \delta^*]$, *for all* $t \geq 0$, *and for* $\nu = 1/\beta$ *and all* $\lambda \in \mathbb{R}$ *with* $\beta < \lambda < \alpha$.
3. *For any* $\mu > 0$, *there exists a* δ_1 *such that*

$$\left\| I + \nu\lambda\hat{F}_\delta \left(sI - \hat{A} - \nu\lambda\hat{B}\hat{F}_\delta \right)^{-1} \hat{B} \right\| \leq 1 + \mu \tag{10.32}$$

for all $\delta \in [0, \delta_1]$ *and for* $\nu = 1/\beta$ *and all* $\lambda \in \mathbb{R}$ *with* $\beta < \lambda$.

Proof The result directly follows from Lemma 4.11. However, note that the notation in that Lemma is different since it relies on an F_δ based on P_δ given by (2.58) instead of P_δ given by (6.22). Therefore the extra factor ν is needed relying on Remark 2.25. ■

Proof of Theorem 10.7 Define

$$\dot{\tilde{\chi}}_i = (A + KC)\tilde{\chi}_i - Ky_i$$
$$\dot{v} = (A + KC)v$$

with

$$v(0) = \frac{1}{N}\sum_{j=1}^{N} \chi_i(0)$$

and

$$\begin{pmatrix} \chi_1(0) \\ \vdots \\ \chi_N(0) \end{pmatrix} = L \begin{pmatrix} \tilde{\chi}_1(0) \\ \vdots \\ \tilde{\chi}_N(0) \end{pmatrix} + \begin{pmatrix} v(0) \\ \vdots \\ v(0) \end{pmatrix}.$$

Then it is easily verified that

$$\chi_i = \sum_{j=1}^{N} \ell_{ij}\tilde{\chi}_j + v.$$

Since $A + KC$ is asymptotically stable and $F_\delta \to 0$ as $\delta \to 0$, we have for any $\mu > 0$

$$s_\delta = \rho F_\delta v$$

is such that $\|s_\delta\|_2 \leq \mu$ and $\|s_\delta\|_\infty \leq \mu$ for any initial condition in the given compact set as $\delta \to 0$. The interconnection of the agents (10.1) and the protocol (10.30) can be written as

$$\begin{cases} \dot{x}_i &= Ax_i + B\mathbf{D}_i\tilde{u}_i + B\mathbf{D}_i s_\delta, \\ \dot{\tilde{\chi}}_i &= (A + KC)\tilde{\chi}_i - KCx_i, \\ \tilde{u}_i &= \rho F_\delta \sum_{j=1}^{N} \ell_{ij}\tilde{\chi}_j \end{cases} \tag{10.33}$$

provided the saturation does not get activated. Note that

$$u_i = \tilde{u}_i + s_\delta.$$

Therefore the saturation does not get activated if we show that $\tilde{u}_i$ is arbitrarily small for sufficiently small δ. Let

$$\hat{x}_i = \begin{pmatrix} \hat{x}_{i,x} \\ \hat{x}_{i,\chi} \end{pmatrix} = \begin{pmatrix} x_i \\ \tilde{\chi}_i \end{pmatrix} - \begin{pmatrix} x_N \\ \tilde{\chi}_N \end{pmatrix},$$

and

$$\hat{A} = \begin{pmatrix} A & 0 \\ -KC & A + KC \end{pmatrix}, \quad \hat{B} = \begin{pmatrix} B \\ 0 \end{pmatrix}, \quad \hat{F}_\delta = \begin{pmatrix} 0 & F_\delta \end{pmatrix}.$$

Then, the overall dynamics of $\hat{x} = \operatorname{col}\{\hat{x}_1, \ldots, \hat{x}_{N-1}\}$ without saturation is as follows:

$$\dot{\hat{x}} = (I_{N-1} \otimes \hat{A})\hat{x} + \rho(T_1\mathbf{D}LT_2 \otimes \hat{B}\hat{F}_\delta)\hat{x} + (I \otimes \hat{B})\tilde{s}_\delta \tag{10.34}$$

with

$$\tilde{s}_\delta = (T_1\mathbf{D}\mathbf{1} \otimes I)s_\delta$$

assuming the saturation is not active. Since $\tau\omega_{\max} < \frac{\pi}{2}$, we can choose ρ such that

$$\rho\beta\cos(\tau\omega_{\max}) > 1. \tag{10.35}$$

Note that this ρ is independent of low-gain parameter δ and condition (10.35) implies that $\rho\beta > 1$. Let this ρ be fixed during the remaining proof. We can now use the same argument as we did in the proof of Theorem 4.4 with the bounds we obtained in Lemma 10.8. That means $\hat{x} \to 0$ and, for any $\nu > 0$, we have $\|(LT_2 \otimes \rho \hat{F}_\delta)\hat{x}\|_\infty \leq \nu$ for any initial conditions in the given compact set as $\delta \to 0$.

We note that

$$\begin{pmatrix} u_1 \\ \vdots \\ u_N \end{pmatrix} = (LT_2 \otimes \rho \hat{F}_\delta)\hat{x} + \begin{pmatrix} s_\delta \\ \vdots \\ s_\delta \end{pmatrix}$$

Since, for sufficiently small δ, both $\|(LT_2 \otimes \rho \hat{F}_\delta)\hat{x}\|_\infty$ and $\|s_\delta\|_\infty$ are arbitrarily small for any initial conditions in the given compact sets, we can indeed avoid saturation. ■

10.3 Semi-Global Synchronization for Discrete-Time MAS

Consider a MAS composed of N identical linear time-invariant agents subject to actuator saturation and unknown nonuniform input delay:

$$\begin{aligned} x_i(k+1) &= Ax_i(k) + B\sigma(u_i(k-\kappa_i)), \\ y_i(k) &= Cx_i(k), \\ x_i(\varsigma) &= \phi_{i,\varsigma}, \quad \varsigma \in [-\bar{\kappa}, 0] \end{aligned} \tag{10.36}$$

for $i = 1, \ldots, N$, where $x_i \in \mathbb{R}^n$, $u_i \in \mathbb{R}^m$, and $y_i \in \mathbb{R}^p$ are, respectively, the state, input, and output vectors of agent i, while $\kappa_i \in [0, \bar{\kappa}]$ is an unknown constant, $\bar{\kappa}$ is a known upper bound, $\phi_i \in c_{\bar{\kappa}}^n$, and $\sigma(\cdot)$ as defined in (4.2). Here c_k^n denotes sequences $(w_1, \ldots, w_k)$ with $w_i \in \mathbb{R}^n$.

We make the following standard assumption for the agent dynamics.

Assumption 10.2 *For each agent $i \in \{1, \ldots, N\}$, we assume that*

- *(A, B) is stabilizable, and (A, C) is detectable;*
- *A is at most weakly unstable.*

The communication network among agents is exactly the same as that in Chap. 3, which satisfies Assumption 3.2 and provides each agent with the quantity

$$\zeta_i(k) = \sum_{j=1}^{N} d_{ij}(y_i(k) - y_j(k)), \quad i = 1, \ldots, N. \tag{10.37}$$

The graph $\mathcal{G}$ describing the communication topology of the network is assumed to be undirected, weighted, and strongly connected. When the communication network is with full-state coupling, i.e., C has full-column rank, without loss of generality $C = I$, the quantity ζ_i becomes

$$\zeta_i(k) = \sum_{j=1}^{N} d_{ij}(x_i(k) - x_j(k)). \tag{10.38}$$

10.3.1 Problem Formulation

As usual, we formulate two state synchronization problems, one for a network with full-state coupling and the other for partial-state coupling.

Problem 10.9 (Full-State Coupling) Consider a MAS described by (10.36) and (10.38) with a given upper bound $\bar{\kappa}$ for the input delay. Let $\mathbf{G}$ be a given set of graphs such that $\mathbf{G} \subseteq \mathbb{G}^N$. The *semi-global state synchronization* problem with a set of network graphs $\mathbf{G}$ is to find, if possible, for any a priori given bounded set of initial conditions $\mathcal{W} \subset c_{\bar{\kappa}}^n$, a parameterized family of linear protocols of the form

$$u_i = F_\delta \zeta_i, \quad (i = 1, \ldots, N) \tag{10.39}$$

where there exists a δ^* such that for all $\delta < \delta^*$, state synchronization among agents is achieved for any graph $\mathcal{G} \in \mathbf{G}$ and for any input delay $\kappa_i \in [0, \bar{\kappa}]$ and any initial conditions $\phi_i \in \mathcal{W}$ for $i = 1, \ldots, N$.

Problem 10.10 (Partial-State Coupling) Consider a MAS described by (10.36) and (10.37) with a given upper bound $\bar{\kappa}$ for the input delay. Let $\mathbf{G}$ be a given set of graphs such that $\mathbf{G} \subseteq \mathbb{G}^N$. The *semi-global state synchronization* problem with a set of network graphs $\mathbf{G}$ is to find, if possible, a positive integer q and for any a priori given bounded set of initial conditions $\mathcal{W} \subset c_{\bar{\kappa}}^n \times \mathbb{R}^q$, a parameterized family of linear dynamic protocols of the form

$$\begin{cases} \chi_i(k+1) = A_{c,\delta}\chi_i(k) + B_{c,\delta}\zeta_i(k), \\ u_i(k) \quad\;\; = C_{c,\delta}\chi_i(k) + D_{c,\delta}\zeta_i(k), \\ \chi_i(0) \quad\;\; = \psi_i, \end{cases} \tag{10.40}$$

for $i = 1, \ldots, N$ with $\chi_i \in \mathbb{R}^q$, where there exists a δ^* such that for all $\delta < \delta^*$, state synchronization among agents is achieved for any graph $\mathcal{G} \in \mathbf{G}$, for any input delay $\kappa_i \in [0, \bar{\kappa}]$ and any initial conditions $(\phi_i, \psi_i) \in \mathcal{W}$ for $i = 1, \ldots, N$.

10.3.2 Protocol Design for MAS with Full-State Coupling

The protocol design is again based on the low-gain methodology, which is slightly different from the Protocol Design 7.4 for MAS with unknown input delay in Sect. 7.3.3.

Protocol design 10.3 *Consider a MAS described by* (10.36) *and* (10.38) *with input saturation. We consider the protocol*

$$u_i = \rho F_\delta \zeta_i, \tag{10.41}$$

where

$$F_\delta = -\frac{1}{1-\beta}(B' P_\delta B + I)^{-1} B' P_\delta A \tag{10.42}$$

with $P_\delta > 0$ being the unique solution of the discrete-time algebraic Riccati equation

$$P_\delta = A' P_\delta A + \delta I - A' P_\delta B (B' P_\delta B + I)^{-1} B' P_\delta A, \tag{10.43}$$

while δ is sufficiently small such that

$$B' P_\delta B < \frac{1-\beta}{2\beta} I,$$

where β is the upper bound of the eigenvalues inside the unit disc for some row stochastic matrix D associated with a graph in a set of graphs $\bar{\mathbb{G}}_\beta^{u,N}$.

Remark 10.11 This feedback (10.42) is actually a special case of the protocol using in Protocol Design 3.2 where we choose $\beta_+ = \sqrt{\beta}$.

The main result based on the above design is stated as follows.

Theorem 10.12 *Consider a MAS described by* (10.36) *and* (10.38) *with an input delay upper bound $\bar{\kappa}$ and input saturation. Let any $0 < \beta < 1$ be given, and hence a set of network graphs $\bar{\mathbb{G}}_\beta^{u,N}$ be defined.*

If (A, B) is stabilizable and the agents are at most weakly unstable, then the semi-global state synchronization problem stated in Problem 10.9 with $\mathbf{G} = \bar{\mathbb{G}}_\beta^{u,N}$ is solvable if

$$\bar{\kappa}\omega_{\max} < \tfrac{\pi}{2}. \tag{10.44}$$

Moreover, for any a priori given compact set of initial conditions $\mathcal{W} \subset c_{\bar{\kappa}}^{n}$, *there exist* $\rho > 0$ *and* $\delta^* > 0$ *such that for this* ρ *and any* $\delta \in (0, \delta^*]$, *the protocol* (10.3) *achieves state synchronization for any graph* $\mathcal{G} \in \bar{\mathbb{G}}_{\beta}^{u,N}$, *for any input delay* $\kappa_i \in [0, \bar{\kappa}]$, *and for any initial condition* $\phi_{i,\varsigma} \in \mathcal{W}$ *for* $i = 1, \dots, N$.

In order to present the proof of the above result, we first need some technical lemmas.

Lemma 10.13 *Suppose* (A, B) *is stabilizable and all the eigenvalues of* A *are within the closed unit disc. Let* F_δ *be designed in* (10.42). *Then, we have the following properties:*

1. *The closed-loop system matrix* $A + (1-\lambda)BF_\delta$ *is Schur stable for all* $\delta > 0$ *and for all* λ *with* $|\lambda| < \beta$.
2. *For any* $\beta > 0$, *there exists a* $\delta^* > 0$ *such that for all* $\delta \in (0, \delta^*]$ *there exist* $r_{1,\delta} > 0$ *and* $0 < \eta_\delta < 1$ *with* $r_{1,\delta} \to 0$ *as* $\delta \to 0$ *such that*
$$\|F_\delta(A + (1-\lambda)BF_\delta)^k\| \leq r_{1,\delta}\eta_\delta^k, \tag{10.45}$$
for all $k \geq 0$ *and for all* $\lambda \in \mathbb{R}$ *with* $|\lambda| < \beta$.
3. *Let* $G_\delta(z) = (1-\lambda)F_\delta(zI - A - (1-\lambda)BF_\delta)^{-1}B$. *Then, for any* $\mu > 0$, *there exists a* $\delta^* > 0$ *such that for all* $\delta \in (0, \delta^*]$
$$\|I + G_\delta\|_\infty \leq 1 + \mu \tag{10.46}$$
for all $\lambda \in \mathbb{R}$ *with* $|\lambda| < \beta$.

Proof Note that the first two properties can be directly obtained from Lemma 3.20 if we choose $\beta_+ = \sqrt{\beta}$ (see also Remark 10.11). We prove it here in a different way since the intermediate steps of this proof will be needed to prove the final property. Consider

$$x(k+1) = (A + (1-\lambda)BF_\delta)x(k),$$

and let $A_f = A + (1-\lambda)BF_\delta$.

It is found that

$$\begin{aligned}
&A'P_\delta A - A'P_\delta B(I + B'P_\delta B)^{-1}B'P_\delta A\\
&= A'P_\delta A - (1-\beta)^2F_\delta'(I + B'P_\delta B)F_\delta\\
&= A_f'P_\delta A_f + (1-\lambda)(1-\beta)F_\delta'F_\delta + (\beta-\lambda)F_\delta'[(1-\beta) - (\beta-\lambda)B'P_\delta B]F_\delta\\
&\geq A_f'P_\delta A_f + (1-\lambda)(1-\beta)F_\delta'F_\delta,
\end{aligned}$$

where the last inequality holds because

$$(\beta - \lambda)B'P_\delta B \leq 2\beta B'P_\delta B \leq (1-\beta)I.$$

Therefore, we obtain

$$P_\delta \geq A_f' P_\delta A_f + \delta I + (1-\lambda)(1-\beta)F_\delta' F_\delta, \tag{10.47}$$

which implies that

$$\begin{aligned} x'(k+1)P_\delta x(k+1) &\leq x'(k)P_\delta x(k) - \delta x'(k)x(k) \\ &\leq \eta_\delta^2 x'(k)P_\delta x(k), \end{aligned}$$

where $\eta_\delta^2 = 1 - \delta \|P_1\|^{-1}$. Hence

$$\|P_\delta^{1/2}x(k)\| \leq \eta_\delta^k \|P_\delta^{1/2}x(0)\|. \tag{10.48}$$

From (10.47) we get

$$(1-\beta)^2 F_\delta' F_\delta \leq P_\delta.$$

Hence

$$\begin{aligned} \|F_\delta(A + (1-\lambda)BF_\delta)^k x(0)\| &= \|F_\delta x(k)\| \\ &\leq (1-\beta)^{-1}\|P_\delta^{1/2}x(k)\|\| \leq (1-\beta)^{-1}\eta_\delta^k \|P_\delta^{1/2}x(0)\|. \end{aligned} \tag{10.49}$$

Since (10.49) is true for all $x(0) \in \mathbb{R}^n$, it follows trivially that

$$\|F_\delta(A + (1-\lambda)BF_\delta)^k\| \leq (1-\beta)^{-1}\|P_\delta^{1/2}\|\eta_\delta^k. \tag{10.50}$$

The proof of the inequality (10.45) is then completed by taking $r_{1,\delta} = (1-\beta)^{-1}\|P_\delta^{1/2}\|$.

It remains to prove the final property of this theorem. Using Lemma 4.20, we can show that this H_∞ norm is bounded by choosing $\beta_+ = \sqrt{\beta}$. However, we need a stronger result here. The inequality (10.47) yields

$$\begin{aligned} (z^{-1}I - A_f)P_\delta(zI - A_f) + A_f' P_\delta(zI - A_f) + (z^{-1}I - A_f)P_\delta A_f \\ \geq \delta I + (1-\lambda)(1-\beta)F_\delta' F_\delta. \end{aligned}$$

Premultiplying it with $\frac{1-\lambda}{1-\beta}B'(z^{-1}I - A_f)^{-1}$ and postmultiplying it with $(zI - A_f)^{-1}B$ yields

$$\begin{aligned}\tfrac{1-\lambda}{1-\beta}B'P_\delta B + \tfrac{1-\lambda}{1-\beta}B'(z^{-1}I - A_f)^{-1}A'_f P_\delta B + \tfrac{1-\lambda}{1-\beta}B'P_\delta A_f(zI - A_f)^{-1}B \\ \geq \delta \tfrac{1-\lambda}{1-\beta}B'(z^{-1}I - A'_f)^{-1}(zI - A_f)^{-1}B \\ + (1-\lambda)^2 B'(z^{-1}I - A_f)^{-1}F'_\delta F_\delta(zI - A_f)^{-1}B. \qquad (10.51)\end{aligned}$$

We have

$$\begin{aligned}\tfrac{1-\lambda}{1-\beta}B'P_\delta A_f &= \tfrac{1-\lambda}{1-\beta}B'P_\delta A - \tfrac{(1-\lambda)^2}{1-\beta}B'P_\delta BF_\delta \\ &= -\left[(I + B'P_\delta B) + \tfrac{1-\lambda}{1-\beta}B'P_\delta B\right](1-\lambda)F_\delta \\ &= -V_\delta(1-\lambda)F_\delta,\end{aligned}$$

where

$$V_\delta = (I + B'P_\delta B) + \tfrac{1-\lambda}{1-\beta}B'P_\delta B.$$

Using this equality, (10.51) yields

$$\tfrac{1-\lambda}{1-\beta}B'P_\delta B - G'_\delta(z^{-1})V_\delta - V_\delta G_\delta(z) \geq G'_\delta(z^{-1})G_\delta(z),$$

which is equivalent to

$$[V_\delta + G'_\delta(z^{-1})][V_\delta + G_\delta(z)] \leq V_\delta^2 + \tfrac{1-\lambda}{1-\beta}B'P_\delta B.$$

Since $V_\delta \to I$ and $B'P_\delta B \to 0$ as $\delta \to 0$, we find that, for any $\mu > 0$, there exists a δ^* such that for any $\delta \in (0, \delta^*]$

$$[I + G'_\delta(z^{-1})][I + G_\delta(z)] \leq (1+\mu)I.$$

This implies that

$$\|I + G_\delta\|_\infty \leq 1 + \mu. \qquad \blacksquare$$

Lemma 10.14 *Assume that D is associated with a strongly connected undirected graph and $\bar{L} = I - D$. If $\bar{L} = R_e J_e R_e'$ with R_e unitary and $J_e = \operatorname{diag}\{J, 0\}$ with J is diagonal, then we have*

$$T_1 \bar{L} T_2 = R J R^{-1}, \tag{10.52}$$

with $R = T_1 R_e T_2$ and $R^{-1} = T_2' R_e' T_2$, where $T_1 \in \mathbb{R}^{(N-1)\times N}$ and $T_2 \in \mathbb{R}^{N\times(N-1)}$ are given by

$$T_1 = \begin{pmatrix} I & -\mathbf{1}_{N-1} \end{pmatrix}, \quad T_2 = \begin{pmatrix} I \\ 0 \end{pmatrix}.$$

Proof This directly follows from Lemma 10.5 since $\bar{L}$ is a Laplacian matrix. ■

Lemma 10.15 *Suppose (A, B) is stabilizable and all the eigenvalues of A are within the closed unit disc. Let F_δ be designed in (10.42). Then, we have*

$$\left\| (J \otimes F_\delta)(I_{N-1} \otimes A + (J \otimes B F_\delta))^k \right\| \leq r_{2,\delta}, \tag{10.53}$$

where J is the Jordan form of matrix $T_1 \bar{L} T_2$ and $\bar{L} = I - D$ with D the row stochastic matrix with its associated graph in $\mathbb{G}_\beta^{u,N}$, and moreover, $r_{2,\delta} \to 0$ as $\delta \to 0$.

Define

$$\bar{G}_\delta(z) = J \otimes F_\delta(zI - (I_{N-1} \otimes A) - (J \otimes B F_\delta))^{-1} B.$$

Then, for any $\mu > 0$, there exists a $\delta^* > 0$ such that for all $\delta \in (0, \delta^*]$

$$\| I + \bar{G}_\delta \|_\infty \leq 1 + \mu. \tag{10.54}$$

Proof We have $J = \operatorname{diag}\{\bar{\lambda}_2, \ldots, \bar{\lambda}_N\}$. Consider the dynamics of η

$$\eta(k+1) = (I_{N-1} \otimes A + J \otimes B F_\delta)\eta(k+1),$$

where

$$\eta = \begin{pmatrix} \eta_2 \\ \vdots \\ \eta_N \end{pmatrix},$$

because we have an undirected network $\bar{L}$ is symmetric and J a diagonal matrix. Hence we find that

$$\eta_i(k+1) = (A + \bar{\lambda}_i B F_\delta)\eta_i(k)$$

for $i = 2, \ldots, N$. Then, following the results of the above Lemma 10.13, we can achieve

$$\|F_\delta \eta_i(t)\| \leq r_{1,\delta}$$

with $r_{1,\delta} \to 0$ as $\delta \to 0$. Then, we have

$$\left\|(J \otimes F_\delta)(I_{N-1} \otimes A + (J \otimes BF_\delta))^k\right\| \leq r_{1,\delta},$$

since J is diagonal. Equation (10.54) can be similarly derived from the final result of Lemma 10.13. This completes the proof. ■

Proof of Theorem 10.12 Let $\mathbf{D}_i$ $(i = 1, \ldots, N)$ be a delay operator for agent i such that $(\mathbf{D}_i u_i)(k) = u_i(k - \kappa_i)$. In the frequency domain, $\hat{\mathbf{D}}_i(\omega) = e^{-j\omega\kappa_i}$. Define $x = \operatorname{col}\{x_i\}$, $u = \operatorname{col}\{u_i\}$ and $\mathbf{D} = \operatorname{diag}\{\mathbf{D}_i\}$ and $\hat{\mathbf{D}}(\omega) = \operatorname{diag}\{\hat{\mathbf{D}}_i(\omega)\}$; the overall dynamics of multi-agent system described by (4.1) and (4.4) can be represented by

$$\begin{cases} x(k+1) = (I_N \otimes A)x(k) + (I_N \otimes B)\sigma(\mathbf{D}u(k)), \\ \quad u(k) \quad = (\bar{L} \otimes \rho F_\delta)x(k). \end{cases} \tag{10.55}$$

If the input $u(k) = (\bar{L} \otimes \rho F_\delta)x(t)$ can be squeezed small enough, i.e., the input can avoid triggering saturation, the overall dynamics (10.55) becomes a system without saturation

$$\begin{cases} x(k+1) = (I_N \otimes A)x(k) + (I_N \otimes B)\mathbf{D}u(k), \\ \quad u(k) \quad = (\bar{L} \otimes \rho F_\delta)x(k). \end{cases} \tag{10.56}$$

The synchronization of (10.56) has been proven in Theorem 7.29 (note that ρ in Chap. 7 is equal to $\rho/(1-\beta)$ in the notation of this chapter), and we can show that synchronization of (10.55) by establishing that the system (10.56) does not saturate provided δ is small enough.

Now define $\bar{x}_i = x_i - x_N$ and $\bar{x} = \operatorname{col}\{\bar{x}_1, \ldots, \bar{x}_{N-1}\}$. Since

$$u_i = \rho F_\delta \sum \bar{\ell}_{ij}(x_j - x_N) = \rho F_\delta \sum \bar{\ell}_{ij}\bar{x}_j,$$

we have

$$u = (\bar{L}T_2 \otimes \rho F_\delta)\bar{x} \tag{10.57}$$

and

$$\bar{x}(k+1) = (I_{N-1} \otimes A)\bar{x}(k) + \rho(T_1\mathbf{D}\bar{L}T_2 \otimes BF_\delta)\bar{x}(k). \tag{10.58}$$

The following step is to show that we can avoid the saturation. Since $\operatorname{Im} \bar{L} \perp \ker T_1$, we only need to show that $(T_1\bar{L}T_2 \otimes \rho F_\delta)\bar{x}(k)$ is sufficiently small. We have

$$\left\|(T_1\bar{L}T_2 \otimes \rho F_\delta)\bar{x}\right\| \leq \left\|(T_1\bar{L}T_2 \otimes \rho F_\delta)\bar{x}\right\|_2,$$

and hence it is sufficient to prove that

$$\left\|(T_1\bar{L}T_2 \otimes \rho F_\delta)\bar{x}\right\|_2 \leq r_{3,\delta}, \tag{10.59}$$

with $r_{3,\delta} \to 0$ as $\delta \to 0$.

Next, we define the linear time-invariant operator $g_\delta : v_\delta \to w_\delta$ with the state space representation

$$\begin{cases} \xi(k+1) = (I_{N-1} \otimes A + (J \otimes BF_\delta))\xi(k) + (I_{N-1} \otimes B)v_\delta(k), \\ w_\delta(k) \quad = (J \otimes F_\delta)\xi(k). \end{cases} \tag{10.60}$$

We also define another linear time-invariant operator ϑ by

$$g(k) = \vartheta(f)(k) = R^{-1}T_1 \begin{pmatrix} h_1 \\ \vdots \\ h_N \end{pmatrix}$$

with

$$h_i(k) = \begin{cases} \left[(\rho \mathbf{D_i} - I)e_i' R_e T_2 \otimes I_m\right] f(k) & \text{if } k \geq \kappa_i \\ \left[-e_i' R_e T_2 \otimes I_m\right] f(k) & \text{otherwise} \end{cases}$$

for $i = 1, \ldots, N$ with e_i denoting the ith unit vector in $\mathbb{R}^N$. We can see that the Fourier transform of these two operators are given by

$$G_\delta(j\omega) = (J \otimes F_\delta)(e^{j\omega}I - (I_{N-1} \otimes A) - (J \otimes BF_\delta))^{-1}(I_{N-1} \otimes B),$$
$$\Delta(j\omega) = R^{-1}T_1(\rho\hat{\mathbf{D}}(\omega) - I)R_eT_2 \otimes I_m.$$

Next, define $\tilde{x} = (R^{-1} \otimes I_n)\bar{x}$. Then, the dynamics of $\tilde{x}$ can be written as

$$\begin{aligned} \tilde{x}(k+1) = (I_{N-1} \otimes A + (J \otimes BF_\delta))\tilde{x}(k) + (I_{N-1} \otimes B)\vartheta((J \otimes F_\delta)\tilde{x}(k)) \\ + (I_{N-1} \otimes B)v_\delta(k), \end{aligned} \tag{10.61}$$

where

$$v_\delta(k) = R^{-1}T_1 \begin{pmatrix} \bar{h}_1 \\ \vdots \\ \bar{h}_N \end{pmatrix}$$

while

$$\begin{cases} \left[\rho \mathbf{D}_i e_i' R_e T_2 J \otimes F_\delta\right]\tilde{x}(k) & k < \kappa_i, \\ 0 & k \geq \bar{\kappa}, \end{cases}$$

for $i = 1, \dots, N$. Note that v_δ vanishes for $k \geq \bar{\kappa}$ and $\tilde{x}(k)$ is bounded for $k \leq \bar{\kappa}$ since the initial conditions are in the bounded set $\mathcal{W}$. Moreover, since $F_\delta \to 0$, we have $\|v_\delta\|_\infty \to 0$ and $\|v_\delta\|_2 \to 0$ as $\delta \to 0$.

From (10.61), we obtain

$$(J \otimes F_\delta)\tilde{x}(k) = (J \otimes F_\delta)(I_{N-1} \otimes A + (J \otimes BF_\delta))^k \tilde{x}(0) + (g_\delta \circ \vartheta)((J \otimes F_\delta)\tilde{x}(k)) \\ + g_\delta(v_\delta)(k)$$

and hence

$$(J \otimes F_\delta)\tilde{x}(k) = (1 - g_\delta \circ \vartheta)^{-1}\left[(J \otimes F_\delta)(I_{N-1} \otimes A + (J \otimes BF_\delta))^k \tilde{x}(0) + g_\delta(v_\delta)(k)\right]. \tag{10.62}$$

According to the definition of operator g_δ, we have $w_\delta = g_\delta(v_\delta)(k)$. Then, $\|w_\delta\|_2 \leq \|G_\delta\|_\infty \|v_\delta\|_2 \leq 2\|v_\delta\|_2$ by choosing $\mu = 1$ in Lemma 10.15. Therefore, for any given initial condition $\phi_i \in \mathcal{W}$ ($i = 1, \dots, N$), $\|w_\delta\|_2 \to 0$ as $\delta \to 0$.

From (10.62), we get

$$\begin{aligned} &\|(J \otimes F_\delta)\tilde{x}(k)\|_2 \\ &\quad \leq \|(I - \Delta G_\delta)^{-1}\|_\infty \left\|(J \otimes F_\delta)(I_{N-1} \otimes A + (J \otimes BF_\delta))^k \tilde{x}(0)\right\|_2 \\ &\qquad + \|(I - \Delta G_\delta)^{-1}\|_\infty \|w_\delta\|_2. \end{aligned}$$

Therefore, we can obtain (10.59) provided that we have $\|(I - \Delta G_\delta)^{-1}\|_\infty$ is bounded independent of δ, using that $(T_1 \bar{L} T_2 \otimes \rho F_\delta)\tilde{x}$ is small if and only if $(J \otimes \rho F_\delta)\tilde{x}$ is small.

Since $\bar{\kappa}\omega_{\max} < \frac{\pi}{2}$, we can choose ρ such that

$$\rho \cos(\bar{\kappa}\omega_{\max}) > 1. \tag{10.63}$$

Note that this ρ is independent of low-gain parameter δ and condition (10.63) implies that $\rho > 1$. Let this ρ be fixed during the remaining proof.

Given (10.63), there exists a $\varpi > 0$ such that

$$\rho \cos(\bar{\kappa}(\omega_{\max} + \varpi)) > 1. \tag{10.64}$$

For $|\omega| < \omega_{\max} + \varpi$, we find that

$$\begin{aligned}
\Delta(j\omega) + \Delta(j\omega)' &= T_2' R_e'(\rho \hat{\mathbf{D}}(\omega) - I) R_e T_2 \otimes I_m + T_2' R_e'(\rho \hat{\mathbf{D}}(\omega)' - I) R_e T_2 \otimes I_m \\
&= T_2' R_e'(\rho \hat{\mathbf{D}}(\omega) + \rho \hat{\mathbf{D}}(\omega)' - 2I) R_e T_2 \otimes I_m \\
&\geq T_2' R_e'(2\rho \cos(\omega \bar{\tau}) - 2) R_e T_2 \otimes I_m \\
&\geq 0,
\end{aligned}$$

because ρ is chosen to satisfy (10.64). Furthermore, we obtain that

$$\begin{aligned}
\|\Delta(j\omega)\|_\infty &= \left\| T_2' R_e'(\rho \hat{\mathbf{D}}(\omega) - I) R_e T_2 \otimes I_m \right\|_\infty \\
&= \left\| (\rho \hat{\mathbf{D}}(\omega) - I) R_e T_2 T_2' R_e' \otimes I_m \right\|_\infty \\
&\leq \left\| (\rho \hat{\mathbf{D}}(\omega) - I) \otimes I_m \right\|_\infty \\
&\leq 1 + \rho,
\end{aligned}$$

since $\|\hat{\mathbf{D}}(\omega)\| \leq 1$. Then, we have that, for $|\omega| < \omega_{\max} + \varpi$

$$\begin{aligned}
\Delta(j\omega)'\Delta(j\omega) &\leq \Delta(j\omega)'\Delta(j\omega) + \Delta(j\omega) + \Delta(j\omega)' \\
&\quad + \left[I - (2+\rho)^{-2}(I + \Delta(j\omega)')(I + \Delta(j\omega)) \right] \\
&\leq \left[1 - (2+\rho)^{-2} \right] (I + \Delta(j\omega)')(I + \Delta(j\omega)),
\end{aligned}$$

which leads to

$$(I + \Delta(j\omega)')^{-1} \Delta(j\omega)' \Delta(j\omega)(I + \Delta(j\omega))^{-1} \leq \left[1 - (2+\rho)^{-2} \right] I.$$

Hence, there exists a $\gamma > 0$ that is independent of parameter δ, such that

$$\|\Delta(j\omega)(I + \Delta(j\omega))^{-1}\| \leq 1 - \gamma.$$

Moreover, we get

$$\begin{aligned}\|(I+\Delta(j\omega))^{-1}\| &= \|I-\Delta(j\omega)(I+\Delta(j\omega))^{-1}\| \\ &\leq 1-\|\Delta(j\omega)(I+\Delta(j\omega))^{-1}\| \\ &\leq \gamma.\end{aligned}$$

Hence

$$\underline{\sigma}(I+\Delta(j\omega)) \geq \frac{1}{\gamma}.$$

On the other hand, from Property 3 in Lemma 10.13 with $\mu = \frac{\gamma}{2-2\gamma}$, we can immediately obtain that

$$\|I+G_\delta(j\omega)\| < 1+\frac{\gamma}{2-2\gamma}.$$

Then, it follows that, for $|\omega| < \omega_{\max}+\varpi$

$$\begin{aligned}\underline{\sigma}\,[I-\Delta(j\omega)G_\delta(j\omega)] &= \underline{\sigma}\,[I+\Delta(j\omega)-\Delta(j\omega)(I+G_\delta(j\omega))] \\ &= \underline{\sigma}(I+\Delta(j\omega))\underline{\sigma} \\ &\quad\times\left[I-(I+\Delta(j\omega))^{-1}\Delta(j\omega)(I+G_\delta(j\omega))\right] \\ &\geq \frac{1}{\gamma}\left(1-(1-\gamma)\left(1+\frac{\gamma}{2-2\gamma}\right)\right)=\frac{1}{2},\end{aligned}$$

for all $\kappa_i \in [0,\bar{\kappa}]$ and all possible D (note that $\bar{L} = I-D$) associated with a network graph in $\mathbb{G}_\beta^{u,N}$. Therefore, we have

$$\left\|(I-\Delta(j\omega)G_\delta(j\omega))^{-1}\right\| \leq 2, \tag{10.65}$$

for all $|\omega| < \omega_{\max}+\varpi$.

For all $|\omega| \geq \omega_{\max}+\varpi$, we know that $\Delta(j\omega)G_\delta(j\omega) \to 0$ as $\delta \to 0$ uniformly in ω. Therefore, for all sufficiently small δ, we also have (10.65) for $|\omega| \geq \omega_{\max}+\varpi$. This completes the proof of Theorem 10.12. ■

10.3.3 Protocol Design for MAS with Partial-State Coupling

Protocol design 10.4 *Consider a MAS described by (10.36) and (10.37) with input saturation.*

We consider the protocol

$$\begin{cases} \chi_i(k+1) = (A+KC)\chi_i(k) - K\zeta_i(k), \\ u_i(k) \quad\;\; = \rho F_\delta \chi_i(k), \end{cases} \tag{10.66}$$

where K is such that $A+KC$ is Schur stable and F_δ is designed as in (10.42), that is

$$F_\delta = -\frac{1}{1-\beta}(B'P_\delta B + I)^{-1}B'P_\delta A$$

with $P_\delta > 0$ is the unique solution of the discrete-time Riccati equation (10.43).

The main result in this subsection is stated as follows.

Theorem 10.16 *Consider a MAS described by (10.36) and (10.37) with an input delay upper bound $\bar{\kappa}$ and input saturation. Let any $0 < \beta < 1$ be given, and hence a set of network graphs $\mathbb{G}_\beta^{u,N}$ be defined.*

If (A, B) is stabilizable, (A, C) is detectable, and A is at most weakly unstable, then the semi-global state synchronization problem stated in Problem 10.10 with $\mathbf{G} = \mathbb{G}_\beta^{u,N}$ is solvable if condition (10.44) is satisfied. In particular, there exists an integer n, and for a priori given compact set of initial conditions $\mathcal{W} \subset \mathcal{L}_\infty^n(\bar{\kappa}) \times \mathbb{R}^n$, there exist $\rho > 0$ and $\delta^* > 0$, such that for this ρ and any $\delta \in (0, \delta^*]$, the protocol (10.66) achieves state synchronization for any graph $\mathcal{G} \in \mathbb{G}_\beta^{u,N}$, for any input delay $\kappa_i \in [0, \bar{\kappa}]$, and for any initial condition $(\phi_i, \psi) \in \mathcal{W}$ for $i = 1, \ldots, N$.

Again, we need the following technical lemmas before we proceed to the main result for MAS with partial-state coupling.

Lemma 10.17 *Consider*

$$\hat{A} = \begin{pmatrix} A & 0 \\ -KC & A+KC \end{pmatrix}, \quad \hat{B} = \begin{pmatrix} B \\ 0 \end{pmatrix}, \quad \hat{F}_\delta = \begin{pmatrix} 0 & F_\delta \end{pmatrix}.$$

Then, for any $0 < \beta < 1$, there exists a δ^* such that we have the following:

1. *The closed-loop system matrix $\hat{A} + (1-\lambda)\hat{B}\hat{F}_\delta$ is Schur stable for all $\delta \in [0, \delta^*]$ and all $|\lambda| < \beta$.*

2. *There exists a* $\hat{r}_{1,\delta}$ *with* $\hat{r}_{1,\delta} \to 0$ *as* $\delta \to 0$ *such that*

$$\left\| \hat{F}_\delta(\hat{A} + (1-\lambda)\hat{B}\hat{F}_\delta)^k \right\| \leq \hat{r}_{1,\delta} \tag{10.67}$$

for all $\delta \in [0, \delta^*]$, *for all* $k \geq 0$, *and all* $\lambda \in \mathbb{R}$ *with* $|\lambda| < \beta$.

3. *Let* $\hat{G}_\delta(z) = (1-\lambda)\hat{F}_\delta(zI - \hat{A} - (1-\lambda)\hat{B}\hat{F}_\delta)^{-1}\hat{B}$. *Then, for any* $\hat{\mu} > 0$, *there exists a* $\delta^* > 0$ *such that for all* $\delta \in (0, \delta^*]$

$$\left\| I + \hat{G}_\delta \right\|_\infty \leq 1 + \hat{\mu} \tag{10.68}$$

for all $\delta \in [0, \delta^*]$ *and all* $\lambda \in \mathbb{R}$ *with* $|\lambda| < \beta$.

Proof Since

$$(1-\lambda)\frac{1}{1-\beta} > (1-\beta)\frac{1}{1-\beta} = 1$$

for any $|\lambda| < \beta$, $\hat{A} + (1-\lambda)\hat{B}\hat{F}_\delta$ is Schur stable, according to [187, Lemma 5]. Note that a coefficient $\frac{1}{1-\beta}$ is added because it does not show up in the F_δ in [187, Lemma 5].

Next, the system with realization $(\hat{A}, \hat{B}, \hat{C})$ with input v and output z is given by

$$\begin{cases} \tilde{x}_1(k+1) = A\tilde{x}_1(k) + (1-\lambda)BF_\delta\tilde{x}_2(k) + Bv(k), \\ \tilde{x}_2(k+1) = (A+KC)\tilde{x}_2(k) - KC\tilde{x}_1(k), \\ \quad z(k) \quad = (1-\lambda)\begin{pmatrix} 0 & F_\delta \end{pmatrix}\tilde{x}_2(k). \end{cases} \tag{10.69}$$

Now, let $x_1 = \tilde{x}_1$ and $x_2 = \tilde{x}_2 - \tilde{x}_1$. Then, we have

$$\begin{cases} x_1(k+1) = (A + (1-\lambda)BF_\delta)x_1(k) + (1-\lambda)BF_\delta x_2(k) + Bv(k), \\ x_2(k+1) = -(1-\lambda)BF_\delta x_1(k) + (A + KC - (1-\lambda)BF_\delta)x_2(k) - Bv(k), \\ \quad z(k) \quad = F_\delta x_1(k) + F_\delta x_2(k). \end{cases} \tag{10.70}$$

Using arguments, which we first used in the proof of Lemma 4.11 for the continuous-time case, we can then show that $\|\hat{G}_\delta - G_\delta\|_\infty$ converges to zero as $\delta \to 0$ with G_δ as defined in Lemma 10.13. Hence, (10.68) can then be obtained from the result of Lemma 10.13. ∎

Proof of Theorem 10.16 Define

$$\tilde{\chi}_i(k+1) = (A+KC)\tilde{\chi}_i(k) - Ky_i(k)$$
$$v(k+1) = (A+KC)v(k)$$

with

$$v(0) = \frac{1}{N}\sum_{j=1}^{N} \chi_i(0)$$

and

$$\begin{pmatrix} \chi_1(0) \\ \vdots \\ \chi_N(0) \end{pmatrix} = L \begin{pmatrix} \tilde{\chi}_1(0) \\ \vdots \\ \tilde{\chi}_N(0) \end{pmatrix} + \begin{pmatrix} v(0) \\ \vdots \\ v(0) \end{pmatrix}.$$

Then it is easily verified that

$$\chi_i = \sum_{j=1}^{N} \ell_{ij}\tilde{\chi}_j + v.$$

Since $A + KC$ is asymptotically stable and $F_\delta \to 0$ as $\delta \to 0$, we have that for any $\mu > 0$

$$s_\delta = \rho F_\delta v$$

is such that $\|s_\delta\|_2 \leq \mu$ and $\|s_\delta\|_\infty \leq \mu$ for any initial condition in the given compact set as $\delta \to 0$.

Assuming that the saturation does not get activated, the closed-loop system of the agent (10.36) and the protocol (10.66) can be written as

$$\begin{cases} x_i(k+1) = Ax_i(k) + B(\mathbf{D}_i\tilde{u}_i)(k) + B(\mathbf{D}_i s)(k), \\ \tilde{\chi}_i(k+1) = (A + KC)\tilde{\chi}_i(k) - KCx_i(k), \\ \quad \tilde{u}_i(k) \quad = \rho F_\delta \sum_{j=1}^{N} \bar{\ell}_{ij}\tilde{\chi}_j. \end{cases} \tag{10.71}$$

Note that

$$u_i = \tilde{u}_i + s_\delta. \tag{10.72}$$

Let

$$\hat{x}_i = \begin{pmatrix} \hat{x}_{i,x} \\ \hat{x}_{i,\chi} \end{pmatrix} = \begin{pmatrix} x_i \\ \tilde{\chi}_i \end{pmatrix} - \begin{pmatrix} x_N \\ \tilde{\chi}_N \end{pmatrix},$$

and

$$\hat{A} = \begin{pmatrix} A & 0 \\ -KC & A + KC \end{pmatrix}, \quad \hat{B} = \begin{pmatrix} B \\ 0 \end{pmatrix}, \quad \hat{F}_\delta = \begin{pmatrix} 0 & F_\delta \end{pmatrix}.$$

Then, the overall dynamics of $\hat{x} = \text{col}\{\hat{x}_1, \ldots, \hat{x}_{N-1}\}$ without saturation is as follows:

$$\hat{x}(k+1) = (I_N \otimes \hat{A})\hat{x}(k) + \rho(T_1\mathbf{D}LT_2 \otimes \hat{B}\hat{F}_\delta)\hat{x}(k) + \hat{B}\tilde{s}(k) \tag{10.73}$$

with

$$\tilde{s} = (T_1\mathbf{D}\mathbf{1} \otimes I)s.$$

Since $\tau\omega_{\max} < \frac{\pi}{2}$, we can choose ρ such that

$$\rho\cos(\tau\omega_{\max}) > 1. \tag{10.74}$$

Note that this ρ is independent of low-gain parameter δ. Let this ρ be fixed during the remaining proof. We can now use similar arguments as used in the proof of Theorem 10.12 with the bounds we obtained in Lemma 10.17 and the fact that $\|\tilde{s}_\delta\|_2 \to 0$ and $\|\tilde{s}_\delta\|_\infty \to 0$ as $\delta \to 0$.

This yields that $\hat{x} \to 0$ and $\|\tilde{u}\|_\infty|$ is arbitrarily small for sufficiently small δ. Given (10.72), $\tilde{u}$ is arbitrary small given that both $\tilde{x}$ and $\tilde{s}_\delta$ are arbitrarily small for all sufficiently small δ. This guarantees that the saturation does not get activated for all sufficiently small δ. Since we already established synchronization for the system without saturation, this completes the proof. ∎

Chapter 11
Synchronization of Continuous-Time Linear Time-Varying MAS

11.1 Introduction

This chapter considers synchronization problems for homogeneous linear continuous-time MAS with time-varying communication networks.

The extension from fixed networks to time-varying networks is generally done in the framework of switching, using the concepts of dwell-time and average dwell-time. A critical assumption in most literature is that the network switches among a finite set of network graphs. For example, see [79] and [91] (full-state coupling) and [118, 134, 146, 172] (partial-state coupling). Also, the time-varying network can be piecewise constant and frequently connected (e.g., [149, 171]), uniformly connected (e.g., [114]), and uniformly connected on average (e.g., [44]). In the case of partial-state coupling, restrictions are always imposed on the agent dynamics. That is, the poles of agent dynamics should be in the closed left half complex plane (see [44, 114]) or the zeros of agent dynamics should be in the closed left half complex plane (see [172]).

We consider two kinds of time-varying networks. One is in the framework of switching, using the concept of dwell-time. In this context, the time-varying network is piecewise constant and strongly connected. The dwell-time indicates the minimum amount between two successive switches in the network. We consider arbitrarily small dwell-time and allow switching in sets of graphs which need not be finite. We show how to design purely decentralized protocols for MAS with switching directed graphs. Multiple Lyapunov functions are used for the stability analysis.

The other case is a general time-varying network that includes time-varying weights which can be varying continuously or in the form of switches. In this case, we do not impose a minimum dwell-time on the switches. We will show how a purely decentralized protocol can be designed for undirected networks. We resort to techniques from robust control to obtain appropriate protocols.

A. Saberi et al., *Synchronization of Multi-Agent Systems in the Presence of Disturbances and Delays*, Systems & Control: Foundations & Applications,
https://doi.org/10.1007/978-3-030-88148-1_11

11.2 Switching Graphs

In this section, we consider switching directed graphs. The time-varying network switches within an infinite set of graphs with some, a priori given, properties and a minimum dwell-time. Multiple Lyapunov functions are used to analyze the stability of the switching systems.

Both full-state coupling and partial-state coupling are considered. In the case of full-state coupling, a high gain-based method is used for the protocol design. In the case of partial-state coupling, a low- and high-gain methodology is utilized for a purely decentralized protocol design given a MAS with minimum-phase agents. When an additional communication channel is allowed, a high-gain theory can be used for protocol design for a MAS even in the case of non-minimum-phase agents.

11.2.1 Problem Formulation

Consider a MAS composed of N identical linear time-invariant agents of the form

$$\begin{aligned} \dot{x}_i &= Ax_i + Bu_i, \\ y_i &= Cx_i, \end{aligned} \qquad (i = 1, \dots, N) \tag{11.1}$$

where $x_i \in \mathbb{R}^n$, $u_i \in \mathbb{R}^m$, $y_i \in \mathbb{R}^p$ are, respectively, the state, input, and output vectors of agent i.

The communication network provides each agent with a linear combination of its own output relative to that of other neighboring agents. In particular, each agent $i \in \{1, \dots, N\}$ has access to the quantity:

$$\zeta_i(t) = \sum_{j=1}^{N} a_{ij}(t)(y_i(t) - y_j(t)), \tag{11.2}$$

where $a_{ij}(t) \geq 0$ and $a_{ii}(t) = 0$ are a piecewise constant and right-continuous function of time t, indicating time-varying communication among agents. This time-varying communication topology of the network can be described by a weighted, time-varying graph $\mathcal{G}_t$ with nodes corresponding to the agents in the network and the weight of edges at time t given by the coefficient $a_{ij}(t)$. Specifically, $a_{ij}(t)$ indicates that, at time t there is an edge with a weight a_{ij} in the graph from agent j to agent i. The Laplacian matrix associated with $\mathcal{G}_t$ is defined as $L_t = [\ell_{ij}(t)]$. In terms of the coefficients of L_t, ζ_i can be rewritten as

$$\zeta_i(t) = \sum_{j=1}^{N} \ell_{ij}(t) y_j(t). \tag{11.3}$$

In the case of full-state coupling, i.e., $C = I$, we have

$$\zeta_i(t) = \sum_{j=1}^{N} a_{ij}(t)(x_i(t) - x_j(t)) = \sum_{j=1}^{N} \ell_{ij}(t)x_j(t). \tag{11.4}$$

We need the following assumption on the network graph.

Assumption 11.1 *At each time t, the graph $\mathcal{G}_t(t)$ describing the communication topology of the network contains a directed spanning tree.*

Based on the above Assumption 11.1, it then follows from [91, Lemma 3.3] that the Laplacian matrix L_t at time t has a simple eigenvalue at the origin, with the corresponding right eigenvector **1** and all the other eigenvalues are in the open right half complex plane. Let $\lambda_{t,1}, \ldots, \lambda_{t,N}$ denote the eigenvalues of L_t such that $\lambda_{t,1} = 0$ and $\text{Re}(\lambda_{t,i}) > 0$, $i = 2, \ldots, N$.

Next, we define a set of time-varying graphs based on some rough information of the graph.

Definition 11.1 For any given real numbers $\alpha, \beta, \tau > 0$, and a positive integer N, let the set of fixed graphs $\mathbb{G}^N_{\alpha,\beta}$ be defined by Definition 1.7. The set $\mathbb{G}^{\tau,N}_{\alpha,\beta}$ is defined as the set of all time-varying graphs $\mathcal{G}_t$ for which

$$\mathcal{G}_t(t) = \mathcal{G}_{\sigma(t)} \in \mathbb{G}^N_{\alpha,\beta}$$

for all $t \in \mathbb{R}$, where $\sigma : \mathbb{R} \to \mathcal{L}_N$ is a piecewise constant, right-continuous function with minimal dwell-time τ, i.e., if σ has discontinuities in t_1 and t_2 then $|t_2 - t_1| > \tau$.

Remark 11.2 Note that the minimal dwell-time is assumed to avoid chattering problems. However, our design is able to obtain a suitable protocol for arbitrarily small $\tau > 0$.

We then formulate two state synchronization problems for a switching network with full-state coupling and partial-state coupling, respectively.

Problem 11.3 (Full-State Coupling) Consider a MAS described by (11.1) and (11.4). For any real numbers $\alpha, \beta, \tau > 0$ and a positive integer N, the *state synchronization* problem via full-state coupling with a set of time-varying network graphs $\mathbb{G}^{\tau,N}_{\alpha,\beta}$ is to find, if possible, a linear static protocol of the form:

$$u_i = F\zeta_i, \tag{11.5}$$

such that, for any time-varying graph $\mathcal{G}_t \in \mathbb{G}^{\tau,N}_{\alpha,\beta}$ and for all the initial conditions of agents, state synchronization among agents can be achieved.

Problem 11.4 (Partial-State Coupling) Consider a MAS described by (11.1) and (11.2). For any real numbers $\alpha, \beta, and \tau > 0$ and a positive integer N, the *state synchronization* problem via partial-state coupling with a set of time-varying network graphs $\mathbb{G}_{\alpha,\beta}^{\tau,N}$ is to find, if possible, a linear time-invariant dynamic protocol of the form

$$\begin{cases} \dot{\chi}_i = A_c\chi_i + B_c\zeta_i, \\ u_i = C_c\chi_i, \end{cases} \tag{11.6}$$

where $\chi_i \in \mathbb{R}^{n_c}$, such that, for any time-varying graph $\mathcal{G}_t \in \mathbb{G}_{\alpha,\beta}^{\tau,N}$ and for all the initial conditions of agents and their protocol, state synchronization among agents can be achieved.

As noted at the beginning of this chapter, some papers looked at time-varying networks which are frequently connected or connected on average (e.g., [44, 149, 171]). The next example shows that if the system is only connected on average, then this imposes additional restrictions on the dynamics of the agents. Basically, it requires that the agents are neutrally stable.

Example 11.5 Consider three agents whose dynamics is given by

$$\dot{x}_i = x_i + u_i, \qquad i = 1, 2, 3,$$

where the time-varying network has a Laplacian matrix which switches every T seconds between the following two matrices:

$$L_1 = \begin{pmatrix} 1 & 0 & -1 \\ 0 & 0 & 0 \\ -1 & 0 & 1 \end{pmatrix}, \qquad L_2 = \begin{pmatrix} 0 & 0 & 0 \\ 0 & 1 & -1 \\ 0 & -1 & 1 \end{pmatrix}.$$

In other words

$$L_t = L_1 \text{ for } t \in [2kT, (2k+1)T) \text{ and } L_t = L_2 \text{ for } t \in [(2k+1)T, (2k+2)T).$$

Clearly this system is not connected at any moment in time but the system is connected on average since

$$L_1 + L_2 = \begin{pmatrix} 1 & 0 & -1 \\ 0 & 1 & -1 \\ -1 & -1 & 2 \end{pmatrix}$$

which is associated to a strongly connected network.

If we define $\bar{x}_1 = x_1 - x_3$ and $\bar{x}_2 = x_2 - x_3$ and we use a protocol

$$u = f\zeta_i,$$

then the overall system switches between

$$\dot{\bar{x}}_1 = \bar{x}_1 - 2f\bar{x}_1$$
$$\dot{\bar{x}}_2 = \bar{x}_2 - f\bar{x}_1,$$

and

$$\dot{\bar{x}}_1 = \bar{x}_1 - f\bar{x}_2$$
$$\dot{\bar{x}}_2 = \bar{x}_2 - 2f\bar{x}_2.$$

We get

$$\begin{pmatrix} \bar{x}_1 \\ \bar{x}_2 \end{pmatrix}(2kT) = e^{A_2T}e^{A_1T}\begin{pmatrix} \bar{x}_1 \\ \bar{x}_2 \end{pmatrix}(2k-2)T)$$

with

$$A_1 = \begin{pmatrix} 1-2f & 0 \\ -f & 1 \end{pmatrix}, \qquad A_2 = \begin{pmatrix} 1 & -f \\ 0 & 1-2f \end{pmatrix}.$$

It is easy to verify that

$$\text{trace}\, e^{A_2T}e^{A_1T} > \tfrac{1}{4}e^{2T}$$

for any choice for f. But then for $T > \ln 2$, we have that the trace is larger than 2 which implies that the absolute value of at least one eigenvalue is larger than 1. This clearly shows that we do not achieve synchronization.

One approach around the issue presented in the above example is to consider the case of fast switching; see, for instance, [44, 123]. As noted before, the other approach is to consider neutrally stable agents. For instance, [72] basically considers agents which are single integrators and hence neutrally stable.

11.2.2 Protocol Design for MAS with Full-State Coupling

In this section, we consider a protocol design for MAS with full-state coupling.

According to [94, Theorem 1], if (A, B) is controllable and B has full-column rank, then there exist nonsingular transformation matrices $T_x \in \mathbb{R}^{n\times n}$ and $T_u \in \mathbb{R}^{m\times m}$ such that

$$x = T_x\hat{x}, \quad u = T_u\hat{u},$$

and the dynamics of x is represented as

$$\dot{\hat{x}} = \hat{A}\hat{x} + \hat{B}\hat{u}, \tag{11.7}$$

where

$$\hat{A} = \begin{pmatrix} A_{1d} & 0 & \cdots & 0 \\ 0 & A_{2d} & \ddots & \vdots \\ \vdots & \ddots & \ddots & 0 \\ 0 & \cdots & 0 & A_{md} \end{pmatrix} + \begin{pmatrix} B_{1d} & 0 & \cdots & 0 \\ 0 & B_{2d} & \ddots & \vdots \\ \vdots & \ddots & \ddots & 0 \\ 0 & \cdots & 0 & B_{md} \end{pmatrix} E \triangleq A_d + B_d E, \tag{11.8}$$

and

$$\hat{B} = \begin{pmatrix} B_{1d} & 0 & \cdots & 0 \\ 0 & B_{2d} & \ddots & \vdots \\ \vdots & \ddots & \ddots & 0 \\ 0 & \cdots & 0 & B_{md} \end{pmatrix} \triangleq B_d,$$

with

$$A_{jd} = \begin{pmatrix} 0 & I_{\rho_j - 1} \\ 0 & 0 \end{pmatrix}, \quad B_{jd} = \begin{pmatrix} 0 \\ 1 \end{pmatrix}, \quad E = \begin{pmatrix} E_{11} & \dots & E_{1m} \\ \vdots & \ddots & \vdots \\ E_{m1} & \dots & E_{mm} \end{pmatrix}.$$

Note that $\sum_{j=1}^{m} \rho_j = n$. Using a high-gain method, we give a protocol design for the MAS.

We choose $\hat{F}_j$ and $P_j > 0$ such that

$$\hat{F}_j = B'_{jd} P_j, \qquad A'_{jd} P_j + P_j A_{jd} - 2\beta P_j B_{jd} B'_{jd} P_j + I_{\rho_j} = 0 \tag{11.9}$$

or

$$\hat{F}_j = \mu \begin{pmatrix} -F_{j1} & -1 \end{pmatrix} \text{ and } P_j = \begin{pmatrix} P_{j1} + F'_{j1} F_{j1} & -F'_{j1} \\ -F_{j1} & 1 \end{pmatrix} \tag{11.10}$$

for μ sufficiently large where

$$A_{jd} = \begin{pmatrix} A_{j11} & A_{j12} \\ 0 & 0 \end{pmatrix}$$

with $A_{j11} \in \mathbb{R}^{(\rho_j-1)\times(\rho_j-1)}$ while F_{j1} and $P_{j1} > 0$ are such that $A_{j11} + A_{j12}F_{j1}$ is asymptotically stable and

$$(A_{j11} + A_{j12}F_{j1})'P_{j1} + P_{j1}(A_{j11} + A_{j12}F_{j1}) + I = 0.$$

Both for (11.9) and (11.10), we have

$$(A_{jd} + \lambda_j B_{jd}\hat{F}_j)'P_j + P_j(A_{jd} + \lambda_j B_{jd}\hat{F}_j) \leq -I, \tag{11.11}$$

for any λ_j such that $\operatorname{Re}\lambda > \beta$ for $j = 1, \ldots m$. Note that for the second design method for $\hat{F}_j$, i.e., (11.10), we need to choose μ sufficiently large (but independent of the λ_i) to guarantee that P_j satisfies (11.11). The proof that this is possible follows the arguments presented in the proof of Theorem 2.13.

Protocol design 11.1 *Consider a MAS described by* (11.4) *and* (11.1). *Let* T_x *and* T_u *be such that the transformation results in a system of the form* (11.7).

For $j = 1, \ldots m$, *choose* $\hat{F}_j$ *and* P_j *via either* (11.9) *or* (11.10) *(with* μ *sufficiently large) such that* (11.11) *is satisfied. Let* $\hat{F} = \operatorname{diag}(\hat{F}_1, \ldots, \hat{F}_m)$.

Define

$$D_\varepsilon = \begin{pmatrix} \varepsilon^{-\rho_1} & 0 & \cdots & 0 \\ 0 & \varepsilon^{-\rho_2} & \ddots & \vdots \\ \vdots & \ddots & \ddots & 0 \\ 0 & \cdots & 0 & \varepsilon^{-\rho_m} \end{pmatrix}, \quad S_\varepsilon = \begin{pmatrix} S_{1\varepsilon} & 0 & \cdots & 0 \\ 0 & S_{2\varepsilon} & \ddots & \vdots \\ \vdots & \ddots & \ddots & 0 \\ 0 & \cdots & 0 & S_{m\varepsilon} \end{pmatrix} \tag{11.12}$$

with $S_{j\varepsilon} = \operatorname{diag}(1, \ldots, \varepsilon^{\rho_j-1})$. *Then, the static protocol can be designed as*

$$u_i = T_u D_\varepsilon \hat{F} S_\varepsilon T_x^{-1} \zeta_i. \tag{11.13}$$

The main result based on the above design is stated as follows.

Theorem 11.6 *Consider a MAS described by* (11.1) *and* (11.4). *Let any real numbers* $\alpha, \beta, \tau > 0$ *and a positive integer* N *be given.*

If (A, B) *is controllable, then the state synchronization problem stated in Problem 11.3 is solvable. In particular, the protocol* (11.13) *with* ε *sufficiently small achieves state synchronization for any time-varying graph* $\mathcal{G}_t \in \mathbb{G}_{\alpha,\beta}^{\tau,N}$.

Before we prove the above theorem we need a preliminary lemma.

Lemma 11.7 *The matrix*

$$\tilde{A}_t = (I_{N-1} \otimes A_d) + (U_t \otimes B_d \hat{F})$$

is asymptotically stable for any upper triangular matrix $U_t \in \mathbb{C}^{(N-1)\times(N-1)}$ *with* $\|U_t\| < \tilde{\alpha}$ *whose eigenvalues satisfy* $\mathrm{Re}(\lambda_i) > \beta$ *for all* $i = 1, \dots, N-1$. *Moreover, there exist a* $\tilde{P} > 0$ *and a small enough* $\mu > 0$ *such that*

$$\tilde{A}_t'\tilde{P} + \tilde{P}\tilde{A}_t \le -\mu\tilde{P} - I \tag{11.14}$$

is satisfied for all possible upper triangular matrices U_t.

Proof We have $P_j > 0$ satisfying (11.11) for $j = 1, \dots m$. We define

$$\tilde{A}_{t,i} = A_d + \lambda_i B_d \hat{F}$$

and

$$\tilde{B} = B_d \hat{F}.$$

Then, we have

$$\tilde{A}_{t,i}' P + P\tilde{A}_{t,i} \le -I,$$

where

$$P = \mathrm{diag}(P_1, \dots, P_m).$$

We define, for $k = 1, \dots, N-1$,

$$\bar{P}_k = \begin{pmatrix} \gamma^{i_1}P & 0 & \cdots & 0 \\ 0 & \ddots & \ddots & \vdots \\ \vdots & \ddots & \ddots & 0 \\ 0 & \cdots & 0 & \gamma^{i_k}P \end{pmatrix}, \qquad \bar{A}_k = \begin{pmatrix} \tilde{A}_{t,1} & \mu_{1,2}\tilde{B} & \cdots & \mu_{1,k}\tilde{B} \\ 0 & \ddots & \ddots & \vdots \\ \vdots & \ddots & \ddots & \mu_{k-1,k}\tilde{B} \\ 0 & \cdots & 0 & \tilde{A}_{t,k} \end{pmatrix},$$

where λ_i are eigenvalues of U_t and $\mu_{i,j} = [U_t]_{ij}$ for $j > i$ are the bounded upper triangular elements of U_t.

We will next use a recursive argument. For $k = 1$ and $\gamma^{i_1} = N$, we have

$$\bar{A}_1'\bar{P}_1 + \bar{P}_1\bar{A}_1 = \gamma^{i_1}(\tilde{A}_{t,1}'P + P\tilde{A}_{t,1}) < -NI.$$

Assume that for $k = j$ we have $\gamma^{i_1}, \dots \gamma^{i_j}$ such that

$$A_{11} = \bar{A}_j'\bar{P}_j + \bar{P}_j\bar{A}_j < -(N - j + 1)I \tag{11.15}$$

We will show that for $k = j + 1$ there exists $\gamma^{i_{j+1}}$ such that

$$\bar{A}_{j+1}'\bar{P}_{j+1} + \bar{P}_{j+1}\bar{A}_{j+1} < -(N - j)I. \tag{11.16}$$

Note that

$$\bar{A}'_{j+1}\bar{P}_{j+1} + \bar{P}_{j+1}\bar{A}_{j+1} = \begin{pmatrix} A_{11} & A_{12} \\ A'_{12} & -\gamma^{i_{j+1}} I \end{pmatrix}$$

where A_{11} is defined by (11.15) while A_{12} is given by

$$A_{12} = \begin{pmatrix} \gamma^{i_1}\mu_{1,j+1} P\tilde{B} \\ \vdots \\ \gamma^{i_j}\mu_{j,j+1} P\tilde{B} \end{pmatrix}$$

Note that the coefficients $\mu_{1,j+1}$ are unknown but bounded since the norm of U is bounded and hence there exists a M such that $\|A_{12}\| < M$. Via Schur complement, it is easy to verify that for the given bound M, there exists a $\gamma^{i_{j+1}}$ sufficiently large such that the matrix

$$\begin{pmatrix} A_{11} & A_{12} \\ A'_{12} & -\mu I \end{pmatrix} < -\gamma^{i_{j+1}} I$$

for all matrices A_{11} and A_{12} such that $A_{11} < -(N - j + 1)I$ and $\|A_{12}\| < M$. This guarantees that (11.16) is satisfied.

Using a recursive argument, we find that there exist $\gamma^{i_1}, \ldots, \gamma^{i_{N-1}}$ such that

$$\tilde{A}'_t \bar{P}_{N-1} + \bar{P}_{N-1}\tilde{A}_t \leq -2I,$$

since $\tilde{A}_t = \bar{A}_{N-1}$. This obviously implies that for μ small enough we have (11.14) for $\tilde{P} = \bar{P}_{N-1}$. ■

Proof of Theorem 11.6 The interconnection of the static protocol (11.5) with

$$F = T_u D_\varepsilon \hat{F} S_\varepsilon T_x^{-1}$$

and the agent (11.1) is written as

$$\dot{x}_i = Ax_i + BF\zeta_i.$$

Define $\bar{x}_i = x_N - x_i$ and $\bar{\ell}_{ij}(t) = \ell_{ij}(t) - \ell_{Nj}(t)$ for $i \in \{1, \ldots, N-1\}$. Then, we get

$$\dot{\bar{x}}_i = A\bar{x}_i + BF\sum_{j=1}^{N-1}\bar{\ell}_{ij}(t)\bar{x}_j \tag{11.17}$$

for $i = 1, \ldots, N-1$. We define $\bar{L} \in \mathbb{R}^{(N-1)\times(N-1)}$ such that $[\bar{L}_t]_{ij} = \bar{\ell}_{ij}$. We have $\bar{L}_t = T_1 L_t T_2$ where T_1 and T_2 are given by (6.91). In the proof of Theorem 6.39, it was shown that the eigenvalues of $\bar{L}_t$ are equal to the nonzero eigenvalues of L_t. Moreover, it is easy to check that $\|L_t\| < \alpha$ implies that $\|\bar{L}_t\| \leq \bar{\alpha} = \sqrt{N}\alpha$.

Next, we prove that the system (11.17) is asymptotically stable for any switching graph $\mathcal{G}_t \in \mathbb{G}_{\alpha,\beta}^{\tau,N}$, which immediately implies that

$$\lim_{t\to\infty} (x_i(t) - x_N(t)) = 0 \quad \text{for } i = 1, \ldots, N-1$$

under any time-varying graph $\mathcal{G}_t \in \mathbb{G}_{\alpha,\beta}^{\tau,N}$. Let

$$\bar{x} = \begin{pmatrix} \bar{x}_1 \\ \vdots \\ \bar{x}_{N-1} \end{pmatrix}.$$

The error dynamics of the whole network can be written as

$$\dot{\bar{x}} = (I_{N-1} \otimes A)\bar{x} + (\bar{L}_t \otimes BF)\bar{x}. \tag{11.18}$$

Define $\bar{U}_t^{-1}\bar{L}_t\bar{U}_t = \bar{S}_t$, where $\bar{S}_t$ is the Schur form of $\bar{L}_t$. In particular, the diagonal elements of $\bar{S}_t$ are the nonzero eigenvalues of the Laplacian matrix $\bar{L}_t$ at time t while $\|\bar{S}_t\| \leq \bar{\alpha}$. Moreover, $\bar{U}_t$ is unitary. Let $\tilde{x}(t) = (\bar{U}_t^{-1} \otimes I)\bar{x}(t)$. Then

$$\dot{\tilde{x}} = (I_{N-1} \otimes A)\tilde{x} + (\bar{S}_t \otimes BF)\tilde{x}. \tag{11.19}$$

Now let $\xi = (I_{N-1} \otimes T_x^{-1})\tilde{x}$. Then

$$\dot{\xi} = (I_{N-1} \otimes \hat{A})\xi + (\bar{S}_t \otimes T_x^{-1}BFT_x)\xi. \tag{11.20}$$

Since $B = T_x\hat{B}T_u^{-1}$ and $F = T_u D_\varepsilon \hat{F} S_\varepsilon T_x^{-1}$, we have $T_x^{-1}BFT_x = \hat{B}D_\varepsilon\hat{F}S_\varepsilon$. Moreover, $\hat{A} = A_d + B_d E$ and $B_d = \hat{B}$, and therefore the dynamics (11.20) can be rewritten as

$$\dot{\xi} = (I_{N-1} \otimes (A_d + B_d E))\xi + (\bar{S}_t \otimes B_d D_\varepsilon \hat{F} S_\varepsilon)\xi. \tag{11.21}$$

Let

$$G_\varepsilon = \begin{pmatrix} \varepsilon^{\rho_m-\rho_1} I_{\rho_1} & 0 & \cdots & 0 \\ 0 & \varepsilon^{\rho_m-\rho_2} I_{\rho_2} & \ddots & \vdots \\ \vdots & \ddots & \ddots & 0 \\ 0 & \cdots & 0 & I_{\rho_m} \end{pmatrix} \tag{11.22}$$

and define $v = (I_{N-1} \otimes S_\varepsilon G_\varepsilon)\xi$. Since $S_{j\varepsilon} A_{jd} S_{j\varepsilon}^{-1} = \varepsilon^{-1} A_{jd}$ and $S_{j\varepsilon} B_{jd} \varepsilon^{\rho_m - \rho_j} = \varepsilon^{\rho_m - 1} B_{jd}$, the dynamics of v can be written as

$$\varepsilon \dot{v} = \left((I_{N-1} \otimes A_d) + (\bar{S}_t \otimes B_d \hat{F})\right) v + (I_{N-1} \otimes B_d E D_\varepsilon^{-1} S_\varepsilon^{-1}) v \tag{11.23}$$

Assume the network graph switches at $t_1, t_2, t_3, \ldots$ with $t_{k+1} - t_k > \tau$, and, for ease of presentation, we set $t_0 = 0$.

Note that $\tilde{x}$ can experience discontinuous jumps when the network graph switches which implies that also v can experience discontinuous jumps when the network graph switches. Denote $W_\varepsilon = I_{N-1} \otimes B_d E D_\varepsilon^{-1} S_\varepsilon^{-1}$, which is $O(\varepsilon)$, and let

$$\tilde{A}_t = (I_{N-1} \otimes A_d) + (\bar{S}_t \otimes B_d \hat{F}).$$

Then we get

$$\varepsilon \dot{v} = \tilde{A}_t v + W_\varepsilon v. \tag{11.24}$$

Using Lemma 11.7, we find $\tilde{P}$ such that (11.14) is satisfied for all possible upper triangular matrices U_t. Define a Lyapunov function $V = \varepsilon v' \tilde{P} v$. It is easy to see that V can have discontinuous jumps when the network graph changes. The derivative of V is bounded by

$$\begin{aligned}
\dot{V} &\leq -\mu \varepsilon^{-1} V - \|v\|^2 + 2\,\mathrm{Re}(v' \tilde{P} W_\varepsilon v) \\
&\leq -\mu \varepsilon^{-1} V - \|v\|^2 + \varepsilon r \|v\|^2 \\
&\leq -\mu \varepsilon^{-1} V,
\end{aligned}$$

for a small enough ε. In the above second inequality, choose r such that $\varepsilon r \geq 2\|\tilde{P} W_\varepsilon\|$.

By integrating both sides, we have

$$V(t_k^-) \leq e^{-\mu \varepsilon^{-1}(t_k - t_{k-1})} V(t_{k-1}^+).$$

There is a potential jump in V at time t_{k-1}. However, we have $V(t_{k-1}^+) \leq m V(t_{k-1}^-)$, where

$$m = \frac{\lambda_{\max}(\tilde{P})}{\lambda_{\min}(\tilde{P})}.$$

Using the fact that $t_k - t_{k-1} > \tau$, there exists a small enough ε such that

$$V(t_k^-) \leq e^{-\mu(t_k - t_{k-1})} V(t_{k-1}^-).$$

Combining these time intervals, we get

$$V(t_k^-) \leq e^{-\mu t_k} V(0).$$

Assuming $t_{k+1} > t > t_k$, we have

$$V(t) \leq m e^{-\mu t} V(0).$$

Hence, $\lim_{t\to\infty} v(t) = 0$ under a time-varying network graph. Since

$$\bar{x}(t) = (\bar{U}_t \otimes T_x G_\varepsilon^{-1} S_\varepsilon^{-1}) v(t),$$

while $\bar{U}_t$ is unitary, we obtain that

$$\lim_{t\to\infty} \bar{x}(t) = 0,$$

which proves the result. ■

11.2.3 Protocol Design for MAS with Partial-State Coupling

In this section, a protocol is designed for linear agents in partial-state coupling networks. When no communication channel is considered, agents are required to be minimum-phase; otherwise, agents can be non-minimum-phase.

11.2.3.1 Minimum-Phase Agents

In this section, we make an extra assumption regarding the agent dynamics given in (11.1).

Assumption 11.2 *(A, B, C) is minimum-phase and invertible with uniform relative degree ρ.*

Remark 11.8 Note that if the system (A, B, C) is minimum-phase and right-invertible, then there always exists a dynamic pre-compensator which results in an agent which is minimum-phase and invertible with uniform relative degree ρ.

We show next that the protocol designed for the fixed network also works for time-varying networks. Given Assumption 11.2, we can write the agent system (11.1) in the SCB form as described in (23.27):

$$\begin{aligned}
\dot{\tilde{x}}_{ia} &= A_a \tilde{x}_{ia} + L_{ad} y_i, \\
\dot{\tilde{x}}_{id} &= A_d \tilde{x}_{id} + B_d(\tilde{u}_i + E_{da}\tilde{x}_{ia} + E_{dd}\tilde{x}_{id}), \\
y_i &= C_d \tilde{x}_{id},
\end{aligned} \tag{11.25}$$

where A_a is asymptotically stable, while A_d, B_d, and C_d are of the form

$$A_d = \begin{pmatrix} 0 & I_{m(\rho-1)} \\ 0 & 0 \end{pmatrix}, \quad B_d = \begin{pmatrix} 0 \\ I_m \end{pmatrix}, \quad C_d = \begin{pmatrix} I_m & 0 \end{pmatrix}.$$

Protocol design 11.2 *We choose an observer gain K such that $A_d - KC_d$ is Hurwitz stable. Next, we consider a feedback gain:*

$$F_\delta = -B_d' P_\delta$$

with $P_\delta > 0$ being the unique solution of the continuous-time algebraic Riccati equation:

$$P_\delta A_d + A_d' P_\delta - \beta P_\delta B_d B_d' P_\delta + \delta I = 0, \tag{11.26}$$

where $\delta > 0$ is a low-gain parameter and $\beta > 0$ is a lower bound on the real part of the nonzero eigenvalues of the Laplacian matrix.

We consider a protocol of the form

$$\begin{aligned} \dot{\hat{x}}_{ia} &= A_a \hat{x}_{ia} + L_{ad} C_d \hat{x}_{id}, \\ \dot{\hat{x}}_{id} &= A_d \hat{x}_{id} + B_d (E_{da} \hat{x}_{ia} + E_{dd} \hat{x}_{id}) + K_\varepsilon (\zeta_i - C_d \hat{x}_{id}), \\ \tilde{u}_i &= F_{\delta\varepsilon} \hat{x}_{id}, \end{aligned} \tag{11.27}$$

where

$$F_{\delta\varepsilon} = \varepsilon^{-\rho} F_\delta S_\varepsilon, \quad K_\varepsilon = \varepsilon^{-1} S_\varepsilon^{-1} K,$$

where ρ is the relative degree of the system, while S_ε is a high-gain scaling matrix defined by

$$S_\varepsilon = \begin{pmatrix} I & 0 & \cdots & 0 \\ 0 & \varepsilon I & \ddots & \vdots \\ \vdots & \ddots & \ddots & 0 \\ 0 & \cdots & 0 & \varepsilon^{\rho-1} I \end{pmatrix}$$

with $\varepsilon > 0$ being a high-gain parameter.

The main result is in the following theorem.

Theorem 11.9 *Consider a multi-agent system described by (11.1) and (11.2) where (A, B) is controllable and (C, A) is observable. Let any real numbers α, β, and $\tau > 0$ and a positive integer N be given.*

Under Assumptions 11.1 and 11.2, the state synchronization problem stated in Problem 11.4 is solvable. In particular, controller (11.27) solves the state synchronization problem under any time-varying graph $\mathcal{G}_t \in \mathbb{G}_{\alpha,\beta}^{\tau,N}$.

We need the following lemma for the proof of the above theorem.

Lemma 11.10 *Consider the matrix*

$$\tilde{A}_{\delta,t} = I_{N-1} \otimes \begin{pmatrix} A_d & 0 \\ 0 & A_d - KC_d \end{pmatrix} + S_t \otimes \begin{pmatrix} B_d F_\delta & -B_d F_\delta \\ B_d F_\delta & -B_d F_\delta \end{pmatrix}.$$

For any small enough δ*, the matrix* $\tilde{A}_{\delta,t}$ *is asymptotically stable for any upper triangular matrix* $S_t \in \mathbb{R}^{(N-1)\times(N-1)}$ *with* $\|S_t\| < \tilde{\alpha}$ *whose eigenvalues satisfy* $\operatorname{Re}\{\lambda_i\} > \beta$ *for all* $i = 1, \ldots, N-1$*. Moreover, there exist* $P_\delta > 0$ *and a small enough* $\nu > 0$ *such that*

$$\tilde{A}_{\delta,t} P_\delta + P_\delta \tilde{A}_{\delta,t}' \leq -\nu P_\delta - 4I \tag{11.28}$$

is satisfied for all possible upper triangular matrices S_t *and such that there exists a* $P_a > 0$ *for which*

$$P_a A_a + A_a' P_a = -\nu P_a - I. \tag{11.29}$$

Proof First note that if ν is small enough such that $A_a + \frac{\nu}{2} I$ is asymptotically stable then there exists a $P_a > 0$ satisfying (11.29).

For the existence of P_δ and the stability of $\tilde{A}_{\delta,t}$ we rely on techniques developed earlier in Theorem 2.34. If we define

$$\bar{A}_{\delta,i} = \begin{pmatrix} A_d + \lambda_i B_d F_\delta & -\lambda_i B_d F_\delta \\ \lambda_i B_d F_\delta & A_d - KC_d - \lambda_i B_d F_\delta \end{pmatrix}$$

and

$$\bar{B} = \begin{pmatrix} B_d F_\delta & -B_d F_\delta \\ B_d F_\delta & -B_d F_\delta \end{pmatrix},$$

then

$$\bar{A}_\delta = \begin{pmatrix} \bar{A}_{\delta,1} & \mu_{1,2}\bar{B} & 0 & \cdots & \mu_{1,N-1}\bar{B} \\ 0 & \bar{A}_{\delta,2} & \ddots & \ddots & \vdots \\ \vdots & \ddots & \ddots & \ddots & 0 \\ \vdots & & \ddots & \ddots & \mu_{N-2,N-1}\bar{B} \\ 0 & \cdots & \cdots & 0 & \bar{A}_{\delta,N-1} \end{pmatrix},$$

where λ_i are the eigenvalues of S_t and $\mu_{i,j} = [S_t]_{ij}$ for $j > i$ are the bounded upper triangular elements of S_t. Define

$$\bar{P}_\delta = \rho \begin{pmatrix} P_\delta & 0 \\ 0 & \sqrt{\|P_\delta\|}P \end{pmatrix},$$

where P_δ is the solution of the Riccati equation (11.26) and P is uniquely defined by the Lyapunov equation:

$$P(A_d - KC_d) + (A_d - KC_d)'P = -I.$$

In the above we choose ρ such that $\rho\delta > 1$ and $\rho\sqrt{\|P_\delta\|} > 2$. As shown in [33] we then have

$$\bar{A}_{\delta,i}'\bar{P}_\delta + \bar{P}_\delta\bar{A}_{\delta,i} \leq -\rho \begin{pmatrix} \delta I & 0 \\ 0 & \frac{1}{2}\sqrt{\|P_\delta\|}I \end{pmatrix} \leq -I,$$

for a sufficiently small δ. Using the above, with the same argument as in Lemma 11.7, we prove that for ν small enough we have (11.28). ■

Proof of Theorem 11.9 For each $i \in \{1, \ldots, N-1\}$, let

$$\bar{x}_i = \begin{pmatrix} \bar{x}_{ia} \\ \bar{x}_{id} \end{pmatrix} := x_i - x_N, \qquad \hat{\bar{x}}_i := \begin{pmatrix} \hat{\bar{x}}_{ia} \\ \hat{\bar{x}}_{id} \end{pmatrix} = \hat{x}_i - \hat{x}_N, \qquad \text{where} \quad \hat{x}_i = \begin{pmatrix} \hat{x}_{ia} \\ \hat{x}_{id} \end{pmatrix}.$$

Then, state synchronization among agents is achieved if $\bar{x}_i \to 0$ for $i = 1, \ldots, N-1$.

Next, define

$$\xi_{ia} = \bar{x}_{ia}, \quad \hat{\xi}_{ia} = \hat{\bar{x}}_{ia}, \quad \xi_{id} = S_\varepsilon \bar{x}_{id}, \quad \text{and} \quad \hat{\xi}_{id} = S_\varepsilon \hat{\bar{x}}_{id}.$$

Then, using the identities

$$S_\varepsilon A_d S_\varepsilon^{-1} = \varepsilon^{-1} A_d, \qquad S_\varepsilon B_d = \varepsilon^{\rho-1} B_d, \qquad \text{and} \qquad C_d S_\varepsilon^{-1} = C_d,$$

we have

$$\begin{aligned}
\dot{\xi}_{ia} &= A_a \xi_{ia} + V_{iad} \xi_{id}, \\
\dot{\hat{\xi}}_{ia} &= A_a \hat{\xi}_{ia} + \hat{V}_{iad} \hat{\xi}_{id}, \\
\varepsilon \dot{\xi}_{id} &= A_d \xi_{id} + B_d F_\delta \hat{\xi}_{id} + V_{ida}^\varepsilon \xi_{ia} + V_{idd}^\varepsilon \xi_{id}, \\
\varepsilon \dot{\hat{\xi}}_{id} &= A_d \hat{\xi}_{id} + \hat{V}_{ida}^\varepsilon \hat{\xi}_{ia} + \hat{V}_{idd}^\varepsilon \hat{\xi}_{id} + \sum_{j=1}^{N-1} \bar{\ell}_{ij} K C_d \xi_{jd} - K C_d \hat{\xi}_{id},
\end{aligned}$$

where

$$\begin{aligned}
V_{iad} &= \hat{V}_{iad} = L_{ad}C_d, \\
V_{ida}^{\varepsilon} &= \hat{V}_{ida}^{\varepsilon} = \varepsilon^{\rho} B_d E_{da}, \\
V_{idd}^{\varepsilon} &= \varepsilon^{\rho} B_d E_{dd} S_{\varepsilon}^{-1}, \\
\hat{V}_{idd}^{\varepsilon} &= \varepsilon^{\rho} B_d E_{dd} S_{\varepsilon}^{-1}.
\end{aligned}$$

Clearly $\|V_{iad}\|$ and $\|\hat{V}_{iad}\|$ are ε-independent, while $\|V_{ida}^{\varepsilon}\|$ and $\|\hat{V}_{ida}^{\varepsilon}\|$ are $O(\varepsilon)$. Moreover,

$$\|\varepsilon^{\rho} B_d E_{dd} S_{\varepsilon}^{-1}\| \leq \varepsilon \|B_d E_{dd}\|.$$

Hence $\|V_{idd}^{\varepsilon}\|$ and $\|\hat{V}_{idd}^{\varepsilon}\|$ are $O(\varepsilon)$. Let

$$\xi_a = \begin{pmatrix} \xi_{1a} \\ \vdots \\ \xi_{(N-1)a} \end{pmatrix}, \quad \hat{\xi}_a = \begin{pmatrix} \hat{\xi}_{1a} \\ \vdots \\ \hat{\xi}_{(N-1)a} \end{pmatrix}, \quad \xi_d = \begin{pmatrix} \xi_{1d} \\ \vdots \\ \xi_{(N-1)d} \end{pmatrix}, \quad \text{and } \hat{\xi}_d = \begin{pmatrix} \hat{\xi}_{1d} \\ \vdots \\ \hat{\xi}_{(N-1)d} \end{pmatrix}.$$

Then,

$$\begin{aligned}
\dot{\xi}_a &= (I_{N-1} \otimes A_a)\xi_a + V_{ad}\xi_d, \\
\dot{\hat{\xi}}_a &= (I_{N-1} \otimes A_a)\hat{\xi}_a + \hat{V}_{ad}\hat{\xi}_d, \\
\varepsilon\dot{\xi}_d &= (I_{N-1} \otimes A_d)\xi_d + (I_{N-1} \otimes B_d F_{\delta})\hat{\xi}_d + V_{da}^{\varepsilon}\xi_a + V_{dd}^{\varepsilon}\xi_d, \\
\varepsilon\dot{\hat{\xi}}_d &= (I_{N-1} \otimes A_d)\hat{\xi}_d + \hat{V}_{da}^{\varepsilon}\hat{\xi}_a + \hat{V}_{dd}^{\varepsilon}\hat{\xi}_d + (\bar{L}_t \otimes KC_d)\xi_d - (I_{N-1} \otimes KC_d)\hat{\xi}_d,
\end{aligned}$$

where

$$V_{ad} = \operatorname{diag}(V_{1ad}, \ldots, V_{(N-1)ad})$$

and $\hat{V}_{ad}$, V_{da}^{ε}, $\hat{V}_{da}^{\varepsilon}$, V_{dd}^{ε}, and $\hat{V}_{dd}^{\varepsilon}$ are similarly defined. Define a unitary $\bar{U}_t$ such that $\bar{U}_t^{-1}\bar{L}_t\bar{U}_t = \bar{S}_t$, where $\bar{S}_t$ is the Schur form of $\bar{L}_t$, and let

$$\begin{aligned}
\nu_a &= \xi_a, \\
\tilde{\nu}_a &= \xi_a - \hat{\xi}_a, \\
\nu_d &= (\bar{S}_t\bar{U}_t^{-1} \otimes I_{m\rho})\xi_d, \\
\tilde{\nu}_d &= \nu_d - (\bar{U}_t^{-1} \otimes I_{m\rho})\hat{\xi}_d.
\end{aligned}$$

Then

$$
\begin{aligned}
\dot{\nu}_a &= (I_{N-1} \otimes A_a)\nu_a + W_{ad,t}\nu_d, \\
\dot{\tilde{\nu}}_a &= (I_{N-1} \otimes A_a)\tilde{\nu}_a + W_{ad,t}\nu_d - \hat{W}_{ad,t}(\nu_d - \tilde{\nu}_d), \\
\varepsilon\dot{\nu}_d &= (I_{N-1} \otimes A_d)\nu_d + (\bar{S}_t \otimes B_d F_\delta)(\nu_d - \tilde{\nu}_d) + W^{\varepsilon}_{da,t}\nu_a + W^{\varepsilon}_{dd,t}\nu_d, \\
\varepsilon\dot{\tilde{\nu}}_d &= (I_{N-1} \otimes A_d)\tilde{\nu}_d + (\bar{S}_t \otimes B_d F_\delta)(\nu_d - \tilde{\nu}_d) + W^{\varepsilon}_{da,t}\nu_a - \hat{W}^{\varepsilon}_{da,t}(\nu_a - \tilde{\nu}_a) \\
&\qquad + W^{\varepsilon}_{dd,t}\nu_d - \hat{W}^{\varepsilon}_{dd,t}(\nu_d - \tilde{\nu}_d) - (I_{N-1} \otimes KC_d)\tilde{\nu}_d,
\end{aligned}
$$

where

$$
\begin{aligned}
W_{ad,t} &= V_{ad}(\bar{U}_t\bar{S}_t^{-1} \otimes I_{m\rho}), \\
\hat{W}_{ad,t} &= \hat{V}_{ad}(\bar{U}_t \otimes I_{m\rho}), \\
W^{\varepsilon}_{da,t} &= (\bar{S}_t\bar{U}_t^{-1} \otimes I_{m\rho})V^{\varepsilon}_{da}, \\
W^{\varepsilon}_{dd,t} &= (\bar{S}_t\bar{U}_t^{-1} \otimes I_{m\rho})V^{\varepsilon}_{dd}(\bar{U}_t\bar{S}_t^{-1} \otimes I_{m\rho}), \\
\hat{W}^{\varepsilon}_{da,t} &= (\bar{U}_t^{-1} \otimes I_{m\rho})\hat{V}^{\varepsilon}_{da}, \\
\hat{W}^{\varepsilon}_{dd,t} &= (\bar{U}_t^{-1} \otimes I_{m\rho})\hat{V}^{\varepsilon}_{dd}(\bar{U}_t \otimes I_{m\rho}).
\end{aligned}
$$

Finally, let N_a and N_d be defined such that

$$
\eta_a := N_a \begin{pmatrix} \nu_a \\ \tilde{\nu}_a \end{pmatrix} = \begin{pmatrix} \nu_{1a} \\ \tilde{\nu}_{1a} \\ \vdots \\ \nu_{(N-1)a} \\ \tilde{\nu}_{(N-1)a} \end{pmatrix}, \qquad \eta_d := N_d \begin{pmatrix} \nu_d \\ \tilde{\nu}_d \end{pmatrix} = \begin{pmatrix} \nu_{1d} \\ \tilde{\nu}_{1d} \\ \vdots \\ \nu_{(N-1)d} \\ \tilde{\nu}_{(N-1)d} \end{pmatrix}.
$$

We get

$$
\begin{aligned}
\dot{\eta}_a &= \tilde{A}_a\eta_a + \tilde{W}_{ad,t}\eta_d, \\
\varepsilon\dot{\eta}_d &= \tilde{A}_{\delta,t}\eta_d + \tilde{W}^{\varepsilon}_{da,t}\eta_a + \tilde{W}^{\varepsilon}_{dd,t}\eta_d,
\end{aligned}
\tag{11.30}
$$

where $\tilde{A}_a = (I_{2(N-1)} \otimes A_a)$ and

$$
\tilde{W}_{ad,t} = N_a \begin{pmatrix} W_{ad,t} & 0 \\ W_{ad,t} - \hat{W}_{ad,t} & \hat{W}_{ad,t} \end{pmatrix} N_d^{-1},
$$

$$
\tilde{W}^{\varepsilon}_{ds,t} = N_d \begin{pmatrix} W^{\varepsilon}_{ds,t} & 0 \\ W^{\varepsilon}_{ds,t} - \hat{W}^{\varepsilon}_{ds,t} & \hat{W}^{\varepsilon}_{ds,t} \end{pmatrix} N_s^{-1}, \quad s \in \{a, d\}.
$$

Note that $\tilde{A}_{\delta,t}$, $\tilde{W}_{ad,t}$, and $\tilde{W}_{ds,t}$ ($s \in \{a,d\}$) are functions of time t, since the Laplacian matrix L_t corresponding to the switching graph is a function of time t and then we have $\bar{U}_t^{-1}\bar{L}_t\bar{U}_t = \bar{S}_t$.

We know from (11.30) that η_d can experience discontinuous jumps when the network graph switches, while η_a is continuous. According to Lemma 11.10, there exist P_δ and P_a satisfying (11.28) and (11.29) respectively.

Define $V_a = \varepsilon^2 \eta_a' P_a \eta_a$ as a Lyapunov function for the dynamics of η_a in (11.30). Similarly, we define $V_d = \varepsilon \eta_d' P_\delta \eta_d$ as a Lyapunov function for the dynamics of η_d in (11.30). It is easy to find that V_d also has discontinuous jumps when the network changes. The derivative of V_a is bounded by

$$\begin{aligned}\dot{V}_a &= -\nu V_a - \varepsilon^2\|\eta_a\|^2 + 2\varepsilon^2\operatorname{Re}(\eta_a' P_a \tilde{W}_{ad,t}\eta_d)\\ &\leq -\nu V_a + \tfrac{\nu}{2\sqrt{\varepsilon}} V_d,\end{aligned} \tag{11.31}$$

where ε is small enough such that

$$2\operatorname{Re}(\eta_a' P_a \tilde{W}_{ad,t}\eta_d) \leq 2r_4\|\eta_a\|\|\eta_d\| \leq \|\eta_a\|^2 + r_4^2\|\eta_d\|^2 \leq \|\eta_a\|^2 + \tfrac{\nu}{2\varepsilon^2\sqrt{\varepsilon}} V_d.$$

Note that we can choose r_4 and, hence, ε independent of the network graph but only depending on our bounds on the eigenvalues and on the norm of our expanded Laplacian $\bar{L}_t$.

Next, the derivative of V_d is bounded by

$$\begin{aligned}\dot{V}_d &= -\nu\varepsilon^{-1} V_d - 4\|\eta_d\|^2 + 2\operatorname{Re}(\eta_d' P_\delta \tilde{W}^{\varepsilon}_{da,t}\eta_a) + 2\operatorname{Re}(\eta_d' P_\delta \tilde{W}^{\varepsilon}_{dd,t}\eta_d)\\ &\leq \tfrac{\nu}{2\sqrt{\varepsilon}} V_a - \nu\varepsilon^{-1} V_d,\end{aligned} \tag{11.32}$$

where $2\operatorname{Re}(\eta_d' P_\delta \tilde{W}^{\varepsilon}_{dd,t}\eta_d) \leq \|\eta_d\|^2$ for a small ε, and

$$2\operatorname{Re}(\eta_d' P_\delta \tilde{W}^{\varepsilon}_{da,t}\eta_a) \leq 2\varepsilon r_1\|\eta_a\|\|\eta_d\| \leq \varepsilon^2 r_1^2\|\eta_a\|^2 + \|\eta_d\|^2 \leq \tfrac{\nu}{2\sqrt{\varepsilon}} V_a + \|\eta_d\|^2,$$

provided r_1 is such that we have $\varepsilon r_1 \geq \|P_\delta \tilde{W}^{\varepsilon}_{da,t}\|$ and ε sufficiently small.

We define V_a^u and V_u^d via

$$\begin{pmatrix}\dot{V}_a^u\\ \dot{V}_d^u\end{pmatrix} = A_e\begin{pmatrix}V_a^u\\ V_d^u\end{pmatrix}, \qquad \begin{pmatrix}V_a^u\\ V_d^u\end{pmatrix}(t_{k-1}^+) = \begin{pmatrix}V_a\\ V_d\end{pmatrix}(t_{k-1}^+) \tag{11.33}$$

where

$$A_e = \nu\begin{pmatrix}-1 & \frac{1}{2\sqrt{\varepsilon}}\\ \frac{1}{2\sqrt{\varepsilon}} & -\varepsilon^{-1}\nu\end{pmatrix}.$$

It is then easy to check that if $V_d^u(t) = V_d(t)$ and $V_a^u(t) \geq V_a(t)$, then $\dot{V}_d^u(t) \geq \dot{V}_d(t)$. Similarly, if $V_a^u(t) = V_a(t)$ and $V_d^u(t) \geq V_d(t)$, then $\dot{V}_a^u(t) \geq \dot{V}_a(t)$. It is then easily verified that

$$V_d^u(t) \geq V_d(t), \quad V_a^u(t) \geq V_a(t)$$

for $t \in [t_{k-1}^+, t_k^-]$.

The matrix A_e has eigenvalues $\hat{\lambda}_1 = -\frac{3}{4}\nu$ and $\hat{\lambda}_2 = -\varepsilon^{-1}\nu - \nu$. We find that

$$e^{A_e t} = \frac{1}{4+\varepsilon} \begin{pmatrix} 4e^{\hat{\lambda}_1 t} + \varepsilon e^{\hat{\lambda}_2 t} & 2\sqrt{\varepsilon}\left(e^{\hat{\lambda}_1 t} - e^{\hat{\lambda}_2 t}\right) \\ 2\sqrt{\varepsilon}\left(e^{\hat{\lambda}_1 t} - e^{\hat{\lambda}_2 t}\right) & 4e^{\hat{\lambda}_2 t} + \varepsilon e^{\hat{\lambda}_1 t} \end{pmatrix}.$$

We find

$$V_a(t_k^-) + V_d(t_k^-) \leq V_a^u(t_k^-) + V_d^u(t_k^-) = \begin{pmatrix} 1 & 1 \end{pmatrix} e^{A_e(t_k - t_{k-1})} \begin{pmatrix} V_a \\ V_d \end{pmatrix} (t_{k-1}^+). \tag{11.34}$$

This yields

$$V_a(t_k^-) + V_d(t_k^-) \leq e^{\hat{\lambda}_3(t_k - t_{k-1})} \left[V_a(t_{k-1}^+) + \sqrt{\varepsilon} V_d(t_{k-1}^+)\right]$$

where $\hat{\lambda}_3 = -\nu/2$ provided ε is small enough.

We have a potential jump at time t_{k-1} in V_d. However, there exists a M such that $V_d(t_{k-1}^+) \leq M V_d(t_{k-1}^-)$. On the other hand, V_a is continuous. We find that

$$\begin{aligned} V_a(t_k^-) + V_d(t_k^-) &\leq e^{\hat{\lambda}_3(t_k - t_{k-1})} \left[V_a(t_{k-1}^-) + M\sqrt{\varepsilon} V_d(t_{k-1}^-)\right] \\ &\leq e^{\hat{\lambda}_3(t_k - t_{k-1})} \left[V_a(t_{k-1}^-) + V_d(t_{k-1}^-)\right] \end{aligned}$$

for ε small enough. Combining these time intervals, we get

$$V_a(t_k^-) + V_d(t_k^-) \leq e^{\hat{\lambda}_3 t_k} \left[V_a(0) + V_d(0)\right].$$

Assume $t_{k+1} > t > t_k$. Since we do not necessarily have $t - t_k > \tau$, we use the bound

$$\begin{aligned} V_a(t) + V_d(t) \leq V_a^u(t) + V_d^u(t) &\leq 2e^{\hat{\lambda}_3(t - t_k)} \left[V_a(t_k^+) + V_d(t_k^+)\right] \\ &\leq 2Me^{\hat{\lambda}_3(t - t_k)} \left[V_a(t_k^-) + V_d(t_k^-)\right]. \end{aligned}$$

Combining all together, we get

$$V(t) \leq 2Me^{\hat{\lambda}_3 t} V(0)$$

where $V = V_a + V_d$. Hence,

$$\lim_{t\to\infty} \eta_a(t) = 0, \text{ and } \lim_{t\to\infty} \eta_d(t) = 0. \tag{11.35}$$

Let $\bar{x}_a = \text{col}\{\bar{x}_{ia}\}$ and $\bar{x}_d = \text{col}\{\bar{x}_{id}\}$. We find that

$$\bar{x}_a = \begin{pmatrix} I_{(N-1)(n-m\rho)} & 0 \end{pmatrix} N_a^{-1} \eta_a = \Theta_a \eta_a,$$

and

$$\bar{x}_d = (I_{N-1} \otimes S_\varepsilon^{-1})(U_t S_t^{-1} \otimes I_{m\rho}) \begin{pmatrix} I_{(N-1)m\rho} & 0 \end{pmatrix} N_d^{-1} \eta_d = \Theta_{d,t} \eta_d,$$

for suitably chosen matrix Θ_a and $\Theta_{d,t}$. Although $\Theta_{d,t}$ is time-varying, it is uniformly bounded, because for graphs in $\mathbb{G}_{\alpha,\beta}^{\tau,N}$ the matrices $\bar{U}_t$ and $\bar{S}_t^{-1}$ are bounded. Therefore, we have

$$\lim_{t\to\infty} \bar{x}_a(t) = 0 \text{ and } \lim_{t\to\infty} \bar{x}_d(t) = 0,$$

which proves the result. ■

11.2.3.2 Partial-State Coupling and Additional Communication

In this section, additional communication of protocol states via the network's communication infrastructure is allowed. In this case, there is no restriction on agents' model, which means that agents can be non-minimum-phase.

To be more specific, we assume that agent i ($i = 1, \ldots, N$) has access to the quantity:

$$\hat{\zeta}_i(t) = \sum_{j=1}^{N} a_{ij}(t)(\xi_i(t) - \xi_j(t)) = \sum_{j=1}^{N} \ell_{ij}(t)\xi_j(t), \tag{11.36}$$

where $\xi_i \in \mathbb{R}^p$ ($i = 1, \ldots, N$) is a variable produced internally by agent i as part of the protocol.

Problem 11.11 (Partial-State Coupling with Extra Communication) Consider a MAS described by (11.1), (11.2), and (11.36). For any real numbers $\alpha, \beta, \tau > 0$ and a positive integer N that define a set of time-varying network graphs $\mathbb{G}_{\alpha,\beta}^{\tau,N}$, the *state synchronization* problem with extra communication and with a set of time-varying network graphs $\mathbb{G}_{\alpha,\beta}^{\tau,N}$ is to find, if possible, a linear time-invariant dynamic protocol of the form

$$\begin{cases} \dot{\chi}_i = A_c \chi_i + B_c \,\text{col}\{\zeta_i, \hat{\zeta}_i\}, \\ u_i = C_c \chi_i + D_c \,\text{col}\{\zeta_i, \hat{\zeta}_i\}, \end{cases} \tag{11.37}$$

for $i = 1, \ldots, N$, where $\chi_i \in \mathbb{R}^{n_c}$, such that, for any time-varying graph $\mathcal{G}_t \in \mathbb{G}_{\alpha,\beta}^{\tau,N}$ and for all initial conditions of agents and their protocol, state synchronization among agents can be achieved.

Define $\chi_i = T x_i$, where

$$T = \begin{pmatrix} C \\ \vdots \\ CA^{n-1} \end{pmatrix}. \tag{11.38}$$

Note that T is not necessarily a square matrix; however, the observability of (C, A) ensures that T is injective, which implies that $T'T$ is nonsingular. Then, we can write the dynamics of χ_i as follows:

$$\begin{aligned} \dot{\chi}_i &= (A_d + \mathcal{L})\chi_i + \mathcal{B}u_i, \quad \chi_i(0) = Tx_i(0), \\ y_i &= C_d \chi_i, \end{aligned} \tag{11.39}$$

where $\mathcal{B} = TB$ and

$$A_d = \begin{pmatrix} 0 & I_{p(n-1)} \\ 0 & 0 \end{pmatrix}, \quad C_d = \begin{pmatrix} I_p & 0 \end{pmatrix}, \quad \mathcal{L} = \begin{pmatrix} 0 \\ L \end{pmatrix},$$

where $L = CA^n(T'T)^{-1}T'$. We design the protocol based on a high-gain method.

Protocol design 11.3 *Consider a MAS described by (11.1), (11.2), and (11.36). Let T be defined in (11.38) and a high-gain matrix S_ε be defined as*

$$S_\varepsilon = \operatorname{diag}(I_p, \ldots, I_p \varepsilon^{n-1}).$$

Choose F such that $A + BF$ is Hurwitz stable, and $P > 0$ is the unique solution of the algebraic equation:

$$A_d P + P A_d' - 2\beta P C_d' C_d P + I = 0, \tag{11.40}$$

where β is the lower bound on the real part of nonzero eigenvalues of L_t. Then, for agent $i \in \{1, \ldots, N\}$, the protocol can be designed as

$$\begin{aligned} \dot{\hat{\chi}}_i &= (A_d + \mathcal{L})\hat{\chi}_i + \mathcal{B}u_i + \varepsilon^{-1} S_\varepsilon^{-1} P C_d' (\zeta_i - \hat{\zeta}_i), \\ u_i &= F(T'T)^{-1}T'\hat{\chi}_i. \end{aligned} \tag{11.41}$$

(continued)

Protocol design 11.3 (continued)
where we choose $\xi_i = C_d\hat{\chi}_i$ *for* $i \in \{1, \ldots, N\}$. *We find that*

$$\hat{\zeta}_i = \sum_{j=1}^{N} \ell_{ij}(t)\xi_j = \sum_{j=1}^{N} \ell_{ij}(t)C_d\hat{\chi}_j.$$

The result based on the above design is given in the following theorem.

Theorem 11.12 *Consider a MAS described by* (11.1) *with communication via* (11.2) *and* (11.36). *Let any real numbers* $\alpha, \beta, \tau > 0$ *and a positive integer* N *be given and hence a set of time-varying network graphs* $\mathbb{G}_{\alpha,\beta}^{\tau,N}$ *be defined.*

If (A, B) *is stabilizable and* (C, A) *is detectable, then the state synchronization problem stated in Problem* 11.4 *is solvable. In particular, the protocol* (11.41) *solves the state synchronization problem for any time-varying graph* $\mathcal{G}_t \in \mathbb{G}_{\alpha,\beta}^{\tau,N}$.

Proof For each $i \in \{1, \ldots, N\}$, let $\bar{\chi}_i = \chi_i - \hat{\chi}_i$. Then

$$\dot{\bar{\chi}}_i = (A_d + \mathcal{L})\bar{\chi}_i - \varepsilon^{-1}S_\varepsilon^{-1}PC_d'(\zeta_i - \hat{\zeta}_i). \tag{11.42}$$

The dynamics (11.42) can be rewritten as

$$\dot{\bar{\chi}}_i = A_d\bar{\chi}_i + \mathcal{L}\bar{\chi}_i - \varepsilon^{-1}S_\varepsilon^{-1}PC_d'C_d\sum_{j=1}^{N} \ell_{ij}(t)\bar{\chi}_j.$$

Define $\xi_i = S_\varepsilon(\bar{\chi}_i - \bar{\chi}_N)$. Then, we get

$$\varepsilon\dot{\xi}_i = A_d\xi_i + \mathcal{L}_\varepsilon\xi_i - PC_d'C_d\sum_{j=1}^{N-1} \bar{\ell}_{ij}(t)\xi_i,$$

where $\bar{\ell}_{ij}(t) = \ell_{ij}(t) - \ell_{Nj}(t)$, while

$$\mathcal{L}_\varepsilon = \begin{pmatrix} 0 \\ \varepsilon^n LS_\varepsilon^{-1} \end{pmatrix}.$$

Let $\xi = \operatorname{col}\{\xi_i\}$ and $\tilde{\mathcal{L}}_\varepsilon = \operatorname{diag}\{\mathcal{L}_\varepsilon\}$. Then, the dynamics of the complete network becomes

$$\varepsilon\dot{\xi} = [I_{N-1} \otimes A_d + \tilde{\mathcal{L}}_\varepsilon - \bar{L}_t \otimes PC_d'C_d]\xi, \tag{11.43}$$

where $\bar{L}_t \in \mathbb{R}^{(N-1)\times(N-1)}$ is such that $[\bar{L}_t]_{ij} = \bar{\ell}_{ij}(t)$. Define $U_t^{-1}\bar{L}_t U_t = S_t$, where S_t is the Schur form of $\bar{L}_t$, and let $v = (U_t^{-1} \otimes I_{pn})\xi$. Then we get

$$\varepsilon\dot{v} = (I_{N-1} \otimes A_d)v + W_\varepsilon v - S_t \otimes (PC_d'C_d)v, \tag{11.44}$$

where $W_\varepsilon = (U_t^{-1} \otimes I_{pn})\tilde{\mathcal{L}}_\varepsilon(U_t \otimes I_{pn})$. Note that when a switching of the network graph occurs, v in most cases experiences a discontinuity (because of a sudden change in $\bar{S}_t$ and $\bar{U}_t$).

Assume the network graph switches at $t_1, t_2, t_3, \ldots$ with $t_{k+1} - t_k > \tau$, and, for ease of presentation, we set $t_0 = 0$. Next, we analyze first the stability of dynamics (11.44) between the graph switches, that is, for time $t \in [t_{k-1}, t_k)$.

Let λ_i be the eigenvalue of the upper triangular matrix $\bar{S}_t$. Since

$$\begin{aligned}
&(A_d - \lambda_i PC_d'C_d)P + P(A_d - \lambda_i PC_d'C_d)' \\
&= -I - 2(\operatorname{Re}(\lambda_i) - \beta)PC_d'C_dP \\
&\leq -I,
\end{aligned}$$

we find that $A_d - \lambda_i PC_d'C_d$ is Hurwitz. Following Lemma 11.7, there exists a $\tilde{P} > 0$ and a small enough μ such that

$$A_{d,t}\tilde{P} + \tilde{P}A_{d,t}' \leq -\nu\tilde{P} - I.$$

Define Lyapunov function $V = \varepsilon v'\tilde{P}v$. The derivative of V is

$$\begin{aligned}
\dot{V} &= -\mu\varepsilon^{-1}V - \|v\|^2 + 2\operatorname{Re}(v'\tilde{P}W_\varepsilon v) \\
&\leq -\mu\varepsilon^{-1}V - \|v\|^2 + \varepsilon r\|v\|^2 \\
&\leq -\mu\varepsilon^{-1}V,
\end{aligned}$$

for a small enough ε. For the second inequality, we choose r such that $\varepsilon r \geq 2\|\tilde{P}W_\varepsilon\|$ which is possible since W_ε is bounded since $\bar{U}_t$ is unitary, and therefore both $\bar{U}_t$ and $\bar{U}_t^{-1}$ are bounded.

Similar to the proof of Theorem 11.6, for a small enough ε, we can achieve that $\lim_{t\to\infty} V(t) = 0$ given switching graphs. This implies that

$$\lim_{t\to\infty} \bar{\chi}_i(t) - \bar{\chi}_N(t) = 0 \tag{11.45}$$

given that U_t is unitary for all t and for any graph in $\mathbb{G}_{\alpha,\beta}^{\tau,N}$.

Next, we plug the controller input $u_i = F(T'T)^{-1}T'\hat{\chi}_i$ into the dynamics (11.1). Then, we get

$$\begin{aligned}\dot{\bar{x}}_i &= A\bar{x}_i + BF(T'T)^{-1}T'(\hat{\chi}_i - \hat{\chi}_N),\\ &= (A+BF)\bar{x}_i - BF(T'T)^{-1}T'(\bar{\chi}_i - \bar{\chi}_N),\end{aligned}$$

with

$$\bar{x}_i = x_i - x_N.$$

Given that $A + BF$ is Hurwitz and (11.45), we find from this differential equation that

$$\lim_{t\to\infty} x_i(t) - x_N(t) = \lim_{t\to\infty} \bar{x}_i(t) = 0,$$

which proves the result. ■

11.2.4 Static Protocol Design

In most cases, we design dynamic protocols for a MAS with partial-state coupling. As we have seen, in many cases we can actually design stable dynamic protocols which are desirable since they do not introduce additional synchronized trajectories. However, even more desirable are static protocols. In this section, we investigate the static protocol design for a MAS with partial-state coupling.

In Sect. 2.5.7, we have seen that static protocols can be designed for the following four classes of agents. For switching graphs we will only consider squared-down minimum-phase agents with relative degree 1. Later in Sect. 11.3.7, for general time-varying graphs, we will also consider the other cases but with the restriction that the graphs are balanced.

If the agents are squared-down minimum-phase with relative degree 1 given a pre-compensator $G_1 \in \mathbb{R}^{m\times q}$ and a post-compensator $G_2 \in \mathbb{R}^{q\times p}$, then the square system (A, BG_1, G_2C) is minimum-phase with relative degree 1. Note that for such a system with input $\hat{u}$ where $u = G_1\hat{u}$ and output $\hat{y} = G_2 y$, there exist nonsingular state transformation matrices T_x and T_u with

$$\bar{x} = \begin{pmatrix}\bar{x}_1\\ \bar{x}_2\end{pmatrix} = T_x x, \qquad \bar{u} = T_u\hat{u}$$

and the dynamics of $\bar{x}$ is represented as

$$\begin{aligned}\dot{\bar{x}}_1 &= A_{11}\bar{x}_1 + A_{12}\bar{x}_2,\\ \dot{\bar{x}}_2 &= A_{21}\bar{x}_1 + A_{22}\bar{x}_2 + \bar{u},\\ \hat{y} &= \bar{x}_2,\end{aligned} \tag{11.46}$$

where $\bar{x}_1 \in \mathbb{R}^{n-m}$ and $\bar{x}_2 \in \mathbb{R}^m$. Moreover, A_{11} is Hurwitz stable.

Next, we can design a protocol for a MAS which is squared-down minimum-phase with relative degree 1.

Protocol design 11.4 *Consider a MAS described by agents (11.1) which are squared-down minimum-phase with relative degree 1 with communication via (11.3). The static protocol is designed as*

$$u_i = -\rho G_1 T_u^{-1} G_2 \zeta_i, \tag{11.47}$$

where ρ is a parameter to be designed.

The main result based on the above design can be stated as follows.

Theorem 11.13 *Consider a MAS described by agents (11.1) and (11.3).*

If (A, B, C) is squared-down minimum-phase with relative degree 1, then the state synchronization problem stated in Problem 11.16 is solvable. In particular, there exists a $\rho^* > 0$ such that for any $\rho > \rho^*$, protocol (11.47) solves the state synchronization problem for any graph $\mathcal{G} \in \mathbb{G}_{t,\beta}^N$.

Before we prove the above theorem, we need a preliminary lemma.

Lemma 11.14 *There exist a P_d and a $\nu > 0$ such that for any upper triangular matrix $U_t \in \mathbb{C}^{(N-1)\times(N-1)}$ with $\|U_t\| < \tilde{\alpha}$ and whose eigenvalues satisfy $\mathrm{Re}(\lambda_i) > \beta$ for all $i = 1, \ldots, N-1$, we have*

$$P_d U_t + U_t' P_d \geq \nu P_d + 4I. \tag{11.48}$$

Proof The proof is similar to the proof of Lemma 11.7. ■

Proof Since all agents are squared-down minimum-phase with relative degree 1, we have

$$\begin{aligned}\dot{\tilde{x}}_{1i} &= A_{11}\tilde{x}_{1i} + A_{12}\tilde{x}_{2i},\\ \dot{\tilde{x}}_{2i} &= A_{21}\tilde{x}_{1i} + A_{22}\tilde{x}_{2i} + \tilde{u}_i,\\ \hat{y}_i &= \tilde{x}_{2i},\end{aligned} \tag{11.49}$$

and a protocol

$$\tilde{u}_i = -\rho \sum_{j=1}^{N} \ell_{ij}(t)\hat{y}_j. \tag{11.50}$$

Here we used that

$$G_2 C T_x^{-1} = \begin{pmatrix} 0 & I \end{pmatrix}.$$

The closed-loop system of (11.49) and (11.50) is written as

$$\begin{aligned} \dot{\tilde{x}}_{1i} &= A_{11}\tilde{x}_{1i} + A_{12}\tilde{x}_{2i}, \\ \dot{\tilde{x}}_{2i} &= A_{21}\tilde{x}_{1i} + A_{22}\tilde{x}_{2i} - \rho \textstyle\sum_{j=1}^{N} \ell_{ij}(t)\tilde{x}_{2j}. \end{aligned} \tag{11.51}$$

Since A_{11} is Hurwitz stable, there exists a $P_a > 0$ and a small enough ν such that

$$P_a A_{11} + A_{11}' P_a = -\nu P_a - I. \tag{11.52}$$

We have $\bar{L}_t = T_1 L_t T_2$ where T_1 and T_2 are given by (6.91). Define $\bar{U}_t^{-1}\bar{L}_t\bar{U}_t = \bar{S}_t$, where $\bar{S}_t$ is the Schur form of $\bar{L}_t$

$$\eta_a = T_1 \begin{pmatrix} \tilde{x}_{11} \\ \tilde{x}_{12} \\ \vdots \\ \tilde{x}_{1N} \end{pmatrix}, \qquad \eta_d = (\bar{U}_t Q_t^{-1} \otimes I) T_1 \begin{pmatrix} \tilde{x}_{21} \\ \tilde{x}_{22} \\ \vdots \\ \tilde{x}_{2N} \end{pmatrix}.$$

We obtain

$$\begin{aligned} \dot{\eta}_a &= (I \otimes A_{11})\eta_a + W_{ad,t}\eta_d, \\ \varepsilon\dot{\eta}_d &= W_{da,t}^{\varepsilon}\eta_a + W_{dd,t}^{\varepsilon}\eta_d \end{aligned}$$

where $\varepsilon = \rho^{-1}$, while

$$\begin{aligned} W_{ad,t} &= \bar{U}_t S_t^{-1} \otimes A_{12} \\ W_{da,t}^{\varepsilon} &= \varepsilon \bar{S}_t \bar{U}_t^{-1} \otimes A_{21} \\ W_{dd,t}^{\varepsilon} &= \varepsilon I \otimes A_{22} - \bar{S}_t \otimes I. \end{aligned}$$

We define

$$V_a = \varepsilon^2 \eta_a' P_a \eta_a, \qquad V_d = \varepsilon \eta_d' P_d \eta_d$$

where P_a and P_d are defined by (11.52) and (11.48), respectively.

The derivative of V_a is bounded by

$$\begin{aligned}\dot{V}_a &= -\nu V_a - \varepsilon^2 \|\eta_a\|^2 + 2\varepsilon^2 \operatorname{Re}(\eta_a' P_a \tilde{W}_{ad,t} \eta_d)\\ &\leq -\nu V_a + \tfrac{\nu}{2\sqrt{\varepsilon}} V_d,\end{aligned} \tag{11.53}$$

where ε is small enough such that

$$2\operatorname{Re}(\eta_a' P_a \tilde{W}_{ad,t} \eta_d) \leq 2r_4 \|\eta_a\|\|\eta_d\| \leq \|\eta_a\|^2 + r_4^2 \|\eta_d\|^2 \leq \|\eta_a\|^2 + \tfrac{\nu}{2\varepsilon^2\sqrt{\varepsilon}} V_d.$$

Note that we can choose r_4 and, hence, ε independent of the network graph but only depending on our bounds on the eigenvalues and on the norm of our expanded Laplacian $\bar{L}_t$.

Next, the derivative of V_d is bounded by

$$\begin{aligned}\dot{V}_d &= -\nu\varepsilon^{-1} V_d - 4\|\eta_d\|^2 + 2\operatorname{Re}(\eta_d' P_d \tilde{W}^{\varepsilon}_{da,t} \eta_a) + 2\varepsilon \operatorname{Re}(\eta_d' P_d (I \otimes A_{22}) \eta_d)\\ &\leq \tfrac{\nu}{2\sqrt{\varepsilon}} V_a - \nu\varepsilon^{-1} V_d,\end{aligned} \tag{11.54}$$

where

$$2\varepsilon \operatorname{Re}(\eta_d' P_d (I \otimes A_{12}) \eta_d) \leq \|\eta_d\|^2$$

for a small enough ε, and

$$2\operatorname{Re}(\eta_d' P_d \tilde{W}^{\varepsilon}_{da,t} \eta_a) \leq 2\varepsilon r_1 \|\eta_a\|\|\eta_d\| \leq \varepsilon^2 r_1^2 \|\eta_a\|^2 + \|\eta_d\|^2 \leq \tfrac{\nu}{2\sqrt{\varepsilon}} V_a + \|\eta_d\|^2,$$

provided r_1 is such that we have

$$r_1 \geq \|P_d(\bar{S}_t \bar{U}_t^{-1} \otimes A_{21})\|,$$

and ε sufficiently small. Note that we can choose r_1 independent of the network graph but only depending on our bounds on the eigenvalues and on the norm of our expanded Laplacian $\bar{L}_t$.

We define V_a^u and V_u^d via

$$\begin{pmatrix} \dot{V}_a^u \\ \dot{V}_d^u \end{pmatrix} = A_e \begin{pmatrix} V_a^u \\ V_d^u \end{pmatrix}, \qquad \begin{pmatrix} V_a^u \\ V_d^u \end{pmatrix}(t_{k-1}^+) = \begin{pmatrix} V_a \\ V_d \end{pmatrix}(t_{k-1}^+) \tag{11.55}$$

where

$$A_e = \nu \begin{pmatrix} -1 & \frac{1}{2\sqrt{\varepsilon}} \\ \frac{1}{2\sqrt{\varepsilon}} & -\varepsilon^{-1}\nu \end{pmatrix}.$$

Similar to the proof of Theorem 11.36, we find

$$V_a(t_k^-) + V_d(t_k^-) \leq e^{\hat{\lambda}_3(t_k - t_{k-1})} \left[V_a(t_{k-1}^+) + \sqrt{\varepsilon} V_d(t_{k-1}^+)\right]$$

where $\hat{\lambda}_3 = -\nu/2$ provided ε is small enough.

We have a potential jump at time t_{k-1} in V_d. However, there exists an M such that $V_d(t_{k-1}^+) \leq M V_d(t_{k-1}^-)$. On the other hand, V_a is continuous. We find that

$$\begin{aligned} V_a(t_k^-) + V_d(t_k^-) &\leq e^{\hat{\lambda}_3(t_k - t_{k-1})} \left[V_a(t_{k-1}^-) + M\sqrt{\varepsilon} V_d(t_{k-1}^-)\right] \\ &\leq e^{\hat{\lambda}_3(t_k - t_{k-1})} \left[V_a(t_{k-1}^-) + V_d(t_{k-1}^-)\right] \end{aligned}$$

for ε small enough. Combining these time intervals, we get

$$V_a(t_k^-) + V_d(t_k^-) \leq e^{\hat{\lambda}_3 t_k} \left[V_a(0) + V_d(0)\right].$$

Assume $t_{k+1} > t > t_k$. Since we do not necessarily have $t - t_k > \tau$, we use the bound

$$\begin{aligned} V_a(t) + V_d(t) \leq V_a^u(t) + V_d^u(t) &\leq 2e^{\hat{\lambda}_3(t - t_k)} \left[V_a(t_k^+) + V_d(t_k^+)\right] \\ &\leq 2Me^{\hat{\lambda}_3(t - t_k)} \left[V_a(t_k^-) + V_d(t_k^-)\right]. \end{aligned}$$

Putting everything together, we get

$$V(t) \leq 2Me^{\hat{\lambda}_3 t} V(0)$$

where $V = V_a + V_d$. Hence,

$$\lim_{t\to\infty} \eta_a(t) = 0, \text{ and } \quad \lim_{t\to\infty} \eta_d(t) = 0. \tag{11.56}$$

Let $\bar{x}_a = \operatorname{col}\{\bar{x}_{ia}\}$ and $\bar{x}_d = \operatorname{col}\{\bar{x}_{id}\}$. We find that

$$\bar{x}_a = \begin{pmatrix} I_{(N-1)(n-m\rho)} & 0 \end{pmatrix} N_a^{-1} \eta_a = \Theta_a \eta_a,$$

and

$$\bar{x}_d = (I_{N-1} \otimes S_\varepsilon^{-1})(U_t S_t^{-1} \otimes I_{m\rho}) \begin{pmatrix} I_{(N-1)m\rho} & 0 \end{pmatrix} N_d^{-1} \eta_d = \Theta_{d,t} \eta_d,$$

for suitably chosen matrix Θ_a and $\Theta_{d,t}$. Although $\Theta_{d,t}$ is time-varying, it is uniformly bounded, because for graphs in $\mathbb{G}_{\alpha,\beta}^{\tau,N}$, the matrices $\bar{U}_t$ and $\bar{S}_t^{-1}$ are bounded. Therefore, we have

$$\lim_{t\to\infty} \bar{x}_a(t) = 0 \text{ and } \quad \lim_{t\to\infty} \bar{x}_d(t) = 0,$$

which proves the result. ■

11.3 Time-Varying Graphs

In this section, we consider time-varying graphs. In other words, we expand switching networks which are piecewise constant to arbitrary time variations. Clearly, this means that the issue of dwell-time is no longer relevant or, one could say, that the dwell-time $\tau = 0$.

11.3.1 Problem Formulation

Consider a MAS composed of N identical linear time-invariant agents of the form

$$\dot{x}_i = Ax_i + Bu_i, \qquad (i = 1, \ldots, N) \tag{11.57}$$

where $x_i \in \mathbb{R}^n$, $u_i \in \mathbb{R}^m$ are, respectively, the state, input, and output vectors of agent i.

The communication network provides each agent with a linear combination of its own output relative to that of other neighboring agents. In particular, each agent $i \in \{1, \ldots, N\}$ has access to the quantity

$$\zeta_i(t) = \sum_{j=1}^{N} a_{ij}(t)(y_i(t) - y_j(t)), \tag{11.58}$$

where $a_{ij}(t) \geq 0$ and $a_{ii}(t) = 0$, are right-continuous functions of time t, indicating time-varying communication among agents. This time-varying communication topology of the network can be described by a weighted, time-varying graph $\mathcal{G}(t)$ with nodes corresponding to the agents in the network and the weight of edges at time t given by the coefficient $a_{ij}(t)$. Specifically, $a_{ij}(t) > 0$ indicates that at time t there is an edge with weight $a_{ij}(t)$ in the graph from agent j to agent i. The Laplacian matrix associated with $\mathcal{G}(t)$ is defined as $L_t = [\ell_{ij}(t)]$. In terms of the coefficients of L_t, ζ_i can be rewritten as

$$\zeta_i(t) = \sum_{j=1}^{N} \ell_{ij}(t) y_j(t). \tag{11.59}$$

In the case of full-state coupling, we have $y_j = x_j$, and we obtain

$$\zeta_i(t) = \sum_{j=1}^{N} \ell_{ij}(t)x_j(t). \tag{11.60}$$

We then formulate two state synchronization problems for a time-varying network with full-state coupling and partial-state coupling, respectively.

Problem 11.15 (Full-State Coupling) Consider a MAS described by (11.57) and (11.60).

The *state synchronization* problem via full-state coupling for time-varying network graphs $\mathbf{G}_t$ is to find, if possible, a linear static protocol of the form

$$u_i = F_i\zeta_i, \tag{11.61}$$

for $i = 1, \ldots, N$, such that, for any time-varying graph $\mathcal{G}_t \in \mathbf{G}_t$ and for all initial conditions of agents, state synchronization among agents is achieved.

Problem 11.16 (Partial-State Coupling) Consider a MAS described by (11.57) and (11.59).

The *state synchronization* problem via partial-state coupling with a set of time-varying network graphs $\mathbf{G}_t$ is to find, if possible, a linear time-invariant dynamic protocol of the form

$$\begin{cases} \dot{\chi}_i = A_c\chi_i + B_c\zeta_i, \\ u_i = C_c\chi_i, \end{cases} \tag{11.62}$$

where $\chi_i \in \mathbb{R}^{n_i}$, such that, for any time-varying graph $\mathcal{G}_t \in \mathbf{G}_t$ and for all the initial conditions of agents and their protocol, state synchronization among agents can be achieved.

In order, to solve this problem for general time-varying graphs, we restrict attention to undirected graphs.

Definition 11.17 For any given real numbers $\alpha, \beta > 0$ and a positive integer N, let the set of undirected fixed graphs $\mathbb{G}^{u,N}_{\alpha,\beta}$ be defined by Definition 1.12. The set $\mathbb{G}^{u,N}_{t,\alpha,\beta}$ is the set of all time-varying graphs $\mathcal{G}(t)$ which satisfy

$$\mathcal{G}(t) \in \mathbb{G}^{u,N}_{\alpha,\beta}$$

for all $t \in \mathbb{R}$. We define

$$\mathbb{G}^{u,N}_{t,\alpha} = \bigcup_{\beta>0} \mathbb{G}^{u,N}_{t,\alpha,\beta}, \qquad \mathbb{G}^{u,N}_{t,\beta} = \bigcup_{\alpha>0} \mathbb{G}^{u,N}_{t,\alpha,\beta}, \qquad \mathbb{G}^{u,N}_{t} = \bigcup_{\substack{\alpha>0\\ \beta>0}} \mathbb{G}^{u,N}_{t,\alpha,\beta}$$

Remark 11.18 Note that for a time-varying graph $\mathcal{G} \in \mathbb{G}_{t,\alpha}^{u,N}$, there exists a $\beta > 0$ such that the real parts of the nonzero eigenvalues of the associated Laplacian L_t are larger than β for all $t > 0$. The intrinsic difference is that this value of β is not known and hence cannot be exploited in the protocol design.

Similar comments apply to $\mathbb{G}_{t,\beta}^{u,N}$ and $\mathbb{G}_{t}^{u,N}$.

Similar to earlier analysis, for instance, for time-delay systems in Chap. 6, we need to connect the properties of L_t to the properties of $\bar{L}_t$ given by

$$\bar{L}_t = T_1 L_t T_2 \tag{11.63}$$

with T_1 and T_2 given by (6.91). We have already seen that the eigenvalues of $\bar{L}_t$ are equal to the nonzero eigenvalues of L_t. However, the fact that L_t is symmetric does not guarantee that $\bar{L}_t$ is symmetric. On the other hand, we do have the following crucial property.

Lemma 11.19 *Consider the set* $\mathbb{G}_{t,\alpha,\beta}^{u,N}$. *For any time-varying Laplacian matrix* L_t *associated to a graph in* $\mathbb{G}_{t,\alpha,\beta}^{u,N}$, *we have that*

$$\tilde{L}_t = V T_1 L_t T_2 V^{-1}$$

is a symmetric matrix whose eigenvalues λ_i *satisfy* $\beta < \lambda_i < \alpha$ *where*

$$V = \left(T_1 T_1'\right)^{-1/2} \tag{11.64}$$

and T_1, T_2 *are defined by* (6.91).

Proof Using (6.92), i.e., $L_t T_2 T_1 = L_t$, we find that the eigenvalues of $\bar{L}_t = T_1 L_t T_2$ are exactly the nonzero eigenvalues of L_t. Clearly, this implies that the eigenvalues of $\tilde{L}_t$ are also equal to the nonzero eigenvalues of L_t. Since L_t is symmetric, it has a basis of orthogonal eigenvectors $x_1, \ldots, x_N$ associated with eigenvalues $\lambda_1, \ldots, \lambda_N$. We have $\lambda_1 = 0$ and $x_1 = \mathbf{1}$. It is easily verified that the eigenvectors of $\tilde{L}_t$ are given by

$$\tilde{x}_2 = V T_1 x_2\,, \ \ldots\ , \ \tilde{x}_N = V T_1 x_N$$

associated to $\lambda_2, \ldots, \lambda_N$, respectively, where we again use $L_t T_2 T_1 = L_t$. We have that

$$T_1' \left(T_1 T_1'\right)^{-1} T_1$$

is the orthogonal projection onto $(\ker T_1)^{\perp}$. Since $\ker T_1 = \{\mathbf{1}\}$ we find that

$$x_i \in (\ker T_1)^{\perp} \text{ for } i = 2, \ldots N$$

and hence

$$T_1' \left(T_1 T_1'\right)^{-1} T_1 x_i = x_i \text{ for } i = 2, \dots N.$$

This implies that

$$\langle \tilde{x}_i, \tilde{x}_j \rangle = \langle V T_1 x_i, V T_1 x_j \rangle = \langle T_1' \left(T_1 T_1'\right)^{-1} T_1 x_i, x_j \rangle = \langle x_i, x_j \rangle = 0$$

for $i, j = 2, \dots, N$ with $i \neq j$. Hence, the eigenvectors of $\tilde{L}_t$ are orthogonal which implies that the matrix $\tilde{L}_t$ is normal. Since, we also know that the eigenvalues are real, we can conclude that $\tilde{L}_t$ is symmetric. ■

Remark 11.20 We should note that the results presented in this section are restricted to undirected graphs. However, from the proof it will become clear that the main restriction we need is that there exist $r > s > 0$ such that

$$\|\tilde{L}_t - rI\| \leq r - s$$

for all $t > 0$. Clearly, also many directed graphs satisfy this condition.

11.3.2 Protocol Design for MAS with Full-State Coupling

Protocol design 11.5 *Consider a MAS described by* (11.57) *and* (11.60) *with an associated set of time-varying network graphs* $\mathbb{G}^{u,N}_{t,\alpha,\beta}$.

For agent $i \in \{1, \dots, N\}$, *we design a controller*

$$u_i(t) = -\tfrac{\alpha+\beta}{\alpha\beta} B' P \zeta_i(t), \tag{11.65}$$

where $P \geq 0$ *is the unique solution of the algebraic Riccati equation*

$$PA + A'P - PBB'P = 0, \tag{11.66}$$

such that

$$A - BB'P$$

is asymptotically stable.

The main result based on the above design is stated in the following theorem.

Theorem 11.21 *Consider a MAS described by (11.57) and (11.60). Let real numbers $\alpha > \beta > 0$ and a positive integer N be given.*

Assume (A, B) is stabilizable. Then the state synchronization problem stated in Problem 11.15 is solvable. In particular, the protocol (11.65) achieves state synchronization for any time-varying graph $\mathcal{G}_t \in \mathbb{G}^{u,N}_{t,\alpha,\beta}$.

Proof of Theorem 11.21 Let $\bar{x}_i = x_i - x_N$. The closed-loop system of (11.57) and (11.61) can be written as

$$\dot{\bar{x}}_i(t) = A\bar{x}_i(t) - \sum_{j=1}^{N-1} \bar{\ell}_{ij}(t) BF\bar{x}_j(t) \tag{11.67}$$

where $\bar{\ell}_{ij} = \ell_{ij} - \ell_{N,j}$. Let

$$\tilde{x}(t) = V \begin{pmatrix} \bar{x}_1(t) \\ \vdots \\ \bar{x}_{N-1}(t) \end{pmatrix}.$$

where V is defined by (11.64). Note that

$$\tilde{L}_t = rI + \tilde{M}_t$$

with $\|\tilde{M}_t\| < r - s$ where

$$r = \tfrac{\alpha+\beta}{2}, \qquad s = \beta.$$

This immediately follows from the fact that $\tilde{L}_t$ is symmetric whose eigenvalues are in the interval (α, β). This implies that $\tilde{M}_t$ has eigenvalues in the interval $(s-r, r-s)$ for the above choices for s and r. This yields the norm bound for $\tilde{M}_t$ since the norm of a symmetric matrix is determined by the largest absolute value of the eigenvalues. We have

$$\dot{\tilde{x}}(t) = (I \otimes A)\tilde{x}(t) + \tilde{L}_t \otimes BF)\tilde{x}(t). \tag{11.68}$$

or

$$\begin{aligned} \dot{\tilde{x}}(t) &= [I \otimes (A + rBF)]\tilde{x}(t) + (I \otimes B)\tilde{w}(t) \\ \tilde{z}(t) &= (I \otimes F)\tilde{x}(t) \end{aligned} \tag{11.69}$$

with

$$\tilde{w}(t) = \left(\tilde{M}_t \otimes I\right)\tilde{z}(t).$$

Since

$$\|\tilde{M}_t \otimes I\| < r - s,$$

we find that if the H_∞ norm of (11.69) with input $\tilde{w}$ and output $\tilde{z}$ is less than $(r - s)^{-1}$ then the system (11.68) is asymptotically stable and hence

$$\tilde{x}(t) \to 0 \quad \text{as} \quad t \to \infty$$

which implies that $x_i(t) - x_N(t) \to 0$ as $t \to \infty$. Hence synchronization is achieved.

In order to verify this H_∞ norm bound, we note that (11.69) has a diagonal structure and hence its H_∞ norm is less than $(r - s)^{-1}$ if we have

$$\|F(sI - A - rBF)^{-1}B\|_\infty < \tfrac{1}{r-s}.$$

From the small gain theorem, this is satisfied when F is such that there exists $\tilde{P}$ such that

$$\tilde{P}(A + rBF) + (A + rBF)'\tilde{P} + (r - s)^2\tilde{P}BB'\tilde{P} + F'F = 0$$

which is satisfied provided

$$F = -\tfrac{1}{2s}B'P \quad \text{and} \quad \tilde{P} = \tfrac{1}{2s(r-s)}P$$

using (11.66). The proof is hence complete. ∎

11.3.3 *Solvability Conditions for MAS with Partial-State Coupling*

The following lemma translates the existence of an appropriate protocol to an H_∞ control problem. This lemma will be our key result to obtain a design for a suitable protocol.

Lemma 11.22 *Consider a MAS described by* (11.57) *and* (11.59). *Let real numbers* $\alpha > \beta > 0$ *and a positive integer* N *be given.*

Assume (A, B) *is stabilizable. Then the protocol* (11.62) *achieves state synchronization for any time-varying graph* $\mathcal{G}_t \in \mathbb{G}^{u,N}_{t,\alpha,\beta}$ *provided that the system*

$$\begin{aligned} \dot{x} &= Ax + rBC_c\chi + Bw \\ \dot{\chi} &= B_cCx + A_c\chi \\ z &= C_c\chi \end{aligned} \tag{11.70}$$

is asymptotically stable and the H_∞ norm from w to z is less than ρ where

$$r = \tfrac{\beta+\alpha}{2}, \qquad \rho = \tfrac{2}{\alpha-\beta}.$$

Proof Let $\bar{x}_i = x_i - x_N$ and $\bar{u}_i = u_i - u_N$. The closed-loop system of (11.57) and (11.62) can be written as

$$\begin{aligned} \dot{\bar{x}}_i(t) &= A\bar{x}_i(t) + B\bar{u}(t) \\ \dot{\bar{\chi}}_i(t) &= A_c\bar{\chi}_i(t) + \textstyle\sum_{j=1}^{N-1} \bar{\ell}_{ij} B_c C\bar{x}_j \\ \bar{u}_i(t) &= C_c\bar{\chi}_i(t) \end{aligned} \tag{11.71}$$

which can be rewritten as

$$\begin{aligned} \dot{\bar{x}}_i(t) &= A\bar{x}_i(t) + B\bar{u}_i \\ \dot{\bar{\xi}}_i(t) &= A_c\bar{\xi}_i(t) + B_c C\bar{x}_i \\ \bar{u}_i(t) &= \textstyle\sum_{j=1}^{N-1} \bar{\ell}_{ij} C_c\bar{\xi}_j(t) \end{aligned} \tag{11.72}$$

with

$$\bar{\zeta}_i = \sum_{j=1}^{N} \bar{\ell}_{ij}\bar{\chi}_i$$

and $\bar{\ell}_{ij} = \ell_{ij} - \ell_{N,j}$. Let

$$\tilde{x}(t) = V\begin{pmatrix} \bar{x}_1(t) \\ \vdots \\ \bar{x}_{N-1}(t) \end{pmatrix}, \qquad \tilde{\xi}(t) = V\begin{pmatrix} \bar{\xi}_1(t) \\ \vdots \\ \bar{\xi}_{N-1}(t) \end{pmatrix}$$

where V is defined by (11.64). We find

$$\begin{aligned} \dot{\tilde{x}}(t) &= (I \otimes A)\tilde{x}_i(t) + (I \otimes B)\tilde{u} \\ \dot{\tilde{\xi}}(t) &= (I \otimes A_c)\tilde{\xi}(t) + (I \otimes B_c C)\tilde{x} \\ \tilde{u}(t) &= (\tilde{L}_t \otimes C_c)\tilde{\xi}(t). \end{aligned} \tag{11.73}$$

As already noted in the full-state coupling case, we have

$$\tilde{L}_t = rI + \tilde{M}_t$$

with $\|\tilde{M}_t\| < r - s$, where

$$r = \tfrac{\beta+\alpha}{2}, \qquad s = \beta.$$

We have

$$\begin{aligned}\dot{\tilde{x}}(t) &= (I \otimes A)\tilde{x}_i(t) + (I \otimes rBC_c\tilde{u} + (I \otimes B\tilde{w}\\ \dot{\tilde{\xi}}(t) &= (I \otimes A_c)\tilde{\xi}(t) + (I \otimes B_cC)\tilde{x}\\ \tilde{z}(t) &= (I \otimes C_c)\tilde{\xi}(t)\end{aligned} \tag{11.74}$$

with

$$\tilde{w}(t) = \left(\tilde{M}_t \otimes I\right)\tilde{z}(t).$$

Since

$$\|\tilde{M}_t \otimes I\| < r - s = \tfrac{\alpha-\beta}{2},$$

we find that if the H_∞ norm of (11.74) with input $\tilde{w}$ and output $\tilde{z}$ is less than $(r - s)^{-1}$ then the system (11.73) is asymptotically stable and hence

$$\tilde{x}(t) \to 0 \quad \text{as} \quad t \to \infty$$

which implies that $x_i(t) - x_N(t) \to 0$ as $t \to \infty$. Hence synchronization is achieved.

In order to verify this H_∞ norm bound, we note that (11.74) has a diagonal structure, and hence its H_∞ norm is less than $(r - s)^{-1}$ if the system (11.70) is asymptotically stable and the H_∞ norm from w to z is less than

$$\tfrac{1}{r-s} = \tfrac{2}{\alpha-\beta}.$$

This completes the proof. ■

Now that we have converted the original design to an H_∞ control problem, we can use similar techniques as used in the proof of Theorem 2.20. These conditions will be expressed in terms of two matrices which we will first define.

- Let $P_0 \geq 0$ be the unique solution; see [160], of the Riccati equation:

$$A'P_0 + P_0A - P_0BB'P_0 = 0 \tag{11.75}$$

  such that $A - BB'P_0$ has all eigenvalues in the closed left half plane.
- Moreover, let $Q_0 \geq 0$ be the solution of the linear matrix inequality; see [116]:

$$\begin{pmatrix} AQ_0 + Q_0A' + BB' & Q_0C' \\ CQ_0 & 0 \end{pmatrix} \geq 0, \tag{11.76}$$

  such that

$$\operatorname{rank}\begin{pmatrix} AQ_0 + Q_0A' + BB' & Q_0C' \\ CQ_0 & 0 \end{pmatrix} = \alpha$$

and

$$\operatorname{rank}\begin{pmatrix} sI - A & AQ_0 + Q_0A' + BB' & Q_0C' \\ -C & CQ_0 & 0 \end{pmatrix} = n + v$$

for all s in the open right half plane where

$$v = \operatorname{normrank} C(sI - A)^{-1}B.$$

Note that these two matrices are always well-defined provided (A, B) is stabilizable and (C, A) is detectable.

The main result based on the above design is stated in the following theorem.

Theorem 11.23 *Consider a MAS described by (11.57) and (11.59). Let real numbers $\alpha > \beta > 0$ and a positive integer N be given.*

Assume that (A, B) is stabilizable. Then the state synchronization problem stated in Problem 11.16 is solvable provided

$$(\alpha - \beta)^2 \lambda_{\max}(P_0 Q_0) < 4\alpha\beta. \tag{11.77}$$

In particular, there exist A_c, B_c, C_c such that the system

$$\begin{aligned} \dot{x} &= Ax + rBC_c\chi + Bw \\ \dot{\chi} &= B_cCx + A_c\chi \\ z &= rC_c\chi \end{aligned} \tag{11.78}$$

is asymptotically stable and the H_∞ norm from w to z is less than $1 + \rho$ where

$$r = \tfrac{\beta+\alpha}{2}, \qquad \rho = \tfrac{2\beta}{\alpha-\beta}. \tag{11.79}$$

For the specific choice of A_c, B_c, C_c, the protocol (11.62) achieves state synchronization for any time-varying graph $\mathcal{G}_t \in \mathbb{G}^{u,N}_{t,\alpha,\beta}$.

Proof According to Lemma 11.22, we need to find a controller of the form

$$\begin{aligned} \dot{\chi} &= A_c\chi + B_cy \\ u &= C_c\chi \end{aligned}$$

for the system

$$\begin{aligned} \dot{x} &= Ax + Bu + Bw \\ y &= Cx \\ z &= u \end{aligned}$$

such that the resulting system has an H_∞ norm strictly less than $1 + \rho$ with ρ given by (11.79).

In order to check whether we can make the H_∞ norm less than $1+\rho$, we use the approach presented in [125]. In this reference it is shown that we can make the H_∞ norm less than 1 if and only if four conditions are satisfied.

Firstly, there must exist a solution of the state feedback Riccati equation:

$$A'P + PA - \nu PBB'P = 0 \tag{11.80}$$

such that $A - \nu BB'P$ has all its eigenvalues in the closed left half plane. Here

$$\nu = \frac{\rho^2 + 2\rho}{(1+\rho)^2} = \frac{4\alpha\beta}{(\alpha+\beta)^2}.$$

We note

$$P = \frac{1}{\nu} P_0.$$

Hence this condition is always satisfied.

We have a singular problem, and hence the second condition is the existence of a solution of a quadratic matrix inequality (which reduces to a linear matrix inequality in this special case):

$$\begin{pmatrix} AQ + QA' + \eta BB' & QC' \\ CQ & 0 \end{pmatrix} \geq 0, \tag{11.81}$$

such that

$$\operatorname{rank} \begin{pmatrix} sI - A & AQ + QA' + \eta BB' & QC' \\ -C & CQ & 0 \end{pmatrix} = n + v, \tag{11.82}$$

for all s in the open right half plane where

$$\eta = \frac{1}{(1+\rho)^2} = \frac{(\alpha-\beta)^2}{(\alpha+\beta)^2}.$$

We find that

$$Q = \eta Q_0.$$

Hence this second condition is also always satisfied.

The third condition is

$$\lambda_{\max}(PQ) < 1. \tag{11.83}$$

We have

$$PQ = \frac{\eta}{\nu} P_0 Q_0,$$

which immediately yields that (11.83) is equivalent to the condition (11.77).

The final condition from [125] is that for any $s_0 \in \mathbb{C}^0$ which is an invariant zero of either $(A, B, C, 0)$ or $(A, B, 0, I)$, there should exist a K such that

$$\left\| KC(s_0 I - A - BKC)^{-1} B \right\| < 1 + \rho.$$

Since $\rho > 0$, obviously it is sufficient to prove that we can find a K such that

$$\|KC(s_0 I - A - BKC)^{-1} B\| \leq 1.$$

The existence of such K has already been established in the proof of Theorem 2.20.

In other words the given synchronization problem is solvable if and only if (11.77) is satisfied. ■

In the above we have obtained conditions for the existence of protocols to achieve synchronization by reducing the problem to an H_∞ control problem. As we have already seen in Chap. 2, this H_∞ control problem is obviously always solvable whenever $P_0 = 0$ or $Q_0 = 0$:

- We have $P_0 = 0$ if and only if A is at most weakly unstable.
- We have $Q_0 = 0$ if and only if (A, B, C) is at most weakly non-minimum-phase.

For these two special cases, we already obtained explicit designs for the protocol in Sects. 2.5.3 and 2.5.4. For at most weakly unstable, we found stable protocols. However, for at most weakly non-minimum-phase agents, we might get unstable protocols. In Sect. 2.5.5 we showed that for minimum-phase agents, we can obtain a stable protocol. When neither of these conditions are satisfied, we have to resort to the design in Sect. 2.5.6 whose structure is not as well understood as in the previous cases. These designs are briefly recapitulated below.

11.3.4 MAS with at Most Weakly unstable agents

We present here two protocol designs for a MAS with at most weakly unstable agents. One is based on the algebraic Riccati equation (ARE), and another is based on the direct eigenstructure assignment method. In each case, we are able to obtain a stable protocol.

11.3.4.1 ARE-Based Method

In this subsection, we design a protocol for a MAS with at most weakly unstable agents. Our design is based on a low-gain feedback with a CSS observer.

Protocol design 11.6 *Consider a MAS described by at most weakly unstable agents* (11.1) *and* (11.3).

We choose an observer gain K such that $A + KC$ is Hurwitz stable. Next, we consider a feedback gain $F_\delta = -B'P_\delta$ where $P_\delta > 0$ is the unique solution of the continuous-time algebraic Riccati equation:

$$A'P_\delta + P_\delta A - \beta P_\delta BB'P_\delta + \delta I = 0, \tag{11.84}$$

where $\delta > 0$ is a low-gain parameter and β is the lower bound of the real part of the nonzero eigenvalues of the Laplacian. This results in the protocol

$$\begin{cases} \dot{\chi}_i = (A + KC)\chi_i - K\zeta_i, \\ u_i = F_\delta \chi_i. \end{cases} \tag{11.85}$$

The main result regarding this design can be stated as follows.

Theorem 11.24 *Consider a MAS described by at most weakly unstable agents* (11.1) *and* (11.3). *Let any $\alpha \geq \beta > 0$ be given.*

If (A, B) is stabilizable and (C, A) is detectable, then the state synchronization problem via a stable protocol stated in Problem 11.16 is solvable. In particular, there exists a $\delta^* > 0$ such that for any $\delta \in (0, \delta^*]$, protocol (11.85) *solves the state synchronization problem for any time-varying graph $\mathcal{G}_t \in \mathbb{G}_{t,\alpha,\beta}^{u,N}$.*

11.3.4.2 Direct Eigenstructure Assignment method

We use the same canonical form as in Sect. 2.5.3.2. If (A, B) is controllable, then there exist a nonsingular state transformation T_x and an input transformation T_u, such that

$$\tilde{A} = T_x^{-1}AT_x = \begin{pmatrix} A_1 & 0 & \cdots & 0 \\ 0 & A_2 & \ddots & \vdots \\ \vdots & \ddots & \ddots & 0 \\ 0 & \cdots & 0 & A_q \end{pmatrix},$$

$$\tilde{B} = T_x^{-1} B T_u = \begin{pmatrix} B_1 & B_{1,2} & \cdots & B_{1,q} & B_{1,q+1} \\ 0 & B_2 & \ddots & \vdots & \vdots \\ \vdots & \ddots & \ddots & B_{q-1,q} & \vdots \\ 0 & \cdots & 0 & B_q & B_{q,q+1} \end{pmatrix},$$

and for $i = 1, 2, \ldots, q$,

$$A_i = \begin{pmatrix} 0 & 1 & 0 & \cdots & 0 \\ \vdots & \ddots & \ddots & \ddots & \vdots \\ \vdots & & \ddots & 1 & 0 \\ 0 & \cdots & \cdots & 0 & 1 \\ -a_{n_i}^i & \cdots & -a_{n_3}^i & -a_{n_2}^i & -a_{n_1}^i \end{pmatrix}, \quad B_i = \begin{pmatrix} 0 \\ \vdots \\ \vdots \\ 0 \\ 1 \end{pmatrix}.$$

Clearly, (A_i, B_i) is controllable. Note that in [106, Section 4.3] and [55], a different canonical form was used where A has an upper triangular structure and B has a diagonal structure. However, as pointed out in [106, Remark 4.79], the resulting low-gain feedback can then exhibit peaking which, in our current context, makes the structure above more suitable.

Protocol design 11.7 *Consider a MAS described by at most weakly unstable agents (11.1) and (11.3). Choose T_x and T_u such that the system is in the canonical form described above.*

For each (A_i, B_i), let $F_{\delta,i} \in \mathbb{R}^{1\times n_i}$ be the unique state feedback gain such that the eigenvalues of $A_i + B_i F_{\delta,i}$ can be obtained from the eigenvalues of A_i by moving any eigenvalue λ_i on the imaginary axis to $\lambda_i - 2\delta$ while all the eigenvalues in the open left half complex plane remain at the same location. Note that $F_{\delta,i}$ can be obtained explicitly in terms of δ. Now define F_δ as

$$F_\delta = T_u \begin{pmatrix} F_{\delta,1} & 0 & \cdots & 0 \\ 0 & F_{\delta,2} & \ddots & \vdots \\ \vdots & \ddots & \ddots & 0 \\ \vdots & & \ddots & F_{\delta,q} \\ 0 & \cdots & \cdots & 0 \end{pmatrix} T_x^{-1}. \tag{11.86}$$

Next, choose K such that $A + KC$ is Hurwitz stable. We then construct the dynamic protocol:

(continued)

Protocol design 11.7 (continued)

$$\begin{cases} \dot{\chi}_i = (A+KC)\chi_i - K\zeta_i, \\ u_i = \frac{2}{\beta} F_\delta \chi_i. \end{cases} \tag{11.87}$$

The following theorem guarantees that the above design has the required properties.

Theorem 11.25 *Consider a MAS described by at most weakly unstable agents* (11.1) *and* (11.3)*. Let any $\alpha \geq \beta > 0$ be given.*

If (A, B) is controllable and (C, A) is detectable, then the state synchronization problem stated in Problem 2.4 *is solvable. In particular, there exists a $\delta^* > 0$ such that for any $\delta \in (0, \delta^*]$, the stable protocol* (11.87) *solves the state synchronization problem for any time-varying graph $\mathcal{G}_t \in \mathbb{G}^{u,N}_{t,\alpha,\beta}$.*

11.3.5 MAS with at Most Weakly Non-minimum-Phase Agents

As indicated before, we will study the problem of synchronization for MAS with weakly non-minimum-phase agents.

Protocol design 11.8 *Consider a MAS described by agents* (11.1) *and* (11.3)*. Assume that the agents are at most weakly non-minimum-phase and left-invertible. For $0 < \rho < 1$, we construct $P \geq 0$ such that*

$$A'P + PA - \nu PBB'P + C'C = 0 \tag{11.88}$$

where

$$\nu = \frac{2\rho + \rho^2}{(1+\rho)^2}.$$

For any $\delta > 0$ let Q_δ be the unique solution of the algebraic Riccati equation:

$$AQ_\delta + Q_\delta A' + (1 - \delta^{-2})Q_\delta C'CQ_\delta + \tfrac{1}{(1+\rho)^2} BB' = 0. \tag{11.89}$$

There exists a δ small enough such that the spectral radius of PQ_δ is less than 1. We then construct the dynamic protocol:

(continued)

Protocol design 11.8 (continued)

$$\begin{cases} \dot{\chi}_i = A\chi_i + \nu B u_i + L_{p,q}(C\chi_i - \zeta_i) \\ u_i = -B'P\chi_i, \end{cases} \tag{11.90}$$

where

$$L_{p,q} = -\delta^{-2}(I - Q_\delta P)^{-1} Q_\delta C'. \tag{11.91}$$

The main result is as follows.

Theorem 11.26 *Consider a MAS described by at most weakly non-minimum-phase, left-invertible agents (11.1) and (11.3). Let any $\alpha \geq \beta > 0$ be given.*

If (A, B) is stabilizable and (A, C) is detectable, then the state synchronization problem stated in Problem 11.16 is solvable.

More specifically, choose ρ such that any $\lambda \in \mathbb{C}$ with $\operatorname{Re}\lambda > \beta$ and $|\lambda| < \alpha$ is in the set Λ_ρ defined in (2.40). Choose δ such that the spectral radius of PQ_δ is less than 1 and then the protocol (11.90) solves the state synchronization problem for any time-varying graph $\mathcal{G}_t \in \mathbb{G}^{u,N}_{t,\alpha,\beta}$.

We assume that the system is left-invertible. As argued in Theorem 2.35, this can always be achieved through a pre-compensator while preserving the property that the system is at most weakly non-minimum-phase.

11.3.6 MAS with Minimum-Phase Agents

Compared to the weakly non-minimum-phase systems, for minimum-phase agents, we can always find stable protocols. Another advantage is that we have much better design methods. As before we distinguish between ARE-based and direct methods.

11.3.6.1 ARE-Based Method

Protocol design 11.9 *Consider a MAS described by agents (2.1) and (2.3). Assume that the agents are minimum-phase and left-invertible. For $0 < \rho < 1$, we construct $P_\rho \geq 0$ such that*

$$A'P_\rho + P_\rho A - \tilde{\nu} P_\rho B B' P_\rho + I = 0 \tag{11.92}$$

(continued)

Protocol design 11.9 (continued)
where

$$\tilde{\nu} = \frac{4\rho + \rho^2}{(2+\rho)^2}.$$

Define

$$F_\rho = -BP_\rho.$$

For any $\varepsilon > 0$*, choose* δ *sufficiently small such that there exists a unique solution* Q_ε *of the algebraic Riccati equation:*

$$AQ_\varepsilon + Q_\varepsilon A' + \varepsilon^{-2} Q_\varepsilon F_\rho{}' F_\rho Q_\varepsilon - \delta^{-2} Q_\varepsilon C'CQ_\varepsilon + BB' = 0. \tag{11.93}$$

We then construct the stable dynamic protocol:

$$\begin{cases} \dot{\chi}_i = A\chi_i - \varepsilon^{-2} Q_\varepsilon C'(C\chi_i - \zeta_i), \\ u_i = F_\rho \chi_i. \end{cases} \tag{11.94}$$

The main result is as follows.

Theorem 11.27 *Consider a MAS described by left-invertible, minimum-phase agents* (11.1) *and* (11.3)*. Let any* $\alpha \geq \beta > 0$ *be given.*

If (A, B) *is stabilizable and* (A, C) *is detectable, then the state synchronization problem stated in Problem* 11.16 *is solvable.*

Then there exists an ε^* *such that the protocol* (11.94) *for any* $\varepsilon < \varepsilon^*$ *solves the state synchronization problem by stable protocol for any time-varying graph* $\mathcal{G}_t \in \mathbb{G}^{u,N}_{t,\alpha,\beta}$.

11.3.6.2 Direct Method

There exist a nonsingular state transformation T_x for the system (2.90) such that

$$\bar{x} = \begin{pmatrix} \bar{x}_1 \\ \bar{x}_2 \end{pmatrix} = T_x x, \tag{11.95}$$

and the dynamics of $\bar{x}$ is represented as

$$\begin{aligned} \dot{\bar{x}}_1 &= \bar{A}_{11}\bar{x}_1 + \bar{A}_{12}\bar{x}_2 \\ \dot{\bar{x}}_2 &= \bar{A}_{21}\bar{x}_1 + \bar{A}_{22}\bar{x}_2 + \bar{B}u + \bar{B}v \end{aligned} \tag{11.96}$$

with $\bar{B}$ invertible, and, since (A, B) is stabilizable, we have that $(\bar{A}_{11}, \bar{A}_{12})$ is stabilizable. We choose F_1 such that $\bar{A}_{11} + \bar{A}_{12}F_1$ is asymptotically stable. In that case, a suitable F_δ is given by

$$F_\delta := \frac{1}{\delta}\bar{B}^{-1}\begin{pmatrix} F_1 & -I \end{pmatrix} T_x, \tag{11.97}$$

for a δ sufficiently small.

For the design of K_ε, we use a different state space transformation. We first transform the system:

$$\begin{aligned} \dot{x} &= Ax + Bu \\ y &= Cx \end{aligned}$$

into the SCB form. We know this system is minimum-phase and left-invertible. Theorem 23.3 then tells us that there exist nonsingular matrices Γ_x, Γ_u, and Γ_y such that the system is in SCB form which, in the compact form presented in (23.30)–(23.32), yields

$$\begin{aligned} \dot{x}_a &= A_{aa}x_a + L_{ad}y_d + L_{ab}y_b, \\ \dot{x}_b &= A_{bb}x_b + L_{bd}y_d, \\ \dot{x}_d &= A_{dd}x_d + B_d(u_d + E_{da}x_a + E_{db}x_b + E_{dd}x_d), \\ y_b &= C_bx_b, \\ y_d &= C_dx_d, \end{aligned}$$

with

$$y = \Gamma_y \begin{pmatrix} y_b \\ y_d \end{pmatrix}, \qquad x = \Gamma_x \begin{pmatrix} x_a \\ x_b \\ x_d \end{pmatrix}, \qquad u = \Gamma_u u_d$$

where

$$A_d = \begin{pmatrix} A_{q_1} & 0 & \cdots & 0 \\ 0 & A_{q_2} & \ddots & \vdots \\ \vdots & \ddots & \ddots & 0 \\ 0 & \cdots & 0 & A_{q_d} \end{pmatrix}, \qquad B_d = \begin{pmatrix} B_{q_1} & 0 & \cdots & 0 \\ 0 & B_{q_2} & \ddots & \vdots \\ \vdots & \ddots & \ddots & 0 \\ 0 & \cdots & 0 & B_{q_d} \end{pmatrix},$$

$$C_d = \begin{pmatrix} C_{q_1} & 0 & \cdots & 0 \\ 0 & C_{q_2} & \ddots & \vdots \\ \vdots & \ddots & \ddots & 0 \\ 0 & \cdots & 0 & C_{q_d} \end{pmatrix},$$

where integers $q_1, \ldots, q_d$ describe the infinite zero structure and $A_q \in \mathbb{R}^{q\times q}$, $B_q \in \mathbb{R}^{q\times 1}$ and $C_q \in \mathbb{R}^{1\times q}$ are given by

$$A_q = \begin{pmatrix} 0 & I_{q-1} \\ 0 & 0 \end{pmatrix}, \quad B_q = \begin{pmatrix} 0 \\ 1 \end{pmatrix}, \quad C_q = \begin{pmatrix} 1 & 0 \end{pmatrix}.$$

Define S_ε and $S_{q,\varepsilon} \in \mathbb{R}^{q\times q}$ as

$$S_\varepsilon = \begin{pmatrix} S_{q_1,\varepsilon} & 0 & \cdots & 0 \\ 0 & S_{q_2,\varepsilon} & \ddots & \vdots \\ \vdots & \ddots & \ddots & 0 \\ 0 & \cdots & 0 & S_{q_d,\varepsilon} \end{pmatrix}, \qquad S_{q,\varepsilon} = \begin{pmatrix} 1 & 0 & \cdots & 0 \\ 0 & \varepsilon & \ddots & \vdots \\ \vdots & \ddots & \ddots & 0 \\ 0 & \cdots & 0 & \varepsilon^{q-1} \end{pmatrix}.$$

We choose K_1 such that $A_{bb} + K_1 C_b$ is asymptotically stable and K_2 equals

$$K_2 = \begin{pmatrix} K_{2,q_1} & 0 & \cdots & 0 \\ 0 & K_{2,q_2} & \ddots & \vdots \\ \vdots & \ddots & \ddots & 0 \\ 0 & \cdots & 0 & K_{2,q_d} \end{pmatrix},$$

where $A_{q_i} + K_{2,q_i} C_{q_i}$ is asymptotically stable for $i = 1, \ldots, d$. We set

$$K_\varepsilon = \Gamma_x \begin{pmatrix} -L_{ab} & -L_{ad} \\ K_1 & -L_{bd} \\ 0 & \varepsilon^{-1} S_\varepsilon^{-1} K_2 \end{pmatrix} \Gamma_y^{-1}. \tag{11.98}$$

Protocol design 11.10 *Consider a MAS described by agents* (11.1) *and* (11.3). *Assume that the agents are minimum-phase and left-invertible.*

We then construct the dynamic protocol:

$$\begin{cases} \dot{\chi}_i = A\chi_i - K_\varepsilon(C\chi_i - \zeta_i), \\ u_i = F_\delta \chi_i, \end{cases} \tag{11.99}$$

(continued)

Protocol design 11.10 (continued)
where F_δ and K_ε are chosen according to (11.97) and (11.98) respectively.

We obtain the following result.

Theorem 11.28 *Consider a MAS described by left-invertible, minimum-phase agents (11.1) and (11.3). Let any $\alpha \geq \beta > 0$ be given.*

If (A, B) is stabilizable and (A, C) is detectable, then the state synchronization problem stated in Problem 11.16 is solvable.

More specifically, there exist ε^* and δ^* such that the protocol (11.99) solves the state synchronization problem by stable protocol for any time-varying graph $\mathcal{G}_t \in \mathbb{G}_{t,\alpha,\beta}^{u,N}$ for all $\varepsilon < \varepsilon^*$ and all $\delta < \delta^*$.

11.3.7 Static Protocol Design

In most cases, we design dynamic protocols for a MAS with partial-state coupling. As we have seen, in many cases we can actually design stable dynamic protocols which are desirable since they do not introduce additional synchronized trajectories. However, even more desirable are static protocols. In this section, we investigate the static protocol design for a MAS with partial-state coupling.

In Sect. 2.5.7, we have seen that static protocols can be designed for the following four classes of agents:

- Squared-down passive
- Squared-down passifiable via static output feedback
- Squared-down passifiable via static input feedforward
- Squared-down minimum-phase with relative degree 1

Each of these cases will be studied here but with time-varying protocols.

Earlier in this section, we considered symmetric protocols. In this subsection, we will see that for the design of static protocol, it is sufficient to consider balanced graphs as defined in Definition 1.9. The time-varying graphs are then defined as follows.

Definition 11.29 For any given real numbers $\alpha, \beta > 0$ and a positive integer N, the set $\mathbb{G}_{t,\alpha,\beta}^{b,N}$ is the set of all time-varying graphs $\mathcal{G}(t)$ which satisfy

$$\mathcal{G}(t) \in \mathbb{G}_{\alpha,\beta}^{b,N}$$

for all $t \in \mathbb{R}$. We define

$$\mathbb{G}_{t,\alpha}^{b,N} = \bigcup_{\beta>0} \mathbb{G}_{t,\alpha,\beta}^{b,N}, \qquad \mathbb{G}_{t,\beta}^{b,N} = \bigcup_{\alpha>0} \mathbb{G}_{t,\alpha,\beta}^{b,N}, \qquad \mathbb{G}_{t}^{b,N} = \bigcup_{\substack{\alpha>0\\ \beta>0}} \mathbb{G}_{t,\alpha,\beta}^{b,N}$$

Lemma 11.30 *Consider the set* $\mathbb{G}_{\alpha,\beta}^{b,N}$. *For any time-varying Laplacian matrix* L_t *associated to a graph in* $\mathbb{G}_{t,\alpha,\beta}^{b,N}$, *we have that*

$$\tilde{L}_t = VT_1L_tT_2V^{-1}$$

is such that

$$\tilde{L}_t + \tilde{L}_t' > 0$$

while the eigenvalues λ_i *of* $\tilde{L}_t + \tilde{L}_t'$ *satisfy* $\beta < \lambda_i < \alpha$ *for all* $t > 0$ *where*

$$V = (T_1T_1')^{-1/2} \tag{11.100}$$

and T_1, T_2 *are defined by* (6.91).

Proof We first note that $L_t\mathbf{1} = 0$ implies that $L_tT_2T_1 = L_t$. Moreover, since L_t is balanced, we find that $L_t'\mathbf{1} = 0$ as well. This in turn yields that $L_t'T_2T_1 = L_t'$.

Let $x_1, \ldots, x_N$ be the orthogonal eigenvectors of $L_t + L_t'$ with associated nonnegative eigenvalues $\lambda_1, \ldots, \lambda_N$. Without loss of generality, we choose $x_1 = \mathbf{1}$ and $\lambda_1 = 0$. In that case, we find that

$$\left[\bar{L}_t + \bar{L}_t'\right] VT_1x_i = VT_1\left[L_t + L_t'\right] T_2T_1x_i = \lambda_i VT_1x_i.$$

Note that $VT_1x_1 = 0$ while $VT_1x_i \neq 0$ for $i = 2, \ldots, N$ (it can actually be easily verified that VT_1x_i and VT_1x_j are orthogonal for $i \neq j$).

Therefore, the eigenvalues of $\bar{L}_t + \bar{L}_t'$ are equal to the nonzero eigenvalues of $L_t + L_t'$. Moreover, $\bar{L}_t + \bar{L}_t' > 0$ since its eigenvalues $\lambda_2, \ldots, \lambda_N$ are all nonnegative. ■

11.3.7.1 Squared-Down Passive Agents

In Sect. 1.3 we have defined the concept of squared-down passivity. In the current subsection, we will show that if the agents are squared-down passive, then we can actually find static protocols in the case of partial-state coupling. Let us first present the associated design.

Protocol design 11.11 *Consider a MAS described by squared-down passive agents* (11.1) *with respect to* G_1 *and* G_2 *with communication via* (11.3). *The static protocol is designed as*

$$u_i = -\rho G_1KG_2\zeta_i, \tag{11.101}$$

(continued)

Protocol design 11.11 (continued)
where $\rho > 0$ and K is any positive definite matrix.

The main result regarding this design can be stated as follows.

Theorem 11.31 *Consider a MAS described by agents (11.1) and (11.3). Assume that (11.1) is squared-down passive with respect to G_1 and G_2 such that (A, BG_1) is controllable and (A, G_2C) is observable.*

In that case the state synchronization problem stated in Problem 11.16 is solvable with a static protocol. In particular, the protocol (11.101) for any $\rho > 0$ and $K > 0$ solves the state synchronization problem for any graph $\mathcal{G}$ which is in $\mathbb{G}_t^{b,N}$ for some N.

Proof Firstly, we need to transform the model of system (11.1) and (11.3). By using (11.101), we have

$$\dot{x}_i = Ax_i - \rho \sum_{j=1}^{N} \ell_{ij}(t) BG_1G_2Cx_j. \tag{11.102}$$

We define $\bar{x}_i = x_i - x_N$ and we obtain

$$\dot{\bar{x}}_i = A\bar{x}_i - \rho \sum_{j=1}^{N-1} \hat{\ell}_{ij}(t) BG_1G_2C\bar{x}_j, \tag{11.103}$$

where $\hat{\ell}_{ij}(t) = \ell_{ij}(t) - \ell_{Nj}(t)$. Define

$$\bar{x} = (V \otimes I) \begin{pmatrix} \bar{x}_1 \\ \bar{x}_2 \\ \vdots \\ \bar{x}_N \end{pmatrix}.$$

Then, the overall dynamics of the network can be written as

$$\dot{\bar{x}} = (I \otimes A)\bar{x} - \rho(\bar{L}_t \otimes BG_1G_2C)\bar{x}, \tag{11.104}$$

where $\bar{L}_t$ is defined by (11.63).

Since (A, BG_1) is controllable and (A, G_2C) is observable, we find that there exists a $P > 0$ satisfying (1.9). We obtain

$$\begin{aligned}\tfrac{d}{dt}\bar{x}'(I \otimes P)\bar{x} &= \bar{x}'\left[I \otimes (A'P + PA)\right]\bar{x} \\ &\quad - \rho\bar{x}'\left[(\bar{L}_t + \bar{L}_t') \otimes C'G_2'G_2C\right]\bar{x} \\ &\leq -\rho\bar{x}'\left[(\bar{L}_t + \bar{L}_t') \otimes C'G_2'G_2C\right]\bar{x}\end{aligned}$$

with $\rho > 0$. Recall that $\bar{L}_t + \bar{L}_t' > \beta I$ for some $\beta > 0$ (even though β is not known). We find that

$$\int_0^\infty \bar{x}'\left[I \otimes C'G_2'G_2C\right]\bar{x}\,\mathrm{d}t < \infty. \tag{11.105}$$

Consider the stabilizable and detectable system

$$(I \otimes A, I \otimes B, I \otimes G_2C)$$

with state $\bar{x}$ and input $-\rho(\bar{L} \otimes G_2C)\bar{x}$. Equation (11.105) implies that the output of this system is in L_2. Since $\bar{L}_t$ is bounded, this also implies that the input is in L_2. But then we immediately find that the state $\bar{x}(t)$ converges to zero and therefore synchronization is achieved. ∎

11.3.7.2 Squared-Down Passifiable via Static Output Feedback

Next we will show that the static protocol (11.5) still works for the MAS where agents are squared-down passifiable via static output feedback, but some of the knowledge of network graphs (that is parameter β) is required. In other words, the problem can be solvable for a set of graphs $\mathbb{G}_{t,\beta}^{b,N}$ via the static protocol (11.5). We design the static protocol as follows.

Protocol design 11.12 *Consider a MAS described agents (11.1) with communication via (11.3). Assume the agents are squared-down passifiable via static output feedback with respect to G_1, G_2, and H. The static protocol is designed as*

$$u_i = -\rho G_1 K G_2 \zeta_i, \tag{11.106}$$

where K is any positive definite matrix, and ρ is a positive parameter to be designed later.

The main result based on the above design can be stated as follows.

Theorem 11.32 *Consider a MAS described by agents (11.1) and communication via (11.3). Assume the agents are squared-down passifiable via static output feedback with respect to G_1, G_2, and H while (A, BG_1) is controllable and (A, G_2C) is observable with BG_1 and G_2C full-column and full-row rank, respectively.*

The state synchronization problem stated in Problem 11.16 but with a static protocol is solvable. In particular, given $\beta > 0$, there exists a ρ^* such that for any $\rho > \rho^*$, protocol (11.106) solves the state synchronization problem for any graph $\mathcal{G} \in \mathbb{G}_{t,\beta}^{b,N}$.

Proof The system is squared-down passifiable via static output feedback with respect to G_1, G_2, and H, and therefore there exists a positive definite matrix P such that (1.11) is satisfied. Moreover, there exists a real number $b > 0$ such that

$$H + H' \leq bI. \tag{11.107}$$

Similar to the proof of Theorem 11.31, we obtain that

$$\begin{aligned}
\tfrac{d}{dt}\bar{x}'(I \otimes P)\bar{x} &= \bar{x}'\left[I \otimes (A'P + PA)\right]\bar{x} \\
&\qquad - \rho\bar{x}'\left[(\bar{L}_t + \bar{L}_t') \otimes C'G_2'G_2C\right]\bar{x} \\
&\leq \bar{x}'\left[I \otimes C'G_2'(H + H')G_2C\right]\bar{x} \\
&\qquad - \rho\bar{x}'\left[(\bar{L}_t + \bar{L}_t') \otimes C'G_2'G_2C\right]\bar{x} \\
&\leq (b - \rho\beta)\bar{x}'\left[I \otimes C'G_2'G_2C\right]\bar{x}.
\end{aligned}$$

Choosing $\rho_* = \frac{b}{\beta}$ and $\rho > \rho^*$ then immediately ensures that synchronization is achieved using the same argument as in the proof of Theorem 11.31. ■

11.3.7.3 Squared-Down Passifiable via Static Input Feedforward

Next we consider agents which are squared-down passifiable via static input feedforward. The crucial difference with the previous subsection is that squared-down passifiability via static output feedback in general requires a high-gain feedback, while squared-down passifiability via input feedforward in general requires a low-gain feedback.

We design the static protocol based on the low-gain methodology as follows.

Protocol design 11.13 *Consider a MAS described by agents (11.1) with communication via (11.3). Assume that the agents are squared-down passi-*

(continued)

Protocol design 11.13 (continued)
fiable via static input feedforward with respect to G_1, G_2, and R. The static protocol is designed as

$$u_i = -\delta G_1 K G_2 \zeta_i, \tag{11.108}$$

where $K > 0$ and the low-gain parameter $\delta > 0$ need to be designed.

The main result based on the above design can be stated as follows.

Theorem 11.33 *Consider a MAS described by agents (2.1) and (2.3). Assume that the agents are squared-down passifiable via static input feedforward with respect to G_1, G_2, and R such that (A, BG_1) is controllable and (A, G_2C) is observable with BG_1 and G_2C full-column and full-row rank, respectively.*

The state synchronization problem stated in Problem 11.16 but with a static protocol is solvable. In particular, given $\alpha > 0$, there exists a δ^* such that for any $\delta < \delta^*$, the protocol (11.108) solves the state synchronization problem for any graph $\mathcal{G} \in \mathbb{G}_{t,\alpha}^{u,N}$.

In order to prove the above theorem, we need the following technical lemma.

Lemma 11.34 *Consider a Laplacian matrix L associated to an undirected graph in $\mathbb{G}_{\alpha}^{u,N}$. In that case any eigenvalue of L satisfies*

$$\bar{L}^2 \leq \alpha \bar{L}.$$

Proof Note that if the graph is undirected and hence L is symmetric, then it is easily verified that also $\bar{L}$ is symmetric. The result is then trivial. ■

Proof of Theorem 11.33 Similar to the proof of Theorem 11.31 but using (1.12), we obtain that

$$\begin{aligned}
\tfrac{d}{dt}\bar{x}'(I \otimes P)\bar{x} &= \bar{x}'\left[I \otimes (A'P + PA)\right]\bar{x} \\
&\qquad - 2v\rho\bar{x}'\left[\bar{L}_t \otimes C'G_2'G_2C\right]\bar{x} \\
&\leq \begin{pmatrix} \bar{x} \\ -\rho(\bar{L}_t \otimes G_2C)\bar{x} \end{pmatrix}' \bar{G}(P) \begin{pmatrix} \bar{x} \\ -\rho(\bar{L}_t \otimes G_2C)\bar{x} \end{pmatrix} \\
&\qquad - 2\rho\bar{x}'\left[\bar{L}_t \otimes C'G_2'G_2C\right]\bar{x} \\
&\qquad + \rho^2\bar{x}'\left[\bar{L}^2(t) \otimes C'G_2'G_2C\right]\bar{x} \\
&\leq (\rho\alpha - 2)\rho\bar{x}'\left[\bar{L}_t \otimes C'G_2'G_2C\right]\bar{x}
\end{aligned}$$

where

$$\bar{G}(P) = \begin{pmatrix} I \otimes (PA + A'P) & I \otimes (PBG_1 - C'G_2') \\ I \otimes (G_1'B'P - G_2C) & -I \otimes (D + D') \end{pmatrix} \leq 0.$$

Choosing $\rho_* = \frac{2}{\alpha}$ and $\rho < \rho^*$ then immediately ensures that synchronization is achieved using the same argument as in the proof of Theorem 11.31. ∎

For balanced graphs we need some additional prior knowledge as clarified in the following theorem.

Theorem 11.35 *Consider a MAS described by agents (2.1) and communication via (2.3). Assume that the agents are squared-down passifiable via static input feedforward with respect to G_1, G_2, and R such that (A, BG_1) is controllable and (A, G_2C) is observable with BG_1 and G_2C full-column and full-row rank, respectively.*

The state synchronization problem stated in Problem 11.16 but with a static protocol is solvable. In particular, given $\alpha > \beta > 0$, there exists a δ^* such that for any $\delta < \delta^*$, the protocol (11.108) solves the state synchronization problem for any graph $\mathcal{G} \in \mathbb{G}_{t,\alpha,\beta}^{b,N}$.

Proof The only difference with the proof of Theorem 11.33 is the first inequalities:

$$\begin{aligned}
\tfrac{d}{dt}\bar{x}'(I \otimes P)\bar{x} &= \bar{x}'\left[I \otimes (A'P + PA)\right]\bar{x} \\
&\qquad - \rho\bar{x}'\left[(\bar{L}_t + \bar{L}_t') \otimes C'G_2'G_2C\right]\bar{x} \\
&\leq \begin{pmatrix} \bar{x} \\ -\rho(\bar{L}_t \otimes G_2C)\bar{x} \end{pmatrix}' \bar{G}(P) \begin{pmatrix} \bar{x} \\ -\rho(\bar{L}_t \otimes G_2C)\bar{x} \end{pmatrix} \\
&\qquad - \rho\bar{x}'\left[(\bar{L}_t + \bar{L}_t') \otimes C'G_2'G_2C\right]\bar{x} \\
&\qquad + \rho^2\bar{x}'\left[\bar{L}_t'\bar{L}_t') \otimes C'G_2'(D + D')G_2C\right]\bar{x} \\
&\leq (\rho b\alpha^2 - \beta)\rho\bar{x}'\left[I \otimes C'G_2'G_2C\right]\bar{x}
\end{aligned}$$

where

$$D + D' \leq bI.$$

Choosing $\rho_* = \frac{\beta}{b\alpha^2}$ and $\rho < \rho^*$ then immediately ensures that synchronization is achieved ∎

11.3.7.4 Squared-Down Minimum-Phase Agent with Relative Degree 1

If the agents are squared-down minimum-phase with relative degree 1 given a pre-compensator $G_1 \in \mathbb{R}^{m \times q}$ and a post-compensator $G_2 \in \mathbb{R}^{q \times p}$, then the square system (A, BG_1, G_2C) is minimum-phase with relative degree 1. Note that for such a system with input $\hat{u}$ where $u = G_1\hat{u}$ and output $\hat{y} = G_2 y$, there exist nonsingular state transformation matrices T_x and T_u with

$$\bar{x} = \begin{pmatrix} \bar{x}_1 \\ \bar{x}_2 \end{pmatrix} = T_x x, \qquad \bar{u} = T_u \hat{u}$$

and the dynamics of $\bar{x}$ is represented as

$$\begin{aligned} \dot{\bar{x}}_1 &= A_{11}\bar{x}_1 + A_{12}\bar{x}_2, \\ \dot{\bar{x}}_2 &= A_{21}\bar{x}_1 + A_{22}\bar{x}_2 + \bar{u}, \\ \hat{y} &= \bar{x}_2, \end{aligned} \tag{11.109}$$

where $\bar{x}_1 \in \mathbb{R}^{n-m}$ and $\bar{x}_2 \in \mathbb{R}^m$. Moreover, A_{11} is Hurwitz stable.

Next, we can design a protocol for a MAS which is squared-down minimum-phase with relative degree 1.

Protocol design 11.14 *Consider a MAS described by agents* (11.1) *which are squared-down minimum-phase with relative degree* 1 *with communication via* (11.3)*. The static protocol is designed as*

$$u_i = -\rho G_1 T_u^{-1} G_2 \zeta_i, \tag{11.110}$$

where ρ is a parameter to be designed.

The main result based on the above design can be stated as follows.

Theorem 11.36 *Consider a MAS described by agents* (11.1) *and* (11.3).

If (A, B, C) is squared-down minimum-phase with relative degree 1*, then the state synchronization problem stated in Problem 11.16 is solvable. In particular, there exists a $\rho^* > 0$ such that for any $\rho > \rho^*$, protocol* (11.110) *solves the state synchronization problem for any graph $\mathcal{G} \in \mathbb{G}_{t,\beta}^{b,N}$.*

Proof Since all agents are squared-down minimum-phase with relative degree 1, we have

$$\begin{aligned} \dot{\tilde{x}}_{1i} &= A_{11}\tilde{x}_{1i} + A_{12}\tilde{x}_{2i}, \\ \dot{\tilde{x}}_{2i} &= A_{21}\tilde{x}_{1i} + A_{22}\tilde{x}_{2i} + \tilde{u}_i, \\ \hat{y}_i &= \tilde{x}_{2i}, \end{aligned} \tag{11.111}$$

and a protocol

$$\tilde{u}_i = -\rho \sum_{j=1}^{N} \ell_{ij}(t)\hat{y}_j. \tag{11.112}$$

Here we used that

$$G_2 C T_x^{-1} = \begin{pmatrix} 0 & I \end{pmatrix}.$$

The closed-loop system of (11.111) and (11.112) is written as

$$\begin{aligned} \dot{\tilde{x}}_{1i} &= A_{11}\tilde{x}_{1i} + A_{12}\tilde{x}_{2i}, \\ \dot{\tilde{x}}_{2i} &= A_{21}\tilde{x}_{1i} + A_{22}\tilde{x}_{2i} - \rho \textstyle\sum_{j=1}^{N} \ell_{ij}(t)\tilde{x}_{2j}. \end{aligned} \tag{11.113}$$

Since A_{11} is Hurwitz stable, there exists a $P_1 > 0$ such that

$$P_1 A_{11} + A_{11}' P_1 = -I.$$

Meanwhile, we choose $b > 0$ such that

$$A_{22} + A_{22}' - bI \leq 0. \tag{11.114}$$

Let $\bar{x}_{1i} = \tilde{x}_{1i} - \tilde{x}_{1N}$ and $\bar{x}_{2i} = \tilde{x}_{2i} - \tilde{x}_{2N}$; we have

$$\begin{aligned} \dot{\bar{x}}_{1i} &= A_{11}\bar{x}_{1i} + A_{12}\bar{x}_{2i}, \\ \dot{\bar{x}}_{2i} &= A_{21}\bar{x}_{1i} + A_{22}\bar{x}_{2i} - \rho \textstyle\sum_{j=1}^{N-1} \hat{\ell}_{ij}(t)\bar{x}_{2j}. \end{aligned} \tag{11.115}$$

Define

$$\bar{x}_1 = V \begin{pmatrix} \bar{x}_{11} \\ \bar{x}_{12} \\ \vdots \\ \bar{x}_{1(N-1)} \end{pmatrix}, \quad \bar{x}_2 = V \begin{pmatrix} \bar{x}_{21} \\ \bar{x}_{22} \\ \vdots \\ \bar{x}_{2(N-1)} \end{pmatrix}.$$

The overall dynamics of the network can then be written as

$$\begin{cases} \dot{\bar{x}}_1 = (I \otimes A_{11})\bar{x}_1 + (I \otimes A_{12})\bar{x}_2 \\ \dot{\bar{x}}_2 = (I \otimes A_{21})\bar{x}_1 + (I \otimes A_{22})\bar{x}_2 - \rho(\bar{L}_{(t)} \otimes I)\bar{x}_2. \end{cases} \tag{11.116}$$

Next, we choose the following candidate Lyapunov function:

$$V = \bar{x}_1'(I \otimes P_1)\bar{x}_1 + \bar{x}_2'\bar{x}_2 \tag{11.117}$$

with $P_1 > 0$. And, then, we calculate the time derivatives of V along the trajectories of system (11.116) as

$$\begin{aligned}\dot{V} \leq & -\bar{x}_1'\bar{x}_1 + 2\bar{x}_1'(I \otimes P_1 A_{12})\bar{x}_2 + 2\bar{x}_2'(I \otimes A_{21})\bar{x}_1 \\ & + \bar{x}_2' I \otimes (A_{22} + A_{22}')\bar{x}_2 - \rho \bar{x}_2' \left[(\bar{L}_t + \bar{L}_t') \otimes I\right] \bar{x}_2 \\ \leq & -\bar{x}_1'\bar{x}_1 + 2\bar{x}_1'(I \otimes P_1 A_{12})\bar{x}_2 + 2\bar{x}_2'(I \otimes A_{21})\bar{x}_1 \\ & + b\bar{x}_2'\bar{x}_2 - \rho\beta\bar{x}_2' I \otimes I \bar{x}_2 \\ \leq & \begin{pmatrix} \bar{x}_1 \\ \bar{x}_2 \end{pmatrix}' (I \otimes \Theta) \begin{pmatrix} \bar{x}_1 \\ \bar{x}_2 \end{pmatrix}\end{aligned}$$

where

$$\Theta = \begin{pmatrix} -I & P_1 A_{12} + A_{21}' \\ A_{12}' P_1 + A_{21} & bI - \rho\beta I \end{pmatrix}. \tag{11.118}$$

Let

$$\rho^* = \frac{1}{\beta} \| P_1 A_{12} + A_{21}' \|^2 + b.$$

Clearly $\Theta \leq 0$ for $\rho > \rho^*$. Therefore for $\rho > \rho^*$, we have

$$\dot{V} < 0.$$

It is then easily verified that we have achieved synchronization. ■

Chapter 12
Synchronization of Continuous-Time Nonlinear Time-Varying MAS

12.1 Introduction

In Chap. 5, we considered a specific class of nonlinear, continuous-time agents connected as a multi-agent system (MAS). We designed protocols for fixed networks with either full- or partial-state coupling.

On the other hand, in Chap. 11 we considered time-varying networks. Specific but important cases, studied in that chapter, are switching networks.

In this chapter, we connect these two chapters by considering a MAS with nonlinear agents that have the structure as studied in Chap. 5 but connected via a time-varying, switching network as studied in Chap. 11. Both full-state coupling and partial-state coupling are considered. The protocol for our nonlinear time-varying agents are based on a high-gain method in the case of full-state coupling, while for networks with partial-state coupling, we design a protocol based on a low- and high-gain method.

12.2 MAS with Full-State Coupling

12.2.1 Problem Formulation

Consider a MAS composed of N identical nonlinear time-varying agents that can be represented in the canonical form

$$\dot{x}_i = A_d x_i + \phi(t, x_i) + B_d(u_i + Ex_i), \quad (i = 1, \dots, N) \tag{12.1}$$

where $x_i \in \mathbb{R}^n$ and $u_i \in \mathbb{R}^m$ are states and inputs of agent i. Assume that the above agents (12.1) have a uniform relative degree ρ, while $A_d \in \mathbb{R}^{n\times n}$ and $B_d \in \mathbb{R}^{n\times m}$

A. Saberi et al., *Synchronization of Multi-Agent Systems in the Presence of Disturbances and Delays*, Systems & Control: Foundations & Applications,
https://doi.org/10.1007/978-3-030-88148-1_12

are defined by

$$A_d = \begin{pmatrix} 0 & I_{(\rho-1)m} \\ 0 & 0 \end{pmatrix}, \quad B_d = \begin{pmatrix} 0 \\ I_m \end{pmatrix}. \tag{12.2}$$

The time-varying nonlinearity ϕ satisfies the following assumption.

Assumption 12.1 *Assume that $\phi(t, x)$ is continuously differentiable and globally Lipschitz continuous with respect to x, uniformly in t and piecewise continuous with respect to t. This implies that there exists a M (independent of t and x) such that*

$$\|\phi(t, x_1) - \phi(t, x_2)\| \leq M\|x_1 - x_2\|$$

for all $t \in \mathbb{R}$ and all $x_1, x_2 \in \mathbb{R}^{\rho m}$.

Moreover, the nonlinearity has a lower-triangular structure in the sense that

$$\frac{\partial \phi_j(t, x_i)}{\partial x_{ik}} = 0, \quad \forall k > j. \tag{12.3}$$

Note that in Sect. 5.2.3, we considered the question which class of agents can be transformed in the specific canonical form (12.1) while satisfying Assumption 12.1.

The communication network among agents is exactly the same as in previous chapters and provides each agent with the quantity

$$\zeta_i = \sum_{j=1}^{N} a_{ij}(x_i - x_j) = \sum_{j=1}^{N} \ell_{ij} x_j, \tag{12.4}$$

for $i = 1, \ldots, N$. We formulate the state synchronization problem for the time-varying network with full-coupled nonlinear time-varying agents as follows.

Problem 12.1 (Full-State Coupling) Consider a MAS described by (12.1) and (12.4). For any real numbers $\alpha, \beta, \tau > 0$ and a positive integer N, the *state synchronization* problem with a set of time-varying network graphs $\mathbb{G}_{\alpha,\beta}^{\tau,N}$ is to find, if possible, a linear static protocol of the form (11.5) such that, for any time-varying graph $\mathcal{G}_t \in \mathbb{G}_{\alpha,\beta}^{\tau,N}$ and for all the initial conditions of agents, state synchronization among agents can be achieved.

12.2.2 Protocol Design

In this section, we show that the Protocol Design 5.1 for the fixed network still works for the time-varying network.

Protocol design 12.1 *Consider a MAS described by* (12.1) *and* (12.4). *We design the protocol*

$$u_i = \varepsilon^{-\rho} F S_\varepsilon \zeta_i, \tag{12.5}$$

where ε *is the high-gain parameter and* S_ε *is the high-gain matrix defined by*

$$S_\varepsilon = \begin{pmatrix} I_m & 0 & \cdots & 0 \\ 0 & \varepsilon I_m & \ddots & \vdots \\ \vdots & \ddots & \ddots & 0 \\ 0 & \cdots & 0 & \varepsilon^{\rho-1} I_m \end{pmatrix}, \tag{12.6}$$

and $F = -B_d' P$ *with* $P = P' > 0$ *the unique solution of the algebraic Riccati equation*

$$A'P + PA - 2\beta PBB'P + Q = 0, \tag{12.7}$$

where $Q > 0$.

The result based on the above design is stated in the following theorem.

Theorem 12.2 *Consider a MAS described by* (12.1) *and* (12.4). *Let any real numbers* $\alpha, \beta, \tau > 0$ *and a positive integer* N *be given.*

If (A, B) *is stabilizable, then under Assumption 5.1 the state synchronization problem is solvable. In particular, there exists an* $\varepsilon^* \in (0, 1]$ *such that for all* $\varepsilon \in (0, \varepsilon^*]$, *the protocol* (12.5) *solves the state synchronization problem for any time-varying graph* $\mathcal{G}_t \in \mathbb{G}_{\alpha,\beta}^{\tau,N}$.

Proof For each $i \in \{1, \ldots, N\}$, let $\bar{x}_i = x_N - x_i$. The state synchronization is achieved if $\bar{x}_i \to 0$ for all $i \in \{1, \ldots, N-1\}$. We have $\bar{L}_t = T_1 L_t T_2$ where T_1 and T_2 are given by (6.91). Define $U_t^{-1} \bar{L}_t U_t = S_t$, where S_t is the Schur form of $\bar{L}_t$.

By following the exact same steps as in the proof of Theorem 5.2, we get

$$\varepsilon \dot{\nu} = \hat{A}_t \nu + \hat{W}_\varepsilon \nu, \tag{12.8}$$

for

$$\nu = (U \otimes I_{p\rho})\xi = (U \otimes I_{p\rho}) \operatorname{diag}(\xi_i)$$

where $\xi_i = S_\varepsilon \bar{x}_i$, while

$$\hat{A}_t = (I_{N-1} \otimes A_d) + (S_t \otimes B_d F) \hat{W}_{t,\varepsilon} = (U_t \otimes I_{p\rho}) W_\varepsilon (U_t^{-1} \otimes I_{p\rho}).$$

In the above, $W_\varepsilon = \mathrm{diag}(W_{i\varepsilon})$ and

$$W_{i\varepsilon} = \varepsilon^\rho B_d E S_\varepsilon^{-1} - \varepsilon S_\varepsilon \Phi_i(t) S_\varepsilon^{-1}.$$

Recall from the proof of Theorem 5.2.

$$\phi(t, x_N) - \phi(t, x_i) = \Phi_i(t)\bar{x}_i,$$

where

$$\Phi_i(t) = \int_0^1 \frac{\partial \phi}{\partial x_i}(t, x_i + p\bar{x}_i)\mathrm{d}p. \tag{12.9}$$

with $\Phi_i(t)$ uniformly bounded, and the lower-triangular structure of the nonlinearity implies that $\Phi_i(t)$ is lower-triangular.

We find that ν has discontinuous jumps when the network graph switches. According to Lemma 11.7, the matrix $\hat{A}_t$ is Hurwitz during a time interval $t \in [t_{k-1}, t_k)$, and there exists a $\hat{P}$ of the form

$$\hat{P} = \mathrm{blkdiag}\{\alpha_1 P, \ldots, \alpha_{N-1} P\},$$

such that

$$\hat{P}\hat{A}_t + \hat{A}_t'\hat{P} \leq -\mu\hat{P} - I,$$

where μ is a small enough real number.

Consider a Lyapunov function $V = \varepsilon\mu'\hat{P}\nu$, for which we have

$$\begin{aligned}\dot{V} &= -\mu\varepsilon^{-1}V - \|\nu\|^2 + 2\,\mathrm{Re}(\nu'\hat{W}_{t,\varepsilon}\hat{P}\nu)\\ &\leq -\mu\varepsilon^{-1}V - \|\nu\|^2 + \varepsilon r_1\|\nu\|^2\\ &\leq -\mu\varepsilon^{-1}V,\end{aligned}$$

for ε small enough such that $\varepsilon r_1 \geq \|\hat{W}_{t,\varepsilon}\hat{P}\|$.

Following the steps in the proof of Theorem 11.6, for a small enough ε, we achieve that $\bar{x}_i \to 0$ for all $i \in \{1, \ldots, N-1\}$. ■

12.3 MAS with Partial-State Coupling

12.3.1 Problem Formulation

Consider a MAS composed of N identical nonlinear time-varying agents that can be represented in the canonical form

$$\begin{aligned} \dot{x}_{ia} &= A_a x_{ia} + L_{ad} y_i, \\ \dot{x}_{id} &= A_d x_{id} + \phi_d(t, x_{ia}, x_{id}) + B_d(u_i + E_{da} x_{ia} + E_{dd} x_{id}), \\ y_i &= C_d x_{id}, \end{aligned} \tag{12.10}$$

where

$$x_i := \begin{pmatrix} x_{ia} \\ x_{id} \end{pmatrix} \in \mathbb{R}^n, \qquad u_i \in \mathbb{R}^m, \qquad y_i \in \mathbb{R}^m$$

are, respectively, states, inputs, and outputs of agent i. The linear parts of the agents have a uniform relative degree denoted by ρ. A_d and B_d have the special form of (12.2), while C_d has the special form

$$C_d = \begin{pmatrix} I_m & 0 \end{pmatrix}. \tag{12.11}$$

We assume that A_a is Hurwitz stable. We impose the following assumption on the time-varying nonlinearity ϕ_d.

Assumption 12.2 *Assume that $\phi_d(t, x_{ia}, x_{id})$ is continuously differentiable and globally Lipschitz continuous with respect to (x_{ia}, x_{id}), uniformly in t, and piecewise continuous with respect to t. This implies that there exists a M (independent of t, x_a and x_d) such that*

$$\|\phi_d(t, x_{a1}, x_{d1}) - \phi_d(t, x_{a2}, x_{d2})\| \leq M\|x_{a1} - x_{a2}\| + M\|x_{d1} - x_{d2}\|$$

for all $t \in \mathbb{R}$ and all $x_{a1}, x_{a2}, x_{d1}, x_{d2}$.

Moreover, the nonlinearity has the lower-triangular structure:

$$\frac{\partial \phi_{dj}(t, x_{ia}, x_{id})}{\partial x_{idk}} = 0, \quad \forall k > j. \tag{12.12}$$

Note that in Sect. 5.3.3, we considered the question which class of agents can be transformed to the specific canonical form (12.10) while satisfying Assumption 12.2.

The communication network among agents is exactly the same as in the previous chapters and provides each agent with the quantity

$$\zeta_i = \sum_{j=1}^{N} a_{ij}(y_i - y_j) = \sum_{j=1}^{N} \ell_{ij} y_j, \tag{12.13}$$

for $i = 1, \ldots, N$. We formulate the state synchronization problem for nonlinear time-varying agents with a time-varying network as follows.

Problem 12.3 (Partial-State Coupling) Consider a MAS described by (12.10) and (12.13). For any real numbers $\alpha, \beta, \tau > 0$ and a positive integer N, the *state synchronization* problem with a set of time-varying network graphs $\mathbb{G}_{\alpha,\beta}^{\tau,N}$ is to find, if possible, a linear time-invariant dynamic protocol of the form (11.6) such that, for any time-varying graph $\mathcal{G}_t \in \mathbb{G}_{\alpha,\beta}^{\tau,N}$ and for all the initial conditions of agents and their protocol, state synchronization among agents can be achieved.

12.3.2 Protocol Design

We will show that the Protocol Design 5.3 for the fixed network still works for the time-varying network.

Protocol design 12.2 *Consider a MAS described by (12.4) and (12.10). We choose K such that $A_d - KC_d$ is Hurwitz stable. Next, we consider a feedback gain:*

$$F_\delta = -B_d' P_\delta$$

with $P_\delta > 0$ being the unique solution of the continuous-time algebraic Riccati equation:

$$P_\delta A_d + A_d' P_\delta - \beta P_\delta B_d B_d' P_\delta + \delta I = 0, \tag{12.14}$$

where $\delta > 0$ is a low-gain parameter. Then, we define

$$F_{\delta\varepsilon} = \varepsilon^{-\rho} F_\delta S_\varepsilon, \quad K_\varepsilon = \varepsilon^{-1} S_\varepsilon^{-1} K.$$

We can then construct the protocol as follows:

(continued)

Protocol design 12.2 (continued)

$$\begin{aligned}\dot{\hat{x}}_{ia} &= A_a\hat{x}_{ia} + L_{ad}C_d\hat{x}_{id},\\ \dot{\hat{x}}_{id} &= A_d\hat{x}_{id} + \phi_d(t,\hat{x}_{ia},\hat{x}_{id}) + K_\varepsilon(\zeta_i - C_d\hat{x}_{id}) + B_d(E_{da}\hat{x}_{ia} + E_{dd}\hat{x}_{id}),\\ \tilde{u}_i &= F_{\delta\varepsilon}\hat{x}_{id}.\end{aligned} \tag{12.15}$$

The main result based on the above design is stated in the following theorem.

Theorem 12.4 *Consider a MAS described by* (12.10) *and* (12.13). *Let any real numbers* $\alpha, \beta, \tau > 0$ *and a positive integer* N *be given.*

If the system is stabilizable and detectable, then under Assumption 12.2 *the state synchronization problem with partial-state coupling is solvable. In particular, there exists a* $\delta^* \in (0,1]$*, such that for each* $\delta \in (0,\delta^*]$*, there exists an* $\varepsilon^*(\delta) \in (0,1]$ *such that for all* $\varepsilon \in (0,\varepsilon^*(\delta)]$*, the protocol* (12.15) *solves the state synchronization problem with a set of time-varying network graphs* $\mathbb{G}^{\tau,N}_{\alpha,\beta}$.

Proof For each $i \in \{1,\ldots,N-1\}$, let

$$\bar{x}_i = \begin{pmatrix}\bar{x}_{ia}\\ \bar{x}_{id}\end{pmatrix} := x_i - x_N, \qquad \hat{\bar{x}}_i := \begin{pmatrix}\hat{\bar{x}}_{ia}\\ \hat{\bar{x}}_{id}\end{pmatrix} = \hat{x}_i - \hat{x}_N, \qquad \text{where} \quad \hat{x}_i = \begin{pmatrix}\hat{x}_{ia}\\ \hat{x}_{id}\end{pmatrix}.$$

Then, state synchronization among agents is achieved if $\bar{x}_i \to 0$ for $i = 1,\ldots,N-1$. Similar to the proof of Theorem 5.2, we can write

$$\phi_d(t,x_{Na},x_{Nd}) - \phi_d(t,x_{ia},x_{id}) = \Phi_{ia}(t)\bar{x}_{ia} + \Phi_{id}(t)\bar{x}_{id},$$

using Taylor's theorem where $\Phi_{ia}(t)$ and $\Phi_{id}(t)$ are given by

$$\begin{aligned}\Phi_{ia}(t) &= \int_0^1 \frac{\partial\phi_d}{\partial x_{ia}}(t, x_{ia} + p\bar{x}_{ia}, x_{id} + p\bar{x}_{id})\mathrm{d}p,\\ \Phi_{id}(t) &= \int_0^1 \frac{\partial\phi_d}{\partial x_{id}}(t, x_{ia} + p\bar{x}_{ia}, x_{id} + p\bar{x}_{id})\mathrm{d}p.\end{aligned}$$

Due to the Lipschitz property of the nonlinearity, the elements of $\Phi_{ia}(t)$ and $\Phi_{id}(t)$ are uniformly bounded, and the lower-triangular structure of the nonlinearity implies that $\Phi_{id}(t)$ is lower-triangular. Moreover, we have

$$\phi_d(t,\hat{x}_{Na},\hat{x}_{Nd}) - \phi_d(t,\hat{x}_{ia},\hat{x}_{id}) = \hat{\Phi}_{ia}(t)\hat{\bar{x}}_{ia} + \hat{\Phi}_{id}(t)\hat{\bar{x}}_{id},$$

for matrices $\hat{\Phi}_{ia}(t)$ and $\hat{\Phi}_{id}(t)$ with the same properties.

Following the proof of Theorem 5.9, we surely obtain

$$\begin{aligned}
\dot{\xi}_{ia} &= A_a\xi_{ia} + V_{iad}\xi_{id},\\
\dot{\hat{\xi}}_{ia} &= A_a\hat{\xi}_{ia} + \hat{V}_{iad}\hat{\xi}_{id},\\
\varepsilon\dot{\xi}_{id} &= A_d\xi_{id} + B_dF_\delta\hat{\xi}_{id} + V^\varepsilon_{ida}\xi_{ia} + V^\varepsilon_{idd}\xi_{id},\\
\varepsilon\dot{\hat{\xi}}_{id} &= A_d\hat{\xi}_{id} + \hat{V}^\varepsilon_{ida}\hat{\xi}_{ia} + \hat{V}^\varepsilon_{idd}\hat{\xi}_{id} + \sum_{j=1}^{N-1}\bar{\ell}_{ij}(t)KC_d\xi_{jd} - KC_d\hat{\xi}_{id},
\end{aligned}$$

where

$$\xi_{ia} = \bar{x}_{ia}, \quad \hat{\xi}_{ia} = \hat{\bar{x}}_{ia}, \quad \xi_{id} = S_\varepsilon\bar{x}_{id}, \quad \text{and} \quad \hat{\xi}_{id} = S_\varepsilon\hat{\bar{x}}_{id},$$

and

$$\begin{aligned}
V_{iad} &= \hat{V}_{iad} = L_{ad}C_d,\\
V^\varepsilon_{ida} &= \varepsilon^\rho B_dE_{da} - \varepsilon S_\varepsilon\Phi_{ia}(t),\\
\hat{V}^\varepsilon_{ida} &= \varepsilon^\rho B_dE_{da} - \varepsilon S_\varepsilon\hat{\Phi}_{ia}(t),\\
V^\varepsilon_{idd} &= \varepsilon^\rho B_dE_{dd}S_\varepsilon^{-1} - \varepsilon S_\varepsilon\Phi_{id}(t)S_\varepsilon^{-1},\\
\hat{V}^\varepsilon_{idd} &= \varepsilon^\rho B_dE_{dd}S_\varepsilon^{-1} - \varepsilon S_\varepsilon\hat{\Phi}_{id}(t)S_\varepsilon^{-1}.
\end{aligned}$$

Define a unitary U such that $U^{-1}\bar{L}U = S$, where S is the Schur form of $\bar{L}$, and let

$$\begin{aligned}
v_a &= (SU^{-1}\otimes I_{n-p\rho})\,\mathrm{vec}(\xi_{ia}),\\
\tilde{v}_a &= v_a - (SU^{-1}\otimes I_{n-p\rho})\,\mathrm{vec}(\hat{\xi}_{ia}),\\
v_d &= (SU^{-1}\otimes I_{p\rho})\,\mathrm{vec}(\xi_{id}),\\
\tilde{v}_d &= v_d - (U^{-1}\otimes I_{p(\rho})\,\mathrm{vec}(\hat{\xi}_{id}).
\end{aligned}$$

Then

$$\begin{aligned}
\dot{v}_a &= (I_{N-1}\otimes A_a)v_a + W_{ad}v_d,\\
\dot{\tilde{v}}_a &= (I_{N-1}\otimes A_a)\tilde{v}_a + W_{ad}v_d - \hat{W}_{ad}(v_d - \tilde{v}_d),\\
\varepsilon\dot{v}_d &= (I_{N-1}\otimes A_d)v_d + (S\otimes B_dF_\delta)(v_d - \tilde{v}_d) + W^\varepsilon_{da}v_a + W^\varepsilon_{dd}v_d,\\
\varepsilon\dot{\tilde{v}}_d &= (I_{N-1}\otimes A_d)\tilde{v}_d + (S\otimes B_dF_\delta)(v_d - \tilde{v}_d) + W^\varepsilon_{da}v_a - \hat{W}^\varepsilon_{da}(v_a - \tilde{v}_a)\\
&\quad + W^\varepsilon_{dd}v_d - \hat{W}^\varepsilon_{dd}(v_d - \tilde{v}_d) - (I_{N-1}\otimes KC_d)\tilde{v}_d,
\end{aligned}$$

where

$$
\begin{aligned}
W_{ad} &= (SU^{-1} \otimes I_{n-m\rho})V_{ad}(US^{-1} \otimes I_{m\rho}),\\
\hat{W}_{ad} &= (SU^{-1} \otimes I_{n-m\rho})\hat{V}_{ad}(U \otimes I_{m\rho}),\\
W^{\varepsilon}_{da} &= (SU^{-1} \otimes I_{m\rho})V^{\varepsilon}_{da}(US^{-1} \otimes I_{n-m\rho}),\\
W^{\varepsilon}_{dd} &= (SU^{-1} \otimes I_{m\rho})V^{\varepsilon}_{dd}(US^{-1} \otimes I_{m\rho}),\\
\hat{W}^{\varepsilon}_{da} &= (U^{-1} \otimes I_{m\rho})\hat{V}^{\varepsilon}_{da}(US^{-1} \otimes I_{n-m\rho}),\\
\hat{W}^{\varepsilon}_{dd} &= (U^{-1} \otimes I_{m\rho})\hat{V}^{\varepsilon}_{dd}(U \otimes I_{m\rho}).
\end{aligned}
$$

Then

$$
\begin{aligned}
\dot{\eta}_a &= \tilde{A}_a\eta_a + \tilde{W}_{ad}\eta_d,\\
\varepsilon\dot{\eta}_d &= \tilde{A}_\delta\eta_d + \tilde{W}^{\varepsilon}_{da}\eta_a + \tilde{W}^{\varepsilon}_{dd}\eta_d,
\end{aligned}
\tag{12.16}
$$

where $\tilde{A}_a = I_{2(N-1)} \otimes A_a$ and

$$
\tilde{A}_\delta = I_{N-1} \otimes \begin{pmatrix} A_d & 0 \\ 0 & A_d - KC_d \end{pmatrix} + S \otimes \begin{pmatrix} B_dF_\delta & -B_dF_\delta \\ B_dF_\delta & -B_dF_\delta \end{pmatrix}.
$$

Moreover,

$$
\begin{aligned}
\tilde{W}_{ad} &= N_a \begin{pmatrix} W_{ad} & 0 \\ W_{ad} - \hat{W}_{ad} & \hat{W}_{ad} \end{pmatrix} N_d^{-1},\\
\tilde{W}^{\varepsilon}_{da} &= N_d \begin{pmatrix} W^{\varepsilon}_{da} & 0 \\ W^{\varepsilon}_{da} - \hat{W}^{\varepsilon}_{da} & \hat{W}^{\varepsilon}_{da} \end{pmatrix} N_a^{-1},\\
\tilde{W}^{\varepsilon}_{dd} &= N_d \begin{pmatrix} W^{\varepsilon}_{dd} & 0 \\ W^{\varepsilon}_{dd} - \hat{W}^{\varepsilon}_{dd} & \hat{W}^{\varepsilon}_{dd} \end{pmatrix} N_d^{-1}.
\end{aligned}
$$

We have now obtained (12.16) which has the same structure as (11.30) in the proof of Theorem 11.9. The rest of the proof follows along the lines of the proof of Theorem 11.9. ∎

Chapter 13
H_∞ and H_2 Almost Synchronization of Continuous-Time Linear MAS

13.1 Introduction

So far we considered several synchronization problems for a MAS. Most research has focused on the idealized case where the agents are not affected by external disturbances. In the literature where external disturbances are considered, γ-suboptimal H_∞ design is developed for a MAS to achieve H_∞ norm from an external disturbance to the synchronization error among agents less than to an a priori given γ. In particular, [49, 194] considered the H_∞ norm from an external disturbance to the output error among agents. Saboori and Khorasani [107] considered the H_∞ norm from an external disturbance to the state error among agents. These papers do not present an explicit methodology for designing protocols. The papers [54] and [59] try to obtain an H_∞ norm from a disturbance to the average of the states in a network of single or double integrators.

By contrast, [84] introduced the notion of H_∞ *almost synchronization* for homogeneous MAS, where the goal is to reduce the H_∞ norm from an external disturbance to the synchronization *error*, to any arbitrary desired level. This work is extended later in [83, 179], and [182]. However, in these works, H_∞ *almost output synchronization* is achieved. Moreover, by restricting to homogeneous networks, more explicit designs can be obtained under weaker conditions.

In this chapter, we study the H_∞ *almost state synchronization* for a MAS with a full-state coupling or a partial-state coupling. We also study the H_2 almost state synchronization, since it is closely related to the problems of H_∞ almost state synchronization. In the H_∞ case, we look at the worst case disturbance with the only constraints being on the power, while in the H_2 we only consider white noise disturbances which is a more restrictive class. In both cases, we only consider process noise and not measurement noise.

The write-up of this chapter is partially based on [132].

A. Saberi et al., *Synchronization of Multi-Agent Systems in the Presence of Disturbances and Delays*, Systems & Control: Foundations & Applications,
https://doi.org/10.1007/978-3-030-88148-1_13

13.2 Problem Formulation

Consider a MAS composed of N identical linear time-invariant agents of the form

$$\begin{aligned} \dot{x}_i &= Ax_i + Bu_i + E\omega_i, \\ y_i &= Cx_i, \end{aligned} \qquad (i = 1, \dots, N) \tag{13.1}$$

where $x_i \in \mathbb{R}^n$, $u_i \in \mathbb{R}^m$, $y_i \in \mathbb{R}^p$ are, respectively, the state, input, and output vectors of agent i and $\omega_i \in \mathbb{R}^q$ is the external disturbances.

The communication network among agents is exactly the same as that in Chap. 2 and provides each agent, in the case of partial-state coupling with the quantity

$$\zeta_i = \sum_{j=1}^{N} a_{ij}(y_i - y_j) = \sum_{j=1}^{N} \ell_{ij} y_j, \quad i = 1, \dots, N. \tag{13.2}$$

If we consider full-state coupling, i.e., $C = I$, then the quantity ζ_i becomes

$$\zeta_i = \sum_{j=1}^{N} a_{ij}(x_i - x_j) = \sum_{j=1}^{N} \ell_{ij} x_j. \tag{13.3}$$

The graph $\mathcal{G}$ describing the communication topology of the network is always assumed to have a directed spanning tree.

Let N be any agent and define $\bar{x}_i = x_N - x_i$ and

$$\bar{x} = \begin{pmatrix} \bar{x}_1 \\ \vdots \\ \bar{x}_{N-1} \end{pmatrix} \quad \text{and} \quad \omega = \begin{pmatrix} \omega_1 \\ \vdots \\ \omega_N \end{pmatrix}.$$

Obviously, synchronization is achieved if $\bar{x} = 0$. That is

$$\lim_{t\to\infty} (x_i(t) - x_N(t)) = 0, \quad \text{for all } i \in \{1, \dots, N-1\}. \tag{13.4}$$

Remark 13.1 The agent N is not necessarily a root agent. Moreover, (13.4) is equivalent to the condition

$$\lim_{t\to\infty} (x_i(t) - x_j(t)) = 0, \quad \text{for all } i, j \in \{1, \dots, N\}.$$

We formulate below four almost state synchronization problems for a network with either H_2 or H_∞ almost synchronization.

Problem 13.2 Consider a MAS described by (13.1) and (13.3). Let $\mathbf{G}$ be a given set of graphs such that $\mathbf{G} \subseteq \mathbb{G}^N$. The $\boldsymbol{H_\infty}$ **almost state synchronization problem**

via full-state coupling with a set of network graphs **G** is to find, if possible, a linear static protocol parameterized in terms of a parameter ε, of the form

$$u_i = F(\varepsilon)\zeta_i, \tag{13.5}$$

such that, for any given real number $r > 0$, there exists an ε^* such that for any $\varepsilon \in (0, \varepsilon^*]$ and for any graph $\mathcal{G} \in \mathbf{G}$, (13.4) is satisfied for all the initial conditions in the absence of disturbances and the closed-loop transfer matrix from ω to $\bar{x}$, denoted by $T_{\omega\bar{x}}$, satisfies

$$\|T_{\omega\bar{x}}\|_\infty < r. \tag{13.6}$$

Problem 13.3 Consider a MAS described by (13.1) and (13.2). Let **G** be a given set of graphs such that $\mathbf{G} \subseteq \mathbb{G}^N$. The ***H*$_\infty$ almost state synchronization problem via partial-state coupling** with a set of network graphs **G** is to find, if possible, a linear time-invariant dynamic protocol parameterized in terms of a parameter ε, of the form

$$\begin{aligned} \dot{\chi}_i &= A_c(\varepsilon)\chi_i + B_c(\varepsilon)\zeta_i, \\ u_i &= C_c(\varepsilon)\chi_i + D_c(\varepsilon)\zeta_i, \end{aligned} \tag{13.7}$$

where $\chi_i \in \mathbb{R}^{n_c}$, such that, for any given real number $r > 0$, there exists an ε^* such that for any $\varepsilon \in (0, \varepsilon^*]$ and for any graph $\mathcal{G} \in \mathbf{G}$, (13.4) is satisfied for all the initial conditions in the absence of disturbances and the closed-loop transfer matrix from ω to $\bar{x}$, denoted by $T_{\omega\bar{x}}$, satisfies (13.6).

Problem 13.4 Consider a MAS described by (13.1) and (13.3). Let **G** be a given set of graphs such that $\mathbf{G} \subseteq \mathbb{G}^N$. The ***H*$_2$ almost state synchronization problem via full-state coupling** with a set of network graphs **G** is to find, if possible, a linear static protocol parameterized in terms of a parameter ε, of the form

$$u_i = F(\varepsilon)\zeta_i, \tag{13.8}$$

such that, for any given real number $r > 0$, there exists an ε^* such that for any $\varepsilon \in (0, \varepsilon^*]$ and for any graph $\mathcal{G} \in \mathbf{G}$, (13.4) is satisfied for all the initial conditions in the absence of disturbances and the closed-loop transfer matrix from ω to $\bar{x}$, denoted by $T_{\omega\bar{x}}$, satisfies

$$\|T_{\omega\bar{x}}\|_2 < r. \tag{13.9}$$

Problem 13.5 Consider a MAS described by (13.1) and (13.2). Let **G** be a given set of graphs such that $\mathbf{G} \subseteq \mathbb{G}^N$. The ***H*$_2$ almost state synchronization problem via partial-state coupling** with a set of network graphs **G** is to find, if possible, a linear time-invariant dynamic protocol parameterized in terms of a parameter ε, of the form

$$\begin{aligned} \dot{\chi}_i &= A_c(\varepsilon)\chi_i + B_c(\varepsilon)\zeta_i, \\ u_i &= C_c(\varepsilon)\chi_i + D_c(\varepsilon)\zeta_i, \end{aligned} \tag{13.10}$$

where $\chi_i \in \mathbb{R}^{n_c}$, such that, for any given real number $r > 0$, there exists an ε^* such that for any $\varepsilon \in (0, \varepsilon^*]$ and for any graph $\mathcal{G} \in \mathbf{G}$, (13.4) is satisfied for all the initial conditions in the absence of disturbances and the closed-loop transfer matrix from ω to $\bar{x}$, denoted by $T_{\omega\bar{x}}$, satisfies (13.9).

Earlier in this book, in particular Sect. 2.5.7, we also considered static protocols for a MAS with partial-state coupling when the agents have a specific structure related to passivity. We will also consider the Problems 13.3 and 13.5 restricted to static protocols.

Note that the problems of H_∞ almost state synchronization and H_2 almost state synchronization are closely related. Roughly speaking, H_2 almost synchronization is easier to achieve than H_∞ almost synchronization. This is related to the fact that in H_∞ we look at the worst case disturbance with the only constraints being the power

$$\limsup_{T\to\infty} \frac{1}{2T} \int_{-T}^{T} \omega_i'(t)\omega_i(t)\mathrm{d}t < \infty,$$

while in H_2 we only consider white noise disturbances which is a more restrictive class.

13.3 Protocol Design for a MAS with Full-State Coupling

In this section, we establish a connection between the almost state synchronization among agents in the network and a robust H_∞ or H_2 almost disturbance decoupling problem via state feedback with internal stability (in short H_∞ or H_2-ADDPSS). Then we design a controller for such robust stabilization problems.

13.3.1 *Necessary and Sufficient Conditions for Almost Synchronization*

We derive here the necessary and sufficient conditions based on the connection between almost state synchronization and H_∞ or H_2 almost disturbance decoupling.

13.3.1.1 H_∞ Almost Synchronization

The MAS system described by (13.1) and (13.3) after implementing the linear static protocol (13.5) is described by

$$\dot{x}_i = Ax_i + BF(\varepsilon)\zeta_i + E\omega_i,$$

for $i = 1, \ldots, N$. Let

$$x = \begin{pmatrix} x_1 \\ \vdots \\ x_N \end{pmatrix}, \qquad \omega = \begin{pmatrix} \omega_1 \\ \vdots \\ \omega_N \end{pmatrix}.$$

Then, the overall dynamics of the N agents can be written as

$$\dot{x} = (I_N \otimes A + L \otimes BF(\varepsilon))x + (I_N \otimes E)\omega. \tag{13.11}$$

We define robust H_∞ almost disturbance decoupling with bounded input as follows. Given $\Lambda \subset \mathbb{C}$, there should exist $M > 0$ such that for any given real number $r > 0$, we can find a parameterized controller

$$u = F(\varepsilon)x \tag{13.12}$$

for the following subsystem:

$$\dot{x} = Ax + \lambda Bu + Bw, \tag{13.13}$$

such that for any $\lambda \in \Lambda$ the following hold:

1. The interconnection of the systems (13.13) and (13.12) is internally stable.
2. The resulting closed-loop transfer function T_{wx} from w to x has an H_∞ norm less than r.
3. The resulting closed-loop transfer function T_{wu} from w to u has an H_∞ norm less than M.

In the above, Λ denotes all the possible locations for the nonzero eigenvalues of the Laplacian matrix L when the graph varies over the set $\mathbf{G}$. It is also important to note that M is independent of the choice of r.

Theorem 13.6 *Let* $\mathbf{G}$ *be a set of graphs such that the associated Laplacian matrices are uniformly bounded and let* Λ *consist of all possible nonzero eigenvalues of Laplacian matrices associated with graphs in* $\mathbf{G}$.

(Necessity) The H_∞ almost state synchronization *problem as defined in Problem 13.2 for the MAS described by* (13.1) *and* (13.3) *given* $\mathbf{G}$ *is solvable by a parameterized protocol* $u_i = F(\varepsilon)\zeta_i$ *only if*

$$\operatorname{Im} E \subset \operatorname{Im} B. \tag{13.14}$$

(Sufficiency) The H_∞ almost state synchronization *problem as defined in Problem 13.2 for the MAS described by* (13.1) *and* (13.3) *for the given* $\mathbf{G}$ *is solved by a parameterized protocol* $u_i = F(\varepsilon)\zeta_i$ *if the robust* H_∞*-ADDPSS with bounded input for the system* (13.13) *with* $\lambda \in \Lambda$ *is solved by the parameterized controller* $u = F(\varepsilon)x$.

Proof Note that L has eigenvalue 0 with associated right eigenvector $\mathbf{1}$. Let

$$L = USU^{-1}, \tag{13.15}$$

with U unitary and S the upper triangular Schur form associated to the Laplacian matrix L such that $S(1, 1) = 0$. Let

$$\eta := (U^{-1} \otimes I_n)x = \begin{pmatrix} \eta_1 \\ \vdots \\ \eta_N \end{pmatrix}, \qquad \bar{\omega} = (U^{-1} \otimes I)\omega = \begin{pmatrix} \bar{\omega}_1 \\ \vdots \\ \bar{\omega}_N \end{pmatrix},$$

where $\eta_i \in \mathbb{C}^n$ and $\bar{\omega}_i \in \mathbb{C}^q$. In the new coordinates, the dynamics of η can be written as

$$\dot{\eta}(t) = (I_N \otimes A + S \otimes BF(\varepsilon))\eta + (U^{-1} \otimes E)\omega, \tag{13.16}$$

which is rewritten as

$$\begin{aligned} \dot{\eta}_1 &= A\eta_1 + \sum_{j=2}^{N} s_{1j} BF(\varepsilon)\eta_j + E\bar{\omega}_1, \\ \dot{\eta}_i &= (A + \lambda_i BF(\varepsilon))\eta_i + \sum_{j=i+1}^{N} s_{ij} BF(\varepsilon)\eta_j + E\bar{\omega}_i, \\ \dot{\eta}_N &= (A + \lambda_N BF(\varepsilon))\eta_N + E\bar{\omega}_N, \end{aligned} \tag{13.17}$$

for $i \in \{2, \ldots, N-1\}$ where $S = [s_{ij}]$. The first column of U is an eigenvector of L associated to the eigenvalue 0 with length 1, i.e., it is equal to $\mathbf{1}/\sqrt{N}$. Using this we obtain

$$\begin{aligned} \bar{x} &= \begin{pmatrix} x_N - x_1 \\ x_N - x_2 \\ \vdots \\ x_N - x_{N-1} \end{pmatrix} = \left(\begin{pmatrix} -1 & 0 & \cdots & 0 & 1 \\ 0 & -1 & \ddots & \vdots & \vdots \\ \vdots & \ddots & \ddots & 0 & \vdots \\ 0 & \cdots & 0 & -1 & 1 \end{pmatrix} \otimes I_n \right) (U \otimes I_n)\eta \\ &= \left(\begin{pmatrix} 0 & V \end{pmatrix} \otimes I_n \right) \eta, \end{aligned}$$

for some suitably chosen matrix V. Therefore we have

$$\bar{x} = (V \otimes I_n) \begin{pmatrix} \eta_2 \\ \vdots \\ \eta_N \end{pmatrix}. \tag{13.18}$$

Note that since U is unitary, also the matrix U^{-1} is unitary, and the matrix V is uniformly bounded. Therefore the H_∞ norm of the transfer matrix from ω to $\bar{x}$ can be made arbitrarily small if and only if the H_∞ norm of the transfer matrix from $\bar{\omega}$ to η can be made arbitrarily small.

In order for the H_∞ norm from $\bar{\omega}$ to η to be arbitrarily small, we need the H_∞ norm from $\bar{\omega}_N$ to η_N to be arbitrarily small. From classical results (see [103, 141]) on H_∞ almost disturbance decoupling, we find that this is only possible if (13.14) is satisfied.

Now suppose $u = F(\varepsilon)x$ solves the simultaneous H_∞-ADDPSS of (13.13) and assume (13.14) is satisfied. We show next that $u_i = F(\varepsilon)\zeta_i$ solves the H_∞ almost state synchronization problem for the MAS described by (13.1) and (13.2). Let X be such that $E = BX$.

The fact that $u = F(\varepsilon)x$ solves the simultaneous H_∞-ADDPSS of (13.13) implies that for small ε we have that $A + \lambda BF(\varepsilon)$ is asymptotically stable for all $\lambda \in \Lambda$. In particular, $A + \lambda_i BF(\varepsilon)$ is asymptotically stable for $i = 2, \ldots, N$ which guarantees that $\eta_i \to 0$ for $i = 2, \ldots, N$ for zero disturbances and all the initial conditions. Therefore we have the state synchronization.

Next, we are going to show that for any $\bar{r} > 0$, we can choose an ε sufficiently small such that the transfer matrix from $\bar{\omega}$ to η_i is less than $\bar{r}$ for $i = 2, \ldots, N$. This guarantees that we can achieve (13.6) for any $r > 0$. We have for a given M and arbitrary small $\tilde{r}$ that for ε small enough that

$$T^{\lambda}_{wx}(s) = (sI - A - \lambda BF(\varepsilon))^{-1}B,$$
$$T^{\lambda}_{wu}(s) = F(\varepsilon)(sI - A - \lambda BF(\varepsilon))^{-1}B$$

satisfies

$$\|T^{\lambda}_{wx}\|_\infty < \tilde{r}, \qquad \|T^{\lambda}_{ux}\|_\infty < M$$

for all $\lambda \in \Lambda$. Denote $\nu_i = F(\varepsilon)\eta_i$.

When $i = N$, it is easy to find that

$$T_{\bar{\omega}\eta_N} = T^{\lambda_N}_{wx} \begin{pmatrix} 0 \cdots 0 \; X \end{pmatrix},$$

and hence

$$\|T_{\bar{\omega}\eta_N}\|_\infty < \bar{r}, \qquad \|T_{\bar{\omega}\nu_N}\|_\infty < \bar{M}_N$$

provided

$$\|X\|\tilde{r} < \bar{r}, \qquad \|X\|M < \bar{M}_N. \tag{13.19}$$

Recall that we can make $\tilde{r}$ arbitrarily small without affecting the bound M. Assume

$$\|T_{\bar{\omega}\eta_j}\|_\infty < \bar{r}, \qquad \|T_{\bar{\omega}v_j}\|_\infty < \bar{M}_j$$

holds for $j = i+1, \ldots, N$. We have

$$T_{\bar{\omega}\eta_i}(s) = T_{wx}^{\lambda_i}(s)\left[e_i \otimes X + \sum_{j=i+1}^{N} s_{ij} T_{\bar{\omega}v_j}(s)\right]$$

where e_i is a row vector of dimension N with elements equal to zero except for the ith component which is equal to 1. Since

$$\left\| e_i \otimes X + \sum_{j=i+1}^{N} s_{ij} T_{\bar{\omega}v_j} \right\|_\infty < \|X\| + \sum_{j=i+1}^{N} |s_{ij}| \bar{M}_j,$$

we find that

$$\|T_{\bar{\omega}\eta_i}\|_\infty < \bar{r}, \qquad \|T_{\bar{\omega}v_i}\|_\infty < \bar{M}_i, \tag{13.20}$$

provided

$$\left(\|X\| + \sum_{j=i+1}^{N} |s_{ij}| \bar{M}_j \right) \tilde{r} < \bar{r}, \qquad \left(\|X\| + \sum_{j=i+1}^{N} |s_{ij}| \bar{M}_j \right) \tilde{M} < \bar{M}_i. \tag{13.21}$$

Note that s_{ij} depends on the graph in $\mathbb{G}$, but since the Laplacian matrices associated to graphs in $\mathbb{G}$ are uniformly bounded, we find that also the s_{ij} are uniformly bounded. In this way we can recursively obtain the bounds in (13.20) for $i = 2, \ldots, N$ provided we choose an ε sufficiently small such that the corresponding $\tilde{r}$ satisfies (13.19) and (13.21) for $i = 2, \ldots, N-1$. Hence, we can choose ε sufficiently small such that the transfer matrix from $\bar{\omega}$ to η_i is less than $\bar{r}$ for $i = 2, \ldots, N$. As noted before this guarantees that we can achieve (13.6) for any $r > 0$. ■

For the case when the set of graphs $\mathbf{G} = \mathbb{G}_{\alpha,\beta}^N$ with given $\alpha, \beta > 0$, we develop necessary and sufficient conditions for the solvability of the H_∞ almost state synchronization problem as follows.

Theorem 13.7 *Consider a MAS described by* (13.1) *and* (13.2) *with an associated graph from the set* $\mathbf{G}$*. Moreover, there exist* $\alpha, \beta > 0$ *such that* $\mathbf{G} = \mathbb{G}^N_{\alpha,\beta}$*.*

Then, the H_∞ *almost state synchronization problem as defined in Problem 13.2 is solvable if and only if* (13.14) *is satisfied and* (A, B) *is stabilizable.*

Proof We have already noted before that (13.14) is actually a necessary condition for the H_∞ almost state synchronization problem. Clearly, also (A, B) is stabilizable as a necessary condition. Sufficiency is a direct result of Theorem 13.10 or Theorem 13.13 for H_∞ almost state synchronization. ∎

13.3.1.2 H_2 Almost Synchronization

We define the robust H_2-ADDPSS with bounded input as follows. Given $\Lambda \subset \mathbb{C}$, there should exist a $M > 0$ such that for any given real number $r > 0$, we can find a parameterized controller (13.12) for the following subsystem, (13.13) such that the following holds for any $\lambda \in \Lambda$:

1. The interconnection of the systems (13.13) and (13.12) is internally stable.
2. The resulting closed-loop transfer function T_{wx} from w to x has an H_2 norm less than r.
3. The resulting closed-loop transfer function T_{wu} from w to u has an H_∞ norm less than M.

In the above, Λ denotes all the possible locations for the nonzero eigenvalues of the Laplacian matrix L when the graph varies over the set $\mathbf{G}$. It is also important to note that M is independent of the choice of r. Note that we need to address two aspects in our controller H_2 disturbance rejection and robust stabilization (because of the uncertain Laplacian). The latter translates in the H_∞ norm constraint from w to u.

Theorem 13.8 *Let* $\mathbf{G}$ *be a set of graphs such that the associated Laplacian matrices are uniformly bounded, and let* Λ *consist of all possible nonzero eigenvalues of Laplacian matrices associated with graphs in* $\mathbf{G}$*.*

(Necessity) The H_2 *almost state synchronization problem, as defined in Problem 13.4, for the MAS described by* (13.1) *and* (13.2) *given* $\mathbf{G}$ *is solvable by a parameterized protocol* $u_i = F(\varepsilon)\zeta_i$ *only if* (13.14) *is satisfied.*

(Sufficiency)The H_2 *almost state synchronization problem, as defined in Problem 13.4, for the MAS described by* (13.1) *and* (13.2) *given* $\mathbf{G}$ *is solvable by a parameterized protocol* $u_i = F(\varepsilon)\zeta_i$ *if the robust* H_2*-ADDPSS with bounded input for the system* (13.13) *with* $\lambda \in \Lambda$ *is solved by the parameterized controller* $u = F(\varepsilon)x$*.*

Proof The proof is similar to the proof of Theorem 13.6. This time we need the H_2 norm from $\bar{\omega}_N$ to η_N to be arbitrarily small, and also H_2 almost disturbance decoupling then immediately yields that we need that (13.14) is satisfied.

The rest of the proof follows the same lines except that we require the H_2 norm from $\bar{\omega}$ to η_j arbitrarily small while we keep the H_∞ norm from $\bar{\omega}$ to ν_j. Recall that for any two stable, strictly proper transfer matrices T_1 and T_2, we have

$$\|T_1T_2\|_2 \leq \|T_1\|_2\|T_2\|_\infty$$

which we need in the modifications of the proof of Theorem 13.6. ∎

For the case when the set of graph $\mathbf{G} = \mathbb{G}^N_{\alpha,\beta}$ with given $\alpha, \beta > 0$, we develop necessary and sufficient conditions for the solvability of the H_2 almost state synchronization problem as follows:

Theorem 13.9 *Consider a MAS described by* (13.1) *and* (13.2) *with an associated graph from the set* $\mathbf{G}$. *Moreover, there exists* $\alpha, \beta > 0$ *such that* $\mathbf{G} = \mathbb{G}^N_{\alpha,\beta}$.

Then, the H_2 *almost state synchronization problem as defined in Problem 13.4 is solvable if and only if* (13.14) *is satisfied and* (A, B) *is stabilizable.*

Proof We have already noted before that (13.14) is actually a necessary condition for the H_2 almost state synchronization problem. Clearly, also (A, B) is stabilizable as a necessary condition. Sufficiency of these conditions for H_2 almost state synchronization is a direct result of either Theorem 13.12 or Theorem 13.14. ∎

We present below two protocol design methods for both H_∞ and H_2 almost state synchronization problems. One relies on an algebraic Riccati equation (ARE), and the other is based on an asymptotic time-scale eigenstructure assignment (ATEA) method.

13.3.2 ARE-Based Method

Using an algebraic Riccati equation, we can design a suitable protocol.

Protocol design 13.1 *Consider a MAS described by* (13.1) *and* (13.3). *Let any real numbers* $\alpha, \beta > 0$ *and a positive integer* N *be given.*

We consider the protocol

$$u_i = \rho F \zeta_i, \tag{13.22}$$

where $\rho = \frac{1}{\varepsilon}$ *and* $F = -B'P$ *with* P *being the unique solution of the continuous-time algebraic Riccati equation:*

$$A'P + PA - 2\beta PBB'P + I = 0. \tag{13.23}$$

The main result regarding H_∞ almost state synchronization problem is stated as follows.

Theorem 13.10 *Consider a MAS described by* (13.1) *and* (13.3) *such that* (13.14) *is satisfied. Let any real numbers* $\alpha, \beta > 0$ *and a positive integer* N *be given.*

If (A, B) *is stabilizable, then the* H_∞ *almost state synchronization problem, as defined in Problem 13.2, with* $\mathbf{G} = \mathbb{G}^N_{\alpha,\beta}$ *is solvable. In particular, for any given real number* $r > 0$*, there exists an* ε^**, such that for any* $\varepsilon \in (0, \varepsilon^*)$*, the protocol* (13.22) *achieves state synchronization and an* H_∞ *norm from* ω *to* $x_i - x_j$ *less than* r *for any* $i, j \in 1, \ldots, N$ *and for any graph* $\mathcal{G} \in \mathbb{G}^N_{\alpha,\beta}$.

Proof Since the graph will be in $\mathbb{G}^N_{\alpha,\beta}$, we know the Laplacian matrix is bounded. Therefore, we can apply Theorem 13.6, and we find that we only need to verify that $u = \rho F x$ solves the robust H_∞ almost disturbance decoupling with bounded input for the system (13.13) with $\lambda \in \Lambda$. Given $\mathcal{G} \in \mathbb{G}^N_\beta$, we know that $\lambda \in \Lambda$ implies that $\operatorname{Re}\lambda \geq \beta$.

Consider the interconnection of (13.13) and $u = \rho F x$. We define

$$V(x) = x'Px$$

and we obtain

$$\begin{aligned}
\dot{V} &= x'(A - \rho\lambda BB'P)'Px + w'B'Px \\
&\qquad + x'P(A - \rho\lambda BB'P)x + x'PBw \\
&= 2\beta x'PBB'Px - x'x - 2\rho\beta x'PBB'Px + 2x'PBw \\
&\leq (2\beta - \tfrac{\beta}{\varepsilon})x'PBB'Px - x'x + \tfrac{\varepsilon}{\beta}w'w \\
&\leq -\tfrac{\beta}{2}\varepsilon u'u - x'x + \tfrac{\varepsilon}{\beta}w'w
\end{aligned}$$

which implies that the system is asymptotically stable and the H_∞ norm of the transfer function from w to x is less than ε/β, while the H_∞ norm of the transfer function from w to u is less than $2/\beta^2$. Therefore, $u = \rho F x$ solves the robust H_∞-ADDPSS with bounded input for the system (13.13) as required. ■

For H_2 we have the following classical result.

Lemma 13.11 *Consider an asymptotically stable system:*

$$\dot{p} = A_1 p + B_1 w.$$

The H_2 *norm from* w *to* p *is less than* ε *if there exists a matrix* Q *such that*

$$AQ + QA' + BB' \leq 0, \qquad Q < \varepsilon I.$$

The main result regarding H_2 almost state synchronization is stated as follows.

Theorem 13.12 *Consider a MAS described by* (13.1) *and* (13.3) *such that* (13.14) *is satisfied. Let any real numbers* $\alpha, \beta > 0$ *and a positive integer* N *be given.*

If (A, B) *is stabilizable, then the* H_2 *almost state synchronization problem, as defined in Problem 13.4, with* $\mathbf{G} = \mathbb{G}^N_{\alpha,\beta}$ *is solvable. In particular, for any given real number* $r > 0$*, there exists an* ε^**, such that for any* $\varepsilon \in (0, \varepsilon^*)$*, the protocol* (13.22) *achieves state synchronization and an* H_2 *norm from* ω *to* $x_i - x_j$ *less than* r *for any* $i, j \in 1, \ldots, N$ *and for any graph* $\mathcal{G} \in \mathbb{G}^N_{\alpha,\beta}$.

Proof Using Theorem 13.8, we know that we only need to verify that $u = \rho F x$ solves the robust H_2-ADDPSS with bounded input for the system (13.13) with $\lambda \in \Lambda$. We use the same feedback as in the proof of Theorem 13.10. In the proof of Theorem 13.10, it is already shown that the closed-loop system is asymptotically stable and the H_∞ norm of the transfer function from w to u is bounded. The only remaining part of the proof is to show that the H_2 norm from w to x can be made arbitrarily small. Using the algebraic Riccati equation, it is easy to see that we have

$$(A - \rho\lambda BB'P)'P + P(A - \rho\lambda BB'P) + \rho\beta PBB'P \leq 0$$

for large ρ. But then we have

$$Q_\varepsilon(A - \rho\lambda BB'P)' + (A - \rho\lambda BB'P)Q_\varepsilon + BB' \leq 0$$

for $Q_\varepsilon = \varepsilon\beta^{-1}P^{-1}$. Then Lemma 13.11 immediately yields that we can make the H_2 norm from w to x arbitrarily small by choosing a sufficiently small ε. ∎

13.3.3 ATEA-Based Method

We use the same structure as in Sect. 2.4.3. There exists a nonsingular transformation matrix $T_x \in \mathbb{R}^{n\times n}$ such that

$$\hat{x} = \begin{pmatrix} \hat{x}_1 \\ \hat{x}_2 \end{pmatrix} = T_x x, \tag{13.24}$$

and the dynamics of $\hat{x}$ is represented as

$$\begin{aligned} \dot{\hat{x}}_1 &= \bar{A}_{11}\hat{x}_1 + \bar{A}_{12}\hat{x}_2, \\ \dot{\hat{x}}_2 &= \bar{A}_{21}\hat{x}_1 + \bar{A}_{22}\hat{x}_2 + \lambda\bar{B}u + \bar{B}\omega, \end{aligned} \tag{13.25}$$

with $\bar{B}$ invertible, and (A, B) is stabilizable implies that $(\bar{A}_{11}, \bar{A}_{12})$ is stabilizable.

We design the protocol as follows.

Protocol design 13.2 *Consider a MAS described by* (13.1) *and* (13.3). *Let any real numbers* $\alpha, \beta > 0$ *and a positive integer* N *be given.*

Choose F_1 *such that* $\bar{A}_{11} + \bar{A}_{12}F_1$ *is asymptotically stable. In that case a suitable protocol for* (13.1) *is*

$$u_i = F_\varepsilon \zeta_i, \tag{13.26}$$

where F_ε *is designed as*

$$F_\varepsilon = \frac{1}{\varepsilon}\bar{B}^{-1}\begin{pmatrix} F_1 & -I \end{pmatrix} T_x. \tag{13.27}$$

The main result for the direct method regarding H_∞ almost state synchronization problem is stated as follows. The result is basically the same as Theorem 13.10 except for a different design protocol.

Theorem 13.13 *Consider a MAS described by* (13.1) *and* (13.3) *such that* (13.14) *is satisfied. Let any real numbers* $\beta > 0$ *and a positive integer* N *be given and hence a set of network graphs* $\mathbb{G}_\beta^N$ *be defined.*

If (A, B) *is stabilizable, then the* H_∞ *almost state synchronization problem, as defined in Problem 13.2, with* $\mathbf{G} = \mathbb{G}_\beta^N$ *is solvable. In particular, for any given real number* $r > 0$, *there exists an* ε^*, *such that for any* $\varepsilon \in (0, \varepsilon^*)$, *the protocol* (13.26) *achieves state synchronization and an* H_∞ *norm from* ω *to* $x_i - x_j$ *less than* r *for any* $i, j \in 1, \ldots, N$ *and for any graph* $\mathcal{G} \in \mathbb{G}_\beta^N$.

Proof Since the graph will be in $\mathbb{G}_{\alpha,\beta}^N$, we know the Laplacian matrix is bounded. Similarly to the proof of Theorem 13.10, we then only need to establish that $u = F_\varepsilon x$ solves the robust H_∞ almosty disturbance decoupling with bounded input for the system (13.13) with $\lambda \in \Lambda$. Given $\mathcal{G} \in \mathbb{G}_\beta^N$, we know that $\lambda \in \Lambda$ implies that $\operatorname{Re}\lambda \geq \beta$.

After a basis transformation, the interconnection of the interconnection of (13.13) and $u = F_\varepsilon x$ is equal to the interconnection of (13.25) and (13.26). We obtain

$$\begin{aligned} \dot{\hat{x}}_1 &= \bar{A}_{11}\hat{x}_1 + \bar{A}_{12}\hat{x}_2, \\ \varepsilon\dot{\hat{x}}_2 &= (\varepsilon\bar{A}_{21} + \lambda F_1)\hat{x}_1 + (\varepsilon\bar{A}_{22} - \lambda I)\hat{x}_2 + \varepsilon\bar{B}w. \end{aligned} \tag{13.28}$$

Define

$$\tilde{x}_1 = \hat{x}_1, \quad \tilde{x}_2 = \hat{x}_2 - F_1\hat{x}_1.$$

Then we can write this system (13.28) in the form

$$\begin{aligned}\dot{\tilde{x}}_1 &= \tilde{A}_{11}\tilde{x}_1 + \tilde{A}_{12}\tilde{x}_2,\\ \varepsilon\dot{\tilde{x}}_2 &= \varepsilon\tilde{A}_{21}\tilde{x}_1 + (\varepsilon\tilde{A}_{22} - \lambda I)\tilde{x}_2 + \varepsilon\bar{B}w,\end{aligned} \tag{13.29}$$

where

$$\tilde{A}_{11} = \bar{A}_{11} + \bar{A}_{12}F_1, \quad \tilde{A}_{12} = \bar{A}_{12},$$

$$\tilde{A}_{21} = \bar{A}_{21} - F_1\bar{A}_{11} + \bar{A}_{22} - F_1\bar{A}_{12}, \quad \tilde{A}_{22} = \bar{A}_{22} - F_1\bar{A}_{12}.$$

In the absence of the external disturbances, the above system (13.29) is asymptotically stable for a small enough ε.

Since $\tilde{A}_{11} = \bar{A}_{11} + \bar{A}_{12}F_1$ is Hurwitz stable, there exists $P = P' > 0$ such that the Lyapunov equation $P\tilde{A}_{11} + \tilde{A}_{11}'P = -I$ holds. For the dynamics $\tilde{x}_1$, we define a Lyapunov function $V_1 = \tilde{x}_1'P\tilde{x}_1$. Then the derivative of V_1 is obtained

$$\begin{aligned}\dot{V}_1 &\le -\|\tilde{x}_1\|^2 + \tilde{x}_2'\tilde{A}_{12}'P\tilde{x}_1 + \tilde{x}_1'P\tilde{A}_{12}\tilde{x}_2\\ &\le -\|\tilde{x}_1\|^2 + 2\operatorname{Re}(\tilde{x}_1'P\tilde{A}_{12}\tilde{x}_2)\\ &\le -\|\tilde{x}_1\|^2 + r_1\|\tilde{x}_1\|\|\tilde{x}_2\|,\end{aligned}$$

where $2\|P\tilde{A}_{12}\| \le r_1$.

Now define a Lyapunov function $V_2 = \varepsilon\tilde{x}_2'\tilde{x}_2$ for the dynamics $\tilde{x}_2$, where $d_2 > 0$ is to be selected. The derivative of V_2 is as follows:

$$\begin{aligned}\dot{V}_2 &\le -2\operatorname{Re}(\lambda)\|\tilde{x}_2\|^2 + 2\varepsilon\operatorname{Re}(\tilde{x}_2'\tilde{A}_{21}\tilde{x}_1)\\ &\qquad + 2\varepsilon\tilde{x}_2'\tilde{A}_{22}\tilde{x}_2 + 2\varepsilon\operatorname{Re}(\tilde{x}_2'\bar{B}w)\\ &\le -2\operatorname{Re}(\lambda)\|\tilde{x}_2\|^2 + \varepsilon r_2\|\tilde{x}_1\|\|\tilde{x}_2\| + \varepsilon r_3\|\tilde{x}_2\|^2\\ &\qquad + \varepsilon r_4\|\omega\|\|\tilde{x}_2\|\\ &\le -\beta\|\tilde{x}_2\|^2 + \varepsilon r_2\|\tilde{x}_1\|\|\tilde{x}_2\| + \varepsilon r_4\|\omega\|\|\tilde{x}_2\|\end{aligned}$$

for a small enough ε, where we choose that $2\|\tilde{A}_{21}\| \le r_2$, $2\|\tilde{A}_{22}\| \le r_3$, and $2\|\bar{B}\| \le r_4$.

Let $V = V_1 + \gamma V_2$ for some $\gamma > 0$. Then, we have

$$\begin{aligned}\dot{V} &\le -\|\tilde{x}_1\|^2 + r_1\|\tilde{x}_1\|\|\tilde{x}_2\| - \gamma\beta\|\tilde{x}_2\|^2\\ &\quad + \varepsilon\gamma r_2\|\tilde{x}_1\|\|\tilde{x}_2\| + \varepsilon\gamma r_4\|\omega\|\|\tilde{x}_2\|.\end{aligned}$$

We have that

$$r_1\|\tilde{x}_1\|\|\tilde{x}_2\| \leq r_1^2\|\tilde{x}_2\|^2 + \tfrac{1}{4}\|\tilde{x}_1\|^2,$$
$$\varepsilon\gamma r_2\|\tilde{x}_1\|\|\tilde{x}_2\| \leq \varepsilon^2\gamma^2 r_2^2\|\tilde{x}_1\|^2 + \tfrac{1}{4}\|\tilde{x}_2\|^2,$$
$$\varepsilon\gamma r_4\|\omega\|\|\tilde{x}_2\| \leq \varepsilon^2\gamma^2 r_4^2\|w\|^2 + \tfrac{1}{4}\|\tilde{x}_2\|^2.$$

Now we choose γ such that $\gamma\beta = 1 + r_1^2$ and $r_5 = \gamma r_4$. Then, we obtain

$$\dot{V} \leq -\tfrac{1}{2}\|\tilde{x}_1\|^2 - \tfrac{1}{2}\|\tilde{x}_2\|^2 + \varepsilon^2 r_5^2\|\omega\|^2$$
$$\leq -\tfrac{1}{2}\|\tilde{x}\|^2 + \varepsilon^2 r_5^2\|w\|^2,$$

for a small enough ε. From the above, we have that $\|T_{w\tilde{x}}\|_\infty < 2\varepsilon r_5$, which immediately leads to $\|T_{wx}\|_\infty < r$ for any real number $r > 0$ as long as we choose ε small enough. On the other hand

$$T_{wu}(s) = -\tfrac{1}{\varepsilon}\begin{pmatrix}0 & \bar{B}^{-1}\end{pmatrix} T_{w\tilde{x}}(s),$$

and hence

$$\|T_{wu}\|_\infty \leq \|\bar{B}^{-1}\| r_5.$$

Therefore, $u = F_\varepsilon x$ solves the robust H_∞-ADDPSS with bounded input for the system (13.13) as required. ■

The main result for the direct method regarding H_2 almost state synchronization problem is stated as follows.

Theorem 13.14 *Consider a MAS described by* (13.1) *and* (13.3) *such that* (13.14) *is satisfied. Let any real numbers $\alpha, \beta > 0$ and a positive integer N be given.*

If (A, B) is stabilizable, then the H_2 almost state synchronization problem, as defined in Problem 13.4*, with $\mathbf{G} = \mathbb{G}_\beta^N$ is solvable. In particular, for any given real number $r > 0$, there exists an ε^*, such that for any $\varepsilon \in (0, \varepsilon^*)$, the protocol* (13.26) *achieves state synchronization and an H_2 norm from ω to $x_i - x_j$ less than r for any $i, j \in 1, \ldots, N$ and for any graph $\mathcal{G} \in \mathbb{G}_\beta^N$.*

Proof Since the graph will be in $\mathbb{G}_{\alpha,\beta}^N$, we know that the Laplacian matrix is bounded. Therefore, we can apply Theorem 13.8, and we know that we only need to verify that the feedback solves the robust H_2 almost disturbance decoupling problem with bounded input for the system (13.13) with $\lambda \in \Lambda$. We use the same feedback as in the proof of Theorem 13.13. In the proof of Theorem 13.13, it is already shown that the closed-loop system is asymptotically stable and the H_∞ norm of the transfer function from w to u is bounded. The only remaining part of the proof is to show that the H_2 norm from w to x can be made arbitrarily small. This clearly is equivalent to showing that the system (13.29) has an arbitrarily small H_2 norm from w to $\tilde{x}_1$ and $\tilde{x}_2$ for a sufficiently small ε. Choose Q such that

$$Q\tilde{A}_{11}' + \tilde{A}_{11}Q = -I$$

In that case we have

$$A_{cl}\begin{pmatrix} \sqrt{\varepsilon}Q & 0 \\ 0 & \sqrt{\varepsilon}I \end{pmatrix} + \begin{pmatrix} \sqrt{\varepsilon}Q & 0 \\ 0 & \sqrt{\varepsilon}I \end{pmatrix} A_{cl}' + \begin{pmatrix} 0 & 0 \\ 0 & \bar{B}\bar{B}' \end{pmatrix}$$
$$\leq \begin{pmatrix} \sqrt{\varepsilon} & \sqrt{\varepsilon}(\tilde{A}_{12} + Q\tilde{A}_{21}') \\ \sqrt{\varepsilon}(\tilde{A}_{12}' + \tilde{A}_{21}Q) & -\frac{\beta}{\sqrt{\varepsilon}}I \end{pmatrix}$$

for a sufficiently small ε where

$$A_{cl} = \begin{pmatrix} \tilde{A}_{11} & \tilde{A}_{12} \\ \tilde{A}_{21} & \tilde{A}_{22} - \frac{\lambda}{\varepsilon}I, \end{pmatrix}$$

and we used that

$$\lambda + \lambda' \geq 2\beta.$$

We then obtain for a sufficiently small ε that

$$A_{cl}\begin{pmatrix} \sqrt{\varepsilon}Q & 0 \\ 0 & \sqrt{\varepsilon}I \end{pmatrix} + \begin{pmatrix} \sqrt{\varepsilon}Q & 0 \\ 0 & \sqrt{\varepsilon}I \end{pmatrix} A_{cl}' + \begin{pmatrix} 0 & 0 \\ 0 & \bar{B}\bar{B}' \end{pmatrix}$$
$$\leq 0.$$

Then Lemma 13.11 immediately yields that we can make the H_2 norm from w to x arbitrarily small by choosing a sufficiently small ε. ■

13.4 Protocol Design for a MAS with Partial-State Coupling

In this section, similar to the approach of the previous section, we show first that the almost state synchronization among agents in the network with partial-state coupling can be solved by equivalently solving a robust H_∞ or H_2 almost disturbance decoupling problem via measurement feedback with internal stability. Then we design a controller for such robust stabilization problems.

13.4.1 *Necessary and Sufficient Condition for Solvability*

We derive necessary and sufficient conditions based on the connection of almost state synchronization and H_∞ or H_2 almost disturbance decoupling.

13.4.1.1 H_∞ Almost Synchronization

The MAS system described by (13.1) and (13.2) after implementing the linear dynamical protocol (13.7) is described by

$$\begin{cases} \dot{\hat{x}}_i = \begin{pmatrix} A & BC_c(\varepsilon) \\ 0 & A_c(\varepsilon) \end{pmatrix} \hat{x}_i + \begin{pmatrix} BD_c(\varepsilon) \\ B_c(\varepsilon) \end{pmatrix} \zeta_i + \begin{pmatrix} E \\ 0 \end{pmatrix} \omega_i, \\ y_i = \begin{pmatrix} C & 0 \end{pmatrix} \hat{x}_i, \\ \zeta_i = \sum_{j=1}^{N} a_{ij}(y_i - y_j), \end{cases} \tag{13.30}$$

for $i = 1, \ldots, N$,where

$$\hat{x}_i = \begin{pmatrix} x_i \\ \chi_i \end{pmatrix}.$$

Define

$$\hat{x} = \begin{pmatrix} \hat{x}_1 \\ \vdots \\ \hat{x}_N \end{pmatrix}, \quad \omega = \begin{pmatrix} \omega_1 \\ \vdots \\ \omega_N \end{pmatrix},$$

and

$$\bar{A} = \begin{pmatrix} A & BC_c(\varepsilon) \\ 0 & A_c(\varepsilon) \end{pmatrix}, \quad \bar{B} = \begin{pmatrix} BD_c(\varepsilon) \\ B_c(\varepsilon) \end{pmatrix}, \bar{E} = \begin{pmatrix} E \\ 0 \end{pmatrix}, \quad \text{and } \bar{C} = \begin{pmatrix} C & 0 \end{pmatrix}.$$

Then, the overall dynamics of the N agents can be written as

$$\dot{\hat{x}} = (I_N \otimes \bar{A} + L \otimes \bar{B}\bar{C})\hat{x} + (I_N \otimes \bar{E})\omega. \tag{13.31}$$

We define robust H_∞ almost disturbance decoupling with measurement feedback and bounded input as follows. Given $\Lambda \subset \mathbb{C}$, there should exist a $M > 0$ such that for any given real number $r > 0$, we can find a parameterized controller

$$\begin{aligned} \dot{\chi} &= A_c(\varepsilon)\chi + B_c(\varepsilon)y, \\ u &= C_c(\varepsilon)\chi + D_c(\varepsilon)y, \end{aligned} \tag{13.32}$$

where $\chi \in \mathbb{R}^{n_c}$, for the following system:

$$\begin{aligned} \dot{x} &= Ax + \lambda Bu + Ew, \\ y &= Cx \end{aligned} \tag{13.33}$$

such that the following holds for any $\lambda \in \Lambda$:

1. The closed-loop system of (13.33) and (13.32) is internally stable.
2. The resulting closed-loop transfer function T_{wx} from w to x has an H_∞ norm less than r.
3. The resulting closed-loop transfer function T_{wu} from w to u has an H_∞ norm less than M.

As before, Λ denotes all possible locations for the nonzero eigenvalues of the Laplacian matrix L when the graph varies over the set $\mathbf{G}$. It is also important to note that M is independent of the choice for r.

Let

$$x_c = \begin{pmatrix} x \\ \chi \end{pmatrix}.$$

Then the closed-loop system of (13.32) and (13.33) is written as

$$\begin{aligned} \dot{x}_c &= (\bar{A} + \lambda \bar{B}\bar{C})x_c + \bar{E}w \\ x &= \begin{pmatrix} I & 0 \end{pmatrix} x_c, \end{aligned} \tag{13.34}$$

where

$$\bar{A} = \begin{pmatrix} A_c(\varepsilon) & B_c(\varepsilon)C \\ 0 & A \end{pmatrix}, \qquad \bar{B} = \begin{pmatrix} 0 \\ B \end{pmatrix}, \qquad \bar{C} = \begin{pmatrix} C_c(\varepsilon) & D_c(\varepsilon) \end{pmatrix}.$$

In order to obtain our main result, we will need the following lemma.

Lemma 13.15 *The H_∞ almost disturbance decoupling with measurement feedback and bounded input for a fixed $\lambda \in \mathbb{C}$ is solvable if and only if:*

1. $\operatorname{Im} E \subset \operatorname{Im} B$.
2. $(A, E, C, 0)$ *is left-invertible.*
3. $(A, E, C, 0)$ *is minimum-phase.*

Proof From [103] we immediately obtain that the H_∞ almost disturbance decoupling with measurement feedback and bounded input is solvable if and only if:

1. $\operatorname{Im} E \subset \operatorname{Im} B$.
2. $(A, E, C, 0)$ is at most weakly non-minimum-phase and left-invertible.
3. For any $\delta > 0$ there exists a matrix K such that

$$\|(sI - A - BKC)^{-1}E\|_\infty < \delta. \tag{13.35}$$

Choose a suitable basis such that

$$A = \begin{pmatrix} A_{11} & A_{12} \\ A_{21} & A_{22} \end{pmatrix}, \quad B = \begin{pmatrix} B_1 \\ B_2 \end{pmatrix}, \quad E = \begin{pmatrix} E_1 \\ E_2 \end{pmatrix}, \quad C = \begin{pmatrix} C_1 & 0 \end{pmatrix}.$$

Assume s_0 is an imaginary axis zero of (A, E, C). In that case the rank of the matrix

$$\begin{pmatrix} sI - A_{11} & -A_{12} & E_1 \\ -A_{21} & sI - A_{22} & E_2 \\ I & 0 & 0 \end{pmatrix}$$

drops for $s = s_0$. This implies the existence of $p \neq 0$ and $q \neq 0$ such that

$$\begin{pmatrix} -A_{12} \\ s_0 I - A_{22} \end{pmatrix} p = \begin{pmatrix} E_1 \\ E_2 \end{pmatrix} q.$$

The final condition for H_∞ almost disturbance decoupling requires for any $\delta > 0$ the existence of a K such that

$$\|(s_0 I - A - BKC)^{-1} E\| < \delta.$$

However

$$\begin{aligned} (s_0 I - A - BKC)^{-1} Eq &= \begin{pmatrix} s_0 I - A_{11} - B_1 K & -A_{12} \\ -A_{21} - B_2 K & s_0 I - A_{22} \end{pmatrix}^{-1} \begin{pmatrix} -A_{12} \\ s_0 I - A_{22} \end{pmatrix} p \\ &= \begin{pmatrix} s_0 I - A_{11} - B_1 K & -A_{12} \\ -A_{21} - B_2 K & s_0 I - A_{22} \end{pmatrix}^{-1} \begin{pmatrix} s_0 I - A_{11} - B_1 K & -A_{12} \\ -A_{21} - B_2 K & s_0 I - A_{22} \end{pmatrix} \\ &\quad \times \begin{pmatrix} 0 \\ p \end{pmatrix} \\ &= \begin{pmatrix} 0 \\ p \end{pmatrix}, \end{aligned}$$

which yields a contradiction if δ is such that

$$\|p\| > \delta \|q\|.$$

Therefore we can't have any invariant zeros on the imaginary axis. In other words, the system $(A, E, C, 0)$ needs to be minimum-phase instead of weakly minimum-phase. Conversely, if $(A, E, C, 0)$ is minimum-phase, it is easy to verify that for any $\delta > 0$ there exists a K such that (13.35) is satisfied. ■

Theorem 13.16 *Consider the MAS described by (13.1) and (13.2) with (A, B) stabilizable and (C, A) detectable.*

(Part I) Let $\alpha, \beta > 0$ be given. Then, the H_∞ almost state synchronization problem, as defined in Problem 13.3 with $\mathbf{G} = \mathbb{G}^N_{\alpha,\beta}$, is solvable by a parameterized protocol (13.7) for any $\alpha > \beta > 0$ if and only if

$$\operatorname{Im} E \subset \operatorname{Im} B \tag{13.36}$$

while $(A, E, C, 0)$ is minimum-phase and left-invertible.

(Part II) Let $\mathbf{G}$ be a set of graphs such that the associated Laplacian matrices are uniformly bounded, and let Λ consist of all possible nonzero eigenvalues of Laplacian matrices associated with graphs in $\mathbf{G}$. Then, the H_∞ almost state synchronization problem for the MAS with any graph $\mathcal{G} \in \mathbf{G}$ is solved by a parameterized protocol (13.7) if the robust H_∞ almost disturbance decoupling problem with measurement feedback and bounded input with bounded input for the system (13.33) with $\lambda \in \Lambda$ is solved by the parameterized controller (13.32).

Proof The structure of the proof is very similar to the proof of Theorem 13.6. We again use the Schur form (2.10) for the Laplacian where U unitary and S upper triangular. Let

$$\eta := (U^{-1} \otimes I_n)x_c = \begin{pmatrix} \eta_1 \\ \vdots \\ \eta_N \end{pmatrix}, \qquad \bar{\omega} = (U^{-1} \otimes I)\omega = \begin{pmatrix} \bar{\omega}_1 \\ \vdots \\ \bar{\omega}_N \end{pmatrix},$$

where $\eta_i \in \mathbb{C}^{n+n_c}$ and $\bar{\omega}_i \in \mathbb{C}^q$. In the new coordinates, the dynamics of η can be written as

$$\dot{\eta}(t) = (I_N \otimes \bar{A} + S \otimes \bar{B}\bar{C}\eta + (U^{-1} \otimes E)\omega, \tag{13.37}$$

which is rewritten as

$$\begin{aligned} \dot{\eta}_1 &= \bar{A}\eta_1 + \sum_{j=2}^{N} s_{1j}\bar{B}\bar{C}\eta_j + \bar{E}\bar{\omega}_1, \\ \dot{\eta}_i &= (\bar{A} + \lambda_i\bar{B}\bar{C})\eta_i + \sum_{j=i+1}^{N} s_{ij}\bar{B}\bar{C}\eta_j + \bar{E}\bar{\omega}_i, \quad (i \in \{2, \ldots, N-1\}), \\ \dot{\eta}_N &= (\bar{A} + \lambda_N\bar{B}\bar{C})\eta_N + \bar{E}\bar{\omega}_N, \end{aligned} \tag{13.38}$$

where

$$\bar{E} = \begin{pmatrix} 0 \\ E \end{pmatrix}, \qquad S = [s_{ij}].$$

As in the case of full-state coupling, we can show that

$$\bar{x} = (V \otimes I_n) \begin{pmatrix} \eta_2 \\ \vdots \\ \eta_N \end{pmatrix}, \tag{13.39}$$

for some suitably chosen matrix V which is uniformly bounded. Therefore the H_∞ norm of the transfer matrix from ω to $\bar{x}$ can be made arbitrarily small if and only if the H_∞ norm of the transfer matrix from $\bar{\omega}$ to η can be made arbitrarily small.

In order for the H_∞ norm from $\bar{\omega}$ to η to be arbitrarily small, we need the H_∞ norm from $\bar{\omega}_N$ to η_N to be arbitrarily small. In other words, the H_∞ ADDPMS has to be solvable for the system

$$\begin{aligned} \dot{x} &= Ax + \lambda Bu + Ew, \\ y &= Cx. \end{aligned}$$

From the results of Lemma 13.15, we find that this is only possible if (13.36) is satisfied, $(A, E, C, 0)$ is left-invertible and minimum-phase.

On the other hand, suppose (13.32) solves the robust H_∞ almost disturbance decoupling problem with measurement feedback and bounded input of (13.33) and assume (13.36) is satisfied. We need to show that (13.7) solves the H_∞ almost state synchronization problem for the MAS described by (13.1) and (13.2). This follows directly from arguments very similar to the approach used in the proof of Theorem 13.6. ■

13.4.1.2 H_∞ Almost Synchronization

The MAS system described by (13.1) and (13.2) after implementing the linear dynamical protocol (13.7) is described by (13.30) for $i = 1, \ldots, N$, and, as before, the overall dynamics of the N agents can be written as

$$\dot{\hat{x}} = (I_N \otimes \bar{A} + L \otimes \bar{B}\bar{C})\hat{x} + (I_N \otimes \bar{E})\omega. \tag{13.40}$$

We define a robust H_2-ADDPMS almost disturbance decoupling with measurement feedback and bounded input as follows. Given $\Lambda \subset \mathbb{C}$, there should exist a $M > 0$ such that for any given real number $r > 0$, we can find a parameterized controller

$$\begin{aligned} \dot{\chi} &= A_c(\varepsilon)\chi + B_c(\varepsilon)y, \\ u &= C_c(\varepsilon)\chi + D_c(\varepsilon)y, \end{aligned} \tag{13.41}$$

where $\chi \in \mathbb{R}^{n_c}$, for the following system

$$\begin{aligned} \dot{x} &= Ax + \lambda Bu + Ew, \\ y &= Cx \end{aligned} \tag{13.42}$$

such that the following holds for any $\lambda \in \Lambda$:

1. The closed-loop system of (13.41) and (13.42) is internally stable.
2. The resulting closed-loop transfer function T_{wx} from w to x has an H_2 norm less than r.
3. The resulting closed-loop transfer function T_{wu} from w to u has an H_∞ norm less than M.

In the above, Λ denotes all the possible locations for the nonzero eigenvalues of the Laplacian matrix L when the graph varies over the set $\mathbf{G}$. It is also important to note that M is independent of the choice for r.

The following lemma provides a necessary condition for the H_2 almost disturbance decoupling with measurement feedback and bounded input.

Lemma 13.17 *The H_2 almost disturbance decoupling with measurement feedback and bounded input for a fixed $\lambda \in \mathbb{C}$ is solvable only if:*

1. $\operatorname{Im} E \subset \operatorname{Im} B$.
2. $(A, E, C, 0)$ *is left-invertible.*
3. $(A, E, C, 0)$ *is at most weakly non-minimum-phase.*

Proof This follows directly from [103]. ∎

Theorem 13.18 *Consider the MAS described by (13.1) and (13.2) with (A, B) stabilizable and (C, A) detectable.*

(Part I) Let $\alpha, \beta > 0$ be given. The H_2 almost state synchronization problem, as defined in Problem 13.5 with $\mathbf{G} = \mathbb{G}^N_{\alpha,\beta}$, is solvable by a parameterized protocol (13.7) for any $\alpha > \beta > 0$ only if

$$\operatorname{Im} E \subset \operatorname{Im} B \tag{13.43}$$

while $(A, E, C, 0)$ is at most weakly non-minimum-phase and left-invertible.

(Part II) Let $\mathbf{G}$ be a set of graphs such that the associated Laplacian matrices are uniformly bounded, and let Λ consist of all the possible nonzero eigenvalues of Laplacian matrices associated with graphs in $\mathbf{G}$. Then, the H_2 almost state synchronization problem for the MAS with any graph $\mathcal{G} \in \mathbf{G}$ is solved by a parameterized protocol (13.7) if the robust H_2 almost disturbance decoupling with measurement feedback and bounded input for the system (13.42) with $\lambda \in \Lambda$ is solved by the parameterized controller (13.41).

Proof Similar, to the proof of Theorem 13.16, the dynamics can be written in the form (13.38).

Using (13.39), we note the H_2 norm of the transfer matrix from ω to $\bar{x}$ can be made arbitrarily small if and only if the H_2 norm of the transfer matrix from $\bar{\omega}$ to η can be made arbitrarily small.

In order for the H_∞ norm from $\bar{\omega}$ to η to be arbitrarily small, we need the H_2 norm from $\bar{\omega}_N$ to η_N to be arbitrarily small. In other words, the H_2 almost disturbance decoupling with measurement feedback and bounded input has to be solvable for the system

$$\begin{aligned}\dot{x} &= Ax + \lambda Bu + Ew,\\ y &= Cx.\end{aligned}$$

From the results of Lemma 13.17, we find that this is only possible if (13.43) is satisfied, and $(A, E, C, 0)$ is left-invertible and at most weakly non-minimum-phase.

On the other hand, suppose (13.41) solves the robust H_2 almost disturbance decoupling with measurement feedback and bounded input of (13.42) and assume (13.43) is satisfied. We need to show that (13.7) solves the H_2 almost state synchronization problem for the MAS described by (13.1) and (13.2). This follows directly from arguments very similar to the approach used in the proof of Theorem 13.6. ■

We present below two protocol design methods for both H_∞ and H_2 almost state synchronization problems based on robust almost disturbance decoupling problems for the case $E = B$, therefore the case where $(A, B, C, 0)$ is minimum-phase. One method relies on an algebraic Riccati equation (ARE), and the other one is based on the direct eigenstructure assignment method.

13.4.2 ARE-Based Method

Using an algebraic Riccati equation, we can design a suitable protocol.

Protocol design 13.3 *Consider a MAS described by* (13.1) *and* (13.2). *Let any real numbers* $\alpha, \beta > 0$ *and a positive integer* N *be given.*

As in the full-state coupling case, we choose $F = -B'P$ *with* P *being the unique solution of the continuous-time algebraic Riccati equation:*

$$A'P + PA - 2\beta PBB'P + I = 0. \tag{13.44}$$

If $(A, B, C, 0)$ *is minimum-phase, then for any* ε *there exists a* δ *small enough such that*

$$A'Q + QA + BB' + \varepsilon^{-4}Q^2 - \delta^{-2}QC'CQ = 0 \tag{13.45}$$

(continued)

Protocol design 13.3 (continued)
has a solution $Q > 0$. We then consider the protocol

$$\begin{aligned}\dot{\chi}_i &= (A + K_\varepsilon C)\chi_i - K_\varepsilon \zeta_i,\\ u_i &= F_\varepsilon \chi_i,\end{aligned} \tag{13.46}$$

where

$$F_\varepsilon = -\tfrac{1}{\varepsilon}B'P, \qquad K_\varepsilon = -\tfrac{1}{\delta^2}QC'.$$

The main result regarding the H_∞ almost state synchronization problem is stated as follows.

Theorem 13.19 *Consider a MAS described by (13.1) and (13.2) with (A, B) stabilizable and (C, A) detectable. Let any real numbers $\alpha, \beta > 0$ and a positive integer N be given.*

The H_∞ almost state synchronization problem, as defined in Problem 13.3, with $\mathbf{G} = \mathbb{G}^N_{\alpha,\beta}$ is solvable. In particular, for any given real number $r > 0$, there exists an ε^*, such that for any $\varepsilon \in (0, \varepsilon^*)$, the protocol (13.46) achieves state synchronization and an H_∞ norm from ω to $x_i - x_j$ less than r for any $i, j \in 1, \ldots, N$ and for any graph $\mathcal{G} \in \mathbb{G}^N_{\alpha,\beta}$.

Proof Using Theorem 13.16, we know that we only need to verify that

$$\begin{aligned}\dot{\chi} &= (A + K_\varepsilon C)\chi - K_\varepsilon y,\\ u &= F_\varepsilon \chi,\end{aligned} \tag{13.47}$$

solves the robust H_∞ almost disturbance decoupling problem via measurement feedback with bounded input for the system (13.33) with $\lambda \in \Lambda$. Given $\mathcal{G} \in \mathbb{G}^N_{\alpha,\beta}$, we know that $\lambda \in \Lambda$ implies that $\operatorname{Re}\lambda \geq \beta$. Obviously $A + BF_\varepsilon$ and $A + K_\varepsilon C$ are both asymptotically stable by construction, and hence the intersection of (13.33) and (13.47) is asymptotically stable. The closed-loop transfer function from w to x is equal to

$$T_{wx} = \begin{pmatrix} I & 0 \end{pmatrix} \begin{pmatrix} I & -T_2 \\ \tilde{T}_3 & I \end{pmatrix}^{-1} \begin{pmatrix} T_1 \\ -T_1 \end{pmatrix},$$

where

$$\begin{aligned} T_1(s) &= (sI - A - \lambda B F_\varepsilon)^{-1} B \\ T_2(s) &= \lambda (sI - A - \lambda B F_\varepsilon)^{-1} B F_\varepsilon \\ \tilde{T}_3(s) &= \lambda (sI - A - K_\varepsilon C + \lambda B F_\varepsilon)^{-1} B F_\varepsilon \\ \tilde{T}_4(s) &= (sI - A - K_\varepsilon C + \lambda B F_\varepsilon)^{-1} B. \end{aligned} \tag{13.48}$$

As argued in the proof of Theorem 13.10, we have

$$\|T_1\|_\infty < \frac{\varepsilon}{\beta}, \qquad \|T_2\|_\infty < \frac{2|\lambda|}{\beta^2} \leq \frac{2\alpha}{\beta^2}.$$

On the other hand, (13.45) implies that

$$\|T_3\|_\infty < \varepsilon^2, \tag{13.49}$$

where

$$T_3(s) = (sI - A - K_\varepsilon C)^{-1} B.$$

Note that

$$\tilde{T}_3 = \lambda (I + \lambda T_3 F_\varepsilon)^{-1} T_3 F_\varepsilon$$

which yields, using (13.49), that

$$\|\tilde{T}_3\|_\infty < (1 - \varepsilon M)^{-1} \varepsilon M_1 < 2\varepsilon M_1$$

for a small ε where M_1 is such that

$$|\lambda| \|B'P\| < \alpha \|B'P\| = M_1.$$

The above yields

$$\|T_{wx}\|_\infty < \varepsilon M_2$$

for some suitable constant M_2. The closed-loop transfer function from w to u is equal to

$$T_{wu} = \begin{pmatrix} F_\varepsilon & F_\varepsilon \end{pmatrix} \begin{pmatrix} I & -T_2 \\ \tilde{T}_3 & I \end{pmatrix}^{-1} \begin{pmatrix} T_1 \\ -T_1 \end{pmatrix},$$

which yields using similar arguments as above that

$$\|T_{wu}\|_\infty < M_3,$$

for some suitable constant M_3 independent of ε. Therefore the H_∞ norm of the transfer matrix T_{wx} becomes arbitrarily small for a sufficiently small ε, while the H_∞ norm of the transfer matrix T_{wu} remains bounded. ■

The main result regarding the H_2 almost state synchronization problem is stated as follows.

Theorem 13.20 *Consider a MAS described by (13.1) and (13.2) with (A, B) stabilizable and (C, A) detectable. Let any real numbers $\alpha, \beta > 0$ and a positive integer N be given.*

The H_2 almost state synchronization problem, as defined in Problem 13.5, with $\mathbf{G} = \mathbb{G}^N_{\alpha,\beta}$ is solvable. In particular, for any given real number $r > 0$, there exists an ε^*, such that for any $\varepsilon \in (0, \varepsilon^*)$, the protocol (13.46) achieves state synchronization and an H_2 norm from ω to $x_i - x_j$ less than r for any $i, j \in 1, \dots, N$ and for any graph $\mathcal{G} \in \mathbb{G}^N_{\alpha,\beta}$.

Proof It is easy to verify using a similar proof that the protocol (13.46) also solves the robust H_2 almost disturbance decoupling problem via measurement feedbac with bounded input and therefore solves H_2 almost state synchronization problem for the MAS. Using the same notation as before, this relies on the fact that we have N_1 such that

$$\|T_1\|_2 < \varepsilon N_1$$

which follows directly from the full-state case. On the other hand, we have N_2 such that

$$\|T_3\|_2 < \varepsilon^2 N_2$$

since $Q \to 0$ for $\delta \to 0$ and

$$(A - KC)Q + Q(A - KC)' + BB' \leq 0.$$

It is then easily shown that

$$\|\tilde{T}_4\|_2 < \varepsilon^2 N_3$$

for some $N_3 > 0$. The rest of the proof is then as in the case of the H_∞ almost state synchronization problem. ■

13.4.3 Direct Method

For ease of presentation, we only consider the case $p = 1$, i.e., the case where we have as scalar measurement. Because we consider left-invertible systems, this immediately implies that the system is actually a SISO system.

We consider the state feedback gain F_ε given by (13.27). In other words we use the same transformation as in Sect. 13.3.3, i.e., we note that there exists a nonsingular transformation matrix $T_x \in \mathbb{R}^{n\times n}$ such that

$$\hat{x} = \begin{pmatrix} \hat{x}_1 \\ \hat{x}_2 \end{pmatrix} = T_x x, \tag{13.50}$$

and the dynamics of $\hat{x}$ is represented as

$$\begin{aligned} \dot{\hat{x}}_1 &= \bar{A}_{11}\hat{x}_1 + \bar{A}_{12}\hat{x}_2, \\ \dot{\hat{x}}_2 &= \bar{A}_{21}\hat{x}_1 + \bar{A}_{22}\hat{x}_2 + \lambda\bar{B}u + \bar{B}\omega, \end{aligned} \tag{13.51}$$

with $\bar{B}$ a nonzero scalar, and (A, B) is stabilizable implies that $(\bar{A}_{11}, \bar{A}_{12})$ is stabilizable.

We design the protocol as follows.

Protocol design 13.4 *Consider a MAS described by (13.1) and (13.2) and a set of network graphs $\mathbb{G}^N_{\alpha,\beta}$ be defined with $\alpha > \beta > 0$.*

Choose F_1 such that $\bar{A}_{11} + \bar{A}_{12}F_1$ is asymptotically stable. In that case, we choose

$$F_\varepsilon = \frac{1}{\varepsilon}\bar{B}^{-1}\left(F_1 \ -I\right)T_x. \tag{13.52}$$

Next, we consider the observer design. Note that the system $(A, B, C, 0)$ is scalar and minimum-phase. In that case there is a nonsingular matrix Γ_x such that, by defining $\bar{x} = \Gamma_x x$, we obtain the system

$$\begin{aligned} \dot{x}_a &= A_a x_a + L_{ad} y, \\ \dot{x}_d &= A_d x_d + B_d(u + w + E_{da}x_a + E_{dd}x_d), \\ y &= C_d x_d, \end{aligned} \tag{13.53}$$

where

$$\bar{x} = \Gamma_x x = \begin{pmatrix} x_a \\ x_d \end{pmatrix},$$

(continued)

Protocol design 13.4 (continued)
with $x_a \in \mathbb{R}^{n-\rho}$ *and* $x_d \in \mathbb{R}^{\rho}$ *and where the matrices* $A_d \in \mathbb{R}^{\rho\times\rho}$, $B_d \in \mathbb{R}^{\rho\times 1}$, *and* $C_d \in \mathbb{R}^{1\times\rho}$ *have the special form*

$$A_d = \begin{pmatrix} 0 & 1 & \cdots & 0 \\ \vdots & \vdots & \ddots & \vdots \\ 0 & 0 & \cdots & 1 \\ 0 & 0 & \cdots & 0 \end{pmatrix}, \quad B_d = \begin{pmatrix} 0 \\ \vdots \\ 0 \\ 1 \end{pmatrix}, \quad C_d = \begin{pmatrix} 1 & 0 & \cdots & 0 \end{pmatrix}. \tag{13.54}$$

Furthermore, the eigenvalues of A_a *are the invariant zeros of* (A, B, C), *and hence* A_a *is asymptotically stable. The transformation* Γ_x *can be calculated using available software, either numerically [61] or symbolically [30].*

Next, define a high-gain scaling matrix

$$S_\varepsilon := \operatorname{diag}(1, \varepsilon^2, \ldots, \varepsilon^{2\rho-2}), \tag{13.55}$$

and define the output injection matrix

$$K_\varepsilon = \Gamma_x \begin{pmatrix} 0 \\ \varepsilon^{-2} S_\varepsilon^{-1} K \end{pmatrix}. \tag{13.56}$$

where K *is such that* $A_d + B_d K$ *is asymptotically stable. We then consider the protocol*

$$\begin{aligned} \dot{\chi}_i &= (A + K_\varepsilon C)\chi_i - K_\varepsilon \zeta_i, \\ u_i &= F_\varepsilon \chi_i. \end{aligned} \tag{13.57}$$

The main result regarding the H_∞ almost state synchronization problem is stated as follows.

Theorem 13.21 *Consider a MAS described by a SISO system* (13.1) *and* (13.2) *with* (A, B) *stabilizable and* (C, A) *detectable. Let any real numbers* $\alpha, \beta > 0$ *and a positive integer* N *be given.*

The H_∞ *almost state synchronization problem, as defined in Problem 13.3, with* $\mathbf{G} = \mathbb{G}^N_{\alpha,\beta}$ *is solvable. In particular, for any given real number* $r > 0$, *there exists an* ε^*, *such that for any* $\varepsilon \in (0, \varepsilon^*)$, *the protocol* (13.57) *achieves state synchronization and an* H_∞ *norm from* ω *to* $x_i - x_j$ *less than* r *for any* $i, j \in 1, \ldots, N$ *and for any graph* $\mathcal{G} \in \mathbb{G}^N_{\alpha,\beta}$.

Proof We use a similar argument as in the proof of Theorem 13.21. We know that we only need to verify that

$$\begin{aligned} \dot{\chi} &= (A + K_\varepsilon C)\chi - K_\varepsilon y, \\ u &= F_\varepsilon \chi, \end{aligned} \tag{13.58}$$

solves the robust H_∞-ADDPMS with bounded input for the system (13.33) with $\lambda \in \Lambda$. Given $\mathcal{G} \in \mathbb{G}^N_{\alpha,\beta}$, we know that $\lambda \in \Lambda$ implies that $\operatorname{Re}\lambda \geq \beta$. Obviously $A + BF_\varepsilon$ and $A + K_\varepsilon C$ are both asymptotically stable by construction, and hence the intersection of (13.33) and (13.47) is asymptotically stable. As in the proof of Theorem 13.19, the closed-loop transfer function from w to x is equal to

$$T_{wx} = \begin{pmatrix} I & 0 \end{pmatrix} \begin{pmatrix} I & -T_2 \\ \tilde{T}_3 & I \end{pmatrix}^{-1} \begin{pmatrix} T_1 \\ -\tilde{T}_4 \end{pmatrix} \tag{13.59}$$

where, as before, we use the definitions in (13.48) but with our modified F_ε and K_ε. As argued in the proof of Theorem 13.13, we have

$$\|T_1\|_\infty < M_1 \varepsilon, \qquad \|T_2\|_\infty < M_2,$$

for suitable constants $M_1, M_2 > 0$. Finally

$$T_3(s) = \Gamma_x \begin{pmatrix} sI - A_a & L_{ad} C_d \\ -B_d E_{da} & Z_1 \end{pmatrix}^{-1} \begin{pmatrix} 0 \\ B_d \end{pmatrix}$$

where

$$Z_1 = sI - A_d - \varepsilon^{-2} S_\varepsilon^{-1} K C_d - B_d E_{dd}.$$

We obtain

$$T_3(s) = \varepsilon^{2n} \Gamma_x \begin{pmatrix} I & 0 \\ 0 & S_\varepsilon^{-1} \end{pmatrix} \begin{pmatrix} sI - A_a & L_{ad} C_d \\ \varepsilon^{2n} B_d E_{da} & Z_2 \end{pmatrix}^{-1} \begin{pmatrix} 0 \\ B_d \end{pmatrix}$$

with

$$Z_2 = sI - \varepsilon^{-2} A_d - \varepsilon^{-2} K C_d + \varepsilon^{2n} B_d E_{dd} S_\varepsilon^{-1},$$

using that

$$\varepsilon^{-2} A_d = S_\varepsilon A_d S_\varepsilon^{-1}, \quad S_\varepsilon B_d = \varepsilon^{2n} B_d \quad \text{and} \quad C_d S_\varepsilon^{-1} = C_d.$$

Note that $E_{d\varepsilon}$ is bounded for $\varepsilon < 1$. Next, we note that

$$X_\varepsilon(s) = \begin{pmatrix} sI - A_a & L_{ad}C_d \\ 0 & sI - \varepsilon^{-2}(A_d + KC_d) \end{pmatrix}^{-1} \begin{pmatrix} 0 \\ B_d \end{pmatrix}$$

$$= \varepsilon^2 \begin{pmatrix} -(sI - A_a)^{-1} L_{ad} \\ I \end{pmatrix} (\varepsilon^2 sI - A_d - KC_d)^{-1} B_d.$$

From the above we can easily conclude that there exists a M such that $\|X_\varepsilon\|_\infty < M\varepsilon^2$. We have

$$T_3 = \varepsilon^{2n}\Gamma_x \begin{pmatrix} I & 0 \\ 0 & S_\varepsilon^{-1} \end{pmatrix} (I + \varepsilon^2 X_\varepsilon E_{d\varepsilon})^{-1} X_\varepsilon$$

where

$$E_{d\varepsilon} = \begin{pmatrix} \varepsilon^{2n-2} E_{da} & \varepsilon^{2n-2} E_{dd} S_\varepsilon^{-1} \end{pmatrix}$$

which is clearly bounded for $\varepsilon < 1$. This clearly implies, using our bounds for X_ε and $E_{d\varepsilon}$, that there exists a $M_3 > 0$ such that

$$\|T_3\| \le \varepsilon^2 M_3$$

for a small ε since

$$\varepsilon^{2n} \begin{pmatrix} I & 0 \\ 0 & S_\varepsilon^{-1} \end{pmatrix} < \varepsilon^2 I.$$

Our bound for T_3 guarantees that

$$\|\tilde{T}_3\|_\infty < \varepsilon M_4, \qquad \|\tilde{T}_4\|_\infty < \varepsilon M_5,$$

for suitable M_4 and M_5. Moreover

$$\|F_\varepsilon\| < \varepsilon^{-1} M_0.$$

Given our bounds, we immediately obtain from (13.59) that there exists M_6 such that

$$\|T_{wx}\|_\infty < M_6\varepsilon.$$

The closed-loop transfer function from w to u is equal to

$$T_{wu} = \begin{pmatrix} F_\varepsilon & F_\varepsilon \end{pmatrix} \begin{pmatrix} I & -T_2 \\ \tilde{T}_3 & I \end{pmatrix}^{-1} \begin{pmatrix} T_1 \\ -\tilde{T}_4 \end{pmatrix}$$

which yields, using similar arguments as above, that

$$\|T_{wu}\|_\infty < M_7$$

for some suitable constant M_7 independent of ε. In other words, the transfer function from w to x is arbitrarily small for a sufficiently small ε, while the transfer function from w to u is bounded which completes the proof. ■

The main result regarding H_2 almost state synchronization problem, as defined in Problem 13.5, is stated as follows.

Theorem 13.22 *Consider a MAS described by a SISO system* (13.1) *and* (13.2) *with (A, B) stabilizable and (C, A) detectable. Let any real numbers $\alpha, \beta > 0$ and a positive integer N be given.*

The H_2 almost state synchronization problem, as defined in Problem 13.5, with $\mathbf{G} = \mathbb{G}_{\alpha,\beta}^N$ *is solvable. In particular, for any given real number $r > 0$, there exists an ε^*, such that for any $\varepsilon \in (0, \varepsilon^*)$, the protocol* (13.57) *achieves state synchronization and an H_2 norm from ω to $x_i - x_j$ less than r for any $i, j \in 1, \ldots, N$ and for any graph $\mathcal{G} \in \mathbb{G}_{\alpha,\beta}^N$.*

Proof It is easy to verify using a similar proof that the protocol (13.57) also solves the robust H_2 almost disturbance decoupling problem via measurement feedback with bounded input and therefore solves the H_2 almost state synchronization problem for the MAS. Using the same notation as before, this relies on the fact that we have aa N_1 such that

$$\|T_1\|_2 < \varepsilon N_1$$

which follows directly from the full-state coupling case. On the other hand, we have N_2 such that

$$\|\tilde{T}_4\|_2 < \varepsilon N_2$$

using that X_ε has an H_2 norm of order ε. The rest of the proof is then as before in the case of H_∞ almost state synchronization problem. ■

13.4.4 Static Protocol Design

In this section, we investigate again the static protocol design for a MAS with partial-state coupling. We first connect the almost state synchronization problem of a MAS with partial-state coupling with a static protocol to a robust H_∞ or H_2 almost state disturbance decoupling problem with bounded input via static output feedback (in short $(H_\infty\text{-ADDPMS})_{\text{static}}$ or $(H_2\text{-ADDPMS})_{\text{static}}$). Then, we design static protocols for three classes of agents. They are squared-down passive agents, squared-down passifiable agents via output feedback, and minimum-phase agents with relative degree 1.

13.4.4.1 Necessary and Sufficient Condition for Solvability

The MAS system described by (13.1) and (13.2) after implementing the linear static protocol (13.5) is described by

$$\begin{aligned} \dot{x}_i &= Ax_i + BF(\varepsilon)\zeta_i + E\omega_i, \\ y_i &= Cx_i \end{aligned}$$

for $i = 1, \ldots, N$ with $\zeta_i = \sum_{j=1}^{N} a_{ij}(y_i - y_j)$. Let

$$x = \begin{pmatrix} x_1 \\ \vdots \\ x_N \end{pmatrix}, \quad \omega = \begin{pmatrix} \omega_1 \\ \vdots \\ \omega_N \end{pmatrix}.$$

Then, the overall dynamics of the N agents can be written as

$$\dot{x} = (I_N \otimes A + L \otimes BF(\varepsilon)C)x + (I_N \otimes E)\omega. \tag{13.60}$$

Firstly, we define the robust $(H_\infty\text{-ADDPMS})_{\text{static}}$ with bounded input as follows. Given $\Lambda \subset \mathbb{C}$, there should exist a parameterized controller

$$u = F(\varepsilon)y \tag{13.61}$$

and $M > 0$ such that, for any given $r > 0$, there exists an $\varepsilon^* > 0$ for which the interconnection of (13.61) and the system

$$\begin{aligned} \dot{x} &= Ax + \lambda Bu + E\omega, \\ y &= Cx, \end{aligned} \tag{13.62}$$

has the property that for any $\lambda \in \Lambda$ and for any $0 < \varepsilon < \varepsilon^*$ we have:

1. The interconnection of the systems (13.61) and (13.62) is internally stable.
2. The resulting closed-loop transfer function from ω to x

$$T_{\omega x}^{\lambda} = (sI - A - \lambda BF(\varepsilon)C)^{-1}E \tag{13.63}$$

   has an H_∞ norm less than r.
3. The resulting closed-loop transfer function from u to x

$$T_{ux}^{\lambda} = (sI - A - \lambda BF(\varepsilon)C)^{-1}B \tag{13.64}$$

   has an H_∞ norm less than r.
4. The resulting closed-loop transfer function from ω to u

$$T_{\omega u}^{\lambda} = F(\varepsilon)C(sI - A - \lambda BF(\varepsilon)C)^{-1}E \tag{13.65}$$

   has an H_∞ norm less than M.
5. The resulting closed-loop transfer function from u to u

$$T_{uu}^{\lambda} = F(\varepsilon)C(sI - A - \lambda BF(\varepsilon)C)^{-1}B \tag{13.66}$$

   has an H_∞ norm less than M.

In the above, Λ denotes all the possible locations for nonzero eigenvalues of the Laplacian matrix L when the graph varies over the set $\mathbf{G}$. It is also important to note that M is independent of the choice for r and independent of $\lambda \in \Lambda$.

The connection of the above problems with the H_∞ almost state synchronization problem as defined in Problem 13.3 but with a static protocol is given below.

Theorem 13.23 *Consider the MAS described by (13.1) and (13.2) with (A, B) stabilizable and (C, A) detectable. Let $\mathbf{G}$ be a set of graphs such that the associated Laplacian matrices are uniformly bounded, and let Λ consist of all the possible nonzero eigenvalues of Laplacian matrices associated with graphs in $\mathbf{G}$.*

(Necessity) The H_∞ almost state synchronization problem, as defined in Problem 13.3, is solvable via a static protocol for the MAS described by (13.1) and (13.2) given $\mathbf{G}$ by a parameterized protocol $u_i = F(\varepsilon)\zeta_i$ only if

$$\operatorname{Im} E \subset \operatorname{Im} B. \tag{13.67}$$

(Sufficiency) The H_∞ almost state synchronization problem, as defined in Problem 13.5 for the MAS described by (13.1) and (13.2) given $\mathbf{G}$, is solvable by a parameterized protocol $u_i = F(\varepsilon)\zeta_i$ if the robust $(H_\infty\text{-ADDPMS})_{static}$ with bounded input for the system (13.62) with $\lambda \in \Lambda$ is solved by the parameterized controller (13.61).

Proof Note that L has eigenvalue 0 with associated right eigenvector $\mathbf{1}$. We again use the Schur form (2.10) for the Laplacian where U unitary and S upper triangular such that $S(1,1) = 0$. Let

$$\eta := (U^{-1} \otimes I_n)x = \begin{pmatrix} \eta_1 \\ \vdots \\ \eta_N \end{pmatrix}, \qquad \bar{\omega} = (U^{-1} \otimes I_r)\omega = \begin{pmatrix} \bar{\omega}_1 \\ \vdots \\ \bar{\omega}_N \end{pmatrix}$$

where $\eta_i \in \mathbb{C}^n$ and $\bar{\omega}_i \in \mathbb{C}^r$. In the new coordinates, the dynamics of η can be written as

$$\dot{\eta}(t) = (I_N \otimes A + S \otimes BF(\varepsilon)C)\eta + (U^{-1} \otimes E)\omega, \tag{13.68}$$

which is rewritten as

$$\begin{aligned} \dot{\eta}_1 &= A\eta_1 + \sum_{j=2}^{N} s_{1j} BF(\varepsilon)C\eta_j + E\bar{\omega}_1, \\ \dot{\eta}_i &= (A + \lambda_i BF(\varepsilon)C)\eta_i + \sum_{j=i+1}^{N} s_{ij} BF(\varepsilon)C\eta_j + E\bar{\omega}_i, \\ \dot{\eta}_N &= (A + \lambda_N BF(\varepsilon)C)\eta_N + E\bar{\omega}_N, \end{aligned} \tag{13.69}$$

for $i \in \{2, \dots, N-1\}$ where $S = [s_{ij}]$. The first column of U is an eigenvector of L associated to eigenvalue 0 with length 1, i.e., it is equal to $\mathbf{1}/\sqrt{N}$. Using this we obtain

$$\bar{x} = \begin{pmatrix} x_N - x_1 \\ x_N - x_2 \\ \vdots \\ x_N - x_{N-1} \end{pmatrix} = \left(\begin{pmatrix} -1 & 0 & \cdots & 0 & 1 \\ 0 & -1 & \ddots & \vdots & \vdots \\ \vdots & \ddots & \ddots & 0 & \vdots \\ 0 & \cdots & 0 & -1 & 1 \end{pmatrix} \otimes I_n \right) (U \otimes I_n)\eta,$$

which yields

$$\bar{x} = \left(\begin{pmatrix} 0 & V \end{pmatrix} \otimes I_n \right) \eta,$$

for some suitably chosen matrix V. Therefore we have

$$\bar{x} = (V \otimes I) \begin{pmatrix} \eta_2 \\ \vdots \\ \eta_N \end{pmatrix}, \qquad \bar{u} = (V \otimes I) \begin{pmatrix} \nu_2 \\ \vdots \\ \nu_N \end{pmatrix} \tag{13.70}$$

where

$$\nu_i = F(\varepsilon)C\eta_i, \text{ for } i = 2, \dots, N.$$

Note that since U is unitary, also the matrix U^{-1} is unitary, and the matrix V is uniformly bounded. Therefore the H_∞ norm of the transfer matrix from ω to $\bar{x}$ can be made arbitrarily small if and only if the H_∞ norm of the transfer matrix from $\bar{\omega}$ to η_i can be made arbitrarily small for $i = 2, \dots, N$. Similarly, the H_∞ norm of the transfer matrix from $\bar{\omega}$ to $\bar{u}$ is bounded if and only if the H_∞ norm of the transfer matrix from $\bar{\omega}$ to ν_i is bounded for $i = 2, \dots, N$.

Now suppose the fact that $u = F(\varepsilon)y$ solves the simultaneous $(H_\infty\text{-ADDPMS})_{\text{static}}$ with bounded input for the system (13.62) and (13.67) is satisfied. We show next that $u_i = F(\varepsilon)\zeta_i$ solves the H_∞ almost state synchronization problem for the MAS described by (13.1) and (13.2). Let X be such that $E = BX$.

The fact that $u = F(\varepsilon)y$ solves the simultaneous H_∞ almost state disturbance decoupling problem with bounded input of (13.62) implies that for small ε we have that $A + \lambda BF(\varepsilon)C$ is asymptotically stable for all $\lambda \in \Lambda$. In particular, $A + \lambda_i BF(\varepsilon)C$ is asymptotically stable for $i = 2, \dots, N$ which guarantees that $\eta_i \to 0$ for $i = 2, \dots, N$ for zero disturbances and for all the initial conditions. Therefore we have the state synchronization.

Next, we are going to show that there exists a $\bar{M} > 0$ such that for any $\bar{r} > 0$, we can choose an ε sufficiently small such that the H_∞ norm of the transfer matrix from $\bar{\omega}$ to η_i is less than $\bar{r}$ and the H_∞ norm of the transfer matrix from $\bar{\omega}$ to ν_i is less than $\bar{M}$ for $i = 2, \dots, N$. This would guarantee that we can find a $M > 0$ such that

$$\|T_{\omega\bar{x}}\|_\infty < r, \qquad \|T_{\omega\bar{u}}\|_\infty < M \tag{13.71}$$

for any $r > 0$ provided ε is small enough.

Since the robust $(H_\infty\text{-ADDPMS})_{\text{static}}$ with bounded input is solved by (13.61), there exists a $\tilde{M}$ such that for any arbitrarily small $\tilde{r}$, we have for an ε small enough that

$$\|T^\lambda_{\omega x}\|_\infty < \tilde{r}, \qquad \|T^\lambda_{ux}\|_\infty < \tilde{r},$$

$$\|T^\lambda_{\omega u}\|_\infty < \tilde{M}, \qquad \|T^\lambda_{uu}\|_\infty < \tilde{M},$$

for all $\lambda \in \Lambda$ where $T^\lambda_{\omega x}$, T^λ_{ux}, $T^\lambda_{\omega u}$, and T^λ_{uu}, are given by (13.63), (13.64), (13.65), and (13.66) respectively.

When $i = N$, it is easy to find that

$$U_{\bar{\omega}\eta_N} = e_N \otimes T^{\lambda_N}_{\omega x}, \qquad T_{\bar{\omega}\nu_N} = e_N \otimes T^{\lambda_N}_{\omega u},$$

where e_i is a row vector of dimension N with elements equal to zero except for the ith component which is equal to 1. Hence

$$\|T_{\bar{\omega}\eta_N}\|_\infty < \bar{r}, \qquad \|T_{\bar{\omega}v_N}\|_\infty < \bar{M}_N$$

provided

$$\tilde{r} < \bar{r}, \qquad \tilde{M} < \bar{M}_N. \tag{13.72}$$

Recall that we can make $\tilde{r}$ arbitrarily small without affecting the bound M. Assume that

$$\|T_{\bar{\omega}\eta_j}\|_\infty < \bar{r}, \qquad \|T_{\bar{\omega}v_j}\|_\infty < \bar{M}_j$$

holds for $j = i+1, \ldots, N$. We have

$$T_{\bar{\omega}\eta_i} = e_i \otimes T_{\omega x}^{\lambda_i} + \sum_{j=i+1}^{N} s_{ij} T_{ux}^{\lambda_i} T_{\bar{\omega}v_j} \tag{13.73}$$

$$T_{\bar{\omega}v_i} = e_i \otimes T_{\omega u}^{\lambda_i} + \sum_{j=i+1}^{N} s_{ij} T_{uu}^{\lambda_i} T_{\bar{\omega}v_j}. \tag{13.74}$$

Since

$$\left\| e_i \otimes T_{\omega x}^{\lambda_i} + \sum_{j=i+1}^{N} s_{ij} T_{ux}^{\lambda_i} T_{\bar{\omega}v_j} \right\|_\infty < \tilde{r} + \sum_{j=i+1}^{N} |s_{ij}| \tilde{r} \bar{M}_j$$

and

$$\left\| e_i \otimes T_{\omega u}^{\lambda_i} + \sum_{j=i+1}^{N} s_{ij} T_{uu}^{\lambda_i} T_{\bar{\omega}v_j} \right\|_\infty < M + \sum_{j=i+1}^{N} |s_{ij}| M \bar{M}_j$$

we find that

$$\|T_{\bar{\omega}\eta_i}\|_\infty < \bar{r}, \qquad \|T_{\bar{\omega}v_i}\|_\infty < \bar{M}_i \tag{13.75}$$

provided

$$\begin{aligned} \tilde{r} + \sum_{j=i+1}^{N} |s_{ij}| \tilde{r} \bar{M}_j &< \bar{r}, \\ M + \sum_{j=i+1}^{N} |s_{ij}| M \bar{M}_j &< \bar{M}_i. \end{aligned} \tag{13.76}$$

Note that s_{ij} depends on the graph in $\mathbb{G}$, but since the Laplacian matrices associated to graphs in $\mathbb{G}$ are uniformly bounded, we find that also the s_{ij} are uniformly bounded. In this way we can recursively obtain the bounds in (13.75) for $i = 2, \ldots, N$ provided we choose an ε sufficiently small such that the corresponding $\bar{r}$ satisfies (13.72) and (13.76) for $i = 2, \ldots, N-1$. Hence, we can choose ε sufficiently small such that the transfer matrix from $\bar{\omega}$ to η_i is less than $\bar{r}$ for $i = 2, \ldots, N$. ■

Next, we consider the H_2 norm instead of the H_∞ norm. First we define the robust (H_2-ADDPMS)$_{\text{static}}$with bounded input as follows. Given $\Lambda \subset \mathbb{C}$, there should exist a parameterized controller (13.61) and a $M > 0$ such that, for any given $r > 0$, there exists an $\varepsilon^* > 0$ for which the interconnection of (13.61) and the system (13.62) has the property that for any $\lambda \in \Lambda$ and for any $0 < \varepsilon < \varepsilon^*$ we have:

1. The interconnection of the systems (13.62) and (13.61) is internally stable.
2. The resulting closed-loop transfer function $T_{\omega x}^{\lambda}$ from ω to x has an H_2 norm less than $\sqrt{r}$.
3. The resulting closed-loop transfer function T_{ux}^{λ} has an H_∞ norm less than r.
4. The resulting closed-loop transfer function $T_{\omega u}^{\lambda}$ from ω to u has an H_2 norm less than $M/\sqrt{r}$.
5. The resulting closed-loop transfer function T_{uu}^{λ} has an H_∞ norm less than M.

In the above, Λ denotes all the possible locations for the nonzero eigenvalues of the Laplacian matrix L when the graph varies over the set $\mathbf{G}$. It is also important to note that M is independent of the choice of r and independent of $\lambda \in \Lambda$.

Note that we need to consider two aspects in our controller, H_2 disturbance rejection and robust stabilization. The latter translates to a problem with the H_∞ norm constraints. Next, we present the H_2 equivalent of Theorem 13.23.

Theorem 13.24 *Consider the MAS described by* (13.1) *and* (13.2) *with* (A, B) *stabilizable and* (C, A) *detectable. Let* $\mathbf{G}$ *be a set of graphs such that the associated Laplacian matrices are uniformly bounded and let* Λ *consist of all the possible nonzero eigenvalues of Laplacian matrices associated with graphs in* $\mathbf{G}$.

(Necessity The H_2 *almost state synchronization problem, as defined in Problem 13.5 for the MAS described by* (13.1) *and* (13.2) *given* $\mathbf{G}$ *is solvable by a parameterized protocol* $u_i = F(\varepsilon)\zeta_i$ *only if*

$$\operatorname{Im} E \subset \operatorname{Im} B. \tag{13.77}$$

(Sufficiency) The H_2 *almost state synchronization problem, as defined in Problem 13.5 for the MAS described by* (13.1) *and* (13.2) *given* $\mathbf{G}$ *is solvable by a parameterized protocol* $u_i = F(\varepsilon)\zeta_i$ *if the robust* (H_2*-ADDPMS*)$_{\text{static}}$ *with bounded input for the system* (13.62) *with* $\lambda \in \Lambda$ *is solved by the parameterized controller* (13.61).

Proof The proof is similar to the proof of Theorem 13.23. The proof has the same structure. There is one step we need to be careful. Recall that for two stable transfer matrices T_1 and T_2 with T_1 strictly proper we have

$$\|T_1T_2\|_2 \le \|T_1\|_2\|T_2\|_\infty, \tag{13.78}$$

and therefore we obtain for (13.73) that

$$\|T_{\bar{\omega}\eta_i}\|_2 \le \|T_{\omega x}^{\lambda_i}\|_2 + \sum_{j=i+1}^{N} s_{ij}\|T_{ux}^{\lambda_i}\|_\infty\|T_{\bar{\omega}v_j}\|_2$$

and

$$\|T_{\bar{\omega}v_i}\|_2 \le \|T_{\omega u}^{\lambda_i}\|_2 + \sum_{j=i+1}^{N} s_{ij}\|T_{uu}^{\lambda_i}\|_\infty\|T_{\bar{\omega}v_j}\|_2.$$

Using this, instead of (13.75) we find constants $\bar{N}_i$ and $\bar{M}_i$ such that

$$\|T_{\bar{\omega}\eta_i}\|_2 < \bar{N}_i\sqrt{r}, \qquad \|T_{\bar{\omega}v_i}\|_2 < \frac{\bar{M}_i}{\sqrt{r}}.$$

■

13.4.4.2 Squared-Down Passive Agents

Next, we can design a protocol provided (A, B, C) is squared-down passive.

Protocol design 13.5 *Consider a MAS described by square down-passive agents (13.1) with communication via (13.2). The static protocol is designed as*

$$u_i = -\frac{1}{\varepsilon}G_1KG_2\zeta_i, \tag{13.79}$$

where $\varepsilon > 0$ is a high-gain parameter to be designed, K is any symmetric and positive definite matrix, and G_1, G_2 are given in (1.9).

The main result regarding the H_∞ almost state synchronization problem with a static protocol design is stated as follows.

Theorem 13.25 *Consider a MAS described by* (13.1) *and* (13.2). *Assume* (A, B, C) *is squared-down passive with respect to* G_1 *and* G_2 *such that* (A, BG_1) *is stabilizable, and* (A, G_2C) *is detectable, while* BG_1 *and* G_2C *have full-column and full-row rank, respectively. Let any real numbers* $\beta > 0$ *and a positive integer* N *be given.*

The H_∞ *almost state synchronization problem, as defined in Problem* 13.3, *is solvable via a static protocol with* $\mathbf{G} = \mathbb{G}_\beta^N$. *In particular, for any given real number* $r > 0$, *there exists an* ε^*, *such that for any* $\varepsilon \in (0, \varepsilon^*)$, *the protocol* (13.79) *achieves state synchronization and an* H_∞ *norm from* ω *to* $x_i - x_j$ *less than* r *for any* $i, j \in 1, \ldots, N$ *and for any graph* $\mathcal{G} \in \mathbb{G}_\beta^N$.

Proof Following Theorem 13.23, it is equivalent to solve the problem of the robust $(H_\infty\text{-ADDPMS})_{\text{static}}$ with bounded input. With the protocol (13.79), the closed-loop system of (13.61) and (13.62) is written as

$$\dot{x} = (A - \frac{\lambda}{\varepsilon} BG_1KG_2C)x + BG_1\omega. \tag{13.80}$$

Define $V(x) = x'Px$, where P is given in (1.9). Then, we obtain

$$\begin{aligned}
\dot{V} &= x'(A - \rho\lambda BG_1KG_2C)'Px + \omega'G_1B'Px \\
&\qquad\qquad + x'P(A - \rho\lambda BG_1KG_2C)x + x'PBG_1\omega \\
&\leq -2\frac{\beta}{\varepsilon}x'PBG_1KG_2Cx + 2x'PBG_1\omega \\
&= -2\frac{\beta}{\varepsilon}x'C'G_2'KG_2Cx + 2x'CG_2'\omega.
\end{aligned}$$

Since K is a symmetric and positive matrix, there exists a nonsingular $\bar{K}$ such that $K = \bar{K}\bar{K}'$. Let $\tilde{x} = \bar{K}'G_2Cx$ and $\tilde{\omega} = \bar{K}^{-1}\omega$. Then we get

$$\begin{aligned}
\dot{V} &\leq -2\frac{\beta}{\varepsilon}x'C'G_2'\bar{K}\bar{K}'G_2Cx + 2x'C'G_2'\bar{K}\bar{K}^{-1}\omega \\
&= -2\frac{\beta}{\varepsilon}\tilde{x}'\tilde{x} + 2\tilde{x}'\tilde{\omega} \\
&\leq -\frac{\beta}{\varepsilon}\tilde{x}'\tilde{x} + \frac{\varepsilon}{\beta}\tilde{\omega}'\tilde{\omega},
\end{aligned}$$

which implies that the system is asymptotically stable and the H_∞ norm of the transfer function from $\tilde{\omega}$ to $\tilde{x}$ is less than ε^2/β^2. Hence, the H_∞ norm of the transfer function from ω to x is less than $r_1\varepsilon^2/\beta^2$ for some real number $r_1 > 0$.

Now let $\tilde{u} = -\frac{1}{\varepsilon}\bar{K}' G_2 C x$. Then, we have

$$\dot{V} \leq -\frac{\beta}{\varepsilon}\tilde{x}'\tilde{x} + \frac{\varepsilon}{\beta}\tilde{\omega}'\tilde{\omega}$$
$$= -\beta\varepsilon\tilde{u}'\tilde{u} + \frac{\varepsilon}{\beta}\tilde{\omega}'\tilde{\omega}$$

which implies that the system is asymptotically stable and the H_∞ norm of the transfer function from $\tilde{\omega}$ to $\tilde{u}$ is less than $1/\beta^2$. Since $u = G_1\tilde{u}$, the H_∞ norm of the transfer function from ω to u is less than r_2/β^2 for some real number $r_2 > 0$. ■

The main result regarding the H_2 almost state synchronization problem with a static protocol is stated as follows.

Theorem 13.26 *Consider a MAS described by* (13.1) *and* (13.2) *such that* (13.14) *is satisfied. Let any real number $\beta > 0$ and a positive integer N be given.*

If (A, B, C) is square down-passive with G_1, G_2 given in (1.9)*, then the H_2 almost state synchronization problem, as defined in Problem 13.5, is solvable via a static protocol with $\mathbf{G} = \mathbb{G}_\beta^N$. In particular, for any given real number $r > 0$, there exists an ε^*, such that for any $\varepsilon \in (0, \varepsilon^*)$, the protocol* (13.79) *achieves state synchronization and an H_2 norm from ω to $x_i - x_j$ less than r for any $i, j \in 1, \dots, N$ and for any graph $\mathcal{G} \in \mathbb{G}_\beta^N$.*

Proof It follows the proof of Theorem 13.12. ■

13.4.4.3 Squared-Down Passifiable Agents

Suppose (A, B, C) is squared-down passifiable via output feedback. The protocol design for squared-down passive agents also works for agents which are squared-down passifiable via output feedback. The result is stated in the following theorem.

The main result for the H_∞ almost state synchronization problem with a static protocol is stated as follows.

Theorem 13.27 *Consider a MAS described by* (13.1) *and* (13.2)*. Assume that (A, B, C) is squared-down passifiable with respect to G_1 and G_2 such that (A, BG_1) is stabilizable, and (A, G_2C) is detectable, while BG_1 and G_2C have full-column- and full-row rank, respectively. Let any real numbers $\alpha, \beta > 0$ and a positive integer N be given, and hence a set of network graphs $\mathbb{G}_{\alpha,\beta}^N$ be defined.*

The H_∞ almost state synchronization problem, as defined in Problem 13.3, with a static protocol with $\mathbf{G} = \mathbb{G}_\beta^N$ is solvable if

$$\operatorname{Im} E \subseteq \operatorname{Im} B. \tag{13.81}$$

In particular, for any given real number $r > 0$, there exists an ε^*, such that for any $\varepsilon \in (0, \varepsilon^*)$, the protocol (13.79) achieves state synchronization and an H_∞ norm from ω to $x_i - x_j$ less than r for any $i, j \in 1, \dots, N$ and for any graph $\mathcal{G} \in \mathbb{G}_{\alpha,\beta}^N$.

Proof Given Theorem 13.23, we only need to verify that $u = -\rho G_1 G_2 y$ where $\rho = \frac{1}{\varepsilon}$ solves the robust H_∞ almost output disturbance decoupling problem with bounded input for the system (13.62) with $\lambda \in \Lambda$. Given $\mathcal{G} \in \mathbb{G}_{\alpha,\beta}^N$, we know that $\lambda \in \Lambda$ implies $\operatorname{Re}\lambda \geq \beta$.

The agents are squared-down passifiable given G_1 and G_2. As noted before, this implies that there exist nonsingular state transformation matrices T_x and T_u with

$$\tilde{x} = \begin{pmatrix} \tilde{x}_1 \\ \tilde{x}_2 \end{pmatrix} = T_x x, \qquad \tilde{u} = T_u \hat{u}$$

such that the dynamics of $\tilde{x}$ is represented by

$$\begin{aligned} \dot{\tilde{x}}_1 &= A_{11}\tilde{x}_1 + A_{12}\tilde{x}_2 + E_1\omega, \\ \dot{\tilde{x}}_2 &= -A_{12}'\tilde{x}_1 + A_{22}\tilde{x}_2 + \lambda\tilde{u} + E_2\omega, \\ \hat{y} &= \tilde{x}_2, \end{aligned}$$

such that $A_{11} + A_{11}' \leq 0$. Using our output feedback, we obtain

$$\begin{aligned} \dot{\tilde{x}}_1 &= A_{11}\tilde{x}_1 + A_{12}\tilde{x}_2 + E_1\omega, \\ \dot{\tilde{x}}_2 &= -A_{12}' = -\tilde{x}_1 + (A_{22} - \lambda\rho)\tilde{x}_2 + E_2\omega, \\ \hat{y} &= \tilde{x}_2. \end{aligned} \tag{13.82}$$

It is easily verified that

$$S(s) = (sI + \rho\lambda I - A_{22})^{-1}$$

satisfies

$$\|S\|_\infty \leq \frac{1}{\beta\rho - \alpha_1} \tag{13.83}$$

for some suitable constant α_1 independent of λ and ρ using that $\operatorname{Re}\lambda > \beta$. Moreover,

$$(sI - A_{11} + \tfrac{1}{s+\rho\lambda}A_{12}A_{12}') + (sI - A_{11} + \tfrac{1}{s+\rho\lambda}A_{12}A_{12}')^* - \tfrac{2}{\rho\beta}A_{12}A_{12}' \geq 0$$

which implies that

$$V(s) = A_{12}'(sI - A_{11} + \tfrac{1}{s+\rho\lambda}A_{12}A_{12}')^{-1}A_{12}$$

satisfies

$$\|V\|_\infty \leq \rho\beta \tag{13.84}$$

Next, consider

$$Z(s) = (sI - A_{22} + \lambda\rho)^{-1} - (sI + \lambda\rho)^{-1}.$$

It is then easily verified that

$$\|Z\|_\infty \leq \frac{\|A_{22}\|}{\beta\rho(\beta\rho - \alpha_1)}.$$

Next, we consider

$$T(s) = A_{12}' \left(sI - A_{11} - A_{12}(sI - A_{22} + \lambda\rho I)^{-1}A_{12}'\right)^{-1} A_{12}.$$

We have

$$T(s) = V(s)\,[I - Z(s)V(s)]^{-1}$$

which, given the bounds obtained earlier, implies that

$$\|T\|_\infty \leq 2\rho\beta$$

for a sufficiently large ρ independent of the value of λ. Then we obtain for

$$T_1(s) = \left(sI - A_{11} - A_{12}(sI - A_{22} + \lambda\rho)^{-1}A_{12}'\right)^{-1} A_{12},$$

that

$$\|T_1\| \leq M_1\rho \tag{13.85}$$

for some suitable constant $M_1 > 0$ with M_1 independent of λ and this bound is satisfied for any ρ sufficiently large. After all, T_1 is the transfer matrix from w to y of the system

$$\begin{aligned}
\dot{x}_1 &= A_{11}x_1 + A_{12}x_2 + A_{12}w,\\
\dot{x}_2 &= -A_{12}'x_1 + (A_{22} - \lambda\rho I)x_2.\\
y &= x_2,
\end{aligned}$$

But then we have

$$\begin{aligned}\dot{x}_1 &= A_{11}x_1 + A_{12}x_2 + A_{12}y + A_{12}w,\\ \dot{x}_2 &= -A_{12}'x_1 + (A_{22} - \lambda\rho I)x_2 + y,\end{aligned}$$

which implies that

$$T_1(s) = (sI - A_{11} + A_{12}A_{12}')^{-1}A_{12}[I + T(s) + S(s)T(s)]$$

which, given our bounds for T and S and the fact that $A_{11} - A_{12}A_{12}'$ is Hurwitz stable, yields (13.85) for a suitable bound M.

Using a dual argument, we can show that our bound for T_1 yields that

$$T_2(s) = \left(sI - A_{11} + A_{12}(sI - A_{22} + \lambda\rho)^{-1}A_{12}'\right)^{-1}$$

satisfies

$$\|T_2\| \leq M_2\rho, \tag{13.86}$$

for some suitable constant $M_2 > 0$ with M_2 independent of λ, and this bound is satisfied for any ρ sufficiently large.

After these preparations, we find that for the system (13.82), the transfer matrix from ω to $\hat{y} = \bar{x}_2$ equals

$$T_{\omega\hat{y}} = ST_1SE_2 + SE_2,$$

while the transfer matrix from ω to $\hat{u}$ equals

$$T_{\omega\hat{u}} = -\rho T_{\omega\hat{y}}$$

which given our bounds yields

$$\|T_{\omega\hat{y}}\|_\infty \leq \frac{M_1\rho}{(\beta\rho - \alpha_1)^2}\|E_2\|, \qquad \|T_{\omega\hat{u}}\|_\infty \leq \frac{M_1\rho^2}{(\beta\rho - \alpha_1)^2}\|E_2\|.$$

Similarly

$$\|T_{\hat{u}\hat{y}}\|_\infty \leq \frac{M_1\rho}{(\beta\rho - \alpha_1)^2}, \qquad \|T_{\hat{u}\hat{u}}\|_\infty \leq \frac{M_1\rho^2}{(\beta\rho - \alpha_1)^2},$$

where

$$\begin{aligned}T_{\hat{u}\hat{y}} &= G_2C(sI - A - \lambda BG_1F_\varepsilon G_2C)^{-1}B,\\ T_{\hat{u}\hat{u}} &= F_\varepsilon G_2C(sI - A - \lambda BG_1F_\varepsilon G_2C)^{-1}B.\end{aligned}$$

This clearly implies that protocol (13.79) solves the robust H_∞ almost disturbance decoupling problem with bounded input for the system (13.62) as required. ∎

The main result regarding the H_2 almost state synchronization problem, as defined in Problem 13.5 with a static protocol design, is stated as follows.

Theorem 13.28 *Consider a MAS described by* (13.1) *and* (13.2). *Let any real numbers* $\alpha, \beta > 0$ *and a positive integer* N *be given.*

If (A, B, C) *is square down passifiable via input feedback with* G_1 *and* G_2 *given in* (1.9), *then the* H_2 *almost state synchronization problem, as defined in Problem 13.5, is solvable via a static protocol with* $\mathbf{G} = \mathbb{G}_{\alpha,\beta}^N$ *when* (13.81) *is satisfied. In particular, for any given real number* $r > 0$, *there exists an* ε^*, *such that for any* $\varepsilon \in (0, \varepsilon^*)$, *the protocol* (13.79) *achieves state synchronization and an* H_2 *norm from* ω *to* $x_i - x_j$ *less than* r *for any* $i, j \in 1, \ldots, N$ *and for any graph* $\mathcal{G} \in \mathbb{G}_{\alpha,\beta}^N$.

Proof It follows the proof of Theorem 13.12. ∎

13.4.4.4 *G*-Minimum-Phase Agents with Relative Degree 1

Note that when agents are G-minimum-phase and with relative degree 1 agents, there exist nonsingular state transformation matrices T_x and T_u such that $\bar{x} = [\bar{x}_1; \bar{x}_2] = T_x x$ and $\bar{u} = T_u u$. Then, the dynamics of $\bar{x}$ is represented as

$$\begin{aligned} \dot{\bar{x}}_1 &= A_{11}\bar{x}_1 + A_{12}\bar{x}_2, \\ \dot{\bar{x}}_2 &= A_{21}\bar{x}_1 + A_{22}\bar{x}_2 + \bar{u}, \\ y &= \bar{x}_2, \end{aligned} \tag{13.87}$$

where $\bar{x}_1 \in \mathbb{R}^{n-m}$ and $\bar{x}_2 \in \mathbb{R}^m$. Moreover, A_{11} is Hurwitz stable.

Next, we can design a protocol provided (A, B, C) is G-minimum-phase agent with relative degree 1.

Protocol design 13.6 *Consider a MAS described by G-minimum-phase and with relative degree* 1 *agents* (13.1) *with communication via* (13.2). *The static protocol is designed as*

$$u_i = -\rho T_u^{-1} G \zeta_i, \tag{13.88}$$

where $\rho = \dfrac{1}{\varepsilon}$ *is a parameter to be designed.*

The main result regarding the H_∞ almost state synchronization problem, as defined in Problem 13.3, with a static protocol is stated as follows.

Theorem 13.29 *Consider a MAS described by* (13.1) *and* (13.2) *such that* (13.14) *is satisfied. Let any real number* $\beta > 0$ *and a positive integer* N *be given and hence a set of network graphs* $\mathbb{G}_\beta^N$ *be defined.*

If (A, B, C) *is* G*-minimum-phase with* G *injective, then the* H_∞ *almost state synchronization problem, as defined in Problem 13.3, is solvable via a static protocol with* $\mathbf{G} = \mathbb{G}_\beta^N$*. In particular, for any given real number* $r > 0$*, there exists an* ε^**, such that for any* $\varepsilon \in (0, \varepsilon^*)$*, the protocol* (13.88) *achieves state synchronization and an* H_∞ *norm from* ω *to* $x_i - x_j$ *less than* r *for any* $i, j \in 1, \ldots, N$ *and for any graph* $\mathcal{G} \in \mathbb{G}_\beta^N$*.*

Proof Following the argument in Theorem 2.50, the robust $(H_\infty\text{-ADDPMS})_{\text{static}}$ with bounded input is defined for the system

$$\begin{aligned} \dot{\bar{x}}_1 &= A_{11}\bar{x}_1 + A_{12}\bar{x}_2, \\ \dot{\bar{x}}_2 &= A_{21}\bar{x}_1 + A_{22}\bar{x}_2 + \lambda\bar{u} + \bar{\omega}, \\ y &= GCT_x^{-1}\bar{x} = [0 \quad I]\bar{x} = \bar{x}_2, \end{aligned} \tag{13.89}$$

via a controller

$$\bar{u} = -\rho y \tag{13.90}$$

for all λ with $\text{Re}(\lambda) > \beta$. In the above, $\bar{\omega} = T_u\omega$.

Then, the closed-loop system of (13.89) and (13.90) is written as

$$\begin{aligned} \dot{\bar{x}}_1 &= A_{11}\bar{x}_1 + A_{12}\bar{x}_2, \\ \varepsilon\dot{\bar{x}}_2 &= \varepsilon A_{21}\bar{x}_1 + (\varepsilon A_{22} - \lambda I)\bar{x}_2 + \varepsilon\bar{\omega}. \end{aligned} \tag{13.91}$$

Obviously, the above system is the same as (13.29). Therefore, the remaining proof follows exactly Theorem 13.13. ■

The main result regarding the H_2 almost state synchronization problem with a static protocol is stated as follows.

Theorem 13.30 *Consider a MAS described by* (13.1) *and* (13.2)*. Let any real numbers* $\beta > 0$ *and a positive integer* N *be given.*

If (A, B, C) *is* G*-minimum-phase with* G *injective, then the* H_2 *almost state synchronization problem, as defined in Problem 13.5, is solvable via a static protocol with* $\mathbf{G} = \mathbb{G}_\beta^N$ *when* (13.81) *is satisfied. In particular, for any given real number* $r > 0$*, there exists an* ε^**, such that for any* $\varepsilon \in (0, \varepsilon^*)$*, the protocol* (13.88) *achieves state synchronization and an* H_2 *norm from* ω *to* $x_i - x_j$ *less than* r *for any* $i, j \in 1, \ldots, N$ *and for any graph* $\mathcal{G} \in \mathbb{G}_\beta^N$*.*

Proof It follows along the lines of the proof of Theorem 13.14. ■

Part II
Synchronization of Heterogeneous Systems

The second part of this book focuses on regulated output synchronization of non-introspective heterogeneous multi-agent systems (heterogeneous MAS). In Chap. 14, we discuss why we will concentrate on regulated output synchronization instead of state synchronization as discussed in the first part of this book.

In the following, Chap. 15 and 16 achieve regulated output synchronization of heterogeneous MAS with linear and nonlinear agents, respectively. Time-varying networks with linear agents are considered in Chap. 17. Various disturbances affecting the agents' dynamics and measurements are addressed in Chaps. 18, 19, and 20. Chapter 18 particularly deals with disturbances with known frequencies, where exact regulated output synchronization can be achieved by including the disturbance dynamics into the exosystem. For the case where the disturbances are bounded in power, we will study H_∞ almost regulated output synchronization in Chap. 19. For the case where the disturbances are stochastic in nature, we will consider H_2 almost regulated output synchronization in Chap. 20.

When the agents are passive, static protocols can be used to achieve output synchronization. Passive agents are considered in Chap. 21 where we also look at the effect of disturbances.

Finally, when agents are introspective, the agents can be homogenized and in this case, we can achieve output synchronization. The approach is quite different from the non-introspective case. Our introspective results can be found in Chap. 22.

Chapter 14
Necessary Conditions for Synchronization of Heterogeneous MAS

14.1 Introduction

In this second part of this book, we will study heterogeneous MAS. In contrast to the first part of this book, the dynamics of each agent can be different. This implies that many aspects of synchronization are going to be intrinsically different.

It should be noted that for homogeneous MAS, output synchronization automatically implies state synchronization (under some very basic observability assumptions). Therefore for homogeneous MAS, there is basically no difference between output and state synchronization. However, for heterogeneous MAS, this situation is intrinsically different. First of all, the internal state of each agent might not be comparable since they need not even have the same dimensions. But even if the dimensions are equal and state synchronization can be defined, it is still much harder to achieve state synchronization compared to output synchronization. Therefore for heterogeneous MAS, we are going to investigate only output synchronization.

We have seen in the first part of this book that for homogeneous MAS, the synchronized trajectory converges to some weighted average of the trajectories of the root agents. See, for instance, (2.12) or (2.32). For heterogeneous MAS we will establish in this chapter that either the synchronized trajectory is going to be zero (in other words, we basically achieve asymptotic stabilization) or the protocols are such that all the agents have some dynamics in common which generates the synchronized trajectory. In [157, 159] some early results were available. This chapter obtains an expanded and corrected version of these results.

Besides the regular communication over the network, there are versions where additional communication is sent over the network such as presented for homogeneous MAS in Sect. 2.5.8. Moreover, in some research/applications, agents have access to their own states or outputs independent of the information received via the network. We will then call the agents *introspective*. Otherwise, the agents are called *non-introspective*. In this chapter, we will investigate the necessity of creating

A. Saberi et al., *Synchronization of Multi-Agent Systems in the Presence of Disturbances and Delays*, Systems & Control: Foundations & Applications,
https://doi.org/10.1007/978-3-030-88148-1_14

these common dynamics for introspective agents with additional communication. The idea is showing that this requirement is necessary independent of whether we have introspective agents or not. Similarly, we can conclude that this additional communication also does not affect this issue.

Because many designs in the literature did not generate these common dynamics, the respective protocols effectively achieved stabilization. This resulted in the situation where many people started investigating regulated synchronization. Here there is an exosystem which shares its output to a subset of the agents, and the objective is to guarantee that output of all agents track the output of the exosystem. This is exactly what we are going to do in the following chapters.

Note that for homogeneous systems we can obviously also introduce regulated synchronization. Many of the designs in the first part of this book can be easily modified to achieve regulated synchronization. This should become obvious from our analysis in the coming chapters.

Note that in [157, 159] the common dynamics generated by the protocols in an heterogenous MAS as mentioned before are called a "virtual" exosystem to show its relationship with regulated synchronization. We have decided not to study output synchronization for heterogeneous MAS since this property tends to be very sensitive to issues like communication delays, and then we again effectively obtain stabilization of the network where everything basically converges to zero.

The write-up of this chapter is partially based on [32].

14.2 Multi-Agent Systems

Consider a MAS composed of N non-identical linear time-invariant continuous-time agents of the form

$$\begin{aligned} \dot{x}_i &= A_i x_i + B_i u_i, \\ y_i &= C_i x_i, \qquad\qquad (i = 1, ..., N) \\ z_{mi} &= C_{mi} x_i \end{aligned} \tag{14.1}$$

where $x_i \in \mathbb{R}^{n_i}$, $u_i \in \mathbb{R}^{m_i}$, and $y_i \in \mathbb{R}^p$ are, respectively, the state, input, and output vectors of agent i. We make the following assumptions for the agent dynamics.

Assumption 14.1 *For each agent $i \in \{1, \ldots, N\}$, we assume that*

- *(A_i, B_i) is stabilizable.*
- *(A_i, C_i) is detectable.*

The output z_{mi} is available for the protocol of agent i. Hence the agents are, in general, introspective. However, by setting $C_{mi} = 0$, we obtain agents which are non-introspective.

The agents obtain information through the network, which is a linear combination of its own output relative to that of other neighboring agents. In particular, each agent $i \in \{1, \ldots, N\}$ has access to the quantity

$$\zeta_i = \sum_{j=1}^{N} a_{ij}(y_i - y_j), \tag{14.2}$$

where $a_{ij} \geq 0$, $a_{ii} = 0$ for $i, j \in \{1, \ldots, N\}$. The topology of the network can be described by a weighted, directed graph $\mathcal{G}$ with nodes corresponding to the agents in the network and the weight of edges given by the coefficients a_{ij}. In particular, $a_{ij} > 0$ indicates that there is an edge in the graph from agent j to agent i. In terms of the Laplacian matrix L associated with this weighted graph $\mathcal{G}$, ζ_i can be rewritten as

$$\zeta_i = \sum_{j=1}^{N} \ell_{ij} y_j. \tag{14.3}$$

We also consider additional information. Assume that (part of) the protocol states can also be communicated using the same network. That is, agent i ($i = 1, \ldots, N$) has access to the quantity

$$\hat{\zeta}_i = \sum_{j=1}^{N} a_{ij}(\xi_i - \xi_j) = \sum_{j=1}^{N} \ell_{ij}\xi_j, \tag{14.4}$$

where $\xi_j \in \mathbb{R}^r$ is a variable produced internally by agent j as part of the protocol.

14.3 Problem Formulation

Since all the agents in the network are non-identical, as mentioned before, we can only pursue the output synchronization rather than the state synchronization among agents. Output synchronization is defined as follows.

Definition 14.1 Consider a heterogeneous MAS described by (14.1), (14.2), and (14.3). The agents in the network achieve output synchronization if

$$\lim_{t\to\infty} \left[y_i(t) - y_j(t)\right] = 0, \qquad \forall i, j \in \{1, \ldots, N\}. \tag{14.5}$$

We will formally define the output synchronization problem as follows.

Problem 14.2 Consider a MAS described by (14.1), (14.2), and (14.3). Let $\mathbf{G}$ be a given set of graphs such that $\mathbf{G} \subset \mathbb{G}^N$. The *output synchronization* problem is to find, if possible, a linear time-invariant dynamic protocol of the form

$$\begin{aligned} \dot{\chi}_i &= A_{ci,1}\chi_i + B_{ci,1}\zeta_i + B_{ci,2}\hat{\zeta}_i + B_{ci,3}z_{mi}, \\ \xi_i &= C_{di}\chi_i, \\ u_i &= C_{ci}\chi_i + D_{ci,1}\zeta_i + D_{ci,2}\hat{\zeta}_i + D_{ci,3}z_{mi}, \end{aligned} \tag{14.6}$$

for $i = 1, \dots, N$ with $\chi_i \in \mathbb{R}^{q_i}$ such that, for any graph $\mathcal{G} \in \mathbf{G}$ and for all the initial conditions of agents and their protocols, output synchronization among agents is achieved.

14.4 Necessary Condition for Output Synchronization

Then, the necessary condition for output synchronization is stated in the following theorem.

Theorem 14.3 *Consider a MAS described by* (14.1), (14.2), *and* (14.3). *Consider a dynamics protocol of the form* (14.6).

Assume, the interconnected system is such that $y_i - y_j \to 0$ and $\xi_i - \xi_j \to 0$ for $i, j \in \{1, \dots, N\}$ as $t \to \infty$ for all intial conditions of the agents and the protocols. In that case, one of the following two scenarios applies:

- *We have that $y_i(t) \to 0$ for all $i \in \{1, \dots, N\}$ and all initial conditions as $t \to \infty$.*
- *There exist a scalar $m \in \mathbb{N}$, matrices $S \in \mathbb{R}^{m\times m}$ and $R \in \mathbb{R}^{q\times m}$, where the eigenvalues of S are in the closed right half complex plane and (S, R) is observable. Moreover for any initial conditions for agents and protocols, there exists $z_0 \in \mathbb{R}^m$ such that*

$$\lim_{t\to\infty} \|y_i(t) - Re^{St}z_0\| = 0$$

for all $i \in \{1, \dots, N\}$.

Moreover, in the second scenario, there exist a scalar $r \in \mathbb{N}$, matrices $S_{21} \in \mathbb{R}^{r\times m}$ and $S_{22} \in \mathbb{R}^{r\times r}$, where the eigenvalues of S_{22} are in the closed right half complex plane, such that, for each $i \in \{1, \dots, N\}$, there exist matrices $\Pi_i \in \mathbb{R}^{n_i\times(m+r)}$ and $\Gamma_i \in \mathbb{R}^{m_i\times(m+r)}$ satisfying

$$\begin{aligned} A_i\Pi_i + B_i\Gamma_i &= \Pi_i\bar{S}, \\ C_i\Pi_i &= \bar{R}, \end{aligned} \tag{14.7}$$

where

$$\bar{S} = \begin{pmatrix} S & 0 \\ S_{21} & S_{22} \end{pmatrix}, \quad \bar{R} = \begin{pmatrix} R & 0 \end{pmatrix}.$$

Remark 14.4 The structure of $\bar{S}$ and $\bar{R}$ is a bit unusual. Note that we can choose $\bar{S} = S$ and $\bar{R} = R$ and dispense with S_{21} and S_{22} provided for each $i = 1, \dots, N$, the pair $(A_i^*, C_{i,1}^*)$, where

$$A_i^* = \begin{pmatrix} A_i + B_i D_{ci,3} C_{mi} & B_i C_{ci} \\ B_{ci,3} C_{mi} & A_{ci} \end{pmatrix}, \quad C_{i,1}^* = \begin{pmatrix} C_i & 0 \end{pmatrix}, \tag{14.8}$$

is detectable. Note that [157, 159] provided the result with $\bar{S} = S$ and $\bar{R} = R$ but missed the additional detectability requirement for (14.8). For details we refer to [32].

Proof The closed-loop system consisting of N agent dynamics (14.1) and the controller (14.6) has the form

$$\begin{aligned} \dot{\tilde{x}} &= A^* \tilde{x} + B_1^* \zeta + B_2^* \hat{\zeta}, \\ y &= C_1^* \tilde{x}, \\ \xi &= C_2^* \tilde{x}, \\ \zeta &= (L \otimes I) y \\ \hat{\zeta} &= (L \otimes I) \xi, \end{aligned}$$

where A^* and C_1^* are block diagonal matrices based on A_i^* and $C_{i,1}^*$ as defined in (14.8). Moreover,

$$B_{i,1}^* = \begin{pmatrix} B_i D_{ci,1} \\ B_{ci,1} \end{pmatrix}, \qquad B_{i,2}^* = \begin{pmatrix} B_i D_{ci,2} \\ B_{ci,2} \end{pmatrix}, \qquad C_{i,2}^* = \begin{pmatrix} 0 & C_{di} \end{pmatrix},$$

while B_1^*, B_2^*, and C_2^* are block diagonal matrices based on $B_{i,1}^*$, $B_{i,2}^*$, and $C_{i,2}^*$, respectively. Note that, we have $\tilde{x} \in \mathbb{R}^{\tilde{n}}$ with $\tilde{n} = n_1 + n_2 + \cdots + n_N$.

Because $y_i - y_j \to 0$ and $\xi_i - \xi_j \to 0$ for $i, j = 1, \dots, N$, the closed-loop system has an asymptotically attractive invariant subspace $\mathcal{S}$ on which $y_i = y_j$ and $\xi_i = \xi_j$ for $i, j = 1, \dots, N$ which implies that $\zeta = (L \otimes I) y = 0$ and $\hat{\zeta} = (L \otimes I) \xi = 0$ on $\mathcal{S}$. Hence, on $\mathcal{S}$, the closed-loop system is described by

$$\dot{\tilde{x}} = A^* \tilde{x}.$$

Without loss of generality, we can assume $\mathcal{S}$ contains no asymptotically stable modes.

If $\mathcal{S} = \{0\}$ then $y_i \to 0$ for all initial conditions. Otherwise, it is nontrivial with dimension $m \in \mathbb{N}$. Therefore, $\mathcal{S}$ can be written as $\mathcal{S} = \operatorname{span} \tilde{\Psi}$ for some matrix $\tilde{\Psi} \in \mathbb{R}^{\tilde{n} \times \tilde{m}}$, and there exists a matrix $\tilde{S} \in \mathbb{R}^{\tilde{m} \times \tilde{m}}$ such that

$$A^* \tilde{\Psi} = \tilde{\Psi} \tilde{S}, \tag{14.9}$$

where $\tilde{S}$ describes the dynamics of the closed-loop system restricted to the subspace $\mathcal{S}$ (which, by construction, does not contain any asymptotically stable modes). Write $\tilde{\Psi}$ as

$$\tilde{\Psi} = \begin{pmatrix} \tilde{\Pi}_1 \\ \tilde{\Sigma}_1 \\ \vdots \\ \tilde{\Pi}_N \\ \tilde{\Sigma}_N \end{pmatrix}$$

with $\tilde{\Pi}_i \in \mathbb{R}^{n_i \times \tilde{m}}$ and $\tilde{\Sigma}_i \in \mathbb{R}^{v_i \times \tilde{m}}$ for $i = 1, \ldots, N$. Considering the block diagonal structure of A^*, this yields

$$A_i \tilde{\Pi}_i + B_i \tilde{\Gamma}_i = \tilde{\Pi}_i \tilde{S},$$

for $i = 1, \ldots, N$ where

$$\tilde{\Gamma}_i = D_{ci,3} C_{mi} \tilde{\Pi}_i + C_{ci} \tilde{\Sigma}_1.$$

Furthermore, since $y_i = y_j$ on $\mathcal{S}$, we have $C_{i,1} \tilde{\Pi}_i = \tilde{R}$ for some $\tilde{R}$.

Let Λ be a nonsingular state transformation taking $(\tilde{S}, \tilde{R})$ to the Kalman observable canonical form. That is,

$$\Lambda^{-1} \tilde{S} \Lambda = \bar{S} = \begin{pmatrix} S & 0 \\ S_{21} & S_{22} \end{pmatrix}, \quad \tilde{R} \Lambda = \bar{R} = \begin{pmatrix} R & 0 \end{pmatrix},$$

for some matrices $S \in \mathbb{R}^{m \times m}$, $S_{21} \in \mathbb{R}^{r \times m}$, $S_{22} \in \mathbb{R}^{r \times r}$, and $R \in \mathbb{R}^{q \times m}$, where $m = \tilde{m} - r$. The output trajectories on the agreement manifold are governed by the observable part of $(\bar{S}, \bar{R})$, given by (S, R), and hence $\lim(y_i(t) - Re^{St} z_0) = 0$ for some $z_0 \in \mathbb{R}^m$. Furthermore, it is easily seen that the regulator equations (14.7) are satisfied with $\Pi_i = \tilde{\Pi}_i \Lambda$ and $\Gamma_i = \tilde{\Gamma}_i \Lambda$. ■

Then, we immediately have the following result.

Theorem 14.5 *Assume that the conditions of Theorem 14.3 hold. For each $i \in \{1, \ldots, N\}$, there exists a $T_i \in \mathbb{R}^{(n_i+v_i) \times (n_i+v_i)}$ and matrices $S_{i21} \in \mathbb{R}^{r_i \times m}$ and $S_{i22} \in \mathbb{R}^{r_i \times r_i}$, where $r_i \leq r$, such that*

$$T_i^{-1} A_i^* T_i = \begin{pmatrix} \bar{S}_i & \star \\ 0 & \star \end{pmatrix}, \quad \tilde{C}_{i,1} T_i = \begin{pmatrix} \bar{R}_i & \star \end{pmatrix},$$

where

$$\bar{S}_i = \begin{pmatrix} S & 0 \\ S_{i21} & S_{i22} \end{pmatrix}, \quad \bar{R}_i = \begin{pmatrix} R & 0 \end{pmatrix}.$$

Proof Referring back to the proof of Theorem 14.3, it is clear that $A^*\Psi = \Psi\bar{S}$, where $\Psi = \tilde{\Psi}\Lambda$. Let

$$\Psi = \begin{pmatrix} \Psi_1 \\ \vdots \\ \Psi_N \end{pmatrix},$$

where $\Psi_i \in \mathbb{R}^{(n_i+v_i)\times\bar{m}}$. Then

$$A_i^*\Psi_i = \Psi_i\bar{S}, \quad C_{i,1}\Psi_i = \bar{R}.$$

Let $\Psi_i = \begin{pmatrix} \Psi_{i1} & \Psi_{i2} \end{pmatrix}$, where $\Psi_{i1} \in \mathbb{R}^{(n_i+v_i)\times m}$ and $\Psi_{i2} \in \mathbb{R}^{(n_i+v_i)\times r}$. Then, Ψ_{i1} is of full column rank m and $\operatorname{Im}\Psi_{i1} \bigcap \operatorname{Im}\Psi_{i2} = 0$. To see this, let

$$O(A, C) = \begin{pmatrix} C \\ \vdots \\ CA^{\bar{m}-1} \end{pmatrix}$$

for any matrices A and C of compatible dimensions, and note that $O(\bar{S}, \bar{R}) = O(\tilde{A}_i, \tilde{C}_i)\Psi_i$. Moreover,

$$O(\bar{S}, \bar{R}) = \begin{pmatrix} O(S, R) & 0 \end{pmatrix},$$

and hence we have

$$O(A_i^*, C_{i,1}^*)\Psi_{i1} = O(S, R) \text{ and } O(A_i^*, C_{i,1})\Psi_{i2} = 0.$$

Since $O(S, R)$ is the observability matrix of the observable pair (S, R), we see that $\operatorname{rank} O(A_i^*, C_i^*)\Psi_{i1} = m$, which implies $\operatorname{rank}\Psi_{i1} = m$. Moreover, $\operatorname{Im}\Psi_{i1} \bigcap \operatorname{Im}\Psi_{i2} = 0$, since otherwise there would be an $\eta_1 \neq 0$ such that $\Psi_{i1}\eta_1 = \Psi_{i2}\eta_2$. This would imply

$$O(A_i^*, C_{i,1})\Psi_{i1}\eta_1 = O(A_i^*, C_{i,1})\Psi_{i2}\eta_2 = 0,$$

which contradicts the fact that $\operatorname{rank} O(A_i^*, C_{i,1})\Psi_{i1} = m$.

Define $T_i = \begin{pmatrix} \Psi_{i1} & \Psi_{i2}V_i & T_{i3} \end{pmatrix}$, where V_i is an injective matrix such that the columns of $\Psi_{i2}V_i$ form a nonsingular basis for $\operatorname{Im}\Psi_{i2}$ and T_{i3} is chosen to make T_i

nonsingular. Then,

$$A_i^* T_i = \begin{pmatrix} A_i^*\Psi_{i1} & A_i^*\Psi_{i2}V_i & A_i^*T_{i3} \end{pmatrix} = \begin{pmatrix} \Psi_{i1}S + \Psi_{i2}S_{21} & \Psi_{i2}S_{22}V_i & A_i^*T_{i3} \end{pmatrix}.$$

Let S_{i21} be defined such that $\Psi_{i2}S_{21} = \Psi_{i2}V_iS_{i21}$, and let S_{i22} be defined such that $\Psi_{i2}S_{22}V_i = \Psi_{i2}V_iS_{i22}$. These exist since $\operatorname{Im}\Psi_{i2}V_i = \operatorname{Im}\Psi_{i2}$. Then,

$$A_i^* T_i = \begin{pmatrix} \Psi_{i1} & \Psi_{i2}V_i & T_{i3} \end{pmatrix} \begin{pmatrix} S & 0 & \star \\ S_{i21} & S_{i22} & \star \\ 0 & 0 & \star \end{pmatrix} = T_i \begin{pmatrix} \bar{S}_i & \star \\ 0 & \star \end{pmatrix}.$$

Furthermore,

$$C_{i,1}T_i = \begin{pmatrix} C_{i,1}\Psi_{i1} & C_{i,1}\Psi_{i2}V_i & C_{i,1}T_{i3} \end{pmatrix} = \begin{pmatrix} R & 0 & 0 \end{pmatrix} = \begin{pmatrix} \bar{R}_i & \star \end{pmatrix}.$$ ■

From the above discussion, it is clear that in order to achieve synchronization, we need at least common dynamics to achieve nontrivial synchronized trajectories. This motivates why in the upcoming chapters the focus is often on achieving some form of homogenization among agents.

Since in the case of heterogeneous MAS, the synchronized trajectory often becomes zero and we effectively achieve stabilization, it makes sense to prescribe the synchronized trajectory to some of the agents. We then obtain the problem of regulated output synchronization which will be formally defined in the next chapter.

Chapter 15
Regulated Output Synchronization of Heterogeneous Continuous-Time Linear MAS

15.1 Introduction

In the first part of this book, we focused on homogeneous MAS. In this second part of the book, we focus on heterogeneous MAS. There is only a limited amount of research on heterogeneous MAS compared to homogeneous MAS.

Ramírez and Femat [89] presented a robust state synchronization design for networks of nonlinear systems with relative degree one, where each agent implements a sufficiently strong feedback based on the difference between its own state and that of a common reference model. In the work of [169], it is assumed that a common Lyapunov function candidate is available, which is used to analyze stability with respect to a common equilibrium point. Depending on the system, some agents may also implement feedbacks to ensure stability, based on the difference between those agents' states and the equilibrium point. Zhao et al. [192] analyzed state synchronization in a network of nonlinear agents based on the network topology and the existence of certain time-varying matrices. Controllers can be designed based on this analysis, to the extent that the available information and actuation allow for the necessary manipulation of the network topology.

The above-cited works focus on synchronizing the agents' internal states. In heterogeneous networks, however, the physical interpretation of one agent's state may be different from that of another agent. Indeed, the agents may be governed by models of different dimensions. In this case, comparing the agents' internal states is not meaningful, and it is more natural to aim for *output synchronization*—that is, agreement on some partial-state output from each agent. Chopra and Spong [16] focused on output synchronization for weakly minimum-phase systems of relative degree one, using a pre-feedback within each agent to create a single-integrator system with decoupled zero dynamics. Pre-feedbacks were also used by Bai et al. [4] to facilitate passivity-based designs.

A. Saberi et al., *Synchronization of Multi-Agent Systems in the Presence of Disturbances and Delays*, Systems & Control: Foundations & Applications,
https://doi.org/10.1007/978-3-030-88148-1_15

Kim et al. [43] studied output synchronization for uncertain single-input, single-output, minimum-phase systems, by embedding an identical model within each agent, the output of which is tracked by the actual agent output. In the previous chapter we showed that a necessary condition for output synchronization in heterogeneous networks is the existence of a *virtual exosystem* that produces a trajectory to which all the agents asymptotically converge. If one knows the model of an observable virtual exosystem without exponentially unstable modes, which each agent is capable of tracking, then it can be implemented within each agent and synchronized via the network. The agent can then be made to track the model with the help of a local observer estimating the agent's states.

The designs mentioned above for heterogeneous networks rely—explicitly or implicitly—on some sort of self-knowledge that is separate from the information transmitted over the network. In particular, the agents may be required to know their own state, their own output, or their own state/output relative to that of a reference trajectory. We refer to agents that possess this type of self-knowledge as *introspective* agents, to distinguish them from *non-introspective* agents—that is, agents that have no knowledge of their own state or output separate from what is received via the network. This distinction is significant because introspective agents have much greater freedom to manipulate their internal dynamics (e.g., through the use of pre-feedbacks) and thus change the way that they present themselves to the rest of the network. The notion of a non-introspective agent is also practically relevant; for example, two vehicles in close proximity may be able to measure their relative distance without either of them having knowledge of their absolute position.

To the authors' knowledge, the first paper that that solves the output synchronization problem for a well-defined class of heterogeneous networks of non-introspective agents is by Zhao et al. [192]. In their work, the only information available to each agent is a linear combination of outputs received over the network. However, the agents are assumed to be passive—a strict requirement that, among other things, requires the agents to be weakly minimum-phase and of relative degree one.

In this chapter, we will consider regulated output synchronization which, as stated in the previous chapter, means that the output of each agent are regulated asymptotically to a reference trajectory generated by an autonomous exosystem. A purely decentralized controller based on a low- and high-gain methodology is designed for each agent, when agents are minimum-phase. However, when the additional communication channel is allowed, agents can be non-minimum-phase.

The write-up of this chapter is partially based on [31] and [33].

15.2 Multi-Agent Systems

Consider a MAS composed of N non-identical linear time-invariant continuous-time agents of the form

$$\begin{aligned} \dot{x}_i &= A_i x_i + B_i u_i, \\ y_i &= C_i x_i, \end{aligned} \qquad (i = 1, \ldots, N) \tag{15.1}$$

where $x_i \in \mathbb{R}_i^n$, $u_i \in \mathbb{R}_i^m$, and $y_i \in \mathbb{R}^p$ are, respectively, the state, input, and output vectors of agent i. We make the following assumptions for the agent dynamics.

Assumption 15.1 *For each agent $i \in \{1, \ldots, N\}$, we assume that*

- *(A_i, B_i) is stabilizable;*
- *(A_i, C_i) is detectable.*

Since the agents are non-introspective, they have no access to part of their states or outputs. The only information that is available for agents is from the network, which is a linear combination of its own output relative to that of other neighboring agents. In particular, each agent $i \in \{1, \ldots, N\}$ has access to the quantity

$$\zeta_i = \sum_{j=1}^{N} a_{ij}(y_i - y_j), \tag{15.2}$$

where $a_{ij} \geq 0$, $a_{ii} = 0$ for $i, j \in \{1, \ldots, N\}$. The topology of the network is described, as before, by a weighted, directed graph $\mathcal{G}$ with nodes corresponding to the agents in the network and the weight of edges given by the coefficients a_{ij}. In terms of the Laplacian matrix L associated with this weighted graph $\mathcal{G}$, ζ_i can be rewritten as

$$\zeta_i = \sum_{j=1}^{N} \ell_{ij} y_j. \tag{15.3}$$

15.3 Problem Formulation

Since all the agents in the network are non-identical, we can only pursue output synchronization, rather than state synchronization among agents. Output synchronization is defined as follows.

Definition 15.1 Consider a heterogeneous MAS described by (15.1) and (15.2). The agents in the network achieve output synchronization if

$$\lim_{t \to \infty} \left[y_i(t) - y_j(t) \right] = 0, \qquad \forall i, j \in \{1, \ldots, N\}. \tag{15.4}$$

We define first the output synchronization problem as follows.

Problem 15.2 Consider a MAS described by (15.1) and (15.2). For a given set of network graphs $\mathbb{G}$, the *output synchronization* problem is to find, if possible, a linear time-invariant dynamic protocol of the form

$$\begin{aligned} \dot{\chi}_i &= A_{ci}\chi_i + B_{ci}\zeta_i, \\ u_i &= C_{ci}\chi_i, \end{aligned} \tag{15.5}$$

for $i = 1, \ldots, N$ where $\chi_i \in \mathbb{R}^{N_{ci}}$, such that, for any graph $\mathbb{G}$ and for all the initial conditions of agents and their protocol, output synchronization among agents is achieved.

It will be shown that the protocol (15.5) requires the communication of protocol states using the communication network among agents. When the communication of protocol states are not allowed, that is, purely distributed protocols are utilized, we will pursue regulated output synchronization. In other words, our goal is to regulate the output of each agent asymptotically toward an a priori specified reference trajectory. The reference trajectories are usually generated by an autonomous exosystem:

$$\begin{aligned} \dot{x}_0 &= A_0 x_0, \quad x_0(0) = x_{00}, \\ y_0 &= C_0 x_0, \end{aligned} \tag{15.6}$$

where $x_0 \in \mathbb{R}^r$ and A_0 has all its eigenvalues on the imaginary axis and (A_0, C_0) is observable.

Then, the regulated output synchronization is defined as follows.

Definition 15.3 Consider a heterogeneous MAS described by (15.1) and (15.2). Suppose the reference trajectories are generated by (15.6). Then, the agents in the MAS achieve regulated output synchronization if

$$\lim_{t\to\infty} [y_i(t) - y_0(t)] = 0, \qquad \forall i \in \{1, \ldots, N\}. \tag{15.7}$$

In order to achieve our goal, it is clear that a non-empty subset of agents must have knowledge of their output relative to the reference trajectory y_0 generated by the exosystem (15.6). We denote such a subset of agents by π. To be specific, each agent $i \in \{1, \ldots, N\}$ has access to the quantity

$$\psi_i = \iota_i(y_i - y_0), \qquad \iota_i = \begin{cases} 1, & i \in \pi, \\ 0, & i \notin \pi. \end{cases}$$

In order to achieve regulated output synchronization for all agents, the following assumption is clearly necessary.

Assumption 15.2 *Every node of the network graph $\mathcal{G}$ is a member of a directed tree which has its root contained in the set π.*

In the following, we refer to the node set π as *root set* in view of Assumption 15.2 (when the network graph $\mathcal{G}$ has a directed spanning tree, the set π may contain one node which is the root of such a directed spanning tree).

Based on the Laplacian matrix L of the network graph $\mathcal{G}$, we define the expanded Laplacian matrix as

$$\bar{L} = L + \text{diag}\{\iota_i\} = [\bar{\ell}_{ij}].$$

Note that $\bar{L}$ is clearly not a Laplacian matrix associated to some graph since it does not have a zero row sum. From [31, Lemma 7], all eigenvalues of $\bar{L}$ are in the open right half complex plane.

Next we define a set of network graphs based on some rough information of the graph.

Definition 15.4 For a given root set π, a positive integer N, and $\alpha, \beta > 0$, the set $\mathbb{G}^N_{\alpha,\beta,\pi}$ is the set of directed graphs composed of N nodes satisfying the following properties:

- The eigenvalues of the associated expanded Laplacian $\bar{L}$, denoted by $\lambda_1, \ldots, \lambda_N$, satisfy $\text{Re}\{\lambda_i\} > \beta$ and $||\bar{L}|| < \alpha$.
- Assumption 15.2 is satisfied.

We define the regulated output synchronization problem as follows.

Problem 15.5 Consider a MAS described by (15.1) and (15.2). For a given root set π, a positive integer N, and $\alpha, \beta > 0$ that define a set of network graphs $\mathbb{G}^N_{\alpha,\beta,\pi}$ and for any reference trajectories given by (15.6), the *regulated output synchronization* problem is to find, if possible, a linear time-invariant dynamic protocol of the form

$$\begin{aligned} \dot{\chi}_i &= A_{ci}\chi_i + B_{ci}\,\text{col}\{\zeta_i, \psi_i\}, \\ u_i &= C_{ci}\chi_i, \end{aligned} \tag{15.8}$$

for $i = 1, \ldots, N$ where $\chi_i \in \mathbb{R}^{N_{ci}}$, such that, for any graph $\mathcal{G} \in \mathbb{G}^N_{\alpha,\beta,\pi}$ and for all the initial conditions of agents and their protocol, the regulated output synchronization among agents is achieved.

15.4 Protocol Design

We present here the protocol design. When the protocol only uses the network information ζ_i and the reference information ψ_i, the solvability of Problem 15.5 requires agents to be minimum-phase, while an additional communication of protocol states via network's communication infrastructure removes this restriction, that is, the agents can be non-minimum-phase.

15.4.1 Minimum-Phase and Right-Invertible Agents

The protocol design for minimum-phase and right-invertible agents is composed of two steps. The first step is to design pre-compensators such that the cascade system of the pre-compensators and the agent dynamics is of uniform rank and contains the exosystem mode. The second step is to design an observer-based controller such that the regulated output synchronization is achieved.

Step 1 There are three pre-compensators to be designed. The first one ensures that all agents are invertible and have uniform rank. The second one ensures that the dynamics of each agent contains a copy of the dynamics of the exosystem while preserving uniform rank and invertibility. The final pre-compensator guarantees that all agents have an equal uniform rank.

Protocol design 15.1 *: Pre-compensator* 1

According to Theorem 24.2 in Chap. 24, we design a pre-compensator of the form

$$\begin{aligned} \dot{p}_{i,1} &= A_{ip,1}p_{i,1} + B_{ip,1}u_i^1, \\ u_i &= C_{ip,1}p_{i,1} + D_{ip,1}u_i^1, \end{aligned} \tag{15.9}$$

such that each interconnection of the agent and the pre-compensator is minimum-phase, invertible, and uniform rank.

Protocol design 15.2 *: Pre-compensator* 2

First consider the case where the agent has no dynamics in common with the exosystem. Suppose that intersection of the agent dynamics with pre-compensator (15.9) is denoted by the triple $(\hat{A}_i, \hat{B}_i, \hat{C}_i)$. Then, $\tilde{\Pi}_i$ and Γ_i are uniquely determined by the so-called regulator equations:

$$\begin{aligned} \hat{A}_i\tilde{\Pi}_i + \hat{B}_i\Gamma_i &= \tilde{\Pi}_i A_0, \\ \hat{C}_i\tilde{\Pi}_i &= C_0. \end{aligned} \tag{15.10}$$

If (Γ_i, A_0) is not observable, we can construct the observable subsystem $(\Gamma_{i,o}, S_{i,o})$; otherwise simply set $\Gamma_{i,o} = \Gamma_i$ and $S_{i,o} = A_0$. Then the pre-compensator for each agent i is designed as

$$\begin{aligned} \dot{p}_{i,2} &= S_{i,o}p_{i,2} + B_{i,2}u_i^2, \\ u_i^1 &= \Gamma_{i,o}p_{i,2}, \end{aligned} \tag{15.11}$$

(continued)

Protocol design 15.2 (continued)
for $i = 1, \ldots, N$, *where* $p_{i,2} \in \mathbb{R}^{n_{pi}}$ *and* $B_{i,2}$ *are chosen to guarantee that no invariant zeros are introduced by the pre-compensator. Moreover, an appropriate choice of* $B_{i,2}$ *can guarantee that the triple* $(S_{i,o}, B_{i,2}, \Gamma_{i,o})$ *is invertible, uniform rank, and minimum-phase (see [60]).*

If the agent has dynamics in common with the exosystem, then we can use the technique outlined in Appendix "Removing Common Modes of Agent i and A_0*" at the end of this chapter to remove these common modes after which the above design is applicable.*

Protocol design 15.3 *: Pre-compensator* 3
We denote by ρ_i *the highest order of the infinite zeros for the interconnection of agent i with pre-compensators* 1 *and* 2. *Choose* $\rho = \max\{\rho_i\}$*:*

$$\begin{aligned} \dot{p}_{i,3} &= K_i p_{i,3} + L_i \tilde{u}_i, \\ u_i^2 &= M_i p_{i,3}, \end{aligned} \tag{15.12}$$

where (K_i, L_i, M_i) *are such that the transfer function of this pre-compensator is equal to* $(s+1)^{\rho_i - \rho} I$.

Using the above design steps, the cascade system of agent (15.1) and pre-compensator (15.9), (15.11), and (15.12) can be represented in the form

$$\begin{cases} \dot{\tilde{x}}_i = \tilde{A}_i \tilde{x}_i + \tilde{B}_i \tilde{u}_i, \\ y_i = \tilde{C}_i \tilde{x}_i, \end{cases} \tag{15.13}$$

where $\tilde{x}_i \in \mathbb{R}^{\tilde{n}_i}$, $\tilde{u}_i \in \mathbb{R}^p$, and $y_i \in \mathbb{R}^p$ are states, inputs, and outputs of the interconnection system. Moreover, we have the following:

- $(\tilde{A}_i, \tilde{B}_i, \tilde{C}_i)$ has uniform rank $\rho \geq 1$.
- $(\tilde{A}_i, \tilde{C}_i)$ contains (A_0, C_0), i.e., there exists a matrix Π_i such that $\Pi_i A_0 = \tilde{A}_i \Pi_i$ and $\tilde{C}_i \Pi_i = C_0$. When (Γ_i, S) is observable, $\Pi_i = \operatorname{col}\{\tilde{\Pi}_i, I\}$, otherwise, the construction of Π_i needs more work and is given in the Appendix "Construction of Π_i" at the end of this chapter.

The cascade system (15.13) has uniform rank and is invertible. Hence the system can be represented by the SCB form as presented in (23.27). In other words, we obtain

$$
\begin{aligned}
\dot{\tilde{x}}_{ia} &= A_{ia}\tilde{x}_{ia} + L_{iad}y_i, \\
\dot{\tilde{x}}_{id} &= A_d\tilde{x}_{id} + B_d(\tilde{u}_i + E_{ida}\tilde{x}_{ia} + E_{idd}\tilde{x}_{id}), \\
y_i &= C_d\tilde{x}_{id},
\end{aligned}
\tag{15.14}
$$

for $i \in \{1, \ldots, N\}$, where $x_{ia} \in \mathbb{R}^{\tilde{n}_i - p\rho}$ represents the finite zero structure, $x_{id} \in \mathbb{R}^{p\rho}$ represents the infinite zero structure, $u_i, y_i \in \mathbb{R}^p$, and A_d, B_d, C_d have the special form

$$
A_d = \begin{pmatrix} 0 & I \\ 0 & 0 \end{pmatrix}, \qquad B_d = \begin{pmatrix} 0 \\ I \end{pmatrix}, \qquad C_d = \begin{pmatrix} I & 0 \end{pmatrix}. \tag{15.15}
$$

The eigenvalues of A_{ia} are the invariant zeros of the triple $(\tilde{A}_i, \tilde{B}_i, \tilde{C}_i)$, which are all in the open left half complex plane, because no unstable invariant zeros are introduced during the pre-compensator design.

Step 2 The observer-based protocol design is given in the following.

Protocol design 15.4 *Consider a MAS described by* (15.1) *and* (15.2). *With the pre-compensator* (15.9), (15.11), *we obtain dynamics* (15.14).

Let $\delta \in (0, 1]$ *and* $\varepsilon \in (0, 1]$ *denote respectively a low-gain and a high-gain parameter. Let therefore* K *be chosen such that* $A_d - KC_d$ *is Hurwitz stable. Furthermore, let* $P_\delta > 0$ *be the solution of the algebraic Riccati equation*

$$
P_\delta A_d + A_d' P_\delta - \beta P_\delta B_d B_d' P_\delta + \delta I = 0, \tag{15.16}
$$

where $\beta > 0$ *is a lower bound on the real part of the nonzero eigenvalues of the expanded Laplacian matrix* $\bar{L}$ *and* $\delta \in (0, 1]$ *is a low-gain parameter. Define* $F_\delta = -B_d' P_\delta$. *Next, define a high-gain scaling matrix*

$$
S_\varepsilon = \operatorname{diag}(I_p, \ldots, \varepsilon^{\rho-1} I_p), \tag{15.17}
$$

where $\varepsilon \in (0, 1]$ *is a high-gain parameter, and define the feedback and output injection matrices:*

$$
F_{\delta\varepsilon} = \varepsilon^{-\rho} F_\delta S_\varepsilon, \quad K_\varepsilon = \varepsilon^{-1} S_\varepsilon^{-1} K.
$$

Now, for each $i \in \{1, \ldots, N\}$, *the linear dynamic protocol is designed as follows:*

(continued)

Protocol design 15.4 (continued)

$$\begin{aligned}\dot{\hat{x}}_{id} &= A_d\hat{x}_{id} + K_\varepsilon(\zeta_i + \psi_i - C_d\hat{x}_{id}),\\ u_i &= F_{\delta\varepsilon}\hat{x}_{id}.\end{aligned} \tag{15.18}$$

The main result based on the above design can be stated as follows.

Theorem 15.6 *Consider a MAS described by (15.1) and (15.2) and any reference trajectories given by (15.6). Let root agent set π, a positive integer N, and $\alpha, \beta > 0$ be given and hence a set of network graphs $\mathbb{G}^N_{\alpha,\beta,\pi}$ be defined.*

Let Assumption 15.1 hold. Moreover, the agents are minimum-phase and right-invertible. Then, the regulated output synchronization problem is solvable. In particular, there exists a $\delta^* \in (0, 1]$, such that for each $\delta \in (0, \delta^*]$, there exists an $\varepsilon^*(\delta) \in (0, 1]$ such that for all $\varepsilon \in (0, \varepsilon^*(\delta)]$, controller (15.18), together with the two pre-compensators (15.9) and (15.11), solves the regulated output synchronization problem.

Proof For each $i \in \{1, \dots, N\}$, let $\bar{x}_i = \tilde{x}_i - \Pi_i x_0$. Then

$$\begin{aligned}\dot{\bar{x}}_i &= \tilde{A}_i\tilde{x}_i - \Pi_i A_0 x_0 + \tilde{B}_i\tilde{u}_i\\ &= \tilde{A}_i\tilde{x}_i - \tilde{A}_i\Pi_i x_0 + \tilde{B}_i\tilde{u}_i\\ &= \tilde{A}_i\bar{x}_i + \tilde{B}_i\tilde{u}_i.\end{aligned}$$

Furthermore, the output synchronization error $e_i = y_i - y_0$ is given by

$$\begin{aligned}e_i &= \tilde{C}_i\tilde{x}_i - C_0x_0\\ &= \tilde{C}_i\tilde{x}_i - \tilde{C}_i\Pi_i x_0\\ &= \tilde{C}_i\bar{x}_i.\end{aligned}$$

Since the dynamics of the $\bar{x}_i$ system with output e_i is governed by the same triple $(\tilde{A}_i, \tilde{B}_i, \tilde{C}_i)$ as the dynamics of agent i, we can decompose it in the same way as in (15.14), by writing

$$\bar{x}_i = \begin{pmatrix}\bar{x}_{ia}\\ \bar{x}_{id}\end{pmatrix}$$

where

$$\begin{aligned}\dot{\bar{x}}_{ia} &= A_{ia}\bar{x}_{ia} + L_{iad}e_i,\\ \dot{\bar{x}}_{id} &= A_d\bar{x}_{id} + B_d(\tilde{u}_i + E_{ida}\bar{x}_{ia} + E_{idd}\bar{x}_{id}),\\ e_i &= C_d\bar{x}_{id}.\end{aligned} \tag{15.19}$$

Define $\xi_{ia} = \bar{x}_{ia}$, $\xi_{id} = S_\varepsilon \bar{x}_{id}$ and $\hat{\xi}_{id} = S_\varepsilon \hat{x}_{id}$. Note that S_ε as defined in (15.17) is the high-gain matrix. Then it is easy to confirm that we can write

$$\begin{aligned}\dot{\xi}_{ia} &= A_{ia}\xi_{ia} + V_{iad}\xi_{id},\\ \varepsilon\dot{\xi}_{id} &= A_d\xi_{id} + B_d F_\delta \hat{\xi}_{id} + V_{ida}^\varepsilon \xi_{ia} + V_{idd}^\varepsilon \xi_{id},\end{aligned} \tag{15.20}$$

where

$$\begin{aligned}V_{iad} &= L_{iad}C_d,\\ V_{ida}^\varepsilon &= \varepsilon^\rho B_d E_{ida},\\ V_{idd}^\rho &= \varepsilon^\rho E_{idd} S_\varepsilon^{-1}.\end{aligned}$$

We also have

$$e_i = C_d\xi_{id}.$$

Furthermore, noting that $\sum_{j=1}^N \ell_{ij} = 0$, we can write

$$\zeta_i + \psi_i = \sum_{j=1}^N \ell_{ij} y_j + \iota_i(y_i - y_0) = \sum_{j=1}^N \ell_{ij}(y_j - y_0) + \iota_i(y_i - y_0) = \sum_{j=1}^N \bar{\ell}_{ij} e_j,$$

where $\bar{\ell}_{ij}$ represents the coefficients of the matrix $\bar{L}$. We therefore have

$$\varepsilon\dot{\hat{\xi}}_{id} = A_d\hat{\xi}_{id} + K\sum_{j=1}^N \bar{\ell}_{ij} C_d \xi_{id} - KC_d\hat{\xi}_{id}.$$

Let

$$\xi_a = \begin{pmatrix}\xi_{1a}\\ \vdots\\ \xi_{Na}\end{pmatrix}, \quad \xi_d = \begin{pmatrix}\xi_{1d}\\ \vdots\\ \xi_{Nd}\end{pmatrix} \quad \text{and} \quad \hat{\xi}_d = \begin{pmatrix}\hat{\xi}_{1d}\\ \vdots\\ \hat{\xi}_{Nd}\end{pmatrix}.$$

Then

$$
\begin{aligned}
\dot{\xi}_a &= \tilde{A}_a\xi_a + V_{ad}\xi_d,\\
\varepsilon\dot{\xi}_d &= (I_N \otimes A_d)\xi_d + (I_N \otimes B_d F_\delta)\hat{\xi}_d + V^\varepsilon_{da}\xi_a + V^\varepsilon_{dd}\xi_d,\\
\varepsilon\dot{\hat{\xi}}_d &= (I_N \otimes A_d)\hat{\xi}_d + (\bar{L} \otimes KC_d)\xi_d - (I_N \otimes KC_d)\hat{\xi}_d,
\end{aligned}
\tag{15.21}
$$

where $\tilde{A}_a = \operatorname{diag}(A_{ia}, \ldots, A_{Na})$, and

$$
V_{ad} = \begin{pmatrix} V_{1ad} & 0 & \cdots & 0\\ 0 & V_{2ad} & \ddots & \vdots\\ \vdots & \ddots & \ddots & 0\\ 0 & \cdots & 0 & V_{Nad} \end{pmatrix}
\quad
V^\varepsilon_{da} = \begin{pmatrix} V^\varepsilon_{1da} & 0 & \cdots & 0\\ 0 & V^\varepsilon_{2da} & \ddots & \vdots\\ \vdots & \ddots & \ddots & 0\\ 0 & \cdots & 0 & V^\varepsilon_{Nda} \end{pmatrix}
$$

$$
V^\varepsilon_{dd} = \begin{pmatrix} V^\varepsilon_{1dd} & 0 & \cdots & 0\\ 0 & V^\varepsilon_{2dd} & \ddots & \vdots\\ \vdots & \ddots & \ddots & 0\\ 0 & \cdots & 0 & V^\varepsilon_{Ndd} \end{pmatrix}
$$

Note that $\|V_{ad}\|$ is ε-independent, whereas $\|V^\varepsilon_{da}\|$ and $\|V^\varepsilon_{dd}\|$ are $O(\varepsilon)$.

Let a unitary $\bar{U}$ be defined such that $\bar{U}^{-1}\bar{L}\bar{U} = \bar{S}$, where $\bar{S}$ is the Schur form of the matrix $\bar{L}$. Define

$$
\begin{aligned}
v_a &= \xi_a,\ v_d = (\bar{S}\bar{U}^{-1} \otimes I_\rho)\xi_d,\\
\tilde{v}_d &= v_d - (\bar{U}^{-1} \otimes I_\rho)\hat{\xi}_d.
\end{aligned}
$$

Then we have

$$
\begin{aligned}
\dot{v}_a &= \tilde{A}_a v_a + W_{ad}v_d,\\
\varepsilon\dot{v}_d &= (I_N \otimes A_d)v_d + (\bar{S} \otimes B_d F_\delta)(v_d - \tilde{v}_d) + W^\varepsilon_{da}v_a + W^\varepsilon_{dd}v_d,\\
\varepsilon\dot{\tilde{v}}_d &= (I_N \otimes A_d)\tilde{v}_d + (\bar{S} \otimes B_d F_\delta)(v_d - \tilde{v}_d) + W^\varepsilon_{da}v_a + W^\varepsilon_{dd}v_d\\
&\qquad - (I_N \otimes KC_d)\tilde{v}_d,
\end{aligned}
\tag{15.22}
$$

where

$$
\begin{aligned}
W_{ad} &= V_{ad}(\bar{U}\bar{S}^{-1} \otimes I_\rho),\\
W^\varepsilon_{da} &= (\bar{S}\bar{U}^{-1} \otimes I_\rho)V^\varepsilon_{da},\\
W^\varepsilon_{dd} &= (\bar{S}\bar{U}^{-1} \otimes I_\rho)V^\varepsilon_{dd}(\bar{U}\bar{S}^{-1} \otimes I_\rho).
\end{aligned}
$$

Finally, let $\eta_a = v_a$, and define η_d such that

$$\eta_d \triangleq N_d \begin{pmatrix} v_d \\ \tilde{v}_d \end{pmatrix} = \begin{pmatrix} v_{1d} \\ \tilde{v}_{1d} \\ \vdots \\ v_{Nd} \\ \tilde{v}_{Nd} \end{pmatrix} \text{ where } N_d = \begin{pmatrix} e_1 & 0 \\ 0 & e_1 \\ \vdots & \vdots \\ e_N & 0 \\ 0 & e_N \end{pmatrix} \otimes I_{p\rho},$$

where $e_i \in \mathbb{R}^N$ is the i'th standard basis vector whose elements are all zero except for the i'th element which is equal to 1. Then, we obtain dynamics of the form

$$\begin{aligned} \dot{\eta}_a &= \tilde{A}_a \eta_a + \tilde{W}_{ad} \eta_d, \\ \varepsilon \dot{\eta}_d &= \tilde{A}_\delta \eta_d + \tilde{W}^{\varepsilon}_{da} \eta_a + \tilde{W}^{\varepsilon}_{dd} \eta_d, \end{aligned} \tag{15.23}$$

where

$$\tilde{A}_\delta = I_{N-1} \otimes \begin{pmatrix} A_d & 0 \\ 0 & A_d - KC_d \end{pmatrix} + \bar{S} \otimes \begin{pmatrix} B_d F_\delta & -B_d F_\delta \\ B_d F_\delta & -B_d F_\delta \end{pmatrix},$$

and

$$\tilde{W}_{ad} = \begin{pmatrix} W_{ad} & 0 \end{pmatrix} N_d^{-1}, \quad \tilde{W}^{\varepsilon}_{da} = N_d \begin{pmatrix} W^{\varepsilon}_{da} \\ W^{\varepsilon}_{da} \end{pmatrix}, \quad \tilde{W}^{\varepsilon}_{dd} = N_d \begin{pmatrix} W^{\varepsilon}_{dd} & 0 \\ W^{\varepsilon}_{dd} & 0 \end{pmatrix} N_d^{-1}.$$

Following the proof of Theorem 5.9, we find that the above dynamics (15.23) is asymptotically stable. Therefore, $\eta_d \to 0$, which immediately implies that $\bar{x}_i \to 0$ for $i = 1, \ldots, N$. Hence, $e_i \to 0$ for $i = 1, \ldots, N$, i.e., the regulated output synchronization is achieved. ■

15.4.2 *Additional Communication*

In this section, an additional communication of protocol states via the network's communication infrastructure is allowed. As analyzed in Sect. 2.5.8 in Chap. 2, the communication of protocol states allows the agents to be **non-minimum-phase**.

The extra communication for agent i ($i = 1, \ldots, N$) is such that the agent has access to the quantity

$$\hat{\zeta}_i = \sum_{j=1}^{N} \bar{\ell}_{ij} \xi_j, \tag{15.24}$$

where $\xi_j \in \mathbb{R}^p$ is a variable produced internally by the agent j as a part of the protocol. Next, the regulated output synchronization problem with additional communication is formulated.

Problem 15.7 Consider a MAS described by (15.1), (15.2), and (15.24). For a given root set π, a positive integer N, $\alpha, \beta > 0$ that define a set of network graphs $\mathbb{G}^N_{\alpha,\beta,\pi}$, and for any reference trajectories given by (15.6), the *regulated output synchronization* problem is to find, if possible, a linear time-invariant dynamic protocol of the form

$$\begin{aligned} \dot{\chi}_i &= A_{ci}\chi_i + B_{ci}\,\mathrm{col}\{\zeta_i, \psi_i, \hat{\zeta}_i\}, \\ u_i &= C_{ci}\chi_i, \end{aligned} \tag{15.25}$$

for $i = 1, \ldots, N$ where $\chi_i \in \mathbb{R}^{N_{ci}}$, such that, for any graph $\mathcal{G} \in \mathbb{G}^N_{\alpha,\beta,\pi}$ and for all the initial conditions of agents and their protocol, the regulated output synchronization among agents is achieved.

The protocol design is composed of two steps. The first step is to design a pre-compensator such that the cascade system of the pre-compensator and the agent dynamics contains the exosystem mode. The second step is to design an observer-based controller such that regulated output synchronization is achieved.

Step 1 The pre-compensator design is similar to the pre-compensator (15.11) in Design 15.2, which is stated in the following design.

Protocol design 15.5 *First consider the case where the agent has no dynamics in common with the exosystem. Then, $\tilde{\Pi}_i$ and Γ_i are uniquely determined by the so-called regulator equations*

$$\begin{aligned} A_i\tilde{\Pi}_i + B_i\Gamma_i &= \tilde{\Pi}_i A_0, \\ C_i\tilde{\Pi}_i &= C_0. \end{aligned} \tag{15.26}$$

If (Γ_i, A_0) is not observable, we can construct the observable subsystem $(\Gamma_{i,o}, S_{i,o})$; otherwise simply set $\Gamma_{i,o} = \Gamma_i$ and $S_{i,o} = A_0$. Then, the pre-compensator for each agent i is designed as

$$\begin{aligned} \dot{p}_{i,1} &= S_{i,o}p_{i,1} + B_{i,1}\tilde{u}_i, \\ u_i &= \Gamma_{i,o}p_{i,1}, \end{aligned} \tag{15.27}$$

for $i = 1, \ldots, N$, where $p_{i,1} \in \mathbb{R}^{n_{pi}}$ and $B_{i,1}$ are chosen to guarantee that no invariant zeros are introduced by the pre-compesator (see [60]).

If the agent has dynamics in common with the exosystem, then we can use the technique outlined in Appendix "Removing Common Modes of Agent i and A_0" to remove these common modes after which the above design is applicable.

The cascade system of agent (15.1) and pre-compensator (15.27) can be represented in the form

$$\begin{cases} \dot{\tilde{x}}_i = \tilde{A}_i\tilde{x}_i + \tilde{B}_i\tilde{u}_i, \\ y_i = \tilde{C}_i\tilde{x}_i, \end{cases} \tag{15.28}$$

where $\tilde{x}_i \in \mathbb{R}^{\tilde{n}_i}$, $\tilde{u}_i \in \mathbb{R}^p$, and $y_i \in \mathbb{R}^p$ are states, inputs, and outputs of the interconnection system. Moreover, we have $(\tilde{A}_i, \tilde{C}_i)$ containing (A_0, C_0), i.e., there exists a matrix Π_i such that $\Pi_i A_0 = \tilde{A}_i\Pi_i$ and $\tilde{C}_i\Pi_i = C_0$.

For each $i \in \{1, \dots, N\}$, let $\bar{x}_i = \tilde{x}_i - \Pi_i x_0$ and $e_i = y_i - y_0$. Then the dynamics of $\bar{x}_i$ with output e_i can be written as

$$\begin{aligned} \dot{\bar{x}}_i &= \tilde{A}_i\bar{x}_i + \tilde{B}_i\tilde{u}_i, \\ e_i &= \tilde{C}_i\bar{x}_i. \end{aligned} \tag{15.29}$$

Let $n \geq \tilde{n}_i$ for $i \in \{1, \dots, N\}$ and define $\chi_i = T_i\bar{x}_i$, where

$$T_i = \begin{pmatrix} \tilde{C}_i \\ \vdots \\ \tilde{C}_i\tilde{A}_i^{n-1} \end{pmatrix}. \tag{15.30}$$

The observability of $(\tilde{A}_i, \tilde{C}_i)$ ensures that T_i is injective, which implies that $T_i'T_i$ is nonsingular and the dynamics of χ_i can be written as

$$\begin{aligned} \dot{\chi}_i &= (A_d + \mathcal{L}_i)\chi_i + \mathcal{B}_i\tilde{u}_i, \quad \chi_i(0) = T_i\bar{x}_i(0), \\ e_i &= C_d\chi_i, \end{aligned} \tag{15.31}$$

where A_d and C_d are in a special form

$$A_d = \begin{pmatrix} 0 & I_{p(n-1)} \\ 0 & 0 \end{pmatrix}, \quad C_d = \begin{pmatrix} I_p & 0 \end{pmatrix}, \tag{15.32}$$

and

$$\mathcal{L}_i = \begin{pmatrix} 0 \\ L_i \end{pmatrix}, \quad \mathcal{B}_i = T_i\tilde{B}_i, \tag{15.33}$$

with $L_i = \tilde{C}_i\tilde{A}_i^n(T_i'T_i)^{-1}T_i'$.

Step 2 The observer-based controller design is given in the following.

Protocol design 15.6 *Consider a MAS described by (15.1) and (15.2). With the pre-compensator (15.9) and state transformation, we obtain dynamics (15.31).*

Using the high-gain matrix $S_\varepsilon = \operatorname{diag}(I_p, \ldots, I_p\varepsilon^{n-1})$*, the observer-based controller for agent* $i \in \{1, \ldots, N\}$ *is designed as*

$$\begin{aligned} \dot{\hat{\chi}}_i &= (A_d + \mathcal{L}_i)\hat{\chi}_i + \mathcal{B}_i\tilde{u}_i + \varepsilon^{-1}S_\varepsilon^{-1}PC_d'(\zeta_i + \psi_i - \hat{\zeta}_i), \\ \tilde{u}_i &= F_i(T_i'T_i)^{-1}T_i'\hat{\chi}_i. \end{aligned} \tag{15.34}$$

The internal variable $\xi_i = C_d\hat{\chi}_i$ *and* $\hat{\zeta}_i = \sum_{j=1}^N \bar{\ell}_{ij}\xi_j = \sum_{j=1}^N \bar{\ell}_{ij}C_d\hat{\chi}_j$. F_i *is chosen such that* $\tilde{A}_i + \tilde{B}_iF_i$ *is Hurwitz, and* $P > 0$ *is the unique solution of the algebraic Riccati equation*

$$A_dP + PA_d' - 2\beta PC_d'C_dP + I = 0,$$

where $\beta > 0$ *is a lower bound on the real part of the nonzero eigenvalues of the expanded Laplacian matrix* $\bar{L}$.

The main result based on the above design is stated as follows.

Theorem 15.8 *Consider a multi-agent system described by (15.1), (15.2), and (15.24). Suppose a reference trajectory is given by (15.6). Let any root set* π *and* $\alpha, \beta > 0$ *be given and hence a set of network graphs* $\mathbb{G}_{\alpha,\beta,\pi}^N$ *be defined.*

Moreover, the agents are right-invertible. Then, the regulated output synchronization problem is solvable. In particular, there exists an $\varepsilon^* \in (0, 1]$ *such that for all* $\varepsilon \in (0, \varepsilon^*]$*, the protocol (15.34) together with the pre-compensator (15.9) solves the regulated output synchronization problem.*

We need the following lemma:

Lemma 15.9 *Consider the matrix* $\tilde{A}_d$ *defined by*

$$\tilde{A}_d = I_N \otimes A_d - S \otimes PC_d'C_d. \tag{15.35}$$

The matrix $\tilde{A}_d$ *is asymptotically stable for any upper triangular matrix* S *whose eigenvalues satisfy* $\operatorname{Re}\{\lambda_i\} > \beta$ *and* $\|S\| < \alpha$. *Moreover, there exists a* $\tilde{P} > 0$ *and a small enough* ν *such that*

$$\tilde{A}_d\tilde{P} + \tilde{P}\tilde{A}_d' \le -\nu\tilde{P} - I \tag{15.36}$$

is satisfied for all possible upper triangular matrices S.

Proof If we define

$$\bar{A}_{d,i} = A_d - \lambda_i P C_d' C_d$$

and

$$\bar{B} = P C_d' C_d$$

then

$$\tilde{A}_d = \begin{pmatrix} \bar{A}_{d,1} & s_{1,2}\bar{B} & \cdots & s_{1,N}\bar{B} \\ 0 & \bar{A}_{d,2} & \ddots & \vdots \\ \vdots & \ddots & \ddots & s_{N-1,N}\bar{B} \\ 0 & \cdots & 0 & \bar{A}_{d,N} \end{pmatrix},$$

where $\lambda_1, \ldots, \lambda_N$ are the eigenvalues of S and $|s_{i,j}| < \alpha$ are elements of the matrix S.

Note that we have

$$\begin{aligned} &\bar{A}_{d,i} P + P \bar{A}_{d,i}' \\ &= (A_d - \lambda_i P C_d' C_d) P + P (A_d - \lambda_i P C_d' C_d)' \\ &= -I - 2(\operatorname{Re}(\lambda_i) - \beta) P C_d' C_d P \\ &\leq -I. \end{aligned}$$

Then, using similar arguments as in Lemma 5.4, we can prove the existence of $\tilde{P}$ that

$$\tilde{A}_d \tilde{P} + \tilde{P} \tilde{A}_d' \leq -2I.$$

This obviously implies that for ν small enough we have (15.36). ■

Proof of Theorem 15.8 For each $i \in \{1, \ldots, N\}$, let $\bar{\chi}_i = \chi_i - \hat{\chi}_i$. Then

$$\begin{aligned} \dot{\bar{\chi}}_i &= (A_d + \mathcal{L}_i)\bar{\chi}_i - \varepsilon^{-1} S_\varepsilon^{-1} P C_d' (\zeta_i + \psi_i - \hat{\zeta}_i), \\ &= (A_d + \mathcal{L}_i)\bar{\chi}_i - \varepsilon^{-1} S_\varepsilon^{-1} P C_d' C_d \sum_{j=1}^{N} \ell_{ij} \bar{\chi}_j, \end{aligned}$$

in light of

$$\zeta_i + \psi_i = \sum_{j=1}^{N} \bar{\ell}_{ij} C_d \chi_j.$$

Define $\xi_i = S_\varepsilon \bar{\chi}_i$. Then, we get

$$\varepsilon \dot{\xi}_i = A_d \xi_i + \mathcal{L}_{i\varepsilon} \xi_i - P C_d' C_d \sum_{j=1}^{N} \bar{\ell}_{ij} \xi_i,$$

where

$$\mathcal{L}_{i\varepsilon} = \begin{pmatrix} 0 \\ \varepsilon^n L_i S_\varepsilon^{-1} \end{pmatrix}.$$

Let $\xi = \operatorname{col}\{\xi_i\}$ and $\mathcal{L}_\varepsilon = \operatorname{diag}\{\mathcal{L}_{i\varepsilon}\}$. Then, the dynamics of the complete network becomes

$$\varepsilon \dot{\xi} = [I_N \otimes A_d + \mathcal{L}_\varepsilon - \bar{L} \otimes P C_d' C_d] \xi. \tag{15.37}$$

Define $\bar{U}^{-1} \bar{L} \bar{U} = \bar{S}$, where $\bar{S}$ is the Schur form of $\bar{L}$, and let $v = (\bar{U}^{-1} \otimes I_{pn}) \xi$. Then we get

$$\varepsilon \dot{v} = (I_N \otimes A_d) v + W_\varepsilon v - \bar{S} \otimes (P C_d' C_d) v, \tag{15.38}$$

where $W_\varepsilon = (\bar{U}^{-1} \otimes I_{pn}) \mathcal{L}_\varepsilon (\bar{U} \otimes I_{pn})$.

We define $\tilde{A}_d$ by (15.35) and choose $\tilde{P}$ and ν such that (15.36) is satisfied via Lemma 15.9.

Define Lyapunov function $V = \varepsilon v' \tilde{P} v$. The derivative of V is

$$\begin{aligned} \dot{V} &= -\nu \frac{V}{\varepsilon} V - \|v\|^2 + 2 \operatorname{Re}(v' \tilde{P} W_\varepsilon v) \\ &\leq -\nu \frac{V}{\varepsilon} V - \|v\|^2 + \varepsilon r \|v\|^2 \\ &\leq -\nu \frac{V}{\varepsilon} V, \end{aligned}$$

for $\varepsilon \leq \frac{1}{r}$. In the second inequality, $\varepsilon r \geq 2\|\tilde{P} W_\varepsilon\|$ given that $\bar{U}$ is unitary. Therefore, we get $\lim_{t\to\infty} \bar{\chi}_i(t) = 0$, i.e., $\lim_{t\to\infty} (\chi_i(t) - \hat{\chi}_i(t)) = 0$.

Next, we plug the controller input $\tilde{u}_i = F_i (T_i' T_i)^{-1} T_i' \hat{\chi}_i$ into the dynamics (15.29). Then, we get

$$\begin{aligned} \dot{\bar{x}}_i &= \tilde{A}_i \bar{x}_i + \tilde{B}_i F_i (T_i' T_i)^{-1} T_i' \hat{\chi}_i, \\ &= (\tilde{A}_i + \tilde{B}_i F_i) \bar{x}_i + \tilde{B}_i F_i ((T_i' T_i)^{-1} T_i' \hat{\chi}_i - \bar{x}_i), \end{aligned}$$

$$= (\tilde{A}_i + \tilde{B}_i F_i)\bar{x}_i + \tilde{B}_i F_i (T_i' T_i)^{-1} T_i' (\hat{\chi}_i - \chi_i), \tag{15.39}$$

which is asymptotically stable, since $\tilde{A}_i + \tilde{B}_i F_i$ is Hurwitz and $\lim_{t\to\infty}(\chi_i - \hat{\chi}_i) = 0$. Hence, $\lim_{t\to\infty}(y_i - y_0) = 0$, which proves the result. ■

15.5 Non-right-Invertible Agents

In the previous sections, we focused on right-invertible agents that could be minimum-phase and non-minimum-phase, where the regulator equations (15.10) and (15.26) are always solvable. On the other hand, if we assume that the regulator equations are solvable, then the right-invertibility condition is not required. In this section, we will study the solvability condition of the regulator equation (15.10) when the agents are non-right-invertible. We show here that the solvability condition depends on the relationship between non-right-invertible dynamics of each agent and the exosystem pair (A_0, C_0).

We make use of the SCB form derived in Theorem 23.3 in compact form given by (23.14)–(23.18). We group the x_a, x_c, and x_d dynamics together (and denote it by x_r) which separates the agent system into right-invertible and non-right invertible dynamics. That is, there exists nonsingular state, input, and output transformations $\Psi_{xi}, \Psi_{ui}, \Psi_{yi}$ such that, by defining

$$x_i = \Psi_{xi} \begin{pmatrix} x_{bi} \\ x_{ri} \end{pmatrix}, \quad y_i = \Psi_{yi} \begin{pmatrix} y_{bi} \\ y_{di} \end{pmatrix}, \quad u_i = \Phi_{ui} u_{ri},$$

the agent system can be described in terms of two subsystems:

$$\Sigma_{bi} : \begin{cases} \dot{x}_{bi} = A_{bi} x_{bi} + L_{bdi} y_{di}, \\ y_{bi} = C_{bi} x_{bi}, \end{cases}$$

$$\Sigma_{ri} : \begin{cases} \dot{x}_{ri} = A_{ri} x_{ri} + B_{ri} u_{ri} + E_{rbi} x_{bi}, \\ y_{di} = C_{di} x_{ri}. \end{cases} \tag{15.40}$$

We denote by n_{bi} and n_{ri} the dimensions of the Σ_{bi} and Σ_{ri} subsystems. Applying the same output transformation

$$y_0 = \Psi_{yi} \begin{pmatrix} y_{bi0} \\ y_{di0} \end{pmatrix}$$

to the exosystem (15.6), we can write it as

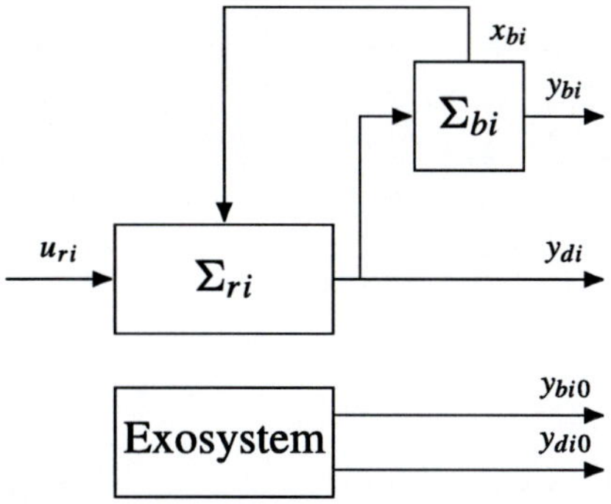

Fig. 15.1 System partitioned into right-invertible and non-right-invertible dynamics

$$\begin{aligned} \dot{x}_0 &= A_0 x_0, \\ y_{bi0} &= C_{bi0} x_0, \\ y_{di0} &= C_{di0} x_0. \end{aligned} \tag{15.41}$$

The goal is now to ensure $(y_{bi} - y_{bi0}) \to 0$ and $(y_{di} - y_{di0}) \to 0$.

Figure 15.1 illustrates how these two subsystems are connected. We see that Σ_{bi} is not directly influenced by the input u_{ri}; it is only indirectly influenced via the output y_{di} from Σ_{ri}. Hence, y_{bi} is entirely dictated by y_{di} together with the initial condition $x_{bi}(0)$. It is evident that if $y_{di}(t) = y_{di0}(t)$, then there is no freedom left to force y_{bi} to track y_{bi0}. Thus, the solvability of the regulation problem depends inherently on the relationship between the non-right-invertible dynamics and the dynamics of the exosystem.

Theorem 15.10 *Define O_i as the observability matrix of the pair*

$$\left[\begin{pmatrix} A_{bi} & L_{bdi} C_{di0} \\ 0 & A_0 \end{pmatrix}, \begin{pmatrix} C_{bi} & -C_{bi0} \end{pmatrix} \right].$$

(Necessity) The regulator equations (15.10) can be solved only if $\operatorname{rank} O_i = n_{bi}$.

(Sufficiency) The regulator equations (15.10) can be solved if $\operatorname{rank} O_i = n_{bi}$ *and, additionally, (A_i, B_i, C_i) have no invariant zeros coinciding with the eigenvalues of A_0.*

Proof (*Sufficiency*) Considering the system equations in (15.40) and (15.41), we find that

$$\Psi_{xi}^{-1} A_i \Psi_{xi} = \begin{pmatrix} A_{bi} & L_{bdi} C_{di} \\ E_{rbi} & A_{ri} \end{pmatrix}, \quad \Psi_{xi}^{-1} B_i \Psi_{ui} = \begin{pmatrix} 0 \\ B_{ri} \end{pmatrix}$$

$$\Psi_{yi}^{-1} C_i \Psi_{xi} = \begin{pmatrix} C_{bi} & 0 \\ 0 & C_{di} \end{pmatrix}, \quad \Psi_{yi}^{-1} C_0 = \begin{pmatrix} C_{bi0} \\ mathbbC_{di0} \end{pmatrix}.$$

Using these expressions, it is easy to verify that (15.10) is equivalent to

$$\begin{aligned} \begin{pmatrix} \Pi_{bi} \\ \Pi_{di} \end{pmatrix} A_0 &= \begin{pmatrix} A_{bi} & L_{bdi}C_{di} \\ E_{rbi} & A_{ri} \end{pmatrix} + \begin{pmatrix} 0 \\ B_{ri} \end{pmatrix} \Gamma_{ri}, \\ \begin{pmatrix} C_{bi} & 0 \\ 0 & C_{di} \end{pmatrix} \begin{pmatrix} \Pi_{bi} \\ \Pi_{ri} \end{pmatrix} &= \begin{pmatrix} C_{bi0} \\ C_{di0} \end{pmatrix}, \end{aligned} \tag{15.42}$$

with

$$\Pi_i = \Psi_{xi} \begin{pmatrix} \Pi_{bi} \\ \Pi_{ri} \end{pmatrix} \quad \text{and} \quad \Gamma_i = \Psi_{ui}\Gamma_{di}.$$

We can write $O_i = \begin{pmatrix} O_{i1} & O_{i2} \end{pmatrix}$, where

$$O_{i1} = \begin{pmatrix} C_{bi} \\ \vdots \\ C_{bi} A_{bi}^{n_{bi}+r-1} \end{pmatrix}.$$

Since (A_{bi}, C_{bi}) is observable, rank $O_{i1} = n_{bi}$. Since rank $O_{i1} = n_{bi}$, we must have

$$\operatorname{Im} O_{i2} \subset \operatorname{Im} O_i = \operatorname{Im} O_{i1}.$$

Hence, there exists a Π_{bi} such that $O_{i1}\Pi_{bi} + O_{i2} = 0$, which means that

$$\begin{pmatrix} \Pi_{bi} \\ I \end{pmatrix}$$

is a basis for the (invariant) unobservable subspace of the matrix pair in the theorem. It follows that there is a V_i such that

$$\begin{aligned} \begin{pmatrix} A_{bi} & L_{bdi}C_{di0} \\ 0 & A_0 \end{pmatrix} \begin{pmatrix} \Pi_{bi} \\ I \end{pmatrix} &= \begin{pmatrix} \Pi_{bi} \\ I \end{pmatrix} V_i, \\ \begin{pmatrix} C_{bi} & -C_{bi0} \end{pmatrix} \begin{pmatrix} \Pi_{bi} \\ I \end{pmatrix} &= 0. \end{aligned} \tag{15.43}$$

It is obvious that we must have $V_i = A_0$ for this to hold, and it then follows that

$$\Pi_{bi} A_0 = A_{bi}\Pi_{bi} + L_{bdi}C_{di0}. \tag{15.44}$$

Consider now the regulator equations

$$\begin{aligned} \Pi_{ri} A_0 &= A_{ri}\Pi_{ri} + E_{rbi}\Pi_{bi} + B_{ri}\Gamma_{ri}, \\ C_{di}\Pi_{ri} &= C_{di0}, \end{aligned} \tag{15.45}$$

where Π_{ri} and Γ_{ri} are the unknowns. Because (A_{ri}, B_{ri}, C_{di}) is right-invertible, the Rosenbrock system matrix

$$\begin{pmatrix} A_{ri} - \lambda I & B_{ri} \\ C_{di} & 0 \end{pmatrix}$$

has normal rank $n_{di} + p$ (see [99, Property 3.1.6]). Since no invariant zeros of (A_{ri}, B_{ri}, C_{di}) coincide with eigenvalues of A_0, the matrix retains its normal rank for all λ that are eigenvalues of A_0. It therefore follows that the regulator equations (15.45) are solvable ([104, Corollary 2.5.1]). From (15.45), we see that $C_{di0} = C_{di}\Pi_{ri}$. Inserting this into (15.44), we have

$$\Pi_{bi}A_0 = A_{bi}\Pi_{bi} + L_{bdi}C_{di}\Pi_{ri}. \tag{15.46}$$

Combining (15.45), (15.46), and the expression $C_{bi}\Pi_{bi} = C_{bi0}$ from (15.43), we see that Π_{bi}, Π_{ri}, and Γ_{ri} are solutions to (15.42), and hence the regulator equations (15.10) are solvable.

(*Necessity*) Note that on the agreement manifold, we must have $y_{di} = y_{di0}$, which means that on this manifold the difference between y_{bi} and y_{bi0} is governed by

$$\begin{aligned} \dot{x}_{bi} &= A_{bi}x_{bi} + L_{bdi}C_{di0}x_0, \\ \dot{x}_0 &= A_0x_0, \\ y_{bi} - y_{bi0} &= C_{bi}x_{bi} - C_{bi0}x_0. \end{aligned} \tag{15.47}$$

This dynamics corresponds precisely to the matrix pair in the statement of the theorem. In order to have $y_{bi} - y_{bi0} = 0$, we must therefore have $O_{i1}x_{bi} + O_{i2}x_0 = 0$. If rank $O_i >$ rank $O_{i1} = n_{bi}$, then this expression can only be satisfied for x_0 in some subspace of dimension lower than r, which means that x_0 must converge to this subspace. However, since the poles of A_0 are all in the closed right half plane, there is no lower-dimensional subspace to which all solutions converge, and hence we must have rank $O_i = n_{bi}$. ■

The following corollary follows immediately from Theorem 15.10.

Corollary 15.11 *The regulator equation* (15.10) *are always solvable if the agent is right-invertible and has no invariant zeros coincide with the poles of* A_0.

Remark 15.12 When the solvability conditions described in Theorem 15.10 are fulfilled, the protocol design for additional communication in Sect. 15.4.2 can be used to achieve regulated output synchronization among agents.

Appendix

Removing Common Modes of Agent i and A_0

We generate here a state transformation such that the common mode of agent i is removed in the exosystem A_0.

From the agent model (15.1) and the exosystem (15.6), the dynamics of e_i is governed by

$$\begin{aligned} \begin{pmatrix} \dot{x}_i \\ \dot{x}_0 \end{pmatrix} &= \begin{pmatrix} A_i & 0 \\ 0 & A_0 \end{pmatrix} \begin{pmatrix} x_i \\ x_0 \end{pmatrix} + \begin{pmatrix} B_i \\ 0 \end{pmatrix} u_i, \\ e_i &= \begin{pmatrix} C_i & -C_0 \end{pmatrix} \begin{pmatrix} x_i \\ x_0 \end{pmatrix}. \end{aligned} \tag{15.48}$$

Now let $\tilde{O}_i$ be the observability matrix corresponding to the pair

$$\left[\begin{pmatrix} A_i & 0 \\ 0 & A_0 \end{pmatrix}, \begin{pmatrix} C_i & -C_0 \end{pmatrix} \right]. \tag{15.49}$$

Let q_i denote the dimension of the null space of matrix $\tilde{O}_i$ and define $r_i = r - q_i$. Next, define $\Lambda_{iu} \in \mathbb{R}^{n_i \times q_i}$ and $\Phi_{iu} \in \mathbb{R}^{r \times q_i}$ such that

$$\tilde{O}_i \begin{pmatrix} \Lambda_{iu} \\ \Phi_{iu} \end{pmatrix} = 0, \quad \operatorname{rank} \begin{pmatrix} \Lambda_{iu} \\ \Phi_{iu} \end{pmatrix} = q_i.$$

Because the pair (A_i, C_i) and (A_0, C_0) are observable, it is easy to show that Λ_{iu} and Φ_{iu} have full-column rank. Suppose that one of the matrices, say Λ_{iu}, has linearly dependent columns. Then there are nonzero vectors $z \in \mathbb{R}^{q_i}$ and $\bar{z} \in \mathbb{R}^r$ such that

$$\begin{pmatrix} \Lambda_{iu} \\ \Phi_{iu} \end{pmatrix} z = \begin{pmatrix} 0 \\ \bar{z} \end{pmatrix} \Rightarrow \tilde{O}_i \begin{pmatrix} 0 \\ \bar{z} \end{pmatrix} = 0 \Rightarrow \begin{pmatrix} C_0 \\ \vdots \\ C_0 A_0^{r-1} \end{pmatrix} \bar{z} = 0.$$

The last statement contradicts with (A_0, C_0) being observable. The definition of Λ_{iu} and Φ_{iu} implies that the columns of $\begin{pmatrix} \Lambda_{iu}' & \Phi_{iu}' \end{pmatrix}'$ span the unobservable subspace of the pair (15.49), which is invariant with respect to blkdiag(A_i, A_0). Hence, there exists a matrix $U_i \in \mathbb{R}^{q_i \times q_i}$ such that

$$\begin{pmatrix} A_i & 0 \\ 0 & A_0 \end{pmatrix} \begin{pmatrix} \Lambda_{iu} \\ \Phi_{iu} \end{pmatrix} = \begin{pmatrix} \Lambda_{iu} \\ \Phi_{iu} \end{pmatrix} U_i, \quad \begin{pmatrix} C_i & -C_0 \end{pmatrix} \begin{pmatrix} \Lambda_{iu} \\ \Phi_{iu} \end{pmatrix} = 0. \tag{15.50}$$

It is easy to get

$$A_i \Lambda_{iu} = \Lambda_{iu} U_i, \quad A_0 \Phi_{iu} = \Phi_{iu} U_i.$$

Now let Λ_{io} and Φ_{io} be defined such that

$$\Lambda_i := \begin{pmatrix} \Lambda_{iu} & \Lambda_{io} \end{pmatrix} \in \mathbb{R}^{n_i \times n_i}, \quad \Phi_i := \begin{pmatrix} \Phi_{iu} & \Phi_{io} \end{pmatrix} \in \mathbb{R}^{r \times r}$$

are nonsingular. Then we can easily derive that there exist matrices Q_i and R_i of the form

$$Q_i = \begin{pmatrix} U_i & Q_{i12} \\ 0 & Q_{i22} \end{pmatrix}, \quad R_i = \begin{pmatrix} U_i & R_{i12} \\ 0 & R_{i22} \end{pmatrix},$$

such that

$$A_i \Lambda_i = \Lambda_i Q_i, \quad A_0 \Phi_i = \Phi_i R_i.$$

Because A_0 is anti-Hurwitz stable and neutrally stable, we know that A_0 is diagonalizable and, hence, R_i is diagonalizable. This implies that R_i has r independent right eigenvectors. Let $v_{i1}, \ldots, v_{ir}$ be r independent right eigenvectors of R_i, such that

$$v_{ij} = \begin{pmatrix} \tilde{v}_{ij} \\ 0 \end{pmatrix}$$

for $j = 1, \ldots, q_i$, where $\tilde{v}_{ij}$ are right eigenvectors of U_i. In that case, we choose $V_{i11} \in \mathbb{R}^{q_i \times q_i}$ such that

$$\operatorname{Im} V_{i11} = \operatorname{span}\{\operatorname{Re} \tilde{v}_{ij}, \operatorname{Im} \tilde{v}_{ij} \mid j = 1, \ldots, q_i\},$$

and we choose $V_{i12} \in \mathbb{R}^{q_i \times r_i}$ and $V_{i22} \in \mathbb{R}^{r_i \times r_i}$ such that

$$\operatorname{Im} \begin{pmatrix} V_{i12} \\ V_{i22} \end{pmatrix} = \operatorname{span}\{\operatorname{Re} v_{ij}, \operatorname{Im} v_{ij} \mid j = q_i + 1, \ldots, r\}.$$

We then construct

$$V_i = \begin{pmatrix} V_{i11} & V_{i12} \\ 0 & V_{i22} \end{pmatrix}.$$

It can be easily verified that $\operatorname{span}\{\operatorname{Re} v_{ij}, \operatorname{Im} v_{ij}\}$ is an invariant subspace of R_i for any $j = 1, \ldots, r$. This implies that

$$R_i V_i = V_i \begin{pmatrix} \Delta_{i1} & 0 \\ 0 & \Delta_{i2} \end{pmatrix}. \tag{15.51}$$

One way of choosing the matrix V_i is choosing

$$\begin{pmatrix} \Delta_{i1} & 0 \\ 0 & \Delta_{i2} \end{pmatrix}$$

to be the real Jordan form of R_i ordered in such a way that Δ_{i1} is the real Jordan form of U_i.

From (15.51), we obtain that

$$V_{i11}^{-1} U_i V_{i11} = \Delta_{i1}, \quad V_{i22}^{-1} R_{i22} V_{i22} = \Delta_{i2},$$

and

$$U_i V_{i12} - V_{i12} \Delta_{i2} = -R_{i12} V_{i22}.$$

We define next

$$\bar{\Phi}_i := \left(\bar{\Phi}_{iu}, \bar{\Phi}_{io}\right) = \Phi_i \begin{pmatrix} I_{q_i} & V_{i12} V_{i22}^{-1} \\ 0 & I_{r_i} \end{pmatrix}. \tag{15.52}$$

Then, we have

$$A_0 \bar{\Phi}_i = \bar{\Phi}_i \begin{pmatrix} U_i & 0 \\ 0 & R_{i22} \end{pmatrix}.$$

We can then define a new state variable $\bar{x}_i \in \mathbb{R}^{n_i + r_i}$ as

$$\bar{x}_i = \begin{pmatrix} \bar{x}_{i1} \\ \bar{x}_{i2} \end{pmatrix} := \begin{pmatrix} x_i - \Lambda_i M_i \bar{\Phi}_i^{-1} x_0 \\ -N_i \bar{\Phi}_i^{-1} x_0 \end{pmatrix},$$

where $M_i \in \mathbb{R}^{n_i \times r}$ and $N_i \in \mathbb{R}^{r_i \times r}$ are defined as

$$M_i = \begin{pmatrix} I_{q_i} & 0 \\ 0 & 0 \end{pmatrix}, \quad N_i = \begin{pmatrix} 0 & I_{r_i} \end{pmatrix}.$$

$$\begin{aligned}
\dot{\bar{x}}_{i1} &= A_i x_i - \Lambda_i M_i \bar{\Phi}_i^{-1} A_0 x_0 + B_i u_i, \\
&= A_i x_i - \Lambda_i M_i \begin{pmatrix} U_i & 0 \\ 0 & R_{i22} \end{pmatrix} \bar{\Phi}_i^{-1} x_0 + B_i u_i, \\
&= A_i x_i - \Lambda_i \begin{pmatrix} U_i & 0 \\ 0 & 0 \end{pmatrix} \bar{\Phi}_i^{-1} x_0 + B_i u_i, \\
&= A_i x_i - \Lambda_i Q_i M_i \bar{\Phi}_i^{-1} x_0 + B_i u_i,
\end{aligned}$$

$$= A_i(x_i - \Lambda_i M_i \bar{\Phi}_i^{-1} x_0) + B_i u_i,$$
$$= A_i \bar{x}_{i1} + B_i u_i.$$

For $\bar{x}_{i2}$ we have

$$\dot{\bar{x}}_{i2} = -N_i \bar{\Phi}_i^{-1} A_0 x_0 = -N_i \begin{pmatrix} U_i & 0 \\ 0 & R_{i22} \end{pmatrix} \bar{\Phi}_i^{-1} x_0 = -R_{i22} N_i \bar{\Phi}_i^{-1} x_0 = R_{i22} \bar{x}_{i2}.$$

Using the equality $C_i \Lambda_{iu} = C_0 \Phi_{iu}$ from (15.50) and $\bar{\Phi}_{iu} = \Phi_{iu}$ from (15.52), we calculate e_i in terms of $\bar{x}_{i1}$ and $\bar{x}_{i2}$:

$$
\begin{aligned}
e_i &= C_i x_i - C_0 x_0, \\
&= C_i x_i - C_0 \begin{pmatrix} \bar{\Phi}_{iu} & \bar{\Phi}_{io} \end{pmatrix} \bar{\Phi}_i^{-1} x_0, \\
&= C_i x_i - \begin{pmatrix} C_i \Lambda_{iu} & C_0 \bar{\Phi}_{io} \end{pmatrix} \Phi_i^{-1} x_0, \\
&= C_i x_i - (C_i \Lambda_i M_i + C_0 \bar{\Phi}_i N_i' N_i) \bar{\Phi}_i^{-1} x_0, \\
&= C_i (x_i - \Lambda_i M_i \bar{\Phi}_i^{-1} x_0) - C_0 \bar{\Phi}_i N_i' N_i \bar{\Phi}_i^{-1} x_0, \\
&= C_i \bar{x}_{i1} + C_0 \bar{\Phi}_i N_i' \bar{x}_{i2}.
\end{aligned}
$$

Therefore, after removing the common mode with agent i, the exosystem is written as

$$
\begin{aligned}
\dot{\bar{x}}_{i0} &= \bar{A}_{i0} \bar{x}_0, \\
\bar{y}_{i0} &= \bar{C}_{i0} \bar{x}_0,
\end{aligned}
\tag{15.53}
$$

where $\bar{A}_{i0} = R_{i22}$ and $\bar{C}_{i0} = -C_0 \bar{\Phi}_i N_i'$.

Construction of Π_i

We construct here a Π_i for the case that (Γ_i, S) is not observable for some $i \in \{1, \ldots, N\}$. There exists a nonsingular matrix Ω_i, such that

$$
\begin{aligned}
\bar{S} &= \Omega_i S \Omega_i^{-1} = \begin{pmatrix} S_{no} & S_{12} \\ 0 & S_o \end{pmatrix}, \\
\bar{\Gamma}_i &= \Gamma_i \Omega_i^{-1} = \begin{pmatrix} 0 & \Gamma_{i,o} \end{pmatrix},
\end{aligned}
$$

where $(\Gamma_{i,o}, S_o)$ is observable, while S_{no} contains the unobservable modes. Choose

$$\Pi_i = \begin{pmatrix} \tilde{\Pi}_i \\ (0 \;\; I)\Omega_i \end{pmatrix}$$

where $(0 \;\; I)$ has the same dimension of $(0 \;\; \Gamma_{i,o})$. Then, we find that

$$\tilde{A}_i \Pi_i = \begin{pmatrix} A_i & B_i \Gamma_{i,o} \\ 0 & S_o \end{pmatrix} \begin{pmatrix} \tilde{\Pi}_i \\ (0 \;\; I)\Omega_i \end{pmatrix} = \begin{pmatrix} A_i \Pi_i + B_i \Gamma_i \\ (0 \;\; S_{ir,o})\Omega_i \end{pmatrix},$$

and

$$\Pi_i S = \begin{pmatrix} \tilde{\Pi}_i \\ (0 \;\; I)\Omega_i \end{pmatrix} S = \begin{pmatrix} \tilde{\Pi}_i S \\ (0 \;\; I)\Omega_i S \end{pmatrix} = \begin{pmatrix} \tilde{\Pi}_i S \\ (0 \;\; S_o)\Omega_i \end{pmatrix}.$$

From the regulation equations (15.10), we have

$$\tilde{A}_i \Pi_i = \Pi_i S.$$

Moreover,

$$\tilde{C}_i \Pi_i = (C_i \;\; 0) \begin{pmatrix} \tilde{\Pi}_i \\ (0 \;\; I)\Omega_i \end{pmatrix} = C_i \tilde{\Pi}_i = R.$$

Chapter 16
Regulated Output Synchronization of Heterogeneous Continuous-Time Nonlinear MAS

16.1 Introduction

The previous chapter considered regulated output synchronization for synchronization problems for heterogeneous continuous-time MAS where the agents are linear. In this chapter, we consider the same problem for nonlinear systems with a specific structure. The class of nonlinear agents we will consider is the same as studied in Chap. 5 for homogeneous MAS. As already noticed in Chap. 5, most of the literature focuses on the concept of passivity for nonlinear systems. We will briefly focus on passivity in Chap. 21.

The write-up of this chapter is partially based on [33].

16.2 Multi-agent Systems

Consider a MAS composed of N non-identical nonlinear time-varying agents that can be represented in the canonical form

$$\begin{aligned} \dot{x}_{ia} &= A_{ia}x_{ia} + L_{iad}y_i, \\ \dot{x}_{id} &= A_{id}x_{id} + \phi_{id}(t, x_{ia}, x_{id}) + B_{id}(u_i + E_{ida}x_{ia} + E_{idd}x_{id}), \\ y_i &= C_{id}x_{id}, \end{aligned} \tag{16.1}$$

where

$$x_i := \begin{pmatrix} x_{ia} \\ x_{id} \end{pmatrix} \in \mathbb{R}^n, \qquad u_i \in \mathbb{R}, \qquad y_i \in \mathbb{R}$$

A. Saberi et al., *Synchronization of Multi-Agent Systems in the Presence of Disturbances and Delays*, Systems & Control: Foundations & Applications,
https://doi.org/10.1007/978-3-030-88148-1_16

are, respectively, states, inputs, and outputs of agent i. We will focus attention on minimum-phase agents where A_{ia} is Hurwitz. For ease of presentation, we consider systems with a scalar input. The ideas of this chapter can easily be extended to a larger class of agents with multiple inputs.

Let the relative degree of the above agent system (16.1) be ρ_i. Then, A_{id}, B_{id}, and C_{id} have the special form

$$A_{id} = \begin{pmatrix} 0 & I_{\rho_i - 1} \\ 0 & 0 \end{pmatrix}, \quad B_d = \begin{pmatrix} 0 \\ 1 \end{pmatrix}, \quad C_d = \begin{pmatrix} 1 & 0 \end{pmatrix}. \tag{16.2}$$

Furthermore, x_{id} and ϕ_{id} have the structure

$$x_{id} = \begin{pmatrix} x_{id1} \\ \vdots \\ x_{id\rho_i} \end{pmatrix}, \quad \phi_{id} = \begin{pmatrix} \phi_{id1} \\ \vdots \\ \phi_{id\rho_i} \end{pmatrix}.$$

We assume that the time-varying nonlinearity $\phi_{id}(t, x_{ia}, x_{id})$ has the following structure.

Assumption 16.1 *Assume that $\phi_{id}(t, x_{ia}, x_{id})$ is continuously differentiable and Lipschitz continuous with respect to (x_{ia}, x_{id}), and it is also Lipschitz continuous uniformly in t and piecewise continuous with respect to t. Moreover, the nonlinearity has the lower-triangular structure:*

$$\frac{\partial \phi_{idj}(t, x_{ia}, x_{id})}{\partial x_{idk}} = 0, \quad \forall k > j. \tag{16.3}$$

The canonical form in (16.1) is similar to various types of chained, lower-triangular canonical forms common in the context of high-gain observer design and output feedback control (see, e.g., [42]. Among the practically relevant types of systems encompassed by this canonical form are mechanical systems with nonlinearities occurring at the acceleration level.

The communication network among agents is exactly the same as that in Chap. 15 and provides each agent with the quantity

$$\zeta_i = \sum_{j=1}^{N} a_{ij}(y_i - y_j) = \sum_{j=1}^{N} \ell_{ij} y_j. \tag{16.4}$$

16.3 Problem Formulation

Similar to Chap. 15, since all the agents are non-identical and non-introspective, we need to pursue regulated output synchronization among agents' outputs, that is, to regulate the outputs of all the agents asymptotically toward a priori specified reference trajectories. The reference trajectories in this section are also generated by (15.6), that is

$$\begin{aligned} \dot{x}_0 &= A_0 x_0, \quad x_0(0) = x_{00}, \\ y_0 &= C_0 x_0. \end{aligned} \tag{16.5}$$

As in Sect. 15.3, a subset of agents, denoted by π, have access to the relative outputs of the reference trajectories, i.e., $\psi_i = \iota_i(y_i - y_0)$ for $i \in \pi$; otherwise, $\psi_i = 0$. Both Assumption 15.2 and Definition 15.4 hold in this Chapter, hence, a set of graphs $\mathbb{G}^N_{\alpha,\beta,\pi}$.

We define the regulated output synchronization problem for a MAS with nonlinear time-varying agents as follows.

Problem 16.1 Consider a MAS described by (16.1) and (16.4). For a given root set π, a positive integer N, and $\alpha, \beta > 0$ that define a set of network graphs $\mathbb{G}^N_{\alpha,\beta,\pi}$ and any reference trajectories given by (16.5), the *regulated output synchronization* problem is to find, if possible, a linear time-invariant dynamic protocol of the form

$$\begin{aligned} \dot{\chi}_i &= A_{c,i}\chi_i + B_{c,i} \operatorname{col}\{\zeta_i, \psi_i\}, \\ u_i &= C_{c,i}\chi_i, \end{aligned} \tag{16.6}$$

for $i = 1, \ldots, N$, such that for any graph $\mathcal{G} \in \mathbb{G}^N_{\alpha,\beta,\pi}$ and for all the initial conditions of agents and their protocol, the regulated output synchronization among agents can be achieved.

16.4 Protocol Design

The protocol design here is similar to the previous chapter. The first step is to design pre-compensators such that the cascade system of the pre-compensators and the agent dynamics have the same relative degree and contain the modes of the exosystem. The second step is to design an observer-based controller such that the regulated output synchronization is achieved.

Step 1 In this chapter, we consider SISO systems, and hence the first pre-compensator used in the previous chapter to make the system invertible with uniform rank is not needed. Two pre-compensators are designed, one for containing the exosystem mode and the other for having the same relative degree.

Protocol design 16.1 : *Pre-compensator* 1

The design is the same as in Design 15.2 with

$$\hat{A}_i = \begin{pmatrix} A_{ia} & L_{iad}C_{id} \\ B_{id}E_{ida} & A_{id} + B_{id}E_{idd} \end{pmatrix}, \quad \hat{B}_i = \begin{pmatrix} 0 \\ B_{id} \end{pmatrix}, \quad \hat{C}_i = \begin{pmatrix} 0 & C_{id} \end{pmatrix}.$$

We obtain

$$\begin{aligned} \dot{p}_{i,1} &= S_{i,o}p_{i,1} + B_{i,1}u_i^1, \\ u_i &= \Gamma_{i,o}p_{i,1}, \end{aligned} \tag{16.7}$$

for $i = 1, \ldots, N$.

Protocol design 16.2 : *Pre-compensator* 2

Let $\tilde{\rho}_i$ *be the relative degree of the interconnection of the system with the first pre-compensator from Design 16.1. Choose* $\rho = \max\{\tilde{\rho}_i\}$, $i = 1, \ldots, N$. *We can add* $\rho - \tilde{\rho}_i$ *integrators for agent* i. *Then, the second pre-compensator is designed as*

$$\begin{aligned} \dot{p}_{i,2} &= \begin{pmatrix} 0 & I_{\rho-\rho_i-1} \\ 0 & 0 \end{pmatrix} p_{i,2} + \begin{pmatrix} 0 \\ 1 \end{pmatrix} \tilde{u}_i, \\ u_i^1 &= \begin{pmatrix} 1 & 0 \end{pmatrix} p_{i,2}, \end{aligned} \tag{16.8}$$

which guarantees that each agent has the relative degree ρ. *Note that we can ignore the nonlinear dynamics in this construction.*

The cascade system of agent (16.1) and pre-compensators (16.7) and (16.8) can be represented in this form

$$\begin{cases} \dot{\tilde{x}}_i = \tilde{A}_i\tilde{x}_i + \tilde{B}_i\tilde{u}_i + \tilde{\phi}_i(t, \tilde{x}_i), \\ y_i = \tilde{C}_i\tilde{x}_i, \end{cases} \tag{16.9}$$

where $\tilde{x}_i \in \mathbb{R}^{\tilde{n}_i}$, $u_i \in \mathbb{R}$, *and* $y_i \in \mathbb{R}$. Moreover, for all $i \in \{1, \ldots, N\}$ we have the following:

- $(\tilde{A}_i, \tilde{C}_i)$ contains (A_0, C_0), i.e., there exists a Π_i such that $\Pi_i A_0 = \tilde{A}_i \Pi_i$ and $\tilde{C}_i \Pi_i = C_0$.
- $(\tilde{A}_i, \tilde{B}_i, \tilde{C}_i)$ has relative degree ρ.

The system $(\tilde{A}_i, \tilde{B}_i, \tilde{C}_i)$ is SISO and hence uniform rank and invertible. Hence the associated system (16.9) can be represented by the SCB form as presented in (23.27). Let

$$\tilde{x}_i = \begin{pmatrix} \tilde{x}_{ia} \\ \tilde{x}_{id} \end{pmatrix},$$

with $\tilde{x}_{ia} \in \mathbb{R}^{\bar{n}_i - \rho}$ representing the finite zero structure and $\tilde{x}_{id} \in \mathbb{R}^{\rho}$ the infinite zero structure, and

$$\begin{cases} \dot{\tilde{x}}_{ia} = \tilde{A}_{ia} x_{ia} + \tilde{L}_{iad} y_i, \\ \dot{\tilde{x}}_{id} = A_d \tilde{x}_{id} + \tilde{\phi}_{id}(t, \tilde{x}_{ia}, \tilde{x}_{id}) + B_d(\tilde{u}_i + \tilde{E}_{ida}\tilde{x}_{ia} + \tilde{E}_{idd}\tilde{x}_{id}), \\ y_i = C_d \tilde{x}_{id}, \end{cases} \tag{16.10}$$

for $i = 1, \ldots, N$, where $\tilde{\phi}_{id}(t, \tilde{x}_{ia}, \tilde{x}_{id})$ is a (possibly time-varying) nonlinearity and

$$A_d = \begin{pmatrix} 0 & I_{\rho-1} \\ 0 & 0 \end{pmatrix}, \qquad B_d = \begin{pmatrix} 0 \\ 1 \end{pmatrix}, \qquad C_d = \begin{pmatrix} 1 & 0 \end{pmatrix}. \tag{16.11}$$

Note that the eigenvalues of $\tilde{A}_{ia}$ are the invariant zeros of the triple $(\tilde{A}_i, \tilde{B}_i, \tilde{C}_i)$, which are all in the open left half complex plane, because no invariant zeros are introduced during the pre-compensator design.

Finally, the following is important to note.

Lemma 16.2 *Consider a system of the form (16.1) with a nonlinearity ϕ_{id} satisfying Assumption 16.1. Design pre-compensators via Design 16.1 and 16.2. In that case, the system we obtain of the form (16.10) is such that $\tilde{\phi}_{id}$ also satisfies Assumption 16.1 (with the obvious notational modifications).*

Proof The proof of this is quite technical and hence the details are omitted. ■

The final step of our design is to derive an observer-based protocol.

Protocol design 16.3 *Let K_ε and $F_{\delta\varepsilon}$ be given in Design 15.4. We then obtain the following observer-based protocol:*

$$\begin{aligned} \dot{\hat{x}}_{id} &= A_d \hat{x}_{id} + K_\varepsilon(\zeta_i + \psi_i - C_d \hat{x}_{id}), \\ \tilde{u}_i &= F_{\delta\varepsilon} \hat{x}_{id}. \end{aligned} \tag{16.12}$$

The main result based on the above design is stated as follows.

Theorem 16.3 *Consider a multi-agent system described by (16.1) and (16.4) and any reference trajectories given by (16.5). Let any root set π, a positive integer N, and $\alpha, \beta > 0$ be given, and hence a set of network graphs $\mathbb{G}^N_{\alpha,\beta,\pi}$ be defined.*

Assume that the agents are minimum-phase. Then, under Assumption 16.1, the regulated output synchronization problem is solvable. In particular, there exists a $\delta^* \in (0, 1]$, such that for each $\delta \in (0, \delta^*]$, there exists an $\varepsilon^*(\delta) \in (0, 1]$ such that for all $\varepsilon \in (0, \varepsilon^*(\delta)]$, protocol (16.12) together with the pre-compensators (16.7) and (16.8) solves the regulated output synchronization problem.

Proof For each $i \in \{1, 2, \ldots, N\}$, let

$$\bar{x}_i = \tilde{x}_i - \Pi_i x_0 = \begin{pmatrix} \tilde{x}_{ia} - \Pi_{i,1}x_0 \\ \tilde{x}_{id} - \Pi_{i,2}x_0 \end{pmatrix},$$

where Π_i is defined to satisfy $\Pi_i A_0 = \tilde{A}_i\Pi_i$ and $\tilde{C}_i\Pi_i = C_0$. Then, the dynamics of $\bar{x}_i$ can be written as

$$\begin{cases} \dot{\bar{x}}_i = \tilde{A}_i\bar{x}_i + \tilde{B}_i\tilde{u}_i + \tilde{\phi}_i(t, \tilde{x}_i), \\ e_i = \tilde{C}_i\bar{x}_i. \end{cases} \tag{16.13}$$

Let $\bar{x}_{ia} = \tilde{x}_{ia} - \Pi_{i,1}x_0$, $\bar{x}_{id} = \tilde{x}_{id} - \Pi_{i,2}x_0$. Then, by Taylor's theorem (see, e.g., [77, Theorem 11.1]), we can write

$$\tilde{\phi}_{id}(t, \tilde{x}_{ia}, \tilde{x}_{id}) = \tilde{\phi}_{id}(t, \Pi_{i,1}x_0, \Pi_{i,2}x_0) + \Phi_{ia}(t)\bar{x}_{ia} + \Phi_{id}(t)\bar{x}_{id},$$

where $\Phi_{ia}(t)$ and $\Phi_{id}(t)$ are given by

$$\Phi_{ia}(t) = \int_0^1 \frac{\partial\phi_{id}}{\partial x_{ia}}(t, s\bar{x}_{ia} + \Pi_{i,1}x_0, s\bar{x}_{id} + \Pi_{i,2}x_0)ds,$$

$$\Phi_{id}(t) = \int_0^1 \frac{\partial\phi_{id}}{\partial x_{id}}(t, s\bar{x}_{ia} + \Pi_{i,1}x_0, s\bar{x}_{id} + \Pi_{i,1}x_0)ds.$$

Due to the global Lipschitz property of the nonlinearity, the elements of $\Phi_{ia}(t)$ and $\Phi_{id}(t)$ are uniformly bounded, and the lower-triangular structure of the nonlinearity implies that $\Phi_{id}(t)$ is lower-triangular. Moreover, because of the Lipschitz property and the fact that both x_0 and $\tilde{\phi}_{id}(t, 0, 0)$ are bounded, we know that $\tilde{\phi}_{id}(t, \Pi_{i,1}x_0, \Pi_{i,2}x_0)$ is uniformly bounded. We obtain

$$\begin{cases} \dot{\bar{x}}_{ia} = \tilde{A}_{ia}\bar{x}_{ia} + \tilde{L}_{iad}e_i, \\ \dot{\bar{x}}_{id} = A_d\bar{x}_{id} + \Phi_{ia}(t)\bar{x}_{ia} + \Phi_{id}(t)\bar{x}_{id} + B_d(\tilde{u}_i + \tilde{E}_{ida}\bar{x}_{ia} + \tilde{E}_{idd}\bar{x}_{id}) \\ e_i = C_d\bar{x}_{id}. \end{cases} \tag{16.14}$$

Define

$$\xi_{ia} = \bar{x}_{ia}, \xi_{id} = S_\varepsilon \bar{x}_{id}$$

$$\hat{\xi}_{id} = S_\varepsilon \hat{x}_{id}.$$

Then we can write

$$\begin{aligned} \dot{\xi}_{ia} &= \tilde{A}_{ia}\xi_{ia} + V_{iad}\xi_{id}, \\ \varepsilon\dot{\xi}_{id} &= A_d\xi_{id} + B_d F_\delta \hat{\xi}_{id} + V^\varepsilon_{ida}\xi_{ia} + V^\varepsilon_{idd}\xi_{id}, \\ \varepsilon\dot{\hat{\xi}}_{id} &= A_d\hat{\xi}_{id} + K\sum_{j=1}^N \bar{\ell}_{ij} C_d \xi_{id} - KC_d\hat{\xi}_{id}, \end{aligned} \tag{16.15}$$

where

$$V_{iad} = \tilde{L}_{iad} C_d,$$

$$V^\varepsilon_{ida} = \varepsilon^\rho B_d E_{da} + \varepsilon S_\varepsilon \Phi_{ia}(t),$$

$$V^\varepsilon_{idd} = \varepsilon^\rho B_d E_{dd} S_\varepsilon^{-1} + \varepsilon S_\varepsilon \Phi_{id}(t) S_\varepsilon^{-1}.$$

It is easily checked that both V^ε_{ida} and V^ε_{idd} are $O(\varepsilon)$. Let

$$\xi_a = \begin{pmatrix} \xi_{1a} \\ \vdots \\ \xi_{Na} \end{pmatrix}, \qquad \xi_d = \begin{pmatrix} \xi_{1d} \\ \vdots \\ \xi_{Nd} \end{pmatrix} \quad \text{and} \quad \hat{\xi}_d = \begin{pmatrix} \hat{\xi}_{1d} \\ \vdots \\ \hat{\xi}_{Nd} \end{pmatrix}.$$

Then

$$\begin{aligned} \dot{\xi}_a &= \tilde{A}_a\xi_a + V_{ad}\xi_d, \\ \varepsilon\dot{\xi}_d &= (I_N \otimes A_d)\xi_d + (I_N \otimes B_d F_\delta)\hat{\xi}_d + V^\varepsilon_{da}\xi_a + V^\varepsilon_{dd}\xi_d, \\ \varepsilon\dot{\hat{\xi}}_d &= (I_N \otimes A_d)\hat{\xi}_d + (\bar{L} \otimes KC_d)\xi_d - (I_N \otimes KC_d)\hat{\xi}_d, \end{aligned} \tag{16.16}$$

where $\tilde{A}_a = \operatorname{diag}(\tilde{A}_{1a}, \ldots, \tilde{A}_{Na})$ and where $V_{ad}, V^\varepsilon_{da}, and V^\varepsilon_{dd}$ are defined in the same way as in the proof of Theorem 15.6.

We find that the above dynamics (16.16) is exactly the same as (15.21). Therefore, in order to show that $\bar{x}_i \to 0$, the remainder of the proof now proceeds in the same way as the proof of Theorem 15.6. This implies in particular that $e_i \to 0$ and therefore regulated output synchronization is achieved. ■

Chapter 17
Regulated Output Synchronization of Heterogeneous Continuous-Time Linear Time-Varying MAS

17.1 Introduction

This chapter considers synchronization problems for heterogeneous continuous-time linear time-varying MAS with non-introspective agents. In the literature, time-varying graphs have been considered in, e.g., [67, 119, 134, 191]. For heterogeneous networks, it is always assumed that the agents are introspective. The main objective of this chapter is to consider the non-introspective case. This chapter is the heterogeneous version of Chap. 11 which considered homogeneous networks.

As in the time-invariant case, when agents are non-introspective, we regulate the output of each agent asymptotically to a reference trajectory generated by an exosystem. A purely decentralized controller based on a low- and high-gain methodology is designed for minimum-phase agents. However, an additional communication channel is allowed, and we will present a design which achieves output synchronization even if the agents are not minimum-phase.

17.2 Multi-Agent Systems

Consider a MAS composed of N non-identical linear time-invariant continuous-time agents of the form

$$\begin{aligned} \dot{x}_i &= A_i x_i + B_i u_i, \\ y_i &= C_i x_i, \end{aligned} \qquad (i = 1, \ldots, N) \tag{17.1}$$

where $x_i \in \mathbb{R}^n$, $u_i \in \mathbb{R}^m$, and $y_i \in \mathbb{R}^p$ are, respectively, the state, input, and output vectors of agent i. We make the following assumptions for the agent dynamics.

A. Saberi et al., *Synchronization of Multi-Agent Systems in the Presence of Disturbances and Delays*, Systems & Control: Foundations & Applications,
https://doi.org/10.1007/978-3-030-88148-1_17

Assumption 17.1 *For each agent $i \in \{1, \ldots, N\}$, we assume that*

- *(A_i, B_i) is stabilizable.*
- *(A_i, C_i) is detectable.*

The time-varying communication network among agents is exactly the same as that in Chap. 11 and provides each agent with the quantity

$$\zeta_i(t) = \sum_{j=1}^{N} a_{ij}(t)(y_i(t) - y_j(t)) = \sum_{j=1}^{N} \ell_{ij}(t) y_j(t), \tag{17.2}$$

for $i = 1, \ldots, N$ where $a_{ij}(t)$ is the time-varying weight in the topology described by graph $\mathcal{G}(t)$ and the Laplacian matrix associated with $\mathcal{G}(t)$ is defined as $L_t = [\ell_{ij}(t)]$.

17.3 Problem Formulation

As before, we pursue here once again the regulated output synchronization among agents. In other words, our goal is to regulate the outputs of all agents asymptotically toward a priori specified reference trajectories. The reference trajectories are generated by

$$\begin{aligned} \dot{x}_0 &= A_0 x_0, \quad x_0(0) = x_{00}, \\ y_0 &= C_0 x_0. \end{aligned} \tag{17.3}$$

As in Sect. 15.3, a subset of agents, denoted by π, has access to the relative output $y_i - y_0$ of the exosystem compared to the output of the agent. To be specific, each agent $i \in \{1, \ldots, N\}$ has access to the quantity

$$\psi_i = \iota_i(y_i - y_0), \quad \iota_i = \begin{cases} 1, & i \in \pi, \\ 0, & i \notin \pi. \end{cases}$$

Based on the Laplacian matrix L_t of the network graph $\mathcal{G}_t$, we define the expanded Laplacian matrix as

$$\bar{L}_t = L_t + \operatorname{diag}\{\iota_i\} = [\bar{\ell}_{ij}(t)].$$

Note that $\bar{L}_t$ is clearly not a Laplacian matrix associated to some graph since it does not have a zero row sum at any time t. From [31, Lemma 7], all eigenvalues of $\bar{L}_t$ at time t are in the open right-half complex plane.

With a set of graph $\mathbb{G}^N_{\alpha,\beta,\pi}$ in Definition 15.4, we define a set of time-varying graphs.

Definition 17.1 For a given root set π, a positive integer N, and $\alpha, \beta, \tau > 0$, the set $\mathbb{G}^{\tau,N}_{\alpha,\beta,\pi}$ is the set of all time-varying graphs $\mathcal{G}_t$ for which

$$\mathcal{G}_t(t) = \mathcal{G}_{\sigma(t)} \in \mathbb{G}^{N}_{\alpha,\beta,\pi}$$

for all $t \in \mathbb{R}$, where $\sigma : \mathbb{R} \to \mathcal{L}_N$ is a piecewise constant, right-continuous function with minimal dwell-time τ.

Remark 17.2 Note that the minimal dwell-time is assumed to avoid chattering problems. However, it can be arbitrarily small.

Problem 17.3 Consider a MAS described by (17.1) and (17.2). For a given root set π, a positive integer N, and $\alpha, \beta, \tau > 0$ that define a set of time-varying network graphs $\mathbb{G}^{\tau,N}_{\alpha,\beta,\pi}$, the *regulated output synchronization* problem is to find, if possible, a linear time-invariant dynamic protocol of the form

$$\begin{aligned} \dot{\chi}_i &= A_{c,i}\chi_i + B_{c,i}\,\mathrm{col}\{\zeta_i, \psi_i\}, \\ u_i &= C_{c,i}\chi_i, \end{aligned} \tag{17.4}$$

for $i = 1, \ldots, N$, where $\chi_i \in \mathbb{R}^{n_{c,i}}$ such that for any time-varying network graph $\mathcal{G}_t \in \mathbb{G}^{\tau,N}_{\alpha,\beta,\pi}$ and for all the initial conditions of agents and their protocol, the regulated output synchronization among agents can be achieved.

17.4 Protocol Design

As in the fixed network graph case in Chap. 15, when the protocol only uses the network information ζ_i and reference information ψ_i, the solvability of Problem 17.3 requires agents to be minimum-phase. This will be discussed in the next subsection. If an additional communication via the network is possible, we will present in the subsection thereafter a design which achieves output synchronization which also works if some of the agents are non-minimum-phase.

17.4.1 Minimum-Phase and Right-Invertible Agents

In this section, the protocol design for minimum-phase and right-invertible agents is exactly the same as that in Sect. 15.4.1. That is, we design pre-compensators using Protocol Designs 15.1, 15.2 and 15.3 such that the resulting interconnection contain all the modes of the exosystem and is invertible and uniform rank. The next step is to design the observer-based controller using Protocol Design 15.4 to achieve output synchronization. We rewrite both protocol designs here.

Protocol design 17.1 *The design of the pre-compensators follows from Protocol Designs 15.1, 15.2, and 15.3. We obtain for $i = 1, \ldots, N$:*

$$\begin{aligned} \dot{p}_{i,1} &= A_{ip,1} p_{i,1} + B_{ip,1} u_i^1, \\ u_i &= C_{ip,1} p_{i,1} + D_{ip,1} u_i^1, \end{aligned} \tag{17.5}$$

$$\begin{aligned} \dot{p}_{i,2} &= S_{i,o} p_{i,2} + B_{i,2} \tilde{u}_i, \\ u_i^1 &= \Gamma_{i,o} p_{i,2}, \end{aligned} \tag{17.6}$$

and

$$\begin{aligned} \dot{p}_{i,3} &= K_i p_{i,3} + L_i \tilde{u}_i, \\ u_i^2 &= M_i p_{i,3}, \end{aligned} \tag{17.7}$$

The cascade system of agent (17.1) and pre-compensators (17.5), (17.6), and (17.7) is uniform rank and invertible. Hence the system can be represented by the SCB form as presented in (23.27). In other words, we obtain

$$\begin{aligned} \dot{\tilde{x}}_{ia} &= A_{ia} \tilde{x}_{ia} + L_{iad} y_i, \\ \dot{\tilde{x}}_{id} &= A_d \tilde{x}_{id} + B_d (\tilde{u}_i + E_{ida} \tilde{x}_{ia} + E_{idd} \tilde{x}_{id}), \\ y_i &= C_d \tilde{x}_{id}, \end{aligned} \tag{17.8}$$

for $i \in \{1, \ldots, N\}$ where A_d, B_d, and C_d are defined in (15.15). The observer-based controller is designed as follows.

Protocol design 17.2 *Let K_ε and $F_{\delta\varepsilon}$ be given in Design 15.4. The observer-based controller is designed as*

$$\begin{aligned} \dot{\hat{x}}_{id} &= A_d \hat{x}_{id} + K_\varepsilon (\zeta_i + \psi_i - C_d \hat{x}_{id}), \\ u_i &= F_{\delta\varepsilon} \hat{x}_{id}. \end{aligned} \tag{17.9}$$

The following theorem shows the main result in this section.

Theorem 17.4 *Consider a MAS described by (17.1) and (17.2) and an exosystem given by (17.3). Let a root agent set π, a positive integer N, and $\alpha, \beta, \tau > 0$ be given and hence a set of time-varying network graphs $\mathbb{G}_{\alpha,\beta,\pi}^{\tau,N}$ be defined.*

Let Assumption 17.1 hold. Moreover, the agents are minimum-phase and right-invertible. Then, the regulated output synchronization problem defined in Problem 17.3 is solvable. In particular, there exists a $\delta^* \in (0, 1]$, such that for each

$\delta \in (0, \delta^*]$, *there exists an* $\varepsilon^*(\delta) \in (0, 1]$ *such that for all* $\varepsilon \in (0, \varepsilon^*(\delta)]$, *the protocol* (17.9) *together with the pre-compensators* (17.5), (17.6), *and* (17.7) *solves the regulated output synchronization problem.*

Proof Following the steps of the proof of Theorem 15.6, we can represent the resulting dynamics in the following form

$$\begin{aligned} \dot{\eta}_a &= \tilde{A}_a \eta_a + \tilde{W}_{ad,t} \eta_d, \\ \varepsilon \dot{\eta}_d &= \tilde{A}_{\delta,t} \eta_d + \tilde{W}^{\varepsilon}_{da,t} \eta_a + \tilde{W}^{\varepsilon}_{dd,t} \eta_d, \end{aligned} \tag{17.10}$$

where $\tilde{A}_a = \operatorname{diag}(A_{1a}, \dots, A_{Na})$,

$$\tilde{A}_{\delta,t} = I_{N-1} \otimes \begin{pmatrix} A_d & 0 \\ 0 & A_d - KC_d \end{pmatrix} + \bar{S}_t \otimes \begin{pmatrix} B_d F_\delta & -B_d F_\delta \\ B_d F_\delta & -B_d F_\delta \end{pmatrix},$$

and

$$\tilde{W}_{ad,t} = \begin{pmatrix} W_{ad,t} & 0 \end{pmatrix} N_d^{-1}, \quad \tilde{W}^{\varepsilon}_{da,t} = N_d \begin{pmatrix} W^{\varepsilon}_{da,t} \\ W^{\varepsilon}_{da,t} \end{pmatrix}, \quad \tilde{W}^{\varepsilon}_{dd,t} = N_d \begin{pmatrix} W^{\varepsilon}_{dd,t} & 0 \\ W^{\varepsilon}_{dd,t} & 0 \end{pmatrix} N_d^{-1}.$$

Note that $\bar{U}_t^{-1} \bar{L}_t \bar{U}_t = \bar{S}_t$, where $\bar{S}_t$ is the Schur form of the matrix $\bar{L}_t$. The remaining proof follows directly the proof of Theorem 11.9 in Chap. 11. ■

17.4.2 Additional Communication

In this section, an additional communication of protocol states is allowed. Hence the agents need not be minimum-phase. As before, the extra communication implies that the quantity

$$\hat{\zeta}_i = \sum_{j=1}^{N} \bar{\ell}_{ij} \xi_j, \tag{17.11}$$

is available to each agent i ($i = 1, \dots, N$) where $\xi_j \in \mathbb{R}^p$ is a variable produced internally by agent j as a part of the protocol. The regulated output synchronization problem with additional communication is formulated as follows.

Problem 17.5 Consider a MAS described by (17.1), (17.2), and (17.11). For a given root set π, a positive integer N, $\alpha, \beta, \tau > 0$ that define a set of time-varying network graphs $\mathbb{G}^{\tau,N}_{\alpha,\beta,\pi}$, and for any reference trajectories given by (17.3), the *regulated output synchronization* problem is to find, if possible, a linear time-invariant dynamic protocol of the form

$$\begin{aligned}\dot{\chi}_i &= A_{ci}\chi_i + B_{ci}\operatorname{col}\{\zeta_i, \psi_i, \hat{\zeta}_i\},\\ u_i &= C_{ci}\chi_i,\end{aligned} \tag{17.12}$$

for $i = 1, \ldots, N$ where $\chi_i \in \mathbb{R}^{N_{ci}}$, such that, for any graph $\mathcal{G}_t \in \mathbb{G}^{\tau,N}_{\alpha,\beta,\pi}$ and for all the initial conditions of agents and their protocol, regulated output synchronization among agents is achieved.

The protocol design is exactly the same as that in Sect. 15.4.2. That is, design the pre-compensator using Design 15.5 to contain the exosystem mode, and then design the observer-based controller using Design 15.6. We also rewrite protocol designs here.

Protocol design 17.3 *The design of the pre-compensator follows Design 15.5 and we obtain*

$$\begin{aligned}\dot{p}_{i,1} &= S_{i,o}p_{i,1} + B_{i,1}\tilde{u}_i,\\ u_i &= \Gamma_{i,o}p_{i,1},\end{aligned} \tag{17.13}$$

for $i = 1, \ldots, N$.

The cascade system of agent (17.1) and pre-compensator (17.13) is of the form (15.28). For each $i \in \{1, \ldots, N\}$, we define $\bar{x}_i = \tilde{x}_i - \Pi_i x_0$ and $e_i = y_i - y_0$. We define $\chi_i = T_i\bar{x}_i$, where T_i is defined by (15.30). Then the dynamics of χ_i with output e_i can be written as

$$\begin{aligned}\dot{\chi}_i &= (A_d + \mathcal{L}_i)\chi_i + \mathcal{B}_i\tilde{u}_i, \quad \chi_i(0) = T_i\bar{x}_i(0),\\ e_i &= C_d\chi_i,\end{aligned} \tag{17.14}$$

where A_d, C_d, $\mathcal{B}_i$, and $\mathcal{L}_i$ are defined in (15.32) and (15.33). Then, the observer-based controller from Design 15.6 has the following form.

Protocol design 17.4 *Following Protocol Design 15.6, the observer-based controller is*

$$\begin{aligned}\dot{\hat{\chi}}_i &= (A_d + \mathcal{L}_i)\hat{\chi}_i + \mathcal{B}_i\tilde{u}_i + \varepsilon^{-1}S_\varepsilon^{-1}PC_d'(\zeta_i + \psi_i - \hat{\zeta}_i),\\ \tilde{u}_i &= F_i(T_i'T_i)^{-1}T_i'\hat{\chi}_i.\end{aligned} \tag{17.15}$$

Then, the result is stated in the following theorem.

Theorem 17.6 *Consider a MAS described by* (17.1), (17.2), (17.11), *and an exosystem given by* (17.3). *Let a root agent set π, a positive integer N, and $\alpha, \beta, \tau > 0$ be given and hence a set of time-varying network graphs $\mathbb{G}^{\tau,N}_{\alpha,\beta,\pi}$ be defined.*

Let Assumption 17.1 be satisfied. Moreover, assume the agents are right-invertible. Then, the regulated output synchronization problem stated in Problem 17.5 is solvable. In particular, there exists an $\varepsilon^* \in (0, 1]$ such that for all $\varepsilon \in (0, \varepsilon^*]$, the protocol (17.15) *together with the pre-compensators from Protocol Design 17.1 solves the regulated output synchronization problem.*

Proof With the same argument as in the proof of Theorem 15.8, we can obtain the dynamics

$$\varepsilon \dot{v} = (I_N \otimes A_d)v + W_{\varepsilon,t}v - \bar{S}_t \otimes (PC_d'C_d)v, \tag{17.16}$$

where $\bar{U}_t^{-1}\bar{L}_t\bar{U}_t = \bar{S}_t$ with $\bar{S}_t$ the Schur form of $\bar{L}_t$ and

$$W_{\varepsilon,t} = (\bar{Q}_t^{-1} \otimes I_{pn})\mathcal{L}_\varepsilon(\bar{U}_t \otimes I_{pn}).$$

Then, following the proof of Theorem 11.12, we can get immediately

$$\lim_{t\to\infty} (\chi_i(t) - \hat{\chi}_i(t)) = 0$$

for any time-varying graph $\mathcal{G}_t \in \mathbb{G}^{\tau,N}_{\alpha,\beta,\pi}$. Rewriting the dynamics in the form (15.39), we can then conclude

$$\lim_{t\to\infty} (y_i(t) - y_0(t)) = 0$$

which proves the result. ■

Chapter 18
Exact Regulated Output Synchronization for Heterogeneous Continuous-Time MAS in the Presence of Disturbances and Measurement Noise with Known Frequencies

18.1 Introduction

The vast majority of the research in MAS has focused on the idealized case where the agents are unaffected by external disturbances. Among the authors that have considered external disturbances, Lin et al. [49] considered the problem of minimizing the $\mathcal{H}_\infty$ norm from an external disturbance to the output of each agent, whereas Lin and Jia [54] and Li et al. [59] considered minimization of the $\mathcal{H}_\infty$ norm from a disturbance to the average of the states in a network of single or double integrators. By contrast, Peymani et al. [83] introduced the notion of $\mathcal{H}_\infty$ *almost synchronization* for a class of heterogeneous networks, where the goal is to reduce the $\mathcal{H}_\infty$ norm from an external disturbance to the synchronization *error*, to any arbitrary desired level.

This chapter, together with the following two chapters, studies the case of disturbances in the context of heterogeneous networks. For homogeneous networks we already considered disturbances in Chap. 13. This chapter considers the case of disturbances and measurement noise with known frequency. Next in Chap. 19, we consider $\mathcal{H}_\infty$ *almost synchronization*, while in Chap. 20 we will consider $\mathcal{H}_\infty$ *almost synchronization*.

As in the previous chapters, we consider agents which are non-introspective and therefore have no direct information about their own states but only information of their output relative to the output of other agents. Also we consider regulated output synchronization where the output of each agent has to converge to a reference trajectory generated by an autonomous exosystem. In the chapter, the disturbances and measurement noise are of known frequency which implies that they are generated by a separate exosystem for which we have no measurements but we do know the model. As mentioned before more general disturbances will be considered in the next two chapters.

A. Saberi et al., *Synchronization of Multi-Agent Systems in the Presence of Disturbances and Delays*, Systems & Control: Foundations & Applications,
https://doi.org/10.1007/978-3-030-88148-1_18

The design in this chapter will be similar to the design philosophy of the previous three chapters but with a modification to ensure disturbance rejection and *exact* regulated output synchronization. We again first consider agents which are minimum-phase. Next, we show how an additional communication channel enables us to drop the requirement of minimum phase. We will also briefly discuss the case of time-varying graphs and show that the results of this chapter can easily be extended to that case.

The write-up of this chapter is partially based on [181].

18.2 Multi-Agent Systems

Suppose a MAS composed of N non-identical linear time-invariant agents of the form

$$\begin{aligned} \dot{x}_i &= A_i x_i + B_i u_i + G_i \omega_i, \\ y_i &= C_i x_i + D_{\omega i} \omega_i, \end{aligned} \qquad (i = 1, \dots, N) \tag{18.1}$$

where $x_i \in \mathbb{R}^{n_i}$, $u_i \in \mathbb{R}^{m_i}$, and $y_i \in \mathbb{R}^p$ are the state, input, and output of agent i. As before we make the following assumption.

Assumption 18.1 *For each agent $i \in \{1, \dots, N\}$, we assume that*

- *(A_i, B_i) is stabilizable.*
- *(A_i, C_i) is detectable.*

The external disturbance ω_i is generated by the exosystem:

$$\begin{aligned} \dot{x}_{\omega i} &= S_i x_{\omega i}, \\ \omega_i &= R_i x_{\omega i}, \end{aligned} \qquad x_{\omega i}(0) = x_{\omega i0} \tag{18.2}$$

where $x_{\omega i} \in \mathbb{R}^{n_{\omega i}}$ and the initial condition $x_{\omega i0}$ can be arbitrarily chosen. It is clear that $G_i \omega_i$ represents the process disturbance, while $D_{\omega i} \omega_i$ represents the measurement noise.

The communication network among agents is exactly the same as in previous chapters and provides each agent with the quantity

$$\zeta_i = \sum_{j=1}^{N} a_{ij}(y_i - y_j) = \sum_{j=1}^{N} \ell_{ij} y_j. \tag{18.3}$$

18.3 Problem Formulation

In this section, as in the previous chapter, we pursue regulated output synchronization among agents' outputs, that is, we regulate the outputs of all agents asymptotically toward a priori specified reference trajectories generated by an exosystem of the form

$$\begin{aligned}\dot{x}_0 &= A_0 x_0, \quad x_0(0) = x_{00},\\ y_0 &= C_0 x_0.\end{aligned} \tag{18.4}$$

As in previous chapters, a subset of agents, denoted by π, has access to the relative output $y_i - y_0$ of the exosystem compared to the output of the agent. To be specific, each agent $i \in \{1, \dots, N\}$ has access to the quantity

$$\psi_i = \iota_i(y_i - y_0), \quad \iota_i = \begin{cases} 1, & i \in \pi,\\ 0, & i \notin \pi.\end{cases} \tag{18.5}$$

We consider a set of graphs $\mathbb{G}^N_{\alpha,\beta,\pi}$ as defined in Definition 15.4.

Note that the regulated output synchronization is achieved if

$$\lim_{t\to\infty} [y_i(t) - y_0(t)] = 0, \quad \forall i \in \{1, \dots, N\}. \tag{18.6}$$

Then, we formulate the exact regulated output synchronization problem for a MAS in the presence of disturbance with known frequencies as follows.

Problem 18.1 Consider a MAS described by (18.1), (18.2), and (18.3). For a given root set π, a positive integer N, and $\alpha, \beta > 0$ that define a set of network graphs $\mathbb{G}^N_{\alpha,\beta,\pi}$ and any reference trajectories given by (18.4), the *exact regulated output synchronization* problem is to find, if possible, a linear time-invariant dynamic protocol of the form

$$\begin{aligned}\dot{\chi}_i &= A_{ci}\chi_i + B_{ci}\,\mathrm{col}\{\zeta_i, \psi_i\},\\ u_i &= C_{ci}\chi_i,\end{aligned} \tag{18.7}$$

for $i = 1, \dots, N$, where $\chi_i \in \mathbb{R}^{N_{ci}}$, such that (18.6) is satisfied for any graph $\mathcal{G} \in \mathbb{G}^N_{\alpha,\beta,\pi}$, for all initial conditions of agents, the protocol and the reference system and for all possible disturbances.

18.4 Protocol Design

There are two protocol designs in this section. When the protocol only uses the network information ζ_i and reference information ψ_i, the solvability of Problem 18.1 requires agents to be minimum-phase. However, as we have seen before, additional communication of protocol states via the network enables us to remove this restriction, that is, the agents can be non-minimum-phase.

18.4.1 Minimum-Phase and Right-Invertible Agents

We start of by building an augmented exosystem for each agent where we combine disturbances ω_i ($i \in \{1, \dots, N\}$) generated by (18.2) with the exosystem (18.4). Let

$$S_{ir} = \text{blkdiag}\{A_0, S_i\}, \qquad R_{ir} = \text{blkdiag}\{C_0, R_i\} \quad \text{and} \quad x_{ir} = \text{col}\{x_0, x_{\omega i}\}.$$

This yields an augmented exosystem which can be written as

$$\begin{aligned} \dot{x}_{ie} &= S_{ie}x_{ie}, \\ \omega_i &= R_{ie}x_{ie}, \qquad x_{ie}(0) = x_{ie0} \\ y_0 &= R_{re}x_{ie}, \end{aligned} \tag{18.8}$$

where

$$S_{ie} = \begin{pmatrix} S_i & 0 \\ 0 & A_0 \end{pmatrix}, \quad R_{ie} = \begin{pmatrix} R_i & 0 \end{pmatrix}, \quad R_{re} = \begin{pmatrix} 0 & C_0 \end{pmatrix},$$

and

$$x_{ie0} = \begin{pmatrix} x_{\omega i0} \\ x_{00} \end{pmatrix}.$$

The protocol design for minimum-phase agents is similar to that in Sect. 15.4.1. That is, we design pre-compensators such that the interconnected system contains all modes of the augmented exosystem and such that all interconnected systems are uniform rank and invertible with identical uniform rank.

Protocol design 18.1 *We first design a pre-compensator similar to Protocol Design 15.1*

$$\begin{aligned} \dot{p}_{i,1} &= A_{ip,1}p_{i,1} + B_{ip,1}u_i^1, \\ u_i &= C_{ip,1}p_{i,1} + D_{ip,1}u_i^1, \end{aligned} \tag{18.9}$$

such that the interconnected system is of the form

(continued)

Protocol design 18.1 (continued)

$$\begin{aligned} \dot{x}_i &= \hat{A}_i\hat{x}_i + \hat{B}_i u_i^1 + \hat{G}_i\omega_i, \\ y_i &= \hat{C}_i x_i + D_{\omega i}\omega_i, \end{aligned} \tag{18.10}$$

and uniform rank and invertible for $i = 1, \ldots, N$. Next, we design pre-compensators such that the interconnected system contains all modes of the augmenten exosystem. This follows Protocol Design 15.2. However, we use the augmented exosystem (18.8) instead of the exosystem (18.4). This implies that the regulator equation (15.10) should be first modified as

$$\begin{aligned} \hat{A}_i\tilde{\Pi}_i + \hat{B}_i\Gamma_i + \hat{G}_i R_{ie} &= \tilde{\Pi}_i S_{ie}, \\ \hat{C}_i\tilde{\Pi}_i + D_{\omega i} R_{ie} &= R_{re}. \end{aligned} \tag{18.11}$$

If (Γ_i, S_{ie}) is not observable, we can construct the observable subsystem $(\Gamma_{i,o}, S_{ie,o})$; otherwise simply set $\Gamma_{i,o} = \Gamma_i$ and $S_{i,o} = S_{ie}$. Then the second pre-compensator for each agent i is designed as

$$\begin{aligned} \dot{p}_{i,2} &= S_{ie,o}p_{i,2} + B_{i,2}u_i^2, \\ u_i^1 &= \Gamma_{i,o}p_{i,2}, \end{aligned} \tag{18.12}$$

for $i = 1, \ldots, N$, where $p_{i,2} \in \mathbb{R}^{n_{pi}}$ and $B_{i,2}$ is chosen to guarantee that no invariant zeros are introduced by the pre-compensator. Moreover, an appropriate choice of $B_{i,2}$ can guarantee that the triple $(S_{ie,o}, B_{i,2}, \Gamma_{i,o})$ is invertible, uniform rank, and minimum-phase (see [60]).

If the agent has dynamics in common with the exosystem, then we can use the technique outlined in Appendix 15.5 to remove these common modes after which the above design is applicable.

The final pre-compensator is designed according to Protocol Design 15.3, and we obtain

$$\begin{aligned} \dot{p}_{i,3} &= K_i p_{i,3} + L_i\tilde{u}_i, \\ u_i^2 &= M_i p_{i,3}, \end{aligned} \tag{18.13}$$

such that the agents interconnected with the three pre-compensators are all uniform rank and invertible with the same uniform rank for all agents.

The cascade system of agent (18.1) and pre-compensators (18.9), (18.12), and (18.13) can be represented in the form

$$\begin{cases} \dot{\tilde{x}}_i = \tilde{A}_i\tilde{x}_i + \tilde{B}_i\tilde{u}_i + \tilde{G}_i\omega_i, \\ y_i = \tilde{C}_i\tilde{x}_i + \tilde{D}_{\omega i}\omega_i, \end{cases} \tag{18.14}$$

where $\tilde{x}_i \in \mathbb{R}^{\tilde{n}_i}$, $\tilde{u}_i \in \mathbb{R}^p$, $and y_i \in \mathbb{R}^p$ are states, inputs, and outputs of the cascade systems. Moreover, there exists $\rho \geq 1$ such that we have

- For the system $(\tilde{A}_i, \tilde{C}_i, \tilde{G}_i, \tilde{D}_{\omega i})$ there exists a matrix Π_i such that

$$\Pi_i S_{ie} = \tilde{A}_i \Pi_i + \tilde{G}_i R_{ie}, \qquad \tilde{C}_i \Pi_i + \tilde{D}_{\omega i} R_{ie} = R_{re}, \tag{18.15}$$

- $(\tilde{A}_i, \tilde{B}_i, \tilde{C}_i)$ has uniform rank ρ.

The cascade system has uniform rank and is invertible. Hence the system can be represented by the SCB form as presented in (23.27). In other words, we obtain

$$\tilde{x}_i = \begin{pmatrix} \tilde{x}_{ia} \\ \tilde{x}_{id} \end{pmatrix} \tag{18.16}$$

with $\tilde{x}_{ia} \in \mathbb{R}^{\tilde{n}_i - p\rho}$ representing the finite zero structure and $\tilde{x}_{id} \in \mathbb{R}^{p\rho}$ the infinite zero structure with the following structure for the dynamics:

$$\begin{cases} \dot{\tilde{x}}_{ia} = A_{ia}\tilde{x}_{ia} + L_{iad} y_i + \tilde{G}_{ia}\omega_i, \\ \dot{\tilde{x}}_{id} = A_d \tilde{x}_{id} + B_d(\tilde{u}_i + E_{ida}\tilde{x}_{ia} + E_{idd}\tilde{x}_{id}) + \tilde{G}_{id}\omega_i, \\ y_i \;= C_d \tilde{x}_{id} + \tilde{D}_{\omega i}\omega_i, \end{cases} \tag{18.17}$$

for $i \in \{1, \ldots, N\}$ where

$$A_d = \begin{pmatrix} 0 & I \\ 0 & 0 \end{pmatrix}, \qquad B_d = \begin{pmatrix} 0 \\ I \end{pmatrix}, \qquad C_d = \begin{pmatrix} I & 0 \end{pmatrix}. \tag{18.18}$$

For each cascade system (18.17), we design an observer-based protocol.

Protocol design 18.2 *We use the same observer-based protocol as in Protocol Design 15.4, that is,*

$$\begin{aligned} \dot{\hat{x}}_{id} &= A_d \hat{x}_{id} + K_\varepsilon(\zeta_i + \psi_i - C_d \hat{x}_{id}), \\ u_i &= F_{\delta\varepsilon}\hat{x}_{id}. \end{aligned} \tag{18.19}$$

The main result of this section is stated in the following theorem.

Theorem 18.2 *Consider a MAS described by* (18.1), (18.2), *and* (18.3), *and any reference system* (18.4). *Let a root set π, a positive integer N, and $\alpha, \beta > 0$ be given and hence a set of network graphs $\mathbb{G}^N_{\alpha,\beta,\pi}$ be defined.*

Let Assumptions 15.1 hold. Moreover, assume the agents are right-invertible. Then the exact regulated output synchronization problem defined in Problem 18.1 is solvable. In particular, for any graph $\mathcal{G} \in \mathbb{G}^N_{\alpha,\beta,\pi}$, *there exists a* δ^* *such that for any* $\delta \in (0, \delta^*]$ *there exists an* ε^* *such that for any* $\varepsilon \in (0, \varepsilon^*]$, *the combination of pre-compensators* (18.9), (18.12), *and* (18.13) *with protocol* (18.19) *solves the exact regulated output synchronization problem.*

Proof Recall that the agents interconnected with the three pre-compensators are of the form (18.14), and there exists Π_i satisfying (18.15). For each $i \in \{1, \dots, N\}$, let $\bar{x}_i = \tilde{x}_i - \Pi_i x_{ie}$. Then we obtain

$$\begin{aligned}\dot{\bar{x}}_i &= \tilde{A}_i\tilde{x}_i - \Pi_i S_{ie}x_{ie} + \tilde{B}_i\tilde{u}_i + \tilde{G}_i\omega_i \\ &= \tilde{A}_i\tilde{x}_i - (\tilde{A}_i\Pi_i + \tilde{G}_i R_{ie})x_{ie} + \tilde{B}_i\tilde{u}_i + \tilde{G}_i R_{ie}x_{ie} \\ &= \tilde{A}_i\bar{x}_i + \tilde{B}_i\tilde{u}_i,\end{aligned}$$

and

$$\begin{aligned}e_i &= y_i - y_0 \\ &= \tilde{C}_i\tilde{x}_i + \tilde{D}_{\omega i}\omega_i - R_{re}x_{ie} \\ &= \tilde{C}_i\tilde{x}_i + \tilde{D}_{\omega i}R_{ie}x_{ie} - (\tilde{C}_i\Pi_i + \tilde{D}_{\omega i}R_{ie})x_{ie} \\ &= \tilde{C}_i\bar{x}_i.\end{aligned}$$

Since the dynamics of the $\bar{x}_i$ system with output e_i is governed by the same dynamics as the dynamics of agent i, we can present $\bar{x}_i$ in the same form as (18.17) but removing the disturbance items. Let

$$\bar{x}_i = \begin{pmatrix}\bar{x}_{ia} \\ \bar{x}_{id}\end{pmatrix}$$

The dynamics of $\bar{x}_i$ is written as

$$\begin{aligned}\dot{\bar{x}}_{ia} &= A_{ia}\bar{x}_{ia} + L_{iad}e_i, \\ \dot{\bar{x}}_{id} &= A_d\bar{x}_{id} + B_d(\tilde{u}_i + E_{ida}\bar{x}_{ia} + E_{idd}\bar{x}_{id}), \\ e_i &= C_d\bar{x}_{id}.\end{aligned}$$

It is clear that the $\bar{x}_i$ dynamic is exactly the same as in (15.19) in Theorem 15.6. Hence the remaining proof follows that in Theorem 15.6. ∎

18.4.2 Additional Communication

In this section, an additional communication of protocol states via the network's communication infrastructure is allowed. From the previous chapters, we know that this enables us to drop the requirement that all agents are minimum-phase.

As before, the extra communication implies that each agent i ($i = 1, \dots, N$) is presumed to have access to the quantity

$$\hat{\zeta}_i = \sum_{j=1}^{N} \bar{\ell}_{ij}\xi_j, \tag{18.20}$$

where $\xi_j \in \mathbb{R}^p$ is a variable produced internally by agent j as the part of the protocol.

We then obtain the following modification of the exact regulated output synchronization problem as defined in Problem 18.1.

Problem 18.3 Consider a MAS described by (18.1), (18.2), (18.3), and (18.20). Consider a set of network graphs $\mathbb{G}^N_{\alpha,\beta,\pi}$ as defined in Definition 15.4. For any reference trajectories given by (18.4), the *exact regulated output synchronization* problem for heterogeneous agents with additional communication is to find, if possible, a linear time-invariant dynamic protocol of the form

$$\begin{aligned} \dot{\chi}_i &= A_{ci}\chi_i + B_{ci}\,\mathrm{col}\{\zeta_i, \psi_i, \hat{\zeta}_i\}, \\ u_i &= C_{ci}\chi_i, \end{aligned} \tag{18.21}$$

for $i = 1, \dots, N$ where $\chi_i \in \mathbb{R}^{N_{ci}}$, such that regulated output synchronization among agents is achieved, for any graph $\mathcal{G} \in \mathbb{G}^N_{\alpha,\beta,\pi}$, for all initial conditions of agents, protocols and exosystem, and for all possible disturbances.

Similar to the minimum-phase case, an augmented exosystem (18.8) is built. From the pre-compensator design of Protocol Design 18.1, we only use the second pre-compensator for agent (18.1) to obtain a cascade system of the form

$$\begin{aligned} \dot{\tilde{x}}_i &= \tilde{A}_i\tilde{x}_i + \tilde{B}_i\tilde{u}_i + \tilde{G}_i\omega_i, \\ y_i &= \tilde{C}_i\tilde{x}_i + \tilde{D}_{\omega i}\omega_i, \end{aligned} \tag{18.22}$$

where $\tilde{x}_i \in \mathbb{R}^{\tilde{n}_i}$ for which there exists a Π_i satisfying (18.15).

Let $\bar{x}_i = \tilde{x}_i - \Pi_i x_{ie}$ and $e_i = y_i - y_0$. Following the same steps as in the proof of Theorem 18.2, we achieve

$$\begin{aligned} \dot{\bar{x}}_i &= \tilde{A}_i\bar{x}_i + \tilde{B}_i\tilde{u}_i, \\ e_i &= \tilde{C}_i\bar{x}_i. \end{aligned} \tag{18.23}$$

Denote $n = \max_i\{\tilde{n}_i\}$, and define a matrix

$$T_i = \begin{pmatrix} \tilde{C}_i \\ \vdots \\ \tilde{C}_i \tilde{A}_i^{n-1} \end{pmatrix},$$

which is the same as in Sect. 15.4.2. Let $\chi_i = T_i \bar{x}_i$. Then, the dynamics of χ_i is written as

$$\begin{aligned} \dot{\chi}_i &= (\mathcal{A} + \mathcal{L}_i)\chi_i + \mathcal{B}_i \tilde{u}_i, \quad \chi_i(0) = T_i \bar{x}_i(0), \\ e_i &= C\chi_i, \end{aligned} \tag{18.24}$$

where $i = 1, \ldots, N$, and

$$\mathcal{A} = \begin{pmatrix} 0 & I_{p(n-1)} \\ 0 & 0 \end{pmatrix}, \quad C = \begin{pmatrix} I_p & 0 \end{pmatrix}, \quad \mathcal{L}_i = \begin{pmatrix} 0 \\ L_i \end{pmatrix}, \quad \mathcal{B}_i = T_i \tilde{B}_i,$$

while $L_i = \tilde{C}_i \tilde{A}_i^n (T_i' T_i)^{-1} T_i'$. We find that the χ_i dynamics is exactly the same as (15.31). Therefore, the observer-based controller (15.34) in Design 15.6 also works in this section, which is rewritten as follows.

Protocol design 18.3 *Following Protocol Design 15.6, the observer-based controller is*

$$\begin{aligned} \dot{\hat{\chi}}_i &= (A_d + \mathcal{L}_i)\hat{\chi}_i + \mathcal{B}_i \tilde{u}_i + \varepsilon^{-1} S_\varepsilon^{-1} P C_d' (\zeta_i + \psi_i - \hat{\zeta}_i), \\ \tilde{u}_i &= F_i (T_i' T_i)^{-1} T_i' \hat{\chi}_i. \end{aligned} \tag{18.25}$$

The main result of this section is stated as follows.

Theorem 18.4 *Consider a MAS described by (18.1), (18.2), (18.3), and (18.20) and any reference system (18.4). Let a root set π, a positive integer N, α, and$\beta > 0$ be given and hence a set of network graphs $\mathbb{G}^N_{\alpha,\beta,\pi}$ be defined.*

Assume the agents are right-invertible. Then the exact regulated output synchronization problem defined in Problem 18.3 is solvable. In particular, for any graph $\mathcal{G} \in \mathbb{G}^N_{\alpha,\beta,\pi}$, there exists an $\varepsilon^* \in (0, 1]$ such that for any $\varepsilon \in (0, \varepsilon^*]$, the combination of pre-compensator (18.12) (with $u_i = u_i^1$ and $\tilde{u}_i = u_i^2$) and protocol (18.25) solves the exact regulated output synchronization problem.

Proof It follows along the same lines as the proof of Theorem 15.8. ∎

18.4.3 Time-Varying Graphs

In this section, we consider a MAS with time-varying graphs. We show here that the protocol design for a fixed graph can be applied directly to a time-varying graph. The time-varying communication network among agents is exactly the same as that in Chap. 11. The network information provided to each agent is modified as

$$\zeta_i = \sum_{j=1}^{N} a_{ij}(t)(y_i - y_j) = \sum_{j=1}^{N} \ell_{ij}(t) y_j. \tag{18.26}$$

The target is also to regulate all agents' outputs to a priori given reference trajectories generated by an exosystem (18.4). The extra information is available to all the agents from the root set π, i.e., each agent $i \in \{1, \ldots, N\}$ has access to the quantity (18.5). Following Chap. 17, we consider a set of time-varying graphs $\mathbb{G}^{\tau,N}_{\alpha,\beta,\pi}$ as defined in Definition 17.1.

We formulate the exact regulated output synchronization problems with and without additional communication under time-varying graphs as follows.

Problem 18.5 Consider a MAS described by (18.1), (18.2), and (18.3). For a given root set π, a positive integer N, and $\alpha, \beta, \tau > 0$ that define a set of network graphs $\mathbb{G}^{\tau,N}_{\alpha,\beta,\pi}$ and any reference trajectories given by (18.4), the *exact regulated output synchronization* problem is to find, if possible, a linear time-invariant dynamic protocol of the form (18.7) such that for any graph $\mathcal{G}_t \in \mathbb{G}^{\tau,N}_{\alpha,\beta,\pi}$ (18.6) is satisfied for all the initial conditions of agents and the reference system.

Problem 18.6 Consider a MAS described by (18.1), (18.2), (18.3), and (18.20). For any given root set π, a positive integer N, $\alpha, \beta, and \tau > 0$ that define a set of network graphs $\mathbb{G}^{\tau,N}_{\alpha,\beta,\pi}$ and exosystem (18.4), the *exact regulated output synchronization* problem for heterogeneous networks with non-introspective agents and with exchange of controller states is to find, if possible, a linear time-invariant dynamic protocol of the form (18.21) such that exact regulated output synchronization among agents is achieved, for any graph $\mathcal{G}_t \in \mathbb{G}^{\tau,N}_{\alpha,\beta,\pi}$, for all the initial conditions of agents, protocols and reference system, and all possible disturbances.

We show here that the protocol design generated for a fixed graph works exactly for a time-varying graph. The result for minimum-phase agent is stated in the following theorem.

Theorem 18.7 *Consider a MAS described by* (18.1), (18.2), (18.3), *and any reference trajectories given by* (18.4). *Consider the set of network graphs* $\mathbb{G}^{\tau,N}_{\alpha,\beta,\pi}$ *and exosystem* (18.4).

Assume that the agents are minimum-phase and right-invertible. Then the exact regulated output synchronization problem defined in Problem 18.5 is solvable. In particular, for any graph $\mathcal{G}_t \in \mathbb{G}^{\tau,N}_{\alpha,\beta,\pi}$, *there exists a* δ^* *such that for any* $\delta \in (0, \delta^*]$

there exists an ε^* *such that for any* $\varepsilon \in (0, \varepsilon^*]$, *the combination of pre-compensators* (18.9), (18.12), *and* (18.13) *with protocol* (18.19) *solves the exact regulated output synchronization problem.*

Proof Combine the proofs of Theorems 18.2 and 17.4. ■

The result for additional communication is stated in the following theorem.

Theorem 18.8 *Consider a MAS described by* (18.1), (18.2), (18.3), (18.20), *and any reference system* (18.4). *Consider the set of network graphs* $\mathbb{G}^{\tau,N}_{\alpha,\beta,\pi}$ *and exosystem* (18.4).

Assume that the agents are right-invertible. Then the exact regulated output synchronization problem defined in Problem 18.6 is solvable. In particular, for any graph $\mathcal{G}_t \in \mathbb{G}^{\tau,N}_{\alpha,\beta,\pi}$, *there exists an* $\varepsilon^* \in (0, 1]$ *such that for any* $\varepsilon \in (0, \varepsilon^*]$, *the combination of pre-compensator* (18.12) *(with* $u_i = u_i^1$ *and* $\tilde{u}_i = u_i^2$*) with protocol* (18.25) *solves the exact regulated output synchronization problem.*

Proof Combine the proofs of Theorems 18.4 and 17.6. ■

Chapter 19
H_∞ Almost Output Synchronization for Heterogeneous Continuous-Time MAS

19.1 Introduction

In the previous chapter, we started looking at regulated synchronization for heterogeneous continuous-time linear MAS, where agents are affected by external disturbances. There, we considered disturbances of known frequency and showed that we could still achieve exact regulated synchronization. In this chapter, we will consider disturbances with bounded power. It is easily established that exact regulated synchronization is no longer possible. We therefore focus on almost regulated synchronization. We will show that for minimum-phase and right-invertible agents, we can reduce the effect of disturbances on the regulated output synchronization error to an arbitrarily small value.

The write-up of this chapter is partially based on [179] and [182].

19.2 Multi-Agent Systems

Consider a MAS composed of N non-identical linear time-invariant agents of the form

$$\begin{aligned} \dot{x}_i &= A_i x_i + B_i u_i + G_i \omega_i, \\ y_i &= C_i x_i, \end{aligned} \qquad (i = 1, \ldots, N) \tag{19.1}$$

where $x_i \in \mathbb{R}^{n_i}$, $u_i \in \mathbb{R}^{m_i}$, and $y_i \in \mathbb{R}^p$ are the state, input, and output of agent i. We make the following standard assumptions for the agent dynamics.

A. Saberi et al., *Synchronization of Multi-Agent Systems in the Presence of Disturbances and Delays*, Systems & Control: Foundations & Applications,
https://doi.org/10.1007/978-3-030-88148-1_19

Assumption 19.1 *For each agent $i \in \{1, \dots, N\}$, we assume that:*

- *(A_i, B_i) is stabilizable.*
- *(A_i, C_i) is detectable.*

The external disturbance $\omega_i \in \mathbb{R}^{\omega_i}$ is either in the set $\Gamma_\kappa^{\text{rms}}$ or in the set Γ_κ^∞ for a given κ as defined below.

Definition 19.1 The set of disturbances with power less than κ is defined as

$$\Gamma_\kappa^{\text{rms}} = \{\omega \in L_{2,\text{loc}} : \|\omega\|_{\text{rms}} \triangleq \limsup_{T\to\infty} \frac{1}{T}\int_0^T \omega(t)'\omega(t)\text{d}t < \kappa^2\}.$$

The set of disturbances which are bounded by κ is defined as

$$\Gamma_\kappa^\infty = \{\omega \in L_\infty : \|\omega\|_\infty < \kappa\}.$$

It is interesting to note that $\Gamma_\kappa^\infty \subset \Gamma_\kappa^{\text{rms}}$ for any κ. The communication network among agents is exactly the same as that in Chap. 15 and provides each agent with the quantity

$$\zeta_i = \sum_{j=1}^N a_{ij}(y_i - y_j) = \sum_{j=1}^N \ell_{ij} y_j. \tag{19.2}$$

19.3 Problem Formulation

In this section, we also pursue regulated output synchronization among the output of the agents, that is, to regulate the outputs of all agents asymptotically toward a priori specified reference trajectories generated by (15.6). We rewrite it here:

$$\begin{aligned} \dot{x}_0 &= A_0 x_0, \quad x_0(0) = x_{00}, \\ y_0 &= C_0 x_0. \end{aligned} \tag{19.3}$$

In the same way as in Chap. 15, a subset of agents, denoted by π, has access to their output relative to the reference trajectory y_0 generated by the exosystem (19.3). To be specific, each agent $i \in \{1, \dots, N\}$ has access to the quantity

$$\psi_i = \iota_i(y_i - y_0), \quad \iota_i = \begin{cases} 1, & i \in \pi, \\ 0, & i \notin \pi. \end{cases}$$

We consider the class of network graphs $\mathbb{G}_{\alpha,\beta,\pi}^N$ as defined in Definition 15.4.

Define $e_{i0} = y_i - y_0$ and

$$\mathbf{e}_0 = \begin{pmatrix} e_{i0} \\ \vdots \\ e_{N0} \end{pmatrix} \quad \text{and} \quad \omega = \begin{pmatrix} \omega_1 \\ \vdots \\ \omega_N \end{pmatrix}.$$

It is obvious that the regulated output synchronization is achieved if $\lim_{t\to\infty} \mathbf{e}_0(t) = 0$, i.e.,

$$\lim_{t\to\infty} (y_i(t) - y_0(t)) = 0, \quad \forall i \in \{1, \dots, N\}. \tag{19.4}$$

We denote by $T_{\omega\mathbf{e}_0}$ the transfer matrix from ω to $\mathbf{e}_0$.

We formulate the $\mathcal{H}_\infty$ almost regulated output synchronization problem for a MAS in the presence of disturbance as follows.

Problem 19.2 Consider a MAS described by (19.1) and (19.2), and exosystem (19.3). For a given root set π, a positive integer N, and $\alpha, \beta > 0$ that define a set of network graphs $\mathbb{G}^N_{\alpha,\beta,\pi}$, the $\mathcal{H}_\infty$ *almost regulated output synchronization* problem is to find, if possible, for any given real number $\gamma > 0$, a linear time-invariant dynamic protocol of the form

$$\begin{aligned} \dot{\chi}_i &= A_{ci}\chi_i + B_{ci}\,\mathrm{col}\{\zeta_i, \psi_i\}, \\ u_i &= C_{ci}\chi_i, \end{aligned} \tag{19.5}$$

for $i = 1, \dots, N$ where $\chi_i \in \mathbb{R}^{N_{ci}}$, such that for any graph $\mathcal{G} \in \mathbb{G}^N_{\alpha,\beta,\pi}$, (19.4) is satisfied, for all initial conditions of agents, protocols, and exosystem, in the absence of disturbances. Moreover, the closed-loop transfer function $T_{\omega\mathbf{e}_0}$ satisfies

$$\|T_{\omega\mathbf{e}_0}\|_\infty \le \gamma. \tag{19.6}$$

19.4 Protocol Design

As mentioned in the beginning, the agents in this chapter are assumed minimum-phase and right-invertible. We will first focus on this case and then illustrate how these technique can still solve the problem for time-varying networks. We will also consider a mix with the techniques from the previous chapter where some disturbances have known frequencies while other disturbances have bounded power.

19.4.1 Minimum-Phase and Right-Invertible Agents

It turns out that we can use techniques very similar to the design presented in Chap. 15. We start by applying the pre-compensators to make the agents invertible with identical uniform rank and containing the dynamics of exosystem 19.3.

Protocol design 19.1 *We design three pre-compensators following Protocol Designs 15.1, 15.2 and 15.3. We obtain for $i = 1, \dots, N$:*

$$\begin{aligned} \dot{p}_{i,1} &= A_{ip,1}p_{i,1} + B_{ip,1}u_i^1, \\ u_i &= C_{ip,1}p_{i,1} + D_{ip,1}u_i^1, \end{aligned} \tag{19.7}$$

and

$$\begin{aligned} \dot{p}_{i,2} &= S_{i,o}p_{i,2} + B_{i,2}u_i^2, \\ u_i^1 &= \Gamma_{i,o}p_{i,2}, \end{aligned} \tag{19.8}$$

and

$$\begin{aligned} \dot{p}_{i,3} &= K_i p_{i,3} + L_i\tilde{u}_i, \\ u_i^2 &= M_i p_{i,3}, \end{aligned} \tag{19.9}$$

The cascade system of agent (19.1) and pre-compensators (19.7), (19.8), and (19.9) can be represented in the form

$$\begin{cases} \dot{\tilde{x}}_i = \tilde{A}_i\tilde{x}_i + \tilde{B}_i\tilde{u}_i + \tilde{G}_i\omega_i, \\ y_i = \tilde{C}_i\tilde{x}_i, \end{cases} \tag{19.10}$$

where $\tilde{x}_i \in \mathbb{R}^{\tilde{n}_i}$, $\tilde{u}_i \in \mathbb{R}^p$, and $y_i \in \mathbb{R}^p$ are states, inputs, and outputs of the cascade system. Moreover, there exists a $\rho > 0$ such that for each agent:

- $(\tilde{A}_i, \tilde{C}_i)$ contains (A_0, C_0), i.e., there exists a matrix Π_i such that $\Pi_i A_0 = \tilde{A}_i\Pi_i$ and $\tilde{C}_i\Pi_i = C_0$.
- $(\tilde{A}_i, \tilde{B}_i, \tilde{C}_i)$ is of uniform rank $\rho \geq 1$.

The cascade system (19.10) has uniform rank and is invertible. Hence the system can be represented by the SCB form as presented in (23.27). In other words, we obtain

$$\begin{cases} \dot{\tilde{x}}_{ia} = A_{ia}\tilde{x}_{ia} + L_{iad}y_i + \tilde{G}_{ia}\omega_i, \\ \dot{\tilde{x}}_{id} = A_d\tilde{x}_{id} + B_d(\tilde{u}_i + E_{ida}\tilde{x}_{ia} + E_{idd}\tilde{x}_{id}) + \tilde{G}_{id}\omega_i, \\ y_i = C_d\tilde{x}_{id}, \end{cases} \tag{19.11}$$

for $i \in \{1, \dots, N\}$, where $\tilde{x}_{ia} \in \mathbb{R}^{\tilde{n}_i - p\rho}$ represents the finite zero structure, $\tilde{x}_{id} \in \mathbb{R}^{p\rho}$ represents the infinite zero structure, $\tilde{u}_i$, $y_i \in \mathbb{R}^p$, $\omega_i \in \mathbb{R}^{m_{\omega_i}}$, and

$$A_d = \begin{pmatrix} 0 & I \\ 0 & 0 \end{pmatrix}, \qquad B_d = \begin{pmatrix} 0 \\ I \end{pmatrix}, \qquad C_d = \begin{pmatrix} I & 0 \end{pmatrix}.$$

Then, for each cascade system (19.11), the observer-based protocol (15.18) from Protocol Design 15.4 can be used, which is rewritten in the following.

Protocol design 19.2 *Following Protocol Design 15.4, the observer-based controller is designed as*

$$\begin{aligned} \dot{\hat{x}}_{id} &= A_d \hat{x}_{id} + K_\varepsilon(\zeta_i + \psi_i - C_d \hat{x}_{id}), \\ u_i &= F_{\delta\varepsilon} \hat{x}_{id}. \end{aligned} \tag{19.12}$$

The main result is stated in the following theorem.

Theorem 19.3 *Consider a MAS described by (19.1) and (19.2), and any reference trajectories given by (19.3). Let a root agent set π, a positive integer N, and $\alpha, \beta > 0$ be given and hence a set of network graphs $\mathbb{G}^N_{\alpha,\beta,\pi}$ be defined.*

Let Assumption 15.1 hold. Moreover, assume the agents are minimum-phase and right-invertible. Then the $\mathcal{H}_\infty$ almost regulated output synchronization problem defined in Problem 19.2 is solvable. In particular, for any given real number $\gamma > 0$, there exists a δ^* such that for any $\delta \in (0, \delta^*]$ there exists an ε^* such that for any $\varepsilon \in (0, \varepsilon^*]$, the combination of pre-compensator (19.7), (19.8), and (19.9) with protocol (19.12) solves the $\mathcal{H}_\infty$ almost regulated output synchronization problem for any graph $\mathcal{G} \in \mathbb{G}^N_{\alpha,\beta,\pi}$.

Lemma 19.4 *Consider the matrix*

$$\tilde{A}_\delta = I_N \otimes \begin{pmatrix} A_d & 0 \\ 0 & A_d - KC_d \end{pmatrix} + S \otimes \begin{pmatrix} B_d F_\delta & -B_d F_\delta \\ B_d F_\delta & -B_d F_\delta \end{pmatrix}. \tag{19.13}$$

For any small enough δ, the matrix $\tilde{A}_\delta$ is asymptotically stable for any upper triangular matrix S whose eigenvalues satisfy $\operatorname{Re} \lambda_i > \beta$ *for all $i = 1, \dots, N-1$ and $\|S\| < \alpha$. Moreover, there exists a $\tilde{P}_\delta > 0$, P_a and a small enough $\nu > 0$ such that*

$$P_a A_a + A_a' P_a = -\nu P_a - 3I. \tag{19.14}$$

and

$$\tilde{A}_\delta \tilde{P}_\delta + \tilde{P}_\delta \tilde{A}_\delta' \leq -\nu \tilde{P}_\delta - 5I \tag{19.15}$$

is satisfied for all possible upper triangular matrices S

Proof First note that if ν is small enough, we know that $A_a + \frac{\nu}{2}I$ is asymptotically stable and hence there exists a $P_a > 0$ satisfying (19.14).

For the existence of $\tilde{P}_\delta$ and the stability of $\tilde{A}_\delta$, we rely on techniques developed earlier in Theorem 5.9. If we define

$$\bar{A}_{\delta,i} = \begin{pmatrix} A_d + \lambda_i B_d F_\delta & -\lambda_i B_d F_\delta \\ \lambda_i B_d F_\delta & A_d - K C_d - \lambda_i B_d F_\delta \end{pmatrix}$$

and

$$\bar{B} = \begin{pmatrix} B_d F_\delta & -B_d F_\delta \\ B_d F_\delta & -B_d F_\delta \end{pmatrix},$$

then

$$\tilde{A}_\delta = \begin{pmatrix} \bar{A}_{\delta,1} & s_{1,2}\bar{B} & \cdots & s_{1,N}\bar{B} \\ 0 & \bar{A}_{\delta,2} & \ddots & \vdots \\ \vdots & \ddots & \ddots & s_{N-1,N}\bar{B} \\ 0 & \cdots & 0 & \bar{A}_{\delta,N} \end{pmatrix},$$

where $\lambda_1, \ldots, \lambda_N$ are the eigenvalues (i.e., the diagonal elements) of S and $|s_{i,j}| < \alpha$ are elements of the matrix S. Define

$$\bar{P}_\delta = \rho \begin{pmatrix} P_\delta & 0 \\ 0 & \sqrt{\|P_\delta\|}P \end{pmatrix},$$

where $P_\delta > 0$ is the solution of the Riccati equation (15.16) and $P > 0$ is uniquely defined by the Lyapunov equation

$$P(A_d - KC_d) + (A_d - KC_d)'P = -6I.$$

In the above we choose ρ such that $\rho\delta > 1$ and $\rho\sqrt{\|P_\delta\|} > 2$. As shown in Theorem 5.9, we then have

$$\bar{A}_{\delta,i}'\bar{P}_\delta + \bar{P}_\delta \bar{A}_{\delta,i} \leq -\rho \begin{pmatrix} \delta I & 0 \\ 0 & \frac{1}{2}\sqrt{\|P_\delta\|}I \end{pmatrix} \leq -I.$$

The remaining proof follows that of Lemma 15.9. ∎

Proof of Theorem 19.3 Note that

$$\tilde{x}_i = \begin{pmatrix} \tilde{x}_{ia} \\ \tilde{x}_{id} \end{pmatrix}$$

and that (19.10) is a shorthand notation for (19.11). For each $i \in \{1, \ldots, N\}$, let $\bar{x}_i = \tilde{x}_i - \Pi_i x_0$, where Π_i is defined to satisfy $\Pi_i A_0 = \tilde{A}_i \Pi_i$, $\tilde{C}_i \Pi_i = C_0$. Then

$$\dot{\bar{x}}_i = \tilde{A}_i \tilde{x}_i - \Pi_i A_0 x_0 + \tilde{B}_i \tilde{u}_i + \tilde{G}_i \omega_i = \tilde{A}_i \bar{x}_i + \tilde{B}_i \tilde{u}_i + \tilde{G}_i \omega_i \tag{19.16}$$

and

$$e_i = y_i - y_r = \tilde{C}_i \tilde{x}_i - C_0 x_0 = \tilde{C}_i \tilde{x}_i - \tilde{C}_i \Pi_i x_0 = \tilde{C}_i \bar{x}_i.$$

Since the dynamics of the $\bar{x}_i$ system with output e_i is governed by the same dynamics as the dynamics of agent i, we can present $\bar{x}_i$ in the same form as (19.11), with

$$\bar{x}_i = \begin{pmatrix} \bar{x}_{ia} \\ \bar{x}_{id} \end{pmatrix}$$

where

$$\begin{aligned} \dot{\bar{x}}_{ia} &= A_{ia} \bar{x}_{ia} + L_{iad} e_i + \tilde{G}_{ia} \omega_i, \\ \dot{\bar{x}}_{id} &= A_d \bar{x}_{id} + B_d(\tilde{u}_i + E_{ida} \bar{x}_{ia} + E_{idd} \bar{x}_{id}) + \tilde{G}_{id} \omega_i, \\ e_i &= C_d \bar{x}_{id}. \end{aligned}$$

Let a unitary $\bar{U}$ be defined such that $\bar{U}^{-1} \bar{L} \bar{U} = \bar{S}$, where $\bar{S}$ is the Schur form of the matrix $\bar{L}$. Then, following exactly the proof of Theorem 15.6, we obtain

$$\begin{aligned} \dot{\eta}_a &= A_a \eta_a + \tilde{W}_{ad} \eta_d + \tilde{G}_a \omega, \\ \varepsilon \dot{\eta}_d &= \tilde{A}_\delta \eta_d + \tilde{W}^\varepsilon_{da} \eta_a + \tilde{W}^\varepsilon_{dd} \eta_d + \varepsilon \tilde{G}^\varepsilon_d \omega, \end{aligned} \tag{19.17}$$

where $\tilde{A}_\delta$ is defined by (19.13) for $S = \bar{S}$ and

$$\tilde{W}_{ad} = \begin{pmatrix} W_{ad} & 0 \end{pmatrix} N_d^{-1}, \qquad \tilde{G}^\varepsilon_d = N_d \begin{pmatrix} \bar{G}^\varepsilon_d \\ \bar{G}^\varepsilon_d \end{pmatrix},$$

$$\tilde{W}^\varepsilon_{da} = N_d \begin{pmatrix} W^\varepsilon_{da} \\ W^\varepsilon_{da} \end{pmatrix}, \qquad \tilde{W}^\varepsilon_{dd} = N_d \begin{pmatrix} W^\varepsilon_{dd} & 0 \\ W^\varepsilon_{dd} & 0 \end{pmatrix} N_d^{-1},$$

where $W_{ad}, W^\varepsilon_{da}, and W^\varepsilon_{dd}$ are given in the proof of Theorem 15.6 and

$$\tilde{G}_a = \text{blkdiag}\{\tilde{G}_{ia}\}, \qquad \bar{G}^\varepsilon_d = (\bar{S}\bar{U}^{-1} \otimes I_{p\rho}) G^\varepsilon_d,$$

where $G_d^\varepsilon = \text{blkdiag}\{G_{id}^\varepsilon\}$ with $G_{id}^\varepsilon = S_\varepsilon \tilde{G}_{id}$.

Next, we make use of Lemma 19.4 which yields an appropriate $\tilde{P}_\delta$, P_a, and ν.

Define $V_a = \varepsilon\eta_a' P_a \eta_a$ as a Lyapunov function for the dynamics of η_a in (19.17). Similarly, we define $V_d = \varepsilon\eta_d' \tilde{P}_\delta \eta_d$ as a Lyapunov function for the dynamics of η_d in (19.17). It is easy to find that V_d also has discontinuous jumps when the network changes. The derivative of V_a is bounded by

$$\begin{aligned}\dot{V}_a &= -\nu V_a - 3\varepsilon\|\eta_a\|^2 + 2\varepsilon\,\text{Re}(\eta_a' P_a \tilde{W}_{ad}\eta_d) + 2\varepsilon\,\text{Re}(\eta_a' P_a \tilde{G}_a \omega) \\ &\leq -2\varepsilon\|\eta_a\|^2 + 2\varepsilon r_1^2\|\eta_d\|^2 + 2\varepsilon r_2^2\|\omega\|^2, \end{aligned} \tag{19.18}$$

where r_1 and r_2 are such that

$$2\,\text{Re}(\eta_a' P_a \tilde{W}_{ad,t}\eta_d) \leq 2r_1\|\eta_a\|\|\eta_d\| \leq \tfrac{1}{2}\|\eta_a\|^2 + 2r_1^2\|\eta_d\|^2,$$

$$2\,\text{Re}(\eta_a' P_a \tilde{G}_a w) \leq 2r_2\|\eta_a\|\|\omega\| \leq \tfrac{1}{2}\|\eta_a\|^2 + 2r_2^2\|\omega\|^2.$$

Note that we can choose r_1 and r_2 independent of the network graph but only depending on our bounds on the matrix $\bar{L}$.

Next, the derivative of V_d is bounded by

$$\begin{aligned}\dot{V}_d &= -\nu\varepsilon^{-1} V_d - 5\|\eta_d\|^2 + 2\,\text{Re}(\eta_d' \tilde{P}_\delta \tilde{W}_{da}^\varepsilon \eta_a) \\ &\qquad + 2\,\text{Re}(\eta_d' \tilde{P}_\delta \tilde{W}_{dd}^\varepsilon \eta_d) + 2\varepsilon\,\text{Re}(\eta_d' \tilde{P}_\delta \tilde{G}_d^\varepsilon \omega) \\ &\leq -2\|\eta_d\| + \varepsilon^2 r_4^2\|\eta_a\|^2 + \varepsilon^2 r_3^2\|\omega\|^2, \end{aligned} \tag{19.19}$$

where

$$2\,\text{Re}(\eta_d' \tilde{P}_\delta \tilde{W}_{dd}^\varepsilon \eta_d) \leq \|\eta_d\|^2$$

for a small ε, and

$$\begin{aligned}2\varepsilon\,\text{Re}(\eta_d' \tilde{P}_\delta \tilde{G}_d^\varepsilon \omega) &\leq 2\varepsilon r_3\|\eta_d\|\|\omega\| \leq \|\eta_d\|^2 + \varepsilon^2 r_3^2\|\omega\|^2, \\ 2\,\text{Re}(\eta_d' \tilde{P}_\delta \tilde{W}_{da}^\varepsilon \eta_a) &\leq 2\varepsilon r_4\|\eta_a\|\|\eta_d\| \leq \varepsilon^2 r_4^2\|\eta_a\|^2 + \|\eta_d\|^2,\end{aligned}$$

provided r_3 and r_4 are such that we have $\varepsilon r_3 \geq \varepsilon\|\tilde{P}_\delta \tilde{G}_d^\varepsilon\|$, $\varepsilon r_4 \geq \|\tilde{P}_\delta \tilde{W}_{da}^\varepsilon\|$.

Let $V = V_a + V_d$. Then, we get

$$\begin{aligned} V &\leq -\varepsilon\|\eta_a\|^2 - \|\eta_d\|^2 + 2\varepsilon r_2^2\|\omega\|^2 + \varepsilon^2 r_3^2\|\omega\|^2 \\ &\leq -\|\eta\|^2 + \varepsilon_1^2 r_5^2\|\omega\|^2, \end{aligned}$$

where $\eta = \text{col}\{\eta_a; \eta_d\}$, $\varepsilon_1^2 = \varepsilon$, and $r_5^2 = \max(2r_2^2, \varepsilon r_3^2)$. So

$$V + \|\eta\|^2 - \varepsilon_1^2 r_5^2 \|\omega\|^2 \le 0. \tag{19.20}$$

From the Kalman–Yakubovich–Popov Lemma (see e.g. [195]), we conclude that (19.20) implies that $\|T_{\omega\eta}\|_\infty \le \varepsilon_1 r_5$. Next, we show that this implies that $\|T_{\omega\mathbf{e}_0}\|_\infty \le \varepsilon_1 r_5'$ for a certain r_5'.

Following the proof of Theorem 15.6, we find that

$$\begin{aligned}\mathbf{e}_0 &= (I_N \otimes C_d)(I_N \otimes S_\varepsilon^{-1})(\bar{U}\bar{S}^{-1} \otimes I_{p\rho}) \left(I_{Np\rho}\ 0\right) N_d^{-1} \left(0\ I_{2Np\rho}\right) \eta \\ &= \Theta\eta,\end{aligned}$$

for a suitably chosen matrix Θ. The norm of Θ is obviously bounded for $C_d S_\varepsilon^{-1} = C_d$. Thus, we obtain

$$\|T_{\omega\mathbf{e}_0}\|_\infty \le \varepsilon_1 \|\Theta\| r_5,$$

which can be made arbitrarily small by an appropriate choice of ε. ■

19.4.2 *Disturbances with and Without Known Frequencies*

We consider here the $\mathcal{H}_\infty$ almost regulated output synchronization problem for a MAS in the presence of two classes of disturbances: disturbances with known frequencies and disturbances with bounded power.

The agent model (19.1) is modified as follows:

$$\begin{aligned}\dot{x}_i &= A_i x_i + B_i u_i + G_i \omega_i + H_{x,i} d_i, \\ y_i &= C_i x_i + H_{y,i} d_i,\end{aligned} \qquad (i = 1, \ldots, N) \tag{19.21}$$

where x_i, u_i and y_i are the same as those in (19.1) where ω_i represents disturbances with bounded power, while $d_i \in \mathbb{R}^{d_i}$ represents external disturbances with known frequencies, which can be generated by the following exosystem:

$$\begin{aligned}\dot{x}_{id} &= S_i x_{id}, \\ d_i &= R_i x_{id},\end{aligned} \qquad x_{id}(0) = x_{id0} \tag{19.22}$$

where $x_{id} \in \mathbb{R}^{n_{d_i}}$ and the initial condition x_{id0} can be arbitrarily chosen.

Then the $\mathcal{H}_\infty$ almost regulated output synchronization problem for a MAS in the presence of disturbance and disturbances with known frequencies is formulated as follows.

Problem 19.5 Consider a MAS described by (19.21), (19.22), and (19.2). For a given root set π, a positive integer N, and $\alpha, \beta > 0$ that define a set of network graphs $\mathbb{G}_{\alpha,\beta,\pi}^N$ and any reference trajectories given by (19.3), the $\mathcal{H}_\infty$ *almost* *regulated output synchronization* problem is to find, if possible, a linear time-invariant dynamic protocol of the form (19.5) such that, for any given real number $\gamma > 0$ and for any graph $\mathcal{G} \in \mathbb{G}^N_{\alpha,\beta,\pi}$, (19.4) is satisfied for all initial conditions for the agents, protocols, and exosystems (19.3) and (19.22) when $\omega_i = 0$, while the closed-loop transfer function $T_{\omega e_0}$ satisfies (19.6).

Note that we have two exosystems (19.3) and (19.22) which generate the reference signal y_0 and the disturbance d_i, respectively. We can unify both in to one exosystem:

$$\begin{aligned} x_{ie} &= S_{ie}x_{ie}, & x_{ie}(0) &= x_{ie0} \\ d_i &= R_{ie}x_{ie}, \\ y_0 &= R_{re}x_{ie}, \end{aligned} \tag{19.23}$$

where

$$S_{ie} = \begin{pmatrix} S_i & 0 \\ 0 & A_0 \end{pmatrix}, \quad R_{ie} = \begin{pmatrix} R_i & 0 \end{pmatrix}, \quad R_{re} = \begin{pmatrix} 0 & C_0 \end{pmatrix}. \tag{19.24}$$

Note that

$$x_{ie0} = \begin{pmatrix} x_{id0} \\ x_{00} \end{pmatrix},$$

and therefore the second part of the initial condition has to be the same for each agent while the first part might be different for each agent.

The protocol design follows Protocol Design 18.1, which is written again as follows.

Protocol design 19.3 *We first design a pre-compensator similar to Protocol Design 15.1*

$$\begin{aligned} \dot{p}_{i,1} &= A_{ip,1}p_{i,1} + B_{ip,1}u_i^1, \\ u_i &= C_{ip,1}p_{i,1} + D_{ip,1}u_i^1, \end{aligned} \tag{19.25}$$

such that the interconnected system is of the form

$$\begin{aligned} \dot{x}_i &= \hat{A}_i\hat{x}_i + \hat{B}_i u_i^1 + \hat{G}_i\omega_i + \hat{H}_{x,i}d_i, \\ y_i &= \hat{C}_i x_i + H_{y,i}d_i, \end{aligned} \tag{19.26}$$

with the subsystem with input u_i^1 to y_i uniform rank and invertible for $i = 1, \ldots, N$. Next, we design pre-compensators such that the interconnected

(continued)

Protocol design 19.3 (continued)
system contains all modes of the augmented exosystem. This follows Protocol Design 15.2. However, we use the augmented exosystem (18.8) instead of the exosystem (18.4). This implies that the regulator equation (15.10) should be first modified as

$$\begin{aligned} \hat{A}_i\tilde{\Pi}_i + \hat{B}_i\Gamma_i + \hat{H}_{x,i}R_{ie} &= \tilde{\Pi}_i S_{ie}, \\ \hat{C}_i\tilde{\Pi}_i + H_{y,i}R_{ie} &= R_{re}. \end{aligned} \tag{19.27}$$

If (Γ_i, S_{ie}) is not observable, we can construct the observable subsystem $(\Gamma_{i,o}, S_{ie,o})$; otherwise simply set $\Gamma_{i,o} = \Gamma_i$ and $S_{i,o} = S_{ie}$. Then the second pre-compensator for each agent i is designed as

$$\begin{aligned} \dot{p}_{i,2} &= S_{ie,o}p_{i,2} + B_{i,2}u_i^2, \\ u_i^1 &= \Gamma_{i,o}p_{i,2}, \end{aligned} \tag{19.28}$$

for $i = 1, \ldots, N$, where $p_{i,2} \in \mathbb{R}^{n_{pi}}$ and $B_{i,2}$ and chosen to guarantee that no invariant zeros are introduced by the pre-compensator. Moreover, an appropriate choice of $B_{i,2}$ can guarantee that the triple $(S_{ie,o}, B_{i,2}, \Gamma_{i,o})$ is invertible, uniform rank, and minimum-phase (see [60]).

If the agent has dynamics in common with the exosystem, then we can use the technique outlined in Appendix 15.5 to remove these common modes after which the above design is applicable.

The final pre-compensator is designed according to Protocol Design 15.3 and we obtain

$$\begin{aligned} \dot{p}_{i,3} &= K_i p_{i,3} + L_i\tilde{u}_i, \\ u_i^2 &= M_i p_{i,3}, \end{aligned} \tag{19.29}$$

such that the agents interconnected with the three pre-compensators are all uniform rank and invertible with the same uniform rank for all agents.

Then, the cascade system of agent (19.21) and pre-compensator (19.25), (19.28), and (19.29) can be represented in the form

$$\begin{cases} \dot{\tilde{x}}_i = \tilde{A}_i\tilde{x}_i + \tilde{B}_i\tilde{u}_i + \tilde{G}_i\omega_i + \tilde{H}_{x,i}d_i, \\ y_i = \tilde{C}_i\tilde{x}_i + \tilde{H}_{y,i}d_i, \end{cases} \tag{19.30}$$

where $\tilde{x}_i \in \mathbb{R}^{n_i}$, $\tilde{u}_i \in \mathbb{R}^p$, and $y_i \in \mathbb{R}^p$ are states, inputs, and outputs of the cascade system. Moreover, we have:

- $(\tilde{A}_i, \tilde{C}_i)$ contains the joint exosystem (19.23), i.e., there exists a matrix Π_i such that $\Pi_i S_{ie} = \tilde{A}_i \Pi_i + \tilde{H}_{x,i} R_{ie}$ and $\tilde{C}_i \Pi_i + \tilde{H}_{y,i} R_{ie} = R_{re}$,
- $(\tilde{A}_i, \tilde{B}_i, \tilde{C}_i)$ is of uniform rank $\rho \geq 1$.

The observer-based controller (19.12) from Protocol Design 19.2 can be designed for each cascade system (19.30).

The main result is stated in the following theorem.

Theorem 19.6 *Consider a MAS described by* (19.21), (19.22), *and* (19.2), *and reference system* (19.3). *Let a root set π, a positive integer N, $\alpha, \beta > 0$ be given, and hence a set of network graphs $\mathbb{G}^N_{\alpha,\beta,\pi}$ be defined.*

Assume the agents are minimum-phase and right-invertible. Then the $\mathcal{H}_\infty$ almost regulated output synchronization problem defined in Problem 19.5 *is solvable. In particular, for any given real number $\gamma > 0$, there exists a δ^* such that for any $\delta \in (0, \delta^*]$ there exists an ε^* such that for any $\varepsilon \in (0, \varepsilon^*]$, the combination of pre-compensators* (19.25), (19.28), *and* (19.29) *with protocol* (19.12) *solves the $\mathcal{H}_\infty$ almost regulated output synchronization problem as defined in Problem* 19.5 *for any graph $\mathcal{G} \in \mathbb{G}^N_{\alpha,\beta,\pi}$.*

Proof Let $\bar{x}_i = \tilde{x}_i - \Pi_i x_{ie}$ for $i \in \{1, \ldots, N\}$, where Π_i satisfies $\Pi_i S_{ie} = \tilde{A}_i \Pi_i + \tilde{H}_{x,i} R_{ie}$ and $\tilde{C}_i \Pi_i + \tilde{H}_{y,i} R_{ie} = R_{re}$. Then

$$\begin{aligned}\dot{\bar{x}}_i &= \tilde{A}_i \tilde{x}_i - \Pi_i S_{ie} x_{ie} + \tilde{B}_i \tilde{u}_i + \tilde{G}_i \omega_i + \tilde{H}_{x,i} d_i \\ &= \tilde{A}_i \bar{x}_i + \tilde{B}_i \tilde{u}_i + \tilde{G}_i \omega_i\end{aligned}$$

and

$$e_i = y_i - y_0 = \tilde{C}_i \tilde{x}_i + \tilde{H}_{y,i} R_{ie} x_{ie} - R_{re} x_{ie} = \tilde{C}_i \tilde{x}_i - \tilde{C}_i \Pi_i x_r = \tilde{C}_i \bar{x}_i.$$

The dynamic $\bar{x}_i$ is exactly the same as (19.16). Then, the remaining proof follows that of Theorem 19.3. ∎

19.4.3 Time-Varying Graphs

In this section, we show that the protocol design for a fixed graph can be applied directly to a time-varying graph. The time-varying communication network among agents is exactly the same as that in Chap. 11. The network information provided to each agent is modified as

$$\zeta_i = \sum_{j=1}^{N} a_{ij}(t)(y_i - y_j) = \sum_{j=1}^{N} \ell_{ij}(t) y_j. \tag{19.31}$$

The target is also to regulate the outputs of all agents to an, a priori given, reference trajectory generated by an exosystem (19.3). In the same way as before, a subset of agents, denoted by π, has access to their output relative to the reference trajectory y_0 generated by the exosystem (19.3). To be specific, each agent $i \in \{1, \ldots, N\}$ has access to the quantity

$$\psi_i = \iota_i(y_i - y_0), \quad \iota_i = \begin{cases} 1, & i \in \pi, \\ 0, & i \notin \pi. \end{cases}$$

We consider the class of network graphs $\mathbb{G}^N_{\alpha,\beta,\pi}$ as defined in Definition 15.4 and a set of time-varying graphs $\mathbb{G}^{\tau,N}_{\alpha,\beta,\pi}$ as defined in Definition 17.1.

Because of the time-varying network, the $\mathcal{H}_\infty$ norm of the closed-loop system does not exist. Instead, the measurement on the regulated output synchronization error $\mathbf{e}_0$ is used for the time-varying case. The problem formulation can be modified as follows.

Problem 19.7 Consider a MAS described by (19.1) and (19.2). For a given root set π, a positive integer N, and $\alpha, \beta, \tau > 0$ that define a set of time-varying network graphs $\mathbb{G}^{\tau,N}_{\alpha,\beta,\pi}$ and any reference trajectories given by (19.3), the *almost regulated output synchronization* problem is to find, if possible, a linear time-invariant dynamic protocol of the form (19.5), such that, for any given real number $\gamma > 0$, for any disturbance bound κ, and for any graph $\mathcal{G}_t \in \mathbb{G}^{\tau,N}_{\alpha,\beta,\pi}$ and for all the initial conditions of agents and reference system, the regulated output synchronization error satisfies the following:

- For all $\omega_i \in \Gamma^\infty_\kappa$, $i = 1, \ldots, N$,

$$\limsup_{t\to\infty} \|\mathbf{e}_0(t)\| < \gamma. \tag{19.32}$$

- For all $\omega_i \in \Gamma^{\text{rms}}_\kappa$, $i = 1, \ldots, N$,

$$\|\mathbf{e}_0\|_{\text{rms}} < \gamma. \tag{19.33}$$

The protocol design generated for a fixed graph works exactly for a time-varying graph. The result is stated in the following theorem.

Theorem 19.8 *Consider a MAS described by* (19.1) *and* (19.2)*, and any reference trajectories given by* (19.3)*. Let a root agent set π, a positive integer N, and $\alpha, \beta, \tau > 0$ be given and hence a set of network graphs $\mathbb{G}^{\tau,N}_{\alpha,\beta,\pi}$ be defined.*

Assume the agents are minimum-phase and right-invertible. Then the almost regulated output synchronization problem defined in Problem 19.7 is solvable. In particular, for any given real number $\gamma > 0$, for any disturbance bound κ, there exists a δ^* such that for any $\delta \in (0, \delta^*]$ there exists an ε^* such that for any $\varepsilon \in (0, \varepsilon^*]$, the combination of pre-compensator (19.7), (19.8), *and* (19.9) *with* *protocol* (19.12) *solves the almost regulated output synchronization problem as defined in Problem 19.7 for any graph* $\mathcal{G} \in \mathbb{G}^N_{\alpha,\beta,\pi}$.

Proof For each $t > 0$, we define a unitary $\bar{U}_t$ such that $\bar{U}_t^{-1}\bar{L}_t\bar{U}_t = \bar{S}_t$, where $\bar{S}_t$ is the Schur form of the matrix $\bar{L}_t$. As in the proof of Theorem 19.3, we can surely get

$$\begin{aligned} \dot{\eta}_a &= A_a\eta_a + \tilde{W}_{ad,t}\eta_d + \tilde{G}_a w, \\ \varepsilon\dot{\eta}_d &= \tilde{A}_{\delta,t}\eta_d + \tilde{W}^{\varepsilon}_{da,t}\eta_a + \tilde{W}^{\varepsilon}_{dd,t}\eta_d + \varepsilon\tilde{G}^{\varepsilon}_{d,t}w, \end{aligned} \tag{19.34}$$

where

$$\tilde{A}_{\delta,t} = I_N \otimes \begin{pmatrix} A_d & 0 \\ 0 & A_d - KC_d \end{pmatrix} + \bar{S}_t \otimes \begin{pmatrix} B_dF_\delta & -B_dF_\delta \\ B_dF_\delta & -B_dF_\delta \end{pmatrix}, \tag{19.35}$$

and

$$\tilde{W}_{ad,t} = \begin{pmatrix} W_{ad,t} & 0 \end{pmatrix} N_d^{-1}, \qquad \tilde{G}^{\varepsilon}_{d,t} = N_d \begin{pmatrix} \bar{G}^{\varepsilon}_{d,t} \\ \bar{G}^{\varepsilon}_{d,t} \end{pmatrix},$$

$$\tilde{W}^{\varepsilon}_{da,t} = N_d \begin{pmatrix} W^{\varepsilon}_{da,t} \\ W^{\varepsilon}_{da,t} \end{pmatrix}, \qquad \tilde{W}^{\varepsilon}_{dd,t} = N_d \begin{pmatrix} W^{\varepsilon}_{dd,t} & 0 \\ W^{\varepsilon}_{dd,t} & 0 \end{pmatrix} N_d^{-1}.$$

Moreover, $\tilde{A}_{\delta,t}$, $\tilde{W}_{ad,t}$, $\tilde{W}^{\varepsilon}_{da,t}$, $\tilde{W}^{\varepsilon}_{dd,t}$, and $\tilde{G}^{\varepsilon}_{d,t}$ are functions of time t, because the Laplacian matrix corresponding to the switching graph is a function of time t.

We know from (19.34) that η_d exhibits discontinuous jumps when the network graph switches, while η_a is continuous. Next, we make use of Lemma 19.4 which yields an appropriate $\tilde{P}_\delta$, P_a, and ν.

Define $V_a = \varepsilon^2\eta_a' P_a\eta_a$ as a Lyapunov function for the dynamics of η_a in (19.34). Similarly, we define $V_d = \varepsilon\eta_d'\tilde{P}_\delta\eta_d$ as a Lyapunov function for the dynamics of η_d in (19.34). It is easy to find that V_d also has discontinuous jumps when the network changes. The derivative of V_a is bounded by

$$\begin{aligned} \dot{V}_a &= -\nu V_a - \varepsilon^2\|\eta_a\|^2 + 2\varepsilon^2\,\mathrm{Re}(\eta_a' P_a\tilde{W}_{ad,t}\eta_d) \\ &\quad + 2\varepsilon^2\,\mathrm{Re}(\eta_a' P_a\tilde{G}_a\omega) \\ &\leq -\nu V_a + \varepsilon c_3 V_d + 2\varepsilon^2 r_5^2\|\omega\|^2, \end{aligned} \tag{19.36}$$

where r_5 and c_3 are such that

$$\begin{aligned} 2\,\mathrm{Re}(\eta_a' P_a\tilde{W}_{ad,t}\eta_d) &\leq 2r_4\|\eta_a\|\|\eta_d\| \\ &\leq \tfrac{1}{2}\|\eta_a\|^2 + 2r_4^2\|\eta_d\|^2 \leq \tfrac{1}{2}\|\eta_a\|^2 + \varepsilon^{-1}c_3 V_d, \\ 2\,\mathrm{Re}(\eta_a' P_a\tilde{G}_a\omega) &\leq 2r_5\|\eta_a\|\|\omega\| \leq \tfrac{1}{2}\|\eta_a\|^2 + 2r_5^2\|\omega\|^2. \end{aligned}$$

Note that we can choose r_4, r_5, and c_3 independent of the network graph but only depending on our bounds on the eigenvalues and condition number of our expanded Laplacian $\tilde{L}(t)$.

Next, the derivative of V_d is bounded by

$$\begin{aligned}\dot{V}_d &= -\nu\varepsilon^{-1}V_d - 4\|\eta_d\|^2 + 2\operatorname{Re}(\eta_d'\tilde{P}_\delta\tilde{W}_{da,t}^{\varepsilon}\eta_a) \\ &\qquad + 2\operatorname{Re}(\eta_d'\tilde{P}_\delta\tilde{W}_{dd,t}^{\varepsilon}\eta_d) + 2\varepsilon\operatorname{Re}(\eta_d'\tilde{P}_\delta\tilde{G}_{d,t}^{\varepsilon}\omega) \\ &\leq c_2V_a - (\nu\varepsilon^{-1} + \nu - \varepsilon^2\frac{c_2c_3}{\nu})V_d + \varepsilon^2r_3^2\|\omega\|^2,\end{aligned} \tag{19.37}$$

where

$$2\operatorname{Re}(\eta_d'\tilde{P}_\delta\tilde{W}_{dd,t}^{\varepsilon}\eta_d) \leq \|\eta_d\|^2$$

for a small ε, and

$$\begin{aligned}2\varepsilon\operatorname{Re}(\eta_d'P_\delta\tilde{G}_{d,t}^{\varepsilon}\omega) &\leq 2\varepsilon r_3\|\eta_d\|\|\omega\| \leq \|\eta_d\|^2 + \varepsilon^2r_3^2\|\omega\|^2, \\ 2\operatorname{Re}(\eta_d'P_\delta\tilde{W}_{da,t}^{\varepsilon}\eta_a) &\leq 2\varepsilon r_1\|\eta_a\|\|\eta_d\| \leq \varepsilon^2r_1^2\|\eta_a\|^2 + \|\eta_d\|^2 \\ &\leq c_2V_a + \|\eta_d\|^2,\end{aligned}$$

provided r_3 and r_1 are such that we have $\varepsilon r_3 \geq \varepsilon\|P_\delta\tilde{G}_{d,t}^{\varepsilon}\|$, $\varepsilon r_1 \geq \|P_\delta\tilde{W}_{da,t}^{\varepsilon}\|$, and c_2 sufficiently large. Then, we get

$$\begin{pmatrix}\dot{V}_a\\ \dot{V}_d\end{pmatrix} \leq A_e\begin{pmatrix}V_a\\ V_d\end{pmatrix} + \begin{pmatrix}2\varepsilon^2r_5^2\|\omega\|^2\\ \varepsilon^2r_3^2\|\omega\|^2\end{pmatrix},$$

where

$$A_e = \begin{pmatrix}-\nu & \varepsilon c_3\\ c_2 & -\varepsilon^{-1}\nu - \nu + \varepsilon^2\frac{c_2c_3}{\nu}\end{pmatrix}.$$

Note that the inequality here is component-wise. We find by integration that

$$\begin{pmatrix}V_a\\ V_d\end{pmatrix}(t_k^-) \leq e^{A_e(t_k-t_{k-1})}\begin{pmatrix}V_a\\ V_d\end{pmatrix}(t_{k-1}^+) + \int_{t_{k-1}}^{t_k} e^{A_e(t_k-s)}\begin{pmatrix}2\varepsilon^2r_5^2\|\omega\|^2\\ \varepsilon^2r_3^2\|\omega\|^2\end{pmatrix}ds \tag{19.38}$$

component-wise. We have a potential jump at time t_{k-1} in V_d. However, there exists an m such that $V_d(t_{k-1}^+) \leq mV_d(t_{k-1}^-)$, while V_a is continuous. Using the explicit expression for e^{A_et} and the fact that $t_k - t_{k-1} > \tau$, we find

$$\begin{pmatrix}1 & 1\end{pmatrix}e^{A_e(t_k-t_{k-1})}\begin{pmatrix}V_a\\ V_d\end{pmatrix}(t_{k-1}^+) \leq e^{\lambda_3(t_k-t_{k-1})}\left[V_a(t_{k-1}^-) + V_d(t_{k-1}^-)\right],$$

where $\lambda_3 = -\nu/2$.

When $\omega_i \in \Gamma_\kappa^\infty$, it can be easily verified that

$$(1\ 1)\int_{t_{k-1}}^{t_k} e^{A_e(t_k-s)}\begin{pmatrix}2\varepsilon^2 r_5\|\omega\|^2\\ \varepsilon^2 r_3^2\|\omega\|^2\end{pmatrix} ds \le r\varepsilon^2\|\omega\|_\infty^2,$$

where r is a sufficiently large constant. We find

$$\left[V_a(t_k^-)+V_d(t_k^-)\right] \le e^{\lambda_3(t_k-t_{k-1})}\left[V_a(t_{k-1}^-)+V_d(t_{k-1}^-)\right]+r\varepsilon^2\|\omega\|_\infty^2. \tag{19.39}$$

Combining these time intervals, we get

$$\left[V_a(t_k^-)+V_d(t_k^-)\right] \le e^{\lambda_3 t_k}\left[V_a(0)+V_d(0)\right]+\frac{r\varepsilon^2}{1-\mu}\|\omega\|_\infty^2,$$

where $\mu < 1$ is such that

$$e^{\lambda_3(t_k-t_{k-1})} \le e^{\lambda_3\tau} \le \mu$$

for all k. Assume $t_{k+1} > t > t_k$. Since we do not necessarily have $t - t_k > \tau$, we use the bound

$$(1\ 1)\, e^{A_e(t-t_k)}\begin{pmatrix}V_a\\ V_d\end{pmatrix}(t_k^+) \le 2me^{\lambda_3(t-t_k)}(V_a+V_d)(t_k^-),$$

where the factor m is due to the potential discontinuous jump. Combining all together, we get

$$V(t) \le 2me^{\lambda_3 t}V(0)+(2m+1)\frac{r\varepsilon^2}{1-\mu}\|\omega\|_\infty^2,$$

where $V = V_a + V_d$. Hence,

$$\limsup_{t\to\infty}\|\eta_d(t)\|^2 \le (2m+1)\frac{r\varepsilon^2}{(1-\mu)\sigma_{\min}(P_\delta)}\kappa. \tag{19.40}$$

On the other hand, for $\omega \in \Gamma_\kappa^{\mathrm{rms}}$, we note that (19.38) implies that

$$\int_{t_{k-1}}^{t_k} V_d(s)ds \le r_6\varepsilon(V_a(t_{k-1}^+)+V_d(t_{k-1}^+))+r_7\varepsilon^2\int_{t_{k-1}}^{t_k}\|\omega(s)\|^2 ds \tag{19.41}$$

for some large enough r_6, r_7.

Similar to (19.39), we get

$$\left[V_a(t_k^-) + V_d(t_k^-)\right] \le e^{\lambda_3 \nu/2}\left[V_a(t_{k-1}^-) + V_d(t_{k-1}^-)\right] + r\varepsilon^2 \int_{t_{k-1}}^{t_k} \|\omega(s)\|^2 \mathrm{d}s.$$

We have

$$\frac{1}{T}\int_0^T \eta_d'(t)\eta_d(t)\,\mathrm{d}t \le \frac{1}{\varepsilon\sigma_{\min}(P_\delta)}\frac{1}{T}\int_0^T V_d(t)\,\mathrm{d}t. \tag{19.42}$$

Combining (19.41), (19.42), and (19.43) and taking the limit as $T \to \infty$, we find

$$\|\eta_d\|_{\text{rms}} \le \varepsilon r_8 \|\omega\|_{\text{rms}}. \tag{19.43}$$

Following the proof of Theorem 15.6, we find that

$$\begin{aligned}\mathbf{e}_0 &= (I_N \otimes C_d)(I_N \otimes S_\varepsilon^{-1})(\bar{U}_t \bar{S}_t^{-1} \otimes I_{p\rho})\begin{pmatrix} I_{Np\rho} & 0\end{pmatrix} N_d^{-1}\eta_d\\ &= \Theta_t \eta_d,\end{aligned}$$

for a suitably chosen matrix Θ_t. Although Θ_t is time-varying, it is uniformly bounded, because for graphs in $\mathbb{G}^N_{\alpha,\beta,\pi}$, the matrix $\bar{L}$ is bounded. Therefore, we have

$$\|\mathbf{e}_0(t)\| = \|\Theta_t \eta_d(t)\| \le \|\Theta_t\|\|\eta_d(t)\|.$$

Using this and (19.40), we can conclude that for $\omega \in \Gamma_\kappa^\infty$, we have (19.32) for any fixed $\gamma > 0$ provided we choose ε small enough. Similarly, using this and (19.43), we can conclude that for $\omega \in \Gamma_\kappa^{\text{rms}}$, we have (19.33) for any fixed $\gamma > 0$ provided we choose ε small enough. ∎

Chapter 20
H_2 Almost Regulated Output Synchronization for Heterogeneous Continuous-Time MAS

20.1 Introduction

In the previous two chapters, we considered regulated synchronization problems for heterogeneous continuous-time linear MAS subject to disturbances. In Chap. 18, we considered disturbances with known frequencies for which we could achieve *exact* regulated synchronization. In Chap. 19, we considered disturbances with bounded power and we showed that we could achieve *almost* regulated synchronization. In this chapter, we consider MAS that are affected by external colored stochastic disturbances. Under the same assumptions as in the previous chapters, we will show that we can again achieve *almost* regulated synchronization. We again use a low-and-high gain methodology to design protocols for each agent such that the $\mathcal{H}_2$ norm of the transfer function from the disturbances to the output synchronization errors can be squeezed arbitrarily small.

The write-up of this chapter is partially based on [183].

20.2 Multi-Agent Systems

Consider a MAS composed of N non-identical linear time-invariant continuous-time agents of the form,

$$\begin{aligned} \dot{x}_i &= A_i x_i + B_i u_i + G_i \breve{\omega}_i, \\ y_i &= C_i x_i, \end{aligned} \qquad (i = 1, \ldots, N) \tag{20.1}$$

where $x_i \in \mathbb{R}^{n_i}$, $u_i \in \mathbb{R}^{m_i}$, and $y_i \in \mathbb{R}^p$ are the state, input, and output of agent i. We make the following standard assumptions for the agent dynamics.

A. Saberi et al., *Synchronization of Multi-Agent Systems in the Presence of Disturbances and Delays*, Systems & Control: Foundations & Applications,
https://doi.org/10.1007/978-3-030-88148-1_20

Assumption 20.1 *For each agent $i \in \{1, \dots, N\}$, we assume that*

- *(A_i, B_i) is stabilizable;*
- *(A_i, C_i) is detectable.*

The external disturbance $\breve{\omega}_i \in \mathbb{R}^{\breve{\omega}_i}$ is the external colored stochastic noise. We assume that $\breve{\omega}_i$ can be (approximately) modeled by a linear model,

$$\begin{aligned} \mathrm{d}\breve{p}_i &= \breve{A}_{\omega i}\breve{p}_i \mathrm{d}t + \breve{G}_{\omega i}\mathrm{d}\omega_i, \qquad \breve{p}_i(0) = \breve{p}_{i0} \\ \breve{\omega}_i &= \breve{C}_{\omega i}\breve{p}_i. \end{aligned} \tag{20.2}$$

Combining (20.1) and (20.2), we get

$$\begin{aligned} \mathrm{d}\breve{x}_i &= \breve{A}_i\breve{x}_i \mathrm{d}t + \breve{B}_i u_i \mathrm{d}t + \breve{G}_i \mathrm{d}\omega_i, \qquad \breve{x}_i(0) = \breve{x}_{i0} \\ y_i &= \breve{C}_i\breve{x}_i, \end{aligned} \tag{20.3}$$

where $\breve{x}_i \in \mathbb{R}^{\breve{n}_i}$, and $\omega = \mathrm{col}\{\omega_i\}$ is a Wiener process (a Brownian motion) with mean 0 and rate Q_0, that is, $\mathrm{Cov}[\omega(t)] = tQ_0$ and the initial condition $\breve{x}_{i0}$ of (20.3) is a Gaussian random vector that is independent of ω_i. Here Q_0 is block-diagonal such that ω_i and ω_j are independent for any $i \neq j$. Its solution $\breve{x}_i$ is rigorously defined through Wiener integrals and is a Gauss-Markov process. See, for instance, [78].

The communication network among agents is exactly the same as that in Chap. 15 and provides each agent with the quantity,

$$\zeta_i = \sum_{j=1}^{N} a_{ij}(y_i - y_j) = \sum_{j=1}^{N} \ell_{ij} y_j. \tag{20.4}$$

20.3 Problem Formulation

In this section, we again pursue regulated output synchronization among agents' outputs, that is, the objective is to regulate the outputs of all agents asymptotically towards an, a priori specified, reference trajectory generated by an exosystem:

$$\begin{aligned} \dot{x}_0 &= A_0 x_0, \quad x_0(0) = x_{00}, \\ y_0 &= C_0 x_0. \end{aligned} \tag{20.5}$$

As in the previous chapters, a subset of agents, denoted by π, has access to the relative output $y_i - y_0$ of the exosystem compared to the output of the agent. To be specific, each agent $i \in \{1, \dots, N\}$ has access to the quantity

$$\psi_i = \iota_i(y_i - y_0), \quad \iota_i = \begin{cases} 1, & i \in \pi, \\ 0, & i \notin \pi. \end{cases} \tag{20.6}$$

We consider a set of graphs $\mathbb{G}^N_{\alpha,\beta,\pi}$ as defined in Definition 15.4.

Define $e_{i0} = y_i - y_0$ and

$$\mathbf{e}_0 = \begin{pmatrix} e_{i0} \\ \vdots \\ e_{N0} \end{pmatrix} \quad \text{and} \quad \omega = \begin{pmatrix} \omega_1 \\ \vdots \\ \omega_N \end{pmatrix}.$$

It is obvious that regulated output synchronization is achieved if $\lim_{t\to\infty} \mathbf{e}_0(t) = 0$. That is,

$$\lim_{t\to\infty} (y_i(t) - y_0(t)) = 0, \quad \forall i \in \{1, \dots, N\}. \tag{20.7}$$

We denote by $T_{\omega\mathbf{e}_0}$ the transfer matrix from ω to $\mathbf{e}_0$.

We formulate the $\mathcal{H}_2$ almost regulated output synchronization problem for a MAS in the presence of a stochastic disturbance as follows.

Problem 20.1 Consider a MAS described by (20.3), (20.4) and the exosystem (20.5). For a given root set π, a positive integer N, and $\alpha, \beta > 0$ that define a set of network graphs $\mathbb{G}^N_{\alpha,\beta,\pi}$, and any reference trajectories given by (20.5), the *$\mathcal{H}_2$ almost regulated output synchronization* problem is to find, if possible, a linear time-invariant dynamic protocol of the form,

$$\begin{aligned} \dot{\chi}_i &= A_{ci}\chi_i + B_{ci}\,\mathrm{col}\{\zeta_i, \psi_i\}, \\ u_i &= C_{ci}\chi_i, \end{aligned} \tag{20.8}$$

for $i = 1, \dots, N$ where $\chi_i \in \mathbb{R}^{N_{ci}}$, such that, for any given real number $\gamma > 0$, and for any graph $\mathcal{G} \in \mathbb{G}^N_{\alpha,\beta,\pi}$, (20.7) is satisfied, for all initial conditions of agents, protocols, and exosystem, in the absence of disturbances. Moreover, the closed-loop transfer function $T_{\omega\mathbf{e}_0}$ satisfies

$$\|T_{\omega\mathbf{e}_0}\|_2 \leq \gamma. \tag{20.9}$$

20.4 Protocol Design

The protocol design from Chap. 15 which was also used in the previous chapters is again applicable. We start by applying the pre-compensators to make the agents invertible with identical uniform rank and containing the dynamics of exosystem (20.5).

Protocol design 20.1 *We design three pre-compensators following Protocol Designs 15.1, 15.2 and 15.3. We obtain for $i = 1, \ldots, N$:*

$$\begin{aligned} \dot{p}_{i,1} &= A_{ip,1} p_{i,1} + B_{ip,1} u_i^1, \\ u_i &= C_{ip,1} p_{i,1} + D_{ip,1} u_i^1, \end{aligned} \tag{20.10}$$

and

$$\begin{aligned} \dot{p}_{i,2} &= S_{i,o} p_{i,2} + B_{i,2} u_i^2, \\ u_i^1 &= \Gamma_{i,o} p_{i,2}, \end{aligned} \tag{20.11}$$

and

$$\begin{aligned} \dot{p}_{i,3} &= K_i p_{i,3} + L_i \tilde{u}_i, \\ u_i^2 &= M_i p_{i,3}, \end{aligned} \tag{20.12}$$

The cascade system of agent (20.3) and the pre-compensators (20.10)–(20.12) can be represented in the form,

$$\begin{cases} \mathrm{d}\tilde{x}_i = \tilde{A}_i \tilde{x}_i \mathrm{d}t + \tilde{B}_i \tilde{u}_i \mathrm{d}t + \tilde{G}_i \mathrm{d}\omega_i, \\ y_i = \tilde{C}_i \tilde{x}_i, \end{cases} \tag{20.13}$$

where $\tilde{x}_i \in \mathbb{R}^{\tilde{n}_i}$, $\tilde{u}_i \in \mathbb{R}^p$, $y_i \in \mathbb{R}^p$ are states, inputs, and outputs of the cascade system (20.13). Moreover, there exists a $\rho > 0$ such that for each agent:

- $(\tilde{A}_i, \tilde{C}_i)$ contains (A_0, C_0), i.e., there exists a matrix Π_i such that $\Pi_i A_0 = \tilde{A}_i \Pi_i$ and $\tilde{C}_i \Pi_i = C_0$,
- $(\tilde{A}_i, \tilde{B}_i, \tilde{C}_i)$ is of uniform rank $\rho \geq 1$.

The cascade system (20.13) has uniform rank and is invertible. Hence, the system can be represented by the SCB form as presented in (23.27). In other words, we obtain:

$$\begin{cases} \mathrm{d}\tilde{x}_{ia} = A_{ia} \tilde{x}_{ia} \mathrm{d}t + L_{iad} y_i \mathrm{d}t + \tilde{G}_{ia} \mathrm{d}\omega_i, \\ \mathrm{d}\tilde{x}_{id} = A_d \tilde{x}_{id} \mathrm{d}t + B_d(\tilde{u}_i + E_{ida} \tilde{x}_{ia} + E_{idd} \tilde{x}_{id}) \mathrm{d}t + \tilde{G}_{id} \mathrm{d}\omega_i, \\ y_i = C_d \tilde{x}_{id}, \end{cases} \tag{20.14}$$

for $i = 1, \ldots, N$, where

$$A_d = \begin{pmatrix} 0 & I \\ 0 & 0 \end{pmatrix}, \qquad B_d = \begin{pmatrix} 0 \\ I \end{pmatrix}, \qquad C_d = \begin{pmatrix} I & 0 \end{pmatrix}. \tag{20.15}$$

Then, the observer-based controller (15.18) in Design 15.4 can be designed for each cascade system (20.14). The protocol design is rewritten here.

Protocol design 20.2 *Following Design 15.4, the observer-based controller is designed as,*

$$\begin{aligned} \dot{\hat{x}}_{id} &= A_d\hat{x}_{id} + K_\varepsilon(\zeta_i + \psi_i - C_d\hat{x}_{id}), \\ u_i &= F_{\delta\varepsilon}\hat{x}_{id}. \end{aligned} \tag{20.16}$$

The main result is stated in the following theorem.

Theorem 20.2 *Consider a MAS described by* (20.3), (20.4), *and reference system* (20.5). *Let a root set* π, *a positive integer* N, $\alpha, \beta > 0$ *be given, and hence a set of network graphs* $\mathbb{G}^N_{\alpha,\beta,\pi}$ *be defined.*

Let Assumption 20.1 hold. Moreover, assume the agents are minimum-phase and right-invertible. Then the $\mathcal{H}_2$ *almost regulated output synchronization problem defined in Problem 20.1 is solvable. In particular, for any given real number* $\gamma > 0$, *there exists a* δ^* *such that for any* $\delta \in (0, \delta^*]$ *there exists an* ε^* *such that for any* $\varepsilon \in (0, \varepsilon^*]$, *the combination of pre-compensators* (20.10)–(20.12) *with protocol* (20.16) *solves the* $\mathcal{H}_2$ *almost regulated output synchronization problem for any graph* $\mathcal{G} \in \mathbb{G}^N_{\alpha,\beta,\pi}$.

Proof Note that

$$\tilde{x}_i = \begin{pmatrix} \tilde{x}_{ia} \\ \tilde{x}_{id} \end{pmatrix}$$

and that (20.13) is a shorthand notation for (20.14). For each $i \in \{1, \ldots, N\}$, let $\bar{x}_i = \tilde{x}_i - \Pi_i x_0$, where Π_i is defined to satisfy $\Pi_i A_0 = \tilde{A}_i\Pi_i$, $\tilde{C}_i\Pi_i = C_0$. Then

$$\mathrm{d}\bar{x}_i = \tilde{A}_i\tilde{x}_i\mathrm{d}t - \Pi_i A_0 x_0\mathrm{d}t + \tilde{B}_i\tilde{u}_i\mathrm{d}t + \tilde{G}_i\mathrm{d}\omega_i = \tilde{A}_i\bar{x}_i\mathrm{d}t + \tilde{B}_i\tilde{u}_i\mathrm{d}t + \tilde{G}_i\mathrm{d}\omega_i \tag{20.17}$$

and

$$e_i = y_i - y_r = \tilde{C}_i\tilde{x}_i - C_0x_0 = \tilde{C}_i\tilde{x}_i - \tilde{C}_i\Pi_i x_0 = \tilde{C}_i\bar{x}_i. \tag{20.18}$$

Since the dynamics of the $\bar{x}_i$ system with output e_i is governed by the same dynamics as the dynamics of (20.13), we can present $\bar{x}_i$ in the same form as (20.14), with

$$\bar{x}_i = \begin{pmatrix} \bar{x}_{ia} \\ \bar{x}_{id} \end{pmatrix}$$

where

$$\begin{aligned} d\bar{x}_{ia} &= A_{ia}\bar{x}_{ia}dt + L_{iad}e_i dt + \tilde{G}_{ia}d\omega_i, \\ d\bar{x}_{id} &= A_d\bar{x}_{id}dt + B_d(\tilde{u}_i + E_{ida}\bar{x}_{ia} + E_{idd}\bar{x}_{id})dt + \tilde{G}_{id}d\omega_i, \\ e_i &= C_d\bar{x}_{id}. \end{aligned}$$

Let a unitary $\bar{U}$ be defined such that $\bar{U}^{-1}\bar{L}\bar{U} = \bar{S}$, where $\bar{S}$ is the Schur form of the matrix $\bar{L}$. Then, following exactly the proof of Theorem 15.6, we obtain

$$\begin{aligned} d\eta_a &= A_a\eta_a dt + \tilde{W}_{ad}\eta_d dt + \tilde{G}_a d\omega, \\ \varepsilon d\eta_d &= \tilde{A}_\delta\eta_d dt + \tilde{W}_{da}^\varepsilon\eta_a dt + \tilde{W}_{dd}^\varepsilon\eta_d dt + \varepsilon\tilde{G}_d^\varepsilon d\omega, \end{aligned} \tag{20.19}$$

where

$$\tilde{A}_\delta = I_N \otimes \begin{pmatrix} A_d & 0 \\ 0 & A_d - KC_d \end{pmatrix} + \bar{S} \otimes \begin{pmatrix} B_d F_\delta & -B_d F_\delta \\ B_d F_\delta & -B_d F_\delta \end{pmatrix}, \tag{20.20}$$

and

$$\tilde{W}_{ad} = \begin{pmatrix} W_{ad} & 0 \end{pmatrix} N_d^{-1}, \qquad \tilde{G}_d^\varepsilon = N_d \begin{pmatrix} \bar{G}_d^\varepsilon \\ \bar{G}_d^\varepsilon \end{pmatrix},$$

$$\tilde{W}_{da}^\varepsilon = N_d \begin{pmatrix} W_{da}^\varepsilon \\ W_{da}^\varepsilon \end{pmatrix}, \qquad \tilde{W}_{dd}^\varepsilon = N_d \begin{pmatrix} W_{dd}^\varepsilon & 0 \\ W_{dd}^\varepsilon & 0 \end{pmatrix} N_d^{-1},$$

where W_{ad}, W_{da}^ε, W_{dd}^ε are given in the proof of Theorem 15.6 and

$$\tilde{G}_a = \text{blkdiag}\{\tilde{G}_{ia}\}, \qquad \bar{G}_d^\varepsilon = (\bar{S}\bar{U}^{-1} \otimes I_{p\rho})\hat{G}_d^\varepsilon,$$

where $\hat{G}_d^\varepsilon = \text{blkdiag}\{G_{id}^\varepsilon\}$ with $G_{id}^\varepsilon = S_\varepsilon\tilde{G}_{id}$.

Following Lemma 19.4, there exists a $P_\delta' = P_\delta > 0$, $P_a' = P_a > 0$, and a small enough $\nu > 0$ such that

$$\tilde{A}_\delta P_\delta + P_\delta\tilde{A}_\delta' \leq -\nu P_\delta - 4I, \tag{20.21}$$

and

$$P_a A_a + A_a' P_a = -\nu P_a - 3I. \tag{20.22}$$

Define $V_a = \varepsilon^2\eta_a' P_a\eta_a$ as a Lyapunov function for the dynamics of η_a in (20.19). Similarly, we define $V_d = \varepsilon\eta_d' P_\delta\eta_d$ as a Lyapunov function for the η_d dynamics in (20.19). The derivative of V_a is bounded by

$$\begin{aligned}
\mathrm{d}V_a &= -\nu V_a \mathrm{d}t - \varepsilon^2 \|\eta_a\|^2 \mathrm{d}t + 2\varepsilon^2 \operatorname{Re}(\eta_a' P_a \tilde{W}_{ad}\eta_d)\mathrm{d}t \\
&\quad + \varepsilon^2 \operatorname{trace}(P_a \tilde{G}_a Q_0 \tilde{G}_a')\mathrm{d}t + 2\varepsilon^2 \operatorname{Re}(\eta_a' P_a \tilde{G}_a)\mathrm{d}\omega \\
&\le -\nu V_a \mathrm{d}t + \varepsilon c_3 V_d \mathrm{d}t + \varepsilon^2 r_5 \operatorname{trace}(Q_0)\mathrm{d}t + 2\varepsilon^2 \operatorname{Re}(\eta_a' P_a \tilde{G}_a)\mathrm{d}\omega, \qquad (20.23)
\end{aligned}$$

where r_5 and c_3 are such that

$$\operatorname{trace}(P_a \tilde{G}_a Q_0 \tilde{G}_a') \le r_5 \operatorname{trace} Q_0,$$

and

$$\begin{aligned}
2\operatorname{Re}(\eta_a' P_a \tilde{W}_{ad}\eta_d) &\le 2r_4 \|\eta_a\| \|\eta_d\| \\
&\le \tfrac{1}{2}\|\eta_a\|^2 + 2r_4^2 \|\eta_d\|^2 \le \tfrac{1}{2}\|\eta_a\|^2 + \varepsilon^{-1} c_3 V_d.
\end{aligned}$$

Note that we can choose r_4, r_5, and c_3 independent of the network graph but only depending on our bounds on the eigenvalues of our matrix $\bar{L}$. Taking the expectation, we get

$$\mathrm{d}\mathbb{E}\, V_a \le -\nu \mathbb{E}\, V_a \mathrm{d}t + \varepsilon c_3 \mathbb{E}\, V_d \mathrm{d}t + \varepsilon^2 r_5 \operatorname{trace}(Q_0)\mathrm{d}t.$$

Next, the derivative of V_d is bounded by

$$\begin{aligned}
\mathrm{d}V_d &= -\nu\varepsilon^{-1} V_d \mathrm{d}t - 4\|\eta_d\|^2 \mathrm{d}t + 2\operatorname{Re}(\eta_d' P_\delta \tilde{W}_{da}^{\varepsilon}\eta_a)\mathrm{d}t \\
&\quad + 2\operatorname{Re}(\eta_d' P_\delta \tilde{W}_{dd}^{\varepsilon}\eta_d)\mathrm{d}t + \varepsilon \operatorname{trace}(P_\delta \tilde{G}_d^{\varepsilon} Q_0 (\tilde{G}_d^{\varepsilon})')\mathrm{d}t \\
&\quad + 2\varepsilon \operatorname{Re}(\eta_d' P_\delta \tilde{G}_d^{\varepsilon})\mathrm{d}\omega \\
&\le c_2 V_a \mathrm{d}t - (\nu\varepsilon^{-1} + \nu - \varepsilon^2 \frac{c_2 c_3}{\nu}) V_d \mathrm{d}t - \|\eta_d\|^2 \mathrm{d}t \\
&\quad + \varepsilon^2 r_3 \operatorname{trace}(Q_0)\mathrm{d}t + 2\varepsilon \operatorname{Re}(\eta_d' P_\delta \tilde{G}_d^{\varepsilon})\mathrm{d}\omega, \qquad (20.24)
\end{aligned}$$

where

$$2\operatorname{Re}(\eta_d' P_\delta \tilde{W}_{dd}^{\varepsilon}\eta_d) \le \|\eta_d\|^2$$

for a small ε and

$$2\operatorname{Re}(\eta_d' P_\delta \tilde{W}_{da}^{\varepsilon}\eta_a) \le \varepsilon r_1 \|\eta_a\| \|\eta_d\| \le \varepsilon^2 r_1^2 \|\eta_a\|^2 + \|\eta_d\|^2 \le c_2 V_a + \|\eta_d\|^2,$$

provided r_1 is such that we have $\varepsilon r_1 \ge \|P_\delta \tilde{W}_{da}^{\varepsilon}\|$ and c_2 sufficiently large. Finally,

$$\operatorname{trace}(P_\delta \tilde{G}_d^\varepsilon Q_0 (\tilde{G}_d^\varepsilon)') \le \varepsilon r_3 \operatorname{trace} Q_0$$

for a suitably chosen r_3. Taking the expectation, we get

$$\mathrm{d}\mathbb{E}\, V_d \le c_2 \mathbb{E}\, V_a \mathrm{d}t - (\nu\varepsilon^{-1} + \nu - \varepsilon^2 \frac{c_2 c_3}{\nu}) \mathbb{E}\, V_d \mathrm{d}t - \mathbb{E}\|\eta_d\|^2 \mathrm{d}t + \varepsilon^2 r_3 \operatorname{trace}(Q_0) \mathrm{d}t.$$

We get

$$\frac{\mathrm{d}}{\mathrm{d}t} \begin{pmatrix} \mathbb{E}\, V_a \\ \mathbb{E}\, V_d \end{pmatrix} \le A_e \begin{pmatrix} \mathbb{E}\, V_a \\ \mathbb{E}\, V_d \end{pmatrix} + \begin{pmatrix} \varepsilon^2 r_5 \operatorname{trace}(Q_0) \\ \varepsilon^2 r_3 \operatorname{trace}(Q_0) \end{pmatrix},$$

where

$$A_e = \begin{pmatrix} -\nu & \varepsilon c_3 \\ c_2 & -\varepsilon^{-1}\nu - \nu + \varepsilon^2 \frac{c_2 c_3}{\nu} \end{pmatrix},$$

with eigenvalues of $\lambda_1 = -\nu + \varepsilon^2 \dfrac{c_2 c_3}{\nu}$ and $\lambda_2 = -\varepsilon^{-1}\nu - \nu$. By integration we find that

$$\begin{pmatrix} \mathbb{E}\, V_a \\ \mathbb{E}\, V_d \end{pmatrix}(t) \le e^{A_e t} \begin{pmatrix} \mathbb{E}\, V_a \\ \mathbb{E}\, V_d \end{pmatrix}(0) + \int_0^t e^{A_e(t-s)} \begin{pmatrix} \varepsilon^2 r_5 \operatorname{trace}(Q_0) \\ \varepsilon^2 r_3 \operatorname{trace}(Q_0) \end{pmatrix} \mathrm{d}s$$

componentwise. Then, for large enough r_6 we have

$$\lim_{t\to\infty} (\mathbb{E}\, V_a(t) + \mathbb{E}\, V_d(t)) \le \varepsilon^2 r_6 \operatorname{trace}(Q_0),$$

which implies that

$$\lim_{t\to\infty} \mathbb{E}\, \eta_d'(t)\eta_d(t) \le \frac{\varepsilon r_6}{\sigma_{\min}(P_\delta)} \operatorname{trace}(Q_0).$$

Following the proof above, we find that

$$\mathbf{e}_0 = (I_N \otimes C_d)(I_N \otimes S_\varepsilon^{-1})(\bar{U}\bar{S}^{-1} \otimes I_{p\rho}) \begin{pmatrix} I_{Np\rho} & 0 \end{pmatrix} N_d^{-1} \eta_d \tag{20.25}$$

$$= (\bar{U}\bar{S}^{-1} \otimes C_d) \begin{pmatrix} I_{Np\rho} & 0 \end{pmatrix} N_d^{-1} \eta_d = \Theta \eta_d, \tag{20.26}$$

for a suitably chosen matrix Θ. The norm of Θ is obviously bounded since $C_d S_\varepsilon^{-1} = C_d$. Therefore, we surely achieve that

$$\lim_{t\to\infty} \mathbb{E}\, \mathbf{e}_0'(t)\mathbf{e}_0(t) \le \frac{\varepsilon r_7}{\sigma_{\min}(P_\delta)} \operatorname{trace}(Q_0),$$

for some suitable chosen r_7, which implies that

$$\|T_{\omega \mathbf{e}_0}\|_2 \leq \frac{\varepsilon r_7}{\sigma_{\min}(P_\delta)}.$$

■

20.5 Disturbances With and Without Known Frequencies

The model (20.2) that we used to generate colored noise clearly cannot incorporate bias terms. In this section, we consider the noise with bias terms as well as other disturbances with known frequency.

The agent model (20.3) is modified as follows:

$$\begin{aligned} \mathrm{d}\breve{x}_i &= \breve{A}_i \breve{x}_i \mathrm{d}t + \breve{B}_i u_i \mathrm{d}t + \breve{G}_i \mathrm{d}\omega_i + \breve{H}_{x,i} d_i \mathrm{d}t, \\ y_i &= \breve{C}_i \breve{x}_i + \breve{H}_{y,i} d_i, \end{aligned} \qquad \breve{x}_i(0) = \breve{x}_{i0} \tag{20.27}$$

where $\breve{x}_i$, u_i, y_i, and ω_i are the same as those in (20.3), while $d_i \in \mathbb{R}^{d_i}$ is an external disturbance with known frequencies, which can be generated by the following exosystem:

$$\begin{aligned} \dot{x}_{id} &= S_i x_{id}, \\ d_i &= R_i x_{id}, \end{aligned} \qquad x_{id}(0) = x_{id0} \tag{20.28}$$

where $x_{id} \in \mathbb{R}^{n_{d_i}}$ and the initial condition x_{id0} can be arbitrarily chosen.

Then the $\mathcal{H}_2$ almost regulated output synchronization problem for a MAS in the presence of stochastic disturbance and disturbances with known frequencies is formulated as follows:

Problem 20.3 Consider a MAS described by (20.27), (20.28), and (20.4). For a given root set π, a positive integer N, and $\alpha, \beta > 0$ that define a set of network graphs $\mathbb{G}^N_{\alpha,\beta,\pi}$, and any reference trajectories given by (20.5), the $\mathcal{H}_2$ *almost regulated output synchronization* problem is to find, if possible, a linear time-invariant dynamic protocol of the form (20.8) such that, for any given real number $\gamma > 0$ and for any graph $\mathcal{G} \in \mathbb{G}^N_{\alpha,\beta,\pi}$, (20.7) is satisfied for all initial conditions of agents, protocols, and exosystems for $\omega = 0$. Moreover, the closed-loop transfer function $T_{\omega \mathbf{e}_0}$ satisfies (20.9).

Note that we have two exosystems, (20.5) and (20.28), which generate the reference signal y_0 and the disturbance d_i, respectively. We can unify them both into one exosystem as

$$\begin{aligned} x_{ie} &= S_{ie} x_{ie}, \qquad x_{ie}(0) = x_{ie0} \\ d_i &= R_{ie} x_{ie}, \\ y_0 &= R_{re} x_{ie}, \end{aligned} \tag{20.29}$$

where

$$S_{ie} = \begin{pmatrix} S_i & 0 \\ 0 & A_0 \end{pmatrix}, \quad R_{ie} = \begin{pmatrix} R_i & 0 \end{pmatrix}, \quad R_{re} = \begin{pmatrix} 0 & C_0 \end{pmatrix}. \tag{20.30}$$

Note that

$$x_{ie0} = \begin{pmatrix} x_{id0} \\ x_{00} \end{pmatrix}$$

and therefore, the second part of the initial condition has to be the same for each agent while the first part might be different for each agent.

The protocol design follows Protocol Design 18.1 in Chap. 18, which is written again as follows:

Protocol design 20.3 *We first design a pre-compensator similar to Protocol Design 15.1*

$$\begin{aligned} \dot{p}_{i,1} &= A_{ip,1} p_{i,1} + B_{ip,1} u_i^1, \\ u_i &= C_{ip,1} p_{i,1} + D_{ip,1} u_i^1, \end{aligned} \tag{20.31}$$

such that the interconnected system is of the form:

$$\begin{aligned} \dot{x}_i &= \hat{A}_i \hat{x}_i + \hat{B}_i u_i^1 + \hat{G}_i \omega_i + \hat{H}_{x,i} d_i, \\ y_i &= \hat{C}_i x_i + H_{y,i} d_i, \end{aligned} \tag{20.32}$$

with the subsystem with input u_i^1 to y_i uniform rank and invertible for $i = 1, \ldots, N$. Next, we design pre-compensators such that the interconnected system contains all modes of the augmenten exosystem. This follows Protocol Design 15.2. However, we use the augmented exosystem (20.29) instead of the exosystem (20.5). This implies that the regulator equation (15.10) should be first modified as

$$\begin{aligned} \hat{A}_i \tilde{\Pi}_i + \hat{B}_i \Gamma_i + \hat{H}_{x,i} R_{ie} &= \tilde{\Pi}_i S_{ie}, \\ \hat{C}_i \tilde{\Pi}_i + H_{y,i} R_{ie} &= R_{re}. \end{aligned} \tag{20.33}$$

If (Γ_i, S_{ie}) is not observable, we can construct the observable subsystem $(\Gamma_{i,o}, S_{ie,o})$; otherwise simply set $\Gamma_{i,o} = \Gamma_i$ and $S_{i,o} = S_{ie}$. Then the second pre-compensator for each agent i is designed as

$$\begin{aligned} \dot{p}_{i,2} &= S_{ie,o} p_{i,2} + B_{i,2} u_i^2, \\ u_i^1 &= \Gamma_{i,o} p_{i,2}, \end{aligned} \tag{20.34}$$

(continued)

Protocol design 20.3 (continued)
for $i = 1, \ldots, N$, where $p_{i,2} \in \mathbb{R}^{n_{pi}}$ and $B_{i,2}$ is chosen to guarantee that no invariant zeros are introduced by the pre-compensator. Moreover, an appropriate choice of $B_{i,2}$ can guarantee that the triple $(S_{ie,o}, B_{i,2}, \Gamma_{i,o})$ is invertible, uniform rank, and minimum-phase (see [60]).

If the agent has dynamics in common with the exosystem, we can use the technique outlined in Appendix 15.5 to remove these common modes after which the above design is applicable.

The final pre-compensator is designed according to Protocol Design 15.3 and we obtain:

$$\begin{aligned} \dot{p}_{i,3} &= K_i p_{i,3} + L_i \tilde{u}_i, \\ u_i^2 &= M_i p_{i,3}, \end{aligned} \tag{20.35}$$

such that the agents interconnected with the three pre-compensators are all uniform rank and invertible with the same uniform rank for all agents.

Then, the cascade system of agent (20.27) and pre-compensators (20.31), (20.34), and (20.35) can be represented in the form,

$$\begin{cases} \mathrm{d}\tilde{x}_i = \tilde{A}_i \tilde{x}_i \mathrm{d}t + \tilde{B}_i \tilde{u}_i \mathrm{d}t + \tilde{G}_i \mathrm{d}\omega_i + \tilde{H}_{x,i} d_i \mathrm{d}t, \\ y_i = \tilde{C}_i \tilde{x}_i + \tilde{H}_{y,i} d_i, \end{cases} \tag{20.36}$$

where $\tilde{x}_i \in \mathbb{R}^{\tilde{n}_i}$, $\tilde{u}_i \in \mathbb{R}^p$, $y_i \in \mathbb{R}^p$ are states, inputs, and outputs of the cascade system (20.36). Moreover, we have

- $(\tilde{A}_i, \tilde{C}_i)$ contains the joint exosystem (20.29), i.e., there exists a matrix Π_i such that $\Pi_i S_{ie} = \tilde{A}_i \Pi_i + \tilde{H}_{x,i} R_{ie}$ and $\tilde{C}_i \Pi_i + \tilde{H}_{y,i} R_{ie} = R_{re}$,
- $(\tilde{A}_i, \tilde{B}_i, \tilde{C}_i)$ is of uniform rank $\rho \geq 1$.

Then, the observer-based controller (20.16) in Design 20.2 can be designed for each cascade system (20.36). The main result is stated in the following theorem.

Theorem 20.4 *Consider a MAS described by* (20.27), (20.28), (20.4), *and reference system* (20.5). *Let a root set π, a positive integer N, $\alpha, \beta > 0$ be given, and hence a set of network graphs $\mathbb{G}^N_{\alpha,\beta,\pi}$ be defined.*

Let Assumption 20.1 hold. Moreover, agents are minimum-phase and right-invertible. Then the $\mathcal{H}_2$ almost regulated output synchronization problem defined in Problem 20.3 is solvable. In particular, for any given real number $\gamma > 0$, there exists a δ^* such that for any $\delta \in (0, \delta^*]$ there exists an ε^* such that for any $\varepsilon \in (0, \varepsilon^*]$, the combination of pre-compensator (20.31), (20.34), *and* (20.35) *with protocol* (20.16) *solves the $\mathcal{H}_2$ almost regulated output synchronization problem for any graph $\mathcal{G} \in \mathbb{G}^N_{\alpha,\beta,\pi}$.*

Proof Let $\bar{x}_i = \tilde{x}_i - \Pi_i x_{ie}$ for $i \in \{1, \ldots, N\}$, where Π_i satisfies $\Pi_i S_{ie} = \tilde{A}_i \Pi_i + \tilde{H}_{x,i} R_{ie}$ and $\tilde{C}_i \Pi_i + \tilde{H}_{y,i} R_{ie} = R_{re}$. Then

$$\begin{aligned}
d\bar{x}_i &= \tilde{A}_i \tilde{x}_i dt - \Pi_i S_{ie} x_{ie} dt + \tilde{B}_i \tilde{u}_i dt + \tilde{G}_i d\omega_i + \tilde{H}_{x,i} d_i dt \\
&= \tilde{A}_i \bar{x}_i dt + \tilde{B}_i \tilde{u}_i dt + \tilde{G}_i d\omega_i,
\end{aligned}$$

and

$$e_i = y_i - y_0 = \tilde{C}_i \tilde{x}_i + \tilde{H}_{y,i} R_{ie} x_{ie} - R_{re} x_{ie} = \tilde{C}_i \tilde{x}_i - \tilde{C}_i \Pi_i x_r = \tilde{C}_i \bar{x}_i.$$

The dynamic $\bar{x}_i$ is exactly the same as (20.17) in Sect. 20.4. Then, the remaining proof follows that of Theorem 20.2. ■

20.6 Time-Varying Graph

The protocol design for a fixed graph can also be applied directly to a time-varying graph. The time-varying communication network among agents is exactly the same as that in Chap. 11. The network information provided to each agent is modified as,

$$\zeta_i = \sum_{j=1}^{N} a_{ij}(t)(y_i - y_j) = \sum_{j=1}^{N} \ell_{ij}(t) y_j. \tag{20.37}$$

In this section, we regulate again all agents' outputs to a priori given reference trajectories generated by an exosystem (20.5). The extra information from the exosystem ψ_i is also obtained for each agent, i.e., $\psi_i = \iota_i(y_i - y_0)$ for $i \in \pi$; otherwise, $\psi_i = 0$. Moreover, Assumption 15.2 and Definition 15.4 hold. Following Chap. 17, a set of fixed graphs $\mathbb{G}^N_{\alpha,\beta,\pi}$ and a set of time-varying graphs $\mathbb{G}^{\tau,N}_{\alpha,\beta,\pi}$ are defined.

Because of the time-varying network, the $\mathcal{H}_2$ norm of the closed-loop system does not exist. Instead, the measurement on the regulated output synchronization error $\mathbf{e}_0$ is used for the time-varying case. Then the problem formulation can be modified as follows:

Problem 20.5 Consider a MAS described by (20.3), (20.37) and exosystem (20.5). For a given root set π, a positive integer N, and $\alpha, \beta, \tau > 0$ that define a set of time-varying network graphs $\mathbb{G}^{\tau,N}_{\alpha,\beta,\pi}$. The *almost regulated output synchronization* problem is to find, if possible, a linear time-invariant dynamic protocol of the form (20.8), such that, for any given real number $\gamma > 0$, for any upper bound for the rate of the stochastic disturbance $\bar{Q}_0$, the regulated output synchronization error satisfies,

$$
\lim_{t\to\infty} \mathbb{E}\,\mathbf{e}_0(t) = 0,
$$
$$
\limsup_{t\to\infty} \mathrm{Var}[\mathbf{e}_0(t)] = \limsup_{t\to\infty} \mathbb{E}\,\mathbf{e}_0'(t)\mathbf{e}_0(t) < \gamma, \tag{20.38}
$$

for any graph $\mathcal{G} \in \mathbb{G}^{\tau,N}_{\alpha,\beta,\pi}$ and for all the initial conditions of agents, protocols, and exosystems and for any $Q_0 \leq \bar{Q}_0$.

Remark 20.6 Clearly, we can also define (20.38) via the expectation of the RMS [8] as

$$
\limsup_{T\to\infty} \frac{1}{T}\mathbb{E} \int_0^T \mathbf{e}(t)'\mathbf{e}(t)\,\mathrm{d}t < \gamma.
$$

Remark 20.7 Note that because of the time-varying graph, the complete system is time-variant and hence the variance of the error signal might not converge as time tends to infinity. Hence, we use in the above a lim sup instead of a regular limit.

The protocol design generated for a fixed graph works exactly for a time-varying graph. The result is stated in the following theorem.

Theorem 20.8 *Consider a MAS described by* (20.3), (20.37)*, and reference system* (20.5)*. Let a root agent set π, a positive integer N, and $\alpha, \beta, \tau > 0$ be given, and hence a set of network graphs $\mathbb{G}^{\tau,N}_{\alpha,\beta,\pi}$ be defined.*

Let Assumption 20.1 hold. Moreover, agents are minimum-phase and right-invertible. Then the almost regulated output synchronization problem defined in Problem 20.5 is solvable. In particular, for any given real number $\gamma > 0$, and for any upper bound for the rate of the stochastic disturbance $\bar{Q}_0$, and for any graph $\mathcal{G}_t \in \mathbb{G}^{\tau,N}_{\alpha,\beta,\pi}$ there exists a δ^* such that for any $\delta \in (0, \delta^*]$ there exists an ε^* such that for any $\varepsilon \in (0, \varepsilon^*]$, the combination of pre-compensator (20.10)–(20.12) *with protocol* (20.16) *solves the almost regulated output synchronization problem.*

Proof For each $t > 0$, we define a unitary $\bar{U}_t$ such that $\bar{U}_t^{-1}\bar{L}_t\bar{U}_t = \bar{S}_t$, where $\bar{S}_t$ is the Schur form of the matrix $\bar{L}_t$. Following the route of proving Theorem 20.2, we can surely get

$$
\begin{aligned}
\mathrm{d}\eta_a &= A_a\eta_a\mathrm{d}t + \tilde{W}_{ad,t}\eta_d\mathrm{d}t + G_a\mathrm{d}\omega,\\
\varepsilon\mathrm{d}\eta_d &= \tilde{A}_{\delta,t}\eta_d\mathrm{d}t + \tilde{W}^{\varepsilon}_{da,t}\eta_a\mathrm{d}t + \tilde{W}^{\varepsilon}_{dd,t}\eta_d\mathrm{d}t + \varepsilon\tilde{G}^{\varepsilon}_{d,t}\mathrm{d}\omega,
\end{aligned} \tag{20.39}
$$

$$
\tilde{A}_{\delta,t} = I_N \otimes \begin{pmatrix} A_d & 0 \\ 0 & A_d - KC_d \end{pmatrix} + \bar{S}_t \otimes \begin{pmatrix} B_dF_\delta & -B_dF_\delta \\ B_dF_\delta & -B_dF_\delta \end{pmatrix}, \tag{20.40}
$$

and

$$
\tilde{W}_{ad,t} = \begin{pmatrix} W_{ad,t} & 0 \end{pmatrix} N_d^{-1}, \qquad \tilde{G}^{\varepsilon}_{d,t} = N_d \begin{pmatrix} \bar{G}^{\varepsilon}_{d,t} \\ \bar{G}^{\varepsilon}_{d,t} \end{pmatrix},
$$

$$\tilde{W}^{\varepsilon}_{da,t} = N_d \begin{pmatrix} W^{\varepsilon}_{da,t} \\ W^{\varepsilon}_{da,t} \end{pmatrix}, \qquad \tilde{W}^{\varepsilon}_{dd,t} = N_d \begin{pmatrix} W^{\varepsilon}_{dd,t} & 0 \\ W^{\varepsilon}_{dd,t} & 0 \end{pmatrix} N_d^{-1}.$$

Moreover, $\tilde{A}_{\delta,t}$, $\tilde{W}_{ad,t}$, $\tilde{W}^{\varepsilon}_{da,t}$, $\tilde{W}^{\varepsilon}_{dd,t}$, and $\tilde{G}^{\varepsilon}_{d,t}$ are functions of time t because the Laplacian matrix corresponding to the switching graph is a function of time t.

Define $V_a = \varepsilon^2 \eta_a' P_a \eta_a$ and $V_d = \varepsilon \eta_d' P_\delta \eta_d$. By using the same arguments as in the proof for the fixed graph, we obtain that

$$\frac{\mathrm{d}}{\mathrm{d}t} \begin{pmatrix} \mathbb{E}\, V_a \\ \mathbb{E}\, V_d \end{pmatrix} \leq A_e \begin{pmatrix} \mathbb{E}\, V_a \\ \mathbb{E}\, V_d \end{pmatrix} + \begin{pmatrix} \varepsilon^2 r_5 \operatorname{trace}(Q_0) \\ \varepsilon r_3 \operatorname{trace}(Q_0) \end{pmatrix}$$

where

$$A_e = \begin{pmatrix} -\nu & \varepsilon c_3 \\ c_2 & -\varepsilon^{-1}\nu - \nu + \varepsilon^2 \frac{c_2 c_3}{\nu} \end{pmatrix}.$$

Note that the inequality here is componentwise. We find by integration that

$$\begin{pmatrix} \mathbb{E}\, V_a \\ \mathbb{E}\, V_d \end{pmatrix} (t_k^-) \leq e^{A_e(t_k - t_{k-1})} \begin{pmatrix} \mathbb{E}\, V_a \\ \mathbb{E}\, V_d \end{pmatrix} (t_{k-1}^+) + \int_{t_{k-1}}^{t_k} e^{A_e(t_k - s)} \begin{pmatrix} \varepsilon^2 r_5 \operatorname{trace}(Q_0) \\ \varepsilon r_3 \operatorname{trace}(Q_0) \end{pmatrix} \mathrm{d}s$$

componentwise. We have a potential jump at time t_{k-1} in V_d. However, there exists a m such that

$$V_d(t_{k-1}^+) \leq m V_d(t_{k-1}^-)$$

while V_a is continuous. Using our explicit expression for $e^{A_e t}$ and the fact that $t_k - t_{k-1} > \tau$ we find that,

$$\left[\mathbb{E}\, V_a(t_k^-) + \mathbb{E}\, V_d(t_k^-)\right] \leq \\ e^{\lambda_3 (t_k - t_{k-1})} \left[\mathbb{E}\, V_a(t_{k-1}^-) + \mathbb{E}\, V_d(t_{k-1}^-)\right] + r\varepsilon^2 \operatorname{trace}(Q_0)$$

where $\lambda_3 = -\nu/2$ and r is a sufficiently large constant. Combining these time-intervals, we get

$$\left[\mathbb{E}\, V_a(t_k^-) + \mathbb{E}\, V_d(t_k^-)\right] \leq e^{\lambda_3 t_k} \left[\mathbb{E}\, V_a(0) + \mathbb{E}\, V_d(0)\right] + \frac{r\varepsilon^2}{1-\mu} \operatorname{trace}(Q_0),$$

where $\mu < 1$ is such that $e^{\lambda_3 (t_k - t_{k-1})} \leq e^{\lambda_3 \tau} \leq \mu$ for all k. Assume $t_{k+1} > t > t_k$. Since we do not necessarily have that $t - t_k > \tau$ we use the bound,

$$\begin{pmatrix}1 & 1\end{pmatrix} e^{A_e(t-t_k)} \begin{pmatrix}\mathbb{E}\, V_a \\ \mathbb{E}\, V_d\end{pmatrix} (t_k^+) \leq 2me^{\lambda_3(t-t_k)}[\mathbb{E}\, V_a + \mathbb{E}\, V_d](t_k^-),$$

where the factor m is due to the potential discontinuous jump. Combining all together, we get

$$[\mathbb{E}\, V_a(t) + \mathbb{E}\, V_d(t)] \leq 2me^{\lambda_3 t}\, [\mathbb{E}\, V_a(0) + \mathbb{E}\, V_d(0)] + (2m+1)\frac{r\varepsilon^2}{1-\mu}\, \text{trace}(Q_0).$$

This implies that

$$\limsup_{t\to\infty} \mathbb{E}\, \eta_d'(t)\eta_d(t) \leq \frac{2m+1}{\sigma_{\min}(P_\delta)} \frac{r\varepsilon}{1-\mu}\, \text{trace}(Q_0)$$

Similarly, we find that

$$\begin{aligned}\mathbf{e}_0 &= (I_N \otimes C_d)(I_N \otimes S_\varepsilon^{-1})(\bar{U}_t \bar{S}_t^{-1} \otimes I_{p\rho}) \begin{pmatrix}I_{Np\rho} & 0\end{pmatrix} N_d^{-1}\eta_d \\ &= (\bar{U}_t \bar{S}_t^{-1} \otimes C_d) \begin{pmatrix}I_{Np\rho} & 0\end{pmatrix} N_d^{-1}\eta_d = \Theta_t \eta_d,\end{aligned}$$

for a suitably chosen matrix Θ_t, which is bounded because of the boundedness of $\bar{L}$ for any graph in $\mathbb{G}^N_{\alpha,\beta,\pi}$. The fact that we can make the asymptotic variance of η_d arbitrarily small immediately implies that the asymptotic variance of $\mathbf{e}_0$ can be made arbitrarily small. Because all agents and protocols are linear, it is obvious that the expectation of $\mathbf{e}_0$ is equal to zero. ■

Chapter 21
Almost Output Synchronization of Heterogeneous Continuous-Time Linear MAS with Passive Agents

21.1 Introduction

In the previous chapters of this book, we often considered systems which either have all poles in the closed left half plane or have all zeros in the open left half plane (at most weakly unstable or minimum phase). For homogeneous MAS, we also considered the case where agents are passive or, at least, squared-down passifiable. One of the advantages of passivity was the fact that we could achieve our objectives by static protocols.

In this chapter, we briefly address heterogeneous MAS with passive agents to illustrate how to exploit the structure of passivity in the heterogenous case. For this purpose, we revisit the H_2 and H_∞ almost synchronization problems for heterogeneous continuous-time linear MAS with passive square agents. We will show that there is an intrinsic difference between H_2 and H_∞ almost synchronization. We will see that H_2 almost synchronization is achievable. However, H_∞ almost synchronization is, in general, not achievable. In addition, expanding from passive agents to squared-down passifiable yields problems as illustrated by a simple example. Note that in this chapter we consider synchronization and not regulated synchronization. The main reason is that we consider static protocols and regulated synchronization will in most cases require a dynamic protocol. However, the insight of this chapter gives us a clear view of what can be done in the case of regulated synchronization.

The write-up of this chapter is partially based on [132].

A. Saberi et al., *Synchronization of Multi-Agent Systems in the Presence of Disturbances and Delays*, Systems & Control: Foundations & Applications,
https://doi.org/10.1007/978-3-030-88148-1_21

21.2 Problem Formulation

Consider a MAS composed of N non-identical linear time-invariant agents of the form,

$$\begin{cases} \dot{x}_i = A_i x_i + B_i u_i + E_i \omega_i, \\ y_i = C_i x_i, \end{cases} \qquad (i = 1, \dots, N) \tag{21.1}$$

where $x_i \in \mathbb{R}^{n_i}$, $u_i \in \mathbb{R}^p$, and $y_i \in \mathbb{R}^p$ are the state, input, and output of agent i while $\omega_i \in \mathbb{R}^q$ is an external disturbance.

We make the following assumption on the agent dynamics.

Assumption 21.1 *For each agent $i \in \{1, \dots, N\}$, we have,*

- *(A_i, B_i) is stabilizable, and (A_i, C_i) is detectable;*
- *(A_i, B_i, C_i) is passive*

Lemma 1.25 and the associated remark 1.26 yield that the above assumption guarantees that there exists a matrix $P_i > 0$ such that

$$\begin{aligned} P_i A_i + A_i' P_i &\le 0, \\ P_i B_i &= C_i'. \end{aligned} \tag{21.2}$$

for $i = 1, \dots N$.

The communication network among agents is the same as in many of the previous chapters and provides each agent with the quantity,

$$\zeta_i = \sum_{j=1}^{N} a_{ij}(y_i - y_j) = \sum_{j=1}^{N} \ell_{ij} y_j, \quad i = 1, \dots, N. \tag{21.3}$$

Let N be any agent and define $\bar{y}_i = y_N - y_i$ and

$$\bar{y} = \begin{pmatrix} \bar{y}_1 \\ \vdots \\ \bar{y}_{N-1} \end{pmatrix} \quad \text{and} \quad \omega = \begin{pmatrix} \omega_1 \\ \vdots \\ \omega_N \end{pmatrix}.$$

Obviously, output synchronization is achieved if $\bar{y}$ coverges to 0 asymptotically. That is,

$$\lim_{t \to \infty} (y_i(t) - y_N(t)) = 0, \quad \text{for all } i \in \{1, \dots, N-1\}. \tag{21.4}$$

Remark 21.1 The agent N is not necessarily a root agent. Moreover, (21.4) is equivalent to the condition,

$$\lim_{t\to\infty}(y_i(t)-y_j(t))=0, \quad \text{for all } i,j\in\{1,\ldots,N\}.$$

We formulate below two almost output synchronization problems for a network with either H_2 or H_∞ almost output synchronization.

Problem 21.2 Consider a MAS described by (21.1) and (21.3). Let $\mathbf{G}$ be a given set of graphs such that $\mathbf{G}\subseteq\mathbb{G}^N$. The **$H_\infty$ almost output synchronization problem with partial-state coupling via a static protocol** with a set of network graphs $\mathbf{G}$ is to find, if possible, a linear static protocol parameterized in terms of a parameter ε, of the form,

$$u_i=F(\varepsilon)\zeta_i, \tag{21.5}$$

such that, for any given real number $r>0$ and for any graph $\mathcal{G}\in\mathbf{G}$, there exists an ε^* such that for any $\varepsilon\in(0,\varepsilon^*]$, synchronization is achieved for all initial conditions in the absence of disturbances, and the closed-loop transfer matrix from ω to $\bar{y}$, denoted by $T_{\omega\bar{y}}$, satisfies

$$\|T_{\omega\bar{y}}\|_\infty<r. \tag{21.6}$$

Problem 21.3 Consider a MAS described by (21.1) and (21.3). Let $\mathbf{G}$ be a given set of graphs such that $\mathbf{G}\subseteq\mathbb{G}^N$. The **$H_2$ almost output synchronization problem with partial-state coupling via a static protocol** with a set of network graphs $\mathbf{G}$ is to find, if possible, a linear static protocol parameterized in terms of a parameter ε, of the form,

$$u_i=F(\varepsilon)\zeta_i, \tag{21.7}$$

such that, for any given real number $r>0$ and for any graph $\mathcal{G}\in\mathbf{G}$, there exists an ε^* such that for any $\varepsilon\in(0,\varepsilon^*]$, synchronization is achieved for all initial conditions in the absence of disturbances, and the closed-loop transfer matrix from ω to $\bar{y}$, denoted by $T_{\omega\bar{y}}$, satisfies

$$\|T_{\omega\bar{y}}\|_2<r. \tag{21.8}$$

21.3 Protocol Design

We first design a static protocol for H_2 almost output synchronization problem with partial-state coupling. Thereafter, we show that H_∞ almost output synchronization problem with partial-state coupling is, in general, not solvable for heterogeneous

MAS with passive agents. The latter is still the case even if we restrict the agents further and assume they are minimum-phase.

21.3.1 H_2 Almost Output Synchronization

We design a static protocol for the H_2 almost output synchronization problem for a heterogeneous MAS with passive agents.

Protocol design 21.1 *Consider a MAS described by (21.1) and (21.3).*
We consider the protocol,

$$u_i = -\rho\zeta_i, \tag{21.9}$$

where $\rho = \frac{1}{\varepsilon}$ with ε a positive parameter.

In this chapter, we revisit the class of network graphs $\mathbb{G}^{b,N}$, i.e., the class of detailed balanced, strongly connected graphs with N nodes, as defined in Definition 4.28.

The main result regarding H_2 almost output synchronization problem via a static protocol is stated as follows:

Theorem 21.4 *Consider a MAS described by (21.1) and (21.3). Let the set of detailed balanced graphs $\mathbb{G}^{b,N}$ be given.*

If (A_i, B_i) is stabilizable and (A_i, C_i) is detectable, then the H_2 almost output synchronization problem, as defined in Problem 21.3, with $\mathbf{G} = \mathbb{G}^{b,N}$ is solvable. In particular, for any given real number $r > 0$ and for any graph $\mathcal{G} \in \mathbb{G}^{b,N}$, there exists an ε^*, such that for any $\varepsilon \in (0, \varepsilon^*)$, the protocol (21.9) achieves output synchronization (without disturbances) and an H_2 norm from ω to $y_i - y_j$ less than r for any $i, j \in \{1, \ldots, N\}$

Proof Note that in the absence of disturbance, output synchronization is achieved for any $\rho > 0$, as shown in [15]. Therefore, we only need to establish the almost output synchronization in the presence of disturbances. Let

$$x = \begin{pmatrix} x_1 \\ \vdots \\ x_N \end{pmatrix}, \quad u = \begin{pmatrix} u_1 \\ \vdots \\ u_N \end{pmatrix}, \quad y = \begin{pmatrix} y_1 \\ \vdots \\ y_N \end{pmatrix}, \quad \zeta = \begin{pmatrix} \zeta_1 \\ \vdots \\ \zeta_N \end{pmatrix}$$

and

$$A = \begin{pmatrix} A_1 & & \\ & \ddots & \\ & & A_N \end{pmatrix}, \qquad B = \begin{pmatrix} B_1 & & \\ & \ddots & \\ & & B_N \end{pmatrix},$$

$$C = \begin{pmatrix} C_1 & & \\ & \ddots & \\ & & C_N \end{pmatrix}, \qquad E = \begin{pmatrix} E_1 & & \\ & \ddots & \\ & & E_N \end{pmatrix}.$$

Then, we have

$$\zeta = (L \otimes I_p)y = (L \otimes I_p)Cx.$$

The closed-loop system can then be written as

$$\begin{aligned} \dot{x} &= \left[A - \rho B(L \otimes I_p)C\right]x + E\omega, \\ y &= Cx. \end{aligned} \tag{21.10}$$

Our first claim is that

$$A_c = A - \rho B(L \otimes I_p)C$$

is neutrally stable. Since our graph is detailed balanced, by Definition 4.28 there exists a positive diagonal matrix

$$Z = \begin{pmatrix} z_1 & & \\ & \ddots & \\ & & z_N \end{pmatrix},$$

such that

$$ZL + L'Z \geq 0.$$

Moreover, our agents are passive, and therefore there exists a P_i satisfying (21.2) for all $i = 1, \ldots, N$. Using this matrix Z and $P = \operatorname{diag}\{P_i\}$ we find:

$$\begin{aligned} (Z \otimes I_n)PA_c + [(Z \otimes I_n)PA_c]' &= -\rho C'\left[(ZL + L'Z) \otimes I_p\right]C \\ &\quad + \begin{pmatrix} z_1(P_1A_1 + A_1'P_1) & & \\ & \ddots & \\ & & z_N(P_NA_N + A_N'P_N) \end{pmatrix} \leq 0. \end{aligned}$$

Next, we consider (21.10) for $\omega = 0$. We obtain

$$x(t)'(Z \otimes I_n)Px(t) \leq x'(0)(Z \otimes I_n)Px(0).$$

Given that

$$x(t) = e^{A_c t}x(0),$$

we find that

$$\left\| [(Z \otimes I_n)P] e^{A_c t} \right\| \leq \|(Z \otimes I_n)P\| .$$

Next, consider (21.10) for $x(0) = 0$ and a disturbance $\omega \in L_1$. We get

$$[(Z \otimes I_n)P]x(t) = \int_0^t [(Z \otimes I_n)P] e^{A_c(t-\tau)} E\omega(\tau)\, d\tau$$

which yields for any $t > 0$ that

$$\|[(Z \otimes I_n)P]x(t)\| \leq \|(Z \otimes I_n)P\| \, \|E\| \|\omega\|_1$$

where $\| \cdot \|_1$ denotes the L_1-norm. We define,

$$V = x'(t)\, [(Z \otimes I_n)P]\, x(t). \tag{21.11}$$

We obtain,

$$\begin{aligned}
\dot{V} &= x'(t) \left[(Z \otimes I_n)PA + A'(Z \otimes I_n)P\right] x(t) \\
&\qquad - 2\rho x'(t)C' \left[(ZL + L'Z) \otimes I_p\right] Cx(t) + 2x'(t)(Z \otimes I_n)PE\omega(t) \\
&\leq -2\rho \left\| \left[(ZL + L'Z) \otimes I_p\right]^{1/2} Cx(t) \right\|^2 + \|(Z \otimes I_n)P\| \, \|E\|^2 \|\omega\|_1 \|\omega(t)\|.
\end{aligned}$$

By integrating this equation, we obtain

$$\left\| \left[(ZL + L'Z) \otimes I_p\right]^{1/2} Cx \right\|_2 \leq \frac{1}{\sqrt{2\rho}} \|(Z \otimes I_n)P\|^{1/2} \|E\| \|\omega\|_1.$$

Define S such that

$$\hat{y} = \begin{pmatrix} y_1 - y_N \\ \vdots \\ y_{N-1} - y_N \end{pmatrix} = Sy.$$

Note that from [178, Lemma 9], we have

$$\ker((ZL + L'Z) \otimes I_p) = \ker(L \otimes I_p) = \ker S,$$

and hence there exists a matrix X such that

$$S = X[(ZL + L'Z) \otimes I_p]^{\frac{1}{2}}.$$

Therefore, we find that

$$\|\hat{y}\|_2 \leq \tfrac{1}{\sqrt{2\rho}} \|X\| \, \|[(Z \otimes I_n)P]\|^{1/2} \, \|E\| \|\omega\|_1.$$

In other words,

$$\|\hat{y}\|_2 \leq \tfrac{M}{\sqrt{\rho}} \|\omega\|_1,$$

for some constant M (independent of ρ). However, this implies that the induced L_1 to L_2 norm from ω to $\hat{y}$ can be made arbitrarily small by choosing ρ sufficiently large.

According to [162] the H_2, norm is less than nN times the induced norm from L_1 to L_2 and hence can be made arbitrarily small by choosing ρ sufficiently large. ■

An interesting question is whether the above result can be extended to a larger class of systems. In particular, the class of agents which are minimum-phase is interesting in this regard because these agents can be made passive through an appropriate output feedback, see [46]. This is effectively done in [16] using an introspective protocol. The following example shows that H_2 almost output synchronization, in general, cannot be achieved for this class of systems.

Example 21.5 Consider a multi-agent system containing two minimum-phase agents with relative degree 1:

$$\begin{aligned} \dot{x}_1 &= x_1 + u_1 + \omega_1, \\ y_1 &= x_1, \end{aligned}$$

and

$$\begin{aligned} \dot{x}_2 &= 2x_2 + u_2 + \omega_2, \\ y_2 &= x_2, \end{aligned}$$

with an associated network graph $\mathcal{G}$ whose Laplacian is given by

$$L = \begin{pmatrix} 1 & -1 \\ -1 & 1 \end{pmatrix}.$$

In that case, if we use a static protocol, we get

$$\begin{aligned} u_1 &= -\rho(x_1 - x_2), \\ u_2 &= -\rho(x_2 - x_1). \end{aligned}$$

But then it is easily seen that the closed-loop transfer matrix from (ω_1, ω_2) to $y_1 - y_2$ is given by

$$\begin{pmatrix} \dfrac{s-2}{s^2 + (2\rho - 3)s + (2 - 3\rho)} & \dfrac{s-1}{s^2 + (2\rho - 3)s + (2 - 3\rho)} \end{pmatrix},$$

which is exponentially unstable for any ρ and therefore we obviously cannot achieve almost output synchronization.

21.3.2 H_∞ Almost Output Synchronization

Two results are provided to show that H_∞ almost output synchronization cannot be achieved for MAS with agents which are passive as well as minimum-phase.

Theorem 21.6 *The H_∞ almost output synchronization problem as stated in Problem 21.2 is, in general, not solvable for a heterogeneous MAS with passive agents.*

Proof Basically, we only need to construct one example for which the H_∞ almost output synchronization problem is not solvable.

Consider a MAS (21.1) consisting of two agents,

$$\begin{aligned} \dot{x}_1 &= u_1 + \omega_1, \\ y_1 &= x_1, \end{aligned}$$

and

$$\begin{aligned} \dot{x}_2 &= \begin{pmatrix} 0 & 1 \\ -1 & 0 \end{pmatrix} x_2 + \begin{pmatrix} 0 \\ 1 \end{pmatrix} u_2 + \begin{pmatrix} 0 \\ 1 \end{pmatrix} \omega_2, \\ y_2 &= \begin{pmatrix} 0 & 1 \end{pmatrix} x_2, \end{aligned}$$

with an associated network graph $\mathcal{G}$ whose Laplacian is given by

$$L = \begin{pmatrix} 1 & -1 \\ -1 & 1 \end{pmatrix}.$$

We use the protocol as given in (21.5),

$$u_i = F(\varepsilon)\zeta_i.$$

Since $F(\varepsilon)$ is scalar, we can, without loss of generality, set $\rho = F(\varepsilon)$. Then the closed-loop transfer matrix $T_{\omega\bar{y}}$ is

$$T_{\omega\bar{y}}(s) = \begin{pmatrix} \frac{s^2+1}{s^3+2\rho s^2+s+\rho} & \frac{s^2}{s^3+2\rho s^2+s+\rho} \end{pmatrix}.$$

The transfer matrix $\|T_{\omega\bar{y}}(s)\|$ at $s = \frac{\sqrt{2}}{2}i$ is equal to 2, which implies that for any $\rho > 0$,

$$\|T_{\omega\bar{y}}\|_{H_\infty} \geq 2.$$

■

Remark 21.7 One might ask whether a heterogeneous protocol might help in the above case. If we consider the same example as in the proof of the above theorem but with a more general protocol,

$$u_1 = -\rho_1\zeta_1, \qquad u_2 = -\rho_2\zeta_1,$$

with ρ_1 different from ρ_2, then it can be shown that we still cannot achieve H_∞ almost output synchronization.

We introduce next an additional assumption on invariant zeros of the agents.

Assumption 21.2 *We assume the agents are minimum-phase.*

Our next result is again a negative result as stated in the following theorem.

Theorem 21.8 *The H_∞ almost output synchronization problem as stated in Problem 21.2, in general, is not solvable for a heterogeneous MAS with passive and minimum-phase agents.*

Proof As before, we only need to construct one example for which the H_∞ almost output synchronization problem is not solvable.

Consider MAS (21.1) with two agents,

$$\begin{aligned} \dot{x}_1 &= \begin{pmatrix} -1 & 0 \\ 2 & -1 \end{pmatrix} x_1 + u_1 + \omega_1, \\ y_1 &= x_1, \end{aligned}$$

and

$$\begin{aligned} \dot{x}_2 &= \begin{pmatrix} -1 & 2 \\ 0 & -1 \end{pmatrix} x_2 + u_2 + \omega_2, \\ y_2 &= x_2. \end{aligned}$$

The Laplacian matrix associated with the communication network is as follows:

$$L = \begin{pmatrix} 1 & -1 \\ -1 & 1 \end{pmatrix}$$

We use the protocol,

$$u_i = F(\varepsilon)\zeta_i.$$

and the following notation:

$$F(\varepsilon) = -\begin{pmatrix} f_{11} & f_{12} \\ f_{21} & f_{22} \end{pmatrix}, \quad \bar{f}_1 = 2f_{11} + 2f_{21}, \quad \bar{f}_2 = 2f_{22} + 2f_{12}.$$

Then the closed-loop transfer matrix $T_{\omega\bar{y}}$ at $s = 0$ is

$$T_{\omega\bar{y}}(0) = \frac{1}{\bar{f}_1 + \bar{f}_2 + 1}\begin{pmatrix} 1 - \bar{f}_2 & -\bar{f}_2 & -\bar{f}_2 - 1 & -\bar{f}_2 - 2 \\ \bar{f}_1 + 2 & \bar{f}_1 + 1 & \bar{f}_1 & \bar{f}_1 - 1 \end{pmatrix},$$

It is easily checked that $T_{\omega\bar{y}}(0)$ is bounded away from zero for any $F(\varepsilon)$, which implies that $\|T_{\omega\bar{y}}\|_{H_\infty}$ is also bounded away from zero for any F and $\rho > 0$. ■

Chapter 22
Synchronization of Heterogeneous Continuous-/and Discrete-Time Linear MAS With Introspective Agents

22.1 Introduction

The initial research on MAS with heterogeneous networks often relied on introspective agents. Whenever agents are introspective, agents can be reshaped as asymptotically identical using a pre-feedback. See for instance the work [4, 16].

In the previous chapters we have focused on non-introspective agents. However, this chapter considers synchronization problems for heterogeneous continuous-/and discrete-time linear MAS with introspective agents. The objective is mainly to show how agents can then be reshaped as asymptotically identical. Moreover, the eigenvalues of the homogenized agents can be placed freely. To ease the protocol design, the eigenvalues are usually placed in the closed-left half complex plane for continuous-time agents (or in the closed unit disc for discrete-time agents).

After these preliminary feedbacks, the agents are basically identical and therefore most of the designs of the first half of this book addressed to homogeneous networks are directly applicable. This is illustrated by showing how to design a protocol to achieve synchronization for continuous-time agents.

This chapter is partially based on the literature [174].

22.2 Multi-Agent Systems

When agents in a heterogeneous network are introspective, i.e., agents have access to parts of their states, pre-compensators can be designed for each agent in the network such that all agents' dynamics are the same asymptotically. We call the heterogeneous network with the pre-compensators almost homogenized.

A. Saberi et al., *Synchronization of Multi-Agent Systems in the Presence of Disturbances and Delays*, Systems & Control: Foundations & Applications,
https://doi.org/10.1007/978-3-030-88148-1_22

Consider a MAS composed of N non-identical linear time-invariant agents of the form,

$$\begin{cases} \rho x_i = A_i x_i + B_i u_i, \\ y_i = C_i x_i, \\ z_{mi} = C_{mi} x_i, \end{cases} \qquad (i = 1, \dots, N) \tag{22.1}$$

where $x_i \in \mathbb{R}^{n_i}$, $u_i \in \mathbb{R}^{m_i}$, and $y_i \in \mathbb{R}^p$ are the state, input, and output of agent i. $z_{mi} \in \mathbb{R}^{p_{mi}}$ is the information that agent i has access to, for agents are introspective. In the above presentation, for continuous-time systems, ρ denotes the time derivative, i.e., $(\rho x_i)(t) = \dot{x}_i$ with $t \in \mathbb{R}$; while for discrete-time systems, ρ denotes the time shift, i.e., $(\rho x_i)(t) = x_i(t+1)$, with $t \in \mathbb{Z}$.

We make the following assumptions on the agent dynamics.

Assumption 22.1 *For each agent $i \in \{1, \dots, N\}$, we have*

- *(A_i, B_i, C_i) is right-invertible.*
- *(A_i, B_i) is stabilizable, and (A_i, C_i) is detectable.*
- *(A_i, C_{mi}) is detectable.*
- *Each agent has the order of infinite zero equal to n_{qi}, $i = 1, \dots, N$.*

The communication network among agents is exactly the same as that in Chap. 2 and provide each agent with the quantity,

$$\zeta_i = \sum_{j=1}^{N} a_{ij}(y_i - y_j) = \sum_{j=1}^{N} \ell_{ij} y_j. \tag{22.2}$$

22.3 Problem Formulation

Since all agents in the network are introspective, we can achieve output synchronization rather than regulated output synchronization among agents. We then define the output synchronization problem as follows.

Problem 22.1 Consider a MAS described by (22.1) and (22.2). Let $\mathbf{G}$ be a given set of graphs such that $\mathbf{G} \subseteq \mathbb{G}^N$. The *output synchronization* problem with a set of network graph $\mathbf{G}$ is to find, if possible, a linear time-invariant dynamic protocol of the form,

$$\begin{aligned} \rho \chi_i &= A_{ci} \chi_i + B_{ci} \operatorname{col}\{\zeta_i, z_{mi}\}, \\ u_i &= C_{ci} \chi_i + D_{ci} \operatorname{col}\{\zeta_i, z_{mi}\}, \end{aligned} \tag{22.3}$$

for $i = 1, \dots, N$ where $\chi_i \in \mathbb{R}^{N_{ci}}$, such that, for any graph $\mathcal{G} \in \mathbf{G}$ and for all the initial conditions of agents and their protocol, output synchronization among agents is achieved.

22.4 Protocol Design

The protocol design is composed of two steps. The first step is to design a pre-compensator for each agent such that the interconnection of the pre-compensator and the agent is asymptotically identical. The second step is to apply the protocol design in the homogeneous section directly on the shaped agents.

22.4.1 Homogenization of the Agents

Since each agent is introspective, we use the local information z_{mi} to manipulate the agent dynamics such that the dynamics for all the agents are almost identical. This is shown in the following lemma.

Lemma 22.2 *Consider a heterogeneous MAS of N agents (22.1) and (22.2). Let Assumption 22.1 hold, and we denote by n_d the maximal order of infinite zeros of (C_i, A_i, B_i), $i \in \{1, \dots, N\}$. Suppose a triple (C, A, B) is given such that*

1. $\operatorname{rank}(C) = p$,
2. *(C, A, B) is invertible, of uniform rank $n_q \geq n_d$, and has no invariant zero.*

Then for each agent $i \in \{1, \dots, N\}$, there exists a pre-compensator of the form,

$$\begin{cases} \rho\xi_i &= A_{i,h}\xi_i + B_{i,h}z_i + E_{i,h}v_i, \\ u_i &= C_{i,h}\xi_i + D_{i,h}v_i, \end{cases} \tag{22.4}$$

such that the interconnection of (22.1) and (22.4) can be written in the form,

$$\begin{cases} \rho\bar{x}_i &= A\bar{x}_i + B(v_i + d_i), \\ y_i &= C\bar{x}_i, \\ \zeta_i &= \sum_{j=1}^{N} \ell_{ij} y_j, \end{cases} \tag{22.5}$$

where d_i is given by

$$\begin{cases} \rho e_i &= A_{i,e}e_i, \\ d_i &= C_{i,e}e_i, \end{cases} \tag{22.6}$$

and $A_{i,e}$ is Hurwitz stable for the continuous-time case, and Schur stable for the discrete-time case.

Remark 22.3 Without loss of generality, we assume that the triple (C, A, B) has the form,

$$A = A_d + B_d H,\ \ A_d := \begin{pmatrix} 0 & I \\ 0 & 0 \end{pmatrix},\ \ B = B_d := \begin{pmatrix} 0 \\ I \end{pmatrix},\ \ C = C_d := \begin{pmatrix} I & 0 \end{pmatrix}, \tag{22.7}$$

where H is such that the matrix $A_d + B_d H$ has desired eigenvalues. The existence of such an H is guaranteed by the fact that (A_d, B_d) is controllable.

Proof The following proof is for continuous-time agents. It is exactly the same for discrete-time agents with obvious changes.

Let $n_q \geq \max_{i=1,\ldots,N}(n_{qi})$. Then, according to Theorem 24.2 in Chap. 24, there exists a pre-compensator to make agent $i \in \{1,\ldots,N\}$ invertible and of equal rank n_q,

$$\begin{cases} \dot{x}^1_{p,i} &= A^1_{p,i}x^1_{p,i} + B^1_{p,i}u^1_i, \\ u_i &= C^1_{p,i}x^1_{p,i}x^1_{p,i} + D^1_{p,i}u^1_i, \end{cases} \tag{22.8}$$

where $u^1_i \in \mathbb{R}^p$. Next, we concentrate on transforming different invertible system dynamics of equal rank to almost identical ones.

Let

$$\tilde{x}_i = \begin{pmatrix} x_i \\ x^1_{p,i} \end{pmatrix}.$$

There always exist nonsingular state transformation $\Gamma_{i,x}$ and input transformation $\Gamma_{i,u}$, see Chap. 23, such that

$$\tilde{x}_i = \Gamma_{i,x}\bar{x}_i, \qquad u^1_i = \Gamma_{i,u}u^2_i. \tag{22.9}$$

where

$$\bar{x}_i := \begin{pmatrix} \bar{x}_{i,a} \\ \bar{x}_{i,d} \end{pmatrix}.$$

Then, the interconnection of (22.1) and (22.8) can be written in the form,

$$\begin{cases} \dot{\bar{x}}_{i,a} &= \bar{A}_{i,a}\bar{x}_{i,a} + \bar{L}_{i,a}y_i, \\ \dot{\bar{x}}_{i,d} &= A_d\bar{x}_{i,d} + B_d(u^2_i + D_{i,a}\bar{x}_{i,a} + D_{i,d}\bar{x}_{i,d}), \\ y_i &= C_d\bar{x}_{i,d}, \end{cases} \tag{22.10}$$

where A_d, B_d, C_d are as defined in (22.7). Note that there is an output injection for the zero dynamics. Therefore, we can have internal stability even if the system is not minimum-phase.

Note that the information

$$\bar{z}_{mi} := \begin{pmatrix} z_{mi} \\ x^1_{p,i} \end{pmatrix}$$

is available for agent i, and $\bar{z}_{mi}$ can be represented in terms of $\bar{x}_{i,a}$, $\bar{x}_{i,d}$ as

$$\bar{z}_{mi} = \bar{C}_{mi}\begin{pmatrix}\bar{x}_{i,a}\\ \bar{x}_{i,d}\end{pmatrix}, \quad \text{where } \bar{C}_{mi} = \begin{pmatrix}C_{mi} & 0\\ 0 & I\end{pmatrix}\Gamma_{i,x}.$$

We define that, for $i = 1, \ldots, N$,

$$\bar{A}_i = \begin{pmatrix}\bar{A}_{i,a} & \bar{L}_{i,a}C_d\\ B_d D_{i,a} & A_d + B_d D_{i,d}\end{pmatrix}, \quad \bar{B}_i = \begin{pmatrix}0\\ B_d\end{pmatrix}.$$

Assumption 22.1 implies that $(C_{m,i}, A_i)$ is observable, which yields that $(\bar{C}_{m,i}, \bar{A}_i)$ is observable. We then design an observer-based pre-compensator for the system (22.10) as

$$\begin{cases}\dot{\hat{\bar{x}}}_i = \bar{A}_i\hat{\bar{x}}_i + \bar{B}_i u_i^2 - \bar{K}_i(\bar{z}_{mi} - \bar{C}_{mi}\hat{\bar{x}}_i),\\ u_i^2 = \begin{pmatrix}-D_{i,a} & H - D_{i,d}\end{pmatrix}\hat{\bar{x}}_i + v_i,\end{cases} \tag{22.11}$$

where $v_i \in \mathbb{R}^p$, $\bar{K}_i$ is chosen such that $\bar{A}_i + \bar{K}_i\bar{C}_{mi}$ is Hurwitz stable, and H is chosen such that $A_d + B_d H$ has desired eigenvalues. Define the observer error $e_i = \bar{x}_i - \hat{\bar{x}}_i$. It is easy to see that the observer error dynamics is asymptotically stable. Moreover, the effect of $\bar{x}_{i,a}$ on the dynamics $\bar{x}_{i,d}$ is asymptotically canceled. Thus, the mapping from the new input v_i to the output y_i is given by

$$\begin{cases}\dot{\bar{x}}_{i,d} = (A_d + B_d H)\bar{x}_{i,d} + B_d(v_i + d_i),\\ y_i = C_d\bar{x}_{i,d},\end{cases} \tag{22.12}$$

where

$$\begin{cases}\dot{e}_i = (\bar{A}_i + \bar{K}_i\bar{C}_{m,i})e_i,\\ d_i = \begin{pmatrix}D_{i,a} & D_{i,d} - \bar{F}_d\end{pmatrix}e_i.\end{cases}$$

■

22.4.2 Protocol Design

We can choose H such that $A_d + B_d H$ has eigenvalues in the closed left-hand half complex plane for continuous-time agents (or in the closed unit disc for discrete-time agents). This means that agents, after homogenization, are represented by (22.5) and (22.6), and are at most weakly unstable. We first show that the output synchronization among almost homogenized agents with partial-state coupling can be solved by equivalently solving a robust stabilization problem. Then, the protocol design in the homogeneous part can be applied for such a robust stabilization problem.

22.4.2.1 Connection to a Robust Stabilization Problem

Suppose the synchronization problem for a continuous-time MAS described by (22.5), (22.6) with $\rho\bar{x}_i = \dot{\bar{x}}_i$ and $\rho e_i = \dot{e}_i$, and (22.2) can be solved by a protocol

$$\begin{cases} \dot{\bar{\chi}}_i = A_c\bar{\chi}_i + B_c\zeta_i, \\ v_i = C_c\bar{\chi}_i, \end{cases} \tag{22.13}$$

with $\bar{\chi}_i \in \mathbb{R}^{n_c}$.

Let

$$\bar{\bar{x}}_i = \begin{pmatrix} \bar{x}_{i,d} \\ \bar{\chi}_i \end{pmatrix}.$$

Then, the closed-loop of each agent can be written as

$$\begin{cases} \dot{\bar{\bar{x}}}_i = \begin{pmatrix} A & BC_c \\ 0 & A_c \end{pmatrix} \bar{\bar{x}}_i + \begin{pmatrix} 0 \\ B_c \end{pmatrix} \zeta_i + \begin{pmatrix} B \\ 0 \end{pmatrix} d_i, \\ y_i = \begin{pmatrix} C & 0 \end{pmatrix} \bar{\bar{\chi}}_i, \\ \zeta_i = \sum_{j=1}^N \ell_{ij} y_i. \end{cases} \tag{22.14}$$

Define

$$\bar{\bar{x}} = \begin{pmatrix} \bar{\bar{x}}_1 \\ \vdots \\ \bar{\bar{x}}_N \end{pmatrix}, \qquad d = \begin{pmatrix} d_1 \\ \vdots \\ d_N \end{pmatrix},$$

and

$$\bar{A} = \begin{pmatrix} A & BC_c \\ 0 & A_c \end{pmatrix}, \quad \bar{B} = \begin{pmatrix} 0 \\ B_c \end{pmatrix}, \quad \bar{C} = \begin{pmatrix} C & 0 \end{pmatrix} \text{ and } \bar{E} = \begin{pmatrix} B \\ 0 \end{pmatrix}.$$

The overall dynamics of the N agents can be written as

$$\dot{\bar{\bar{x}}} = \left[I_N \otimes \bar{A} + L \otimes \bar{B}\bar{C} \right] \bar{\bar{x}} + (I_N \otimes \bar{E})d. \tag{22.15}$$

Similar to the observation in the case of partial-state coupling in Chap. 2, synchronization for the system (22.15) is equivalent to the asymptotic stability of the following $N - 1$ subsystems,

$$\dot{\tilde{\eta}}_i = (\bar{A} + \lambda_i \bar{B}\bar{C})\tilde{\eta}_i, \quad i = 2, \ldots, N, \tag{22.16}$$

where λ_i for $i = 2, \ldots, N$ are the non-zero eigenvalues of the Laplacian matrix L. This is formalized in the following theorem.

Theorem 22.4 *The MAS* (22.15) *achieves output synchronization if and only if the system* (22.16) *is globally asymptotically stable for* $i = 2, \ldots, N$.

Proof As we did in the proof of Theorem 2.15, we define the Jordan form (2.10) of the Laplacian. Let

$$\eta := \left(T^{-1} \otimes I_{n_q+n_c}\right) \bar{\bar{x}} = \begin{pmatrix} \eta_1 \\ \vdots \\ \eta_N \end{pmatrix},$$

where $\eta_i \in \mathbb{C}^{n_q+n_c}$. In the new coordinates, the dynamics of η can be written as

$$\dot{\eta} = \left[I_N \otimes \bar{A} + J_L \otimes \bar{B}\bar{C}\right]\eta + (T \otimes \bar{E})d.$$

Following the proof of Theorem 2.15, we obtain that the MAS (22.15) achieves output synchronization if and only if $\eta_i \to 0$ for $i = 2, \ldots, N$. Next, we show that $\eta_i \to 0$ for $i = 2, \ldots, N$ if and only if the system (22.16) is globally asymptotically stable for $i = 2, \ldots, N$.

Let

$$J_L = \begin{pmatrix} 0 & 0 & 0 & \ldots & 0 \\ 0 & \lambda_2 & \mu_2 & \ddots & \vdots \\ \vdots & \ddots & \ddots & \ddots & 0 \\ \vdots & & \ddots & \ddots & \mu_{N-1} \\ 0 & \ldots & \ldots & 0 & \lambda_N \end{pmatrix},$$

and

$$T = \begin{pmatrix} w_1 \\ w_2 \\ \vdots \\ w_N \end{pmatrix}$$

with $\mu_i \in \{0, 1\}$ for $i = 2, \ldots, N$.

Then, for $i = 2, \ldots, N$, we have

$$\dot{\eta}_i = \left(\bar{A} + \lambda_i \bar{B}\bar{C}\right)\eta_i + \mu_i \eta_{i+1} + (w_i \otimes \bar{E})d. \tag{22.17}$$

Since the signal d goes to zero, the stability of system (22.17) is equivalent to that of

$$\dot{\bar{\eta}}_i = \left(\bar{A} + \lambda_i \bar{B}\bar{C}\right) \bar{\eta}_i + \mu_i \bar{\eta}_{i+1}, \tag{22.18}$$

which proves the results. ∎

Remark 22.5 Similar to what we did in Theorem 2.17, we consider η_1 defined in the proof of Theorem 22.4 and show that it satisfies,

$$\dot{\eta}_1 = \bar{A}\eta_1 + (w_1 \otimes \bar{E})d, \quad \eta_1(0) = (w_1 \otimes I_{n_p+n_c})\bar{\bar{x}}(0), \tag{22.19}$$

where w_1 is the first row of T^{-1}, i.e., the normalized left eigenvector associated with the zero eigenvalue.

Note that the system (22.16) is exactly the same as the system (2.30) in Sect. 2.5.1, Chap. 2. Therefore, the protocol design in the homogeneous chapters can be applied here. As an example, we apply an ARE-based Protocol 2.4 from Sect. 2.5.3 for a continuous-time heterogeneous MAS after homogenization which has the form of (22.5) and (22.6) with $\rho\bar{x}_i = \dot{\bar{x}}_i$ and $\rho e_i = \dot{e}_i$.

22.4.2.2 ARE-Based Method

We use Protocol Design 2.4 in Sect. 2.5.3.1, which is rewritten here.

Protocol design 22.1 *Consider a MAS described by non-identical agents (22.1) and (22.2). Suppose a pre-compensator (22.4) is designed such that agents are almost homogenized, and the interconnection is given by (22.5) and (22.6).*

We choose an observer gain K such that $A + KC$ is Hurwitz stable. Next, we consider a feedback gain $F_\delta = -B'P_\delta$ where $P_\delta > 0$ is the unique solution of the continuous-time algebraic Riccati equation,

$$A'P_\delta + P_\delta A - \beta P_\delta BB'P_\delta + \delta I = 0, \tag{22.20}$$

where $\delta > 0$ is a low-gain parameter and β is the lower bound of the real part of the non-zero eigenvalues of the Laplacian. This results in the protocol,

$$\begin{cases} \dot{\bar{\chi}}_i = (A + KC)\bar{\chi}_i - K\zeta_i, \\ v_i = F_\delta \bar{\chi}_i. \end{cases} \tag{22.21}$$

The main result regarding this design can be stated as follows.

Theorem 22.6 *Consider a MAS described by non-identical agents* (22.1) *and* (22.2). *Let any $\alpha \geq \beta > 0$ be given, and hence a set of network graphs $\mathbb{G}^N_{\alpha,\beta}$ be defined.*

If (A_i, B_i) is stabilizable, (C_i, A_i) is detectable, (C_{mi}, A_i) is detectable, and (A_i, B_i, C_i) is right-invertible, then the output synchronization problem via a dynamic protocol stated in Problem 22.3 *with $\mathbf{G} = \mathbb{G}^N_{\alpha,\beta}$ is solvable. In particular, there exists a $\delta^* > 0$ such that for any $\delta \in (0, \delta^*]$, the combination of pre-compensator* (22.4) *and protocol* (22.21) *solves the output synchronization problem for any graph $\mathcal{G}$ which is in $\mathbb{G}^N_{\alpha,\beta}$ for some N. Moreover, the synchronized trajectory is given by* (22.19).

Proof Using Theorem 22.4, we only need to prove that the system

$$\begin{cases} \dot{x} = Ax - \lambda BB'P_\delta \chi, \\ \dot{\chi} = (A + KC)\chi - KCx, \end{cases} \tag{22.22}$$

is asymptotically stable for any λ that satisfies $\mathrm{Re}(\lambda) \geq \beta$ and $|\lambda| < \alpha$. The remaining proof follows exactly that of Theorem 2.27. ∎

Chapter 23
A Special Coordinate Basis (SCB) of Linear Multivariable Systems

23.1 Introduction

What is called the special coordinate basis (SCB) of a multivariable linear time-invariant system plays a dominant role throughout this book; hence, its clear understanding is essential. The purpose of this chapter is to recall the SCB as well as its properties pertinent to this book. The SCB as originated in [109, 111, 112] was crystallized for strictly proper systems in [110] and for proper systems in [98]. Our presentation of the SCB here omits all the proofs that can be found in the literature.

What is the SCB? It is a fine-grained structural decomposition of a multivariable linear time-invariant system. It partitions a given system into separate but interconnected subsystems that reflect the architectural mapping of inner workings of the system. By this we mean that the SCB identifies all pertinent structural elements of a system and their functions, and most significantly, it also displays the interconnections among all such structural elements. In doing so, the SCB representation explicitly reveals the system's finite and infinite zero structure, and invertibility properties. Since its introduction, the SCB has been used in a large body of research, on topics including loop transfer recovery, time-scale assignment, disturbance rejection, H_2 control, and H_∞ control. It has also been used as a fundamental tool in the study of linear systems theory. For details on these topics, we refer to the books [12, 14, 99, 102, 105] all of which are based on the SCB, and the references therein. Other topics include multi-variable root loci [109, 111], decoupling theory [110], factorization of linear systems [13], squaring down of non-square systems [95, 98, 110], and model order reduction [82]. As will be clear to the readers, the influence of the SCB will be felt amply throughout this book and often from different angles.

A. Saberi et al., *Synchronization of Multi-Agent Systems in the Presence of Disturbances and Delays*, Systems & Control: Foundations & Applications,
https://doi.org/10.1007/978-3-030-88148-1_23

23.2 The Special Coordinate Basis

We present the SCB in this section. For readers unfamiliar with the topic, the complexities of the SCB may initially appear overwhelming. This is only a reflection, however, of the inherent complexities that exist in general multivariable linear time invariant systems. In the following exposition, significant complexity is added to accommodate general non-strictly proper multivariable systems. To get an overview of the SCB of progressively complex systems, we recommend first reading the SCB of uniform rank systems [109], the SCB of invertible systems [111], the SCB of strictly proper systems [110], and then the SCB of general multivariable systems as presented shortly. Also, the complexities encountered can be dissipated by carefully following various notations used.

We consider a linear time-invariant system Σ_* characterized by a quadruple (A, B, C, D). Let the dynamic equations of Σ_* be

$$\Sigma_* : \begin{cases} \rho x = Ax + Bu \\ y = Cx + Du, \end{cases} \tag{23.1}$$

where ρ is an operator indicating the time derivative $\frac{d}{dt}$ for continuous-time systems and a forward unit time shift for discrete-time systems. Also, $x \in \mathbb{R}^n$ is the state, $u \in \mathbb{R}^m$ is the control, and $y \in \mathbb{R}^p$ is the output. Without loss of generality, we assume that $(B' \; D')'$ and $(C \; D)$ have full rank.

Next, it is simple to verify that nonsingular transformations $\tilde{U}$ and $\tilde{V}$ exist such that

$$\tilde{U} D \tilde{V} = \begin{pmatrix} I_{m_0} & 0 \\ 0 & 0 \end{pmatrix}, \tag{23.2}$$

where m_0 is the rank of matrix D. Hence, hereafter, without loss of generality, it is assumed that the matrix D has the form given on the right-hand side of (23.2). As such, without loss of generality, we can focus on a system Σ_* of the form,

$$\Sigma_* : \begin{cases} \rho x = Ax + \begin{pmatrix} B_0 & \hat{B}_1 \end{pmatrix} \begin{pmatrix} u_0 \\ \hat{u}_1 \end{pmatrix} \\ \begin{pmatrix} y_0 \\ \hat{y}_1 \end{pmatrix} = \begin{pmatrix} C_0 \\ \hat{C}_1 \end{pmatrix} x + \begin{pmatrix} I_{m_0} & 0 \\ 0 & 0 \end{pmatrix} \begin{pmatrix} u_0 \\ \hat{u}_1 \end{pmatrix}, \end{cases} \tag{23.3}$$

where the matrices B_0, $\hat{B}_1$, C_0, and $\hat{C}_1$ have appropriate dimensions.

By nonsingular transformation of the state, output, and input, the system (23.3) can be transformed to the SCB. We use the symbols $\tilde{x}$, $\tilde{y}$, and $\tilde{u}$ to denote the state, output, and input of the system transformed to the SCB form. The transformations between the original system (23.3) and the SCB are called Γ_s, Γ_y, and Γ_u, and we write $x = \Gamma_s \tilde{x}$, $y = \Gamma_y \tilde{y}$, and $u = \Gamma_u \tilde{u}$.

The state $\tilde{x}$, output $\tilde{y}$, and input $\tilde{u}$ are partitioned as

$$\tilde{x} = \begin{pmatrix} x_a \\ x_b \\ x_c \\ x_d \end{pmatrix}, \qquad \tilde{y} = \begin{pmatrix} y_0 \\ y_d \\ y_b \end{pmatrix}, \qquad \tilde{u} = \begin{pmatrix} u_0 \\ u_d \\ u_c \end{pmatrix}.$$

Each component of state $\tilde{x}$ represents a particular subsystem described in the next section. In the partition of output $\tilde{y}$, y_0 is the original part of output as given in (23.3), y_d is the output from the x_d subsystem, and y_b is the output from the x_b subsystem. Similarly, in the partition of input $\tilde{u}$, u_0 is the original part of input as given in (23.3), u_d is the input to the x_d subsystem, and u_c is the input to the x_c subsystem. Because u_0 appears first in both u and $\tilde{u}$, Γ_u is of the form $\operatorname{diag}(I_{m_0}, \bar{\Gamma}_u)$, for some nonsingular $\bar{\Gamma}_u$.

23.2.1 Structure of the SCB

Consider first the case when the general system Σ_* as given in (23.1) is strictly proper, that is, the matrix $D = 0$, and consequently u_0 and y_0, do not exist in (23.3). The meaning of the four subsystems can be explained as follows:

- The x_a subsystem represents the zero dynamics. This part of the system is not directly affected by any inputs, nor does it affect any outputs directly. It may be affected, however, by the outputs y_b and y_d from x_b and x_d subsystems.
- The x_b subsystem has a direct effect on the output y_b, but it is not directly affected by any inputs. It may be affected, however, by the output y_d from the x_d subsystem. The x_b subsystem is observable from the output y_b. The existence of x_b subsystem renders the given system Σ_* non-right invertible. That is, Σ_* is right invertible whenever x_b subsystem is nonexistent.
- The x_c subsystem is directly affected by the input u_c, but it does not have a direct effect on any outputs. It may also be affected by the outputs y_b and y_d from the x_b and x_d subsystems, as well as the state x_a. However, the influence from x_a is matched with the input u_c (that is, the influence from x_a is additive to that of the input u_c). The x_c subsystem is controllable from the input u_c. The existence of x_c subsystem renders the given system Σ_* non-left invertible. That is, Σ_* is left invertible whenever x_c subsystem is nonexistent.
- The x_d subsystem represents the infinite zero structure. This part of the system is directly affected by the input u_d, and it also affects the output y_d directly. The x_d subsystem can be further partitioned into m_d single-input single-output (SISO) subsystems with states x_i and outputs y_i for $i = 1, \ldots, m_d$. Each of these subsystems consist of a chain of integrators of length q_i, from the i'th element of u_d (denoted by u_i) to the i'th element of y_d (denoted by y_i). Each integrator chain may be affected at each stage by the output y_d from the x_d subsystem, and

at the lowest level of the integrator chain (where the input appears), it may be affected by all the states of the system. The x_d subsystem is observable from y_d, and controllable from u_d.

The structure of strictly proper SCB systems is summarized in Table 23.1. For non-strictly proper systems the structure is the same, except for the existence of the direct-feedthrough output y_0, which is directly affected by the input u_0, and can be affected by any of the states of the system. It can also affect all the states of the system.

A block diagram of the SCB is given in Figs. 23.1, 23.2, 23.4, and 23.4. Figure 23.1 expresses the zero dynamics. Figure 23.2 represents the dynamics that is present if and only if the system Σ_* is not right invertible. Figure 23.3 represents the dynamics that is present if and only if the system Σ_* is not left invertible. Finally, the dynamics in Fig. 23.4 is related to the infinite zero structure. In this last figure, a signal given by a double-edged arrow is some linear combination of outputs y_i, $i = 0$ to m_d, whereas the signal, given by the double-edged arrow with a solid dot in it, is some linear combination of all states.

Definition 23.1 A linear time-invariant dynamics (A, B, C, D) is left invertible if, given zero initial conditions, there exists only one input signal u, namely, $u = 0$, that ensures $y(t) = 0$ for all $t \geq 0$. For single-input-single-output system, a system is left invertible if and only if its transfer function is nonzero.

Table 23.1 Summary of strictly proper SCB structure. The *Interconnections* column indicates influences from other subsystems.

Subsystem	Input	Output	Interconnections
x_a	–	–	y_b, y_d
x_b	–	y_b	y_d
x_c	u_c	–	y_b, y_d, $x_a^\star$
x_d	u_d	y_d	$x_a^\star$, $x_b^\star$, $x_c^\star$

*Matched with input

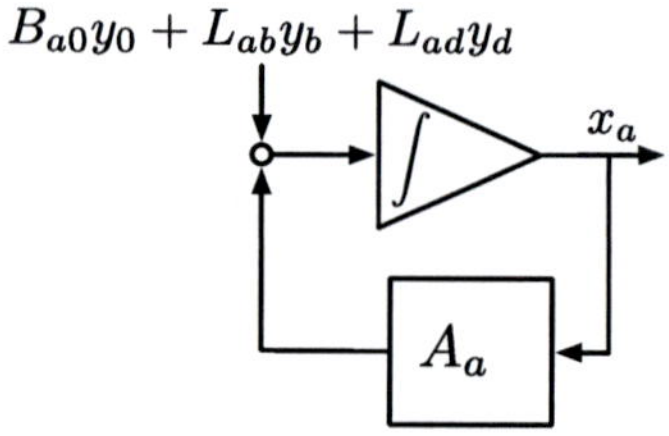

Fig. 23.1 Zero dynamics

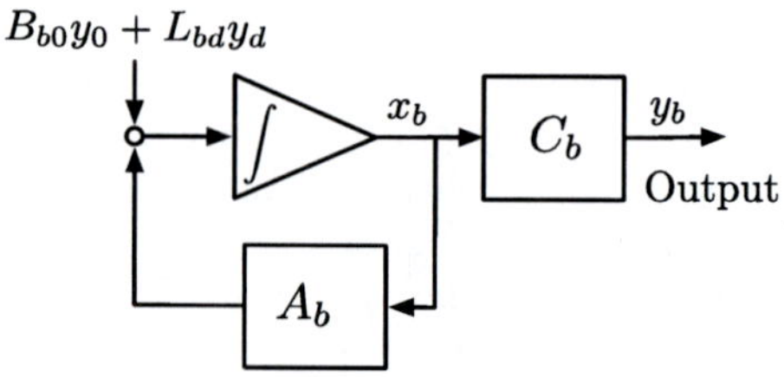

Fig. 23.2 Non-right-invertible dynamics

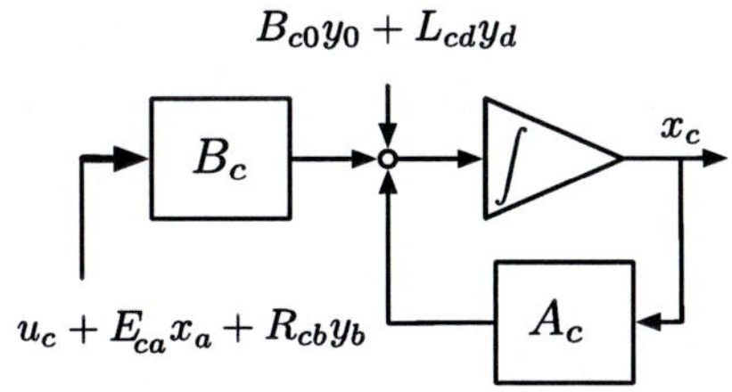

Fig. 23.3 Non-left-invertible dynamics

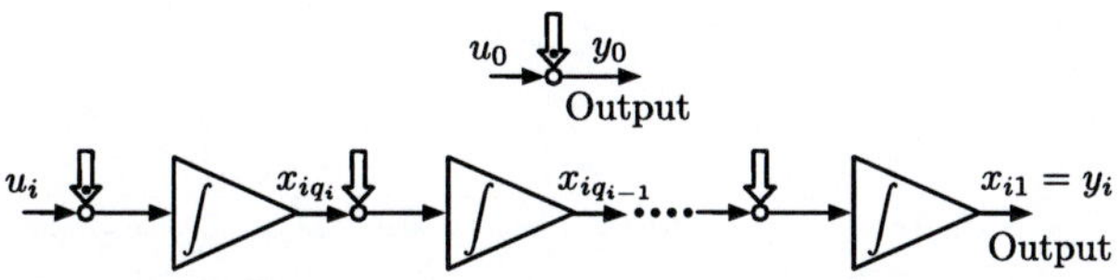

Fig. 23.4 Infinite zero structure

Definition 23.2 A linear time-invariant dynamics (A, B, C, D) is right invertible if, given a smooth reference output y_r, there exists an initial condition $x(0)$ and an input u that ensures $y(t) = y_r(t)$ for all $t \geq 0$. For single-input-single-output system, a system is right invertible if and only if its transfer function is nonzero.

23.2.2 SCB Equations

The SCB representation of the system Σ_* as given in (23.3) is articulated by the following theorem.

Theorem 23.3 *For any given system Σ_* characterized by the matrix quadruple (A, B, C, D), there exist*

1. *unique coordinate free non-negative integers,* $n_a(\Sigma_*)$, $n_b(\Sigma_*)$, $n_c(\Sigma_*)$, $n_d(\Sigma_*)$, $m_d \leq m - m_0$, *and* q_i, $i = 1, \ldots, m_d$, *and*
2. *nonsingular state, output, and input transformations Γ_s, Γ_y, and Γ_u that take the given Σ_* into the SCB that displays explicitly both the finite and the infinite zero structures of Σ_* as well as its invertibility properties.*

The SCB is described by the following set of equations:

$$\rho x_a = A_a x_a + B_{a0} y_0 + L_{ad} y_d + L_{ab} y_b, \tag{23.4}$$

$$\rho x_b = A_b x_b + B_{b0} y_0 + L_{bd} y_d, \tag{23.5}$$

$$\rho x_c = A_c x_c + B_{c0} y_0 + L_{cd} y_d + B_c(E_{ca} x_a + R_{cb} y_b) + B_c u_c, \tag{23.6}$$

$$y_0 = C_{0a}x_a + C_{0b}x_b + C_{0c}x_c + C_{0d}x_d + u_0, \tag{23.7}$$

$$y_b = C_b x_b, \tag{23.8}$$

and for each $i = 1, \ldots, m_d$,

$$\rho x_i = A_{q_i}x_i + L_{i0}y_0 + L_{id}y_d + B_{q_i}\left(u_i + E_{ia}x_a + E_{ib}x_b + E_{ic}x_c + \sum_{j=1}^{m_d} E_{ij}x_j\right), \tag{23.9}$$

$$y_i = C_{q_i}x_i. \tag{23.10}$$

Here the states x_a, x_b, x_c, *and* x_d *have dimension* $n_a(\Sigma_*)$, $n_b(\Sigma_*)$, $n_c(\Sigma_*)$, *and* $n_d(\Sigma_*) = \sum_{i=1}^{m_d} q_i$, *respectively, whereas* x_i *is of dimension* q_i *for each* $i = 1, \ldots, m_d$. *The control vectors* u_0, u_d, *and* u_c *have, respectively, dimensions* m_0, m_d, *and* $m_c = m - m_0 - m_d$, *whereas the output vectors* y_0, y_d, *and* y_b *have, respectively, dimensions* $p_0 = m_0$, $p_d = m_d$, *and* $p_b = p - p_0 - p_d$. *Also, we have*

$$x = \Gamma_s\tilde{x}, \qquad y = \Gamma_y\tilde{y}, \qquad u = \Gamma_u\tilde{u},$$

$$\tilde{x} = \begin{pmatrix} x_a \\ x_b \\ x_c \\ x_d \end{pmatrix}, \qquad x_d = \begin{pmatrix} x_1 \\ x_2 \\ \vdots \\ x_{m_d} \end{pmatrix},$$

$$\tilde{y} = \begin{pmatrix} y_0 \\ y_d \\ y_b \end{pmatrix}, \qquad y_d = \begin{pmatrix} y_1 \\ y_2 \\ \vdots \\ y_{m_d} \end{pmatrix},$$

$$\tilde{u} = \begin{pmatrix} u_0 \\ u_d \\ u_c \end{pmatrix}, \qquad u_d = \begin{pmatrix} u_1 \\ u_2 \\ \vdots \\ u_{m_d} \end{pmatrix}.$$

The x_i $(i = 1, \dots, m_d)$ together form x_d and, similarly, the y_i $(i = 1, \dots, m_d)$ together form y_d, and

$$y_d = C_d x_d, \quad \text{where} \quad C_d = \begin{pmatrix} C_{q_1} & 0 & \cdots & 0 \\ 0 & \ddots & \ddots & \vdots \\ \vdots & \ddots & \ddots & 0 \\ 0 & \cdots & 0 & C_{q_{m_d}} \end{pmatrix}. \tag{23.11}$$

The matrices A_{q_i}, B_{q_i}, and C_{q_i} have the following form:

$$A_{q_i} = \begin{pmatrix} 0 & I_{q_i - 1} \\ 0 & 0 \end{pmatrix}, \quad B_{q_i} = \begin{pmatrix} 0 \\ 1 \end{pmatrix}, \quad C_{q_i} = \begin{pmatrix} 1 & 0 \end{pmatrix}. \tag{23.12}$$

(Obviously, for the case when $q_i = 1$, we have $A_{q_i} = 0$, $B_{q_i} = 1$, and $C_{q_i} = 1$). Clearly, (A_{q_i}, B_{q_i}) and (C_{q_i}, A_{q_i}) form, respectively, controllable and observable pairs. This implies that all the states x_i are both controllable and observable. Assuming that the x_i are arranged such that $q_i \le q_{i+1}$, the matrix L_{id} has the particular form

$$L_{id} = \begin{pmatrix} L_{i1} & L_{i2} & \cdots & L_{i\,i-1} & 0 & 0 & \cdots & 0 \end{pmatrix}.$$

The last row of each L_{id} is identically zero. Furthermore, the pair (A_c, B_c) is controllable, and the pair (C_b, A_b) is observable.

23.2.3 A Compact Form

We can rewrite the SCB given by Theorem 23.3 in a more compact form as a system characterized by the quadruple $(\tilde{A}, \tilde{B}, \tilde{C}, \tilde{D})$,

$$\begin{cases} \rho\tilde{x} = \tilde{A}\tilde{x} + \tilde{B}\begin{pmatrix} y_0 \\ u_d \\ u_c \end{pmatrix} \\ \tilde{y} \ = \tilde{C}\tilde{x} + \tilde{D}\tilde{u}, \end{cases} \tag{23.13}$$

where

$$\tilde{A} := \Gamma_s^{-1}(A - B_0 C_0)\Gamma_s = \begin{pmatrix} A_a & L_{ab}C_b & 0 & L_{ad}C_d \\ 0 & A_b & 0 & L_{bd}C_d \\ B_c E_{ca} & B_c R_{cb} C_b & A_c & L_{cd}C_d \\ B_d E_{da} & B_d E_{db} & B_d E_{dc} & A_d \end{pmatrix}, \tag{23.14}$$

and where

$$A_d = \begin{pmatrix} A_{q_1} & 0 & \cdots & 0 \\ 0 & \ddots & \ddots & \vdots \\ \vdots & \ddots & \ddots & 0 \\ 0 & \cdots & 0 & A_{q_{m_d}} \end{pmatrix} + L_{dd}C_d + B_d E_{dd}, \tag{23.15a}$$

$$B_d = \begin{pmatrix} B_{q_1} & 0 & \cdots & 0 \\ 0 & \ddots & \ddots & \vdots \\ \vdots & \ddots & \ddots & 0 \\ 0 & \cdots & 0 & B_{q_{m_d}} \end{pmatrix}, \tag{23.15b}$$

$$\tilde{B} := \Gamma_s^{-1} \left(B_0 \ \hat{B}_1\right) \Gamma_u = \begin{pmatrix} B_{a0} & 0 & 0 \\ B_{b0} & 0 & 0 \\ B_{c0} & 0 & B_c \\ B_{d0} & B_d & 0 \end{pmatrix}, \tag{23.16}$$

$$\tilde{C} := \Gamma_y^{-1} \begin{pmatrix} C_0 \\ \hat{C}_1 \end{pmatrix} \Gamma_s = \begin{pmatrix} C_{0a} & C_{0b} & C_{0c} & C_{0d} \\ 0 & 0 & 0 & C_d \\ 0 & C_b & 0 & 0 \end{pmatrix}, \tag{23.17}$$

and

$$\tilde{D} := \Gamma_y^{-1} D \Gamma_u = \begin{pmatrix} I_{m_0} & 0 & 0 \\ 0 & 0 & 0 \\ 0 & 0 & 0 \end{pmatrix}. \tag{23.18}$$

In the above equations, if one needs expanded expressions for the matrices E_{da}, E_{db}, E_{dc} and E_{dd}, they can easily be obtained from (23.9). Note that we always have that (A_c, B_c) and (A_d, B_d) are controllable while (C_b, A_b) is observable.

Admittedly, the SCB of Theorem 23.3 looks complicated with all its innate decompositions of state, output, and input variables. However, as will be evident throughout the book, the SCB of a linear system clearly displays its underlying structure. In fact, the proofs of several theorems, lemmas, and properties stated in later chapters will be hard to follow without having endured the intricacies of the SCB.

23.3 Properties of the SCB

The SCB is closely related to the canonical form of Morse [73], which is obtained through transformations of the state, input, and output spaces, and the application of state feedback and output injection. A system in the canonical form of Morse consists of four decoupled subsystems that reflect essential geometric properties of the original system. The SCB form of a system largely reflects the same properties; however, the SCB is obtained through transformations of the state, input, and output spaces alone, without the application of state feedback and output injection. Thus, the SCB is merely a representation of the original system in a different coordinate basis, and it can therefore be used directly for design purposes.

Some properties of the SCB, which correspond directly to the properties of the canonical form of Morse, are first summarized as follows:

- The invariant zeros of the system Σ_* given in (23.1) are the eigenvalues of the matrix A_a.
- The system Σ_* given in (23.1) is right invertible if, and only if, the subsystem x_b is nonexistent.
- The system Σ_* given in (23.1) is left invertible if, and only if, the subsystem x_c is nonexistent.
- The system Σ_* given in (23.1) is invertible if, and only if, both the subsystem x_b and the subsystem x_c are nonexistent.
- The system Σ_* given in (23.1) has m_0 infinite zeros of order 0 and $i\bar{q}_i$ infinite zeros of order i, where $\bar{q}_i$ is the number integrator chains of length i in the x_d subsystem.

By studying the dynamics of the x_a subsystem and its connections to the rest of the system, one obtains a precise description of the invariant zero dynamics of the system and the classes of input signals that may be blocked by these zeros. The information thus obtained goes beyond what can be obtained through the notions of state and input *pseudo zero directions* (see [65, 100]).

The representation of the infinite zero structure through integrator chains in the x_d subsystem allows for the explicit construction of high-gain controllers and observers in a general multiple-input multiple-output setting (see, e.g., [97]). This removes unnecessary restrictions of square-invertibility and uniform relative degree that are found in much of the high-gain literature.

We next describe in detail the pertinent properties of the SCB as summarized above; each main property is stated in a subsection devoted to it. The properties discussed below are true for both continuous- and discrete-time systems. However, sometimes, for convenience of writing, we use the notations commonly used for continuous-time systems. The reader can easily decipher the corresponding notations for discrete-time systems. For clarity, whenever it is needed, we repeat our discussion for discrete-time systems.

23.3.1 Observability (Detectability) and Controllability (Stabilizability)

In this subsection, we examine the issues related to observability, detectability, controllability, and stabilizability of a system via its SCB. Note that we simply use detectability and stabilizability, which for continuous-time systems refers to $\mathbb{C}^-$-detectability and $\mathbb{C}^-$-stabilizability, whereas for discrete-time systems, this refers to $\mathbb{C}^\ominus$-detectability and $\mathbb{C}^\ominus$-stabilizability.

We have the following property.

Property 23.4 We note that (C_b, A_b) and (C_{q_i}, A_{q_i}) form observable pairs. Unobservability can arise only in the variables x_a and x_c. In fact, the given system Σ_* is observable (detectable) if and only if $(C_{\rm obs}, A_{\rm obs})$ is observable (detectable), where

$$A_{\rm obs} = \begin{pmatrix} A_a & 0 \\ B_c E_{ca} & A_c \end{pmatrix}, \quad C_{\rm obs} = \begin{pmatrix} C_{0a} & C_{0c} \\ B_d E_{da} & B_d E_{dc} \end{pmatrix},$$

Similarly, (A_c, B_c) and (A_{q_i}, B_{q_i}) form controllable pairs. Basically, the variables x_a and x_b determine the controllability of the system. In fact, Σ_* is controllable (stabilizable) if and only if $(A_{\rm con}, B_{\rm con})$ is controllable (stabilizable), where

$$A_{\rm con} = \begin{pmatrix} A_a & L_{ab} C_b \\ 0 & A_b \end{pmatrix}, \quad B_{\rm con} = \begin{pmatrix} B_{a0} & L_{ad} \\ B_{b0} & L_{bd} \end{pmatrix},$$

23.3.2 Left and Right Invertibility

In this subsection, we examine the invertibility properties of Σ_* via its SCB. Let us first recall from [74] the definition of right and left invertibility.

Definition 23.5 Consider a linear system Σ_*.

- Let u_1 and u_2 be any inputs to the system Σ_*, and let y_1 and y_2 be the corresponding outputs (for the same initial conditions). Then Σ_* is said to be **left invertible**, if $y_1(t) = y_2(t)$ for all $t \geq 0$ implies that $u_1(t) = u_2(t)$ for all $t \geq 0$.
- The system Σ_* is said to be **right invertible** if, for any $y_{\rm ref}(t)$ defined on $[0, \infty)$, an input u and a choice of $x(0)$ exist such that $y(t) = y_{\rm ref}(t)$ for all $t \in [0, \infty)$.
- The system Σ_* is said to be invertible if the system is both left and right invertible.

Remark 23.6 One can easily deduce the following:

1. Σ_* is right invertible if and only if its transfer function matrix is a surjective rational matrix.

2. Σ_* is right invertible if and only if the rank of $P_{\Sigma_*}(s) = n + p$ for all but finitely many $s \in \mathbb{C}$, where the polynomial matrix $P_{\Sigma_*}(s)$ is the Rosenbrock system matrix of Σ_* defined as

$$P_{\Sigma_*}(s) := \begin{pmatrix} sI - A & -B \\ C & D \end{pmatrix}.$$

3. Σ_* is left invertible if and only if its transfer function matrix is an injective rational matrix.
4. Σ_* is left invertible if and only if the rank of $P_{\Sigma_*}(s) = n + m$ for all but finitely many $s \in \mathbb{C}$.

We have the following property connecting these properties to the SCB:

Property 23.7 The given system Σ_* is right invertible if and only if x_b and hence y_b are nonexistent ($n_b = 0$, $p_b = 0$), left invertible if and only if x_c and hence u_c are nonexistent ($n_c = 0$, $m_c = 0$), and invertible if and only if both x_b and x_c are nonexistent. Moreover, Σ_* is called degenerate if it is neither left nor right invertible.

23.3.3 *Finite Zero Structure*

In this subsection, we recall first the definition of invariant zeros of a system and their generalized associated right state and input zero direction chains, and then we discuss how SCB exhibits them in its structure.

The invariant zeros of a system Σ_* that is characterized by (A, B, C, D) are defined via the Smith canonical form of the Rosenbrock system matrix $P_{\Sigma_*}(s)$. Let us first briefly recall the Smith canonical form for any polynomial matrix $P(s) \in \mathbb{R}^{n\times m}[s]$. It is well known, see e.g., [28], that for any $P(s) \in \mathbb{R}^{n\times m}[s]$, there exist unimodular[1] matrices $U(s) \in \mathbb{R}^{n\times n}[s]$, $V(s) \in \mathbb{R}^{m\times m}[s]$ and $\Psi(s) \in \mathbb{R}^{n\times m}[s]$ with the latter of the form

$$\Psi(s) = \begin{pmatrix} \psi_1(s) & 0 & \cdots & & \cdots\cdots & 0 \\ 0 & \ddots & \ddots & & & \vdots \\ \vdots & & \ddots & \psi_r(s) & \ddots & \vdots \\ \vdots & & & \ddots & 0 & \ddots \vdots \\ \vdots & & & & \ddots & \ddots 0 \\ 0 & \cdots & \cdots & \cdots & 0 & 0 \end{pmatrix},$$

[1] A polynomial matrix in $\mathbb{R}^{n\times m}[s]$ that is invertible with a polynomial inverse is called unimodular.

such that

$$P(s) = U(s)\Psi(s)V(s).$$

Here $\Psi(s)$ is called the Smith canonical form of $P(s)$ when the $\psi_i(s)$ are monic polynomials with the property that $\psi_i(s)$ divides $\psi_{i+1}(s)$ for $i = 1, \ldots, r-1$, and r is the normal rank of the matrix $P(s)$. The polynomials $\psi_i(s)$ are called the invariant factors of $P(s)$. Their product $\psi(s) = \psi_1(s)\psi_2(s)\cdots\psi_r(s)$ is called the zero polynomial of $P(s)$. Each invariant factor $\psi_i(s)$, $i = 1, 2, \ldots, r$, can be written as a product of linear factors

$$\psi_i(s) = (s - \lambda_{i1})^{\alpha_{i1}}(s - \lambda_{i2})^{\alpha_{i2}} \cdots (s - \lambda_{ik_i})^{\alpha_{ik_i}}, \quad i = 1, 2, \ldots, r,$$

where $\lambda_{ik} \neq \lambda_{i\ell}$ ($k \neq \ell$) are complex numbers and α_{ik} ($k, \ell \in \{1, \ldots, k_i\}$) are positive integers. Then the complete set of factors, $(s - \lambda_{ik})^{\alpha_{ik}}$, $k = 1, 2, \ldots, k_i$, and $i = 1, 2, \ldots, r$, are called the *elementary divisors* of the polynomial matrix $P(s)$.

Now we are ready to recall the definition of the invariant zeros [65] of Σ_*.

Definition 23.8 The roots of the zero polynomial $\psi(s)$ of the (Rosenbrock) system matrix $P_{\Sigma_*}(s)$ are called the **invariant zeros** of Σ_*.

Remark 23.9 It is obvious from the above definition that an alternative way of defining an invariant zero of Σ_* is as follows: $\lambda \in \mathbb{C}$ is called an invariant zero of Σ_* if the rank of $P_{\Sigma_*}(\lambda)$ is strictly smaller than the normal rank of $P_{\Sigma_*}(s)$. Note that the normal rank is defined as the rank of a polynomial or rational matrix in all but finitely many $s \in \mathbb{C}$.

The SCB of Theorem 23.3 shows explicitly the invariant zeros of the system. To be more specific, we have the following property.

Property 23.10 The invariant zeros of Σ_* are the eigenvalues of A_a.

For continuous-time systems, if all invariant zeros of a system Σ_* are in $\mathbb{C}^-$, then we say Σ_* is minimum-phase; otherwise, Σ_* is said to be non-minimum-phase. Those invariant zeros that are in $\mathbb{C}^-$ are called the stable invariant zeros. Also, those that are not in $\mathbb{C}^-$ are called the unstable invariant zeros. Analogously, for discrete-time systems, if all the invariant zeros of a system Σ_* are in $\mathbb{C}^\ominus$, then we say Σ_* is minimum-phase; otherwise, Σ_* is said to be of non-minimum-phase. Those invariant zeros that are in $\mathbb{C}^\ominus$ are called the stable invariant zeros. Also, those that are not in $\mathbb{C}^\ominus$ are called the unstable invariant zeros.

The following definition introduces the notions of algebraic and geometric multiplicities [39] of an invariant zero and its multiplicity structure.

Definition 23.11 The **algebraic multiplicity** ρ_z of an invariant zero z is defined as the degree of the product of the elementary divisors of $P_{\Sigma_*}(s)$ corresponding to z. Likewise, the **geometric multiplicity** ν_z of an invariant zero z is defined as the number of the elementary divisors of $P_{\Sigma_*}(s)$ corresponding to z. Moreover, the invariant zero z is said to have a **semi-simple structure** if its algebraic and geometric multiplicities are equal. Otherwise, it is referred to as an invariant zero with **nonsimple** structure.

Given an invariant zero z, let $n_{z,i}$ be the degree of $(s - z)$ in the invariant factor $\Psi_i(s)$ of the Rosenbrock system matrix. Then the multiplicity structure of an invariant zero is defined as

$$S_z^* = \{n_{z,1}, n_{z,2}, \ldots, n_{z,\nu_z}\}. \tag{23.19}$$

If $n_{z,1} = n_{z,2} = \cdots = n_{z,\nu_z} = 1$, then we say z is a semi-simple invariant zero of the given system Σ_*. It is called a simple invariant zero if $\nu_z = 1$ and $n_{z,1} = 1$.

We discuss next the invariant zeros together with their multiplicity structure of the system Σ_* as displayed by the SCB.

Property 23.12 Consider the system Σ_* with its associated SCB. Then, z is an invariant zero of Σ_* with multiplicity structure S_z^* if and only if z is an eigenvalue of A_a with multiplicity structure S_z^*.

We need to recall next the notion of the right state and input zero directions and left state and input zero directions [39] associated with an invariant zero of a system. We focus first on the right state and input zero directions associated with an invariant zero for a left invertible system Σ_* (left invertibility is discussed in Definition 23.5).

Definition 23.13 Consider an invariant zero z with a semi-simple structure of a left invertible system Σ_*. Then the associated **right state and input zero directions,** $x_z \neq 0$ and u_z, of Σ_* are defined as those that satisfy the condition

$$P_{\Sigma_*}(z)\begin{pmatrix} x_z \\ u_z \end{pmatrix} = \begin{pmatrix} zI - A & -B \\ C & D \end{pmatrix}\begin{pmatrix} x_z \\ u_z \end{pmatrix} = 0.$$

Some papers in the literature extend the above definition to non-left-invertible systems. This is incorrect as argued in [100].

Whenever an invariant zero has a nonsimple multiplicity structure, a concept of generalized right state and input zero direction[2] chain associated with that invariant zero exist. A proper definition of this is given in [100]. At this time, we would like to point out that although some researchers (e.g., see [65] and [122] among others) define the generalized right state and input zero direction chains as x_{R}^j and w_{R}^j, $j = 1, \cdots, \rho_z - \sigma_z$, satisfying

$$P_{\Sigma_*}(z)\begin{pmatrix} x_{\mathrm{R}}^j \\ w_{\mathrm{R}}^j \end{pmatrix} = -\begin{pmatrix} x_{\mathrm{R}}^{j-1} \\ 0 \end{pmatrix}, \quad j = 1, \ldots, \rho_z - \sigma_z. \tag{23.20}$$

However, this is also incorrect for non-left-invertible systems as argued in [100].

[2] Generalized right state and input zero directions are also called pseudo-right state and input zero directions.

In the following, we identify the right state and input zero direction chain associated with an invariant zero z of a general system whether it is left invertible or not, and whether z has a semi-simple multiplicity structure or not. However, as discussed, in the absence of a precise definition that is not based on any special coordinate basis, we caution that the Property 23.14 can be viewed either as a definition or as a property.

Let us start by defining the eigenvector chain associated with an eigenvalue of the matrix A_a. Given an invariant zero z of the system Σ_* (i.e., the eigenvalue z of the matrix A_a), for each $i = 1$ to ν_z, a set of vectors in $\mathbb{R}^{n_a}$ that satisfies the following condition (23.21) is the eigenvector chain of A_a associated with the invariant zero z:

$$A_a x_{i,1}^z = z x_{i,1}^z, \quad \text{and} \quad (A_a - zI_{n_a}) x_{i,j+1}^z = x_{i,j}^z, \quad j = 1, \ldots, n_{z,i} - 1. \tag{23.21}$$

We have the following property regarding the right state and input zero direction chain associated with an invariant zero of a system.

Property 23.14

1. For each $i \in \{1, \ldots, \nu_z\}$, a set of vectors in $\mathbb{R}^n$ given in (23.22) is the **generalized right state zero direction chain** of Σ_* associated with the invariant zero z

$$x_{ij}^z = \Gamma_s \begin{pmatrix} x_{ij}^z \\ 0 \\ \vdots \\ 0 \end{pmatrix} \qquad j \in \{1, \ldots, \nu_{z,i}\}. \tag{23.22}$$

   Also, x_{i1}^z is the right state zero direction of Σ_* associated with z.
2. For each $i \in \{1, \ldots, \nu_z\}$, a set of vectors w_{ij}^z, $j = 1$ to $n_{z,i}$, in $\mathbb{R}^m$ as given in (23.23) is a **generalized right input zero direction chain** of Σ_* associated with the invariant zero z:

$$w_{ij}^z = -\Gamma_y \begin{pmatrix} E_{da} \\ E_{dc} \end{pmatrix} x_{ij}^z, \tag{23.23}$$

   where E_{da} is as defined in Property 23.4. Also, w_{i1}^z is said to be a right input zero direction of Σ_* associated with z.

The following property gives a dynamical interpretation of finite zero structure of a system. It is formulated for continuous-time systems. An analogous formulation is valid for discrete-time systems as well.

Property 23.15 (Dynamical Interpretation of Finite Zero Structure) For a system Σ_* that is not necessarily left invertible, given that the initial condition, $x(0) = x_{i\alpha}^z$

for any $\alpha \le n_{z,i}$ and the input

$$u = \sum_{j=1}^{\alpha} \frac{w_{ij}^{z} t^{\alpha-j} \exp(zt)}{(\alpha - j)!} \text{ for all } t \ge 0, \tag{23.24}$$

where z is any invariant zero of the system and $n_{z,i} \in S_z^*$, we have

$$y \equiv 0$$

and

$$x(t) = \sum_{j=1}^{\alpha} \frac{x_{ij}^{z} t^{\alpha-j} \exp(zt)}{(\alpha - j)!} \text{ for all } t \ge 0. \tag{23.25}$$

One can define the left state and input zero direction chain associated with an invariant zero of Σ_* as follows.

Definition 23.16 The **left state and input zero direction chain** associated with each invariant zero of Σ_* are defined as the corresponding right state and input zero direction chain of the dual system Σ_{*d}.

Remark 23.17 We can connect the list of structural invariant indices due to Morse [73] to the SCB. In particular, the list $\mathfrak{I}_1$ of Morse is exactly equal to the invariant factors of A_a.

Next, we would like to recall the definition of the input decoupling zeros and the output decoupling zeros of a system.

Definition 23.18 The zeros of the matrix pencil

$$\begin{pmatrix} \lambda I - A & -B \end{pmatrix},$$

i.e., the values of λ for which the above pencil loses rank, are called the **input decoupling zeros** of Σ_*. They are also referred to as the input decoupling zeros of the pair (A, B).

The zeros of the matrix pencil

$$\begin{pmatrix} \lambda I - A \\ C \end{pmatrix},$$

i.e., the values of λ for which the above pencil loses rank, are called the output decoupling zeros of Σ_*. They are also referred to as the output decoupling zeros of the pair (C, A).

Remark 23.19 In the literature, input decoupling zeros are also referred to as uncontrollable eigenvalues, whereas the output decoupling zeros are referred to as unobservable eigenvalues.

Note that as we have done for invariant zeros, we can also associate a multiplicity structure with an input or output decoupling zero. The precise definition should be obvious from the above and hence is not included here.

The following property shows how input decoupling zeros and output decoupling zeros of Σ_* are displayed by the SCB.

Property 23.20 Consider a system Σ_* with corresponding special coordinate basis. Define A_{con}, B_{con}, A_{obs}, and C_{obs} according to Property 23.4.

1. The input decoupling zeros of Σ_* are the input decoupling eigenvalues of the pair $(A_{\text{con}}, B_{\text{con}})$. Also, input decoupling zeros of (A_b, L_{bd}) are contained in the set of input decoupling zeros of Σ_*. Some of the input decoupling zeros of Σ_* could be contained among its invariant zeros.
2. The output decoupling zeros of Σ_* are the output decoupling eigenvalues of the pair $(C_{\text{obs}}, A_{\text{obs}})$. Also, output decoupling zeros of $(B_d E_{dc}, A_c)$ are contained in the set of output decoupling zeros of $(C_{\text{obs}}, A_{\text{obs}})$. Some of the output decoupling zeros of Σ_* could be contained among its invariant zeros.

Remark 23.21 As it is obvious from Property 23.20, it is crucial to realize that the input decoupling zeros and the output decoupling zeros of a system need not be invariant zeros.

Finally,

Definition 23.22 A continuous-time system is called minimum-phase if all invariant zeros are in $\mathbb{C}^-$. A system is called weakly minimum-phase if all invariant zeros are in $\mathbb{C}^- \cap \mathbb{C}^0$ and all invariant zero in $s \in \mathbb{C}^0$ are semi-simple. A system is called weakly non-minimum-phase if all invariant zeros are in $\mathbb{C}^- \cap \mathbb{C}^0$ and there exists an invariant zero which is not semi-simple. A system is called at most weakly non-minimum-phase if all invariant zeros are in $\mathbb{C}^- \cap \mathbb{C}^0$.

For a discrete-time system, the above definitions still hold with $\mathbb{C}^-$ and $\mathbb{C}^0$ replaced by $\mathbb{C}^{\ominus}$ and $\mathbb{C}^{\circ}$, respectively.

23.3.4 Infinite Zero Structure

In this subsection, we examine the infinite zero structure of a system and how it is displayed by the SCB. Let us first recall some pertinent information from the literature. Infinite zeros are defined either in association with root-locus theory or as Smith–McMillan zeros of the transfer function at infinity.

Let us first view the infinite zeros from the viewpoint of root-locus theory. For this, consider a strictly proper system Σ_* subject to a high-gain feedback $u = \tilde{\rho} y$

for a scalar gain $\tilde{\rho}$. It can then be shown (see, e.g., Hung and MacFarlane [38]) that the unbounded closed-loop poles of the feedback system can be listed as

$$s_{j\ell}(\tilde{\rho}) = \tilde{\rho}^{1/\nu_j}\eta_{j\ell} + \zeta_{j\ell}(\tilde{\rho}) \quad \text{for} \quad \ell = 1, \dots, \nu_j, \quad j = 1, \dots, m,$$

where

$$\lim_{\tilde{\rho}\to\infty} \tilde{\rho}^{-1/\nu_j}\zeta_{j\ell}(\tilde{\rho}) = 0.$$

Here $s_{j\ell}$ is termed an infinite zero of order ν_j. Actually, until recently, the infinite zeros defined this way were considered to be fictitious objects introduced for the convenience of visualization.

Let us next consider the infinite zeros from the viewpoint of Smith-McMillan theory. To define the zero structure of the system Σ_* at infinity, one can use the familiar Smith–McMillan description of the zero structure at finite frequencies of the corresponding transfer matrix H, which need neither to be square nor strictly proper. A rational matrix $H(s)$ possesses an infinite zero of order k when $H(1/z)$ has a finite zero of precisely that order at $z = 0$; see [20, 87, 92], and [147].

The number of zeros at infinity together with their orders indeed defines an infinite zero structure. It is important to note that for strictly proper transfer matrices, the above two definitions of the infinite zeros and their structure are consistent.

Owens [81] related the orders of the infinite zeros of the root-loci of a square system with a nonsingular transfer function matrix to the C^* structural invariant indices list $\mathfrak{I}_4$ of Morse [73]. This connection reveals that the *structure at infinity is in fact the topology of inherent integrations between the input and the output variables.* The SCB of Theorem 23.3 explicitly shows this topology of inherent integrations. The following property pinpoints this.

Property 23.23 Let $\overline{q}_0 = m_0$. Let $\overline{q}_j$ be the integer such that exactly $\overline{q}_j$ elements of $\{q_1, \dots, q_{m_d}\}$ are equal to j. Also, let σ be the smallest integer such that $\overline{q}_j = 0$ for all $j > \sigma$. Then there are $\overline{q}_0$ infinite zeros of order 0, and $j\overline{q}_j$ number of infinite zeros of order j, for $j = 1, \dots, \sigma$. Moreover, the C^* structural invariant indices list $\mathfrak{I}_4$ of Morse is given by

$$\mathfrak{I}_4 = \{\overbrace{0, 0, \dots, 0}^{\overline{q}_0},\ \overbrace{1, 1, \dots, 1}^{\overline{q}_1},\ \dots,\ \overbrace{\sigma, \sigma, \dots, \sigma}^{\overline{q}_\sigma}\}.$$

Remark 23.24 The state vector x_d in the SCB of a system is nonexistent if and only if the given system does not have infinite zeros of order greater than or equal to one.

Remark 23.25 A particular subclass of a general MIMO system termed as a *uniform rank system* plays a significant role in designing MIMO systems. For strictly proper systems, a uniform rank system is square, invertible, and more importantly has only one C^* structural invariant indices list $\mathcal{I}_4$ of Morse as given by

$$\mathcal{I}_4 = \{\overbrace{r, r, \ldots, r}^{n_r}\},$$

for some integers r and n_r. That is, the infinite zero structure of a uniform rank system contains n_r integers all equal to r. This is unlike several integers repeating themselves many times for a general MIMO system as discussed in Property 23.23.

As depicted in Fig. 23.5, an invertible uniform rank system Σ_w in its SCB format consists of two systems Σ_{w0} and Σ_{wr} of the form,

$$\Sigma_{w0} : \{ \; \dot{w}_0 = A_0 w_0 + A_{0w} y_w, \quad \text{(invariant zero dynamics)} \tag{23.26a}$$

$$\Sigma_{wr} : \begin{cases} \dot{w}_j = w_{j+1}, \;\; \text{for } \; j = 1, 2, \; \cdots \;, r-1, \\ \dot{w}_r = g(w) + u_w, \quad (r\text{-th order infinite zero dynamics}) \end{cases} \tag{23.26b}$$

with $y_w := w_1$ considered as its output. Let us emphasize that the integer r here denotes the order of infinite zeros of Σ_w. Also, $w_i, i = 0, \; 1, \; 2, \; \ldots \; r$ are the states, y_w is the output, and u_w is the input of Σ_w. Both the input u_w and the output y_w have the same dimension. Furthermore, $g(w)$ denotes a linear function of all the state variables $w_i, i = 0, \; 1, \; 2, \; \ldots \; r$. Thus, in the format of SCB, Σ_w consists of Σ_{w0} representing its finite zero structure and Σ_{wr} representing its infinite zero structure. We note further that Σ_{w0} is fed only by the output of Σ_{wr}. For clarity, let us also point out that the infinite zero structure Σ_{wr} consists of r integrator blocks forming a chain; the output of the integrator block on the extreme right (see Fig. 23.5) is its output as well as the output of Σ_w, while the extreme left integrator block is fed by a linear combinations of all the states and the input of Σ_w. We note also that each integrator block in the chain contains n_r scalar integrators. It is easy to recognize that the system Σ_w is square and invertible.

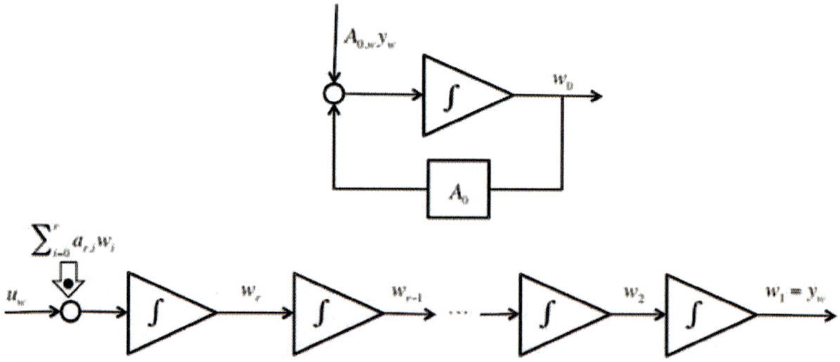

Fig. 23.5 Uniform rank system—SCB format

Given the above structure we can write invertible, uniform rank systems (using a suitable basis for the state space) in the form:

$$\begin{aligned} \dot{x}_a &= A_a x_a + L_{ad} y, \\ \dot{x}_d &= A_d x_d + B_d(u + E_{da} x_a + E_{dd} x_d), \\ y &= C_d x_d, \end{aligned} \tag{23.27}$$

where the eigenvalues of A_a are the invariant zeros of the system while A_d, B_d and C_d are of the form

$$A_d = \begin{pmatrix} 0 & I_{m(r-1)} \\ 0 & 0 \end{pmatrix}, \quad B_d = \begin{pmatrix} 0 \\ I_m \end{pmatrix}, \quad C_d = \begin{pmatrix} I_m & 0 \end{pmatrix}.$$

with m is the dimension of u.

Remark 23.26 The class of SISO systems is a particular subclass of a general uniform rank system. Let r be the relative degree of a SISO system. If the system is strictly proper, then the C^* structural invariant indices list $\mathcal{I}_4$ of Morse is given by

$$\mathcal{I}_4 = \{r\}.$$

23.3.5 Geometric Subspaces

Geometric theory is concerned with the study of subspaces of the state space with certain invariance properties, for example, A-invariant subspaces (which remain invariant under the unforced motion of the system), (A, B)-invariant subspaces (which can be made invariant by the proper application of state feedback), and (C, A) invariant subspaces (which can be made invariant by the proper application of output injection) (see, e.g., [142, 163]). Prominent examples of A-invariant subspaces are the controllable subspace (i.e., the image of the controllability matrix) and the unobservable subspace (the kernel of the observability matrix).

The development of geometric theory has in large part been motivated by the challenge of decoupling disturbance inputs from the outputs of a system, either exactly or approximately. Toward this end, a number of subspaces have been identified, which can be related to the different partitions in the SCB. Of particular importance in the context of control design for exact disturbance decoupling are the *weakly unobservable subspace*, which, by the proper selection of state feedback, can be made not to affect the outputs; and the *controllable weakly unobservable subspace*, which has the additional property that the dynamics restricted to this subspace is controllable. Of particular importance in the context of observer design for exact disturbance decoupling are the *strongly controllable subspace*, which, by the proper selection of output injection, is such that its quotient space can be rendered unaffected by the system inputs; and the *distributionally weakly unobserv-*

able subspace, which has the additional property that the dynamics restricted to its quotient space is observable.

We now proceed to connect some classic subspaces from the geometric theory of linear systems to SCB. That is, in the following, we show certain interconnections between the decomposition of the state space as done by the SCB, and various invariant subspaces from the geometric theory. To do so, we recall the first two subspaces. The subspaces $\mathcal{V}_g(\Sigma_*)$ and $\mathcal{S}_g(\Sigma_*)$ are classic subspaces and are crucial elements of geometric theory of linear systems. Also, later on we recall two more subspaces, $\mathcal{V}_\lambda(\Sigma_*)$ and $\mathcal{S}_\lambda(\Sigma_*)$, which are recently introduced in the context of H_∞ theory.

Definition 23.27 Consider a linear system Σ_* characterized by the matrix quadruple (A, B, C, D). Then,

1. The $\mathbb{C}_g$**-stabilizable weakly unobservable subspace** $\mathcal{V}_g(\Sigma_*)$ is defined as the largest subspace of $\mathbb{R}^n$ for which a matrix F exists such that the subspace is $(A+BF)$-invariant, contained in $\ker(C+DF)$, whereas the eigenvalues of $(A+BF)|\mathcal{V}_g$ are contained in $\mathbb{C}_g \subseteq \mathbb{C}$.
2. The $\mathbb{C}_g$**-detectable strongly controllable subspace** $\mathcal{S}_g(\Sigma_*)$ is defined as the smallest subspace of $\mathbb{R}^n$ for which a matrix K exists such that the subspace is $(A+KC)$-invariant, contains $\operatorname{Im}(B+KD)$, and is such that the eigenvalues of the map that is induced by $(A+KC)$ on the factor space $\mathbb{R}^n/\mathcal{S}_g$ are contained in $\mathbb{C}_g \subseteq \mathbb{C}$.

For the case when $\mathbb{C}_g = \mathbb{C}$, $\mathcal{V}_g$ and $\mathcal{S}_g$ are, respectively, denoted by $\mathcal{V}^*$ and $\mathcal{S}^*$.

Finally, let a $\mathbb{C}_g$ be chosen such that it has no common elements with the set of invariant zeros of Σ_*. Then the corresponding $\mathcal{V}_g(\Sigma_*)$, which is always independent of the particular choice of such a $\mathbb{C}_g$ is referred to as the strongly controllable subspace $\mathcal{R}^*(\Sigma_*)$.

Remark 23.28 We note that $\mathcal{V}_g(\Sigma_*)$ and $\mathcal{S}_g(\Sigma_*)$ are dual in the sense that

$$\mathcal{V}_g(\Sigma_{*d}) = \mathcal{S}_g(\Sigma_*)^\perp,$$

where Σ_{*d} is the dual system of Σ_*.

Moreover, it can be shown that $\mathcal{R}^*(\Sigma_*)$ equals $\mathcal{V}^*(\Sigma_*) \cap \mathcal{S}^*(\Sigma_*)$.

Remark 23.29 It is easy to observe that $\mathcal{V}_g(\Sigma_*)$ and $\mathcal{S}_g(\Sigma_*)$ are invariant under state feedback and output injection.

Remark 23.30 We should note that if (A, B) is $\mathbb{C}_g$-stabilizable, then for $\mathcal{V}_g(\Sigma_*)$, a matrix F exists that satisfies the conditions stated in Definition 23.27 and, moreover, $A+BF$ is $\mathbb{C}_g$-stable. An analogous comment can be made for $\mathcal{S}_g(\Sigma_*)$.

Remark 23.31 It is easily shown that the subspaces $\mathcal{V}_g(\Sigma_*)$ and $\mathcal{S}_g(\Sigma_*)$ satisfy the following:

$$\begin{pmatrix} A \\ C \end{pmatrix} \mathcal{V}_g(\Sigma_*) \subseteq \left(\mathcal{V}_g(\Sigma_*) \oplus \{0\}\right) + \operatorname{Im} \begin{pmatrix} B \\ D \end{pmatrix} \tag{23.28}$$

and

$$\ker \begin{pmatrix} C & D \end{pmatrix} \cap \begin{pmatrix} A & B \end{pmatrix} \left(\mathcal{S}_g(\Sigma_*) \oplus \mathbb{R}^m\right) \subseteq \mathcal{S}_g(\Sigma_*). \tag{23.29}$$

Remark 23.32 In the paper [70, 71], two geometric subspaces, namely, maximum uncontrollable subspace (MUCS) and maximum unobservable subspace (MUS) of Σ_*, are defined as follows: Consider the controllability matrix $Q_c(L)$ of Σ_* with output injection matrix L,

$$Q_c(L) = \begin{pmatrix} B & (A+LC)B & \dots & (A+LC)^{n-1}B \end{pmatrix}.$$

Let L be chosen such that the null space of $Q_c(L)'$ is of maximal order. Such a null space is called MUCS of Σ_*. Similarly, consider the observability matrix $Q_o(F)$ of Σ_* with state feedback gain F,

$$Q_o(F) = \begin{pmatrix} C \\ C(A+BF) \\ \vdots \\ C(A+BF)^{n-1} \end{pmatrix}.$$

Let F be chosen such that the null space of $Q_o(F)$ is of maximal order. Such a null space is called the MUS of Σ_*.

We observe that MUS equals $\mathcal{V}^*$ while MUCS is the orthogonal complement of $\mathcal{S}^*$.

We define below the notion of strong (also called perfect or ideal) controllability.

Definition 23.33 A given system Σ_* is said to be **strongly controllable** if it is controllable under any arbitrary output injection matrix L, i.e., the pair $(A + LC,\ B + LD)$ is controllable for every L of appropriate dimensions.

Analogously, we define below the notion of strong (also called perfect or ideal) observability.

Definition 23.34 A given system Σ_* is said to be **strongly observable** if it is observable under any arbitrary state feedback gain F, i.e., the pair $(A + BF,\ C + DF)$ is observable for every F of appropriate dimensions.

The above properties can be characterized in terms of the geometric subspaces defined above:

Theorem 23.35 *Consider a linear system Σ_* characterized by the matrix quadruple (A, B, C, D). Then,*

- *The system is strongly controllable if and only if $\mathcal{V}(\Sigma_*) = \mathbb{R}^n$.*
- *The system is strongly observable if and only if $\mathcal{V}(\Sigma_*) = \{0\}$.*

By now it is clear that the SCB decomposes the state space into several distinct parts. In fact, the state space $\mathcal{X}$ is decomposed as

$$\mathcal{X} = \mathcal{X}_a \oplus \mathcal{X}_b \oplus \mathcal{X}_c \oplus \mathcal{X}_d\,.$$

Here $\mathcal{X}_a$ is related to the invariant zeros, i.e., the eigenvalues of A_a are the invariant zeros of Σ. On the other hand, $\mathcal{X}_b$ is related to right invertibility, i.e., the system is right invertible if and only if $\mathcal{X}_b = \{0\}$, whereas $\mathcal{X}_c$ is related to left invertibility, i.e., the system is left invertible if and only if $\mathcal{X}_c = \{0\}$. The latter two equivalence are true provided that, as assumed before, $(B' \quad D')'$ and $(C \quad D)$ are of full rank. Finally, $\mathcal{X}_d$ is related to zeros of Σ at infinity.

We focus next on certain interrelationships between the SCB and some basic ingredients of the geometric control theory.

Property 23.36 Consider a system Σ_* that has already been transformed in the special coordinate basis.

1. $\mathcal{X}_a \oplus \mathcal{X}_c$ is equal to $\mathcal{V}^*(\Sigma_*)$.
2. $\mathcal{X}_c \oplus \mathcal{X}_d$ is equal to $\mathcal{S}^*(\Sigma_*)$.
3. $\mathcal{X}_c$ is equal to $\mathcal{R}^*(\Sigma_*)$.

Remark 23.37 In view of Property 23.7, it is obvious that Σ_* is left invertible if and only if

$$\mathcal{R}^*(\Sigma_*) = 0 \text{ and } \begin{pmatrix} B \\ D \end{pmatrix} \text{ is injective}$$

or equivalently

$$\mathcal{V}^*(\Sigma_*) \cap B \ker D = 0 \text{ and } \begin{pmatrix} B \\ D \end{pmatrix} \text{ is injective.}$$

Similarly, Σ_* is right invertible if and only if

$$\mathcal{V}^*(\Sigma_*) + \mathcal{S}^*(\Sigma_*) = \mathbb{R}^n \text{ and } \begin{pmatrix} C & D \end{pmatrix} \text{ is surjective}$$

or equivalently

$$\mathcal{S}^*(\Sigma_*) + C^{-1}\operatorname{Im} D = \mathbb{R}^n \text{ and } \begin{pmatrix} C & D \end{pmatrix} \text{ is surjective.}$$

23.3.6 Miscellaneous Properties of the SCB

Several properties of linear multivariable time-invariant systems can trivially be visualized using the SCB of Theorem 23.3. We give below some of these properties.

Property 23.38 The normal rank of Σ_* is equal to $m_d + m_0$. Moreover, it is easy to see the following:

1. normrank $P_{\Sigma_*}(s) = n + \text{normrank}[C(sI - A)^{-1}B + D]$.
2. The normal rank of Σ_* is equal to p if and only if Σ_* is right invertible.
3. The normal rank of Σ_* is equal to m if and only if Σ_* is left invertible.

Property 23.39

1. Σ_* is right invertible, and minimum-phase $\Rightarrow$ Σ_* is stabilizable.
2. Σ_* is left invertible, and minimum-phase $\Rightarrow$ Σ_* is detectable.
3. Σ_* is invertible, and minimum-phase $\Rightarrow$ Σ_* is stabilizable and detectable.
4. Σ_* is right invertible, and the invariant zeros are disjoint from the eigenvalues (or unstable eigenvalues) of $A \Rightarrow \Sigma_*$ is controllable (stabilizable).
5. Σ_* is left invertible, and its invariant zeros are disjoint from the eigenvalues (or unstable eigenvalues) of $A \Rightarrow \Sigma_*$ is observable (detectable).
6. Σ_* is invertible, and its invariant zeros are disjoint from the eigenvalues (or unstable eigenvalues) of $A \Rightarrow \Sigma_*$ is controllable and observable (stabilizable and detectable).
7. The feedthrough matrix D in Σ_* is injective $\Rightarrow$ Σ_* is left invertible and has no infinite zeros of order greater than or equal to one.
8. The feedthrough matrix D in Σ_* is surjective $\Rightarrow$ Σ_* is right invertible and has no infinite zeros of order greater than or equal to one.

We connected in Sects. 23.3.3 and 23.3.4 the lists $\mathfrak{I}_1$ and $\mathfrak{I}_4$ of Morse [73] to SCB. The following property connects the lists $\mathfrak{I}_2$ and $\mathfrak{I}_3$ of Morse to SCB.

Property 23.40

$$\text{The list } \mathfrak{I}_2 \text{ of Morse} = \text{The controllability indices of the pair } (A_c, B_c).$$

$$\text{The list } \mathfrak{I}_3 \text{ of Morse} = \text{The observability indices of the pair } (C_b, A_b).$$

We have also the following remark.

Remark 23.41 The integers

$$n_a(\Sigma_*), n_b(\Sigma_*), n_c(\Sigma_*), n_d(\Sigma_*), m_d,$$
$$\text{and } q_i (i = 1, \ldots, m_d)$$

are structurally invariant with respect to state feedback and output injection. Moreover, the integers $n_a(\Sigma_*)$, $n_b(\Sigma_*)$, $n_c(\Sigma_*)$, and $n_d(\Sigma_*)$ are, respectively, equal to the number of elements in the lists $\mathfrak{I}_1$, $\mathfrak{I}_2$, $\mathfrak{I}_3$, and $\mathfrak{I}_4$ of Morse. For further details, one can refer to [73] and [101].

23.3.7 Additional Compact Form of the SCB

Finally, let us observe that depending on some specific properties a given system satisfies, SCB can be written compactly in different formats. For convenience, we present below one specific format that we have used in this book.

If the given system Σ_* is left invertible, then the decomposition in (23.15)–(23.18) simplifies because x_c is no longer present (see Property 23.7), and we obtain the following structure:

$$\Gamma_s^{-1}(A - B_0 C_0)\Gamma_s = \begin{pmatrix} A_a & 0 & L_{ab}C_b & L_{ad}C_d \\ 0 & A_b & L_{bd}C_d & \\ B_d E_{da} & B_d E_{db} & A_d & \end{pmatrix}, \tag{23.30}$$

$$\Gamma_s^{-1}\left(B_0 \ \hat{B}_1\right)\Gamma_u = \begin{pmatrix} B_{a0} & 0 \\ B_{b0} & 0 \\ B_{d0} & B_d \end{pmatrix}, \tag{23.31}$$

$$\Gamma_y^{-1}\begin{pmatrix} C_0 \\ \hat{C}_1 \end{pmatrix}\Gamma_s = \begin{pmatrix} C_{0a} & C_{0b} & C_{0d} \\ 0 & 0 & C_d \\ 0 & C_b & 0 \end{pmatrix}, \tag{23.32}$$

and

$$\Gamma_y^{-1}\begin{pmatrix} I_{m_0} & 0 \\ 0 & 0 \end{pmatrix}\Gamma_u = \begin{pmatrix} I_{m_0} & 0 \\ 0 & 0 \\ 0 & 0 \end{pmatrix}. \tag{23.33}$$

23.4 Software Packages to Generate SCB

While the SCB provides a fine-grained decomposition of multivariable linear time invariant systems, transforming an arbitrary system to the SCB is a complex operation. A constructive algorithm for strictly proper systems is provided in [110] based on a modified Silverman algorithm [120]. This algorithm is lengthy and involved, and includes repeated rank operations and construction of non-unique transformations to divide the state space. Thus, the algorithm can realistically be executed by hand only for very simple systems.

To automate the process of finding transformations to the SCB, numerical algorithms were developed and implemented as part of the *Linear Systems Toolkit* for *Matlab*. The first among such work is by Lin et al. [57] and the resulting software was commercialized in early 1990s [56]. The Toolkit was revised subsequently twice each time with further improvements, the first revision was reported in [14, 18] and the second revision in [61]. Although these numerical algorithms are invaluable in practical applications, engineers also often operate on systems where some or all of the elements of the system matrices have a symbolic representation. There are obvious and definite advantages in being able to obtain symbolic representation of SCB, without having to insert numerical values in place of symbolic variables. Furthermore, the numerical algorithms are based on inherently inaccurate floating-point operations that make them prone to numerical errors. Ideally, if the elements of the system matrices are represented by symbols and exact fractions, one would be able to obtain an exact SCB representation of that system, also represented by symbols and exact fractions. To address these issues, Grip and Saberi [30] recently developed a procedure for symbolic transformation of multivariable linear time invariant systems to the SCB, using the commercial mathematics software suite '*Maple*'. The procedure is based on the modified Silverman algorithm from [110], with a modification to achieve a later version of the SCB that includes an additional structural property (see, e.g., [100]), and an extension to SCB for non-strictly proper systems [98]. Symbolic transformations are useful complement to available numerical tools [14, 18]. Also, symbolic transformation to the SCB makes it possible to work directly on the SCB representation of a system without first inserting numerical values, thereby removing an obstacle to more widespread use of SCB such as squaring down of non-square systems and asymptotic time-scale assignment and other topics where the SCB has previously been applied, and to some other topics where the SCB has not yet been applied.

The '*Maple*' software code is the work of Grip and Saberi and can be found in [30].

Chapter 24
Squaring Down of General MIMO Systems to Invertible Uniform Rank Systems via Pre- and/or Post-compensators

24.1 Introduction and Problem Statement

As the title of the chapter indicates, the goal of this chapter is to transform a given not necessarily square and not necessarily invertible MIMO system to a particular subclass of a square invertible MIMO system termed as a *uniform rank system*, by utilizing only pre-compensators, and without using any state feedback as depicted in Fig. 24.1. Traditionally and historically, the term "squaring down" is used when a non-square and non-invertible MIMO system is transformed to a square invertible MIMO system [95]. Hence, such a process of transforming is termed as a squaring down process.

One should observe some constraints in pre-compensator design in accomplishing the above goal of squaring down. The required compensators for squaring down are often dynamic and as such induce additional finite zeroes. It is desirable that the additional finite zeros induced can be placed at desired locations in the open left half complex plane. Also, it is desirable that the compensators by themselves are asymptotically stable, and thus do not deteriorate the robustness and performance of an eventual feedback design. Moreover, it is desirable that the compensator design preserve the stabilizability, detectability, and minimum-phase properties of the original system.

To accomplish the above goal of squaring down with added constraints, let us first recall a fundamental difference between general MIMO systems, uniform rank systems, and SISO systems. As pointed out in Chap. 23, the dynamics associated with finite zeros does not play any role in distinguishing uniform rank systems from general MIMO systems. However, the dynamics associated with infinite zeros, particularly the C^* structural invariant indices list $\mathcal{I}_4$ of it plays a profound role. This list $\mathcal{I}_4$ is invariant with respect to output injection, state feedback, and nonsingular transformations of state, input, and output. For a general strictly proper **square invertible** MIMO system, the list $\mathcal{I}_4$ contains several subsets, a subset of n_1 integers

A. Saberi et al., *Synchronization of Multi-Agent Systems in the Presence of Disturbances and Delays*, Systems & Control: Foundations & Applications,
https://doi.org/10.1007/978-3-030-88148-1_24

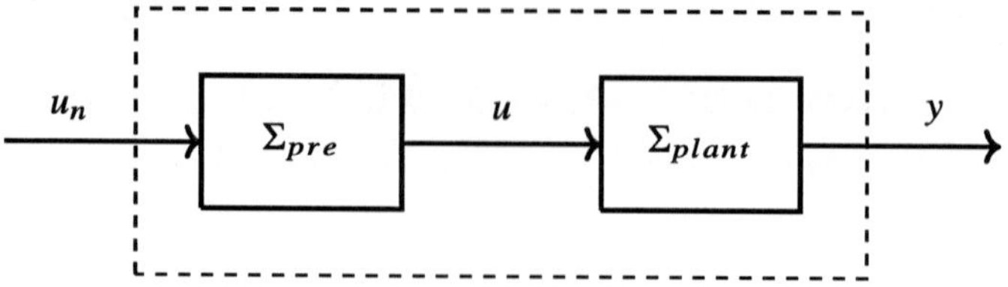

Fig. 24.1 Squaring down via pre-compensators

all equal to 1, a subset of n_2 integers all equal to 2, and so on, ending with a subset of n_k integers all equal to k for a certain k. That is,

$$\mathcal{I}_4 = \{\overbrace{1, 1, \ldots, 1}^{n_1},\ \overbrace{2, 2, \ldots, 2}^{n_2},\ \cdots,\ \overbrace{k, k, \ldots, k}^{n_k}\}.$$

The total number of integers in $\mathcal{I}_4$ as expressed above equals the number of inputs (which is equal to the number of outputs).[1] In contrast to the above, the $\mathcal{I}_4$ for a uniform rank system is of the form,

$$\mathcal{I}_4 = \{\overbrace{r, r, \ldots, r}^{n_r}\},$$

for some integers r and n_r, whereas the $\mathcal{I}_4$ for a SISO system is a single integer, namely, its relative degree r,

$$\mathcal{I}_4 = \{r\}.$$

That is, the infinite zero structure of a uniform rank system contains n_r integers all equal to r, while $\mathcal{I}_4$ of a SISO system is only a single integer. Thus, SISO system is a special case of a uniform rank system.

Before we proceed further, let us understand the significance of squaring down. Obviously, the structural properties of SISO systems are easy to decipher in comparison with MIMO systems. Also, it is common knowledge that MIMO systems are structurally much more complex than SISO systems. Because of this, it is of immense interest to simplify the complex structure of a general MIMO system by transforming it to a uniform rank system which is structurally similar to a SISO system and hence simpler for an eventual feedback design.

Now that the structural differences between a MIMO system and a uniform rank system are clear and transparent, we can formally state the general squaring down

[1] For a non-square system, the total number of integers in $\mathcal{I}_4$ is less than or equal to the number of inputs or the number of outputs whichever is smaller.

problem as described below. Consider a general linear time invariant strictly proper MIMO system,

$$\Sigma_{plant} : \begin{cases} \dot{\tilde{x}} = \tilde{A}\tilde{x} + \tilde{B}\tilde{u} \\ \tilde{y} = \tilde{C}\tilde{x}, \end{cases} \tag{24.1}$$

where $\tilde{x} \in \mathcal{R}^n$, $\tilde{u} \in \mathcal{R}^m$, and $\tilde{y} \in \mathcal{R}^p$ are respectively state, input, and output of the system Σ_{plant}.

We can now state the following formal problem statement. Consider a general MIMO system Σ_{plant} as described in (24.1). Let k be the highest order of infinite zeros of Σ_{plant}, that is, the list $\mathcal{I}_4$ is

$$\mathcal{I}_4 = \{\overbrace{1, 1, \dots, 1}^{n_1}, \ \overbrace{2, 2, \dots, 2}^{n_2}, \ \cdots, \ \overbrace{k, k, \dots, k}^{n_k}\}.$$

Consider a goal as follows:

Problem 24.1 Design a pre-compensator Σ_{pre} whenever possible such that the cascade connection of Σ_{pre} and Σ_{plant} in that order as depicted in Fig. 24.1 results in an invertible, uniform rank system with the list $\mathcal{I}_4$ as

$$\mathcal{I}_4 = \{k, k, \dots, k\},$$

and that the squared down system contains the invariant zeros of the given system Σ_{plant} and possibly additional invariant zeros which can be freely assigned in the open left half complex plane.

In this chapter we only consider pre-compensator design because that is what we need for this book. The general results, including post-compensator design and combined pre- and post-compensator designs, can be found in [113].

It is clear that the process of squaring down a general MIMO system to a uniform rank system requires a tool that can carefully and structurally examine the infinite zero structure of the given MIMO system. We utilize here the powerful tool of SCB that captures all the inherent structures of a MIMO system including the infinite zero structure.

The algorithmic procedures developed here to square-down a given system to a uniform rank system are presented for continuous-time systems. However, the same procedures can be utilized with appropriate straightforward modifications for discrete-time systems as well.

The write-up of this chapter is partially based on [113].

24.2 Results

We have the following results.

Theorem 24.2 *Consider a general right invertible MIMO system Σ_{plant} as described in (24.1) with the highest order of its infinite zeros equal to k. Let Σ_{plant} be stabilizable and detectable.*

There exists a pre-compensator that squares down the system Σ_{plant} to an invertible uniform rank system with its order of infinite zeros equal to k satisfying the requirements presented in Problem 24.1.

Remark 24.3 We consider here only strictly proper systems. The results can be extended trivially to the non-strictly proper system.

Remark 24.4 In addition to being stabilizable and detectable, if the given system Σ_{plant} is minimum-phase, then the squared down system, the cascade of Σ_{pre} and Σ_{plant} is, besides of uniform rank and square, also stabilizable, detectable, and of minimum-phase.

The proof of the above theorem is by explicit construction of pre-compensators that accomplish the tasks itemized in the theorem. There are two phases to such a construction. In the first phase, a given general linear time-invariant strictly proper MIMO system is squared down to an invertible square MIMO system in Sect. 24.3. In the second phase, an invertible square MIMO system is squared down to an invertible uniform rank system by utilizing a pre-compensator in Sect. 24.4.

24.3 Transformation to an Invertible Square MIMO System

This section, based on [95], considers the first phase in which a given general right invertible linear time-invariant strictly proper MIMO system is squared down to an invertible square MIMO system. We will see that a general right invertible system Σ_{plant} which is stabilizable and detectable can be squared down to an invertible square system via a pre-compensator. The squaring down procedure introduces extra invariant zeros in addition to the invariant zeros of the original system Σ_{plant}. However, all such additional invariant zeros can be freely assigned in the open left half s plane.

We utilize here the powerful tool of SCB to do so. This requires writing the SCB equations for strictly proper systems in a compact form by suppressing the unwanted variables The next subsection presents such SCB equations.

24.3.1 *Squaring-Down to a Square Invertible System*

We will first present the use of a post-compensator for a left invertible system. The required pre-compensator for right invertible systems is then obtained via dualization.

24.3.1.1 Left Invertible Systems

To simplify the presentation, we first consider the squaring down problem for a left invertible system Σ_{plant}. We utilize SCB as a basis to design a post compensator. For a left invertible system, in the format of SCB of Σ_{plant}, as illustrated in Sect. 23.3.7, the variables x_c and u_c are nonexistent. Also, the control variable u_d and one of the output variables, namely, y_d have the same dimension m_d. The other output variable y_b has the dimension p_b. We need to square down the output variables y_d and y_b by designing a post compensator Σ_{post} in such a way that the additional invariant zeros induced by it are within a desirable set $\mathbb{C}_d$, a subset of the open left half complex plane $\mathbb{C}^-$.

Let us first attempt a static post compensator K_{post}. In this case, in the format of SCB we need to select a K_{post} to generate a new output y_{new} by a linear combination of the outputs y_d and y_b as

$$y_{new} = K_{post} \begin{pmatrix} y_d \\ y_b \end{pmatrix} = K_d y_d + K_b y_b.$$

When K_d is a singular matrix, it can be easily shown that the infinite zero structure of the resulting system is different from that of the given system. Hence, to preserve the infinite zero structure, we take K_d as a non-singular matrix. Without loss of generality, K_d is then taken as an identity matrix. The matrix K_b in general cannot be taken as a zero matrix, since it amounts to completely neglecting y_b as an output variable. Also, in the case when $K_b = 0$, the status of state x_b is the same as the status of state x_a implying that $\lambda(A_{bb})$ (the eigenvalues of A_{bb}) are additional invariant zeros. In general, $\lambda(A_{bb})$ need not be within the desired set $\mathbb{C}_d$, and hence our design problem is to choose K_b such that the induced additional invariant zeros are within $\mathbb{C}_d$. Choosing

$$y_{new} = K_{post} \begin{pmatrix} y_d \\ y_b \end{pmatrix} = y_d + K_b y_b, \quad K_{post} = \begin{pmatrix} I_{p_d} & K_b \end{pmatrix}, \tag{24.2}$$

and letting

$$\bar{x}_a = \begin{pmatrix} x_a \\ x_b \end{pmatrix},$$

can be written as

$$\dot{\bar{x}}_a = \bar{A}_{aa}\bar{x}_a + \bar{A}_{ad}\, y_{new} \tag{24.3}$$

where

$$\bar{A}_{aa} = \begin{pmatrix} A_{aa} & L_{ab}C_b - L_{ad}K_bC_b \\ 0 & A_{bb} - L_{bd}K_bC_b \end{pmatrix}, \quad \bar{A}_{ad} = \begin{pmatrix} L_{ad} \\ L_{bd} \end{pmatrix}. \tag{24.4}$$

Equation (24.3) together with

$$\dot{x}_i = A_{q_i}x_i + L_{id}y_d + B_{q_i}\left(u_i + E_{ia}x_a + E_{ib}x_b + \sum_{j=1}^{m_d} E_{ij}x_j\right),$$
$$y_i = C_{q_i}x_i.$$

for $i = 1, \ldots, m_d$, can be considered as the new dynamic equations. The new invariant zeros can be easily calculated if these equations are re-written in the form of an SCB. It can be easily shown that there exist matrices M_{ib}, $i = 1$ to m_d, such that

$$\bar{x}_i = x_i + M_{ib}x_b,$$

and

$$\begin{aligned} \dot{\bar{x}}_i &= A_{q_i}\bar{x}_i + \bar{L}_{id}\bar{y}_d \\ &\quad + B_{q_i}\left(u_i + \bar{E}_{ia}x_a + \bar{E}_{ib}x_b + \sum_{j=1}^{m_d} \bar{E}_{ij}\bar{x}_j\right), \end{aligned} \tag{24.5}$$

$$y_{new} = \bar{y}_d = C_d\bar{x}_d, \tag{24.6}$$

where

$$\bar{x}_d = \begin{pmatrix} \bar{x}_1 \\ \bar{x}_2 \\ \vdots \\ \bar{x}_{m_d} \end{pmatrix}.$$

Equations (24.3), (24.5), and (24.6) are in the form of SCB and indeed correspond to the new squared down system in the format of SCB. Hence, the new invariant zeros are given by $\lambda(\bar{A}_{aa})$. Examination of $\lambda(\bar{A}_{aa})$ shows that $\lambda(A_{aa})$ and $\lambda(A_{bb} - L_{bd}K_bC_b)$ are the new invariant zeros. This discussion leads to the next theorem.

Theorem 24.5 *The post compensator K_{post} as in (24.2) induces additional set of invariant zeros in the new squared down system comprising of equations (24.3), (24.5), and (24.6) where y_{new} is the new output. These additional set of invariant zeros are $\lambda(A_{bb} - L_{bd}K_bC_b)$.*

Since the additional set of invariant zeros are $\lambda(A_{bb} - L_{bd}K_bC_b)$, the problem of additional invariant zero assignment by an appropriate choice of K_b is the same as the problem of pole assignment by output feedback of a system characterized by the matrix triple (A_{bb}, L_{bd}, C_b). Thus, as well known, it is not always possible to place the additional invariant zeros by output feedback within the desired set $\mathbb{C}_d$. This implies that squaring down with a constraint of placing additional invariant zeros within a desirable set $\mathbb{C}_d$ cannot in general be accomplished by a static compensator alone. We are thus lead to the use of dynamic compensators.

We can pursue several different approaches to design dynamic compensators. We use here observer theory. We recall the following lemma for linear state function observers which is an excerpt from [80]

Lemma 24.6 *Consider a stabilizable and detectable system,*

$$\dot{x} = Ax + Bu, \quad y = Cx,$$

where x, u, and y are of dimensions n, m, and p, respectively. Also, let $u = -Fx + u_0$ be a desired state feedback control law and $\lambda(A - BF)$ belongs to $\mathbb{C}^-$. Here u_0 is a new open-loop control. A minimal order observer is characterized by matrices W, N, and M of dimensions $(n-p) \times n$, $(n-p) \times (n-p)$, and $(n-p) \times p$, respectively, such that

- $WA = NW + MC$
- $\operatorname{rank}\begin{pmatrix} W \\ C \end{pmatrix} = n$
- $\lambda(N) \subseteq \mathbb{C}^-$.

Then, the observer-based compensator realizing the state feedback control law $u = -Fx + u_0$ is given by

$$\begin{aligned} \dot{z} &= (N - WBG)z + (M - WBJ)y + WBu_0 \\ u &= -Gz - Jy + u_0 \end{aligned}$$

where G and J are given by

$$\begin{pmatrix} G \\ J \end{pmatrix} = F \begin{pmatrix} W \\ C \end{pmatrix}^{-1}.$$

Remark 24.7 Note that the eigenvalues of the matrix N can be arbitrarily designed as long as we guarantee that they contain the unobservable eigenvalues of the given system.

We now proceed to the design of dynamic post-compensators. We assume that the given system Σ_{plant} is stabilizable and detectable. Then, from Property 23.4 of SCB, the pair $(A_{bb},\ L_{bd})$ is stabilizable and the pair $(A_{bb},\ C_b)$ is observable. Let F and N be such that $\lambda(A_{bb} - L_{bd}F)$ and $\lambda(N)$ are in $\mathbb{C}_d$. We note that, since $(A_{bb},\ L_{bd})$ is in general only stabilizable, those $\lambda(A_{bb})$ which are stable but not controllable must be included in $\mathbb{C}_d$. Now employing Lemma 24.6 and using (A_{bb}, L_{bd}, C_b) in place of (A, B, C), evaluate the matrices W, M, G, and J. Finally, determine the dynamic post-compensator $\tilde{\Sigma}_{post}$ as

$$\dot{z} = (N - WL_{bd}G)z + (M - WL_{bd}J)y_b + WL_{bd}y_d. \tag{24.7}$$

$$y_{new} = y_d + Jy_b + Gz \tag{24.8}$$

Here z is a $(n_b - p_b)$ dimensional state which represents the dynamics of $\tilde{\Sigma}_{post}$. One can re-write (24.7) by substituting (24.8) in it,

$$\dot{z} = Nz + My_b + WL_{bd}y_d. \tag{24.9}$$

Because of the resulting cancellation of terms when (24.8) is substituted in (24.7), the resulting dynamic system (24.9) turns out to be asymptotically stable as $\lambda(N)$ are selected to be in $\mathbb{C}_d$. This is important to note since in general one cannot always design stable compensators [148]. Thus, since our compensator is asymptotically stable, in view of the results of [27] and [108], we can point out that our compensator does not deteriorate the robustness and performance (e.g., sensor noise attenuation) of an eventual feedback design. The transfer function of the compensator $\tilde{\Sigma}_{post}$ as given by (24.8) and (24.9) is

$$\tilde{H}_{post} = G(sI - N)^{-1}\begin{pmatrix} WL_{bd} & M \end{pmatrix} + \begin{pmatrix} I_{p_d} & J \end{pmatrix}.$$

Then, in view of the output transformation $\tilde{y} = \Gamma_y y$, the compensator to be implemented is given by

$$\Sigma_{post} = \tilde{\Sigma}_{post}\Gamma_y^{-1}, \tag{24.10}$$

and has the transfer function,

$$H_{post} = \tilde{H}_{post}\Gamma_y^{-1}.$$

We have the following theorem.

Theorem 24.8 *Consider a left invertible, stabilizable, and detectable plant Σ_{plant} with transfer function H_{plant}. Define a square system Σ_{sq} with transfer function H_{sq} as*

$$H_{sq} = H_{post}H_{plant}.$$

Note that Σ_{sq} is formed by cascading Σ_{plant} and Σ_{post} as in Fig. 24.1. Then, Σ_{sq} has the following properties:

- *Σ_{sq} is invertible.*
- *The invariant zeros of Σ_{sq} are the invariant zeros of Σ_{plant}, $\lambda(N)$ and the eigenvalues of $A_{bb} - L_{bd}F$. Thus, the number of invariant zeros of Σ_{sq} is equal to $n_a + 2n_b - p_b$.*
- *The dynamic order of Σ_{sq} is $n + n_b - p_b$.*
- *The structure at infinity of Σ_{sq} is the same as that of Σ_{plant}.*
- *The poles of Σ_{sq} are the poles of Σ_{plant} together with $\lambda(N)$.*

Proof Let

$$e = Wx_b - z, \quad \bar{x}_a = \begin{pmatrix} x_a \\ x_b \\ e \end{pmatrix}$$

Then, it is straightforward to write the dynamic equations of Σ_{sq},

$$\begin{aligned} \dot{e} &= Ne \\ \dot{\bar{x}}_a &= \bar{A}_{aa}\bar{x}_a + \bar{L}_{ad}y_{new}, \end{aligned} \tag{24.11}$$

and for each $i = 1, \ldots, m_d$,

$$\dot{x}_i = A_{q_i}x_i + L_{id}y_d + B_{q_i}\left(u_i + \bar{E}_{ia}\bar{x}_a + \sum_{j=1}^{m_d} E_{ij}x_j\right), \tag{24.12}$$

$$y_{new} = y_d + Jy_b + GWx_b - Ge, \tag{24.13}$$

where

$$\bar{A}_{aa} = \begin{pmatrix} A_{aa} & L_{ab}C_b - L_{ad}F & L_{ad}G \\ 0 & A_{bb} - L_{bd}F & L_{bd}G \\ 0 & 0 & N \end{pmatrix}, \quad \bar{L}_{ad} = \begin{pmatrix} L_{ad} \\ L_{bd} \\ 0 \end{pmatrix}, \tag{24.14}$$

and where $\bar{E}_{ia}$ is appropriately defined. While (24.11) conforms to the form required for SCB, (24.12) and (24.13) do not. However, as in Appendix 2 of [110], a new variable $\bar{x}_i$ and its dynamic equation can be defined to conform with the format of the SCB. In this process, the matrices $\bar{A}_{aa}$ and $\bar{L}_{ad}$ are preserved as in (24.14). Thus, $\lambda(\bar{A}_{aa})$ are the new invariant zeros. The fact that $\lambda(N)$ are uncontrollable in Σ_{sq} follows easily by observing that

$$\dot{e} = Ne$$

is one of the dynamic equations of Σ_{sq}. The other properties of Theorem 24.8 are now obvious. ■

Remark 24.9 If K_b exists such that $\lambda(A_{bb} - L_{bd}K_bC_b)$ are at the desired locations in $\mathbb{C}^-$, then $F = K_bC_b$, $J = K_b$, and $G = 0$. Thus, in this case, the dynamic compensator Σ_{post} reduces to a static compensator K_{post} because then (24.8) does not require the state z of (24.7).

Remark 24.10 As mentioned earlier, Σ_{post} by itself is an asymptotically stable system since its eigenvalues $\lambda(N)$ are selected to be in the desired locations in $\mathbb{C}^-$. Note that $\lambda(N)$ are among the invariant zeros as well as among the eigenvalues of the squared system Σ_{sq}, and hence are uncontrollable in Σ_{sq} although they might be observable in Σ_{sq}. The fact that $\lambda(N)$ are uncontrollable in Σ_{sq} is not detrimental to the overall system design as $\lambda(N)$ have already been placed at desired locations in $\mathbb{C}^-$.

Remark 24.11 If Σ_{plant} is stabilizable and detectable, and of minimum -phase, so is the squared system Σ_{sq}. This can be seen easily since Σ_{plant} is of minimum-phase implies that its invariant zeros are in in $\mathbb{C}^-$. This means that by our design all the invariant zeros of Σ_{sq} are in $\mathbb{C}^-$ implying that Σ_{sq} is of minimum-phase, and is also stabilizable and detectable.

We consider next an example to illustrate the design of a post-compensator.

Example 24.12 Let the given system be

$$\begin{aligned} \dot{x}_{b1} &= x_{b2}, \quad \dot{x}_{b2} = x_{b3} + x_d, \quad \dot{x}_{b3} = -x_{b1} \\ \dot{x}_d &= x_{b1} + u_d, \quad y_d = x_d, \quad y_b = x_{b1}. \end{aligned} \tag{24.15}$$

This system is already in the form of an SCB, and is controllable, observable, left invertible, and minimum-phase (as it does not have any invariant zeros). However, it is not square. Here $n = 4$, $n_a = 0$, $n_b = 3$, $n_d = 1$, $p_b = 1$, and $p_d = 1$. A static controller alone cannot place all the induced invariant zeros in $\mathbb{C}^-$. We need a second order dynamic compensator. For the squared system, the number of invariant zeros will be $n_a + 2n_b - p_b = 5$. Let -1, -2, -3, -4, and -5 be the required invariant zeros. Let -1 and -2 be $\lambda(N)$, and -3, -4, and -5 be $\lambda(A_{bb} - L_{bd}F)$. Then, we can select

$$F = \begin{pmatrix} 47 & 12 & -59 \end{pmatrix}, \quad N = \begin{pmatrix} 0 & -2 \\ 1 & -3 \end{pmatrix}, \quad W = \begin{pmatrix} -2 & 0 & 1 \\ -3 & 1 & 0 \end{pmatrix},$$

$$M = \begin{pmatrix} -7 \\ -7 \end{pmatrix}, \quad G = \begin{pmatrix} -59 & 12 \end{pmatrix}, \quad \text{and } J = -35.$$

The Σ_{post} is defined by

$$\begin{aligned} \dot{z}_1 &= -2z_2 - 7y_b \\ \dot{z}_2 &= z_1 - 3z_2 - 7y_b + y_d, \\ y_{new} &= y_d - 35y_b - 59z_1 + 12z_2 \end{aligned} \tag{24.16}$$

and has a transfer function,

$$H_{post} = \frac{1}{s^2 + 3s + 2}\left(s^2 + 15s + 120 \; -35s^2 + 224s + 259\right). \tag{24.17}$$

Note that Σ_{post} defined by (24.16) is asymptotically stable. Defining

$$e_1 = -2x_{b1} + x_{b3} - z_1, \quad e_2 = -3x_{b1} + x_{b2} - z_2,$$

the squared system Σ_{sq} with the input u_d and the output y_{new} is described by

$$\dot{x}_{b1} = -x_{b2},$$

$$\dot{x}_{b2} = -47x_{b1} - 12x_{b2} + 60x_{b3} - 59e_1 + 12e_2 + y_{new}$$

$$\dot{x}_{b3} = -x_{b1}, \quad \dot{e}_1 = -2e_2, \quad \dot{e}_2 = e_1 - 3e_2$$

$$\dot{\tilde{x}}_d = -504x_{b1} - 97x_{b2} + 720x_{b3} - 720e_1 + 62e_2 + 12\tilde{x}_d + u_d, \quad y_{new} = \tilde{x}_d.$$

One can verify the properties of Theorem 24.8 from the above equations.

24.3.1.2 Right Invertible Systems

We now consider a right invertible system Σ_{plant}. In this case, in the format of SCB of Σ_{plant}, the state x_b and hence the output y_b is nonexistent. However, both the inputs u_d and u_c exist. To square down these input variables, we need to design a pre-compensator Σ_{pre} with a new input u_{new} with dimension m_d as its input and u_d and u_c as its outputs in the format of SCB.

One expects that the two problems of designing a post-compensator Σ_{post} for left invertible systems and designing a pre-compensator Σ_{pre} for right invertible systems are algebraically dual to each other. This implies that one can design a pre-compensator Σ_{pre} by the following steps:

- Transpose the given right invertible system Σ_{plant} to obtain its dual Σ^d_{plant} which then is inherently left invertible.
- Obtain a new SCB for Σ^d_{plant}.
- Following the previous development, design a post-compensator Σ^d_{post} that squares down Σ^d_{plant}.

- Transpose thus designed post-compensator Σ_{post}^{d} for the left invertible system Σ_{plant}^{d} to obtain the pre-compensator Σ_{pre} for the given right invertible system Σ_{plant}.

Remarks analogous to Remarks 24.9–24.11 can be made to emphasize the properties of thus designed Σ_{pre}. In particular, let us emphasize that Σ_{pre} is asymptotically stable since its eigenvalues are placed in a desired set $\mathbb{C}_d$, a subset of open left hand complex plane $\mathbb{C}^-$. Also, it can be seen that the eigenvalues of Σ_{pre} are among both the eigenvalues and invariant zeros of the resulting squared system Σ_{sq}, and that they are unobservable but nevertheless are placed in the desired set $\mathbb{C}_d$.

We consider next an example taken from [21] to illustrate the design of a pre-compensator.

Example 24.13 Let the system be given by:

$$\dot{\bar{x}}_1 = \bar{x}_2 + \bar{u}_2, \quad \dot{\bar{x}}_2 = \bar{x}_3, \quad \dot{\bar{x}}_3 = \bar{x}_4 + \bar{u}_1, \quad \dot{\bar{x}}_4 = \bar{u}_2, \quad \bar{y} = \bar{x}_1. \tag{24.18}$$

This system is controllable, observable, right invertible, and of minimum-phase as it does not have any invariant zeros. Davison shows that a static pre-compensator that squares down the above system while placing the induced invariant zeros at desirable locations does not exist. Hence, we pursue here the design of a dynamic pre-compensator that places the induced invariant zeros at desirable locations. We do so by constructing a dual system as discussed above. The dual system of (24.18) is

$$\begin{aligned}
&\dot{\bar{x}}_{d1} = \bar{u}_d, \quad \dot{\bar{x}}_{d2} = \bar{x}_{d1}, \quad \dot{\bar{x}}_{d3} = \bar{x}_{d2}, \quad \dot{\bar{x}}_{d4} = \bar{x}_{d3}, \\
&\bar{y}_{d1} = \bar{x}_{d3}, \quad \bar{y}_{d2} = \bar{x}_{d1} + \bar{x}_{d4}.
\end{aligned}$$

The SCB of the above dual system can be constructed by the coordinate transformation,

$$\bar{x}_d = \Gamma_s x, \quad \bar{y}_d = \Gamma_y \begin{pmatrix} y_d \\ y_b \end{pmatrix}, \quad \bar{u}_d = \Gamma_u u_d,$$

where

$$\bar{x}_d = \begin{pmatrix} \bar{x}_{d1} \\ \bar{x}_{d2} \\ \bar{x}_{d3} \\ \bar{x}_{d4} \end{pmatrix}, \quad x = \begin{pmatrix} x_{b1} \\ x_{b2} \\ x_{b3} \\ x_d \end{pmatrix},$$

$$\Gamma_s = \begin{pmatrix} 0 & 0 & 1 & 1 \\ 0 & 1 & 0 & 0 \\ 1 & 0 & 0 & 0 \\ 0 & 0 & -1 & 0 \end{pmatrix}, \quad \Gamma_y = \begin{pmatrix} 0 & 1 \\ 1 & 0 \end{pmatrix}, \quad \Gamma_u = 1.$$

The SCB has the dynamic equations,

$$\begin{aligned} &\dot{x}_{b1} = x_{b2}, \quad \dot{x}_{b2} = x_{b3} + x_d, \quad \dot{x}_{b3} = -x_{b1}, \quad \dot{x}_d = x_{b1} + u_d, \\ &y_d \ = x_d, \quad y_b = x_{b1}. \end{aligned} \tag{24.19}$$

Note that the above system is exactly the same as the system (24.15) considered in Example 24.12. As in Example 24.12, let -1, -2, -3, -4, and -5 be the required invariant zeros. Then, the post compensator $\tilde{\Sigma}^d_{post}$ which squares down the above dual system in SCB format is given by (24.16) whose transfer function is (24.17). Then, Σ^d_{post} for the system (24.19) is

$$\Sigma^d_{post} = \tilde{\Sigma}^d_{post}\Gamma_s^{-1}.$$

Thus, Σ_{pre} for the given system Σ_{plant} has the transfer function

$$H_{pre} = (H^d_{post})' = \frac{1}{s^2+3s+2}\begin{pmatrix} -35s^2+224s+259 \\ s^2+15s+120 \end{pmatrix}.$$

The overall squared down system Σ_{sq} is described by the given system (24.18), and the pre-compensator equations as given by

$$\begin{aligned} \dot{z}_1 &= z_2 - 59u_{new}, \\ \dot{z}_2 &= -2z_1 - 3z_2 + 12u_{new} \\ \bar{u}_1 &= -7z_1 - 7z_2 - 35u_{new}, \quad \bar{u}_2 = z_2 + u_{new}. \end{aligned}$$

It is straightforward to verify that Σ_{sq} is stabilizable and detectable, and moreover -1 and -2 are among both the invariant zeros and eigenvalues of it. These modes -1 and -2 are not observable, but they are obviously detectable.

24.4 Transformation to a Uniform Rank System

The goal of this section is to construct a pre-compensator that squares down a general linear invertible MIMO system to a uniform rank system. Our development to do so is divided into several subsections. Section 24.4.1 details the description of a general square invertible system in an SCB format with a convenient notation useful here, while Sect. 24.4.2 does the same for a uniform rank system. Section 24.4.3 describes how the construction of a pre-compensator can be sub-divided into the construction of a number of simple pre-compensators, each one of which can be constructed by an algorithm termed as $\mathcal{P}$-algorithm given in Sect. 24.4.4. The data required for the construction of each of simple pre-compensator is obtained in

Sect. 24.4.5. Finally, Sect. 24.4.6 presents a numerical example illustrating the pre-compensator design methodology.

24.4.1 An Invertible System in a Particular SCB Format

It is clear from the development so far that squaring down requires a careful structural decomposition of finite and infinite zero structures of the given MIMO system. Evidently, such a decomposition is captured by the special coordinate basis (SCB). We recall below the SCB for square invertible systems by considering a system Σ with the highest order of infinite zeros equal to k. Such a system Σ can be decomposed into $k+1$ systems, Σ_0 and Σ_i, $i = 1$ to k. Here Σ_0 represents the zero dynamics, while Σ_i, $i = 1$ to k represents the dynamics of i-th order infinite zeros. Thus, the dynamics of the given squared system Σ in an SCB format that is suitable for our development here can be described as follows where $g_i(x)$ for an integer i denotes linear functions of all state variables x:
Invariant zero dynamics:

$$\Sigma_0 : \left\{ \dot{x}_0 \;=\; A_0 x_0 + \Sigma_{i=1}^{k} A_{0,i}\, y_i, \right. \tag{24.20a}$$

First order infinite zero dynamics:

$$\Sigma_1 : \left\{ \dot{x}_{1,1} = g_1(x) + u_1, \right. \tag{24.20b}$$

Second order infinite zero dynamics:

$$\Sigma_2 : \begin{cases} \dot{x}_{2,1} = x_{2,2} \;+ v_{2,1,1}\, x_{1,1} \\ \dot{x}_{2,2} = g_2(x) + u_2, \end{cases} \tag{24.20c}$$

Third order infinite zero dynamics:

$$\Sigma_3 : \begin{cases} \dot{x}_{3,1} = x_{3,2} \;+ v_{3,1,1}\, x_{1,1} + v_{3,1,2}\, x_{2,1} \\ \dot{x}_{3,2} = x_{3,3} \;+ v_{3,2,1}\, x_{1,1} + v_{3,2,2}\, x_{2,1} \\ \dot{x}_{3,3} = g_3(x) + u_3, \end{cases} \tag{24.20d}$$

which continues until the final one containing the kth order infinite zeros:

$$\Sigma_k : \begin{cases} \dot{x}_{k,1} & = x_{k,2} + \Sigma_{j=1}^{k-1} v_{k,1,j}\, x_{j,1} \\ \dot{x}_{k,2} & = x_{k,3} + \Sigma_{j=1}^{k-1} v_{k,2,j}\, x_{j,1} \\ \vdots & \\ \dot{x}_{k,k-1} & = x_{k,k} + \Sigma_{j=1}^{k-1} v_{k,k-1,j}\, x_{j,1} \\ \dot{x}_{k,k} & = g_k(x) + u_k. \end{cases} \tag{24.20e}$$

The states of the above SCB system are defined as

$$x = \begin{pmatrix} x_0 \\ x_1 \\ x_2 \\ \vdots \\ x_k \end{pmatrix}, \quad x_1 = x_{1,1}, \quad x_2 = \begin{pmatrix} x_{2,1} \\ x_{2,2} \end{pmatrix}, \quad x_i = \begin{pmatrix} x_{i,1} \\ x_{i,2} \\ \vdots \\ x_{i,i} \end{pmatrix}, \quad x_k = \begin{pmatrix} x_{k,1} \\ x_{k,2} \\ \vdots \\ x_{k,k} \end{pmatrix}.$$

Notationally, the states x_0, and x_i, $i = 1$ to k pertain respectively to the dynamics of invariant zeros, and i-th order infinite zeros. Their dimensions are respectively n_0, and in_i, $i = 1$ to k. Each state x_i has i components denoted by $x_{i,j}$, $j = 1$ to i. The dimension of each $x_{i,j}$ is n_i. In the notation $x_{i,j}$, the first subscript i always refers to the order of infinite zeros, and the second subscript j refers to a particular component of x_i with $j = i$ being its last component. A word about the subscript notation of the coefficient matrix $v_{i,j,m}$ is in order. In general, the subscripts i and j in $v_{i,j,m}$ refer to the dynamic equation of the state $x_{i,j}$ in which $v_{i,j,m}$ appears, and the subscript m refers to the output component y_m to which the coefficient belongs. The input u and the output y are defined as

$$u = \begin{pmatrix} u_1 \\ u_2 \\ \vdots \\ u_k \end{pmatrix}, \quad y = \begin{pmatrix} y_1 \\ y_2 \\ \vdots \\ y_k \end{pmatrix} = \begin{pmatrix} x_{1,1} \\ x_{2,1} \\ \vdots \\ x_{k,1} \end{pmatrix}.$$

Notationally, u_i is the input to the dynamics of i-th order infinite zeros, while the first component $x_{i,1}$ in x_i is the output y_i of i-th order infinite zero chain. The dimension of both u_i and y_i is n_i.

Remark 24.14 We observe that the dynamics associated with different orders of infinite zeros are interwoven. Such an interweaving occurs in two ways. The first one is output injection through the coefficients of the type $v_{i,j,m}$ in Eq. (24.20), while the second one is through the linear functions of the type $g_i(x)$ which are additive to the input u_i for $i = 1$ to k.

24.4.2 An Invertible Uniform Rank System in a Suitable SCB Format

As already discussed in Sect. 23.3.4, an invertible uniform rank system Σ_w in its SCB format consists of two systems Σ_{w0} and Σ_{wk} of the form,

$$\Sigma_{w0} : \{ \; \dot{w}_0 = A_0 w_0 + A_{0y} y_w, \quad \text{(invariant zero dynamics)} \tag{24.21a}$$

$$\Sigma_{wk} : \begin{cases} \dot{w}_j = w_{j+1}, \quad \text{for } j = 1, 2, \cdots, k-1, \\ \dot{w}_k = g(w) + u_w, \quad \text{(k-th order infinite zero dynamics)} \end{cases} \tag{24.21b}$$

with $y_w := w_1$ considered as its output. The integer k here denotes the order of infinite zeros of Σ_w. Also, w_i, $i = 0, 1, 2, \cdots, k$ are the states, y_w is the output, and u_w is the input of Σ_w. Furthermore, $g(w)$ denotes a linear function of all state variables w_i, $i = 0, 1, 2, \cdots, k$.

We observe that, in the format of SCB, Σ_w consists of Σ_{w0} representing its finite zero structure and Σ_{wk} representing its infinite zero structure. We note further that Σ_{w0} is fed only by the output of Σ_{wk}. We also note that the infinite zero structure Σ_{wk} consists of k integrator blocks forming a chain; the output of first integrator is its output as well as the output of Σ_w, while the last integrator is fed by a linear combinations of all the states and the input of Σ_w. We also note that each integrator block in the chain contains n_k scalar integrators.

24.4.3 Transforming a System to a Corresponding Uniform Rank System in SCB Format

In what follows, our goal is to design a pre-compensator, denoted by $\mathcal{P}_i$, to reduce Σ_i, $i = 1$ to $k - 1$, to a uniform rank system Σ_{wi} in SCB format of order k.

There are two phases in accomplishing our goal. We observe from Eq. (24.20) that the dynamic equations of Σ_i, for each $i = 2$ to k, are not in the form of a uniform rank system in SCB format as given in Eq. (24.21) in Sect. 24.4.2 because of the presence of output injection through the coefficients of the type $\nu_{i,j,m}$. The first phase consists of rewriting such dynamic equations in the SCB format by an appropriate selection of state variables. A successful completion of this phase gives rise to a uniform rank system $\check{\Sigma}_{i,i}$ in SCB format whose order of infinite zeros is the same as that of the given system Σ_i, namely, i.

The second phase consists of designing the pre-compensator $\mathcal{P}_i$ that enables squaring down $\check{\Sigma}_{i,i}$ to Σ_{wi} having k-th order infinite zeros. This implies that the pre-compensator $\mathcal{P}_i$ must inherently be a dynamic system.

Before we proceed to a general method of designing $\mathcal{P}_i$, we consider below a simple example to bring forward the concepts involved. The purpose of the example is to show that just using integrators or integrator chains alone as pre-compensators cannot always accomplish the task, and certain inputs or feedback loops between integrators or within integrator chains must be designed appropriately. The fundamental concept behind such a design is introduced in the example.

Example 24.15 We consider here a system with highest order of infinite zeros equal to three. Let $g_1(x) = \alpha_{1,1,3}x_{3,3}$, $g_2(x) = 0$, $g_3(x) = 0$, $\nu_{2,1,1} = 0$, $\nu_{3,1,1} = 0$, $\nu_{3,1,2} = 0$, $\nu_{3,2,1} = 0$, and $\nu_{3,2,2} = 0$. Let us also assume that x_0 is nonexistent implying that there are no invariant zeros. Then, the given system is described by the following equations:

$$\Sigma_1 : \{\dot{x}_{1,1} = \alpha_{1,1,3}x_{3,3} + u_1, \qquad \text{(First order infinite zero dynamics)}$$

$$\Sigma_2 : \{\dot{x}_{2,1} = x_{2,2}, \quad \dot{x}_{2,2} = u_2, \qquad \text{(Second order infinite zero dynamics)}$$

$$\Sigma_3 : \{\dot{x}_{3,1} = x_{3,2}, \quad \dot{x}_{3,2} = x_{3,3}, \quad \dot{x}_{3,3} = u_3. \qquad \text{(Third order infinite zero dynamics)}$$

We observe that Σ_3 is already in the form of a uniform rank system with infinite zeros of order 3. To conform with the notation we follow in general, we simply change the notation of state variables as

$$w_{3,1} = x_{3,1}, \quad w_{3,2} = x_{3,2}, \quad w_{3,3} = x_{3,3}.$$

Thus, we have Σ_{w3} same as Σ_3 except for notation,

$$\Sigma_{w3} : \{\dot{w}_{3,1} = w_{3,2}, \ \dot{w}_{3,2} = w_{3,3}, \ \dot{w}_{3,3} = u_3. \ \text{(Third order infinite zero dynamics)}$$

Since Σ_2 has only infinite zeros of order 2, we add an integrator as a part of pre-compensator (see Fig. 24.2). Let $w_{2,1} = x_{2,1}, \ w_{2,2} = x_{2,2}$ and $w_{2,3} = u_2 = x_{p_2,2}$, then we have

$$\Sigma_{w2} : \{\dot{w}_{2,1} = w_{2,2}, \dot{w}_{2,2} = w_{2,3}, \dot{w}_{2,3} = u_{new,2}. \text{(Third order infinite zero dynamics)}$$

Now we proceed to transform Σ_1 to a uniform rank system with infinite zeros of order 3 by trying to add a chain of two integrators as a pre-compensator (see Fig. 24.3). Let $w_{1,1} = x_{1,1}$ and $w_{1,2} = \alpha_{1,1,3}x_{3,3} + u_1$, then we have

$$\dot{w}_{1,1} = w_{1,2}, \quad \dot{w}_{1,2} = \alpha_{1,1,3}u_3 + x_{p_1,2}.$$

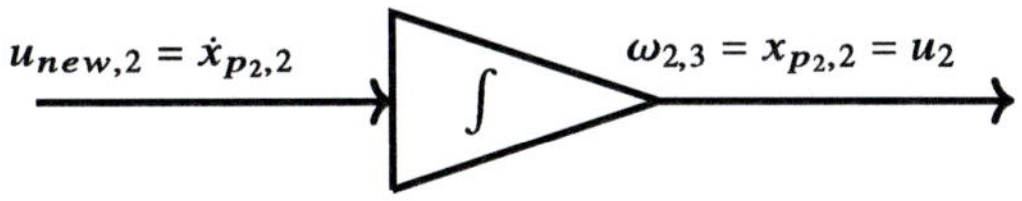

Fig. 24.2 A pre-compensator $\mathcal{P}_2$ which is a simple integrator that transforms Σ_2 to Σ_{w2}

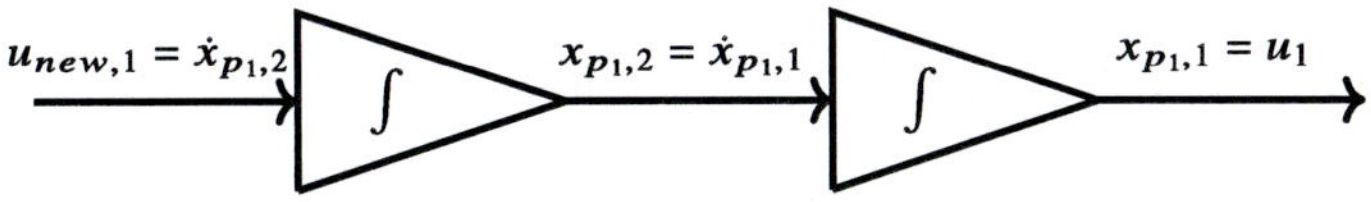

Fig. 24.3 A chain of two integrators being tried as a pre-compensator to transform Σ_1 to Σ_{w1}

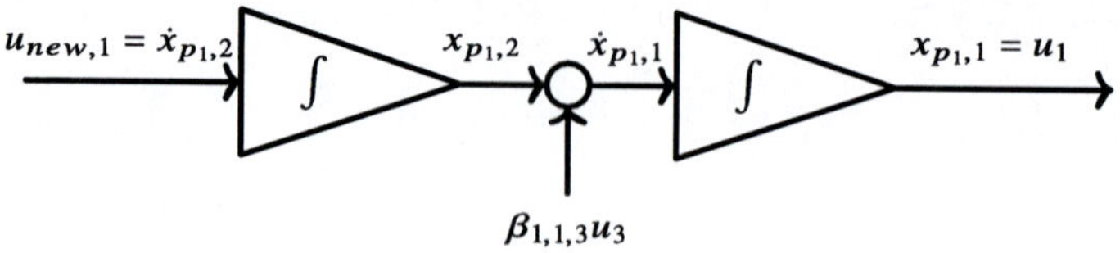

Fig. 24.4 A pre-compensator $\mathcal{P}_1$ which is a chain of two integrators with a feedback signal that transforms Σ_1 to Σ_{w1}

Let us define $w_{1,3} = \alpha_{1,1,3}u_3 + w_{p_1,2}$ so that $\dot{w}_{1,2} = w_{1,3}$ as required by the uniform rank system. Then we get

$$\dot{w}_{1,3} = \alpha_{1,1,3}\dot{u}_3 + u_{new,1}.$$

The presence of $\dot{u}_3$ in the above equation implies that the above set of equations do not conform to a uniform rank system with infinite zeros of order 3, also we cannot supply $\dot{u}_3$ to form $\dot{w}_{1,3}$. This in turn implies that just using integrators or simple integrator chains without any other feedback other than just the input $u_{new,1}$ as a pre-compensator will not lead to a uniform rank system. To rectify the above situation, any pre-compensator that is designed appropriately must eliminate the presence of $\dot{u}_3$ in the above equation.

Let us consider the pre-compensator depicted in Fig. 24.4. Choose $w_{1,1} = x_{1,1}$ and $w_{1,2} = \alpha_{1,1,3}x_{3,3} + u_1$ where $u_1 = x_{p_1,1}$. Then we have,

$$\begin{aligned}\dot{w}_{1,1} &= w_{1,2},\\ \dot{w}_{1,2} &= \alpha_{1,1,3}u_3 + x_{p_1,2} + \beta_{1,1,3}u_3.\end{aligned} \tag{24.22}$$

By selecting

$$\beta_{1,1,3} = -\alpha_{1,1,3},$$

we can eliminate the presence of u_3 in 24.22. We emphasize that we can supply u_3 easily to form $\dot{w}_{1,2}$. Now we can let $w_{1,3} = x_{p_1,2}$, and this leads to

$$\Sigma_{w1} : \left\{\dot{w}_{1,1} = w_{1,2},\quad \dot{w}_{1,2} = w_{1,3},\quad \dot{w}_{1,3} = u_{new,1}.\right. \text{(Third order infinite zero dynamics)}$$

We now summarize the pre-compensator equations:

$$\begin{aligned}&\mathcal{P}_1 : \left\{\dot{x}_{p_1,1} = x_{p_1,2} - \alpha_{1,1,3}u_3,\quad \dot{x}_{p_1,2} = u_{new,1},\right.\\ &\mathcal{P}_2 : \left\{\dot{x}_{p_2,2} = u_{new,2},\right.\end{aligned} \tag{24.23}$$

where $u_1 = x_{p_1,1}$ and $u_2 = x_{p_2,2}$.

Remark 24.16 Example 24.15 reveals the following. A pre-compensator cannot be a simple chain of integrators in all situations. Moreover, a chain of integrators with appropriate feedback signals as inputs to one integrator or more in the chain can be used as a pre-compensator to transform a system with a lower order infinite zeros to another system which is of uniform rank and has higher order infinite zeros. The key concept here is the introduction of appropriately designed feedback signals to eliminate the presence of given control variables in the dynamic equations of the resulting system other than the last equation while increasing the order of infinite zeros. If the given control variables in the dynamic equations other than the last equation are not eliminated, differential coefficients of such given control variables will be present in the last dynamic equation of the resulting system, and such differential coefficients are not allowed in any uniform rank system.

Remark 24.17 The key concept of eliminating control variables as discussed in Remark 24.16 is the basis of the so called $\mathcal{P}$-algorithm developed subsequently (see also Fig. 24.5). The $\mathcal{P}$-algorithm can be used over and over repetitively in transforming a general square invertible system to a uniform rank system.

We now proceed to design $\mathcal{P}_i$ that enables squaring down $\breve{\Sigma}_{i,i}$ for $i = 2$ to $k-1$ having i-th order infinite zeros to Σ_{wi} having k-th order infinite zeros. This is done step by step. That is, we decompose $\mathcal{P}_i$ as a cascade of $k-i$ simple pre-compensators as,

$$\mathcal{P}_i = \mathcal{P}_{i,k-1}\,\mathcal{P}_{i,k-2}\,\cdots\,\mathcal{P}_{i,i+1}\,\mathcal{P}_{i,i}. \tag{24.24}$$

Each simple pre-compensator $\mathcal{P}_{i,j}$, $j = i$ to $k-1$, is designed by following what is termed as *$\mathcal{P}$-algorithm* that is developed soon in Sect. 24.4.4. The goal of each simple pre-compensator $\mathcal{P}_{i,j}$ is to enable increasing the order of infinite zeros of a certain given system by a value 1. To be precise, let us define

$$\breve{\mathcal{P}}_{i,i} = \mathcal{P}_{i,i}, \tag{24.25}$$

and for $i+1 \le j \le k-1$,

$$\breve{\mathcal{P}}_{i,j} = \mathcal{P}_{i,j}\,\mathcal{P}_{i,j-1}\,\cdots\,\mathcal{P}_{i,i+1}\,\mathcal{P}_{i,i}. \tag{24.26}$$

Let us also denote, for $i+1 \le j \le k-1$,

$$\breve{\Sigma}_{i,j+1} = \breve{\mathcal{P}}_{i,j}\,\breve{\Sigma}_{i,i}. \tag{24.27}$$

The above notations allow us to explain the step-by-step design procedure we envision. We design first $\mathcal{P}_{i,i}$ such that the system

$$\breve{\Sigma}_{i,i+1} := \breve{\mathcal{P}}_{i,i}\,\breve{\Sigma}_{i,i} := \mathcal{P}_{i,i}\,\breve{\Sigma}_{i,i}$$

is a uniform rank system in SCB format with its order of infinite zeros equal to $i+1$. We design next $\mathcal{P}_{i,i+1}$ such that the system

$$\breve{\Sigma}_{i,i+2} := \breve{\mathcal{P}}_{i,i+1}\breve{\Sigma}_{i,i} := \mathcal{P}_{i,i+1}\mathcal{P}_{i,i}\,\breve{\Sigma}_{i,i}$$

is a uniform rank system in SCB format with its order of infinite zeros equal to $i+2$. We continue this iterative procedure until we design the $\mathcal{P}_{i,k-1}$ such that $\breve{\Sigma}_{i,k}$ is a uniform rank system in SCB format with its order of infinite zeros equal to k. By our design, $\breve{\Sigma}_{i,k}$ is indeed Σ_{wi} and $\mathcal{P}_i$ is indeed $\breve{\mathcal{P}}_{i,k-1}$.

24.4.4 A Building Block of Pre-compensator Design

In this section, we design each $\mathcal{P}_{i,j}$ given in (24.24) step by step starting from $j = i$. To do so, an algorithm termed as *$\mathcal{P}$-algorithm* is presented next.

As a preliminary to the constructive details of *$\mathcal{P}$-algorithm*, we state first its objective. Consider the last equation of $\breve{\Sigma}_{i,j}$ for $i = 1$ to $k-1$ and $i \le j \le k-1$,

$$\dot{w}_{i,j} = g_{i,j}(x) + f_{i,j}(w) + x_{pi,j}, \tag{24.28}$$

where $g_{i,j}(x)$ is a linear function of all states of x, $x_{pi,j}$ denotes an input variable which turns out to be the state of a simple pre-compensator $\mathcal{P}_{i,j}$, $w_{i,j}$ is a substate variable, and the function $f_{i,j}(w)$ has a certain structure, namely, it is a linear function of all variables $w_{p,q}$ for $p < i, q \le j$.

Remark 24.18 As will be evident shortly and in subsequent sections, the function $f_{i,j}(w)$ is identically zero for $i = 1$. For the sake of generality of the algorithm, we just keep the presence of $f_{i,j}(w)$ in the algorithm. The expression for $f_{i,i}(w)$, and the expression for each $g_{i,i}(x)$ are derived in the subsequent section that pertains to phase 1. Based on $g_{i,i}(x)$ and $f_{i,i}(w)$, the algorithm developed below will generate recursively $g_{i,j}(x)$ and $f_{i,j}(w)$ for $i < j \le k-1$.

Define $w_{i,j+1}$ as

$$w_{i,j+1} := \dot{w}_{i,j} = g_{i,j}(x) + f_{i,j}(w) + x_{pi,j}. \tag{24.29}$$

Given $g_{i,j}(x)$ and $f_{i,j}(w)$, the goal of *$\mathcal{P}$-algorithm* is to design the *simple pre-compensator $\mathcal{P}_{i,j}$* containing one block of integral operation such that the cascade connection of $\mathcal{P}_{i,j}$ and the subsystem with its dynamic equation as (24.28) is of the form,

$$\begin{aligned} \dot{w}_{i,j} &= w_{i,j+1} \\ \dot{w}_{i,j+1} &= g_{i,j+1}(x) + f_{i,j+1}(w) + x_{pi,j+1}, \end{aligned} \tag{24.30}$$

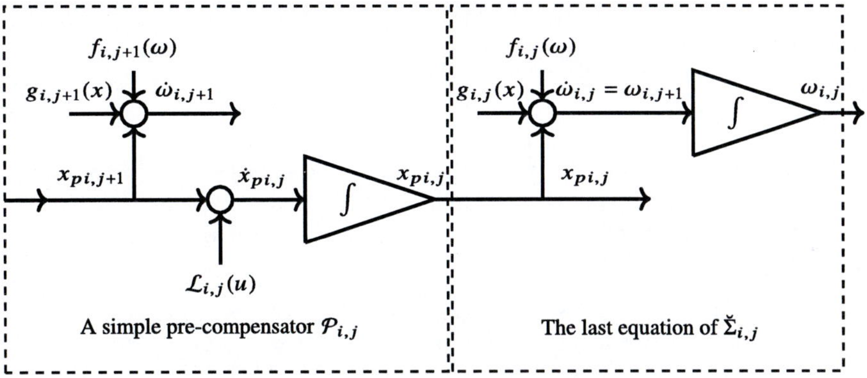

Fig. 24.5 A cascade of simple pre-compensator $\mathcal{P}_{i,j}$ with a part of $\check{\Sigma}_{i,j}$, $i \leq j \leq k-1$

for an appropriately defined linear functions $g_{i,j+1}(x)$ and $f_{i,j+1}(w)$, and a new input variable $x_{pi,j+1}$.

Remark 24.19 Equations (24.28), (24.29), and (24.30) imply that the goal of the $\mathcal{P}$*-algorithm* developed below that constructs the pre-compensator $\mathcal{P}_{i,j}$ is twofold, to generate the input variable $x_{pi,j}$ to $\dot{w}_{i,j}$ as shown in (24.29) and to generate the linear functions $g_{i,j+1}(x)$ and $f_{i,j+1}(w)$ which define the dynamics of $\dot{w}_{i,j+1}$ as shown in (24.30) where $x_{pi,j+1}$ can be considered as a new input variable to the pre-compensator $\mathcal{P}_{i,j}$. Towards accomplishing such a dual goal, we envision the structure of pre-compensator $\mathcal{P}_{i,j}$ as shown in Fig. 24.5.

Construction of $\mathcal{P}$-Algorithm That Designs $\mathcal{P}_{i,j}$

As input data to this algorithm, we are given the linear functions $g_{i,j}(x)$ and $f_{i,j}(w)$. We need to design the input variable $x_{pi,j}$ such that we get the system in (24.30). We do so in three stages:

Stage a In this stage, we decompose the given linear function $g_{i,j}(x)$ into two appropriately defined linear functions. Let us first define the following:

$$\check{x}_0 = x_0, \quad x_r = \begin{pmatrix} \check{x}_r \\ x_{r,r} \end{pmatrix}, \quad \check{x}_r = \begin{pmatrix} x_{r,1} \\ x_{r,2} \\ \vdots \\ x_{r,r-1} \end{pmatrix} \text{ for } 2 \leq r \leq k\,, \quad \check{x} = \begin{pmatrix} \check{x}_0 \\ \check{x}_2 \\ \vdots \\ \check{x}_k \end{pmatrix}.$$

For each $1 < r \leq k$, we observe that $\check{x}_r$ is obtained from x_r by omitting its last element, namely, $x_{r,r}$. Note that $\check{x}_1$ does not exist. Then, it is easy to see that

$$g_{i,j}(x) = \check{g}_{i,j}(\check{x}) + \Sigma_{r=1}^{k} \alpha_{i,j,r} x_{r,r}, \tag{24.31}$$

for an appropriately defined linear function $\breve{g}_{i,j}(\breve{x})$ and coefficients $\alpha_{i,j,r}$. We can then rewrite the dynamic equation (24.28) as

$$\dot{w}_{i,j} = \breve{g}_{i,j}(\breve{x}) + \Sigma_{r=1}^{k}\alpha_{i,j,r}x_{r,r} + f_{i,j}(w) + x_{pi,j} := w_{i,j+1}. \tag{24.32}$$

In this step, we isolated $x_{r,r}$ for $r = 1$ to k in the expression for $\dot{w}_{i,j}$. We note that $\dot{x}_{r,r}$ will have the given control variable u_r in its expression. This presence of u_r in $\dot{x}_{r,r}$ is to be eliminated by an appropriate feedback signal while designing the pre-compensator as pursued in Stages b and c below.

Stage b In this stage, we construct the variable $x_{pi,j}$ via a *simple pre-compensator* $\mathcal{P}_{i,j}$ with an integral operation as shown in Fig. 24.5. That is, we construct $x_{pi,j}$ as

$$x_{pi,j} := \mathcal{I}_1\left[x_{pi,j+1} + \mathcal{L}_{i,j}(u)\right], \tag{24.33}$$

where the operator $\mathcal{I}_1[h]$ denotes the integral of $h(t)$,

$$\mathcal{I}_1[h] = \int h(\tau)d\tau.$$

In Eq. (24.33), $x_{pi,j+1}$ is the input to the *simple pre-compensator*, and the linear function $\mathcal{L}_{i,j}(u)$ is defined as

$$\mathcal{L}_{i,j}(u) = \Sigma_{r=1}^{k}\beta_{i,j,r}\, u_r, \tag{24.34}$$

where the coefficients $\beta_{i,j,r}$, $1 \le r \le k$ are yet unknown. The above two equations lead to

$$\dot{x}_{pi,j} := x_{pi,j+1} + \Sigma_{r=1}^{k}\beta_{i,j,r}\, u_r. \tag{24.35}$$

Stage c In this stage, we determine the unknown coefficients $\beta_{i,j,r}$ for $1 \le r \le k$. To do so, let us differentiate $w_{i,j+1}$ as defined in (24.32) with respect to time to obtain,

$$\begin{aligned}\dot{w}_{i,j+1} = \bar{g}_{i,j+1}(x) + \Sigma_{r=1}^{k}\alpha_{i,j,r}\left[g_r(x) + u_r\right] + f_{i,j+1}(w)\\ + \Sigma_{r=1}^{k}\beta_{i,j,r}\, u_r + x_{pi,j+1},\end{aligned} \tag{24.36}$$

where the linear function $\bar{g}_{i,j+1}(x) := \breve{g}_{i,j}(\dot{\breve{x}})$, $f_{i,j+1}(w) := f_{i,j}(\dot{w})$.

Remark 24.20 Clearly, in view of (24.20), $\dot{\breve{x}}$ contains only x variables and no input variables, thus $\breve{g}_{i,j}(\dot{\breve{x}})$ is of the form $\bar{g}_{i,j+1}(x)$. Similarly, since $i \le j \le k-1$ and since $f_{i,j}(w)$ is a linear function of only variables $w_{p,q}$ for $p < i$ and $q \le j$, we observe that $f_{i,j+1}(w) := f_{i,j}(\dot{w})$ is a linear function of only variables $w_{p,q}$ for $p < i$ and $q \le j+1$.

We select

$$\beta_{i,j,r} = -\alpha_{i,j,r} \tag{24.37}$$

to cancel all the input variables in (24.36). This completely specifies $x_{pi,j}$ given in (24.33). In other words, by the above selection of $\mathcal{L}_{i,j}(u)$, we complete the task of designing the *simple pre-compensator* $\mathcal{P}_{i,j}$ shown in Fig. 24.5.

All the above development, lets us rewrite $\dot{w}_{i,j+1}$ as

$$\begin{aligned} \dot{w}_{i,j+1} &= \bar{g}_{i,j+1}(x) + \Sigma_{r=1}^{k}\alpha_{i,j,r}g_r(x) + f_{i,j+1}(w) + x_{pi,j+1} \\ &= g_{i,j+1}(x) + f_{i,j+1}(w) + x_{pi,j+1}, \end{aligned} \tag{24.38}$$

where

$$g_{i,j+1}(x) = \bar{g}_{i,j+1}(x) + \Sigma_{r=1}^{k}\alpha_{i,j,r}g_r(x).$$

At this time, it is wise to pause and reflect on the above constructional details. Equations (24.33)–(24.37) are crucial aspects of the algorithm. Expressing $x_{pi,j}$ as an integral of some terms as in (24.33) enables the cancellation of unwanted input variables in $\dot{w}_{i,j+1}$ as seen in (24.36) and (24.37). Although the mechanics of writing these equations are straightforward, the underlying concepts as to why such cancellations are done are profound. These concepts enable us to move the unwanted input variables from one step to another until they are accumulated in the last step where they are permissible.

We now summarize different notations and inputs and outputs of the *$\mathcal{P}$-algorithm* whose main task is to design the *simple pre-compensator* $\mathcal{P}_{i,j}$ which is essentially prescribed by prescribing the linear function $\mathcal{L}_{i,j}(u)$ that cancels some unwanted variables. We encounter here the following terminology:

$g_{i,j}(x)$	Input data to the algorithm
$g_{i,j+1}(x)$	Output data of the algorithm
$f_{i,j}(w)$	Input data to the algorithm
$f_{i,j+1}(w)$	Output data of the algorithm
$\mathcal{L}_{i,j}(u)$	Output data of the algorithm
$\beta_{i,j,r}$	Coefficients that define $\mathcal{L}_{i,j}(u)$
$\alpha_{i,j,r}$	Internal coefficients used to rewrite a part of $g_{i,j}(x)$
$x_{pi,j}$	Output variable and state of *simple pre-compensator* $\mathcal{P}_{i,j}$

$x_{pi,j+1}$	Input variable to *simple pre-compensator* $\mathcal{P}_{i,j}$
$w_{i,j+1}$	Prescribed variable ahead of the algorithm
$\dot{w}_{i,j+1}$	A variable which is the output of the algorithm

Remark 24.21 It is evident how we can make use of $\mathcal{P}$-*algorithm* repetitively to design all the *simple pre-compensators* $\mathcal{P}_{i,j}$, $i = 1$ to $k-1$ and $j = i$ to $k-1$. To be explicit, we narrate the design procedure here. For each i, $i = 1$ to $k-1$, we start with the design of $\mathcal{P}_{i,i}$. This requires $g_{i,i}(x)$ and $f_{i,i}(w)$ as inputs to the $\mathcal{P}$-*algorithm*. We will show in the subsequent section how to generate these inputs. This involves transforming each Σ_i to a uniform rank system $\check{\Sigma}_{i,i}$ in SCB format having the same order of infinite zeros as Σ_i does, namely, i. For our present discussion, we assume that both $g_{i,i}(x)$ and $f_{i,i}(w)$ are known and can be fed as input data to the $\mathcal{P}$-*algorithm*. The algorithm not only designs $\mathcal{P}_{i,i}$ (and thus $x_{pi,i} = u_i$), it also generates the linear functions $g_{i,i+1}(x)$ and $f_{i,i+1}(w)$. By feeding $g_{i,i+1}(x)$ and $f_{i,i+1}(w)$ as input data to the $\mathcal{P}$-*algorithm*, we can design $\mathcal{P}_{i,i+1}$ and at the same time generate the linear functions $g_{i,i+2}(x)$ and $f_{i,i+2}(w)$. In turn, $g_{i,i+2}(x)$ and $f_{i,i+2}(w)$ can be used as input data to the $\mathcal{P}$-*algorithm* to design $\mathcal{P}_{i,i+2}$ and to generate the linear functions $g_{i,i+3}(x)$ and $f_{i,i+3}(w)$. Thus, successively and recursively we can design all $\mathcal{P}_{i,j}$, $j = i$ to $k-1$ once we have $g_{i,i}(x)$ and $f_{i,i}(w)$.

Remark 24.22 It is critical to understand the structure of the pre-compensator $\mathcal{P}_{i,j}$ shown in Fig. 24.5. Clearly, $x_{pi,j}$ represents the state of $\mathcal{P}_{i,j}$. Also, as emphasized in Remark 24.19, the goal of $\mathcal{P}_{i,j}$ is to generate as its outputs, $x_{pi,j}$ which is an input to $\dot{w}_{i,j}$, and the linear functions $g_{i,j+1}(x)$ and $f_{i,j+1}(w)$ which define the dynamics of $\dot{w}_{i,j+1}$. To accomplish its goal, $\mathcal{P}_{i,j}$ requires two types of input data. One of them is $x_{pi,j+1}$ which indeed is a new input variable to the pre-compensator $\mathcal{P}_{i,j}$. Besides this input, as displayed clearly in Fig. 24.5, the linear function $\mathcal{L}_{i,j}(u)$ is also an input to $\mathcal{P}_{i,j}$. Equation (24.34) shows that $\mathcal{L}_{i,j}(u)$ is a linear function of the input variables of the given system Σ, namely, u_r, $r = 1$ to k. However, each $u_r := x_{pr,r}$ for $r = 1$ to $k-1$, is generated by the simple pre-compensator $\mathcal{P}_{r,r}$, $r = 1$ to $k-1$. This leads us to emphasize two aspects of pre-compensator design, (1) all simple pre-compensators $\mathcal{P}_{\ell,m}$, $\ell = 1$ to $k-1$, $m = \ell$ to $k-1$ are linked via linear functions of the type $\mathcal{L}_{\ell,m}(u)$. When all the pre-compensators $\mathcal{P}_{\ell,m}$ are successfully and iteratively designed, they can all be put together as a single over-all pre-compensator whose new inputs are $x_{pi,k}$ which can be designated as $u_{new,i}$, $i = 1$ to $k-1$ and one of the original inputs to the given system Σ, namely, u_k which can be called $u_{new,k}$. The outputs of the over-all pre-compensator are the variables $u_r := x_{pr,r}$ for $r = 1$ to $k-1$ to the given system Σ. This is illustrated via an example in Sect. 24.4.6 by presenting the dynamic equations of each simple pre-compensator and the over all pre-compensator. Moreover, the over-all pre-compensator, by generating the linear functions of the type $g_{i,j+1}(x)$ and $f_{i,j+1}(w)$, enables us to write the entire squared down uniform rank system in the format of SCB as given in Eq. (24.21) in Sect. 24.4.2.

Remark 24.23 Once all the subsystems Σ_i, $i = 1$ to k are transformed to uniform rank systems, they can all be put together easily as the system Σ_w given in (24.21). We define the output y_w whose components $y_{w,i}$ are outputs of individual subsystems Σ_i, $i = 1$ to k, and define the input u_{new} to consist of individual inputs $u_{new,i}$, $i = 1$ to k. The design of each pre-compensator does not affect the finite zero structure. All we need to do is define $A_{0,w}y_w = \Sigma_{i=1}^{k} A_{0,i}y_{w,i}$.

24.4.5 Transformation to a Uniform Rank System

As we discussed in Sect. 24.4.3, squaring down each Σ_i, $i = 1$ to $k-1$, to a uniform rank system Σ_{wi} is done in two phases, phase 1 during which Σ_i is transformed to a uniform rank system $\breve{\Sigma}_{i,i}$ in SCB format whose order of infinite zeros is the same as that of Σ_i, and phase 2 during which the pre-compensator $\mathcal{P}_i$ is designed that squares down $\breve{\Sigma}_{i,i}$ to a uniform rank system Σ_{wi} having its order of infinite zeros equal to k. Clearly, phase 2 can be accomplished by the $\mathcal{P}$*-algorithm* as explained in Remark 24.21. What remains is the determination of the expressions for $g_{i,i}(x)$ and $f_{i,i}(w)$ that are required to get started for phase 2. This remaining task is what is pursued below by successfully completing phase 1.

In the processing of determining $\breve{\Sigma}_{i,i}$, we introduce variables $w_{i,j}$ for $1 \le i \le k$ and $1 \le j \le i$ such that there exists a recursive relationship between these variables,

$$\dot{w}_{i,j} = w_{i,j+1} \text{ for } 1 < i \le k \text{ and } 1 \le j < i,$$

and

$$\dot{w}_{i,i} = g_{i,i}(x) + f_{i,i}(w) + u_i \text{ for } 1 \le i \le k.$$

Also, $\dot{w}_{i,j}$ for $1 < i \le k$ and $1 \le j < i$ can be expressed in terms of x and certain known variable w. In fact, $\dot{w}_{i,j}$ in this case is of the form,

$$\dot{w}_{i,j} = x_{i,j+1} + f_{i,j}(w), \tag{24.39}$$

where $f_{i,j}(w)$ is a function of w with a certain structure, namely, it is a linear function of all variables $w_{p,q}$ for $p < i, q \le j$. The form of Eq. (24.39) is depicted in Fig. 24.6. We observe that Fig. 24.6 compliments Fig. 24.5 to show the structure of each $\dot{w}_{i,j}$ as related to x and w for all possible indices i and j.

In what follows, expressions for all $f_{i,j}(w)$ are obtained recursively leading to the determination of $g_{i,i}(x)$ and $f_{i,i}(w)$. We illustrate this at first by considering specific values of i, namely, $i = 1$ to 4, before considering a general index i.

Fig. 24.6 Structure of $\dot{w}_{i,j}$ for $1 < i \leq k$ and $1 \leq j < i$

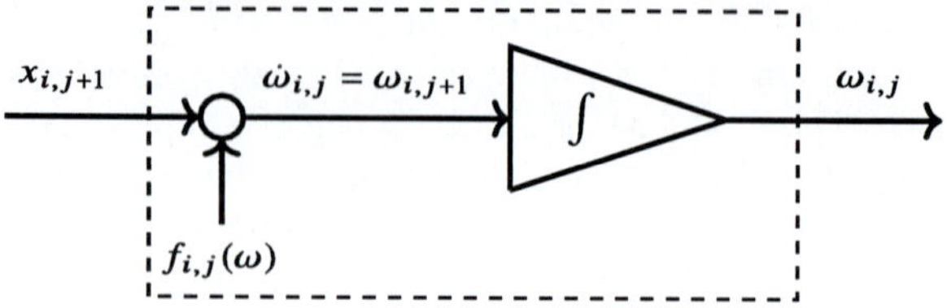

24.4.5.1 Determination of $g_{1,1}(x)$ and $f_{1,1}(w)$

As alluded to, our goal here is to transform Σ_1 to $\breve{\Sigma}_{1,1}$, and in that process determine the expressions for $g_{1,1}(x)$ and $f_{1,1}(w)$.

Let us recall the dynamic equation of Σ_1,

$$\Sigma_1 : \left\{ \dot{x}_{1,1} = g_1(x) + u_1. \right. \tag{24.40}$$

Clearly, this is already in the form of a uniform rank system with its order of infinite zeros equal to 1. Nevertheless, for uniformity of notations to follow, we let

$$x_{1,1} = w_{1,1},$$

and re-write Eq. (24.40) as

$$\breve{\Sigma}_{1,1} : \left\{ \dot{w}_{1,1} = g_1(x) + u_1. \right.$$

Comparing the above equation with the Eq. (24.28) for $i = 1$ and $j = 1$, we find that

$$g_{1,1}(x) = g_1(x) \quad \text{and} \quad f_{1,1}(w) = 0.$$

This completes the task of this subsection.

24.4.5.2 Determination of $g_{2,2}(x)$ and $f_{2,2}(w)$

To start with, we assume that the pre-compensator $\mathcal{P}_1$ is designed that squares down $\breve{\Sigma}_{1,1}$ to a uniform rank system Σ_{w1} by utilizing appropriately the *$\mathcal{P}$-algorithm*, and its notations are consistent with those in Figs. 24.5 and 24.6.

We recall next the dynamic equations of Σ_2 as

$$\Sigma_2 : \begin{cases} \dot{x}_{2,1} = x_{2,2} + v_{2,1,1}\, w_{1,1} \\ \dot{x}_{2,2} = g_2(x) + u_2, \end{cases}$$

where $y_1 = x_{1,1} = w_{1,1}$ by the notation introduced in the previous subsection. We denote

$$w_{2,1} = x_{2,1} \quad \text{and} \quad \dot{w}_{2,1} = w_{2,2} = x_{2,2} + \nu_{2,1,1} w_{1,1}.$$

In view of the notations introduced in Eq. (24.39) and in Fig. 24.6, we note that

$$f_{2,1}(w) = \nu_{2,1,1} w_{1,1}.$$

We have

$$\dot{w}_{2,2} = g_2(x) + u_2 + \nu_{2,1,1} \dot{w}_{1,1} = g_2(x) + \nu_{2,1,1} w_{1,2} + u_2.$$

Comparing the expression for $\dot{w}_{2,2}$ with the Eq. (24.28) for $i = 2$ and $j = 2$, we find that

$$g_{2,2}(x) = g_2(x) \quad \text{and} \quad f_{2,2}(w) = \nu_{2,1,1} w_{1,2}.$$

This lets us rewrite the above system as

$$\breve{\Sigma}_{2,2} : \begin{cases} \dot{w}_{2,1} = w_{2,2} := x_{2,2} + f_{2,1}(w) \\ \dot{w}_{2,2} = g_{2,2}(x) + f_{2,2}(w) + u_2. \end{cases}$$

This completes the task of this subsection.

24.4.5.3 Determination of $g_{3,3}(x)$ and $f_{3,3}(w)$

To start with, we assume that the pre-compensators $\mathcal{P}_1$ and $\mathcal{P}_2$ are designed that square down $\breve{\Sigma}_{1,1}$ and $\breve{\Sigma}_{2,2}$ to uniform rank systems Σ_{w1} and Σ_{w2} by utilizing appropriately the *$\mathcal{P}$-algorithm*, and the notations are consistent with those in Figs. 24.5 and 24.6.

We recall below the dynamic equations of Σ_3 as

$$\Sigma_3 : \begin{cases} \dot{x}_{3,1} = x_{3,2} + \nu_{3,1,1}\, w_{1,1} + \nu_{3,1,2}\, w_{2,1} \\ \dot{x}_{3,2} = x_{3,3} + \nu_{3,2,1}\, w_{1,1} + \nu_{3,2,2}\, w_{2,1} \\ \dot{x}_{3,3} = g_3(x) + u_3, \end{cases}$$

where $y_1 = x_{1,1} = w_{1,1}$ and $y_2 = x_{2,1} = w_{2,1}$ by the notations introduced in the previous two subsections.

We denote

$$w_{3,2} = x_{3,2} + \nu_{3,1,1}\, w_{1,1} + \nu_{3,1,2}\, w_{2,1}.$$

This yields

$$\dot{w}_{3,1} = w_{3,2} = x_{3,2} + f_{3,1}(w),$$

where

$$f_{3,1}(w) = v_{3,1,1}\, w_{1,1} + v_{3,1,2}\, w_{2,1},$$

consistent with the notations introduced in Fig. 24.6. We have

$$\dot{w}_{3,2} = x_{3,3} + f_{3,2}(w),$$

where

$$f_{3,2}(w) = v_{3,2,1}\, w_{1,1} + v_{3,2,2}\, w_{2,1} + v_{3,1,1}\, w_{1,2} + v_{3,1,2}\, w_{2,2},$$

once again notationally consistent with those introduced in Fig. 24.6.

We denote

$$w_{3,3} := \dot{w}_{3,2} = x_{3,3} + f_{3,2}(w),$$

and obtain

$$\dot{w}_{3,3} = g_3(x) + u_3 + f_{3,2}(\dot{w}) = g_{3,3}(x) + f_{3,3}(w) + u_3, \tag{24.41}$$

where $g_{3,3}(x) = g_3(x)$ and

$$f_{3,3}(w) = f_{3,2}(\dot{w}) = v_{3,2,1}\, w_{1,2} + v_{3,2,2}\, w_{2,2} + v_{3,1,1}\, w_{1,3} + v_{3,1,2}\, w_{2,3}.$$

We observe that all $f_{3,j}(w)$ has a definite structure, namely, $f_{3,j}(w)$ is a function of only $w_{p,q}$ for $p < 3$ and $q \le j$. We can easily see that

$$\breve{\Sigma}_{3,3} : \begin{cases} \dot{w}_{3,1} = w_{3,2} \\ \dot{w}_{3,2} = w_{3,3} \\ \dot{w}_{3,3} = g_{3,3}(x) + f_{3,3}(w) + u_3. \end{cases}$$

This completes the task of this subsection.

24.4.5.4 Determination of $g_{4,4}(x)$ and $f_{4,4}(w)$

To start with, once again we assume that the pre-compensators $\mathcal{P}_1$ to $\mathcal{P}_3$ are designed that square down $\breve{\Sigma}_{1,1}$ to $\breve{\Sigma}_{3,3}$ to uniform rank systems Σ_{w1} to Σ_{w3} by utilizing appropriately the $\mathcal{P}$*-algorithm*, and the notations are consistent with those in Figs. 24.5 and 24.6.

We recall next the dynamic equations of Σ_4 as given,

$$\Sigma_4 : \begin{cases} \dot{x}_{4,1} = x_{4,2} + \Sigma_{j=1}^{3} \nu_{4,1,j}\, x_{j,1} \\ \dot{x}_{4,2} = x_{4,3} + \Sigma_{j=1}^{3} \nu_{4,2,j}\, x_{j,1} \\ \dot{x}_{4,3} = x_{4,4} + \Sigma_{j=1}^{3} \nu_{4,3,j}\, x_{j,1} \\ \dot{x}_{4,4} = g_4(x) + u_4, \end{cases}$$

where $y_i = x_{i,1} = w_{i,1}$, $i = 1$ to 3 by the notations introduced in the previous two subsections.

We denote

$$w_{4,2} = \dot{x}_{4,1} = x_{4,2} + \Sigma_{j=1}^{3} \nu_{4,1,j}\, w_{j,1}.$$

This yields

$$\dot{w}_{4,1} = w_{4,2} = x_{4,2} + \Sigma_{j=1}^{3} \nu_{4,1,j}\, w_{j,1} = x_{4,2} + f_{4,1}(w),$$

where

$$f_{4,1}(w) = \Sigma_{j=1}^{3} \nu_{4,1,j}\, w_{j,1}.$$

We have

$$\dot{w}_{4,2} = x_{4,3} + f_{4,2}(w),$$

where

$$f_{4,2}(w) = \Sigma_{j=1}^{3} \nu_{4,2,j}\, w_{j,1} + f_{4,1}(\dot{w}) = \Sigma_{m=1}^{2} \Sigma_{j=1}^{3} \nu_{4,3-m,j}\, w_{j,m}.$$

We denote

$$w_{4,3} = x_{4,3} + f_{4,2}(w),$$

and obtain

$$\dot{w}_{4,3} = x_{4,4} + f_{4,3}(w),$$

where

$$f_{4,3}(w) = \Sigma_{m=1}^{3} \Sigma_{j=1}^{3} \nu_{4,4-m,j}\, w_{j,m}.$$

By denoting

$$w_{4,4} = x_{4,4} + f_{4,3}(w),$$

we obtain

$$\dot{w}_{4,4} = g_{4,4}(x) + f_{4,4}(w) + u_4,$$

where $g_{4,4}(x) = g_4(x)$ and

$$f_{4,4}(w) = f_{4,3}(\dot{w}) = \Sigma_{m=1}^{3} \Sigma_{j=1}^{3} \nu_{4,4-m,j}\, w_{j,m+1}.$$

As in the previous subsection, we observe that all $f_{4,j}(w)$, $j = 1$ to 3, are consistent with (24.39) and the notations of Fig. 24.6. Also, all $f_{4,j}(w)$, $j = 1$ to 4, have a definite structure, namely, $f_{4,j}(w)$ is a linear function of only $w_{p,q}$ for $p < 4$ and $q \leq j$. Moreover, we can easily see that

$$\breve{\Sigma}_{4,4} : \begin{cases} \dot{w}_{4,1} = w_{4,2} \\ \dot{w}_{4,2} = w_{4,3} \\ \dot{w}_{4,3} = w_{4,4} \\ \dot{w}_{4,4} = g_{4,4}(x) + f_{4,4}(w) + u_4. \end{cases}$$

This completes the task of this subsection.

24.4.5.5 Determination of $g_{i,i}(x)$ and $f_{i,i}(w)$

The previous subsections lead the way to generalize the recursive procedure of determining $g_{i,i}(x)$ and $f_{i,i}(w)$ for any $i = 2$ to $k - 1$, and determining $\breve{\Sigma}_{i,i}$ for any $i = 2$ to k. This is pursued in this subsection. A good grasp of notations of Figs. 24.5 and 24.6 facilitates an easy understanding of what follows.

We first recall below the dynamic equations of Σ_i as given,

$$\Sigma_i : \begin{cases} \dot{x}_{i,1} \ \ = x_{i,2} + \Sigma_{j=1}^{i-1} \nu_{i,1,j}\, x_{j,1} \\ \dot{x}_{i,2} \ \ = x_{i,3} + \Sigma_{j=1}^{i-1} \nu_{i,2,j}\, x_{j,1} \\ \ \vdots \\ \dot{x}_{i,i-1} = x_{i,i} \quad + \Sigma_{j=1}^{i-1} \nu_{i,i-1,j}\, x_{j,1} \\ \dot{x}_{i,i} \ \ = g_i(x) + u_i. \end{cases}$$

We denote

$$w_{i,1} = x_{i,1} \ \text{ and } \ w_{i,2} = x_{i,2} + \Sigma_{j=1}^{i-1} \nu_{i,1,j}\, w_{j,1}.$$

We have

$$\dot{w}_{i,1} = w_{i,2} := x_{i,2} + f_{i,1}(w),$$

where $f_{i,1}(w) = \Sigma_{j=1}^{i-1} \nu_{i,1,j}\, w_{j,1}$.

If $i = 2$, we have

$$\dot{w}_{2,2} = g_2(x) + u_2 + f_{2,1}(\dot{w}) = g_{2,2}(x) + f_{2,2}(w) + u_2,$$

where $g_{2,2}(x) = g_2(x)$ and

$$f_{2,2}(w) = f_{2,1}(\dot{w}) = \nu_{2,1,1}\, w_{1,2}.$$

If $i \neq 2$, but $3 \leq i \leq k$, we have

$$\dot{w}_{i,2} = x_{i,3} + \Sigma_{j=1}^{i-1} \nu_{i,2,j}\, w_{j,1} + f_{i,1}(\dot{w}) := x_{i,3} + f_{i,2}(w),$$

where

$$f_{i,2}(w) = \Sigma_{j=1}^{i-1} \nu_{i,2,j}\, w_{j,1} + \Sigma_{j=1}^{i-1} \nu_{i,1,j}\, w_{j,2} = \Sigma_{m=1}^{2} \Sigma_{j=1}^{i-1} \nu_{i,3-m,j}\, w_{j,m}.$$

We denote $w_{i,3} = \dot{w}_{i,2}$. If $i = 3$, we have

$$\dot{w}_{3,3} = g_3(x) + u_3 + \Sigma_{m=1}^{2} \Sigma_{j=1}^{i-1} \nu_{i,3-m,j}\, \dot{w}_{j,m} = g_{3,3}(x) + f_{3,3}(w) + u_3,$$

where $g_{3,3}(x) = g_3(x)$ and

$$f_{3,3}(w) = f_{3,2}(\dot{w}) = \Sigma_{m=1}^{2} \Sigma_{j=1}^{2} \nu_{3,3-m,j}\, w_{j,m+1}.$$

If $i \neq 3$, but $4 \leq i \leq k$, we have

$$\dot{w}_{i,3} = x_{i,4} + \Sigma_{j=1}^{i-1} \nu_{i,3,j}\, w_{j,1} + f_{i,2}(\dot{w}) := x_{i,4} + f_{i,3}(w),$$

where

$$f_{i,3}(w) = \Sigma_{m=1}^{3} \Sigma_{j=1}^{i-1} \nu_{i,4-m,j}\, w_{j,m}.$$

We denote $w_{i,4} = \dot{w}_{i,3}$. If $i = 4$, we have

$$\dot{w}_{4,4} = g_4(x) + u_4 + f_{4,3}(\dot{w}) = g_{4,4}(x) + f_{4,4}(w) + u_4,$$

where $g_{4,4}(x) = g_4(x)$ and

$$f_{4,4}(w) = f_{4,3}(\dot{w}) = \Sigma_{m=1}^{3} \Sigma_{j=1}^{3} \nu_{4,4-m,j}\, w_{j,m+1}.$$

We emphasize that all the development above is consistent with (24.39) and the notations of Fig. 24.6. Proceeding recursively, for an integer ℓ such that $\ell < i \leq k$, we observe that, once again consistent with (24.39) and the notations of Fig. 24.6,

$$\begin{aligned} \dot{w}_{i,\ell} &= x_{i,\ell+1} + \Sigma_{j=1}^{i-1} \nu_{i,\ell,j}\, x_{j,1} + \Sigma_{m=1}^{\ell-1} \Sigma_{j=1}^{i-1} \nu_{i,\ell-m,j}\, \dot{w}_{j,m} \\ &= x_{i,\ell+1} + f_{i,\ell}(w) := w_{i,\ell+1}, \end{aligned} \tag{24.42}$$

where

$$f_{i,\ell}(w) = \Sigma_{m=1}^{\ell} \Sigma_{j=1}^{i-1} \nu_{i,\ell+1-m,j}\, w_{j,m}. \tag{24.43}$$

Remark 24.24 For phase 2, as needed by *$\mathcal{P}$-algorithm*, it is important to recognize the structure of $f_{i,\ell}(w)$; it is a function of only $w_{p,q}$ for $p < i$ and $q \leq \ell$.

If $i = \ell + 1$, we observe that

$$\dot{w}_{i,i} = g_{i,i}(x) + f_{i,i}(w) + u_i,$$

where $g_{i,i}(x) = g_i(x)$ and

$$f_{i,i}(w) = f_{i,i-1}(\dot{w}) = \Sigma_{m=1}^{i-1} \Sigma_{j=1}^{i-1} \nu_{i,i-m,j}\, w_{j,m+1}.$$

We emphasize that the structure of $f_{i,i}(w)$ conforms to what is discussed in Remark 24.24.

Finally, the above development yields for all $2 \leq i \leq k$,

$$\breve{\Sigma}_{i,i} : \begin{cases} \dot{w}_{i,1} = w_{i,2} \\ \dot{w}_{i,2} = w_{i,3} \\ \quad \vdots \\ \dot{w}_{i,i} = g_{i,i}(x) + f_{i,i}(w) + u_i. \end{cases} \tag{24.44}$$

This completes the task of this subsection.

24.4.5.6 Conversion of $x_{i,j}$, to Corresponding $w_{i,j}$

In phase 1 of each stage i of transforming Σ_i to $\breve{\Sigma}_{i,i}$, $i = 1$ to k, we define $w_{i,j}$, $i = 1$ to k and $j = 1$ to i, in terms of $x_{i,j}$, $i = 1$ to k and $j = 1$ to i. After all these stages are complete, we can solve for all $x_{i,j}$, $i = 1$ to k in terms of $w_{i,j}$. From the notations summarized in Fig. 24.6 and from equations (24.42) and (24.43), we can easily obtain such relationships. In fact, we have

$$\begin{aligned} x_{i,1} &= w_{i,1} \text{ for all } i = 1 \text{ to } k, \\ x_{i,j} &= w_{i,j} - f_{i,j-1}(w) \\ &= w_{i,j} - \Sigma_{m=1}^{j-1} \Sigma_{p=1}^{i-1} v_{i,j-m,p}\, w_{p,m} \text{ for all } 1 < i \le k \text{ and } 1 < j \le i. \end{aligned} \tag{24.45}$$

This enables all linear functions $g_{i,j}(x)$ to be re-written as $g_{i,j}(w)$.

Remark 24.25 We observe that all our transformations in all stages never touch Σ_0, and hence the finite zero structure is preserved by the design of a pre-compensator.

24.4.6 Numerical Example

Example 24.26 To illustrate the construction of a pre-compensator, we present here an example with the highest order of infinite zeros equal to 4. For simplicity, we set all coefficients equal to one. That is, we set $g_i(x) = g_{i,i}(x) = \sum_{i=0}^{4} \sum_{j=0}^{i} x_{i,j}$ where $x_{0,0} = x_0$. Also, we set $v_{m,i,j} = 1$ for all possible m, i, j. The dimensions n_i for all $i = 1$ to 4 are assumed to be equal to one. Moreover, we set $A_{0,i} = 1, i = 1, \ldots, 4$ and $A_{aa} = -1$.

It is relatively easy to transform each infinite zero subsystem Σ_i to its corresponding uniform rank system $\breve{\Sigma}_i$ whose order of infinite zeros equals t i. We have

$$\breve{\Sigma}_{1,1}: \quad \left\{ \dot{w}_{1,1} = g_{1,1}(x) + f_{1,1}(w) + u_1, \right.$$

$$\breve{\Sigma}_{2,2}: \quad \begin{cases} \dot{w}_{2,1} = w_{2,2}, \\ \dot{w}_{2,2} = g_{2,2}(x) + f_{2,2}(w) + u_2, \end{cases}$$

$$\breve{\Sigma}_{3,3}: \quad \begin{cases} \dot{w}_{3,1} = w_{3,2}, \quad \dot{w}_{3,2} = w_{3,3}, \\ \dot{w}_{3,3} = g_{3,3}(x) + f_{3,3}(w) + u_3, \end{cases}$$

$$\breve{\Sigma}_{4,4}: \quad \begin{cases} \dot{w}_{4,1} = w_{4,2}, \quad \dot{w}_{4,2} = w_{4,3}, \quad \dot{w}_{4,3} = w_{4,4}, \\ \dot{w}_{4,4} = g_{4,4}(x) + f_{4,4}(w) + u_4, \end{cases}$$

where

$$g_{1,1}(x) = \sum_{i=0}^{4} \sum_{j=0}^{i} x_{i,j}, \qquad f_{1,1}(w) = 0,$$

$$g_{2,2}(x) = \sum_{i=0}^{4} \sum_{j=0}^{i} x_{i,j}, \qquad f_{2,2}(w) = w_{1,2},$$

$$g_{3,3}(x) = \sum_{i=0}^{4} \sum_{j=0}^{i} x_{i,j}, \qquad f_{3,3}(w) = \sum_{i=1}^{2} \sum_{j=1}^{2} w_{j,i+1},$$

$$g_{4,4}(x) = \sum_{i=0}^{4} \sum_{j=0}^{i} x_{i,j}, \qquad f_{4,4}(w) = \sum_{i=1}^{3} \sum_{j=1}^{3} w_{j,i+1}.$$

In what follows we use the notation,

$$x = (x_0,\ x_1,\ x_2,\ x_3,\ x_4),$$

where

$$x_1 = x_{1,1}, \quad x_2 = (x_{2,1},\ x_{2,2}), \quad x_3 = (x_{3,1},\ x_{3,2},\ x_{3,3}), \quad x_4 = (x_{4,1},\ x_{4,2},\ x_{4,3},\ x_{4,4}).$$

We now proceed to design pre-compensators $\mathcal{P}_i$ for $i = 1$ to 3. To do so, we follow the structure and notation of a pre-compensator $\mathcal{P}_{i,j}$ as shown in Fig. 24.5.

Construction of $\mathcal{P}_1$ For subsystem $\check{\Sigma}_{1,1}$, we need to construct $\mathcal{P}_1 = \mathcal{P}_{1,3}\mathcal{P}_{1,2}\mathcal{P}_{1,1}$ to transform $\check{\Sigma}_{1,1}$ to Σ_{w1} which is of uniform rank with order of infinite zeros equal to 4.

Construction of $\mathcal{P}_{1,1}$ Let $i = 1$ and $j = 1$ in Fig. 24.5. The state-variable as well as the output of $\mathcal{P}_{1,1}$ is $x_{p_1,1}$. We let $u_1 = x_{p_1,1}$. Also, let

$$w_{1,2} = g_{1,1}(x) + f_{1,1}(w) + x_{p_1,1}$$

so that $\dot{w}_{1,1} = w_{1,2}$. The dynamic equation of $\mathcal{P}_{1,1}$ is

$$\dot{x}_{p_1,1} = x_{p_1,2} + \Sigma_{r=1}^{4} \beta_{1,1,r}\, u_r \tag{24.46}$$

where $\beta_{1,1,r} = -1$ for $r = 1$ to 4. By using such values for $\beta_{1,1,r}$, we eliminate the presence of u_i, $i = 1$ to 4 in $\dot{w}_{1,2}$ given below. We emphasize that this is a key step in designing $\mathcal{P}_{1,1}$. We have

$$\dot{w}_{1,2} = g_{1,2}(x) + f_{1,2}(w) + x_{p_1,2} \tag{24.47}$$

where

$$g_{1,2}(x) = 3x_0 + 11x_1 + (10\ 5)\, x_2 + (8\ 5\ 5)\, x_3 + (5\ 5\ 5\ 5)\, x_4, \tag{24.48}$$

while $f_{1,2}(w) = 0$, and $x_{p_1,2}$ is the input to $\mathcal{P}_{1,1}$ and is also the output of $\mathcal{P}_{1,2}$ which is yet to be designed.

Construction of $\mathcal{P}_{1,2}$ Let $i = 1$ and $j = 2$ in Fig. 24.5 which prompts us to define

$$w_{1,3} = \dot{w}_{1,2} = g_{1,2}(x) + f_{1,2}(w) + x_{p_1,2}.$$

The dynamic equation of $\mathcal{P}_{1,2}$ is

$$\dot{x}_{p_1,2} = x_{p_1,3} + \Sigma_{r=1}^{4}\beta_{1,2,r}\, u_r \tag{24.49}$$

where

$$\beta_{1,2} = \begin{pmatrix}\beta_{1,2,1} & \beta_{1,2,2} & \beta_{1,2,3} & \beta_{1,2,4}\end{pmatrix} = \begin{pmatrix}-11 & -5 & -5 & -5\end{pmatrix}.$$

By using such values for $\beta_{1,2,r}$, we eliminate the presence of u_i, $i = 1$ to 4 in $\dot{w}_{1,3}$ given below. Again, this is a key step in designing $\mathcal{P}_{1,2}$. We have

$$\dot{w}_{1,3} = g_{1,3}(x) + f_{1,3}(w) + x_{p_1,3}, \tag{24.50}$$

where

$$g_{1,3}(x) = 23x_0 + 67x_1 + \begin{pmatrix}57 & 36\end{pmatrix} x_2 \\ + \begin{pmatrix}44 & 34 & 31\end{pmatrix} x_3 + \begin{pmatrix}29 & 31 & 31 & 31\end{pmatrix} x_4,$$

while $f_{1,3}(w) = 0$, and $x_{p_1,3}$ is the input to $\mathcal{P}_{1,2}$ and is also the output of $\mathcal{P}_{1,3}$ which is yet to be designed.

Construction of $\mathcal{P}_{1,3}$ Let $i = 1$ and $j = 3$ in Fig. 24.5 which prompts us to define

$$w_{1,4} = \dot{w}_{1,3} = g_{1,3}(x) + f_{1,3}(w) + x_{p_1,3}.$$

The dynamic equation of $\mathcal{P}_{1,3}$ is

$$\dot{x}_{p_1,3} = x_{p_1,4} + \Sigma_{r=1}^{4}\beta_{1,3,r}\, u_r \tag{24.51}$$

where

$$\beta_{1,3} = \begin{pmatrix}\beta_{1,3,1} & \beta_{1,3,2} & \beta_{1,3,3} & \beta_{1,3,4}\end{pmatrix} = \begin{pmatrix}-67 & -36 & -31 & -31\end{pmatrix}.$$

By using such values for $\beta_{1,3,r}$, we eliminate the presence of u_i, $i = 1$ to 4 in $\dot{w}_{1,4}$ given below. Again, this is a key step in designing $\mathcal{P}_{1,3}$. We have

$$\dot{w}_{1,4} = g_{1,4}(x) + f_{1,4}(w) + x_{p_1,4}, \tag{24.52}$$

where

$$g_{1,4}(x) = 142x_0 + 414x_1 + \begin{pmatrix}357 & 222\end{pmatrix} x_2 + \begin{pmatrix}279 & 209 & 199\end{pmatrix} x_3 + \begin{pmatrix}188 & 194 & 196 & 196\end{pmatrix} x_4$$

while $f_{1,4}(w) = 0$, and $x_{p_1,4}$ is the input to $\mathcal{P}_{1,3}$. This is the final step in designing $\mathcal{P}_1$, and thus $x_{p_1,4}$ is indeed $u_{new,1}$.

Construction of $\mathcal{P}_2$ For subsystem $\breve{\Sigma}_{2,2}$, we need to construct $\mathcal{P}_2 = \mathcal{P}_{2,3}\mathcal{P}_{2,2}$ to transform $\breve{\Sigma}_{2,2}$ to Σ_{w2}, a uniform rank system with order of infinite zeros equal to 4.

Construction of $\mathcal{P}_{2,2}$ Let $i = 2$ and $j = 2$ in Fig. 24.5. The state-variable as well as the output of $\mathcal{P}_{2,2}$ is $x_{p_2,2}$. We let $u_2 = x_{p_2,2}$. Also, let

$$w_{2,3} = g_{2,2}(x) + f_{2,2}(w) + x_{p_2,2}$$

so that $\dot{w}_{2,2} = w_{2,3}$. The dynamic equation of $\mathcal{P}_{2,2}$ is

$$\dot{x}_{p_2,2} = x_{p_2,3} + \Sigma_{r=1}^{4} \beta_{2,2,r}\, u_r \tag{24.53}$$

where $\beta_{2,2,r} = -1$ for $r = 1$ to 4. By using such values for $\beta_{2,2,r}$, we eliminate the presence of u_i, $i = 1$ to 4 in $\dot{w}_{2,3}$ given below. We emphasize that this is a key step in designing $\mathcal{P}_{2,2}$. We have

$$\dot{w}_{2,3} = g_{2,3}(x) + f_{2,3}(w) + x_{p_2,3} \tag{24.54}$$

where

$$g_{2,3}(x) = 3x_0 + 11x_1 + \begin{pmatrix}10 & 5\end{pmatrix} x_2 + \begin{pmatrix}8 & 5 & 5\end{pmatrix} x_3 + \begin{pmatrix}5 & 5 & 5 & 5\end{pmatrix} x_4, \tag{24.55}$$

while

$$f_{2,3}(w) = w_{1,3}, \tag{24.56}$$

$x_{p_2,3}$ is the input to $\mathcal{P}_{2,2}$ and is also the output of $\mathcal{P}_{2,3}$ which is yet to be designed.

Construction of $\mathcal{P}_{2,3}$ Let $i = 2$ and $j = 3$ in Fig. 24.5. The state-variable as well as the output of $\mathcal{P}_{2,3}$ is $x_{p_2,3}$. Let

$$w_{2,4} = g_{2,3}(x) + f_{2,3}(w) + x_{p_2,3}$$

so that $\dot{w}_{2,3} = w_{2,4}$. The dynamic equation of $\mathcal{P}_{2,3}$ is

$$\dot{x}_{p_2,3} = x_{p_2,4} + \Sigma_{r=1}^{4} \beta_{2,3,r}\, u_r \tag{24.57}$$

where

$$\beta_{2,3} = \begin{pmatrix}\beta_{2,3,1} & \beta_{2,3,2} & \beta_{2,3,3} & \beta_{2,3,4}\end{pmatrix} = \begin{pmatrix}-11 & -5 & -5 & -5\end{pmatrix}.$$

By using such values for $\beta_{2,3,r}$, we eliminate the presence of u_i, $i = 1$ to 4 in $\dot{w}_{2,4}$ given below. We emphasize that this is a key step in designing $\mathcal{P}_{2,3}$. We have

$$\dot{w}_{2,4} = g_{2,4}(x) + f_{2,4}(w) + x_{p_2,4} \tag{24.58}$$

where

$$g_{2,4}(x) = 23x_0 + 67x_1 + \begin{pmatrix}57 & 36\end{pmatrix} x_2 + \begin{pmatrix}44 & 34 & 31\end{pmatrix} x_3 + \begin{pmatrix}29 & 31 & 31 & 31\end{pmatrix} x_4, \tag{24.59}$$

while

$$f_{2,4}(w) = w_{1,4}, \tag{24.60}$$

and $x_{p_2,4}$ is the input to $\mathcal{P}_{2,3}$. This is the final step in designing $\mathcal{P}_2$, and thus $x_{p_2,4}$ is indeed $u_{new,2}$.

Construction of $\mathcal{P}_3$ For subsystem $\breve{\Sigma}_{3,3}$, we need to construct $\mathcal{P}_3 = \mathcal{P}_{3,3}$ to transform $\breve{\Sigma}_{3,3}$ to Σ_{w3} which is of uniform rank with order of infinite zeros equal to 4.

Construction of $\mathcal{P}_{3,3}$ Let $i = 3$ and $j = 3$ in Fig. 24.5. The state-variable as well as the output of $\mathcal{P}_{3,3}$ is $x_{p_3,3}$. We let $u_3 = x_{p_3,3}$. Also, let

$$w_{3,4} = g_{3,3}(x) + f_{3,3}(w) + x_{p_3,3}$$

so that $\dot{w}_{3,3} = w_{3,4}$. The dynamic equation of $\mathcal{P}_{3,3}$ is

$$\dot{x}_{p_3,3} = x_{p_3,4} + \Sigma_{r=1}^{4} \beta_{3,3,r}\, u_r \tag{24.61}$$

where $\beta_{3,3,r} = -1$ for $r = 1$ to 4. By using such values for $\beta_{3,3,r}$, we eliminate the presence of u_i, $i = 1$ to 4 in $\dot{w}_{3,4}$ given below although it is not essential. We have

$$\dot{w}_{3,4} = g_{3,4}(x) + f_{3,4}(w) + x_{p_3,4} \tag{24.62}$$

where

$$g_{3,4}(x) = 3x_0 + 11x_1 + \begin{pmatrix}10 & 5\end{pmatrix} x_2 + \begin{pmatrix}8 & 5 & 5\end{pmatrix} x_3 + \begin{pmatrix}5 & 5 & 5 & 5\end{pmatrix} x_4, \tag{24.63}$$

while

$$f_{3,4}(w) = w_{1,4} + w_{2,4} + w_{1,3} + w_{2,3}. \tag{24.64}$$

This is the final step in designing $\mathcal{P}_3 = \mathcal{P}_{3,3}$, and thus $x_{p_{3},4}$ is indeed $u_{new,3}$.

By now the design of all simple pre-compensators is complete. We can convert all $x_{i,j}$, $i = 1$ to 4 and $j = 1$ to i, to corresponding $w_{i,j}$. In view of (24.45), we have

$$\begin{aligned} x_{i,1} &= w_{i,1} \quad \text{for all } i = 1 \text{ to } 4, \\ x_{i,j} &= w_{i,j} - \Sigma_{m=1}^{j-1} \Sigma_{p=1}^{i-1} w_{p,m} \quad \text{for all } 1 < i \leq 4 \text{ and } 1 < j \leq i. \end{aligned}$$

This in turn lets us express,

$$\begin{aligned} g_{w1}(w) = g_{1,4}(x) &= 142w_0 + \begin{pmatrix}-802 & -591 & -196 & 0\end{pmatrix} w_1 \\ &+ \begin{pmatrix}-637 & -369 & -196 & 0\end{pmatrix} w_2 \\ &+ \begin{pmatrix}-307 & -183 & 3 & 0\end{pmatrix} w_3 + \begin{pmatrix}188 & 194 & 196 & 196\end{pmatrix} w_4 \\ g_{w2}(w) = g_{2,4}(x) + f_{2,4}(w) &= 23w_0 + \begin{pmatrix}-127 & -93 & -31 & 1\end{pmatrix} w_1 \\ &+ \begin{pmatrix}-101 & -57 & -31 & 0\end{pmatrix} w_2 \\ &+ \begin{pmatrix}-49 & -28 & 0 & 0\end{pmatrix} w_3 + \begin{pmatrix}29 & 31 & 31 & 31\end{pmatrix} w_4 \\ g_{w3}(w) = g_{3,4}(x) + f_{3,4}(w) &= 3w_0 + \begin{pmatrix}-19 & -15 & -4 & 1\end{pmatrix} w_1 \\ &+ \begin{pmatrix}-15 & -10 & -4 & 1\end{pmatrix} w_2 \\ &+ \begin{pmatrix}-7 & -5 & 0 & 0\end{pmatrix} w_3 + \begin{pmatrix}5 & 5 & 5 & 5\end{pmatrix} w_4 \\ g_{w4}(w) = g_{4,4}(x) + f_{4,4}(w) &= w_0 + \begin{pmatrix}-5 & -2 & 0 & 1\end{pmatrix} w_1 \\ &+ \begin{pmatrix}-4 & -1 & 0 & 1\end{pmatrix} w_2 \\ &+ \begin{pmatrix}-2 & 0 & 1 & 1\end{pmatrix} w_3 + + \begin{pmatrix}1 & 1 & 1 & 1\end{pmatrix} w_4, \end{aligned}$$

where $w_i = \text{col}(w_{i,1}, w_{i,2}, w_{i,3}, w_{i,4})$. The above expressions lead to the squared down uniform rank system in SCB format for $i = 1$ to 4 as

$$\breve{\Sigma}_{i,4}: \begin{cases} \dot{w}_{i,1} = w_{i,2} \\ \dot{w}_{i,2} = w_{i,3} \\ \dot{w}_{i,3} = w_{i,4} \\ \dot{w}_{i,4} = g_{wi}(w) + u_{new,i}, \end{cases}$$

where $u_{new,i}$, $i = 1$ to 3 are as defined earlier, and $u_{new,4} := u_4$.

We can now put together the dynamic equations of all simple pre-compensators of the type $\mathcal{P}_{i,j}$ to obtain the dynamic equations of over all pre-compensator Σ_{pre},

$$\Sigma_{pre}: \begin{pmatrix} \dot{x}_{p1,1} \\ \dot{x}_{p1,2} \\ \dot{x}_{p1,3} \\ \dot{x}_{p2,2} \\ \dot{x}_{p2,3} \\ \dot{x}_{p3,3} \end{pmatrix} = \begin{pmatrix} -1 & 1 & 0 & -1 & 0 & -1 \\ -11 & 0 & 0 & -5 & 0 & -5 \\ -67 & 0 & 0 & -36 & 0 & -31 \\ -1 & 0 & 0 & -1 & 1 & -1 \\ -11 & 0 & 0 & -5 & 0 & -5 \\ -1 & 0 & 0 & -1 & 0 & -1 \end{pmatrix} \begin{pmatrix} x_{p1,1} \\ x_{p1,2} \\ x_{p1,3} \\ x_{p2,2} \\ x_{p2,3} \\ x_{p3,3} \end{pmatrix} + \begin{pmatrix} 0 & 0 & 0 & -1 \\ 0 & 0 & 0 & -5 \\ 1 & 0 & 0 & -31 \\ 0 & 0 & 0 & -1 \\ 0 & 1 & 0 & -5 \\ 0 & 0 & 1 & -1 \end{pmatrix} \begin{pmatrix} u_{new,1} \\ u_{new,2} \\ u_{new,3} \\ u_{new,4} \end{pmatrix}.$$

We observe that the output of Σ_{pre} is the input to the given system Σ,

$$u = \text{col}(x_{p1,1}, x_{p2,2}, x_{p3,3}, u_{new,4}) = \text{col}(u_1, u_2, u_3, u_4),$$

while the input of Σ_{pre} is

$$u_{new} = \text{col}(u_{new,1}, u_{new,2}, u_{new,3}, u_{new,4}),$$

where $u_{new,4} = u_4$.

References

1. Agaev, R.P., Chebotarev, P.Yu.: The matrix of maximum out forests of a digraph and its applications. Autom. Remote Control **61**(9), 424–1450 (2000)
2. Anderson, B.D.O.: A system theory criterion for positive real matrices. SIAM J. Control **5**(2), 171–182 (1967)
3. Arcak, M.: Passivity as a design tool for group coordination. IEEE Trans. Autom. Control **52**(8), 1380–1390 (2007)
4. Bai, H., Arcak, M., Wen, J.: Cooperative Control Design: A Systematic, Passivity-Based Approach. Communications and Control Engineering. Springer, Berlin (2011)
5. Berman, A., Varga, R.S., Ward, R.C.: ALPS: matrices with nonpositive off-diagonal entries. Linear Algebra Appl. **21**(3), 233–244 (1978)
6. Bliman, P., Ferrari-Trecate, G.: Average consensus problems in networks of agents with delayed communications. Automatica **44**(8), 1985–1995 (2008)
7. Blondel, V., Gevers, M.: Simultaneous stabilizability of three linear systems is rationally undecidable. Math. Control Signals Syst. **6**(2), 135–145 (1993)
8. Boyd, S.P., Barratt, C.H.: Linear Controller Design: Limits of Performance. Information and System Sciences. Prentice Hall, Englewood Cliffs (1991)
9. Boyd, S.P., Doyle, J.C.: Comparison of peak and RMS gains for discrete-time systems. Syst. Control Lett. **9**(1), 1–6 (1987)
10. Cao, Y., Yu, W., Ren, W., Chen, G.: An overview of recent progress in the study of distributed multi-agent coordination. IEEE Trans. Ind. Inf. **9**(1), 427–438 (2013)
11. Chen, B.M.: On properties of the special coordinate basis of linear systems. Int. J. Control **71**(6), 981–1003 (1998)
12. Chen, B.M.: Robust and H_∞ Control. Communication and Control Engineering Series. Springer, Berlin (2000)
13. Chen, B.M., Saberi, A., Sannuti, P.: Explicit expressions for cascade factorization of general nonminimum phase systems. IEEE Trans. Autom. Control **37**(3), 358–363 (1992)
14. Chen, B.M., Lin, Z., Shamash, Y.: Linear Systems Theory: A Structural Decomposition Approach. Birkhäuser, Boston (2004)
15. Chopra, N.: Output synchronization on strongly connected graphs. IEEE Trans. Autom. Control **57**(1), 2896–2901 (2012)
16. Chopra, N., Spong, W.: Output synchronization of nonlinear systems with relative degree one. In: Blondel, V.D., Boyd, S.P., Kimura, H. (eds.). Recent Advances in Learning And Control. Lecture Notes in Control and Information Sciences, vol. 371, pp. 51–64. Springer, London (2008)

A. Saberi et al., *Synchronization of Multi-Agent Systems in the Presence of Disturbances and Delays*, Systems & Control: Foundations & Applications,
https://doi.org/10.1007/978-3-030-88148-1

17. Chopra, N., Spong, W.: Passivity-based control of multi-agent systems. In: Kawamura, S., Svinin, M. (eds.). Advances in Robot Control: From Everyday Physics to Human-Like Movements, pp. 107–134. Springer, Heidelberg (2008)
18. Chu, D., Liu, X., Tan, C.E.: On the numerical computation of structural decomposition in systems and control. IEEE Trans. Autom. Control **47**(11), 1786–1799 (2002)
19. Chu, H., Yuan, J., Zhang, W.: Observer-based consensus tracking for linear multi-agent systems with input saturation. IET Control Theory Appl. **9**(14), 2124–2131 (2015)
20. Commault, C., Dion, J.M.: Structure at infinity of linear multivariable systems: a geometric approach. IEEE Trans. Autom. Control **27**(3), 693–696 (1982)
21. Davison, E.J.: Some properties of minimum phase systems and 'squared-down' systems. IEEE Trans. Autom. Control **28**(2), 221–222 (1983)
22. Doyle, J., Chu, C.-C.: Robust control of multivariable and large scale systems. Tech. Report, Honeywell systems and research center (1986)
23. Doyle, J.C., Glover, K., Khargonekar, P.P., Francis, B.A.: State space solutions to standard H_2 and H_∞ control problems. IEEE Trans. Autom. Control **34**(8), 831–847 (1989)
24. Eichler, A., Werner, H.: Closed-form solution for optimal convergence speed of multi-agent systems with discrete-time double-integrator dynamics for fixed weight ratios. Syst. Control Lett. **71**, 7–13 (2014)
25. Fisher, M.E., Fuller, A.T.: On the stabilization of matrices and the convergence of linear iterative processes. Proc. Cambridge Philos. Soc. **54**, 417–425 (1958)
26. Fradkov, A.: Passification of nonsquare linear systems. In: Proceedings of European Control Conference, Porto, pp. 3338–3343 (2001)
27. Freudenberg, J.S., Looze, D.P.: Right half plane poles and zeros and design tradeoffs in feedback systems. IEEE Trans. Autom. Control **30**(6), 555–565 (1985)
28. Gantmacher, F.R.: The Theory of Matrices. Chelsea, New York (1959)
29. Golub, G.H., van Loan, C.F.: Matrix Computations, 2nd edn. John Hopkins University Press, Baltimore (1989)
30. Grip, H.F., Saberi, A.: Structural decomposition of linear multivariable systems using symbolic computations. Int. J. Control **83**(7), 1414–1426 (2010)
31. Grip, H.F., Yang, T., Saberi, A., Stoorvogel, A.A.: Output synchronization for heterogeneous networks of non-introspective agents. Automatica **48**(10), 2444–2453 (2012)
32. Grip, H.F., Saberi, A., Stoorvogel, A.A.: On the existence of virtual exosystems for synchronized linear networks. Automatica **49**(10), 3145–3148 (2013)
33. Grip, H.F., Saberi, A., Stoorvogel, A.A.: Synchronization in networks of minimum-phase, non-introspective agents without exchange of controller states: homogeneous, heterogeneous, and nonlinear. Automatica **54**, 246–255 (2015)
34. Hadjicostis, C.N., Charalambous, T.: Average consensus in the presence of delays in directed graph topologies. IEEE Trans. Autom. Control **59**(3), 763–768 (2014)
35. Hatanaka, T., Chopra, N., Fujita, M., Spong, M.W.: Passivity-Based Control and Estimation in Networked Robotics. Springer, Heidelberg (2015)
36. Hengster-Movric, K., You, K., Lewis, F.L., Xie, L.: Synchronization of discrete-time multi-agent systems on graphs using Riccati design. Automatica **49**(2), 414–423 (2013)
37. Hitz, L., Anderson, B.D.O.: Discrete positive-real functions and their application to system stability. Proc. IEE **116**(1), 153–155 (1969)
38. Hung, Y.S., MacFarlane, A.G.J.: On the relationships between the unbounded asymptote behavior of multivariable root loci, impulse response and infinite zeros. Int. J. Control **34**(1), 31–69 (1981)
39. Karcanias, N., Kouvaritakis, B.: The output zeroing problem and its relationship to the invariant zero structure: a matrix pencil approach. Int. J. Control **30**(2), 395–415 (1979)
40. Katayama, T.: On the matrix Riccati equation for linear systems with a random gain. IEEE Trans. Autom. Control **21**(5), 770–771 (1976)
41. Khalil, H.K.: On the robustness of output feedback control methods to modeling errors. IEEE Trans. Autom. Control **26**(2), 524–526 (1981)

42. Khalil, H.K., Praly, L.: High–gain observers in nonlinear feedback control. Int. J. Robust Nonlinear Control **24**(6), 993–1015 (1987)
43. Kim, H., Shim, H., Seo, J.H.: Output consensus of heterogeneous uncertain linear multi-agent systems. IEEE Trans. Autom. Control **56**(1), 200–206 (2011)
44. Kim, H., Shim, H., Back, J., Seo, J.: Consensus of output-coupled linear multi-agent systems under fast switching network: averaging approach. Automatica **49**(1), 267–272 (2013)
45. Lafferriere, G., Williams, A., Caughman, J., Veerman, J.J.P.: Decentralized control of vehicle formations. Syst. Control Lett. **54**(9), 899–910 (2005)
46. Larsen, M., Kokotovic, P.V.: On passivation with dynamic output feedback. IEEE Trans. Autom. Control **46**(6), 962–967 (2001)
47. Lee, J., Kim, J., Shim, H.: Disc margins of the discrete-time LQR and its application to consensus problem. Int. J. Syst. Sci. **43**(10), 1891–1900 (2012)
48. Li, T., Zhang, J.: Consensus conditions of multi-agent systems with time-varying topologies and stochastic communication noises. IEEE Trans. Autom. Control **55**(9), 2043–2057 (2010)
49. Li, Z., Duan, Z., Chen, G., Huang, L.: Consensus of multi-agent systems and synchronization of complex networks: a unified viewpoint. IEEE Trans. Circ. Syst.-I Regul. Pap. **57**(1), 213–224 (2010)
50. Li, Y., Xiang, J., Wei, W.: Consensus problems for linear time-invariant multi-agent systems with saturation constraints. IET Control Theory Appl. **5**(6), 823–829 (2011)
51. Li, Z., Duan, Z., Chen, G.: Consensus of discrete-time linear multi-agent system with observer-type protocols. Discrete Continuous Dynam. Syst. B **16**(2), 489–505 (2011)
52. Lin, Z.: Low Gain Feedback. Lecture Notes in Control and Information Sciences. Springer, Berlin (1998)
53. Lin, P., Jia, Y.: Consensus of second-order discrete-time multi-agent systems with nonuniform time-delays and dynamically changing topologies. Automatica **45**(9), 2154–2158 (2009)
54. Lin, P., Jia, Y.: Robust H_∞ consensus analysis of a class of second-order multi-agent systems with uncertainty. IET Control Theory Appl. **4**(3), 487–498 (2010)
55. Lin, Z., Saberi, A.: Low-and-high gain design technique for linear systems subject to input saturation – a direct method. Int. J. Robust Nonlinear Control **7**(12), 1071–1101 (1997)
56. Lin, Z., Saberi, A., Chen, B.M.: Linear Systems Toolbox. A.J. Controls, Seattle. Washington State University Technical Report No. EE/CS 0097 (1991)
57. Lin, Z., Saberi, A., Chen, B.M.: Linear systems toolbox: system analysis and control design in the Matlab environment. In: Proceedings of 1st IEEE Conference on Control Applications, Dayton, pp. 659–664 (1992)
58. Lin, P., Jia, Y., Du, J., Yuan, S.: Distributed consensus control for second-order agents with fixed topology and time-delay (in Chinese). In: Chinese Control Conference, Hunan, pp. 577–581 (2007)
59. Lin, P., Jia, Y., Li, L.: Distributed robust H_∞ consensus control in directed networks of agents with time-delay. Syst. Control Lett. **57**(8), 643–653 (2008)
60. Liu, X., Chen, B.M., Lin, Z.: On the problem of general structure assignments of linear systems through sensor/actuator selection. Automatica **39**(2), 233–241 (2003)
61. Liu, X., Chen, B.M., Lin, Z.: Linear systems toolkit in Matlab: structural decompositions and their applications. J. Control Theory Appl. **3**(3), 287–294 (2005)
62. Liu, Z., Saberi, A., Stoorvogel, A.A., Zhang, M.: Passivity-based state synchronization of homogeneous multiagent systems via static protocol in the presence of input saturation. Int. J. Robust Nonlinear Control **28**(7), 2720–2741 (2018)
63. Liu, Z., Zhang, M., Saberi, A., Stoorvogel, A.A.: State synchronization of multi-agent systems via static or adaptive nonlinear dynamic protocols. Automatica **95**, 316–327 (2018)
64. Ma, C.Q., Zhang, J.F.: Necessary and sufficient conditions for consensusability of linear multi-agent systems. IEEE Trans. Autom. Control **55**(5), 1263–1268 (2010)
65. MacFarlane, A.G.J., Karcanias, N.: Poles and zeros of linear multivariable systems: a survey of the algebraic, geometric and complex-variable theory. Int. J. Control **24**(1), 33–74 (1976)
66. Macon, N., Spitzbart, A.: Inverses of Vandermonde matrices. Am. Math. Month. **65**(2), 95–100 (1958)

67. Meng, Z., Yang, T., Dimarogonas, D.V., Johansson, K.H.: Coordinated output regulation of multiple heterogeneous linear systems. In: Proceedings of 52nd Conference on Decision and Control (CDC), Florence, pp. 2175–2180 (2013)
68. Meng, Z., Zhao, Z., Lin, Z.: On global leader-following consensus of identical linear dynamic systems subject to actuator saturation. Syst. Control Lett. **62**(2), 132–142 (2013)
69. Mesbahi, M., Egerstedt, M.: Graph Theoretic Methods in Multiagent Networks. Princeton University Press, Princeton (2010)
70. Mita, T.: On maximal unobservable subspace, zeros, and their applications. Int. J. Control **25**(6), 885–899 (1977)
71. Mita, T.: On the synthesis of an unknown observer for a class of multi-input/output systems. Int. J. Control **26**(6), 841–851 (1977)
72. Moreau, L.: Stability of continuous-time distributed consensus algorithms. In Proceedings of 43rd Conference on Decision and Control (CDC). The Bahamas, pp. 3998–4003 (2004)
73. Morse, A.S.: Structural invariants of linear multivariable systems. SIAM J. Control Opt. **11**(3), 446–465 (1973)
74. Moylan, P.: Stable inversion of linear systems. IEEE Trans. Autom. Control **22**(1), 74–78 (1977)
75. Münz, U., Papachristodoulou, A., Allgöwer, F.: Delay robustness in consensus problems. Automatica **46**(8), 1252–1265 (2010)
76. Münz, U., Papachristodoulou, A., Allgöwer, F.: Delay robustness in non-identical multi-agent systems. IEEE Trans. Autom. Control **57**(6), 1597–1603 (2012)
77. Nocedal, J., Wright, S.J.: Numerical Optimization. Springer, New York (1999)
78. Øksendal, B.: Stochastic Differential Equations: An Introduction with Applications. Universitext, 6th edn. Springer, Berlin (2003)
79. Olfati-Saber, R., Murray, R.M.: Consensus problems in networks of agents with switching topology and time-delays. IEEE Trans. Autom. Control **49**(9), 1520–1533 (2004)
80. O'Reilly, J.: Observers for Linear Systems. Academic Press, London (1983)
81. Owens, D.H.: Invariant zeros of multivariable systems: a geometric analysis. Int. J. Control **26**(4), 537–548 (1977)
82. Ozcetin, H.K., Saberi, A., Sannuti, P.: Special coordinate basis for order reduction of linear multivariable systems. Int. J. Control **52**(1), 191–226 (1990)
83. Peymani, E., Grip, H.F., Saberi, A., Wang, X., Fossen, T.I.: H_∞ almost ouput synchronization for heterogeneous networks of introspective agents under external disturbances. Automatica **50**(4), 1026–1036 (2014)
84. Peymani, E., Grip, H.F., Saberi, A.: Homogeneous networks of non-introspective agents under external disturbances - H_∞ almost synchronization. Automatica **52**, 363–372 (2015)
85. Pogromsky, A.Y., Santoboni, G., Nijmeijer, H.: Partial synchronization: from symmetry towards stability. Phys. D **172**(1–4), 65–87 (2002)
86. Proskurnikov, A.V., Mazo Jr., M.: Simple synchronization protocols for heterogeneous networks: beyond passivity. In: Proceedings of the 20th IFAC World Congress, Toulouse, pp. 9426–9431 (2017)
87. Pugh, A.C., Ratcliffe, P.A.: On the zeros and poles of a rational matrix. Int. J. Control **30**(2), 213–227 (1979)
88. Qu, Z.: Cooperative Control of Dynamical Systems: Applications to Autonomous Vehicles. Spinger, London (2009)
89. Ramírez, J.G.B., Femat, R.: On the controlled synchronization of dynamical networks with non identical agents. In International IEEE Conference on Physics and Control, Potsdam, pp. 1253–1257 (2007)
90. Ren, W., Atkins, E.: Distributed multi-vehicle coordinate control via local information. Int. J. Robust Nonlinear Control **17**(10–11), 1002–1033 (2007)
91. Ren, W., Beard, R.W.: Consensus seeking in multiagent systems under dynamically changing interaction topologies. IEEE Trans. Autom. Control **50**(5), 655–661 (2005)
92. Rosenbrock, H.H.: State-Space and Multivariable Theory. Wiley, New York (1970)
93. Rudin, W.: Real and Complex Analysis, 3rd edn. McGraw-Hill, New York (1987)

94. Saberi, A.: Decentralization of large-scale systems: a new canonical form for linear multivariable systems. IEEE Trans. Autom. Control **30**(11), 1120–1123 (1985)
95. Saberi, A., Sannuti, P.: Squaring down by static and dynamic compensators. IEEE Trans. Autom. Control **33**(4), 358–365 (1988)
96. Saberi, A., Sannuti, P.: Time-scale structure assignment in linear multivariable systems using high-gain feedback. Int. J. Control **49**(6), 2191–2213 (1989)
97. Saberi, A., Sannuti, P.: Observer design for loop transfer recovery and for uncertain dynamical systems. IEEE Trans. Autom. Control **35**(8), 878–897 (1990)
98. Saberi, A., Sannuti, P.: Squaring down of non-strictly proper systems. Int. J. Control **51**(3), 621–629 (1990)
99. Saberi, A., Sannuti, P., Chen, B.M.: H_2 Optimal Control. Prentice Hall, Englewood Cliffs (1995)
100. Saberi, A., Chen, B.M., Sannuti, P.: Theory of LTR for non-minimum phase systems, recoverable targets, and recovery in a subspace. Part 1: analysis. Int. J. Control **53**(5), 1067–1115 (1991)
101. Saberi, A., Ozcetin, H., Sannuti, P.: New structural invariants of linear multivariable systems. Int. J. Control **56**(4), 877–900 (1992)
102. Saberi, A., Chen, B.M., Sannuti, P.: Loop Transfer Recovery: Analysis and Design. Springer, Berlin (1993)
103. Saberi, A., Lin, Z., Stoorvogel, A.A.: H_2 and H_∞ almost disturbance decoupling problem with internal stability. Int. J. Robust Nonlinear Control **6**(8), 789–803 (1996)
104. Saberi, A., Stoorvogel, A.A., Sannuti, P.: Control of Linear Systems with Regulation and Input Constraints. Communication and Control Engineering Series. Springer, Berlin (2000)
105. Saberi, A., Stoorvogel, A.A., Sannuti, P.: Filtering Theory: With Applications to Fault Detection, Isolation, and Estimation. Birkhäuser, Boston (2007)
106. Saberi, A., Stoorvogel, A.A., Sannuti, P.: Internal and External Stabilization of Linear Systems with Constraints. Birkhäuser, Boston (2012)
107. Saboori, I., Khorasani, K.: H_∞ consensus achievement of multi-agent systems with directed and switching topology networks. IEEE Trans. Autom. Control **59**(11), 3104–3109 (2014)
108. Safonov, M.G., Chen, B.S.: Multivariable stability-margin optimisation with decoupling and output regulation. IEE Proc. D **129**(6), 276–282 (1982)
109. Sannuti, P.: A direct singular perturbation analysis of high-gain and cheap control problems. Automatica **19**(1), 41–51 (1983)
110. Sannuti, P., Saberi, A.: Special coordinate basis for multivariable linear systems – finite and infinite zero structure, squaring down and decoupling. Int. J. Control **45**(5), 1655–1704 (1987)
111. Sannuti, P., Wason, H.S.: A singular perturbation canonical form of invertible systems: determination of multivariable root-loci. Int. J. Control **37**(6), 1259–1286 (1983)
112. Sannuti, P., Wason, H.S.: Multiple time-scale decomposition in cheap control problems – singular control. IEEE Trans. Autom. Control **30**(7), 633–644 (1985)
113. Sannuti, P., Saberi, A., Zhang, M.: Squaring down of general MIMO systems to invertible uniform rank systems via pre- and/or post-compensators. Automatica **50**(8), 2136–2141 (2014)
114. Scardovi, L., Sepulchre, R.: Synchronization in networks of identical linear systems. Automatica **45**(11), 2557–2562 (2009)
115. Schenato, L., Sinopoli, B., Franceschetti, M., Poolla, K., Sastry, S.S.: Foundations of control and estimation over lossy networks. Proc. IEEE **95**(1), 163–187 (2007)
116. Schumacher, J.M.: The role of the dissipation matrix in singular optimal control. Syst. Contr. Lett. **2**(5), 262–266 (1983)
117. Seo, J.H., Shim, H., Back, J.: Consensus of high-order linear systems using dynamic output feedback compensator: low gain approach. Automatica **45**(11), 2659–2664 (2009)
118. Seo, J.H., Back, J., Kim, H., Shim, H.: Output feedback consensus for high-order linear systems having uniform ranks under switching topology. IET Control Theory Appl. **6**(8), 1118–1124 (2012)

119. Shi, G., Johansson, K.H.: Robust consensus for continuous-time multi-agent dynamics. SIAM J. Control Opt. **51**(5), 3673–3691 (2013)
120. Silverman, L.M.: Inversion of multivariable linear systems. IEEE Trans. Autom. Control **14**(3), 270–276 (1969)
121. Sinopoli, B., Schenato, L., Franceschetti, M., Poolla, K., Jordan, M.I., Sastry, S.S.: Kalman filtering with intermittent observations. IEEE Trans. Autom. Control **49**(9), 1453–1464 (2004)
122. Soroka, E., Shaked, U.: The properties of reduced order minimum variance filters for systems with partially perfect measurements. IEEE Trans. Autom. Control **33**(11), 1022–1034 (1988)
123. Stillwell, D.J., Bolt, E.M., Roberson, D.G.: Sufficient conditions for fast switching synchronization in time-varying network topologies. SIAM J. Appl. Dynam. Syst. **5**(1), 140–156 (2006)
124. Stoorvogel, A.A.: The H_∞ Control Problem: A State Space Approach. Prentice-Hall, Englewood Cliffs (1992)
125. Stoorvogel, A.A.: The H_∞ control problem with zeros on the boundary of the stability domain. Int. J. Control **63**(6), 1029–1053 (1996)
126. Stoorvogel, A.A., Saberi, A.: Consensus in the network with nonuniform constant input delay. In: American Control Conference, Chicago, pp. 4106–4111 (2015)
127. Stoorvogel, A.A., Saberi, A., Chen, B.M.: The discrete-time H_∞ control problem with measurement feedback. Int. J. Robust Nonlinear Control **4**(4), 457–480 (1994)
128. Stoorvogel, A.A., Zhang, M., Saberi, A., Grip, H.F.: Synchronization in networks of weakly-non-minimum-phase, non-introspective agents without exchange of controller states. In: American Control Conference, Portland, pp. 3548–3552 (2014)
129. Stoorvogel, A.A., Saberi, A., Zhang, M.: Synchronization for heterogeneous networks of weakly-non-minimum-phase, non-introspective agents without exchange of controller states. In: American Control Conference, Boston, pp. 5187–5192 (2016)
130. Stoorvogel, A.A., Saberi, A., Zhang, M.: Solvability conditions and design for state synchronization of multi-agent systems. Automatica **84**, 43–47 (2017)
131. Stoorvogel, A., Saberi, A., Zhang, M., Liu, Z.: Solvability conditions and design for synchronization of discrete-time multi-agent systems. Int. J. Robust Nonlinear Control **28**(4), 1381–1401 (2018)
132. Stoorvogel, A.A., Saberi, A., Zhang, M., Liu, Z.: Solvability conditions and design for H_∞ and H_2 almost state synchronization of homogeneous multi-agent systems. Eur. J. Control **46**, 36–48 (2019)
133. Su, H., Chen, M.Z.Q.: Multi-agent containment control with input saturation on switching topologies. IET Control Theory Appl. **9**(3), 399–409 (2015)
134. Su, Y., Huang, J.: Stability of a class of linear switching systems with applications to two consensus problem. IEEE Trans. Autom. Control **57**(6), 1420–1430 (2012)
135. Su, H., Chen, M.Z.Q., Lam, J., Lin, Z.: Semi-global leader-following consensus of linear multi-agent systems with input saturation via low gain feedback. IEEE Trans. Circ. Syst.-I Regul. Pap. **60**(7), 1881–1889 (2013)
136. Su, H., Chen, M.Z.Q., Chen, G.: Robust semi-global coordinated tracking of linear multi-agent systems with input saturation. Int. J. Robust Nonlinear Control **25**(14), 2375–2390 (2015)
137. Su, H., Jia, G., Chen, M.Z.Q.: Semi-global containment control of multi-agent systems with intermittent input saturation. J. Franklin Inst. **352**(9), 3504–3525 (2015)
138. Takaba, K.: A dynamic protocol for local synchronization of linear multi-agent systems subject to input saturation. In: Proceedings of 54th Conference on Decision and Control, Osaka (2015), pp. 4923–4927
139. Tian, Y.-P., Liu, C.-L.: Consensus of multi-agent systems with diverse input and communication delays. IEEE Trans. Autom. Control **53**(9), 2122–2128 (2008)
140. Tian, Y.-P., Liu, C.-L.: Robust consensus of multi-agent systems with diverse input delays and asymmetric interconnection perturbations. Automatica **45**(5), 1347–1353 (2009)
141. Trentelman, H.L.: Almost Invariant Subspaces and High Gain Feedback. CWI Tracts, Amsterdam, vol. 29 (1986)

142. Trentelman, H.L., Stoorvogel, A.A., Hautus, M.L.J.: Control Theory for Linear Systems. Communication and Control Engineering Series, Springer, London (2001)
143. Tuna, S.E.: LQR-based coupling gain for synchronization of linear systems. arXiv:0801.3390v1 (2008)
144. Tuna, S.E.: Synchronizing linear systems via partial-state coupling. Automatica **44**(8), 2179–2184 (2008)
145. Tuna, S.E.: Conditions for synchronizability in arrays of coupled linear systems. IEEE Trans. Autom. Control **55**(10), 2416–2420 (2009)
146. Vengertsev, D., Kim, H., Shim, H., Seo, J.: Consensus of output-coupled linear multi-agent systems under frequently connected network. In: Proceedings of 49th Conference on Decision and Control (CDC), Atlanta, pp. 4559–4564 (2010)
147. Verghese, G.: Infinite frequency behavior in generalized dynamical systems, Ph.D. Thesis, Department of Electrical Engineering, Stanford University (1978)
148. Vidyasagar, M.: Control System Synthesis: A Factorization Approach. MIT Press, Cambridge (1985)
149. Wang, J., Cheng, D., Hu, X.: Consensus of multi-agent linear dynamic systems. Asian J. Control **10**(2), 144–155 (2008)
150. Wang, X., Saberi, A., Stoorvogel, A.A., Grip, H.F., Yang, T.: Synchronization for heterogenous networks of introspective right-invertible agents with uniform constant communication delay. in Proceedings of 52nd Conference on Decision and Control (CDC), Firenze, pp. 5198–5203 (2012)
151. Wang, X., Saberi, A., Grip, H.F., Stoorvogel, A.A.: Control of open-loop neutrally stable systems subject to actuator saturation and external disturbances. Int. J. Robust Nonlinear Control **23**(2), 229–240 (2013)
152. Wang, X., Saberi, A., Stoorvogel, A.A.: Stabilization of linear system with input saturation and unknown constant delays. Automatica **49**(12), 3632–3640 (2013)
153. Wang, X., Saberi, A., Stoorvogel, A.A., Grip, H.F., Yang, T.: Consensus in the network with uniform constant communication delay. Automatica **49**(8), 2461–2467 (2013)
154. Wang, X., Saberi, A., Yang, T.: Synchronization in heterogeneous networks of discrete-time introspective right-invertible agents. Int. J. Robust Nonlinear Control **24**(18), 3255–3281 (2013)
155. Wang, X., Saberi, A., Stoorvogel, A.A., Grip, H.F., Yang, T.: Synchronization in a network of identical discrete-time agents with uniform constant communication delay. Int. J. Robust Nonlinear Control **24**(18), 3076–3091 (2014)
156. Wang, X., Su, H., Wang, X., Chen, G.: An overview of coordinated control for multi-agent systems subject to input saturation. Perspect. Sci. **7**, 133–139 (2016)
157. Wieland, P., Allgöwer, F.: An internal model principle for consensus in heterogeneous linear multi-agent systems. In: Proceedings of 1st IFAC Workshop on Estimation and Control of Networked Systems, Venice, pp. 7–12 (2009)
158. Wieland, P., Kim, J.S., Scheu, H., Allgöwer, F.: On consensus in multi-agent systems with linear high-order agents. In: Proceedings of the 17th IFAC World Congress, Seoul, pp. 1541–1546 (2008)
159. Wieland, P., Sepulchre, R., Allgöwer, F.: An internal model principle is necessary and sufficient for linear output synchronization. Automatica **47**(5), 1068–1074 (2011)
160. Willems, J.C.: Least squares stationary optimal control and the algebraic Riccati equation. IEEE Trans. Autom. Control **16**(6), 621–634 (1971)
161. Willems, J.C.: Dissipative dynamical systems part II: linear systems with quadratic supply rates. Arch. Rational Mech. Anal. **45**(5), 352–393 (1972)
162. Wilson, D.A.: Convolution and Hankel operator norms for linear systems. IEEE Trans. Autom. Control **34**(1), 94–97 (1989)
163. Wonham, W.M.: Linear Multivariable Control: A Geometric Approach, 3rd edn. Springer, New York (1985)
164. Wu, C.W.: Synchronization in Complex Networks of Nonlinear Dynamical Systems. World Scientific, Singapore (2007)

165. Wu, C.W., Chua, L.O.: Application of graph theory to the synchronization in an array of coupled nonlinear oscillators. IEEE Trans. Circ. Syst.-I Fundam. Theory Appl. **42**(8), 494–497 (1995)
166. Wu, C.W., Chua, L.O.: Application of Kronecker products to the analysis of systems with uniform linear coupling. IEEE Trans. Circ. Syst.-I Fundam. Theory Appl. **42**(10), 775–778 (1995)
167. Xia, T., Scardovi, L.: Output-feedback synchronizability of linear time-invariant systems. Syst. Control Lett. **94**, 152–158 (2016)
168. Xia, T., Scardovi, L.: Synchronization of linear time-invariant systems on rooted graphs. In: Proceedings 55th Conference on Decision and Control (CDC), Las Vegas, pp. 4376–4381 (2016)
169. Xiang, J., Chen, G.: On the V-stability of complex dynamical networks. Automatica **43**(6), 1049–1057 (2007)
170. Xiao, F., Wang, L.: Asynchronous consensus in continuous-time multi-agent systems with switching topology and time-varying delays. IEEE Trans. Autom. Control **53**(8), 1804–1816 (2008)
171. Yang, T., Roy, S., Wan, Y., Saberi, A.: Constructing consensus controllers for networks with identical general linear agents. Int. J. Robust Nonlinear Control **21**(11), 1237–1256 (2011)
172. Yang, T., Grip, H.F., Saberi, A., Zhang, M., Stoorvogel, A.A.: Synchronization in time-varying networks of non-introspective agents without exchange of controller states. In American Control Conference, Portland, pp. 1475–1480 (2014)
173. Yang, T., Meng, Z., Dimarogonas, D.V., Johansson, K.H.: Global consensus for discrete-time multi-agent systems with input saturation constraints. Automatica **50**(2), 499–506 (2014)
174. Yang, T., Saberi, A., Stoorvogel, A.A., Grip, H.F.: Output synchronization for heterogeneous networks of introspective right-invertible agents. Int. J. Robust Nonlinear Control **24**(13), 1821–1844 (2014)
175. Yao, J., Guan, Z.-H., Hill, D.J.: Passivity-based control and synchronization of general complex dynamical networks. Automatica **45**(9), 2107–2113 (2009)
176. You, K., Xie, L.: Network topology and communication data rate for consensusability of discrete-time multi-agent systems. IEEE Trans. Autom. Control **56**(10), 2262–2275 (2011)
177. Zhang, J.R., Knospe, C.R., Tsiotras, P.: Stability of linear time-delay systems: a delay-dependent criterion with a tight conservatism bound. In: American Control Conference, Chicago, pp. 1458–1462 (2000)
178. Zhang, H., Lewis, F.L., Qu, Z.: Lyapunov, adaptive, and optimal design techniques for cooperative systems on directed communication graphs. IEEE Trans. Indus. Electron. **59**(7), 3026–3041 (2012)
179. Zhang, M., Saberi, A., Grip, H.F., Stoorvogel, A.A.: "$\mathcal{H}_\infty$ almost output synchronization for heterogeneous networks without exchange of controller states. IEEE Trans. Control Netw. Syst. **2**(4), 348–357 (2015)
180. Zhang, M., Saberi, A., Stoorvogel, A.A.: Synchronization in a network of identical discrete-time agents with unknown, nonuniform constant input delay. In: Proceedings of 54th Conference on Decision and Control (CDC), Osaka, pp. 7054–7059 (2015)
181. Zhang, M., Saberi, A., Stoorvogel, A.A.: Regulated output synchronization for heterogeneous time-varying networks with non-introspective agents in presence of disturbance and measurement noise with known frequencies. In: American Control Conference, Chicago, pp. 2069–2074 (2015)
182. Zhang, M., Saberi, A., Stoorvogel, A.A., Sannuti, P.: Almost regulated output synchronization for heterogeneous time-varying networks of non-introspective agents and without exchange of controller states. In: American Control Conference, Chicago, pp. 2735–2740 (2015)
183. Zhang, M., Stoorvogel, A.A., Saberi, A.: Stochastic almost regulated output synchronization for heterogeneous time-varying networks with non-introspective agents and without exchange of controller states. In: American Control Conference, Chicago, pp. 775–780 (2015)
184. Zhang, L., Chen, M.Z.Q., Su, H.: Observer-based semi-global consensus of discrete-time multi-agent systems with input saturation. Trans. Inst. Meas. Control **38**(6), 665–674 (2016)

185. Zhang, M., Saberi, A., Stoorvogel, A.A.: Synchronization in the presence of unknown, nonuniform and arbitrarily large communication delays. Eur. J. Control **38**, 63–72 (2017)
186. Zhang, M., Saberi, A., Stoorvogel, A.A.: Semiglobal state synchronization for continuous- or discrete-time multiagent systems subject to actuator saturation. Int. J. Robust Nonlinear Control **28**(16), 4966–4980 (2018)
187. Zhang, M., Saberi, A., Stoorvogel, A.A.: Synchronization in a network of identical continuous- or discrete-time agents with unknown nonuniform constant input delay. Int. J. Robust Nonlinear Control **28**(13), 3959–3973 (2018)
188. Zhang, M., Saberi, A., Stoorvogel, A.A.: Synchronization in a network of identical continuous-or discrete-time agents with unknown nonuniform constant input delay. Int. J. Robust Nonlinear Control **28**(13), 3959–3973 (2018)
189. Zhang, M., Saberi, A., Stoorvogel, A.: Semi-global state synchronization for discrete-time multi-agent systems subject to actuator saturation and unknown nonuniform input delay. Eur. J. Control **54**, 12–21 (2020)
190. Zhang, M., Saberi, A., Stoorvogel, A.A.: Semi-global state synchronization for multi-agent systems subject to actuator saturation and unknown nonuniform input delay. In: American Control Conference, Denver, pp. 3319–3324 (2020)
191. Zhao, J., Hill, D.J., Liu, T.: Synchronization of complex dynamical networks with switching topology: a switched system point of view. Automatica **45**(11), 2502–2511 (2009)
192. Zhao, J., Hill, D.J., Liu, T.: Passivity-based output synchronization of dynamical network with non-identical nodes. In: Proceedings of 49th Conference on Decision and Control (CDC), Atlanta, pp. 7351–7356 (2010)
193. Zhao, H., Park, J., Zhang, Y., Shen, H.: Distributed output feedback consensus of discrete-time multi-agent systems. Neurocomputing **138**, 86–91 (2014)
194. Zhao, Y., Duan, Z., Wen, G., Chen, G.: Distributed H_∞ consensus of multi-agent systems: a performance region-based approach. Int. J. Control **85**(3), 332–341 (2015)
195. Zhou, K., Doyle, J.C., Glover, K.: Robust and Optimal Control. Prentice Hall, Upper Saddle River (1996)
196. Zhou, B., Xu, C., Duan, G.: Distributed and truncated reduced-order observer based output feedback consensus of multi-agent systems. IEEE Trans. Autom. Control **59**(8), 2264–2270 (2014)

Index

A. Saberi et al., *Synchronization of Multi-Agent Systems in the Presence of Disturbances and Delays*, Systems & Control: Foundations & Applications,
https://doi.org/10.1007/978-3-030-88148-1